Parasitism

INTERSPECIFIC INTERACTIONS
A Series Edited by John N. Thompson

Claude Combes

Parasitism

The Ecology and Evolution of Intimate Interactions

Translated by Isaure de Buron
and Vincent A. Connors

With a New Foreword by Daniel Simberloff

The University of Chicago Press
Chicago and London

CLAUDE COMBES is professor of animal biology and director of the Centre de Biologie et Écologie Tropicale et Méditerranéenne at the Université de Perpignan, France. He is coauthor of *L'Homme et l'Animal.*

The University of Chicago Press, Chicago 60637
The University of Chicago Press, Ltd., London

Printed in the United States of America

10 09 08 07 06 05 04 03 02 01 1 2 3 4 5
ISBN: 0-226-11445-7 (cloth)
ISBN: 0-226-11446-5 (paper)

Originally published as *Interactions durables: Écologie et évolution du parasitisme*
© Masson, 1995

Library of Congress Cataloging-in-Publication Data

Combes, Claude.
[Interactions durables. English]
Parasitism : the ecology and evolution of intimate interactions ; translated by Isaure de Buron and Vincent A. Connors ; foreword by Daniel Simberloff.
p. cm. — (Interspecific interactions)
Includes bibliographical references (p.).
ISBN 0-226-11445-7 (cloth : alk. paper) — ISBN 0-226-11446-5 (pbk. : alk. paper)
1. Parasitism. I. Title. II. Series.

QL757 .C61513 2001
577.8′57—dc21

2001017140

♾ The paper used in this publication meets the minimum requirements of the American National Standard for Information Sciences—Permanence of Paper for Printed Library Materials, ANSI Z39.48-1992.

To my mentor, Louis Euzet

Contents

Foreword

Parasites do not fit comfortably into the modern array of biological specialties. Nonmedical parasitology is a small field; many university biology programs lack even one such specialist. Ecologists all recognize that certain parasites can control the distribution and abundance of certain host species, but a small group of examples dominates textbooks, and few ecologists have much sense of the scope and diversity of parasite biology. Similarly, any evolutionist can cite examples of exquisitely evolved parasites, often highly degenerate ones that have lost entire systems (e.g., nerves, digestive tract) no longer needed in a specialized environment like a vertebrate gut. Yet few evolutionists grasp the diversity of adaptations to a parasitic life or consider how an understanding of this diversity might inform larger issues in evolution. Ethologists all know life cycles of a few parasites in which remarkably complex and stylized behavior allows the parasite to find the host, or at least the host's habitat, at a propitious moment. More recently, some have followed parts of the growing literature on how some parasites modify the behavior of their hosts so as to favor the parasite's transmission. But for most ethologists, as for most ecologists and evolutionists, such examples are just that—isolated, spectacular examples of some natural process taken to an extreme level by a particularly intense interaction between two species.

Medical parasitology, of course, is a large field, especially if one combines bacterial and viral "microparasites" with macroorganisms like tapeworms, trematodes, and ticks. Evolutionists and ecologists have recently made a major effort to inculcate evolutionary thinking into medical parasitology, introducing new perspectives on pathogenicity, epidemiology, and other aspects of the literature on infectious disease (e.g., Anderson and May 1991). But the great majority of clinical specialists in infectious disease, as well as most laboratory researchers, still have not delved into the rich world of parasitic relationships much beyond human pathogens. Nor have ecologists, evolutionists, and ethologists integrated the recent advances in evolutionary medical parasitology into their respective fields.

In short, the biology of parasites remains rather a stepchild in the biological sciences. Claude Combes's *Parasitism: The Ecology and Evolution of*

Intimate Interactions is a landmark attempt to raise the profile of parasites among the literate scientific public, and among biologists in particular. He offers two unifying insights. First, the interaction between parasite and host is often so intimate and persistent that they may together be seen as a sort of superorganism with a "supergenome": each species' genome is expressed within the context of the other species' phenotype. Attempts to understand either component without the other are doomed to failure. Second, parasite evolution, ecology, and behavior can be graphically and elegantly studied in the comparative mode that has served evolutionists so well, through the metaphors of the encounter filter and the compatibility filter. The former is a set of features in the potential parasite and potential host that determine how much individuals of the two species actually come into contact with one another, while the latter is a set of traits that determines whether a pair of species that are in contact will form an intimate, persistent, interacting system. These theoretical insights combine with Combes's breathtaking erudition to present a comprehensive yet easily grasped overview of parasite biology in the framework of ecology, evolution, and ethology. That one author can draw so easily and appropriately on examples from plants and animals, microorganisms and macroorganisms, genes and ecosystems, and the medical and molecular literatures is the mark of a virtuoso scientist. That much of the most interesting material is from works in languages other than English is a bonus.

In fact, anyone who reads *Parasitism: The Ecology and Evolution of Intimate Interactions* may wonder why parasites have received such short shrift among most biologists. Combes speculates that the very subtlety of most parasitic interactions (compared with those between predator and prey) may account for their relative neglect. In parasitism one of the protagonists is often nearly invisible to the other, and perhaps to humans as well. Having so successfully mastered a key trick of their trade—stealth—parasites seem to fly beneath everyone's radar screen. Yet there may also be a different reason we often overlook them. Their very intimacy and subtlety as they produce a devastating impact on the host leads to an almost visceral reaction among those who observe them. Who among us, seeing masses of trematodes embedded in mammalian organs or fifty wasp pupae protruding from the surface of a shriveled caterpillar, can avoid a shudder? Predation may be brutal and epitomize "nature, red in tooth and claw," but at least it is swift and some of its practitioners are the most admired of animals. Athletic teams' sobriquets—lions, tigers, hawks, and rattlers—connote admiration. No school has a tapeworm, trematode, or fungus as an emblem. The use of "parasite" as a derogatory epithet is similarly telling. In short, perhaps the very details of parasitic interactions are so off-putting that they have discouraged close study by all but a few brave souls.

If so, we owe a double debt to Combes for having rendered this vast subject not only comprehensible but palatable. His clear, almost conversational style is leavened by frequent understated humor. For instance, his observation that impacts of the protozoan *Toxoplasma gondii* on human behavior include loss of self-control and increased dogmatism, with a report of a survey of its presence among university department heads, casts a new light on academic governance. In sum, he illuminates the sweeping theoretical and practical implications of a largely misunderstood field.

Daniel Simberloff

Acknowledgments

In addition to the persons I have thanked in the French edition, it is my pleasure to renew my gratitude to the faculty, researchers, engineers, and technicians from the Laboratoire de Biologie Animale at the Université de Perpignan. Additionally, I thank all the colleagues who assisted in untold ways during the preparation of this work, including searching out and finding information and documents for this English-language edition: Greta Aeby, Guennady Ataev, Pierre Bartoli, Ian Beveridge, Christian Biémont, Michel Boulétreau, Jacques Brodeur, Yves Carton, Lawrence Curtis, Domitien Debouzie, Alain Dessein, Claude Gabrion, Kirill Galaktionov, Alexander Granovitch, Michael Hochberg, Pierre Juchault, Finn Kjellberg, Jacob Koella, Armand Kuris, Kevin Lafferty, Catherine Moulia, Teresa Pojmanska, Robert Poulin, Andrew Read, Heinz Richner, Klaus Rohde, Arne Skorping, Louise Taylor, Louis-Albert Tchuem-Tchuente, Richard Tinsley, Tellervo Valtonen, and Ian Whittington. In this area I am grateful to A. Ducreux for her excellent work as librarian in my laboratory.

I would like to particularly thank Isaure de Buron and Vincent Connors, who together not only accepted the difficult task of translating this book into English but provided numerous suggestions and comments that have improved its contents. Heinz Richner kindly provided the photograph for the jacket.

Important parts of the text are descriptions of the adaptations found in parasite-host systems, which tend to lend themselves easily to graphic and pictorial anthropomorphic expressions. However, the use of these terms and descriptors does not imply an anthropomorphic interpretation on my part.

Last, I hope this book will contribute, even modestly, to the intent expressed by Robert Poulin in his *Evolutionary Ecology of Parasites:* "Nothing can be more rewarding to a student of parasitism than the sudden and recent popularity of parasites among ecologists and evolutionary biologists."

Part 1

Who in Whom: The Diversity of Durable Interactions

1 The Universe of Parasites

Parasites affect the life and death of practically every other organism.
Price 1980

The Notion of the Durable Interaction

Living organisms produce four types of wealth: their matter (body, organs, etc.); their metabolism (chemical reactions, enzymes, etc.); their work (movement, care of young, etc.); and the products of their work (nests, houses). These assets are coded for, either directly or indirectly but without exception, by the organism's genome and constitute, in its broadest sense, the phenotype (Dawkins 1982).

As in all cases of riches and wealth, other living organisms often lust after these assets, the most sought after being those stored in the form of the organisms' bodies: each second on earth, in prairies and forests, in soils, in lakes and rivers, and in the deepest seas and oceans, billions of living organisms are preyed on by others who covet energy in this form.[1] These are, of course, the predator-prey systems so often described in ecology texts.

Besides the energy exchange of predator-prey systems, however, energy exchanges also occur because of parasite-host systems: just as the predator takes energy from its prey, the parasite takes energy from its host. But one major difference between predator-prey and parasite-host systems is that in the former the actors are usually visible to each other, with scenes happening before our eyes even when played out at such vastly different scales as a lion capturing a buffalo or a spider trapping a fly. In parasite-host systems one of the protagonists is, at best, difficult to observe. Although there are a few exceptions, parasites are in general

---

This chapter is an introduction to the study of parasites and their mode of cycling in ecosystems. Examples have been limited to include only some major groups, such as the protozoa, helminths, and insects, and are sometimes complemented by referring or alluding to some other types of parasites.

1. "Evolution has always sacrificed the individual for the species, and the species for life. Death was necessary for life to progress. This may be the reason why the human conscience, which appeared at the level of the individual, rejects among evil forces processes that have prepared it during 4 billion years" (Combes 1987).

actors that play in the shadows—perhaps the reason they have been ignored until so recently by mainstream ecologists.

Yet other differences between predator-prey and parasite-host systems become visible when one looks at the level of individuals and how they interact. In predator-prey systems the interaction between the predator and its prey is instantaneous or nearly so (e.g., when a cat catches a mouse, when a fox eats a hen, or when a frog snatches an insect). At most the interaction is only as long as the time of pursuit or waiting and a period of digestion. Moreover, and in virtually all cases, the predator exploits only the first of the four energetic assets of its prey: its matter. In a parasite-host system, that is, when an organism lives on or in another (a tick on a dog, a fluke in a sheep's liver, etc.), the interaction is prolonged and often broken only by the death of one of the protagonists. Because of the length of the interaction, the parasite can exploit not only the material assets of the host but also the other three types of wealth produced by the host's genome: its metabolism, work, and products. Not only is host matter exploited as an energy source, it also becomes a habitat for the parasite.

Although at the level of the biosphere the global transfer of energy from hosts to parasites is negligible when compared with that of prey to predators, the parasite-host interaction may have consequences that reach far beyond the scale of the individual host because of the prolonged and *durable* nature of the interaction. Parasites, more subtly but not less significantly, can affect host populations in profound and significant ways (Anderson and May 1978a, 1978b; May and Anderson 1978).

Although at first one might see little or nothing in common between a virus, a mitochondrion, a trypanosome, a fluke, a tick, and a cuckoo's chick, what unites these organisms and gives them cohesion is the durable nature of the interaction between the one called the parasite and the one called the host. This is true no matter what the systematic position of the partners, no matter the way the interaction is borne out, and no matter even that the benefits and costs of the association may be unequally shared.

Because of the different nature of these prolonged energetically based associations, I have previously proposed the term "durable interactions" (Combes 1995) to describe all such associations between genomes that, either directly or through the intermediary of the phenotype, last for extended periods.

Figure 1.1 symbolizes the expression of the parasite's genome within the context of the host's phenotype and the expression of the host's genome within the context of the parasite's phenotype. Significantly, and because of the prolonged interaction of the protagonists, we can view the parasite-host association as a *system*—meaning the association takes on

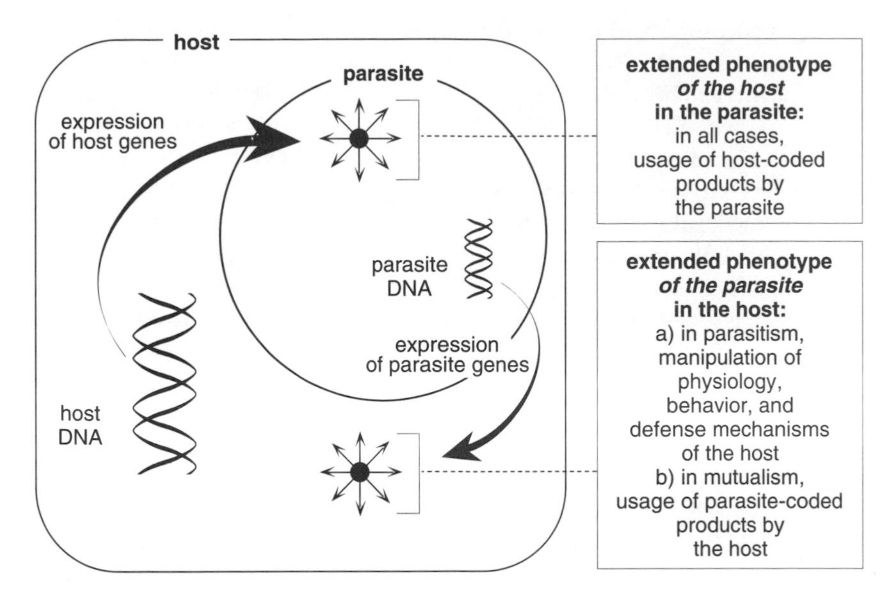

Figure 1.1. The extended phenotype of the host in the parasite and the parasite in the host.

novel characteristics of its own that are *not just the simple sum of its component parts.* In short, these new characteristics are due to the interaction of the *crossed phenotypes* of the parasite and host, each of which may be interfering with the other and each of which is being expressed under the influence of the other's (Combes 2000).

Consequences of the Durable Interaction

The durable interaction has four major consequences, each linked to the fact that there are, side by side, two genomes. This fact makes it, in a sense, a *superorganism* possessing a "*supergenome.*"

First Consequence: The Parasite's Genome Is Expressed in the Context of the Host's Phenotype

To communicate with each other the different parts of a living organism, as well as the different individuals of the same species, use signals (molecules coded directly or indirectly by genes) that are recognized by receptors (other molecules coded for by other genes). All organisms, from the simplest to the most complex in organization and function, use the same fundamental constitutive molecules, metabolic processes, information storage processes, and chemical systems in this communication, and an exchange of signals is made possible between the parasite and the host because of this remarkable unity of life.

In a parasite-host association, the signals produced by the genome of one of the partners may act on the phenotype of the other, thus crossing the species barrier and inducing morphological, anatomical, physiological, or behavioral changes in the recipient. The expression and extension of the parasite's genome into its host's in this way gave Dawkins (1982) one of the most striking illustrations of the concept of the *extended phenotype,* which is a phenotype envisioned to occur beyond the physical limits of the organism to which the genes belong. Among other examples of the extended phenotype, Dawkins cites galls—growths on plants within which certain insect larvae develop. These structures are the result of selection, in the insect genome, of genes able to manipulate the growth of the plant host tissues in their surroundings.

The possibility that a parasite's genome may influence the host phenotype, as in the case of galls, is not necessarily restricted to the action of signaling molecules, however, and in certain "sensitive organs," like those within the nervous system of animals, the parasite's physical presence alone may modify the physiology or behavior of the host.

The influences of a parasite's genes on host physiology and behavior in these ways are sometimes both profound and obvious, and they may favor either host-to-host transmission, the exploitation of the host, or escape from the host's immune system.[2] In some cases, however, the parasite's influence is not very visible, and as a result its consequences may be underestimated. Given that all organisms are parasitized at some time during their life spans, one may wonder if in fact there is even one durable interaction that does not imply a manipulation of hosts, however subtle, by parasites. (Is there even one free-living animal that owes its physiology and behaviors solely to its own genes?)

Second Consequence: The Host Genome Is Expressed within the Context of the Parasite's Phenotype

As soon as a parasite-host system forms and lasts, there is a high probability that some of the genes (or gene combinations) making up the "supergenome" will be duplicated. This does not mean that such genes have an identical nucleotide sequence but rather means that they have similar

---

2. Thompson and Kavaliers (1994) even consider that the study of interactions between parasites and hosts may be very helpful in understanding the relation between animal physiology and behavior on a larger scale.

or overlapping functions. When such *double-usage genes* exist there is a tendency for one of the two genes to stop being expressed over the course of evolution. This may happen either because mutations abrogate the function of one member of the pair, with the system surviving because of the second gene, or more simply because of the cessation of expression of one of the genes in one of the partners. Double-usage genes lost in this way may become pseudogenes (sequences not translated into proteins), or they may ultimately be deleted.

In general parasites and not hosts lose genes, which explains the observed regression of functions otherwise ensured by the host genome in many parasites. That is, the direct consequence of the extension of the host's phenotype into the parasite is the exploitation of the host genome. This is attested to by the fact that all parasite descriptions emphasize morphological simplifications and that parasites save on numerous morphological and functional accessories, such as organs of locomotion, digestive and enzymatic systems, and sense organs because their hosts can take care of the corresponding functions of obtaining food, digestion, and avoiding enemies. An example of organs being lost through this process is the disappearance of the digestive tract in cestodes, while an example of the resulting functional dependence of parasites on their hosts is the lack of the enzymes necessary for purine metabolism in all known parasitic protozoa and some parasitic metazoa, such as the schistosomes. In all these organisms purine metabolism must be ensured by the hosts' enzymes.

What happened to the genes that coded for digestive systems or the enzymes of purine metabolism in the ancient ancestors of current parasites is still an unanswered question. Nevertheless, when such genes are truly lost by the parasite, the survival of the species depends, from then on, strictly on the host. The parasite is "linked" to the host genome and can no longer survive without that host.

Third Consequence: Innovative Genes

Although parasites often depend on host genes for important functions, nothing prevents the host from exploiting the parasite's genome if similar or better genes exist in the parasites. In this case the loss of a double-usage gene may happen in the host, which then becomes as dependent for survival on the presence of the parasite as the parasite is on the host. In this case the parasite is referred to as a mutualist, and the association constitutes *mutualism.*

Thus parasitism and mutualism can be differentiated not only by energetic constraints but also by the fate of double-usage genes: if gene expression stops only on one side, there is parasitism; if different genes on each side stop being expressed, there is mutualism. Cheng (1991) expressed this difference nicely by emphasizing a unilateral physiological dependence in the case of parasitism and a bilateral dependence in mutualism. This distinction was further clarified by Smith (1992): "In parasitism, hosts are exploited by parasites; in symbiosis, hosts always exploit their symbionts, although simultaneous exploitation of host by symbiont may occur in certain associations."[3]

An advantageous situation for the host occurs when the parasite brings not simply a "better" gene but a "new" gene or genes that code for a novel function, such as a new synthetic or metabolic pathway. In acquiring such "innovative" parasite genes the host makes "an evolutionary leap" in the sense that the new function is gained virtually at once and without patient selection and adjustment. Viewed in this way, parasitism is certainly not a marginal process in biological evolution (see Szathmary and Maynard Smith 1995). At least in the first stages of life, and as evidenced by the mitochondrion—the central piece of eukaryotic cell oxidative metabolism and itself an "ancient bacterium"—the acquisition of innovative genes via such a durable interaction likely has played an essential role in the process resulting in the evolution of the more complex life forms themselves.

Obviously, innovative genes coding for novel functions may also have appeared via mutation in one of the partner species after the parasite-host relationship was initiated. In this case the gene may be exploited by both species.

Regardless of the source of the exploited genes, however, selection will not occur under the same conditions in parasitic and mutualistic associations (fig. 1.2).

In a parasitic association the system receives resources from the outside, and both partners are in competition for these resources. This competition puts very strong selective pressures on both the parasite and host, with each constituting separate units of selection. These selective pressures can be far stronger than those pressures between the association and the other units of selection affecting it in the ecosystem.

In a mutualistic association, on the other hand, the selective pressures between the partners becomes less important (although not nonexistent), and the association itself forms the major unit of selection (see Maynard Smith 1989c). In this case the most notable pressure results from the unit's interacting with other entities of selection in the ecosystem.

---

3. Traditionally the European term "symbiosis" is interpreted by Americans to denote "mutualism," and this is the sense Smith uses here.

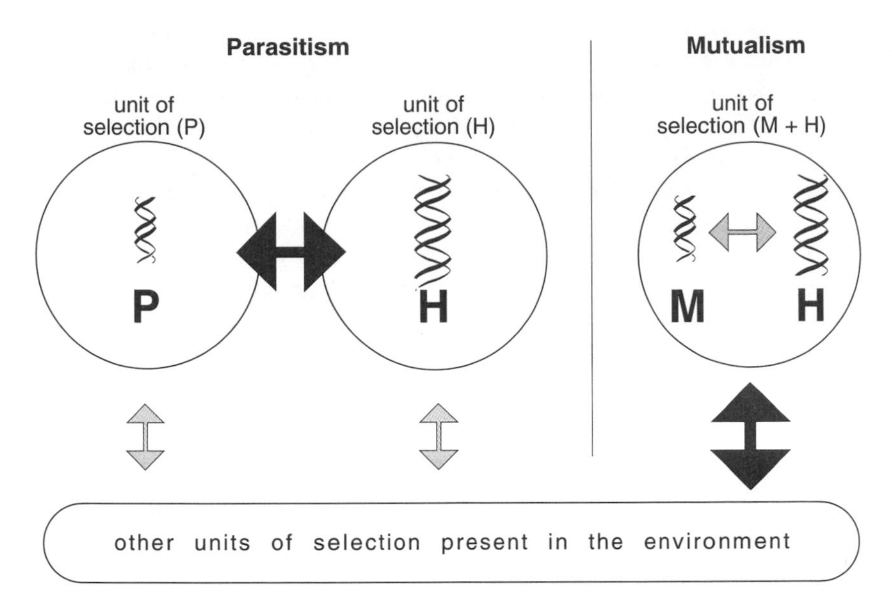

Figure 1.2. Selection in a parasitic (left) and a mutualist (right) system. In mutualism, the host-mutualist system (M + H) constitutes a unique unit of selection. The black arrows indicate the strongest selective pressures. P = parasite; H = host; M = mutualist.

Fourth Consequence: Gene Exchange

The prolonged association between two organisms with different genomes provides an ideal mechanism for the exchange of genes between associated members. Although we know they exist, however, little is known about such transfers in nature. At least in durable interactions of very ancient origin (such as the ones linking mitochondria and chloroplasts to eukaryotic cells), some exchanges of genes have occurred during evolution. It is thought that such exchanges may involve mobile DNA elements, viruses, bacteria such as *Wolbachia,* or still other as yet unidentified mechanisms, and it is suspected that these exchanges may have played a major role in numerous evolutionary processes, one of the major ones being the evolution of the mechanisms by which parasites escape host immune systems.

The Frontiers of the Durable Interaction

Although some workers spend a good bit of time debating whether specific organisms should or should not be classified as parasites, the more encompassing concept of a durable interaction removes all ambiguity and allows us to bring together, alongside parasitism and mutualism, associations such as inquilism (obtaining habitat without taking food), phoresy (obtaining transportation for a set period), and parasitoidism (a durable interaction terminated by the assassination of the host). This concept also allows one to exclude without hesitation organisms whose interaction exists only for the short time of a meal. It is not reasonable to classify as parasitic organisms such as mosquitoes or tsetse flies just because, like parasites, they consume a fraction of an organism without killing it. If we accept mosquitoes as parasites, then we should call the cow that eats grass without killing it, the sparrow that eats cherries without swallowing the entire cherry tree, and all other herbivores, parasites as well. Mosquitoes, tsetse flies, and other insects find their place in parasitology textbooks as vectors of parasites, not as parasites themselves, and it is only when some such species succumbs to the attraction of a habitat provided by another organism, or to taking food exclusively from the same host, or to obtaining a convenient mode of transportation that it would acquire the rights to the parasitic world. Ticks, for example, spend enough time on their hosts so that a typical durable interaction is established—they make molecules that affect the host phenotype by rendering the blood meal both easier and longer lasting: "Tick saliva contains a cocktail of pharmacologically active compounds (e.g. immuno-suppressants, analgesics, anticoagulants, antiplatelet aggregatory compounds)" (Bowman, Dillwith, and Sauer 1996).

As always in biology, however, there are marginal situations that are difficult to classify. Although it is clear that a bee that gathers pollen from flower to flower has nothing to do with parasitism, it is not unreasonable to consider as parasitic the caterpillar of a butterfly who spends its entire larval life on the same nourishing plant.

We should perhaps add another comment about parasitoids, which, after a phase of "classical parasitism" when genomes interact, then kill their host. The best known are insect parasitoids of other insects, but we must also include in this group some viruses (bacteriophages) that, after an initial interaction period, kill their bacterial hosts. Although the end of the association resembles that of predator and prey, the relationship between parasitoids and their victims must be classified among durable interactions because the period of interaction shows numerous facets of the extended phenotype, most notably the forced insidious entry of the parasite genome into the host phenotype.

Last, what fate should we reserve for the word "symbiosis," which literally signifies "life together" and therefore notably applies to the concept of a durable interaction? The word was created at the end of the nineteenth century by the German botanist Heinrich Anton de Bary to designate all associations of two species living together, including parasitism, mutualism, inquilism, and so forth. This is the sense that is always used by certain authors, such as Cheng (1970) and other American workers. Often, however, the appellation is more restrictive and applied only to associations said to be "with reciprocal benefits," here called mutualism. Because of this dichotomous usage, I will make limited use of the term symbiosis, which is also the choice made by Bronstein (1994a): "To reduce semantic confusion, I will use the term mutualism exclusively."

The associations that lead to durable interactions are infinitely numerous in the living world. In fact no organism exists, and none may ever have existed (except at the very beginning of life), that does not establish or is not subjected to a durable interaction with at least one other organism. All living organisms are involved in parasitism, either as hosts or as parasites themselves (Price 1980), and these interactions continue to play an essential and major role in biological evolution and the functioning of the biosphere as we know it.

Why Parasites?

In the "beginning," the ancestors of organisms currently living in association with each other were free-living and not connected. Because in evolution changes most often relate to cost-benefit relationships, we must therefore consider two things: why once free-living organisms found more advantages than disadvantages in adopting the parasitic mode of life, and why free-living organisms that became hosts either were unable to eliminate the parasites or found advantage in such a situation themselves.

To understand why a free-living organism acquires a parasitic lifestyle via selection, one can first consider two free-living species that share the same environment at a given time, then wonder about the advantage that some individuals of one of the two species (call it A) have in taking the second species (B) as their living habitat. It is not enough to say that individuals of species B constitute environments; one must also demonstrate that individuals of species A that possess genes inducing the choice of B as a living milieu benefit from an increased fitness compared with those that remain free-living.

There are three main advantages that individuals of B can give to individuals of A (fig. 1.3): habitat, motility, and energy.

Figure 1.3. What is the advantage of a parasitic life?

Habitat

As long as a species is free-living its members are subject to fluctuations in the environment and to the aggression of other free-living organisms. If certain individuals of this species take individuals of another species for their habitat, we may speculate that they might immediately benefit in two ways: they would occupy a more stable milieu than the exterior environment, and they would be sheltered from the predators and competitors they might otherwise face while free-living.

One cannot contest the fact that a living environment is stable. Living organisms possess mechanisms to ensure homeostasis, or the consistency of a number of physicochemical parameters. Blood composition, for example, is maintained with amazingly little variation, and in many organisms, such as birds or mammals, temperature remains constant. Therefore it is more "comfortable" to live inside a polar bear or desert rat than in the environments where the polar bear and rat live. This may be one reason most parasites have shortened their stay in the external environment as much as possible and also why some have rid themselves of this phase entirely.

That a living environment provides protection from predators, though, deserves more discussion. First of all, this is not necessarily true so long as individuals of A remain outside individuals of B—if they are ectoparasites. It is well known that cleaner fish pick off and eat the ectoparasites of other fish and that cleaner birds peck ectoparasites from the fur of mammals. Whittington (1996) showed that some fish ectoparasites even select for camouflage (i.e., for a transparent body and colored spots) that provides some protection from cleaner fish, and Euzet and Combes (1998) have suggested that the pressures exerted by such predation may in some cases have induced the passage from ecto- to meso- or even endoparasitism in monogeneans.[4] Relative to these worms' habitat these authors write: "If, in a population of a given species, there is a polymorphism of habitat preference, some individuals may remain on the skin while others tend to penetrate inside the body. A process of natural selection can then give rise either to a gradual change in the habitat of the whole population or to two different populations (which may later evolve into separate species)."

In short, then, protection from predators is provided only if individuals of species A find refuge inside individuals of species B. Further, and in an amazing turn but with very few exceptions, parasites are protected from predation in another way: they do not tend to eat each other. Parasites live in a miraculously pacifist world where there are no prey and no predators and where, at the most, bacteria may be absorbed by parasites much bigger than they are. In other words, all parasites are at the same trophic level.[5] As soon as an organism lives on the gill of a fish, in the bladder of a frog, the duodenum of a mammal, or in the tissues of a seed, no other organism will pursue or devour it. In evolutionary terms, it is free to leave aside the apparatus, including sensorial and locomotor organs, that allowed its free-living ancestors to detect and flee enemies. To illustrate the protection given to parasites by this mode of existence, Domitien Debouzie (personal communication) gives an estimation of the mortality rate in the chestnut weevil: as long as these parasitic insects are inside the chestnut the mortality rate is low, but as soon as they reach the ground to complete their development they fall victim to all sorts of predators and mortality increases dramatically (fig. 1.4).

Overall, then there are two clear advantages to having a living habitat: the stability of the environment and protection against predators.

There are, however, two objections to this assessment. First, although the living environment is stable it is also mortal; second, even though the parasite no longer has specific predators, it nevertheless inherits the predators of the host (all the parasites of a mouse should logically be digested when eaten by a cat). There are two responses to these objections:

---

4. Euzet (1989) distinguished the ectoparasites, which are those in direct contact with the outside such that their propagules exit directly to the outside; the mesoparasites, which are those in an organ with communication with the outside so that propagules are released through direct pathways; and the endoparasites, which are in an organ, tissue, or cell without contact with the outside such that the release of propagules is traumatic or problematic.

5. Except, however, for the hyperparasites, which are parasites of parasites.

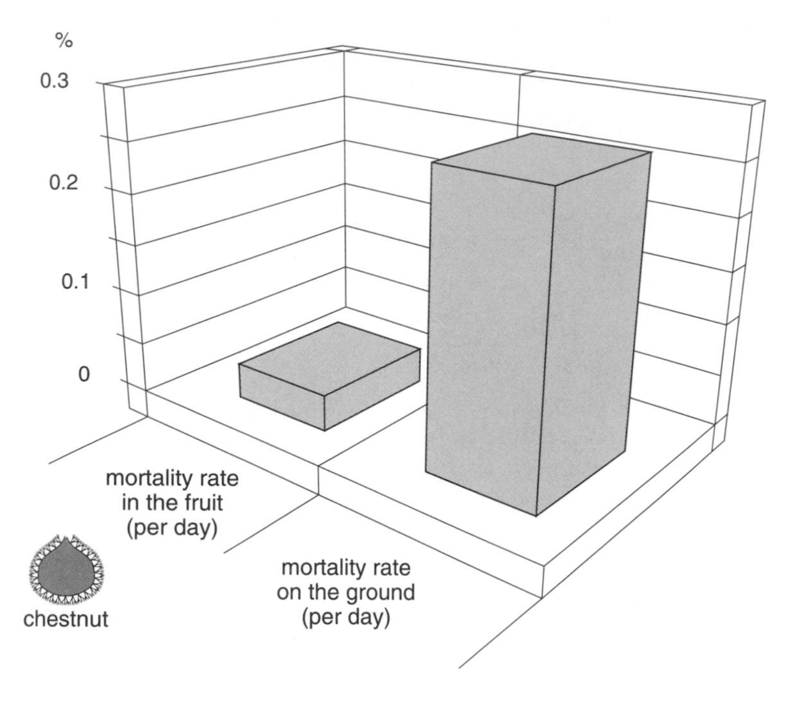

Figure 1.4. Estimated mortality rate of chestnut weevils during their parasitic and free-living stages (original figure based on the data of Domitien Debouzie, personal communication).

Relative to the first objection, to live in a mortal habitat may not be as much of a disadvantage as it at first seems. Parasites themselves are mortal, and therefore establishing a durable interaction would not yield an infinite interaction for the individual. The problem is avoided by parasites' having life cycles that allow their descendants to infect the descendants of their hosts. All living organisms disperse their descendants (seeds get carried away by the wind, birds leave their nests, etc.), and for parasites dispersion out of the founding host is precisely the way to escape the dilemma of host death.

As for the second objection, it is true that virtually all host species (except top consumers) have predators. In each case, however, mechanisms have been selected so that predation does not usually imperil the host species, and parasites are in danger only if the host species is itself in peril. In fact there is even another possibility—that the parasite may take advantage of its host's being eaten by a predator (a mouse eaten by a cat, for example) by surviving in the predator. The parasite in this case wins twice, since it can successively exploit both the prey and the predator (the mouse and then the cat). Evolutionarily, it is then in the interest of the parasite that the cat eats the mouse. Certain parasites have used this as a means to navigate up the food chain and reproduce only in the top consumer, that is, in the host that has the fewest or no predators.

Still another argument in favor of the "security" advantage obtained by occupying a host as a habitat is that in nature any shelter is rapidly exploited by living organisms. For example, rocky coasts and coral reefs harbor a multitude of fixed or not very mobile animals such as sponges, corals, ascidians, and mollusks. These organisms increase the "porosity" of the physical environment, and these new niches are themselves colonized by an impressive list of tenants such as protozoans, planarians, copepods, amphipods, and annelids. Relative to cool ocean waters, the hosts of parasites are luxurious shelters that bring not just safety but also room and board.

Mobility

In many cases (almost all when the habitat is an animal) a living habitat is mobile. Based on this fact we can then ask, Could mobility itself have been the initial advantage obtained by acquiring a living environment?

The dispersion of propagules (a general term designating all forms, such as eggs and larvae, that serve in dissemination) is one of the major parameters of success of living organisms. If in a population certain individuals have a greater capacity to disperse than others, then their or their offspring's chances of survival may be increased, either by an increase in the probability of finding a suitable habitat, by the conquering of new areas, or by the reduction of intraspecific competition. Mobility in this sense is very important for a species, and the less mobile adults are, the more likely it is that some elaborate means of dispersion for its propagules has been selected. For example, plants or fixed aquatic animals ensure the dispersion of their gametes or planktonic larvae by developing various appendages and lift devices, such as bristles, blades, or umbrellas, that allow them to be carried away by the wind or to drift with the currents. The classical examples of such dispersal mechanisms are of course plants, which ensure the dispersion of their pollen by insect pollinators and their seeds by vertebrates (i.e., by hooking on to the fur or passing through the digestive system). When an organism is not very mobile it can improve the dispersion of its propagules by making them more mobile than itself.

Associating oneself durably to a mobile organism, as opposed to associating one's pollen or seeds, amounts to selecting an even more efficient mechanism of dispersion: instead of hooking propagules on to a vehicle, their producer (the adult organism) attaches itself to the vehicle and thus disperses its young throughout the host's travels. All parasites in the digestive tract of a mammal have their eggs or larvae dispersed more or less constantly in this manner as the living vehicle travels: all bird parasites fly with their hosts, and all fish parasites swim with theirs. This could explain why entire groups of parasites are restricted to mobile animal hosts and why they tend to disdain immobile plant hosts. When parasites do exploit immobile hosts such as plants or sedentary animals, their dispersion is ensured by different processes according to the group. For example, in insects parasitic of other insects and crustaceans from the monstrillid group, only the larvae are parasitic: in these cases dispersion is ensured by mobile adults, whereas in helminths with heteroxenous life cycles one of the hosts is often not very mobile and another acts as a vector to ensure dispersion.

The advantage of host mobility for dissemination and the resultant outcrossing of parasites is testified to by observations such as those of Bursten, Kimsey, and Owings (1997), who showed that male fleas prefer to parasitize very young squirrels *(Spermophilus beecheyi)* that will soon be dispersing away from their maternal nest.

Energy

Paradoxically, only prokaryotes are true autotrophs—making complex biomolecules either by capturing the energy of the sun (photosynthesis) or by harnessing it from naturally occurring mineral compounds in a process called chemosynthesis. When eukaryotes are photosynthetic, they owe their ability to synthesize their own food to their *association* with chloroplasts, themselves almost certainly distant descendants of once free-living prokaryotic organisms.

When one is big and not autotrophic, the most natural way to obtain energy is to swallow something smaller than oneself. When one is small and also not autotrophic, one may certainly look to devour something smaller than oneself, but a more elegant solution is to eat, without destroying it completely, something bigger.

Parasites either hijack part of the food of their host or feed on their host's nutritious tissues and fluids, which allows the parasite to decrease the allocation of its own resources to nutritional functions and may lead, for example, to a simplification or loss of the digestive system. In short, *the parasite can be simple because it exploits complicated hosts.*

The Sum

The question stated earlier was, What adaptive advantage did free-living organisms obtain by engaging in a parasitic lifestyle? Furthermore, between habitat, mobility, and energy, is it really possible to make such a choice?

In the beginning of a durable association, habitat, mobility, and energy may each have, in theory, represented a distinct advantage. Later one or both of the other advantages may have been added to the package, with most parasites today taking advantage of all three at the same time, courtesy of their hosts. Taking food becomes, for example, an almost immediate consequence of the association simply because the usual prey of the newly associated organism becomes less accessible or even unavailable and because the new habitat provides a readily available source of food that does not require spending energy to forage.

Parasitism may even be ineluctable, as suggested by the works of the computer scientist Tom Ray.[6] Ray built a series of programs that compete with each other, that reproduce, and that are subject to random mutations in each generation just as in living organisms. Since there was competition in the system, the "best" programs prospered at the expense of others. In an amazing but not at all surprising way, the system quickly gave rise to true parasites—including shortened programs that borrowed missing instructions from others. Even more remarkable, the programs that remained normal (the hosts) rapidly developed new instructions protecting them from the parasitic programs, which in turn modified themselves to elude this host strategy. Thus all the ingredients for a true parasite-host association occurred, including reciprocal selective pressures. This leads one to wonder if parasitism, as much as reproduction or diversity, belongs to the definition of life. Competition for resources, whether matter, energy, space, power, or even computerized information, incites parasitism.

How Parasites?

If the conveniences of a parasitic life seem multiple, the inconveniences are also real. Nevertheless, what we sometimes see as difficulty in the parasitic lifestyle may be an anthropomorphic and naive vision of the problem. There is no doubt, however, that whatever the apparent disadvantages of a parasitic lifestyle, the cost-benefit ratio of a successful interaction is such that the advantages have won out over the disadvantages. That is, an organism becomes parasitic the same way others become marine, terrestrial, or cave dwelling—by providing, generation after generation, the possibility for natural selection to sort the best adapted from the less fit.

---

6. Ray's work, done at the University of Delaware, is summarized by Ridley (1994).

For parasites to adapt, they must respond principally to four essential characteristics of the hosts: their discontinuity in space (host individuals are equivalent to hospitable islands dispersed over an inhospitable ocean); their discontinuity in time (hosts are mortal, as though the islands were to sink); their hostility (hosts attack parasites, most notably with their immune system); and their evolution (host populations modify their own nature over time).

Discontinuity in Space (the Host as a Fragmented Habitat)

The overall population of a given parasite is divided into as many fragments as there are infected individuals, with each host serving as a habitat distinct from other host individuals. As a result, individuals of a given parasite species form an "infrapopulation" within each infected host. Life in such a fragmented and discontinuous habitat is not peculiar to parasites, and many places such as groves, hedges, springs, and caves form similar discontinuous populations between which the exchange of individuals (and therefore genes) is more or less frequent. Between two groves that shelter populations of tits, for example, exchange may occur because of "emancipated" juvenile or adult tits. Exchanges between parasite infrapopulations also occur by such emigrations and immigrations. But in this case propagules, not migrations of juveniles or adults, ensure host-to-host transmission. Interestingly, it is unusual for a parasite that has acquired maturity in an individual host to leave and parasitize another. This happens only with certain ectoparasites and, in rare cases, where parasites survive host cannibalism (Buron and Maillard 1987; Maillard and Aussel 1988).

Although hosts clearly represent discontinuous resources, such a discontinuity in space nevertheless comes with a certain degree of both reproducibility and variability. Reproducibility occurs because different individuals of the same host species have important resemblances, such as their "topography" (organ placement and relative dimensions). Such reproducibility led Holmes (1987) to state that "animals belonging to the same host species provide the most similar replicates of habitat that a field ecologist is likely to encounter." Similarly, Price (1987) concluded that "the host microcosm is replicated in time and space much more than habitats for most other organisms." Variability of habitat occurs because, on a finer scale (most notably biochemical) no individual host is exactly like another (save for rare exceptions where host populations are clones). This variability plays a large role in the overall distribution of parasites, since not much difference is required for an individual host to be more susceptible to parasitism than its neighbor.

As a general rule the multiplication of the infrapopulation on site is normal for viruses, bacteria, and protozoans. It is rare for metazoans, however, and in this case recruitment—the acquisition of new individuals for the infrapopulation—is usually done from the outside.

Discontinuity in Time (Hosts Are Mortal)

Discontinuity in time is added to the discontinuity in space because hosts are mortal, with even those individual hosts that have high longevity (humans, for example) being at best only temporary habitats. Like life existing in fragmented environments, life in such temporary habitats is not peculiar to parasites. Ponds that appear in the rainy season in tropical areas, the microaquariums that form in tree hollows, and "cow pies" are also habitats with limited life spans. Like free-living organisms in such ephemeral habitats, parasite populations survive the death of their individual habitats by dispersing their progeny to the outside, with descendants being "responsible" for finding similar host habitats for themselves. This is in fact the origin of parasitic life cycles, a frequent characteristic of which is their complexity. The parasitic organism can often find a milieu similar to that of its direct ancestor only after passing through and exploiting a series of habitats.

Hostility of the Milieu (Hosts Attack Their Parasites)

From the point of view of an immunologist or a medical practitioner, a parasite is an aggressor against which the host must defend itself. The host acts against this aggressor by using its defensive systems, which without a doubt have evolved under the direct pressure of pathogenic organisms.

From an ecological or a zoological perspective, however, the parasite does not attack its host, and in fact the opposite occurs. That the parasite takes its food on site and induces some pathogenicity does not make it a killer. In this sense the parasite exploits its milieu without "knowing" that the milieu is alive, and it has no other "goal" than to reproduce. In fact a parasite is counterselected each time the cost of its pathogenicity decreases the success of transmission because the host dies (see chapter 6). The parasite is therefore a killer only if the death of the host favors, either directly or indirectly, parasite fecundity and transmission. The host also has no other goal than reproduction, and if the presence of the parasite decreases its fitness it is in its interest to kill the parasite. However, mechanisms of parasite destruction will not be selected for in hosts in two cases: if the host can take advantage of the intruder's presence, or if defending itself is more expensive than tolerating the infection. Apart from these exceptions, the host attacks if it can damage the parasite.

A parasite's search for a host is, then, nothing other than a particular case of the general rule that all living organisms seek a habitat to establish themselves in. The difference, as noted by Hanley et al. (1996), is that "in parasite settlement, the habitat (host) is able to fight back." Parasites are therefore obliged to defend themselves and as a result have often selected elaborate strategies that differ according to species and taxa, all usually grouped under the term "immune evasion."

Evolution of the Milieu (the Hosts Modify Themselves)

Any environment where life is found changes with time: lakes, forests, and caves do not remain the same indefinitely. On a global scale, the planet has evolved: the chemical composition of the atmosphere and that of the oceans are not the same today as they were 3.5 billion years ago when life appeared. Important climatic cycles such as those that caused glaciations in the Quaternary period also occur over time, and it is the fate of all species to adapt to such changes or perish.

Hosts for their part have evolved constantly because of such things as the occupation of diverse but limited habitats and competition with other species. Over time some phyla have diversified greatly; species have appeared and others (sometimes entire groups) have died out. As a result parasites have also been confronted with changes of different kinds over time. For instance: the host species they inhabit may have been subjected to modifications in anatomy, physiology, behavior, and life history traits (longevity, fecundity, etc.); their host species may have exploded into several genetically distinct populations and eventually into several reproductively isolated species; and changes in the composition of the fauna linked to the evolution of other phyla or to migrations may make local species of hosts occasionally inaccessible or, on the contrary, offer an opportunity to conquer a new host.

In response to these various changes in their living habitats, parasites (like all other life forms) have no "choice" other than to adapt or disappear, and they have apparently adapted more often than they have disappeared. To adapt they themselves have evolved and, far from being at a disadvantage, often pull some benefit from the new possibilities presented by the evolution of their surroundings. The recent evolution of the hominids, for example, has offered extraordinary opportunities of conquest to the parasite world.

Poulin (1996) attempted a first synthesis of the evolution of life history strategies in parasitic organisms by analyzing those host-related factors (age, size, diet, immune status, etc.) thought to affect parasite traits (longevity, size, fecundity, etc.). In effect, "given that hosts provide the principal environmental features parasites have to deal with, and given that these features are well characterized by the life history of the host, we may expect natural selection to result in covariation between parasite and host life history" (Sorci, Morand, and Hugot 1997).

Comparative analysis has opened a new area of research to explore this issue. Probably the most interesting relationships to study are those that link longevity, fecundity, and size. The small body size of parasites usually leads to high fecundity, but the benefit of increased fecundity is expressed only if adult mortality is low, and adult mortality often depends on host longevity. Like Brooks and McLennan (1993) before him, Poulin (1996) denounced some "myths" about parasites (for example, "parasites are smaller and simpler than their free-living ancestors"), and the work of each has shown that parasites, like free-living organisms, have adapted via natural selection only in ways that maximize their own fitness.

Who Are Parasites?

Parasite Groups

The parasitic DNA sequences observed in all genomes and—on the opposite end of the scale—the social parasitism observed in various groups of metazoans, are demonstrations that parasitism addresses all levels of life, from molecules through ecosystems.[7] This recognition has led Anderson and May (1979) to distinguish the "microparasites" (viruses, bacteria, fungi, protists) from the "macroparasites" (helminths, arthropods, and other metazoans).

In general, *microparasites* are small; multiply within the host; induce a durable immunity; have unstable populations; and are responsible for epidemic diseases. *Macroparasites,* on the other hand, are large; do not as a rule multiply within their host; do not induce a durable immunity; have stable populations; and are responsible for endemic diseases.[8]

---

7. The most recent discovery of a new phylum concerns a parasite (Funch and Kristensen 1995). This phylum is the Cycliophora, which is represented by *Symbion pandora* (a small organism less than 1 mm long that lives on the mouthparts of the Norwegian lobster, *Nephrops norvegicus*). The parasite, which could not be placed in any other known phylum, forms by asexual budding larvae that are fixed in place. Sexual reproduction also occurs, and fertilized females release numerous larvae that are good swimmers and seek new hosts.

8. Anderson and May's exact definitions are as follows: "*Microparasites* (viruses, bacteria, protozoans) are characterised by small size, short generation times, extremely high rates of direct reproduction within the host, and a tendency to induce immunity to reinfection in those hosts that survive the initial onslaught. The duration of infection is typically short in

Thus parasites run the gamut from molecules through the most complicated of organisms and are represented in every known taxonomic group on the planet. One may then ask, What relationship exists between parasitism and the global structure of the living world?

One aspect of this relationship concerns the ratio of parasitic to free-living species within groups. For instance, there are taxonomic groups of organisms that are composed totally of parasites, as with viruses, sporozoans, and acanthocephalans; groups almost totally lacking any parasitic forms (for example, the echinoderms);[9] and groups constituted in major part by free-living forms and in minor part by parasitic forms (extreme cases include the cniderians, which are all free-living except in rare cases such as *Polypodium hydriforme,* which parasitizes sturgeon eggs and thus constitutes a serious threat to caviar producers—see Raikova 1994). There are also groups constituted in large part by parasitic forms and in small part by free-living forms. The Platyhelminthes are classically cited in this regard because, as a group, they include several classes of only parasitic forms (monogeneans, trematodes, cestodes) and one class (Turbellaria) of both free-living and parasitic forms. Last, there are groups constituted of both numerous parasitic and numerous free-living forms, as is the case in the bacteria, fungi, and nematodes.

Another aspect of the relationship is that there are also cases of "reciprocity." For example, there are bivalves that parasitize fish and fish that parasitize bivalves: the larvae (glochidia) of freshwater mussels in the family Unionidae develop on the gills of cyprinids, attaching themselves by valves specially adapted for that function. On the other hand, some fish parasitize, or at least exploit, the shelter that mollusks provide: examples include pearlfish, which live in pearl oysters; *Liparis,* which lives in scallops during its preadult phase; and *Rhodeus,* which lays eggs in freshwater mussels.

Parasitism of a Close Relative

Parasitism between members of the same taxonomic group is also not at all exceptional, with parasitic associations between related angio-

---

relation to the expected lifespan of the host, and therefore is of a transient nature (there are, of course, many exceptions, of which the slow viruses are particularly remarkable). *Macroparasites* (parasitic helminths and arthropods) tend to have much longer generation times than microparasites, and direct multiplication within the host is either absent or occurs at a low rate. The immune responses elicited by these metazoans generally depend on the number of parasites present in a given host, and tend to be of relatively short duration. Macroparasitic infections therefore tend to be of a persistent nature, with hosts being continually reinfected."

9. However, some species of ophiurids have larvae that are incubated by other species of ophiurids.

sperms,[10] medusae, crustaceans, and insects being common. Some surprising examples include the red algae, of which about a fourth of the known species parasitize other red algae. In this case the cells of the parasitic algae inject their nuclei into the cytoplasm of the host cells, and the parasitic genome replaces the host genome. Studies of both nuclear and mitochondrial DNA suggest that the parasitic algae are close relatives to their red algae hosts and that they may even be their direct descendants (Goff 1991). This relationship is in fact so narrow that it has been proposed that it could only be a perversion of fertilization, with the male nucleus parasitic of the female cytoplasm.

Another such example includes fungal parasites of other fungi, and as in the red algae, kinship between the parasite and host is often very close here as well. This is particularly intriguing, moreover, because it seems that autoparasitism may occur even within the same fungi, with some mycelial filaments behaving like parasites of other filaments of the same "mycelium individual."[11] Similarly, lichens parasitize other lichens, bringing together up to four associated genomes at one time.

Yet another example of "like parasitizing like" includes fish that are parasites of other fish. Here small fish from the family Trichomycteridae live in the branchial cavities of other fish where, like vampires, they absorb blood from their hosts' branchial filaments. These fish are also occasionally parasitic of humans; they are attracted by currents and penetrate the urethra when humans urinate while swimming. Most surprising, however, is the catfish *Synodontys multipunctatus* from Lake Tanganyika. The fry of this fish develop in the mouth of various cichlids that practice buccal incubation. The parasitic fish somehow places its eggs in the mouth of the host fish along with the host's own eggs. Then the parasitic fry get rid of the host's young by eating them and grow in their place (Sato 1986).

Last, some sharks may harbor a parasitic eel in their hearts, at least in a transitory way. Caira et al. (1997) reported that a shortfin mako *(Iurus oxyrhynchus)* weighing 395 kilograms, which landed at Montauk, New York, in 1992, harbored "two dead, but otherwise healthy appearing pugnose eels, *Simenchelys parasiticus,* within the lumen of its heart." These eels were each about 25 centimeters long, and although no lesion of penetration was obvious, Caira's team hypothesized that the eels might have pen-

10. But there is only one known case of a gymnosperm parasitizing another gymnosperm (see Woltz et al. 1994).

11. In some cases such an effect may be in relation to the enormous dimensions a fungus may attain. For instance, molecular techniques have shown that the mycelium of one individual *Armillaria bulbosa* may cover up to fifteen hectares, weigh ten tons, and be over fifteen hundred years old (Smith et al. 1992).

etrated at the level of the gill apparatus. The exclusive presence of blood in the eels' stomachs and the existence of a pathogenic effect in the heart suggest that *S. parasiticus,* which is sometimes captured free-living in the deep seas, also sometimes acts as a facultative parasite.

Parasitism is also not rare in insect societies, and here we often find very close kinship among the protagonists. For example, colonies of ants of the genus *Lasius* are parasitic of colonies of other species of *Lasius,* and colonies of *Plagiolepsis* are parasitic of other *Plagiolepsis* colonies (see Jaisson 1993).

The biology of the hymenopteran parasitoids of the family Aphelinidae (see Walter 1983) is particularly intriguing in that male eggs develop only if laid in female larvae of the same species, which themselves are parasitoids of homopterans. Female larvae are devoured alive by the male larvae, implying the following scenario: aphelinid females lay their female eggs (the fertilized ones) in the larvae of homopterans and then lay the male eggs (nonfertilized) in the female larvae of their own species. Obviously the system works only because not all the female larvae are sacrificed to the benefit of the males.

Parasitic Molecules

Some of the parasitic associations just described, far from being curiosities, nourish serious and difficult reflections about evolutionary processes. It is even more so concerning parasitic DNA. That molecules of nucleic acids can be parasites, as much so as a liver fluke but at a different scale, is not a familiar concept for either molecular biologists or parasitologists. To understand the notion of parasitic DNA one must first recognize that there are "naked" nucleic acid molecules that are not integrated into host genomes. This is, for example, the case of plasmids—small extrachromosomal circular nucleic acids harbored by bacteria that often carry genes conferring bacterial antibiotic resistance and that, at the same time, reduce their hosts' reproductive efficiency in their absence. One must also cite the viroids, which are naked RNA molecules a few hundred bases long that cause various diseases in plants, and the pseudochromosomes (or B chromosomes) that occur in the cells of numerous host organisms. These pseudochromosomes sometimes interfere with real chromosomes (or A chromosomes) and impair both fecundity and survival.[12]

12. B chromosomes differ from A chromosomes because they are not necessary for the survival of the individual or needed for transmission of genetic information, and because their number is inconsistent in individuals of the same species. Such B chromosomes are frequent and may parasitize up to 15% of all known eukaryotic organisms.

Although the DNA molecule was thought in the 1960s to be composed of genes nicely lined up one after the other with each coding for a well-defined protein, research soon showed that part of the DNA present in the genome of eukaryotes had two worrisome characteristics: an instability of position and an absence of obvious function. These aberrations, originally interesting only to molecular biologists, little by little piqued the curiosity of parasitologists, ecologists, and evolutionary biologists alike. Could these DNAs be, like some living organisms, simple in their structure but obeying the rules of durable interactions? Would some carry advantages for the host molecules and others carry disadvantages? Could they be *parasites* whose benefits would be the exploitation of the replication machinery of the host genome? Such suspected parasitic DNAs are of concern for humans because of their abundance and because evidence has accumulated that they may be responsible for several serious health problems.

Three categories of DNA molecules suspected to be parasitic can be distinguished: mobile elements, repeated sequences, and pseudogenes. Mobile elements are fragments of DNA that can move or transpose themselves into various locations in the genome, and several dozen families of these elements can be counted: they constitute about 10% of the total genome in fruit flies and are widely represented in humans.

The evidence that genomes are not stable and that they contain such *mobile elements* began to accumulate in the 1940s with the work of Barbara McClintock at a time when molecular biology was not yet possible. Working with corn, she showed that the results obtained when some strains were crossed could be explained only if some DNA fragments "jumped" from one place to another in the genome, modifying the expression of the genes affected. More than thirty years later the confirmation of the existence of such "mobile elements" or "transposable elements" (small DNA fragments capable of detaching and reinserting themselves) quickly raised the question, Could these little fragments be authentic parasites, multiplying at the expense of the host genome?

Mobile elements present themselves as DNA sequences that range in size from three hundred to a few thousand base pairs long, and their transposition can occur in two ways: either as DNA (transposons) or by way of an RNA intermediate (retrotransposons). Retrotransposons differ from transposons because retrotransposons contain a gene for a molecule, reverse transcriptase, whose job it is to reverse transcribe RNA (itself originally transcribed from the retrotransposon) back into a double-stranded version of the element for insertion elsewhere in the genome. Two types of insertion exist in eukaryotes, differing from each other by the presence or absence of direct (noninverted) long terminal repeats (LTRs) at each end. Those elements that have LTRs are termed viral

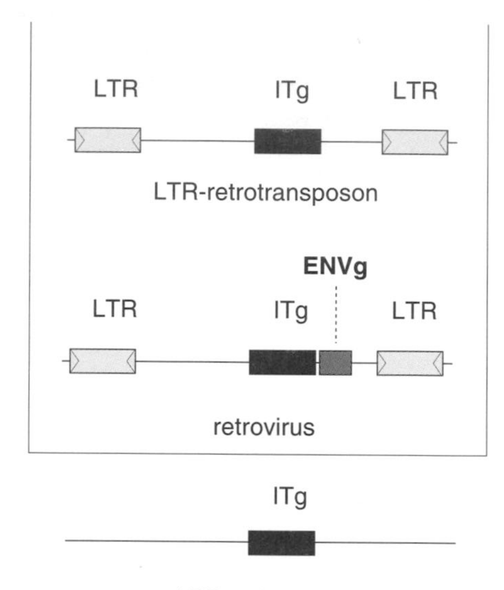

Figure 1.5. Diagram showing the important similarities existing between retroviruses, which are authentic genome parasites, and a retrotransposon (mobile element), which differs from retroviruses only by the absence of a coding sequence for a viral envelope. ENVg: gene for the envelope; ITg: gene for reverse transcriptase; LTR: long terminal repeats.

retrotransposons for their distinct structural relationship to retroviral DNA (fig. 1.5), while those that lack LTRs are denoted as nonviral retrotransposons (Lewin 2000).

Because retrotransposons first produce an obligatory copy of RNA from the element using RNA polymerase and from this RNA then make a copy of DNA (using reverse transcriptase), the multiplication of these mobile elements is indirect. Retrotransposons are known from numerous eukaryotes (animals and plants) but not from prokaryotes, and it is important to note that nonviral retrotransposons (those without LTRs) are found in thousands of copies in the human genome.

Transposons, on the other hand, do not have reverse transcriptase and are characterized by inverted terminal repeats at both extremities. This mobile element codes instead for a transposase, which allows it to detach and then reinsert itself somewhere else. This transposition can occur in two ways. Either the element can leave its place on the DNA and reinsert itself at another location in a process called excision (nonreplicative or conservative transposition), or it can copy itself and insert the copy elsewhere (replicative transposition). The multiplication of transposons is therefore direct and in the form of DNA without passage through a copy of RNA. Transposons are frequent in prokaryotes but are also frequent in eukaryotes (both animals and plants).

LTR retrotransposons show many similarities to retroviruses, which are considered without doubt to be parasitic. Viruses differ from LTR retrotransposons, however, because they have genes that code for proteins that form viral capsides (fig. 1.5). Retrotransposons, on the other hand, differ from retroviruses because they have no known "viral" forms able to ensure their transport from cell to cell. This is so true that a sequence initially identified as "gypsy" was found to be, in fact, a retrovirus (Kim et al. 1994). It is almost certain that retroviruses derive from retrotransposons or, vice versa, that retrotransposons derive from retroviruses.

The presence or absence of a gene for reverse transcriptase in parasitic DNA corresponds to an important functional difference: when the mobile element can make reverse transcriptase, there is an intermediate RNA that is reverse transcribed into DNA before insertion.

The occurrence of *repeated sequences,* however, presents a different problem for the parasitologist. Molecular biologists use the term "C value" to refer to the size of the haploid genome of a living species. The C value is characteristic of each species but varies widely from one species to another, even within a given taxonomic group. The C value is measured either in the number of base pairs or in kilodaltons, with one kilodalton being equal to 1.62 base pairs.

To give an idea of the size of these values, here are some measurements presented by Li and Graur (1991): the bacterial genome has between 650,000 and 13 million base pairs (4 million for *Escherichia coli*); that of invertebrates has between 50 million and 22 billion (180 million for *Drosophila melanogaster*); and that of vertebrates has between 400 million and 140 billion (3.4 billion for *Homo sapiens*).

Inexplicably, the organisms that appear to have the highest C values are certain amoebas (with up to 670 billion base pairs, or two hundred times the number found in humans). This simple example shows that the size of the genome is not proportional to the organisms' degree of complexity. In this context parasites seem "normal": about 30 million base pairs in *Plasmodium falciparum* and 270 million in *Schistosoma mansoni.*

The huge difference between the genetic information supposedly necessary to build an organism and the size of its genome is referred to as the paradox of the C value. It is reasonable to assume that the number of genes present in an organism is more or less correlated with the complexity of the species. The amoeba, however, probably does not need two hundred times as much genetic information as the human to function and build itself. The only explanation for the paradox of the C value is that there is noncoding DNA that does not contain genetic information and that the abundance of this type of DNA varies considerably according to species.

In fact, such noncoding DNA may constitute between 30% and 100% of the genome of some organisms (Cavalier-Smith 1985).

The term "repeated sequences" defines all sequences of a genome found in more than one copy. There are two very distinct subensembles of such sequences: on the one hand, numerous functional genes are present in several copies, thus assuring the cell an abundant production of their corresponding messenger RNAs. On the other hand, however, there are repeated sequences that are not transcribed into RNA and therefore not translated into proteins. These are repetitions, sometimes very numerous, of generally uniform "patterns" that may range in length from only two or three base pairs to over several hundred. For example, the sequences AAG or TTAGGG and ACACAGCGGG are repeated 2.4, 2.2, and 1.2 billion times, respectively, in the genome of the kangaroo rat, *Dipodomys ordi* (see Li and Graur 1991).

These "nonsense" sequences either are placed one behind another (in tandem repetition) or are scattered throughout the genome. As noted earlier, these sequences may constitute a large proportion of the genome.

Pseudogenes are the last category of nucleic acids that may be suspected of being parasitic. These are unexpressed genes or copies of genes that thus escape selective constraint and that therefore may accumulate mutations and drift considerably from their initial sequence. These copies of genes may have different destinies: they may all stay functional and identical to themselves, remain functional but diverge in their composition and function and thus form "gene families," or cease being functional and thus become inactive pseudogenes. It is the third eventuality that we are most interested in here, since there are two possible explanations for pseudogenes depending on the origin and, in fact, the quality of the copy. The first explanation is based on the fact that any gene may be duplicated and the copy will in general be very accurate. Therefore there is no a priori reason for the copy not to be expressed. However, there are other events (mutations, deletions) that may induce these copies to be aberrant, and from then on their sequences will only degrade. The second possible origin of a pseudogene is via the reverse transcription of messenger RNA. In this case it is not the DNA template that duplicates. Rather, a copy of DNA is made from RNA using reverse transcriptase. Such copies are usually of bad quality and become "processed pseudogenes," easily identified because the sequences corresponding to the introns of the initial genes are excised during RNA processing. In both cases careful comparison of numerous sequences will identify the functional gene that is at the origin of the pseudogene, so long as any subsequent drift has not been too great.

Most curiously, it is among the "processed pseudogenes" that we find certain sequences becoming extremely numerous. For example, in the human genome a sequence named *Alu* (after the restriction site *Alu1* contained within it) is three hundred base pairs long and is repeated 500,000 times in each cell of our bodies. All in all, *Alu* may constitute more than 5% of the total human genome. The sequence *Alu* is therefore a pseudogene multiplied almost to infinity and is derived from a gene coding for a cytoplasmic RNA that intervenes in the orientation of proteins toward the endoplasmic reticulum.

Reverse transcription is therefore potentially dangerous for genomes, because imperfect copies of functional genes may be made any time reverse transcriptase is present and because these copies may then "fall back" anywhere within the genome.

Parasitic Transmission

Biological Cycles and Their Elucidation

The propagules of a living organism survive only if, either actively or passively, they find or end up in a favorable habitat. What characterizes parasites is that this accomplishment often requires several changes of habitat.

A change in habitat over the lifetime of an organism is not unique to parasites. An anuran amphibian lives successively first in water as a "tadpole" and then as a terrestrial "tetrapod." Similarly, a mosquito is first aquatic and then aerial, and even mammals, including humans, live successively first in the aquatic milieu of the uterus, where the fetus behaves like a parasite of the mother, and then in the terrestrial habitat.

It seems difficult, however, to beat parasites for the complexity of the cyclical changes of milieu they undergo (which may include successive hosts, the transfer between successive sites within hosts, and passage through the exterior environment as well). In most parasites the same genome must therefore be capable of building a series of six or seven "successive organisms," each different from the others and each able to *discover* and *exploit* different habitats: "There is a difficult task of switching off and turning on certain metabolic activities in a parasite upon being transmitted from one host to another" (Wang 1991).

Today we know most of the modes of transmission used by the various groups of parasites. However, we still have major gaps at the level of the species simply because the experimental transmission of biological cycles is almost always a difficult, if not impossible, enterprise. How can one elucidate, for example, the life cycles of the parasites of a large marine cetacean that cannot easily be kept in an aquarium?

The techniques of molecular biology allow a new approach to this problem, the only constraint being the initial building of a limited number of hypotheses based on other available data. Such starting hypotheses may be elaborated from taxonomic and biological data obtained from known cycles of closely related species. An excellent example of this approach is given by Jousson et al. (1998), who comparatively analyzed the DNA of a series of adult trematodes of the family Mesometridae from the digestive tract of marine fish and the DNA of a series of cercariae showing the general characteristics of the family Mesometridae (top of fig. 1.6).

In this study the length of the ITS1 + 5.8 S + ITS2 fragment obtained via polymerase chain reaction (PCR) allowed an initial sorting between the adult mesometridean parasites of fish and the larval-stage parasites from the molluskan intermediate hosts (center frame of fig. 1.6). The sequencing of the ITS1 then showed that two species of adult trematodes *(Mesotrema brachycoelia* and *M. orbicularis)* did not have cercariae in the initial sample, that the worm *Wardula capitellata* was obtained from cercariae given off by *Barleeia rubra,* and that *Centroderma spinosissima* was coming from cercariae given off by three species of *Rissoa* (bottom frame of fig. 1.6).

Transmission by Contact

Some parasites take advantage of contacts between hosts of the same species to pass from infected to noninfected individuals. Among the numerous parasites that may be cited in this context are the head lice *(Pediculus capitis)* that pass first from one child to another at school and then from child to family at home; the scabies mite *(Sarcoptes scabiei)* that excavates galleries beneath the skin of numerous mammals; and *Gyrodactylus,* a group of fish monogeneans that have no swimming larvae.

Parasites that are transmitted during host mating also take advantage of direct contact. For example, juveniles of the monogenean *Isancistrum subulatae* (a parasite of squids) are born on an infected squid and transferred to the sexual partner when their tentacles entangle during copulation (Llewellyn 1984). Similarly, the L3 larvae (those that have already gone through two molts) of the nematode *Nemhelix bakeri,* which as adults parasitize the genital tracts of the snail *Helix aspersa,* transfer into the genital tract of the sexual partner during copulation (Morand 1988). The mite *Demodex flagellurus* lives in the sebaceous glands of the male mouse's prepuce and female mouse's clitoris (Bukva 1985) and infects the partner during copulation. In humans the same type of contact is used by viruses such as HIV, the causative agent of AIDS, by protozoans such as *Trichomonas vaginalis* (causative agent of trichomoniasis), and by arthropods such as the body louse *Phthirus inguinalis.*

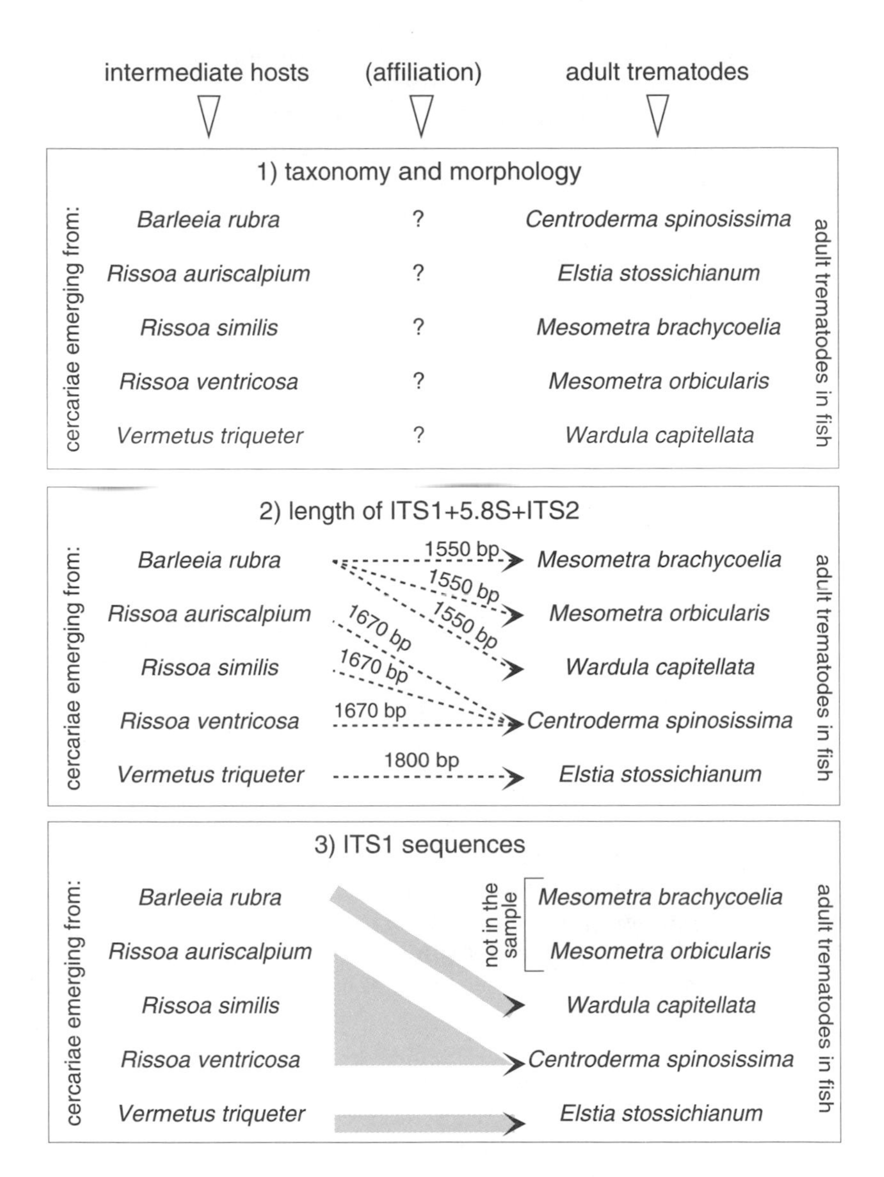

Figure 1.6. Elucidation of some trematode life cycles using the tools of molecular biology. Using data on the morphology of cercariae and adults (top frame), the length of a particular region of the genome allows a preliminary sorting of the possible host sequence (center frame). Further sequencing refines the affiliations (bottom frame); bp = base pairs.

The selection of infection by such a sexual pathway is sometimes unexpected, since it can concern parasites that do not inhabit either the skin or the genital tract. For example, *Trypanosoma equiperdum,* a trypanosome of horses, is transmitted during coitus, although most other trypanosomes (such as the human parasites *T. brucei gambiense* or *T. b. rhodesiense* in Africa and *T. cruzi* in South America) are transmitted by insect bites. Perhaps the most original variation of direct transfer from host to host is that of "vertical" transmission from parent to offspring.[13] In mammals infection from mother to young during either gestation or lactation is used by a variety of protozoans and helminths (see Baer 1972; Shoop 1991). In endocellular parasites, such as some viruses, bacteria, and protozoans, the passage through oocyte cytoplasm (transovarian transmission) is also extremely common (see Smith and Dunn 1991).

Transmission by Consumption

Transmission by consumption is widely used by parasites that are integrated into a food web leading to the host. Such parasites may settle in the digestive tract and its immediate attachments or may migrate to more farflung organs after ingestion.

The infective stage of these parasites may be ingested by the host either in an "isolated" form or in a form that is integrated into the body proper of a prey item. Ingestion of isolated infective stages may be quasi-accidental if, for example, the parasite (which may or may not be encysted) can be found on dirty hands or unclean food. Since using the hands to grasp food is principally a human behavior, the "dirty hand" has probably been a lucky and major opportunity for a variety of parasites. For instance, the human dysentery amoeba, *Entamoeba histolytica,* uses this pathway, and the cysts evacuated from the digestive tract induce infection after ingestion. Infection in this case depends on a lack of hygiene, and other modes of infection include contaminated vegetables or fecally contaminated water.

The ingestion of isolated infective stages is also used by numerous nematodes: the eggs of *Haemonchus contortus,* which causes a widely distributed and serious disease in cattle, are disseminated with the feces and hatch into larvae. These larvae remain in the pasture, and the infective L3 stage is then ingested as the animals graze.

Most often the infective stage is tightly integrated with the food (e.g.,

---

13. The term "vertical transmission" denotes all transmission from parents to their direct descendants, even if it is not transovarian. Any parasite transmission occurring between individual hosts that are not parents and offspring, however, is qualified as "horizontal transmission."

grass or prey) of the host in which it must end up. For example, such an association by consumption is a characteristic of the toxoplasmid *Toxoplasma gondii,* which is ingested by eating meat infected with cysts; the liver fluke *Fasciola hepatica,* whose infective stages (metacercariae) are stuck to plants grazed by ruminants; the Chinese liver fluke, *Clonorchis sinensis* (a cousin of *F. hepatica),* whose metacercariae are harbored by freshwater fish and ingested by fish-eating mammals, including humans; and the taenias or tapeworms of humans, whose infective stages are either in the muscles of pigs *(Taenia solium)* or beef *(Taenia saginata)* and whose adults occur in our digestive tracts.

Examples of transmission patterns similar to the preceding ones would fill dozens of pages. Eating raw or undercooked food is almost always the origin of such infections in humans, and it is interesting that mastering fire during the evolution of humankind and the subsequent cooking of food have probably limited the impact of several major parasitic diseases and thus favored the demographic expansion of our species.

In the very special case of the trichinellids (*Trichinella spiralis* and related species), transmission by consumption proceeds indefinitely without the passage of larval stages outside of the host. These viviparous nematodes, whose adults live in the digestive tract of rats, pigs, humans, or other mammals, disperse their larvae directly into the muscles of the host harboring the adults. A new individual is infected only by eating flesh harboring encysted L3 larvae, which then become adults in the gut. Rats become infected as they eat each other, pigs (and sometimes horses) by eating rats or rat cadavers, and humans by eating undercooked contaminated pork or horsemeat. Once in humans, however, the worms are at a parasitic dead end, since humans generally do not eat each other (see chapter 14).

In some cycles the infective stage survives the death of its host (the intermediate host) and is then ingested by the definitive host. Adults of the protostrongylid nematode *Muellerius capillarius* and several other related species are parasites of sheep and goats. The infective L3 larval stages are found in terrestrial pulmonate snails that have ingested L1 larvae released with sheep or goat feces. Small infected mollusks are ingested by the definitive hosts along with the grass, but larger infected mollusks, such as slugs, are not. Instead, when these slugs die the cycle continues because the L3 larvae, which can survive and remain infective for many months, spread over the grass and await ingestion (Cabaret 1990).

Transmission by Biting Vector

In addition to transmission by contact and by ingestion, parasites often use transmission by biting vectors. In much the same way that HIV can travel by needle, some parasites (viruses as well as protozoans and metazoans) use the living and flying needles that are mosquitoes, blackflies (simulians), sandflies (phlebotomes), tsetse flies, and others to get from one host to another. In this mode of transmission the biting arthropod takes up a parasite during a blood meal and either injects it directly into another host at a later time or leaves it close to the site of a bite. Most often the parasite is subject to a developmental phase in the vector before transmission.

This mode is, for example, used by the myxoma virus, which is transmitted to rabbits by either mosquitoes or fleas depending on the geographic region; by the bacterium *Yersinia pestis* (causative agent of plague), which is usually transmitted to rats or human by fleas; by the causative agent of sleeping sickness (protozoans of the genus *Trypanosoma*), which are transmitted by tsetse flies *(Glossina)*; by the causative agent of malaria (protozoans of the genus *Plasmodium*), which are transmitted by mosquitoes *(Anopheles)*; by the causative agent of onchocerciasis or river blindness (nematodes of the genus *Onchocerca*), which are transmitted by blackflies *(Simulium)*; by the causative agent of elephantiasis (nematodes of the genus *Wuchereria*), which are also transmitted by mosquitoes *(Aedes)*; and by the causative agent of loiasis, transmitted by horseflies of the genus *Chrysops.*

Regardless of whether the biting vector is an arthropod, it is nearly always a hematophagous organism. For example, some leeches act as vectors for fish and amphibian trypanosomes.

With the only condition being that the vector take several blood meals during its lifetime, this mode of transmission ensures the parasite's entry to and exit from the proper host. Sometimes, however, the process may differ for entry and exit. For example, and contrary to the African trypanosomes *T. b. gambiense* and *T. b. rhodesiense,* which are transmitted by biting both during entry and exit, *T. cruzi* (the causative agent of Chagas' disease) uses kissing bugs that ingest the parasite during the bite but "leave" the infective stages with their feces. In this case the parasites penetrate on their own if in direct contact with soft mucosal tissues (lips, eyelids) or through a lesion caused by scratching at the site of the bite.

Sometimes, however, the vector takes the parasite during a bite but does not really "reinject" it at all. This is the case of *Wuchereria bancrofti,* a nematode parasite of humans. As adults these parasites live in the deep lymphatic vessels of humans and produce multitudes of larvae (microfilariae) that circulate in the peripheral blood. These are taken in during the bite of a mosquito, within which they transform into infective stages. When the mosquito bites again the larvae drop onto the victim's skin and penetrate through the hole left by the bite after the mosquito has gone.

Similarly, some ticks (acarids) may also vector microfilariae, which after breaking out of the hemocoel penetrate the host's skin using the hole the tick pierced during its blood meal.

The injection of the parasite is not required when the parasite remains "outside" the host. This is the case for the worms that cause thelaziosis (nematodes of the genus *Thelazia,* which live beneath the eyelid of wild and domestic ungulates). Here larvae are transmitted by flies that land on the eyes, and in this case the parasite's larvae are simply lapped up by the flies and later deposited on another host's eyes by the buccal apparatus without biting. However, and despite this apparently simple process, the nematodes still must undergo maturation in the fly.

Transmission by Active Free-Living Stage

An active infective stage that fixes itself to the host or that crosses the host's tegument is characteristic of many ectoparasites. The classic examples are fish monogeneans, whose swimming larvae, the oncomiracidium, attaches to the host's gills or skin. However, some internal parasites also infect their host using free-living stages that have organs to ensure their penetration. In humans the nematode *Ancylostoma duodenale* (there are also dog ancylostomes such as *A. caninum* and *A. brasiliense*) has larvae that penetrate the skin after contact with contaminated ground. Another example includes trematodes of the genus *Schistosoma,* whose cercariae pierce the tegument while the host is swimming. As a general rule such infections are followed by migration of the larval stages through the organism, since the parasite usually does not initially penetrate the organ needed for its development (see chapter 2).

The Sequence of Hosts

Several species of hosts, most often not related, may be *successively* exploited as the parasite develops over its life cycle. I have previously cited the hypothetical example of a parasite that develops successively in the mouse and then in the cat. In fact this is a real case: by eating a mouse, a cat can become infected with any number of different organisms, including protozoans (*Toxoplasma gondii, Sarcocystis muris,* etc.), nematodes (*Spirura rytipleurites, Aelurostrongylus abstrusus,* etc.), and cestodes (*Taenia taeniaeformis,* etc.).

The *sequence of hosts* constitutes the series of hosts utilized during the life cycle and is sometimes referred to as the longitudinal component of the cycle. This sequence involves both an ecological dimension (the succession of environments) and an ontogenetic dimension (the various developmental stages of the parasite). Each time the parasite changes host some genes are turned off while others are activated, forming an organism that is morphologically, anatomically, physiologically, and ethologically distinct. That is, the parasites undergo an authentic *metamorphosis* at each change of host.

Little is known about the signals parasites perceive when they arrive at a new host, which may be of a physical nature (for example, temperature if the host is homeothermic) or, more commonly, chemical.

The simplest sequence has only one obligatory host species, while the most complex ones have up to four. This diversity of sequences in parasites may be illustrated by examples drawn from the helminths (fig. 1.7).

The One-Host Sequence

Adults of the nematode *Ascaris lumbricoides,* the causative agent of ascariasis, live in the duodenum of humans, where they lay eggs. These eggs are shed with the feces, where they begin their maturation, and the larvae then undergo two molts. When the eggs contain the L3 larval stage they are infective, and once swallowed by a human they hatch and become adults following an internal migration through the heart and lungs.

The host sequence of the nematode *A. lumbricoides* therefore includes only *one host species* (the human) and only *one transmission event* (human to human).

The Two-Host Sequence

Adults of the cestode *Taenia taeniaeformis* live in the duodenum of cats. Segments of the worm (proglottids) are packed with eggs and are passed with the feces. Each egg contains an embryo that, when swallowed by a rodent (most notably mice), develops into a larval stage called a cysticercus in the mouse's muscle. Once a cat eats the infected rodent the cysticerci then develop into adults in the cat's gut.

The host sequence for *T. taeniaeformis* therefore includes *two species of hosts* (the cat and the rodent); and *two transmission events* (from the cat to the mouse and from the mouse to the cat).

The Three-Host Sequence

Adults of the trematode *Clonorchis sinensis,* the Chinese liver fluke, live in the bile ducts of various mammals, predominantly in the Far East, including humans. Eggs are evacuated with the feces and ingested by a freshwater gastropod mollusk (a dozen species have been reported as vectors, the most frequent being *Parafossarulus manchouricus*). Numerous swimming larvae (cercariae) are produced by asexual multiplication and then leave the mollusk. The cercariae penetrate a variety of fish species

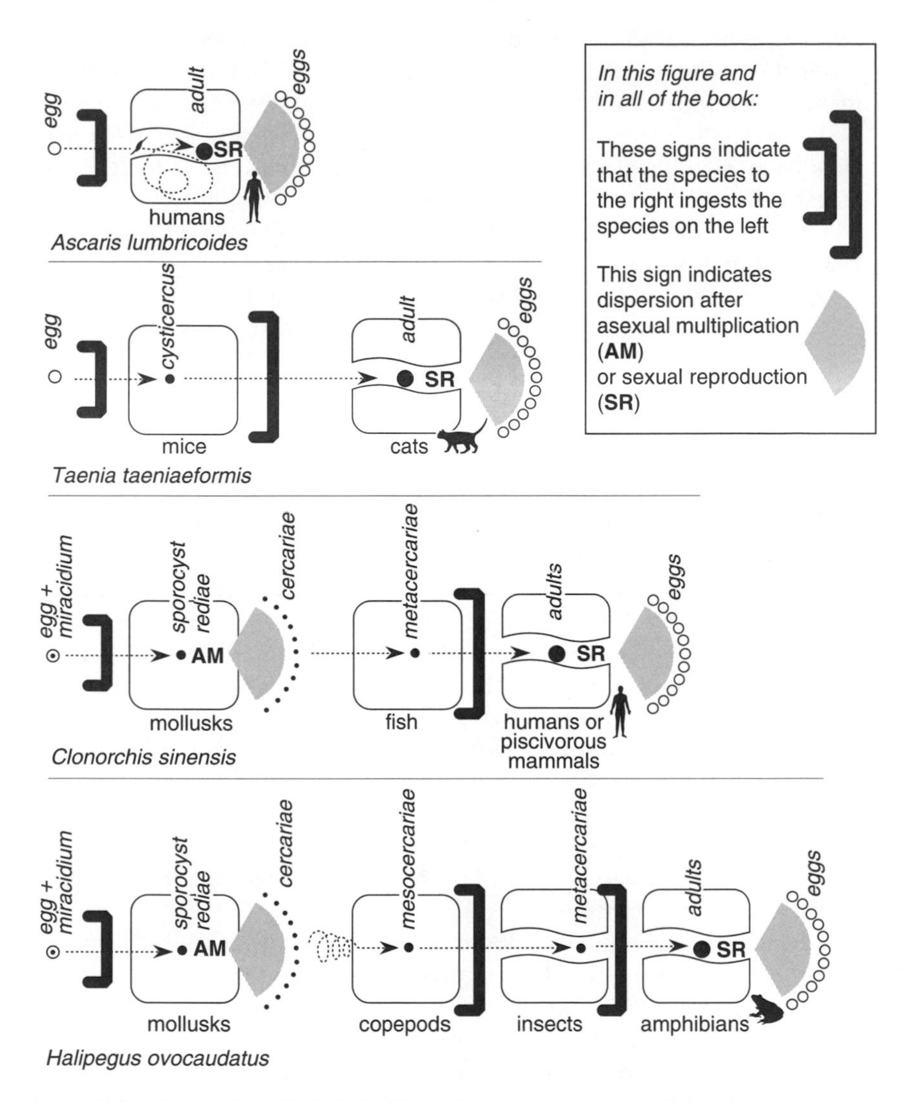

Figure 1.7. Four cycles of helminths illustrating one, two, three, and four host sequences in life cycles. Note the internal migration of *Ascaris*. AM = asexual multiplication; SR = sexual reproduction. Dotted arrow indicates route of ingestion or migration. Open circles = eggs; solid circles = larvae or adults.

(about one hundred), where they then encyst in the muscle as metacercariae. Once the metacercariae are eaten with the fish they "excyst" in the mammal's stomach and mature, after a brief migration, in the bile ducts of the liver (this trematode may cause serious disease, with loads of over 25,000 parasites per patient, and each worm may measure up to 2.5 cm in length).

The host sequence for *C. sinensis* therefore includes *three species of hosts* (the mammal, the mollusk, and the fish); and *three transmission events* (from human to mollusk, mollusk to fish, and fish to human).

The Four-Host Sequence

Cycles with a sequence of four hosts are rare, and no human parasite is known to have one. However, the cycle of the trematode *Halipegus ovocaudatus,* a parasite of amphibians, has been shown by Kechemir (1978) to include adult parasites that live and reproduce in the oral cavity of green frogs. The eggs are evacuated into the water, and the miracidia hatching from them then penetrate a mollusk, where asexual multiplication results in cercarial production. Cercariae then exit the mollusk, reach the water, and lie still on the bottom of a pond until ingested by a copepod. Once eaten by a copepod the cercariae actively inject themselves into the hemocoel. Once the copepod is eaten by a dragonfly larva the parasite settles in the digestive tract, and if the dragonfly is then eaten by a green frog, the parasite settles in the mouth cavity, where it becomes an adult.

The host sequence for *H. ovocaudatus* therefore includes *four host species* (the amphibian, mollusk, crustacean, and insect); and *four transmission events* (from amphibian to mollusk, mollusk to crustacean, crustacean to insect, and insect to amphibian).

All the stages of the life cycle of *Halipegus ovocaudatus* are obligatory, and the events are spread over two to three months. In discussing *Halipegus* Moore (1993b) wrote: "Such an animal is worthy of our liveliest curiosity, if not our frank admiration." Based on our current knowledge there appear to be only a few four-host cycles. However, there may be more than we know of, simply because they are so difficult to demonstrate, particularly in marine environments.

A Few Definitions

The preceding examples raise the need for a few definitions concerning host sequences before we continue discussing transmission: a cycle with a one-host sequence is said to be *holoxenous* (or monoxenous); a cycle with a sequence of more than one host is said to be *heteroxenous;* the definitive host is that host where the parasite reproduces sexually; the intermediate host is an obligatory host that shelters a stage that does not reproduce or that reproduces asexually and within which some development of the larva occurs. For each event of the cycle there is a host "from which" the parasite comes and a host "to which" it goes; the former is the *upstream host,* and the latter the *downstream host.*[14] Three remarks must be made concerning these definitions.

We have a tendency to identify the human as the definitive host in human parasitic infections, but this is not always true. In the case of *Plasmodium,* for example, sexual reproduction is in the mosquito and asexual multiplication in the human. Therefore the mosquito is, technically speaking, the definitive host and the human the intermediate host. Nevertheless, the mosquito is usually termed the *vector,* probably because it is an easily identifiable direct source of infection for humans. "Vector" has, then, a strong medical and anthropocentric connotation that does not always fit with the technical definition of the term.

We often designate as a vector any intermediate host or physical entity that transmits the parasite to the definitive host. The term therefore should normally be used only for arthropods that transmit parasites by deposition or injection and for prey that transmit the parasite when ingested. However, it is also commonly used to describe the molluskan intermediate hosts of trematodes, where the infection is the result of contact with stages produced in the mollusks that swim freely in the water for a time. Logically speaking, then, the water should be considered the disease vector in this case.

In the very peculiar cycle of *Trichinella* (see above), there is first sexual reproduction in the digestive tract of the host (pig, rat, etc.), dispersal of the larvae into the muscles, and then infection of another mammal that ingests the muscle-dwelling larvae. The same individual that is the definitive host therefore also acts as the intermediate host.

The Shortening of the Life Cycle

A parasite's life cycle may be shortened by several distinct processes (see Grabda-Kazubska 1976; Monod 1977), each of which can again be illustrated by trematodes. Note, however, that here we are talking about *optional shortenings* and not about shortenings that may occur on an evolutionary scale.

The usual life cycle of a trematode (fig. 1.8A) normally includes first a

14. More precise definitions of these terms are given in Combes (1991d).

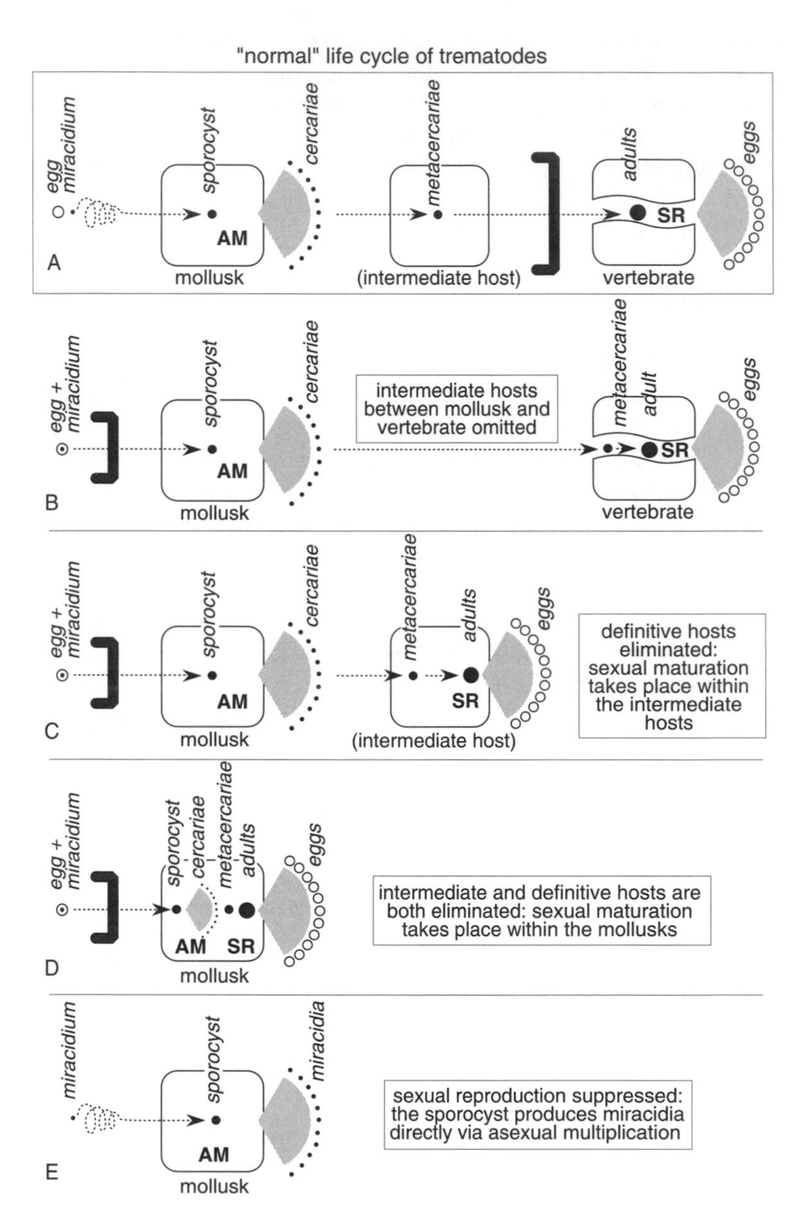

Figure 1.8. Optional ways helminth life cycles can be shortened. (A) "Normal" trematode cycle; (B) skipping a host: *Opisthioglyphe ranae* "skips" the intermediate host; (C) anticipated maturation: *Paralepoderma cloacicola* becomes gravid in the second intermediate host; (D) more precocious anticipated maturation of *P. cloacicola:* gravid in the first intermediate host; (E) shortcut of the cycle: *Mesostephanus haliasturis* shortcuts both the cercarial stage and sexual reproduction. Open circles = eggs; dotted circles = larvae; solid circles = encysted stage or adult.

molluskan host (from which cercariae escape) and then a second intermediate host, whereupon the cercaria encysts to form a metacercaria whose nature is variable. This is followed by a definitive host, where the metacercaria becomes an adult. The body of the cercaria, the metacercaria, and the adult parasite are in fact one single object that ultimately acquires sexual maturity after a series of developmental shifts.

The first way of shortening the life cycle consists of the parasite's "skipping" a host (fig. 1.8B). In the typical life cycle of *Opisthioglyphe ranae* the cercariae classically transform into metacercariae in tadpoles, and adult frogs get infected when they eat their own young during metamorphosis, which is common in amphibians. The originality in this case is that the cercariae can also enter adult frogs directly via the oral cavity, go through a transient metacercarial stage lasting about twenty-four hours, excyst, and transform directly into an adult in the digestive tract. There is, then, a normal cycle with three hosts and an abbreviated cycle with two hosts, and it appears that both coexist within the same parasite population (Grabda-Kazubska 1976).

Another way of shortening the life-cycle consists of acquiring early sexual maturation. The normal host sequence of *Paralepoderma cloacicola* includes the mollusk *Planorbis planorbis,* amphibian tadpoles from a variety of frog species, and a snake, either *Natrix natrix* or *N. maura.* Monod (1977) showed that within the same population the parasite, normally adult in the reptile's intestine, may also become gravid in the tadpole (fig. 1.8C) and even in the mollusk (fig. 1.8D), even though each occupies a different habitat. When maturation occurs in the mollusk, the cercariae surprisingly transform "on site" into metacercariae, with eggs appearing in the uterus of these metacercariae shortly thereafter. As a result, transmission from mollusk to mollusk may occur indefinitely in this cycle.

A last way of shortening the life cycle consists of getting rid of one of the reproductive phases. In both of the preceding cases shortening the life cycle involved a telescoping event in that the stages normally present in several types of hosts followed one after the other in only one host and the developing stages themselves were not modified. A more surprising shortening is illustrated by another trematode, *Mesostephanus haliasturis,* which during its "normal" cycle successively parasitizes the mollusk *Melanoides tuberculata* (normal host of the sporocysts that then produce cercariae), followed by "rainbow" fish of the genus *Melanotaenia* (the normal hosts of the metacercariae), and then cormorants or other piscivorous marine birds (the normal hosts of adults). In parallel with this cycle, which is in every respect a classical trematode life cycle, sporocysts can also produce miracidia directly and thus shortcut not only the sequence of hosts but also the sequence of developmental stages (fig. 1.8E). The results in this case mean that there is no more sexual reproduction (Barker and Cribb

1993) and that there are not only fewer hosts but also fewer ontogenetic stages as well.

The Lengthening of the Life Cycle

The life cycle can be lengthened by incorporating additional hosts termed “transport” or “paratenic” hosts. Cycles with such hosts exist in numerous groups of helminths, most notably the cestodes and acanthocephalans.

The tapeworm *Diphyllobothrium latum,* commonly known as the broadfish tapeworm, is one well-known example. In this case the adult parasite is found in the intestines of humans and various other fish-eating mammals. To go from human to human the parasite must pass successively through both an obligatory freshwater copepod and an obligatory fish host. It is by eating undercooked infected fish that humans become infected. Thus the life cycle consists of three obligatory hosts: humans, crustaceans, and fish. However, if a big fish eats a small fish harboring the parasite, the larvae may survive and remain able to infect a mammal. The process may repeat itself (if a very big fish eats the big fish), resulting in the accumulation of a large number of infective larvae in the biggest fish.

The host sequence of *Diphyllobothrium latum* therefore includes a minimum of three obligatory hosts (the mammal, the copepod, the fish) and can be extended by incorporating nonobligatory paratenic fish hosts (fig. 1.9).

Significantly to us, there is a paratenic host in yet another human parasitic disease, paragonimiasis. This disease is caused by a variety of species of trematodes in the genus *Paragonimus.* As in all trematodes, the cercariae are produced in a mollusk. These cercariae then encyst in fresh-

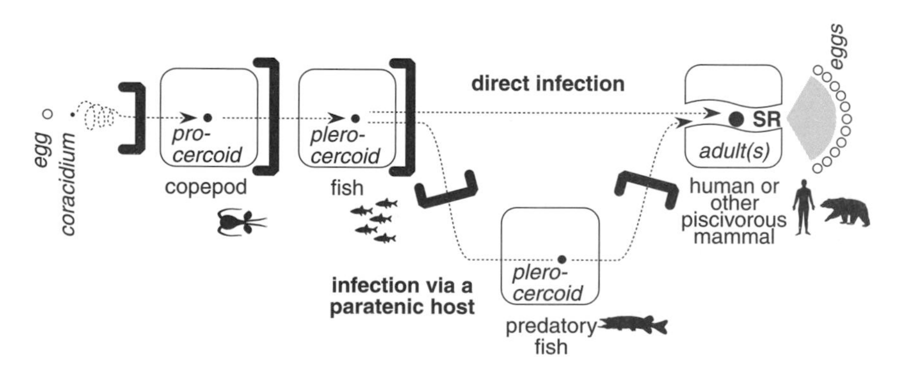

Figure 1.9. Cycle of the cestode *Diphyllobothrium latum* showing the possible integration of a paratenic host (here a carnivorous fish). SR = sexual reproduction.

water crustaceans (crabs) to form metacercariae, and for a long time it was thought that humans became infected by eating undercooked crustaceans. This mode of infection is in fact possible, although in the 1980s it was recognized that there is also a different route. Mammals of several species, including wild pigs, eat infected crustaceans, and the metacercariae, which are neither digested nor capable of developing to adulthood in the pig, go to the muscles, where they survive. Humans therefore are probably most often infected by eating undercooked wild pig meat (paragonimiasis is a serious disease in humans because adult worms are large—up to several centimeters long—and locate themselves in the lungs, where they are highly pathogenic).

Very often paratenic hosts appear as "concentrators" of infective stages and, far from complicating transmission, may in fact promote it. For example, such hosts may ensure the infection of age-specific definitive hosts by a distinct pathway. In this case younger age-classes become infected "directly" by eating small intermediate hosts, whereas older hosts become infected by eating paratenic hosts that are bigger. This occurs in the life cycle of the nematode *Anguillicola crassus,* where small eels are infected by eating the small copepod intermediate host. Big eels, however, become infected by eating paratenic fish hosts, notably bream (Guillaume Blanc, personal communication).

Last, there are entire groups, such as some families of acanthocephalans, where transmission, although physiologically possible without paratenic hosts, can really continue only through a paratenic host. The cycle of the acanthocephalan *Corynosoma semerme,* for instance, includes a copepod first host within which the cystacanth, the infective stage to the seal that is the definitive host, matures. However, direct transmission to the seal via the microscrustacean is highly improbable, and the cycle is completed because paratenic hosts (various fish species) first ingest the copepods and accumulate the cystacanths before being eaten by the seals.

Polymorphism in Life Cycles

There are parasite life cycles that vary depending on characteristics such as the sex or age of the host. An example is the intestinal nematode *Toxocara canis,* a dog parasite already mentioned. The life cycle of *T. canis* proceeds as follows.

The eggs are evacuated outside with the dog's feces. Once outside, they mature to give an embryo and then, within a few weeks, pass through two successive larval stages, L1 and L2, while still protected in the egg. When the egg contains the L2 larva it is "infective" and can survive up to two years in the environment if the temperature and humidity are favorable. Once a dog swallows infective eggs, the L2 larvae are released and cross the intestinal wall, pass into the circulatory system, arrive at the right side of the heart, and pass to the lungs via the pulmonary artery. At this point the L2 larvae then have a different fate depending on what animal has ingested the eggs: if it is a puppy younger than five weeks of age the L2 larvae molt to give L3 larvae in the lung tissue. These larvae then migrate up into the trachea, are swallowed, reach the stomach, and ultimately molt into adults in the duodenum. The entire event lasts about five weeks in total. If the puppy is older than five weeks, however, the L2 larvae instead pass into the trachea, are carried off by the circulation, and locate in the muscles of the growing pup. If the puppy is a female that later becomes pregnant, the L2 larvae "wake up" on the forty-second day of gestation, migrate to the circulatory system, cross the placenta, and reach the liver of the developing fetus through the umbilical vein. The L2 larvae then complete their migration, previously interrupted for months in the mother, in newborn pups by passing through the usual L3 and L4 stages in the lungs and digestive tract. Last, if a "nondog"—for example, a rodent or a bird—were to ingest the eggs, the L2 larvae would then distribute themselves in the muscles as in the case of puppies older than five weeks. The rodent or the bird then becomes a paratenic host. If it is eaten by a dog, the migration and settlement in the lungs of larvae then occurs normally. Moreover, if the rodent or bird is eaten by a predator other than a dog, the larvae may encyst in this new paratenic host, and transmission may still be possible if this host is then later eaten by a dog.

Toxocara cati is a related parasite found in the cat. The general host sequences of the life cycles are comparable, but the mother cat transmits the L2 larvae to the kittens through the milk. *T. canis* and *T. cati* can be distinguished morphologically with scanning electron microscopy (Dorchies and Guitton 1993).

In general, then, it is important to recognize that the alteration in the physiological and morphological characteristics of parasites that occurs during their life cycles, as well as their ability to alter their sequence of hosts, is due to the *phenotypic plasticity* of parasite species. Given that parasitism as a lifestyle is the most common mode of existence on the planet, such plasticity has obviously stood the test of evolutionary time very well.

2 Specialization in Parasites

The literature of parasitology contains no general, heuristic, testable, and nonteological explanation for why some parasite species are highly discriminatory and others are not.
Janovy, Clopton, and Percival 1992

Why are there no parasite species exploiting all the members of large taxa such as mammals or birds?
Timms and Read 1999

Two Scales of Specialization

The ecological niche represents the ensemble of all variables that characterize the portion of the environment where a living species fits (MacArthur 1958). It is defined by an *n*-dimensional space representing as many variables as one can consider or measure, although in practice the variables defining the niche of a free-living species usually boil down to space and resources. It is no different for a parasite, and the simple mention of the host species and microhabitat occupied are usually considered sufficient to give a satisfying approximation of the parasite's niche. These are, then, the scales on which the degree of specialization of parasites is generally evaluated: the host species and the part of the host occupied.

Using the host species to characterize the parasite's environment is so familiar to parasitologists that the host is almost automatically integrated into the description of the parasite species. "This ecological particularity has always influenced the study of parasites; particularly, it never seemed useless to have the parasite name followed by the host name" (Euzet and Combes 1980). A parasite species is therefore associated with a spectrum of hosts, that is, with a list of taxa (usually species, but sometimes other levels) whose range allows the definition of the degree of specificity of the parasite.

In 1980 Euzet and Combes proposed the concept of "filters" to symbolize the mechanisms responsible for the formation of host ranges. Holmes (1987) later used this idea under the name "screens." Figure 2.1 illustrates the four mechanisms of sorting that allow one to go from the long list of theoretically available host species to the more restricted list of the real range of hosts occupied.

Arrows 1–4 in the figure point to the parameters that restrict the range

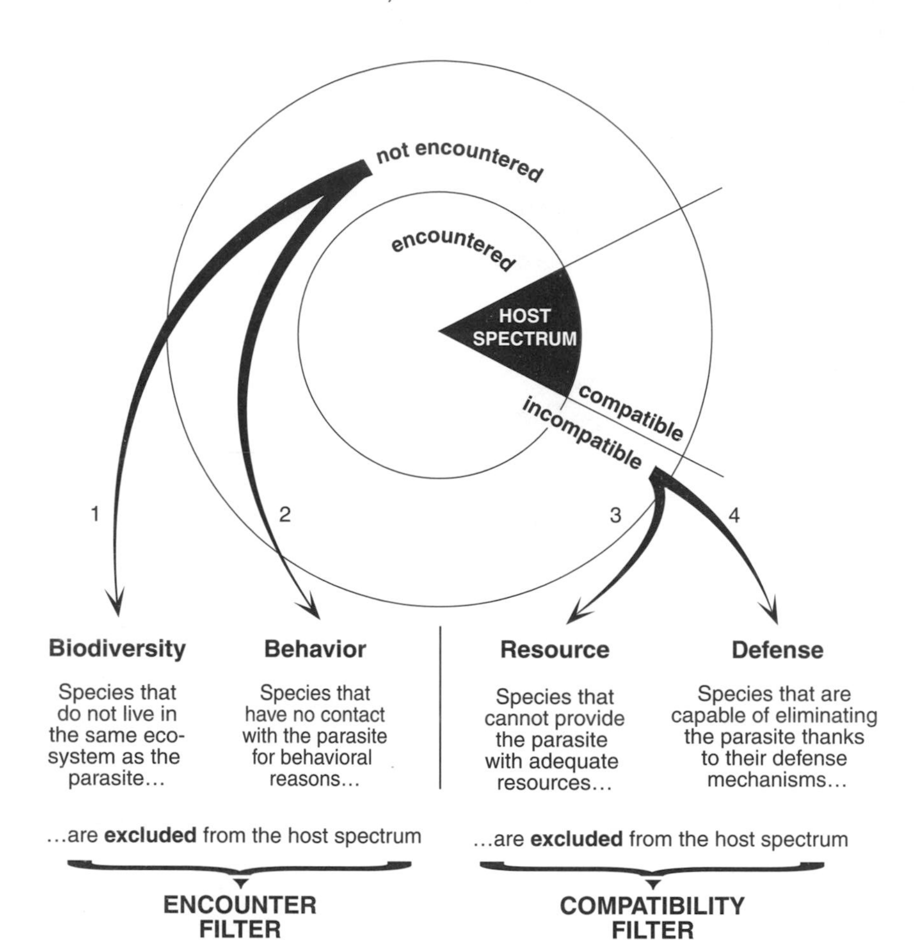

Figure 2.1. "Formation" of the host spectrum (based on the proposal of Euzet and Combes 1980). The intersection of compatible hosts encountered constitutes the host spectrum (in black). 1 and 2: species not encountered because they are not found together (1) or because host behavior prevents encounter (2); 3 and 4: species the parasite does not develop in because of incompatibility owing to lack of resources (3) or host defense mechanisms (4).

of potential hosts. These include (1) a *"biodiversity" parameter*—species that do not live in the same ecosystem as the parasite are excluded from the host spectrum; (2) a *"behavior" parameter*—species with behaviors rendering contact with parasite infective stages impossible are excluded from the spectrum; (3) a *"resource" parameter*—species that do not support the spatial or metabolic resources necessary for the parasite are excluded from the spectrum of hosts; and (4) a *"defense" parameter*—species that can destroy the parasite via their immunodefense systems or other mechanisms are excluded from the host spectrum.

Mechanisms 1 and 2 define the probability of contact between the parasite and a potential host and together constitute what Euzet and Combes (1980) termed the *encounter filter.* Mechanisms 3 and 4, on the other hand, delimit the probability of the parasite and potential host living together durably after encounter and constitute what Euzet and Combes termed the *compatibility filter.*

All degrees of opening or closing of the two filters are possible regardless of the level—species, populations, or individuals—addressed. However, one can simplify the model for heuristic purposes by supposing that each filter may be either open or closed. In this case there are four possible combinations:

1. Both the encounter filter and the compatibility filter are closed. Here the parasite-host association cannot exist. This is, for example, the situation that occurs in the case of schistosome cercariae (schistosomes are parasites of mammals) and earthworms, crickets, or lizards. No encounter and compatibility are possible between this parasite and these organisms.

2. The encounter filter is opened, but the compatibility filter is closed. In this case the parasite-host association does not exist, although selective pressure is on the parasite to open the filter of compatibility and any mutant capable of doing so may be selected for. This is, for example, the situation faced by the cercariae of bird schistosomes that penetrate swimming humans—they are not able to develop because cercarial penetration through a human's skin induces a severe immunologic response ("swimmer's itch") that destroys the invading parasite. That is, there is encounter but not compatibility.

3. The encounter filter is closed but the compatibility filter is open. The parasite-host association does not exist, although a change in the composition of the ecosystem or in the behavior of the organisms themselves may open the encounter filter and result in association. This is, for example, the case of the cercariae of human schistosomes and mice. The encounter filter is closed (the mouse does not normally enter water and is therefore not normally infected in nature). The compatibility filter is open however, and in the laboratory most human schistosomes are maintained in mice. That is, the researcher opens the encounter filter.

4. Both the encounter filter and the compatibility filter are open. This is, for example, what happens in the actual bird and human schistosome systems that exist in nature. In each case, the appropriate cercariae en-

counter and penetrate compatible hosts and develop into adults. There are, at the same time, both encounter and compatibility, and the association exists.

The concept of encounter and compatibility as a sequence of filters allows a more complete understanding of the parasite-host relationship than that provided by delineation of the host spectrum alone because it recognizes the role natural selection plays in the system. As shown in figure 2.2, the different filters can be represented as camera diaphragms that can be opened and closed to varying degrees. Selection on the parasites provides pressure to open filters while selection on the hosts tends to

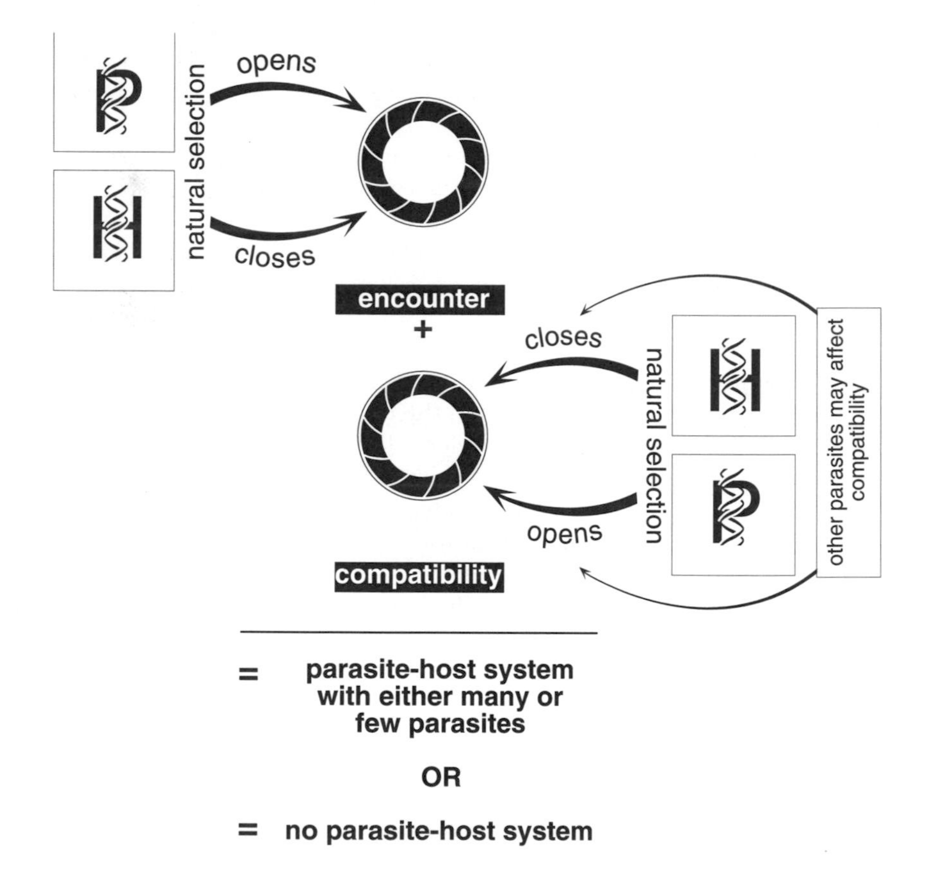

Figure 2.2. Encounter and compatibility filters (modified from Combes 1987). The degrees of opening or closing of each filter together add up to the occurrence or nonoccurrence of a parasite-host system. As will be seen later, the opening of the filters may be limited by the cost of virulence to the parasite, while their closing may be limited by the cost of resistance in the host.

close them. The notion that parasites struggle to increase their abundance in hosts, and thus to increase their virulence, arises directly from the interplay of these counterpressures, although later in this book we will see that this basic scheme needs to be adjusted on only one point: increased parasite virulence may in some cases be disadvantageous to the parasite, and selection may then have a tendency to close the filters for the parasites as well.

Parasite-host systems are only one level in the hierarchy and organization of parasitism. It is not the host that constitutes the habitat of the parasite, however, but rather a fraction of the host: an organ, a part of an organ, a type of cell, a part of a cell.

In practice authors often restrict the definition of microhabitat in parasitology to several easily measured and defined spatial dimensions. This approach is in part justified because the organs or tissues of individual hosts often show marked gradations on a number of parameters (structure, metabolites, waste products, pH, water currents, etc.), and the location of the parasite often spans a wide range along each gradient making up its immediate environment (see Holmes 1990b). This is the obvious case, for example, within the digestive tracts of animals and on the gills of fish, two favorite and convenient environments for parasitologists to study.

In fact, however, such a topographic delineation of the microhabitat tends to ignore the actual resources exploited by the parasites occupying them, and two parasites of different species may live in the same portion of the host without necessarily belonging to the same guild or ensemble of species that share the same resources (see Crompton 1973). This is the case in the digestive tract of a vertebrate, where one may distinguish guilds of absorbing parasites that feed on intestinal contents via the transport of nutrients across their tegument; guilds of swallowers that ingest particles; guilds of hematophagous parasites that consume blood; and guilds of grazers that actively ingest fragments of epithelial or mucosal tissue.

For instance, Radomski and Pence (1995) showed that the eight most common species of intestinal helminths in coyotes *(Canis latrans)* in Texas belong to four guilds, each distinguishable by the development of its intestinal tract. Along the same lines Bansemir and Sukhdeo (1994) showed that it is possible to precisely determine the guild intestinal parasites belong to by having the host ingest fluorescent dyes that are selectively localized in the host's ingesta, blood, or intestinal cells. That is, one cannot forget that identifying the microhabitat alone, as well as identifying the host alone, gives only partial information on the parasite's ecology.

There are not many, if any, organs or tissues in organisms that are totally protected from parasites, and it is in microhabitats other than the skin and digestive tract that the pathogenic effects of parasites seem strongest. Malaria, leishmaniasis, trypanosomiasis, filariasis, and schistosomiasis, which together still constitute the major obstacles of modern humanity to good health, are caused by parasites found mainly in the circulatory system. Moreover, the most "vital" microhabitats, such as those in the nervous system, are usually exploited by parasites only when transmission needs to provoke the death of the host or otherwise alter its behavior to ensure success.

Specialization for the Host Species

A Preamble: The Partners' Identity

Analyzing the host spectrum makes sense only if two conditions are met: a satisfactory inventory from field investigations exists; and the parasite and host species have been correctly identified.

According to May (1986), the 1.5 million or so identified species alive today may represent only 10% of the actual number of species present in the biosphere. There is no doubt that parasites, because of their "hidden" lifestyle, are the least known of all extant species and that there is an unimaginable number of them remaining to be discovered and described. The result is that known host spectra are often far from reality because of nonexistent or insufficient study. This is testified to by the positive correlations that exist between the number of investigations on a parasite (sampling effort) and the broadness of its host spectrum (Poulin 1992b, 1998a). The situation is further modified by the recognition of morphologically indistinct species, and it is here that modern biomolecular techniques have allowed substantial progress over the two past decades, most notably through DNA sequencing and allozyme analysis. According to Anderson, Blouin, and Beech (1998), "Genetic markers are invaluable tools for differentiating between morphologically identical sibling species, for detecting hybridization, for identifying eggs or larval forms and for quantifying patterns of host affiliation and specificity."

In short, allozyme analysis has been the method of choice to ensure that two morphologically indistinct organisms represent reproductively isolated species. Allozymes are slightly different forms of the same enzyme, coded by different alleles, and their amino acid composition differs without, in most cases, modifying the activity of the enzyme.

This method works because it can detect reproductive isolation. If for example, there are two alleles (a and b) at an enzyme locus in a diploid population, then there are three possible genotypes: homozygous aa, homozygous bb, and heterozygous ab. If there are no heterozygotes in a population it almost always means there is reproductive isolation between aa individuals and bb individuals. Such reproductive isolation between ensembles of individuals living in sympatry indicates that they belong to different species. In this technique each individual's genotype is revealed by electrophoresis, and because alleles of the same locus are codominant, this technique allows almost immediate detection of heterozygotes. However, the method is applicable only if the organisms involved reproduce sexually.

When there are morphologically indistinct parasites in two animals of different host species living in sympatry, two hypotheses may be presented a priori: the parasite may show little specificity and can be found indifferently in the two hosts; or there are two distinct species of parasites, each one specific to its host but not distinguishable using classical morphological methods.

Determining which of these two hypotheses is correct is possible because of allozyme analysis: if there are no heterozygotes for alleles at the same locus, then the two parasites under study do not mate and are of different species. This locus is therefore qualified as being diagnostic, and except in cases involving the total counterselection of heterozygotes, one such diagnostic locus is (in principle) enough to confirm reproductive isolation between two species. Analyzing genomic sequences also gives precious clues about the degree of divergence or similarity between the populations (or ensembles) under study.

In most cases population analysis by such molecular methods has shown that species of parasites once thought to be generalists were in reality complexes of several specialist species. This has been shown to be the case so often that nowadays generalist parasites are considered with suspicion. This is true even though in other cases the opposite has happened: in certain groups researchers have had a historical tendency to describe new species of parasites each time they examined a new species of host, basing their judgment solely on the dogma of narrow specificity rather than on distinctive characteristics.

Many host spectra have been clarified and further elucidated using the available molecular methodologies, as in the following examples (fig. 2.3).

1. The acanthocephalan *Acanthocephaloides propinquus,* historically considered a generalist because of the number of hosts it has been reported to parasitize, has been revealed to be a complex of species. Using both morphological and allozyme analyses, Buron, Renaud, and Euzet (1986) identified those *A. propinquus* occurring in a variety of host fish in Mediterranean lagoons and those along the coast as two sympatric species, one parasitizing mostly gobies in lagoons *(Gobius niger* and

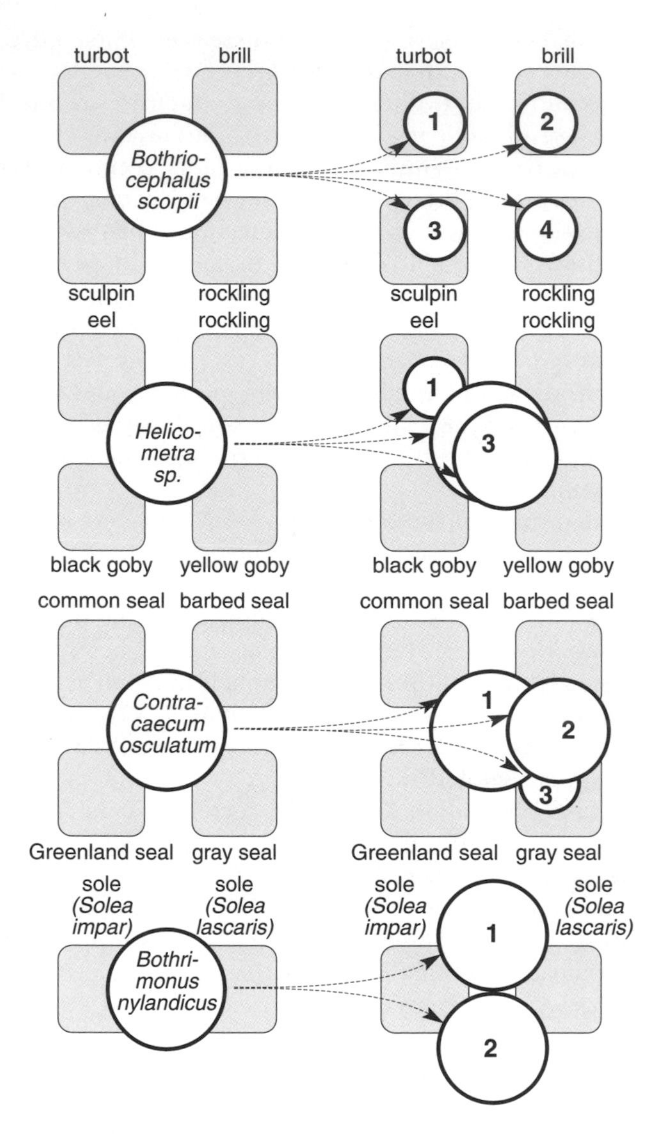

Figure 2.3. Some examples of the discovery of cryptic species of parasitic helminths (from various authors; see text). Open circles denote cryptic species distinguished from previously described named species. Rectangles denote host species. Intersections of circles and rectangles denote parasite-host distribution.

Zosterisessor ophiocephalus) and the other parasitizing the scaldfish along the coast.

2. Until 1983 a common bothriocephalid cestode designated *Bothriocephalus scorpii* was reported to occur in a variety of marine fish, including brill *(Scophthalmus rhombus)*, turbot *(Psetta maxima)*, bull-rout *(Myoxocephalus scorpius)*, and the five-bearded rockling *(Ciliata mustela)*. Allozyme analysis then revealed that there were diagnostic loci for the worms found in each fish, with each population having an allele characteristic for each host species without the occurrence of heterozygotes. Thus all four species of hosts harbored cryptic populations of bothriocephalids that were reproductively isolated from the others (Euzet, Renaud, and Gabrion 1984). Looking more closely at the parasites, very small morphological differences were then observed between the four types (size, organ disposition, etc.), but these differences on their own were too small to justify recognizing them as separate species. Since *B. scorpii* has been reported from a variety of fish families, it is probable that the real number of species in the *Bothriocephalus* complex may be higher.

3. Three species of trematodes of the genus *Helicometra* are harbored by a total of four fish species *(Symphodus cinereus, Zosterisessor ophiocephalus, Gobius niger,* and *Anguilla anguilla)* in Mediterranean lagoons, but the host spectra are now recognized to be unequal: one species of *Helicometra* is found in only one of the fish, another in the three others, and the last one in all four (Reversat, Renaud, and Maillard 1989).

4. The seal nematode known as *Contracaecum osculatum* is in fact a complex of three reproductively isolated and morphologically indistinguishable cryptic species. One of them (A) parasitizes both the bearded seal, *Erignathus barbatus,* and the gray seal, *Halichoerus grypus.* The second (B) has been found once in the bearded seal and normally parasitizes only the gray seal, the Greenland seal *(Phoca groenlandica)*, and the common seal *(Phoca vitulina)*. The third species (C) has been found only in the gray seal (Nascetti et al. 1993).

5. The cestode known as *Bothrimonus nylandicus,* a parasite of sole *(Solea lascaris* and *S. impar)* is in fact composed of two reproductively isolated "twin taxa," both parasitizing both species of sole. In this case research increased the number of known species but did not change host specificity (Renaud and Gabrion 1988).

6. The use of molecular methods has allowed the recognition of genetically distinct species or populations of medically and economically important parasites on numerous occasions. For example, the cestode *Monieza benedeni,* which for a long time was thought to be parasitic of both bovids and sheep, is in fact a complex of two cryptic species, one specific to bovids and the other to sheep (Ba et al. 1993). Similarly, the common

horse ascarid is also composed of two cryptic species, *Parascaris equorum* and *P. univalens* (Bullini et al. 1978). The problem posed for the longest time by the amoeba *Entamoeba histolytica,* which seemed to present both pathogenic and nonpathogenic "forms," was resolved by distinguishing two species, with *E. histolytica* being pathogenic and *E. dispar* being non-pathogenic or only slightly pathogenic (see Clark and Diamond 1994). In *Toxoplasma gondii* DNA analysis using restriction fragment length polymorphism (RFLP) has also shown that virulent strains are genetically distinct (Sibley and Boothroyd 1992). Similarly *Pneumocystis carinii,* an organism that is related to fungi and develops in vertebrate lungs, "was considered for almost one century a widely spread, euxenic, unique species. At present, this concept is changing . . . the scientific community is progressively accepting that the terminology '*P. carinii*' is hiding a heterogeneous group of microorganisms" Mazars and Dei Cas (1998). In *Trypanosoma* allozyme analysis has allowed considerable progress toward the distinction of zymodemes, or ensembles characterized by common allozyme profiles (Tibayrenc and Ayala 1988). Although it is not formally recognized that the cestode *Echinococcus granulosus* (causative agent of hydatidosis in its vertebrate intermediate hosts) is a complex of species, the genetic distances between the three lineages corresponding to sheep, cow, and horse are on the order of those recognized for other well-established species (Bowles, Blair, and McManus 1995b).

7. Molecular techniques have reduced the abusive synonymizing of species distinguished only according to hosts or other uncertain characters. Blair and coworkers (Blair, Agatsuma, and Watanobe 1997; Blair et al. 1997) have analyzed the ITS2 region (internal transcribed spacer number 2 of the ribosomal transcription subunit) and CO1 region (cytochrome c oxydase subunit 1) in three species of the trematode genus *Paragonimus: P. ohirai, P. sadoensis,* and *P. iloktsuenensis.* Based on the fact that the three species have identical ITS2 sequences and almost identical CO1 sequences, these authors conclude that this evidence is "an indication that they are conspecific or at least capable of exchanging genes." Therefore the three must be considered to belong to only one species (named *P. ohirai* because of the priority of the name). Conversely, studies of the same DNA sequences have shown that two species are confounded under the name *Paragonimus westermani,* with the parasites of the China-Japan-Korea-Taiwan ensemble distinguishing themselves clearly from parasites of the Malaysia-Thailand-Philippines ensemble.

8. Last, Clayton, Gregory, and Price (1992) suspect that the phthiropterous insects of bird feathers, often cited as examples of narrow specificity, may include numerous nonvalid cryptic species such that these ectoparasites might in fact be generalists.

Such taxonomic difficulties concern not just parasites but also hosts, most notably vectors (see Crampton 1994; Tabachnick and Black 1995). For example, among the anopheles that transmit malaria there are several complexes (*Anopheles gambiae, A quadrimaculatus, A. freeborni,* etc.) that are each composed of closely related but reproductively isolated species: of sixty-six species considered capable of playing a role in malaria transmission, more than half belong to complexes of cryptic species (Besansky and Collins 1992). It is necessary to know these cryptic host species in order to correctly evaluate the host spectrum of the parasites, particularly in light of the growing drive for the development of bioengineered resistant vectors to control these devastating diseases. Obviously, failure to recognize cryptic vector species and their separate roles in transmission would seriously undermine any such effort at control.

The Host Spectrum

The General Characteristics of the Host Spectrum

Once the problem of identifying partners is resolved, it is possible to build an array of hosts for each species of parasite. This array, or host spectrum, is the list of species the parasite exploits at each particular stage of its development.

The host spectrum for a parasite generally exhibits four characteristics: (1) at any given stage of its life cycle a parasite usually only uses a few host species (and sometimes only one); (2) there is almost always a marked difference in the number of parasites that pass through the various species of the spectrum, with some host species playing a reduced or even negligible role relative to others in the cycle; (3) there is important geographic variability in the host spectra: they may be very different from one region to another because of genetic variation and local adaptation (see Kaltz and Shykoff 1998); and last (4) specialization for the host and specialization for the habitat are not linked.

Parasitologists have developed a precise vocabulary to characterize the parasite's degrees of requirement for a host. The following distinctions were made by Euzet and Combes (1980) relative to host spectra: *oioxenous species* live only in one host species (fig. 2.4A); *stenoxenous species* are found in a small group of related host species, usually in the same genus or family ("sister groups" in the terms of phylogeny; see fig. 2.4B); and *euryxenous species,* which live in unrelated host species. According to Janovy, Clopton, and Percival (1992) "the occupied hosts do not appear to form a monophyletic group" but are convergent in their ecology and ethology (fig. 2.4C).

Recognizing a parasite as stenoxenous or euryxenous is sometimes

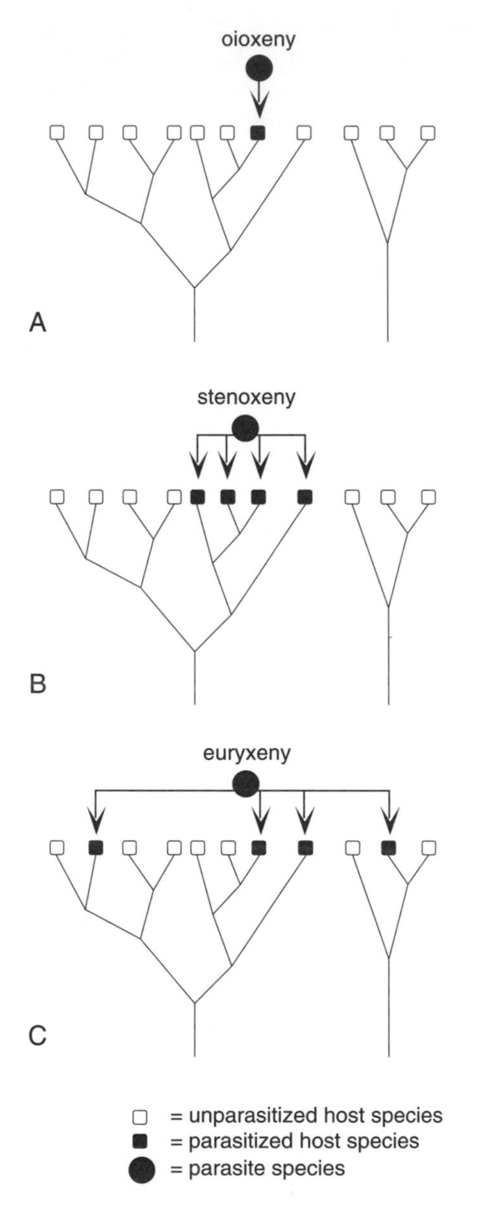

Figure 2.4. Distinction between oioxenous, stenoxenous, and euryxenous parasite species.

difficult and may require estimating the relative evolutionary proximity of the hosts if it is not already known with precision.

The terms "specialists" and "generalists" as used by parasitologists are borrowed from general ecology and designate parasites with, respectively, oioxenous/stenoxenous and euryxenous tendencies. It is also said that specificity is narrow when the host spectrum contains only a few species and wide in the opposite case. Importantly, the host spectrum should not be interpreted to demonstrate the relative tolerance of the parasite species for environmental conditions, for two reasons.

First, we generally know nothing about the "ecological distance" between any two host species; that is, we usually know little about the real differences in environment that these host species offer to the parasite. However, specificity results from the confrontation between tolerance of the parasite and variation in the available habitat. As such, "specialization is not favoured unless habitat differences are large in relation to the tolerance of the phenotype" (Adamson and Caira 1994). Thus two species of taxonomically distant hosts may offer convergent conditions relative to the habitat provided and two taxonomically close species may offer widely different ones. Moreover, when the number of species in a host's phylum is small the parasite may have little choice: it is difficult to evaluate the adaptive potential of parasitic organisms when evolution has left only a small number of available hosts in a phylum (Combes 1968).

The second reason the host spectrum does not truly demonstrate the parasite's tolerance of environmental conditions is that the spectrum is often deduced from a series of investigations carried out on the same parasite but in different locations. This approach may have the additive effect of making the list of host species longer, but it says nothing about the real needs of the parasite at a given point in its distribution. A parasite with extreme stringency for a particular host may appear as a generalist if we look only at the list of host species given in many parasitology textbooks.

In the following examples it is sometimes possible to invoke either the encounter or the compatibility filter, or sometimes both, to "explain" the host spectrum, whereas in other cases any such supposition is impossible.

The "Yes/No" Aspect of the Host Spectra

Specialization is a "normal" characteristic of the living world—no one finds it surprising that the same plant does not live on beaches, in humid prairies, and on the tundra. In parasitism, however, we are often surprised that certain species show a marked specialization for one host or a very few hosts, particularly if the ecosystem they live in seems to offer numerous taxonomically related species to exploit.

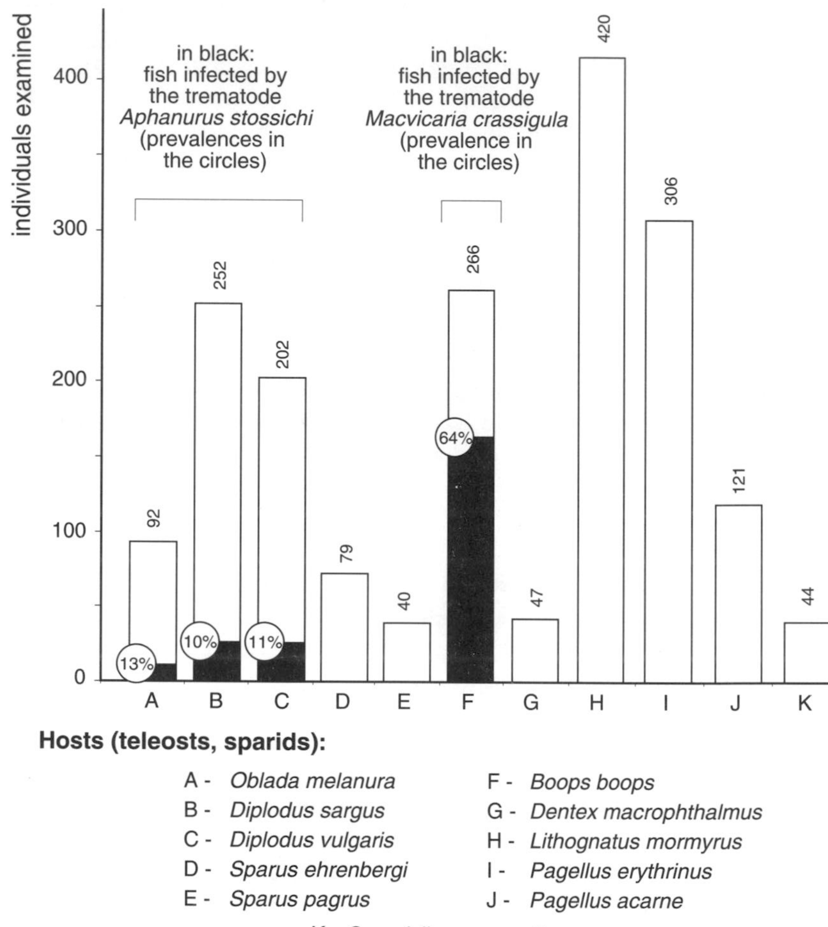

Figure 2.5. Investigation of parasitism in fish (Sparidae) in the same geographical region showing that each parasite species is harbored by only part of the host species theoretically available to it (data from Saad-Fares 1985).

Possible Explanation: The Encounter Filter. Saad-Fares (1985) studied intestinal trematodes of about a dozen marine fish, all belonging to the family Sparidae, along the eastern coasts of the Mediterranean. Although numerous trematodes were identified, none were found to occur in all eleven species of fish examined, even though large numbers were sampled. Figure 2.5 indicates, for each fish species, the size of the sample and the number of species that harbored two trematodes, *Aphanurus stossichi* and *Macvicaria crassigula*. We can see that three species of Sparidae harbored *A. stossichi*, only one harbored *M. crassigula*, and the others harbored neither.[1] In this case it is possible to imagine that the worms are excluded from the other possible hosts by the encounter filter. If the trematodes were transmitted by different vectors, not eating certain prey would protect specific host species from infection by the juvenile parasites within it. However, the compatibility filter could also be involved if, for example, certain intestinal environments do not fit some parasites.

Possible Explanation: The Compatibility Filter. Emson and Mladenov (1987) analyzed the distribution of the parasitic copepod *Ophyopsyllus reductus*, among seven species of closely related ophiurids (brittle stars) living together in algal clumps along the coast of Jamaica. Their results are summarized in figure 2.6. We can see that despite large sample sizes for most of the "potential" hosts, the copepod was found on only one species, *Ophiocomella ophiactoides*. Since *O. reductus* may easily be detached and reattached from its host, the researchers could experimentally offer the parasites the choice of different species of ophiurids. In all cases the parasites attached only to *O. ophiactoides*. This investigation shows the extreme but not exceptional case of a spectrum narrowed to one host species even though apparently closely related species coexist in the environment—a strong indication that it is the compatibility filter and not the encounter filter that is closed for the nonhost species of ophiurids.

The best demonstration of the consequences of the "yes/no" aspect of the compatibility filter is probably that of Tinsley and his collaborators, who worked on parasitism by several species of monogeneans of the genus *Protopolystoma* in African amphibians of the genus *Xenopus* (Tinsley and Jackson 1998a, 1998b). This genus of hosts includes one diploid species and a series of polyploid hybrids. If two diploid species, say AA and BB, give rise to a hybrid AABB by polyploidism, the hybrid would then have the genetic information of both AA and BB side by side. If the parent AA harbors parasite P1 and the parent BB parasite P2, it is expected that the hybrid AABB would be compatible to both P1 and P2. Tinsley's group showed that this hypothesis is in fact correct. Figure 2.7 shows (center) the two species, *Xenopus fraseri* and *X. laevis*, which both have thirty-six chromosomes. Above and below these two are *X. wittei* and *X. vestitus*, which are both considered hybrids of the two previous species based on morphological, biochemical, and molecular data. *Xenopus wittei* and *X. vestitus* are allopolyploids with

1. Specificity of fish trematodes is not always so narrow (see, for example, Barker et al. 1994).

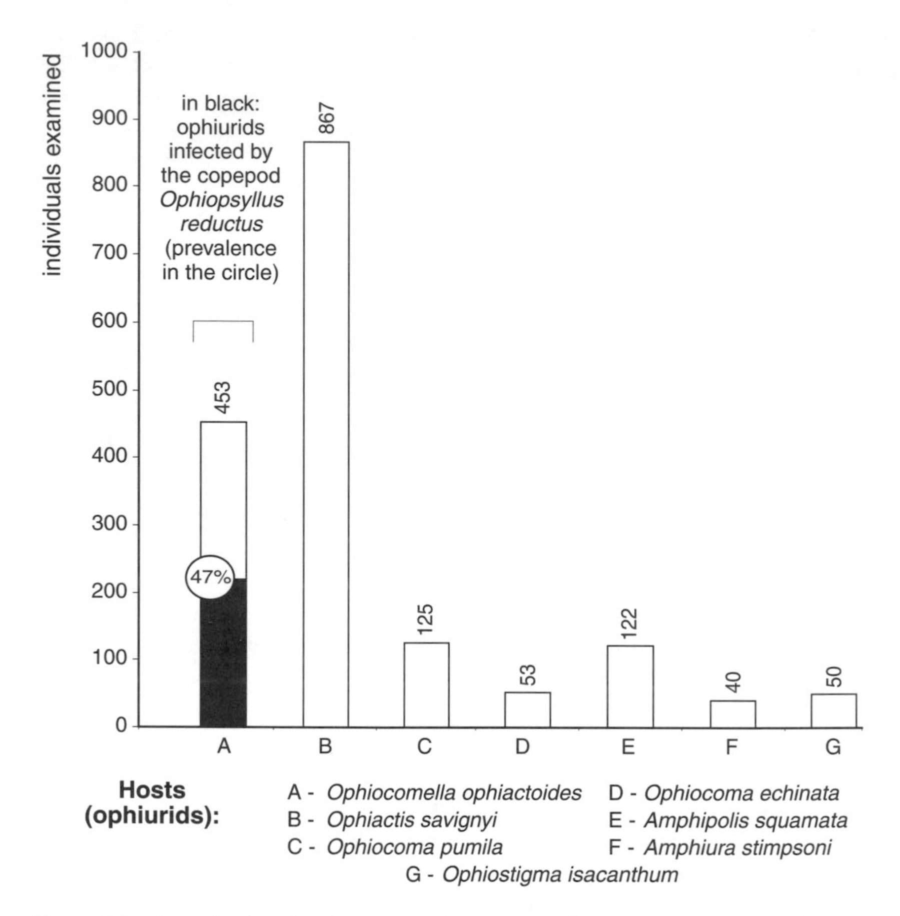

Figure 2.6. Investigation showing one species of copepod exploiting only one of the seven species of ophiurids available in the same environment (after Emson and Mladenov 1987).

seventy-two chromosomes each and thus, in principle, contain the information of both *X. fraseri* and *X. laevis*. The parasite *Protopolystoma simplicis* "follows" the genetic information of *X. laevis,* since it is found in *X. laevis* and also in *X. wittei* and *X. vestitus*. The parasite *P. fissilis* "follows" the information of *X. fraseri,* since it is found in *X. fraseri* as well as *X. wittei*.[2]

---

2. *Protopolystoma fissilis* would be expected to also be found in *Xenopus vestitus*. Tinsley and Jackson suggest that its absence may be due to either a secondary loss or a failure to colonize *X. vestitus* despite the presumed genetic links. Similarly, *X. fraseri* and *X. laevis* each have a specific *Protopolystoma* that has not colonized the hybrids.

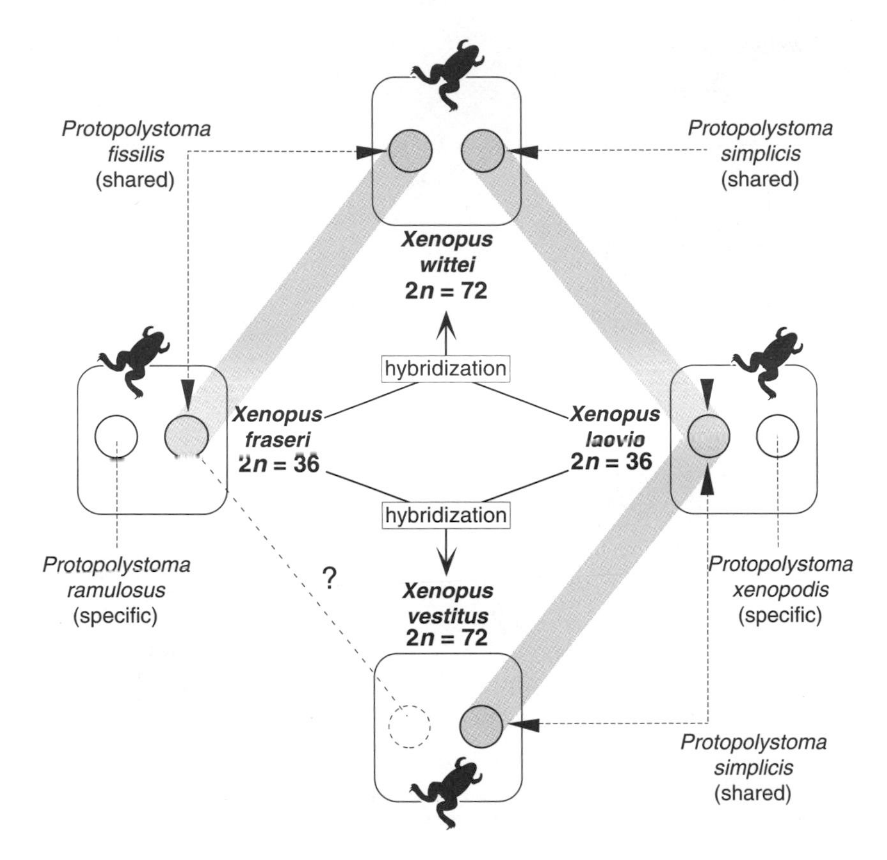

Figure 2.7. Specificity of *Protopolystoma* for xenopids: hybrid polyploid xenopids harbor parasites of both parent species (after Tinsley and Jackson 1998a, 1998b).

The Nuances of the Host Spectrum

When a parasite can infect several species of hosts, the rule is that the different hosts share the parasite population unequally. The probability that the encounter and compatibility filters will be opened exactly to the same degree in two or more host species is very small. Many examples demonstrating this could be cited; only a few are presented here.

Possible Explanation: The Encounter Filter. *Diplozoon gracile* is a gill monogenean that is found on four species of freshwater fish in the south of France: *Gobio gobio, Telestes soufia, Phoxinus phoxinus,* and *Barbus meridionalis.* Interestingly, however, another species of barbel, *Barbus barbus,* is never found to be infected. The host spectrum therefore does not respect

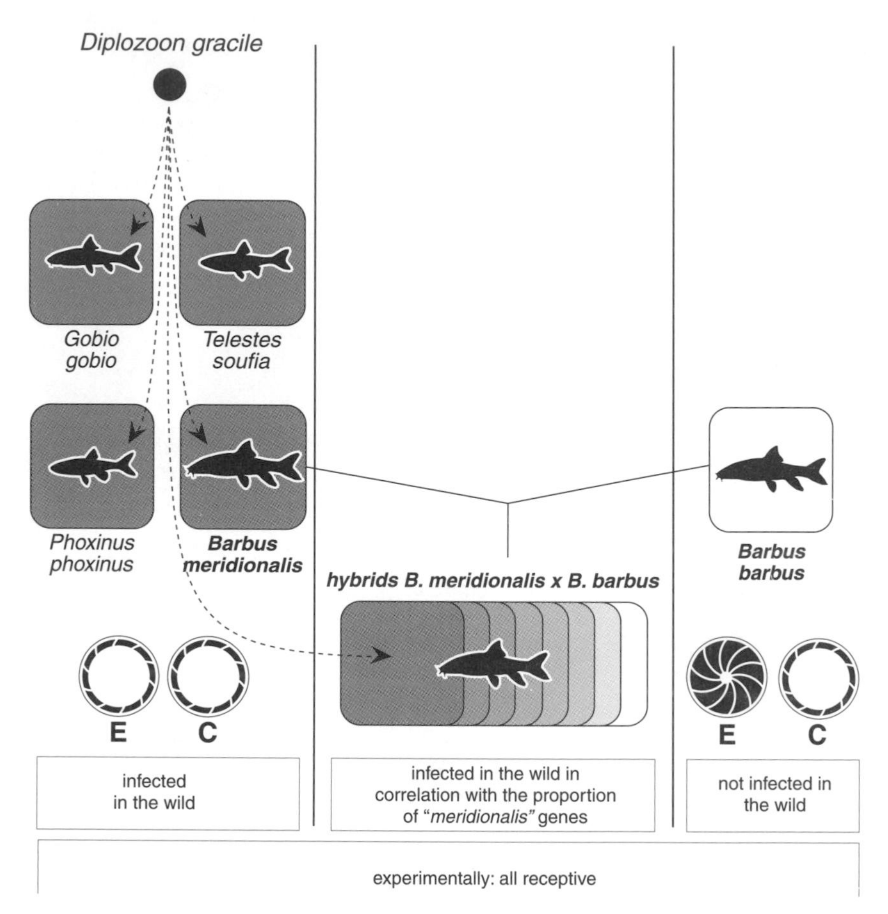

Figure 2.8. Host spectrum of *Diplozoon gracile* in rivers of the south of France. The compatibility filter (C) is open for the five species of freshwater fish that are represented. The encounter filter (E) is closed for *Barbus barbus* but is open for hybrid *Barbus meridionalis* (mult} *Barbus barbus* in proportion to the extent that "*meridionalis*" genes are present in the hybrids. Shaded squares denote degree of infection (dark = heavily infected; light = lightly infected; white = not infected.

the usual rule of a phyletic relationship between host species, since logically both barbels should be within the host spectrum.

Research by Le Brun, Renaud, and Lambert (1988) has shown how important the encounter filter is for the specificity of *D. gracile* (fig. 2.8). *Barbus meridionalis* essentially lives in small tributaries, while *B. barbus* lives in the main rivers. There is, however, an intermediate area where the two species hybridize. Experimentally, *B. meridionalis* and *B. barbus* become infected to the same degree with *D. gracile,* meaning that in this case *the compatibility filter is not closed.* In *G. gobio,* which is present over the entire area studied, the proportion of individual hosts infected by *D. gracile* is about equal in the different types of waters. Knowing that *B. barbus* lives in the central part of the rivers where the water column is maximal while *B. meridionalis* and the other fish hosts (with the exception of *G. gobio*) live close to the edges—and knowing that the infective larvae of *D. gracile* rarely swim in the open water but stay close to the bottom—it is reasonable to assume that the more a host hybrid has *B. meridionalis* genes the more it frequents the edge zones where the probability of encountering the *Diplozoon*'s larvae is greatest. That is, there is a direct correlation between the introgression of *B. meridionalis* genes and the degree of opening of the encounter filter. This is, in fact, the situation in nature: the level of parasitism is directly correlated with the introgression of *B. meridionalis* genes into the *Barbus* genome as determined by measurement of the allelic frequencies of several enzymes. Significantly, this work demonstrates how essential studies associating both genetics and behavior are to understanding the circulation of parasites in ecosystems.

A different opening of the encounter filter probably explains the differences in parasitism observed by Fiorillo and Font (1996) in four species of *Lepomis* (Centrarchidae) in Louisiana. These fish species are exploited in different ways by the acanthocephalan *Leptorhynchoides thecatus.* The prevalence, average intensity, and maximum intensity are, respectively: 3%, 1, and 1 in *Lepomis macrochirus;* 24%, 2.4, and 5 in *L. punctatus;* 100%, 9.9, and 31 in *L. megalotis;* and 100%, 31.6, and 88 in *L. microlophus.* The differences are explained by the diets of the four lepomids and therefore by the degree of opening of the encounter filter for the intermediate hosts that harbor the infective stages.

Possible Explanation: The Superposition of the Two Filters. The two following examples allow us to separate, up to a certain point, the role of the encounter and compatibility filters.

1. Metacercariae of the trematode *Labratrema minimus* parasitize several species of fish in Mediterranean lagoons (fig. 2.9). These metacercariae develop into adults when one of these fish is eaten by the piscivorous fish *Dicentrarchus labrax.* The fish that harbor metacercariae are atherines *(Atherina boyeri)*, gobies *(Pomatoschistus microps)*, gilthead seabreams *(Sparus aurata)*, and mullets *(Liza ramada)*. Faliex and Morand (1994) have shown that it is possible to estimate, if not the exact proportion of parasites that reach the definitive host via different intermediate hosts, then at least the relative importance of each. The flux of metacercariae carried by atherines into *D. labrax* is the largest, next that of gobies, and much less so that of the seabreams and mullets.

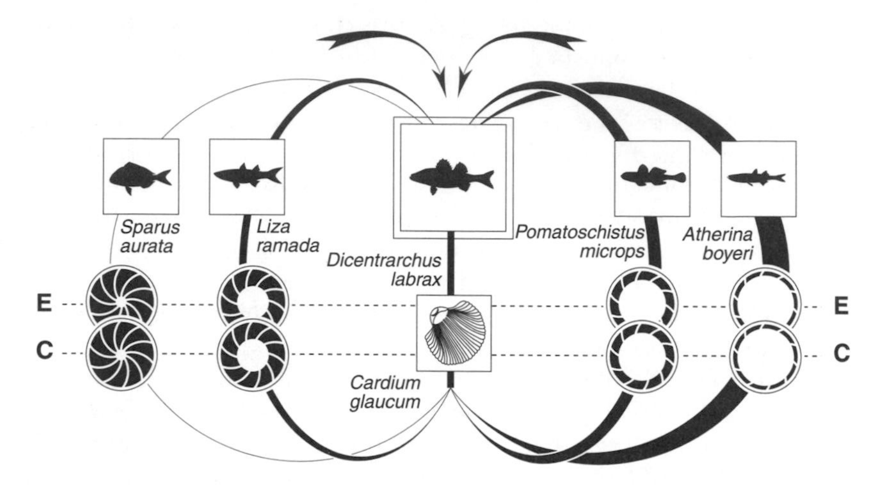

Figure 2.9. The transmission of the trematode *Labratrema minimus* in a Mediterranean lagoon, illustrating its use of four species of fish intermediate hosts for metacercariae (after data communicated by E. Faliex). The thickness of connecting lines denotes degree of use as determined by the relative opening of the encounter (E) and compatibility (C) filters.

In this parasite-host system the encounter between cercariae and fish happens by simple contact between the parasite and a fish. After emergence from the bivalve *Cardium glaucum,* the cercariae float passively in the water column by spreading two adhesive "tail-like" elongations, and the fish simply knock into them as they swim. Almost all the cercariae attach to the pectoral, pelvic, and caudal fins, after which they penetrate the tegument and encyst mainly in the liver. The processes described so far do not suggest that contact is easier for one fish species than for another. However, an essential difference does exist: only atherines and gobies are present year round in the lagoon where the mollusks giving off the cercariae are found. Therefore the encounter filter is, over time, more open for the atherines and gobies than for the other fish.

Compatibility in this system, which is estimated by the degree of histologic reaction in the liver and the rapidity of metacercarial maturation, is clearly different according to the fish species. Only atherines show no accumulation of cells, such as granulocytes and macrophages, around the hepatic metacercariae.

Differential specificity in this system therefore results from *both* the encounter filter and the compatibility filter.

2. A comparable example involves the distribution of the metacercariae of several trematodes in odonate nymphs found in a lake in North Carolina (Wetzel and Esch 1996). This work shows several marked corre-

lations between important parameters of parasitism (prevalence and intensity) and the ecologic, ethologic, and taxonomic characteristics of the odonate hosts (six anisopterans and three zygopteran species). Three correlations are particularly clear.

The host species, *Gomphus exilis,* a forager, is the least parasitized by the three trematodes studied (fig. 2.10A). The suggested explanation is that the nymph of this species, by circulating in the sediment, is partially protected from being infected by cercariae that are emitted into the open water. The encounter filter is therefore partially closed.

Erythemis simplicicollis, the smallest of all the odonates studied, is the most highly parasitized by *Halipegus occidualis* (fig. 2.10B). The explanation is that this small nymph most likely consumes the small crustaceans (ostracods) that are the source of infection for this worm. In this case the encounter filter is therefore relatively wide open.

Last, nymphs of the six species of anisopterans transmit all three species of trematodes, while the three species of zygopterans studied never transmit one of the species, *Haematoloechus longiplexus* (fig. 2.10C). In this case the compatibility filter, as opposed to the encounter filter, is likely closed, because the anisopterous nymphs surely meet the cercariae of the various trematodes.

Even if not all the factors influencing specificity can be identified in this example, one can still see that the trematodes present in odonate nymphs are influenced by characteristics linked to encounter, such as foraging behavior, and in the case of *H. longiplexus,* also linked to compatibility.

Possible Explanation: Unclear. If the clutch size of the great tit, *Parus major,* is experimentally increased, the males but not the females increase their feeding of the chicks. These males are about twice as heavily parasitized by some avian *Plasmodium* as are those that do not have an increased clutch size (Richner, Christe, and Oppliger 1995). Similar results have been obtained relative to infection of the same bird by species of *Leucocytozoon* (Norris, Anwar, and Read 1995). Right away we see that there are two possible explanations: the increased number of young in the nest requires the males to make more frequent feeding trips, thus exposing them to more bites of the insect vectors (i.e., the encounter filter is open wider), or the increased activity imposed on them decreases their allocation of resources for immune defense, thus increasing the success of infection (i.e., the compatibility filter is open wider).

Similarly, a several-year study of the lizard *Sceloporus occidentalis* has shown that, independent of location, year, and size, males are more parasitized by *Plasmodium mexicanum* than are females (Schall and Marghoob 1995). Such a result leaves open whether this is because the males are

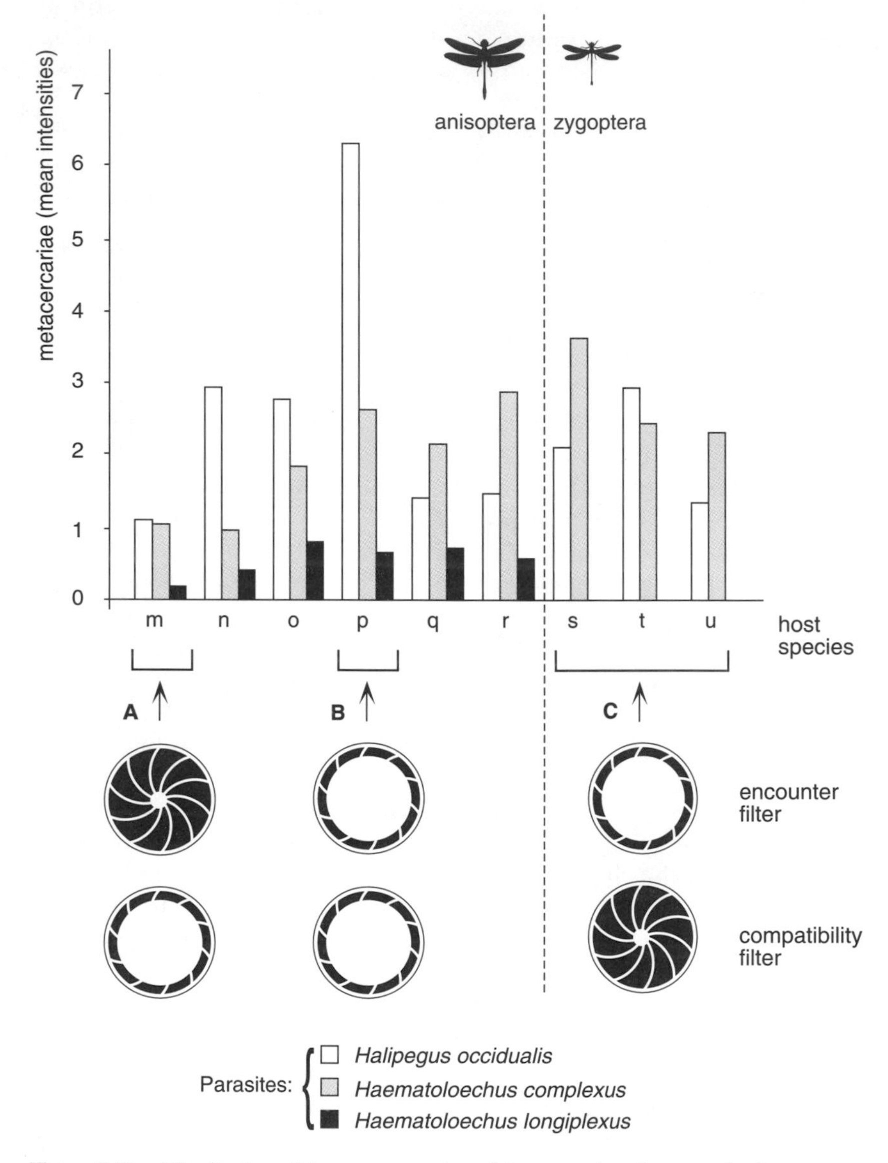

Figure 2.10. Distribution of the metacercariae of three species of trematodes in a population of odonate nymphs. A designates the foraging species, which is clearly less parasitized than the benthic species; B designates the species that consume small crustaceans, which are highly parasitized; and C designates the three species of zygopterans that never transmit the trematode *Haematoloechus longiplexus* (after Wetzel and Esch 1996).

more exposed to infection as the result of behaviors different from those of females or whether it results from increased compatibility because of, for example, the depressing effect of male hormones on the host immunodefense system.

Variability in Host Spectra

One aspect of the variable nature of specificity appears when we compare apparently related species. A well-known example involves the human schistosomes: *Schistosoma haematobium* parasitizes only humans; *S. mansoni* parasitizes humans and numerous rodents; and *S. japonicum* parasitizes dozens of mammalian species, including humans. It may also be, however, that such asymmetries, particularly the striking example of *S. japonicum,* in fact reflect insufficient taxonomic data and not true variability in specificity. The studies of Bayssade-Dufour (1980) on the localization of the sensorial setae in the cercarial tegument have shown that *S. japonicum* may be related to the schistosomatids of birds. These bird schistosomes have a very broad specificity, with, for example, *Austrobilharzia variglandis* being able to parasitize dozens of marine bird species. Maybe the broad host spectrum of *S. japonicum* in mammals can be explained by such a relationship.

A second aspect of variability in specificity occurs within species and depends on the geographic location of the populations of both the intermediate and the definitive hosts. That is, the compatibility filter of a parasite may be opened or closed for hosts of different scaldfish species according to the geographic area considered. This is well demonstrated by examining the intermediate hosts of schistosomes. Véra (1991) has been able to partition the genetic diversity of the parasite and that of its snail vectors in West Africa. This author has studied experimentally the compatibility of two populations of schistosomes, one from western Niger (here called NIG) and the other from the northern area of Ivory Coast (here called IVO), with the two major species of the snail host *(Bulinus)* in these regions (Bertrand Sellin, Philippe Brémond, and Charles Véra, personal communication).

Cross-infections in the laboratory have shown (fig. 2.11) that *S. haematobium* NIG is strongly compatible with *B. truncatus,* whatever their origin (NIG or IVO); that *S. haematobium* NIG shows little or no compatibility with *B. globosus* NIG (the local mollusks) but is compatible with *B. globosus* IVO; that *S. haematobium* IVO is strongly compatible with local *B. globosus* IVO but has little compatibility with *B. globosus* NIG; and that *S. haematobium* IVO has little compatibility with *B. truncatus,* whatever their origin (IVO or NIG).

At first these results seem complicated, but the take-home message is in fact quite simple: west of Niger, *S. haematobium* uses *B. truncatus* as

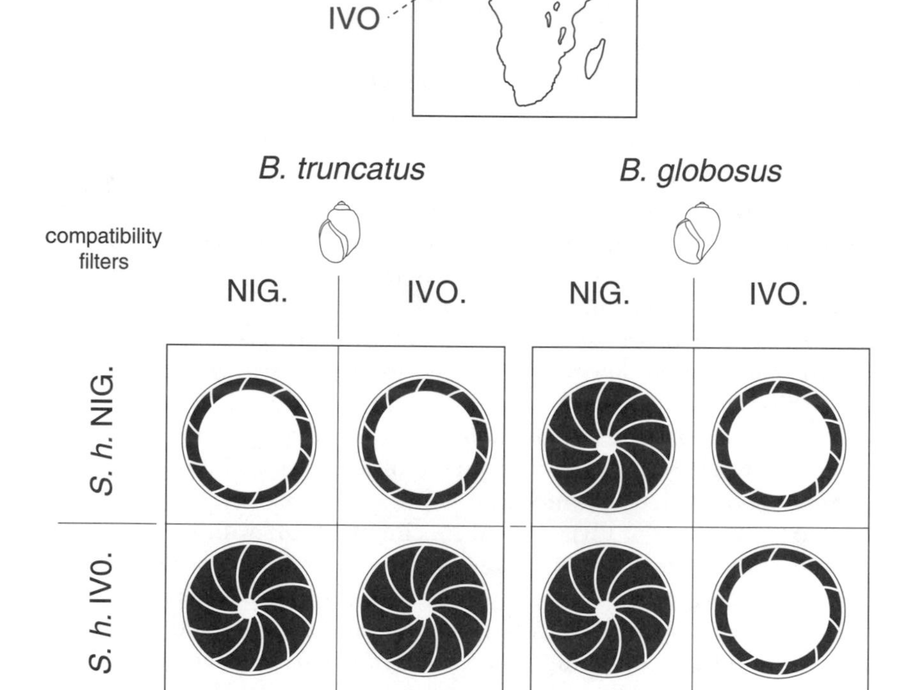

Figure 2.11. Opening of the compatibility filters for the two populations of *Schistosoma haematobium* in local and foreign populations of *Bulinus* (after the experiments of C. Véra). IVO. = Ivory Coast; NIG. = Niger; *S.h.* = *Schistosoma haematobium.*

its principal vector even though *B. globosus* is present; in Ivory Coast, *S. haematobium* uses *B. globosus* as its principal vector, although *B. truncatus* is also present. Based on this information, we can formulate the following hypotheses.

1. Relative to the difference of compatibility with *B. globosus,* the limiting factor seems to be the species of mollusk, since *S. haematobium* NIG as well as IVO experimentally accepts *B. globosus* IVO but not *B. globosus* NIG (see the opening of the filters symbolized on the right side of fig. 2.11).

2. For the differences in compatibility with *B. truncatus* it is the parasites that seem to be the limiting factor, since *S. haematobium* NIG is compatible with *B. truncatus* wherever it comes from, while *S. haematobium*

IVO is incompatible with *B. truncatus* wherever it comes from (see the opening of the filters symbolized on the left side of fig. 2.11).

This research shows that, depending on the geographic region, the distribution of parasites in hosts may be very different and is related to the genetic diversity of both the parasite and the host. Preston and Southgate (1994) consider that, in the case of schistosomes, the geographic variation in compatibility depends at least in part on the diversity of proteins expressed in the molluskan intermediate hosts' hemolymph as well as those on the surface of the miracidia. For example, molluskan agglutinins may play an important role in recognizing the parasites, and their variation could come from their being glycosylated in different ways.

Numerous situations analogous to the mollusk just discussed are also known for insects. Mosquitoes from the genera *Anopheles, Aedes,* and *Mansonia* (as well as others) transmit a wide array of causative agents of diseases including viruses (yellow fever, dengue), protozoans (various species of *Plasmodium*), and nematodes (*Wucheria bancrofti* in humans and *Dirofilaria immitis* in the dog). Depending on the region, the percentage of mosquitoes of a given species susceptible to a given population of pathogenic agent may vary from 0 to 100%.

Determinants of Host Spectra

Although the variable nature of the width of the host spectrum may leave one perplexed, it does not prevent us from relating this diversity to particular characteristics of parasite-host systems in general. That is, various hypotheses of causation can be formulated either by comparing a small number of species or by using information available in data banks.

Hypothesis 1: The host spectrum is characteristic of the taxonomic group of parasites

This hypothesis is based on the supposition that there is a phyletic constraint that durably influences specificity. The hypothesis is verified in numerous cases, since parasite groups are often characterized by their specificities. For some time now, parasitologists have recognized highly specific groups of parasites: among them are the monogenean parasites of fish, where oioxeny is usually the rule (Baer and Euzet 1961); the cestodes, with the most clear-cut cases being those of cartilaginous fish (elasmobranchs) (Euzet 1956) and birds (Baer 1957); and the insects, such as the anoplurans (lice) and siphonopterans (fleas).

Based on investigations made in both Canada and the Mediterranean, Sasal, Desdevises, and Morand (1998) confirm that there is such a "phylogenetic conservatism" in specificity. Note, however, that there are taxonomic groups for which it is impossible to attribute a particular degree of specificity. Such is the case with the schistosomes, for which, as we have seen, there are oioxenous *(Schistosoma haematobium)*, stenoxenous *(S. mansoni)*, and euryxenous *(S. japonicum)* species.

Although there have been many discussions about the reason certain groups of parasites are more specific than others, no conclusion with general consequences has ever been reached on the subject. To say that some parasites have stricter requirements for exploiting space (attachment to the host) or resources (metabolism) only displaces the question, and Adamson and Caira (1994) suggest that the observed specificity of the "current" parasites may also reflect some trace of the predation habits of their remote free-living ancestors.

Hypothesis 2: The width of the host spectrum is correlated with the age of the parasite group on an evolutionary scale

This hypothesis suggests that specificity is narrower or wider depending on the parasite's being more or less "young" on the global scale of biological evolution. The simple comparison of prokaryotes, eukaryotic protozoans, and metazoan eukaryotes, which all contain both specialist and generalist parasites, indicates that this hypothesis, contrary to the preceding one, is not justified. It could, however, have great value at a less ambitious scale. For example, among the Platyhelminthes cestodes are usually considered more specific than trematodes relative to their vertebrate definitive hosts. However, the very narrow adaptation to parasitism by the cestodes (attested by the total disappearance of the digestive tract), and also the fact that they are very abundant in the elasmobranchs ("primitive" vertebrates), most likely makes them "older" parasites than the trematodes. Thus in this case there would be a correlation between narrow specificity and the older age of the group that is explainable by ever increased adaptation and specialization over a longer period (see Price 1977).

Hypothesis 3: The host spectrum depends on the taxonomic group of hosts being exploited

To test this hypothesis one must first compare specificity within the same group of parasites found in different groups of hosts. Setting apart viruses and bacteria, this is possible for the trematodes and cestodes, which are both found in all five vertebrate classes (fish, amphibians, reptiles, birds, and mammals). As already noted, the cestodes of elasmobranchs and of birds are particularly specific.

There is an interesting track to explore within the framework of this hypothesis, and that is to determine if the groups of hosts of numerous and related species have parasites with wide host spectra that could be explained simply by the proximity and similarity of the environment these species offer the parasites.

Poulin (1992b), by comparing a series of investigations made on the copepod and helminth parasites of freshwater fish, showed that this proposition is at least partially supported. A parasite that has as its host a family of fish containing few species is in general narrower in specificity than a parasite whose host family has numerous species. Sasal, Desdevises, and Morand (1998) support this hypothesis in their previously discussed analysis of parasitism in fish from Canada and the Mediterranean.

Hypothesis 4: The host spectrum differs depending on the mode of infection

It has been said that parasites that infect their hosts via a free-living active stage may be more specific than those ingested or transmitted by a biting vector (Noble, Noble, and Schad 1989). This has certainly not always been supported: cestodes, which are transmitted to their definitive hosts by ingestion, and monogeneans, which infect fish by an actively swimming free stage, both are characterized by narrow specificities.

Hypothesis 5: The host spectrum differs depending on the complexity of the cycle

It does not seem that parasites with heteroxenous life cycles show a difference in specificity, at least for their definitive hosts, compared with those that have holoxenous cycles.

Hypothesis 6: The host spectrum differs depending on the microhabitat

Adamson and Caira (1994) consider that intestinal parasites that do not induce lesions should be less specific than parasites that penetrate tissues. However, this does not seem compatible with the case of intestinal cestodes, which exhibit little or no intrusiveness and are, overall, highly specific. A different hypothesis has been suggested by John C. Holmes (personal communication), who thinks parasites that are found in the sites richest in nutrients and that are the most stable (for example, the middle portion of the intestine) are more specific than those found in poorer and more unstable habitats (for example, the anterior portion of the intestine and the rectum). However, this could also be an indirect effect of interspecific competition, which is stronger in the most favorable habitats.

Hypothesis 7: The host spectrum is influenced by the more or less random nature of transmission

Holmes (1990a) studied parasites of the marine fish *Sebastes nebulosus,* which exists in small populations that correspond to rocky emergences more or less isolated from one another. Such a fragmented host population results in the sporadic appearance and disappearance of some parasites because of the "ups and down" associated with transmission, with repopulation occurring when infected fish of various species just pass by. Holmes noticed that almost all parasites of *S. nebulosus* were not specific.

From this arises his remark, "Perhaps periodic local extinctions mitigate the development of strict host specificity." Pushing the hypothesis to its fullest (Combes 1995), we may suppose that a narrow specificity can be selected for only in an environment sufficiently stable to warrant that the cycle is not periodically interrupted, and only if the hosts form abundant and relatively unfragmented populations.

In their analysis of monogenean parasitism in Mediterranean fish, Sasal, Niquil, and Bartoli (1999) support the hypothesis that generalist parasites would be the only ones able to adjust to host species whose populations are not very predictable. One may also conclude that selection for nonspecific species happens when transmission conditions become difficult, with generalist species favored relative to specialist species in such circumstances.

Hypothesis 8: The host spectrum is characteristic of the level of the cycle being considered

It has been emphasized for some time that specificity is rather wide for hosts in which the durable interaction implies only a limited metabolic exchange (for example, the hosts of encysted stages such as metacercariae), whereas specificity is much narrower when such exchanges are not as limited (e.g., when there is sexual or asexual multiplication of the parasite). Other explanations still need to be found for those cases, such as the trematodes, where specificity for the mollusk intermediate host (where asexual multiplication occurs) is almost always narrower than specificity for the vertebrate host (where sexual reproduction occurs).

Hypothesis 9: The host spectrum is wider in restricted geographic areas

This hypothesis, due to Freeland (1983), rests on the following considerations: over a large geographic area (a continent or large island) a narrow specificity is established because only a few host species provide a favorable habitat for a given parasite. Individuals that develop in less favorable hosts transmit fewer of their genes and therefore are counterselected.

On a smaller or more restricted geographic scale (i.e, a small island), free-living species occur in limited numbers and are subject to random extinctions. Under these conditions the parasites that are successful are those that can invade several species of hosts. Islands, then, should logically be incubators of parasite faunas consisting primarily of generalists. The work of Thomas (1953) on the nematodes and trematodes of the small mammals of the Inner Hebrides seems to refute this hypothesis, whereas support for it comes from Mas-Coma and Montoliu (1978, 1987), who studied the life cycle of the trematode *Dollfusinus frontalis,* a parasite of the frontal sinuses of mammals. Although *D. frontalis* is typically a parasite of insectivores (the hedgehog, *Erinaceus europaeus*) in continental

Europe, it invades populations of smaller rodents (the black rat, *Rattus rattus,* and the garden dormouse, *Eliomys quercinus*) on the small Mediterranean island of Formentera, thus supporting a widening of the host spectrum on islands.

Abundance-Specificity Relationships

It is generally admitted that there is a correlation between the local abundance of free-living species and their regional distributions (Brown 1995). In other words, species that are abundant in patches have a tendency to be observed frequently in the patches. This also occurs for parasitic species. Barker, Marcogliese, and Cone (1996) showed, for example, that there was a positive correlation between the local abundance of eel parasites and the number of localities where they were encountered in Nova Scotia.

Poulin (1998b) asked two other questions relative to abundance and specificity:

1. If host individuals are considered to be the patches (fig. 2.12, top), are the most abundant parasitic species in individuals also the most frequent? That is, is there a relationship between intensity and prevalence? The answer is yes, and it has been demonstrated in a number of investigations.

2. Now, if host species are considered to be patches (fig. 2.12, bottom), are the most abundant parasitic species in populations (high prevalence and intensities) also more frequent in the host species? That is, do they have the widest host spectra?

Poulin makes the prediction that the relationship must be reversed (that is, the higher the abundance, the narrower the specificity) in the following terms: "Species that occur in large numbers in many host individuals . . . are likely to occur in fewer patches, or host species. The logic behind this prediction is that adaptations to combat host immune defenses are costly, and that parasites specializing on very few hosts may achieve greater abundance in those hosts than if they had to invest in a wider range of evasive mechanisms to infect a wider array of host species. . . . There may thus be a trade-off between host specificity and the average abundance parasites can achieve in their hosts."

Poulin demonstrates this by a comparative analysis of the helminths of freshwater fish in Canada and suggests "a continuum from high abundance in few host species to low abundance in many host species." However, he also shows (Poulin 1999b), based on data obtained in Azerbaijan by Vaidova (1978), that the relationship between the abundance and distribution of helminth parasites in birds is different from that found in fish. In birds the ability to exploit several host species and the potential for heavy parasite loads in individual hosts are positively linked. The explanation Poulin suggests is that birds are exposed to a wider range of

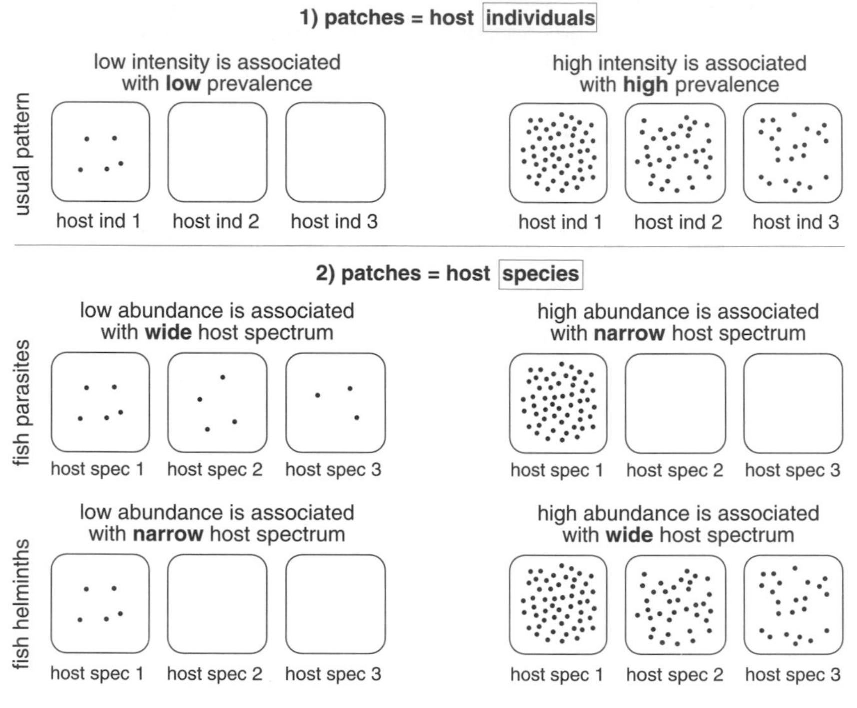

Figure 2.12. Top: at the level of the individual, a parasite may in general be both abundant and frequent. Bottom: at the level of the species, an abundant parasite is not necessarily a generalist (after Poulin 1998b): ind = individual; spec = species.

helminth species than fish and also offer a greater variety of microhabitats. A definitive explanation of the relationship between parasite abundance and host specificity, however, awaits further research.

I Love You Too[3]

As discussed earlier, four factors are involved in the formation of a host spectrum: biodiversity, behavior, resources, and host resistance.

Among these four factors, the two that determine incompatibility are *resources* and *resistance*. These two factors are particularly difficult to distinguish, whereas *biodiversity* and *behavior*, which constitute the encounter filter, are easier. The question is whether parasites that cannot develop in a host "suffer either from the lack of necessary factors, or from the presence of adverse factors" (Cioli, Knopf, and Senft 1977).

3. Translation of "Je t'aime, moi non plus," title of a song by S. Gainsbourg and J. Birkin that denotes a two-way love-hate relationship loaded with conflicts between the protagonists.

Figure 2.13 explains the dilemma. Here the same spectrum of hosts can be obtained by a balanced dosage of the respective roles of resources and resistance (A), by a dominant role for resources (B), or by a dominant role for resistance (C).

To understand this situation, we must first clarify some vocabulary. After a period of confusion, most authors have adopted the following definitions.

An organism in which the given parasite completely develops (until the transmission of its genes) is called *compatible* (also often referred to as *susceptible, permissive,* or *receptive*). When not compatible, the host may either be *resistant* or *unsuitable.* A *resistant* organism is one in which a parasite cannot develop because of host defense mechanisms. An organism in which a parasite cannot develop for reasons other than those related to immunodefense is often called *refractory.* However, there is a problem with using the term "refractory" to describe the failure of a parasite to develop because "refractory" has also been used in the past to denote resistance. Therefore here I will use the word "unsuitable," which better expresses the inadequacy of the host for the parasite, noting that the unsuitable character of a host may arise from factors not related to the host itself, such as the presence of competing parasites.

Anthropomorphically speaking, parasites can "perceive" other living organisms as belonging to one of three categories (fig. 2.14):[4] those they "love" and infect; those they would love to live in but cannot infect; and those they do not like and do not infect.

This is not very different from the perception that predators have of other organisms, among which are those they "love" and can capture; those they would love to consume but who run faster than they do; and those they do not like and do not pursue.

Various indexes lead parasitologists to conclude that host defense mechanisms represent one of the "dominant" parameters making up the compatibility filter and thus that situation C of fig. 2.13 is the most frequently observed. That is, *a nonsusceptible host would more often be resistant than unsuitable.* This is supported by several considerations, five of which are discussed here.

1. I shall cite first the experiments carried out by Cioli and Dennert (1976) on the development of *Schistosoma mansoni* in immunosuppressed white rats *(Rattus norvegicus).* In this system control rats normally eliminate schistosomes between the fourth and fifth week after exposure to cercariae, and very few adult schistosomes are found after the fifth week.

---

4. See the acknowledgments to this book to prevent any anthropomorphic interpretation of these imagined illustrations.

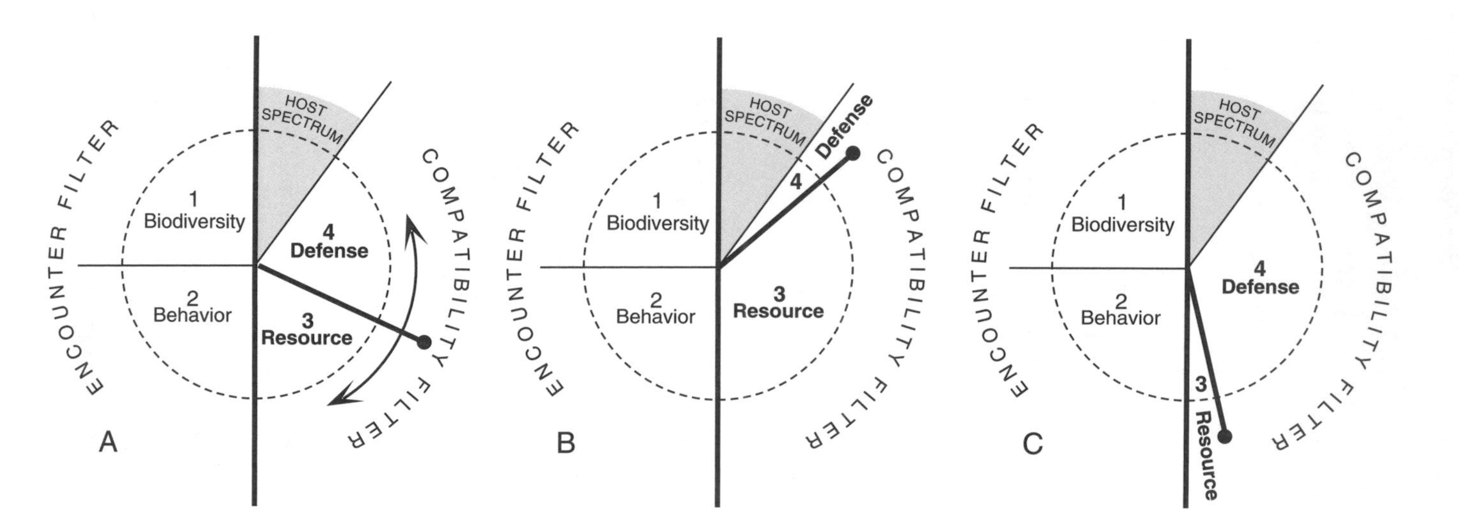

Figure 2.13. Diagram representing the limiting factors of the host spectrum in the compatibility filter. In A, limitation is due in part to resources and in part to resistance; in B, it is due mainly to resources; and in C, mainly to resistance.

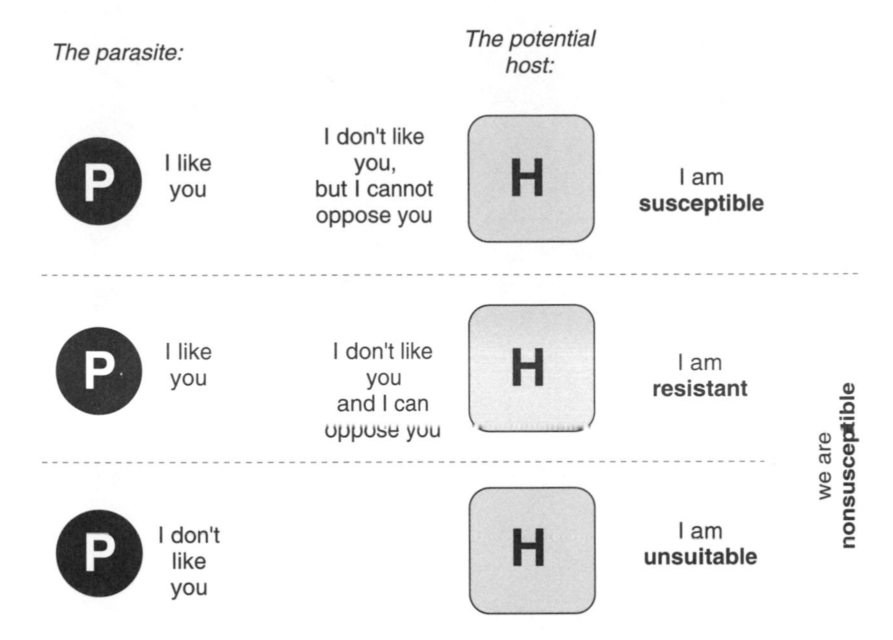

Figure 2.14. Susceptibility and nonsusceptibility; unsuitability and resistance.

In immunosuppressed rats, however, the number of schistosomes recovered after the fifth week is considerable. Figure 2.15, built from the data of Cioli and Dennert, leaves little doubt that the differences between control and immunosuppressed animals are highly significant ($p < 0.005$). However, the experimenters failed to obtain either the same recovery rate or the same level of maturity of worms relative to those obtained in a normally receptive animal, such as the black rat, *Rattus rattus*. As such, increased survival in the noncompatible immunosuppressed host indicates that though immunity is strongly involved in determining specificity, it is not the only causative factor.

2. Research done in nature, or that involving experimental infections as well as in vitro manipulation, confirms that parasites develop easily in a normally noncompatible host if something decreases the level of defensive response. For example, there are several cases where certain trematodes develop in specific species of mollusks only if a different species of trematode has prepared the way. This is the case for *Schistosoma bovis,* which seems to be incapable of infecting healthy *Bulinus tropicus*. However, if *B. tropicus* is already infected by worms of the genus *Calicophoron,* development of *S. bovis* is possible (Southgate et al. 1989). Similarly, Walker (1979) has shown that *Austrobilharzia terrigalensis* is able to

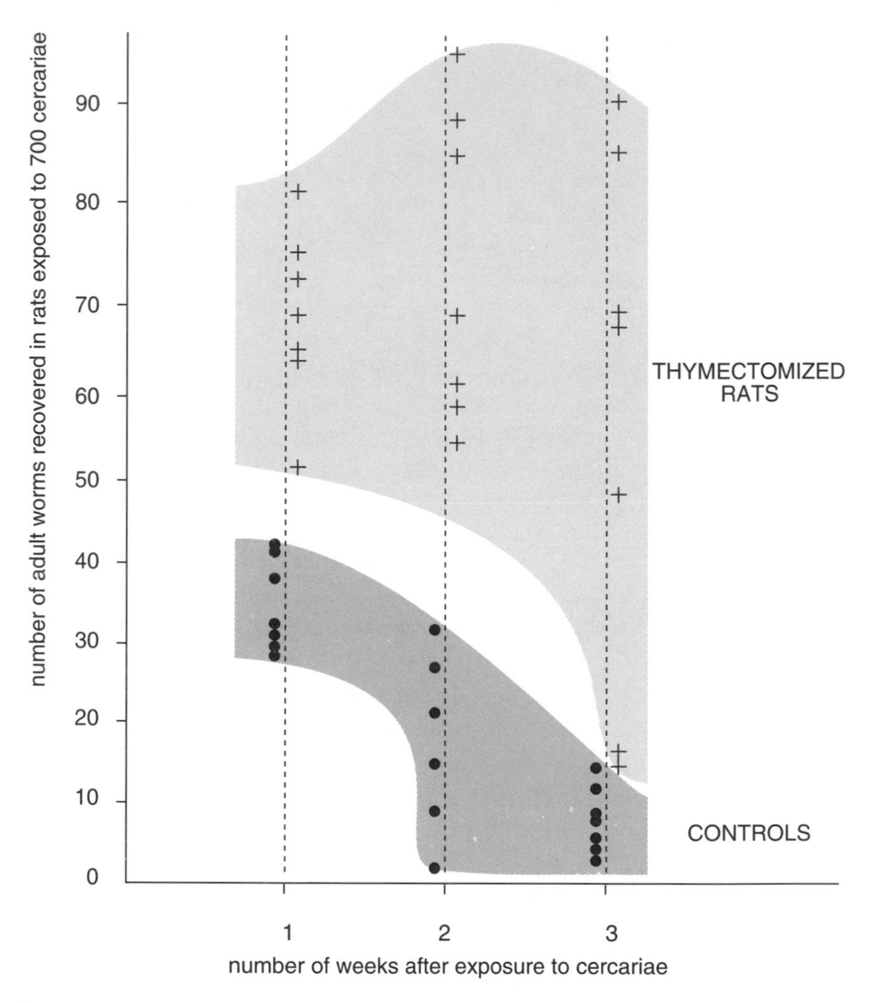

Figure 2.15. Experiments of Cioli and Dennert showing the numbers of recovered schistosomes in rats. Rats that normally reject schistosomes become receptive if they are experimentally immunosuppressed (after the data of Cioli and Dennert 1976).

parasitize the mollusk *Velacumentus australis* only if it already harbors a different species of trematode. This has led Kuris (1990) to conclude that other species of *Australobilharzia* may also be "obligatory secondary invaders."

Similarly, laboratory experiments show that a mollusk normally resistant to a parasite may become susceptible to that parasite if it is already parasitized by another. This is the case with *Echinostoma paraensei,* which renders a strain of *Biomphalaria glabrata* that is normally resistant to *Schistosoma mansoni* susceptible to it (Lie and Richards 1977). In vitro experiments on this system have shown that the sporocysts (intramolluskan larval stages) of *E. paraensei* produce an as yet to be identified substance that prevents the hemocytes (macrophagelike defensive cells of the invertebrate) from adhering to *S. mansoni* sporocysts and killing them (Loker, Cimino, and Hertel 1992).

Last, some parasitic diseases affect humans only after a splenectomy causes immunodeficiency. This is the case in babesiosis, which is caused by a protozoan in the genus *Babesia* transmitted by ticks—human cases involve splenectomized persons.

The most convincing demonstration of the role of defensive mechanisms in restricting the spread of the host spectrum is, however, probably given by the numerous "opportunistic" parasitoses observed in persons with AIDS. Immunodeficient people may be victims not only of the uncontrolled development of parasites that are otherwise normally contained (such as the fungus *Candida albicans*, the protozoans, *Cryptosporidium parvum, Toxoplasma gondii,* and *Pneumocystis carinii,* and numerous bacteria and viruses) but also parasites that are rarely found in humans, such as several species of microsporidians (see Canning and Hollister 1987).

In each of the cases above it could be imagined that one or the other of the parasites had modified a factor within their host other than immunity. It is, however, more logical to conclude that *S. bovis* in the first example, and *S. mansoni* in the two others, takes advantage of an immunosuppression induced by the first invader (Loker 1994). This is obviously not contestable in the cases of those opportunistic parasitoses linked to AIDS. Since we are used to the great ability of parasites to take advantage of situations favorable to them, it should not be surprising to see some of them use the process just described (obtaining a "helping hand" from another) to establish themselves (see chapter 10).

3. In phytopathogen-plant relationships the narrow specificity of bacteria, such as *Pseudomonas* and *Xanthomonas,* is usually related to a host's defensive mechanisms and, in particular, to the plant's capacity to recognize the pathogen. Van der Ackerveken and Bonas (1997) write: "One very important component of successful infection is to avoid recognition by the host. In fact, the narrow host range of these pathogens is thought to result from the ability of the majority of plants (the so-called 'non-host' plants) to recognize the bacteria at an early stage of infection and to mount a battery of defense responses."

4. It may be interesting to compare the specificity in strict parasitic systems (where the parasite does not bring any advantage to its host) with

that in more mutualistic systems (where the parasite brings in innovative genes that benefit the host). Indeed, one may think that defense against the parasite weakens or even disappears in mutualistic relationships. Such mutualistic systems should logically be characterized by a broadening of the host spectrum, since immune avoidance would then be useless. Classical texts explain that this is the case, and it is thus another argument to support the causative relationship between immunity and specificity.

5. Research on hybrid compatibility supplies even more information in support of resistance as the major player in nonsusceptibility: in Cameroon, *Schistosoma haematobium* uses the vector *Bulinus rohlfsi,* while *S. intercalatum* uses *B. forskalii.* Experimental research (Southgate, Van Wijk, and Wright 1976) has shown than hybrids of *S. haematobium* × *S. intercalatum* are capable of developing in both species of mollusks, just as if the hybrids had molecules that allow them to avoid the defense mechanisms of both mollusks at the same time.

Coustau, Renaud, Maillard, et al. (1991) have shown that the trematode *Prosorhynchus squamatus* shows a very different affinity for the northern mussel *Mytilus edulis* (susceptible) and the southern mussel *M. galloprovincialis* (nonsusceptible). In zones where the two bivalves hybridize, it is the introgression of *M. edulis* genes that allows the parasite to develop in the hybrids. The logical interpretation is therefore that *M. edulis* does not possess genes that allow it to recognize or destroy *P. squamatus.*

In both cases the results show that there is heritability in the characteristics linked to compatibility and suggest that these characteristics may be linked to the host's resistance and not to its being unsuitable.

A theoretical consideration: Living matter shows great continuity across life. For example, the enzymes that break down rice protein and those that break down human protein of the same type need not be *very* different. In fact, with the amino acids from ingested rice proteins, humans easily build their own such proteins. If we limit ourselves to one taxonomic group—for example, the mammals—such molecular similarities are even more common. As such, it is not a stretch to imagine a parasite that could develop, if not in all living organisms, at least in thousands or even tens of thousands of species. That this does not happen is probably due to the huge diversity of host molecules that can be involved in defense. In humans, for example, tens of millions of different antibodies may exist. Although it is theoretically possible for a parasite to colonize any living organism, *it is also probably impossible for it to fight against every one.*

The privileged role I have assigned to internal defense systems in the mechanism of compatibility implies that hosts play a major role in configuring the filter of compatibility and thus also in the determination of parasite specificity. This point of view is not unanimously accepted, and the important role given to defense systems here does not mean that other major factors affecting specificity should be ignored. Enzyme-substrate relationships, for example, could play a role as soon as an enzyme of the parasite must link to a particular substrate of the host. Foley and Tilley (1995) proposed that susceptibility or resistance of mammals for *Plasmodium falciparum* depends on the structure of a membrane protein found on red blood cells. This protein would normally be cleaved by a parasite protease, whereas in resistant species cleavage would be impossible because of the spatial conformation of the protein.

There is, in fact, a factor that has been left aside up to now for simplification. We have considered situations in which a host species is associated with only a single parasite species. This is a caricature of the real situation in nature because a host rarely or never harbors only one species of parasite. Within the host, every parasite must deal with other parasite species, each of which can help create a more hostile milieu. When we look at competition later in the book we shall see that initial infection by certain parasites, not just the host, can cause the closing of the filter of compatibility to other species. In a way this is nothing other than host immunity being coded for by the parasites' genes.

Between Encounter and Compatibility: The Identification of the Host Species

The Risk of Getting Lost

When infection involves consumption or the intervention of a biting vector, parasite identification of the host can only be "postinvasive." That is, the parasite is not itself directly responsible for ending up in an adequate host. Much as windborne seeds may fall on good soil, a few parasites carried by vectors may end up in an adequate host while many others simply get lost. If a prey has a "spectrum of predators" more open than that of the host spectrum of the larval parasite it harbors, then each parasite that ends up in a nonhost predator is lost. For example, if a frog parasite is transmitted by an insect, the parasite is lost each time the insect is eaten by a bird, a bat, or a spider.

Similarly, when a biting arthropod harboring a parasite bites an organism that is not one where the parasite can develop, the infective stages are cheated out of their chance to make it and are also lost. An experiment by Philippon (1977) is significant in this regard: in the African savanna, simulian (blackfly) vectors of human onchocerciasis bite humans, donkeys, cows, sheep, and goats, although the parasite it harbors (the nematode, *Onchocerca volvulus*) develops only in humans. Over a six-day period Philippon captured the blackflies on a man and a donkey placed side by side (fig. 2.16). The results showed that, in proportion to surface area, the blackflies were indifferent about which host they bit; 423 blackflies (of which 39 harbored the parasite) attacked the man, and 2,918 (of which 204 were parasitized) attacked the donkey (which did not become infected, since *O. volvulus* is human specific).

When the parasite infects its host in the form of an active free-living stage, it should logically be able to identify substances—for example, those of the host tegument—that trigger attachment and penetration. This is, in fact, true for many parasites, where selection for mechanisms of host identification may prevent the loss of part of the infective stages. We can therefore expect that the process of host identification is under strong selective constraint, since any parasite that does not know how to identify a compatible host loses an opportunity to transmit its genes and any parasite that wrongly takes an organism as compatible pays for this mistake with its life and does not transmit its genes.

It is possible to incorporate identification into the model of encounter and compatibility filters by simply imagining that the compatibility filter has both an identification and an installation component. Intuition suggests that the identification component should be open when the installation component is open and closed in the opposite case. In other words, the two components should form a unique filter that would be useless to subdivide. There are, however, cases where identification appears to be too precise and other cases where it appears to be too imprecise, just as one would expect in a constantly evolving system.

"Too Precise" an Identification

Can a host that is at the same time encountered and compatible be "ignored" by a parasite?

De Meeus, Renaud, and Gabrion (1990) studied the specificity of the ectoparasitic copepods *Lepeophtheirus thompsoni* and *L. europaensis* and showed that under natural conditions the former is found only in turbot *(Psetta maxima)* whereas the latter is found both on the brill *(Scopthalmus rhombus)* and flounder *(Platichthys flesus)*. They also showed that under experimental conditions it is possible to obtain development until maturity on "nonnatural" hosts, for example, *L. europaensis* (the parasite of the brill and flounder) on the turbot.

The combination *L. europaensis*/turbot, which is not found in nature, is a paradox, since experimental results demonstrate that the compatibility filter is open. Moreover, the encounter filter is also open, since all the organisms are found living side by side in the same ecosystem. In this case it is difficult to understand why the identification component stays closed.

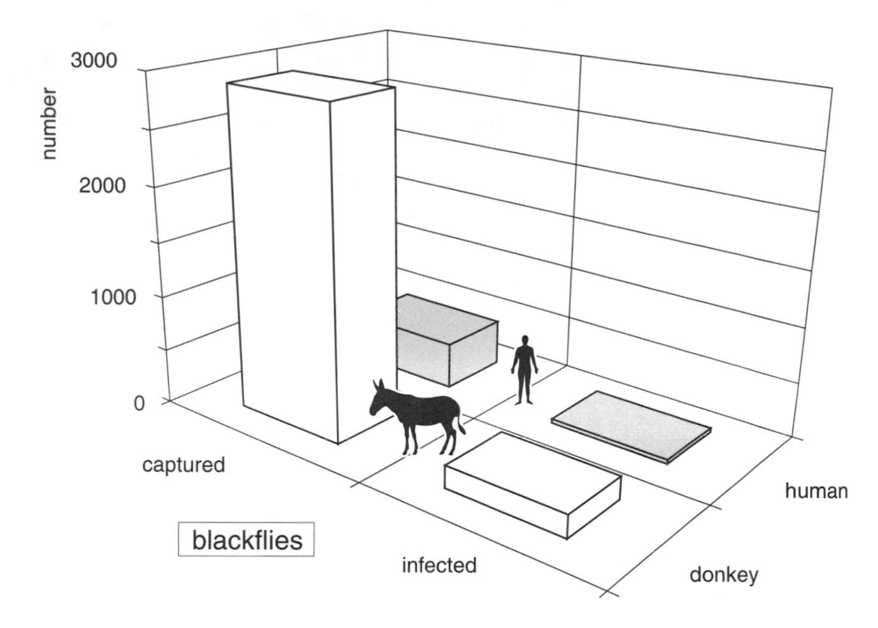

Figure 2.16. One donkey and one man set side by side attract the same proportion of parasitized blackflies, vectors of human onchocerciasis, relative to their surface area. All infective stages injected into the donkey by the flies are lost (from Philippon 1977).

The paradox is explained, however, by the fact that compatibility, as I have emphasized, involves many subtle nuances. On the one hand, the percentage of survival of the copepods was slightly better on the natural host than on the experimental host. On the other hand, the presence of the "other" copepod decreases the reproductive success on the "other" species of fish. This means the parasite is compatible with several hosts, but since one is more compatible than the other, it has selected a mechanism of identification for the best one.

The copepods of the turbot and the brill support the idea that a narrow host specificity characterizes many parasites because it is more advantageous to be well adapted to a species of host than to get dispersed onto hosts for which one is less well suited.

"Too Imprecise" an Identification

Trematodes have in their life cycle two free-living infective stages: the miracidium, which penetrates a mollusk, and the cercariae, which penetrates either a second intermediate host or a definitive host. The mechanisms of identification have been well studied in these organisms, and overall the quality of host identification is better in the miracidium than in cercariae (see Haas et al. 1995; Haas and Haberl 1997). Identification is, however, not entirely precise for either of them.

For miracidia the recognition of high molecular weight glycoconjugates (more than thirty kilodaltons) emitted by mollusks allows them to be responsive to other aquatic organisms such as leeches or fish but triggers a more rigorous response to host mollusks than to nonhost mollusks. For example, miracidia of *Trichobilharzia ocellata* orient themselves in water conditioned by their target mollusk, *Lymnaea stagnalis,* but not in water conditioned by *L. truncatula* (Kalbe, Haberl, and Haas 1997). Nevertheless, numerous experiments show that despite these mechanisms important losses still occur. This imprecision in identification, known as the *decoy effect,* leads miracidia to penetrate noncompatible mollusks where they have no possibility of escape and are destroyed. In some cases just attempting penetration of the "wrong" mollusk is enough to make the miracidium inept at later penetration of the "right" mollusk (Moné and Combes 1986; Combes and Moné 1987). That these results are contradicted by the results of Kalbe, Haberl, and Haas (1997) shows that there is no general rule and that natural selection works differently in different lineages of trematodes. It is thus likely that losses due to the decoy effect vary greatly between species. These losses have rarely been quantified, although Touassem and Combes (1985) showed that out of a thousand miracidia of *Schistosoma bovis* confronted with mollusks of different species, three-quarters penetrated vectors and one-quarter penetrated nonvectors.

As with miracidia, cercariae that penetrate the "wrong" host are destroyed and lost from transmission. Körner and Haas (1998) showed that cercariae respond to the overall concentration of weak molecular weight excretory/secretory products coming from the host but not to the specific composition of the mixture, which leads to only a weakly specific response. These authors make us notice that, surprisingly, host search behavior has apparently evolved in different ways in miracidia and cercariae. For example, cercariae of *Echinostoma revolutum* are capable of orienting themselves in a concentration gradient of a mixture of amino acids but cannot distinguish between mixtures of different compositions (Körner and Haas 1998). As noted earlier, cercariae of bird schistosomes that normally must penetrate a specific bird species also regularly target humans when they swim, causing swimmer's dermatitis, or "swimmer's itch."

The significance of the decoy effect in nature is not known (see Combes 1991a; Haas and Haberl 1997). Preserving the possibility of conquering new hosts is probably not sufficient justification for the effect, however, since natural selection likely would not retain characteristics whose advantages would be expressed only ten thousand or one million years in the future. A more likely explanation is that better precision in host selection did not evolve because the cost would have been greater that the benefit obtained.

Specialization for the Microhabitat

If a parasite could "see" and "think," once installed in its host it would have no other horizon and subject of meditation than its immediate environment. An adult cestode would see only the intestinal microvilli and lumen of the vertebrate, and *Plasmodium* only human cell membranes and cytoplasm. The existence of such different sites is the result of the heterogeneity of the genes expressed in different organs and tissues in each living organism—the result being that the habitat offered to parasites is not ecologically the same in different parts of the same host. This heterogeneity is true even at the scale of the cell, within which the cytoplasm, nucleus, DNA molecules, and so on represent distinct environments. Such heterogeneity becomes even greater at the scale of tissues and organs in multicellular hosts, and it is almost trivial to note that each organ is characterized by numerous and specific morphoanatomical parameters (shape, toughness, consistency of the boundaries) and physiochemical parameters (pH, osmolarity, solute makeup). The synthesis done by Crompton (1973) on the vertebrate digestive tract shows, for example, how numerous and different the sites are in alimentary tracts according to taxonomic group. Although there are some parasite generalists relative to microhabitat, such as *Toxoplasma gondii* (which invades almost all types of nucleated cells; see Kasper and Mineo 1994), the rule is a very fine discrimination for the quality of the microhabitat itself.

Parasites are sensitive to variations in the milieu that seem minuscule or even inaccessible to the casual observer. One would think, for example, that the gill system of a fish constitutes a global microhabitat because the conditions of the milieu do not seem, a priori, very different at the top and bottom of the gill arch or at the proximal and distal ends of a gill filament. However, this is not the case, and monogenean species occupy very precise and limited sites on the gills, demonstrating that this system offers more diversity in habitat than a superficial analysis suggests.

Certain parasites also show intermicrohabitat migrations, the most studied case being the cestode *Hymenolepis diminuta* in the rat intestine. This species undergoes a circadian migration related to host feeding and digestion. Outside the period of digestion, cestodes are in the posterior region of the small intestine. When the content of the stomach is dumped into the intestine, however, the worms rapidly migrate toward the anterior region of the duodenum. Based on experimental results, this migration is not an artifact linked to contractions of the intestine as the result of host death and autopsy (Read and Kilejian 1969). Various hypotheses have been proposed about signals that might be detected by *H. diminuta* (see Crompton 1973), and it appears that comparable migrations also occur for other parasites of digestive systems, such as trematodes of the genus *Podocotyle* in flounders *(Platichthys flesus)* (MacKenzie and Gibson cited in Crompton 1973).

To identify the appropriate microhabitat, parasites have surface molecules that recognize host molecules in the same way that hormones, growth factors, or neurotransmitters recognize receptors. Intracellular parasites such as viruses, bacteria, and protozoans most often recognize proteoglycans, which are molecules that have both protein and polysaccharide regions. Very slight modifications of the proteoglycans give these molecules a huge diversity in form (see Frevert et al. 1993), and such a "lock-and-key" system theoretically allows a given stage of a parasite to penetrate only an adequate cell. For example, the system that allows CD4 molecules and associated coreceptors to be recognized by the GP120 molecule of HIV is one of these lock-and-key systems known to the public. Additionally, the same parasite often must recognize different microhabitats at different stages of its life cycle. For example, *Plasmodium* must sometimes recognize human hepatocytes, sometimes human red blood cells, and sometimes the digestive cells of mosquitoes. When the intracellular parasite is not specific for a cell type *(Toxoplasma gondii)*, intuition suggests that it recognizes a common nonspecific protein on the membrane. Such knowledge of the mechanisms of identification of the microhabitat obviously offers numerous avenues for chemotherapy.

Advantages of Specialization

The Classic Advantage of the Specialist

Parasites must invest in numerous and complex adaptations in order to survive and reproduce, and specialization for a particular microhabitat may be explained primarily by the difficulty and expense of exploiting more heterogeneous environments. It is difficult for a parasite to invest in adaptations (for example, attachment structures) that would allow it to settle in several differently structured organs. Whittington et al. (2000) suggest that a specific recognition or reaction occurs between secretions produced by monogeneans and components of fish host mucus, which could explain the narrow host specificity of these parasites. Moreover, parasites have a vital need for the expression of host genes and can survive only in the organs, organ parts, tissues, or cells where these genes are expressed.

Specialization for the host species is explained in a similar way. To live in two or more host species, a parasite must be capable of encountering organisms that have different behaviors, exploiting different resources, successfully confronting different competitors, and avoiding different immune systems.

Thus far I have emphasized that the pressures exerted by hosts' defensive mechanisms necessitate important investments on the part of parasites and as a result limit the way parasites can respond. These costs probably increase very quickly when the genetic distance between hosts increases, and it should not be forgotten that the pressures exerted by the host's immunodefense system do not have a true equivalent in free-living organisms.

A convincing argument for the difficulty and cost of avoiding the defense mechanisms of several host species is exemplified in the following observations.

Leptopilina boulardi is a small parasitoid hymenopteran that develops in the larvae of the fruit fly, *Drosophila melanogaster.* Carton and Nappi (1991) and Dupas and Boscaro (1999) showed experimentally that in most of the populations studied in Europe and Africa host resistance (meaning hosts that can theoretically encapsulate the parasite) is generally of little use in limiting infection. In this case parasite "virulence" seems to win out, although in a particular population in the Brazzaville region *D. melanogaster* are much more successful at encapsulating *L. boulardi* larvae than in other regions. Significantly, in this area the hymenopteran attacks not one species, as in the other localities, but five species, among them *D. melanogaster.* In short, there is a good correlation between the efficacy of the host's immunodefense abilities and narrow specificity in this system. Roughly translated, the French saying "he who kisses too many does not embrace well" is well adapted to this situation.

The advantage of a specialist compared with a generalist is predicted by optimal foraging theory, which stipulates that individuals of an extant species that capture their food while spending the least energy are selected for in the short term: "Optimal foraging models predict that specialization may arise simply in response to differences in the quality and abundance of different kinds of food" (Futuyma 1986).

The Limitation of Interspecific Competition

The more "generalist" living organisms are, the more intense is the competition between them. In the particular case of parasitism, an overall widening of the host spectrum increases the probability that any two species of parasites will encounter each other in the same host simply because the number of available host species is not infinite. If competition occurs in the world of parasites, one would think that selection would favor a given parasite species living in only the few host species within which it would be the most competitive. The result would be a reduction of its spectrum.

Ecologists in general consider that an aggregated distribution of individuals in a fragmented environment increases intraspecific competition while decreasing interspecific competition ("the aggregation model of coexistence"; see Sevenster 1996). One may thus consider specialization for certain species of hosts analogous to the aggregation of individuals in a fragmented environment, which tends to support the hypothesis that specialization similarly decreases competition between parasite species while increasing parasite diversity.

By restricting the dimensions of their microhabitat, parasites would tend to limit competition with other species sharing the finite space of the host. This idea is supported because, at least in certain hosts, one may show that parasites not only have reduced the spatial dimensions of their microhabitat but have also been displaced in order to reduce interspecific competition. Nevertheless, the question of competition among parasites is still debated, and certain authors allow it only a very limited role in the structuring of parasite communities (see chapter 13).

The Maintenance of Genetic Exchanges

The more the parasites of the same species parasitize a small number of hosts, the higher their probability of encountering a mate. Although one might object that many parasites are hermaphroditic, the argument is not valid. In fact all evidence indicates that hermaphroditic parasites practice outcrossing whenever possible to avoid a reduction in genetic diversity as the result of too frequent self-fertilizations (Nollen 1983, 1993). At the microhabitat scale specialization thus increases the probability of encountering a mate—something that is difficult to accomplish if there is too large a dispersion in the host (Rohde 1979).

Rohde's hypothesis is original and comes from observations on the niches of fish monogeneans. In this system there are certain species of monogeneans that occupy only a limited space on the gill even though no other species may be competing for the location. Rohde believes the necessity of maintaining genetic exchanges via mating could be an explanation for such microhabitat restriction. The more the parasites that infect a host individual are grouped onto a small surface, the easier mating is (that is, the probability of encountering a mate becomes higher, particularly when parasite abundance is small).

Lebedev (1978) expresses reservations about this hypothesis ("An increase in the chance of mating . . . will be a consequence of site selection but not its reason") even though supportive arguments seem numerous (Rohde 1994). In short: the niche of mature parasites is often more restricted than that of immature parasites; species whose members have limited mobility often have more restricted niches than mobile species; species that are not abundant have more restricted niches than those of abundant species; and niches become smaller during mating for those parasites that have a particular mating period.

The Reduction of the Pathogenic Effect

One may suppose that parasites in some cases reduce their pathogenic effect by limiting the exploitation of the host to well-defined areas. However, this hypothesis is not well supported because a parasite whose infrapopulation is centered in too small a location can also more seriously alter a particular function of the host by being concentrated. This is so true that, as we will see later, some parasites use concentration in a microhabitat to weaken their host. This happens, for instance, when transmission requires that the host harboring the parasites, the intermediate host, be eaten by a predator (the next host; see chapter 7).

Constraints of Specialization

The Limitation of Geographic Extension

"Those parasites with the most generalized requirements are the most likely to disperse" (Stromberg and Crites 1973). This very simple idea, which implies that specialized parasites have less chance to conquer new areas, was put forth some years ago and has been illustrated since then by numerous examples.

For instance, when heteroxenous parasites were confronted with human migrations, those most specialized for their vectors could not travel with their human hosts as they penetrated regions where the vectors did not exist. This is the main reason exotic parasitic diseases do not occur in temperate regions, although to this restriction must be added social and behavioral constraints as well. This is also the reason urinary schistosomiasis has not established itself in South America—the mollusks *(Bulinus)* that are necessary for transmission do not exist in the New World (see fig. 20.9). This might also be why, when prehistoric men left tropical regions, there was a decrease in the pressure of parasitism and why the size of the human population then increased.

The narrow specialization of parasites, particularly for their vectors, is therefore a constraint that is not arguable, and generalist species are beyond a doubt better able to colonize new areas than are specialists.

The Limitation of Resources

This constraint is an extension of the previous one in the sense that in an ecosystem the fewer the possible hosts, the less chance the infective stages have to encounter their targets and the less chance the parasite has to reproduce. On first analysis it seems it would be to a parasite's advantage to inhabit as variable a number of hosts as possible, which would then guarantee a greater chance of multiplication. The only objection that can be given to this way of looking at specialization is that it does not take into account the individual abundances of different species in the ecosystem. In terms of resources it is probably better to parasitize only one abundant species rather than several rare species. Sadly, however, we do not yet know enough to appreciate the value of this dilemma for parasites, although intuitively it would probably be advantageous for a parasite to be narrowly specific in some cases and less specific in others.

Restricting the habitat to a reduced portion of the host has consequences comparable to those of being limited to parasitizing only one or a few species: by restricting itself to a specific microhabitat, the parasite limits the quantity of both the space and the resources available to it. Therefore it must be that obstacles to a broader exploitation place important constraints on the distribution within the host (membrane receptors preventing parasite penetration, structures of the organs preventing attachment, etc.). Added to this is the fact that when parasitic species have a narrow specificity for a particular microhabitat they can then infect only host species in which that microhabitat is present. For example, nematodes of the *Subulura* group are specialized for the intestinal ceca of mammals and birds, and they are not found in hosts that do not have well-developed ceca.

The Risk of Extinction

The most serious problem for parasites is that if they are specific for a particular host species, the disappearance of the host leads to their own disappearance. Given the rapid reduction in species diversity around the globe today, one must admit it is highly probable that a number of parasite species have effectively disappeared in this way. This situation can be compared to a company that does not diversify its business enough and goes bankrupt the day consumers no longer buy its only product.

In a sense, generalist parasites may be thought of as species confronted with a certain amount of instability in both time and space relative to their hosts. In this case it is better to be able to exploit several different hosts: if one host is temporarily absent they can survive in another. On the other hand, it may also be supposed that the character "specialist" is selected for when the continued presence of the only host (or of a very small number of host species) is perfectly predictable (see "Determinants of Host Spectra" above).

Difficulties in Reaching the Microhabitat

It is rare for the initial encounter with a host individual to happen exactly where the parasite's microhabitat will be. The invading parasite must detect signals or signal gradients and move to reach the microhabitat. That is, the parasite must "know where it is." Sukhdeo and Sukhdeo (1994) note that when a parasite penetrates a host it enters an environment that has numerous highly predictable characteristics. It is this predictability that allows parasites to "memorize" in their genes a *map of the host* and to select the sensory organs that detect appropriate signals and gradients. The migratory pathway is itself predictable because of the defined physical and chemical "topography" of a host, and the more complex the migration, the more the process resembles a modern missile that is able to locate the target by constantly acquiring new information relative to its present position. There are powerful selective pressures for the discovery of the correct habitat, because parasites that localize outside the "normal" microhabitat would likely not transmit their genes—either they die because the resources are not correct or they survive but cannot correctly disperse their offspring.

Nevertheless, the passage along a more or less specific migration route also often results in numerous losses in the stock of invading infective stages. The following examples show that this is even more so when specialization for the microhabitat is narrow.

Metacercariae of the liver fluke *Fasciola hepatica*, a parasite of sheep, bovids, and humans, release the juveniles in the host's duodenum. To reach the proper microhabitat for the adult (the bile ducts), the young fluke must cross first the intestinal wall, then the peritoneum, then the capsule that surrounds the liver and enter the liver parenchyma; finally, it must then perforate the wall of the bile duct, where it will settle. The smaller liver fluke *Dicrocoelium dendriticum* that parasitizes the same animals arrives in the organism in the same form (metacercariae) and goes to the same microhabitat (the bile ducts), but it takes the more logical although no less perilous route up the common bile duct.

Metacercariae of the trematode *Nephrotrema truncatum* (a parasite of the insectivorous mammals *Talpa, Neomys, Sorex,* and *Crocidura*) are ingested by their intermediate hosts, which are freshwater annelids. Juveniles are freed in the duodenum of the insectivore, perforate the intestinal wall, and migrate to the liver. From there they go to the kidney, always passing through tissues. Since the liver is directly underneath the right kidney, it is always in the right kidney that adult *N. truncatum* are found (Jourdane 1974).

The cercariae of schistosomes penetrate anywhere on the skin, but adults are found only in the veins surrounding the intestine *(Schistosoma mansoni* and *S. japonicum)* or the bladder *(S. haematobium).* For these worms a long migration occurs from the region of skin penetration to these microhabitats (see chapter 20).

During each of the migrations mentioned above, juvenile energy expenditure is high either because the parasite moves actively, because it must defend itself, or because it has to perforate various barriers. In many instances during these migrations the parasites are also not protected from the host's immune system, and despite both the signal detection systems discussed earlier and assorted defense mechanisms, numerous juveniles get lost on their way to the target.

Difficulties in Leaving the Microhabitat

The position of the microhabitat in a complex organism influences not only the internal migration to the site but also the exiting of the propagules (eggs or larvae). As a general rule, if a parasite is inside a tissue or in a closed system (e.g., the circulatory system of vertebrates), its propagules can exit only by some form of breakout. There are two main solutions: a biting vector takes up the parasite (a virus, protozoan, or filaria), or the parasite makes its own way out. Examples of the latter process include the filarial worm *Dracunculus medinensis,* which provokes cutaneous ulcers to shed its offspring, and the schistosomes, whose eggs must pass through organ walls to leave the host. Schistosomes lay their eggs in blood vessels as close as possible to an exit route, and most of them expel their eggs through the intestinal tract. However, one species parasitic of humans in Africa *(Schistosoma haematobium)* exits by the urinary tract, and a species parasitic of ungulates in India *(S. nasale)* uses the original pathway of laying eggs in vessels surrounding the nasal sinuses (fig. 2.17). The genus *Trichobilharzia,* closely related to *Schistosoma,* parasitizes birds, and as in the case of *Schistosoma,* most species have their eggs released in the host's feces. However, eight species are known to settle in nasal areas and to have their eggs expelled through the nasal cavity, while one, *T. regenti,* has been

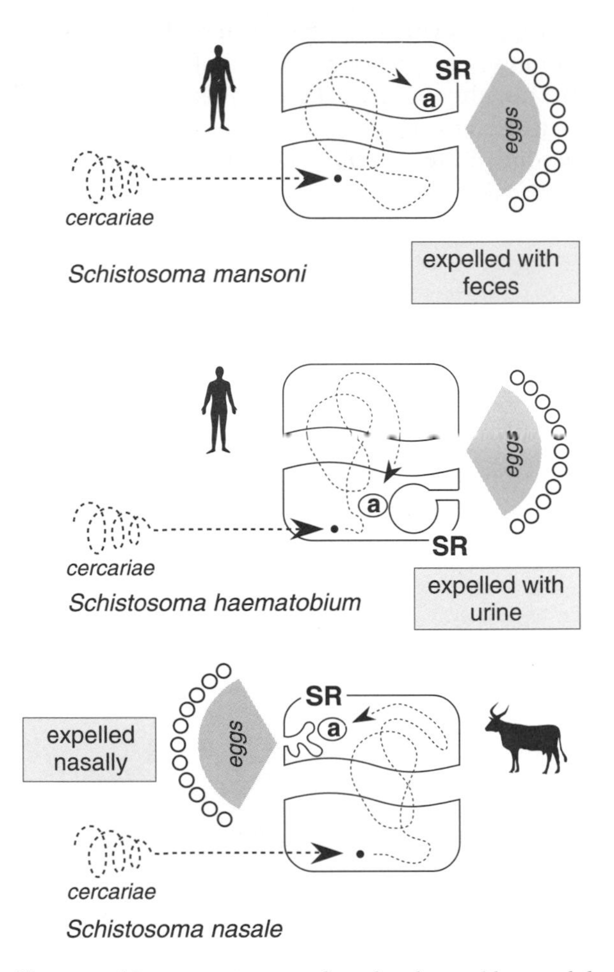

Figure 2.17. The ways schistosome eggs are released to the outside: a = adults; SR = sexual reproduction.

observed to produce free miracidia in mucus (Horak et al. 1999). How do "nasal" schistosomes get to their microhabitat? Little is known. According to Horak et al. (1999), *T. regenti* schistosomula exit blood vessels and migrate through nervous tissues in experimentally infected ducks.

Obviously, the way propagules are released to the outside influences the parasite's pathogenicity in that host. For example, while most of the trematodes living in the digestive tract or its extensions (mesoparasites) have little to no pathogenicity because of their eggs, the pathogenicity of schistosomes occurs principally as the result of the eggs. This is due to the multiple perforations they cause in the wall of the organs of expulsion or, in contrast, because they become trapped in large numbers within the organism when they fail to exit.

As for the scope of the microhabitat, the release of propagules therefore also is a constraint for parasites restricted to areas without easy access to the outside.

3 The Fragmented Nature of Parasite Populations and Communities

The study of communities of parasites has much to offer to community ecology in general.
Holmes 1987

At each stage of the cycle, selection has favoured different carriers of the parasite's genes.
Poulin 1998a

Parasite Populations

A population is an ensemble of individuals of the same species that have about equal probabilities of interbreeding with each other (panmixia). This ensemble is distinct from other such groups of individuals of the same species in that the exchange of genes between them is usually limited if not nonexistent. Populations are the result of the fragmentation of the total number of individuals in the species because of the heterogeneity of the environment: cows are concentrated where there are pastures, birds where there are trees, and frogs where pond and river edges exist. The fragmentation of a parasite population is more complex than that observed for free-living organisms, however, because a particular parasite species may be fragmented along three scales: space, host species, and host individual.

Scale: Space

Like all other species in the biosphere, parasite species form populations that are isolated to one degree or another because of the heterogeneity of their environment. While it can be difficult to trace the limits of populations of a free-living species, it is even harder for parasites. To understand this, one can imagine a pond populated with green frogs harboring parasite P. If the pond is far from neighboring ponds, it is justifiable to consider that the ensemble of green frogs in this pond constitutes a population, since genetic exchange occurs primarily between individuals within the pond. If P has a holoxenous life cycle, then there is a good chance that the limits of its population coincide in space with the limits of its host population simply because its migration depends on the frogs' migration. However, if the life cycle of P is heteroxenous, P may be transported each generation from pond to pond by vectors such as dragonflies or other mobile insects commonly eaten by frogs. In this case the population of parasite P can be spread over the ensemble of ponds and not just one pond.

Scale: The Host Species

Now suppose that among the frogs in the pond there is also a population of toads, and suppose that parasite P can infect both frogs and toads. Two hypothetical scenarios are possible. If in each generation the stages of transmission outside the amphibians mix randomly, then the parasites of the frog and those of the toad constitute a single panmictic population. But if the frogs' parasites, for whatever reasons, have a tendency to be found in frogs during the next generation (and if the same is true for toad parasites and their respective hosts), we cannot conclude that parasite P of the frogs and parasite P of the toads constitute one and only one population because gene exchange between the frogs' parasites and the toads' parasites is limited. Similarly, and even if at each generation the larval parasites from frogs and toads mix, the fecundity of the parasites, their survival, or even their mode of selection may not necessarily be identical in the two hosts.

In short, the broader the host spectrum of a parasite, the more one should be concerned by the level of fragmentation exhibited by the host species. When we consider a human parasite that also develops in animals, it is obviously important in terms of disease control to know if the parasites from the humans and those from the animals constitute a sole panmictic population or if there is a more or less marked isolation of the two. When the host spectrum comprises several species for a particular life-cycle stage, the hypothesis of a random mixing of the transmission stages at each generation should not be considered proved.

Scale: The Host Individual

Without exception, the population of a parasite is "exploded" into as many elementary subensembles as there are parasitized individuals. This fragmentation by individual host itself shows three distinct characteristics.

1. The fragmentation is *temporary* and does not last more than the life span of the host, which is most often measured in months or years and sometimes even less. Except in extraordinary cases, the propagules coming from different individuals within the same host population mix randomly in each generation. That is, isolation by individual host is not comparable to the isolation observed in fragmented habitats such as islands, within which isolation is usually prolonged for numerous generations.

2. In many cases there is *no reproduction on site,* and the recruitment of new individual parasites by an individual host is completely dependent on exogenous sources.

3. Host individuals are sometimes places where a parasite stays only a short time and where *parasite turnover* may be rapid. Some trematodes of migratory birds are often cited in this regard: here some worms can complete their life cycle during a stopover before the birds continue their trip. Galaktionov (1996) notes, "The birds feed on the mollusks, become infected and subsequently disperse the eggs; this is promoted by the rapid development of the adult flukes, 3 or 4 days." For other parasites such turnover is slower, and this is an area of parasite ecology about which, particularly for human parasites, very little is known at present.

Terminology

How, then, do we distinguish the different levels of fragmentation observed in parasite populations without ambiguity?

The following vocabulary is often used (see Margolis et al. 1982; Sousa 1994): the *infrapopulation:* the ensemble of individuals inhabiting an individual host of species X (all parasites of species P of an individual frog in our previous imaginary example); the *metapopulation:* the ensemble of individuals inhabiting a population of hosts of species X (all parasites of species P in the frogs' population); and the *suprapopulation:* the ensemble of individuals inhabiting populations of hosts of species X, Y, etc. (all parasites of species P for the population of frogs plus those of the population of toads).

But these terms—infrapopulation, metapopulation, and suprapopulation—are satisfying only in part. Although "infrapopulation" is widely used and does not pose a problem, "metapopulation" does and is not acceptable. A metapopulation is defined by population geneticists as an ensemble of interconnected populations, meaning that migrating individuals cause genetic exchange between fragments that, although reduced, are not nonexistent. For example, blue tits live within a metapopulation divided among a series of forest fragments between which there is genetic exchange when certain individuals, most notably juveniles, migrate from one forest fragment to another. Often such a metapopulation results from the subdivision of an initial population into smaller populations that then maintain limited exchanges among themselves. Metapopulation structure influences genetic change within species and is dependent on the number of smaller populations forming the metapopulation, the importance of migratory exchanges between the population fragments, and the degree of inequality in the exchange of genes between the groups. In a metapopulation certain populations become extinct, but new populations are established at unoccupied sites (see Hanski 1998). The notion of a metapopulation is vital today because humans, by fractionating habitats (through the construction of highways, deforestation, building in littoral zones), transform originally vast populations into smaller and smaller fragments that are more and more distant from one another, with each subject to extinction.

Relative to durably interacting organisms, metapopulations of parasites exist just as do those of free-living species. For example, Théron et al. (1992; André Théron, personal communication) showed that schistosomes in a series of palustrian mangrove fragments (such as those observed in the Caribbean) constitute a metapopulation. Each fragment has particular frequencies of certain alleles but is not totally isolated from other fragments making up the metapopulation. Extinctions and repopulations occur periodically in these habitats, leading to *instability at the local population level* but global *stability at the metapopulation level* (fig. 3.1).

The epidemiology of the cestode *Echinococcus multilocularis* in some European foci is another example of how metapopulations function: the life cycle of this parasite involves a rodent (vole) (it is also sometimes found in humans) and a carnivore (fox). Depending on the climatic conditions, the parasite may disappear for one or more years in some localities and then reappear via colonization from other parts where the life cycle had been maintained (Giraudoux 1991).[1] The concept of a metapopulation is likely applicable to numerous parasite species whose populations are maintained because of the alternation of local extinction and colonization. That is, they may exist temporarily only in certain environments and come back to empty ones as conditions change. Parasitologists are used to those parasites that seem to disappear from some locales (sometimes for years) and then suddenly reappear for no apparent reason. Often such disappearances occur only in particular places, and parasites survive in others that might not be included in surveys or investigations.

It is thus not a good idea to use the term "metapopulation" for a level of fragmentation hierarchically below that of the population. This does not mean the genetic exchanges occurring between infrapopulations do not resemble those that characterize true metapopulations. If we were to define the metapopulation only by the fact that it regroups interconnected ensembles of individuals, any fragmentation of a population could qualify as a metapopulation.

Last, use of the term "suprapopulation" is debatable for a different reason: it designates in ecology what everyone already calls a population and therefore is redundant and potentially confusing.

To eliminate the difficulties caused by these terms, I have previously

---

1. P. Giraudoux (personal communication) believes that certain metapopulations of *Echinococcus multilocularis* (for instance, in China) may be composed of populations whose intermediate host is different. As such, rodent communities may sometimes recontaminate other communities from which the parasite had disappeared.

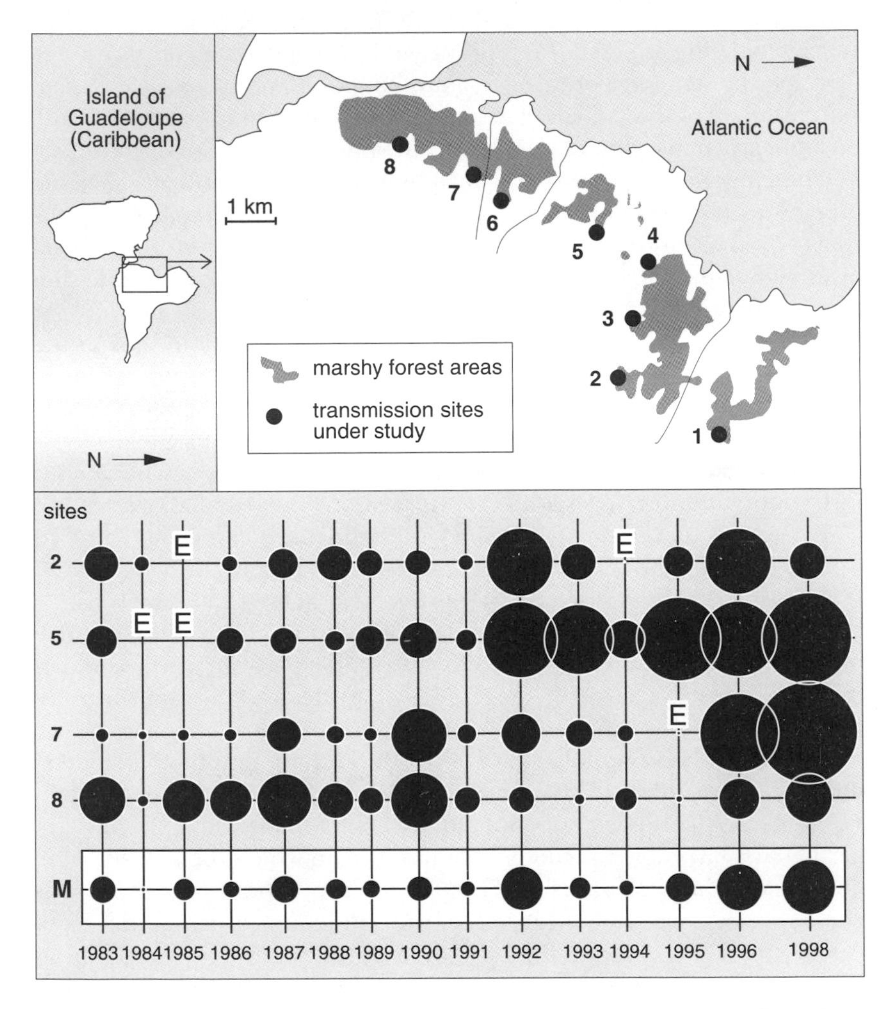

Figure 3.1. An example of a true metapopulation of parasites: the fragmented population of *Schistosoma mansoni* in the palustrian forests of Guadeloupe. The bottom part of the figure summarizes part of the observations (sites 2, 5, 7, and 8) showing that each population is ephemeral, with some extinctions (E) occurring over time even though the metapopulation (M, white area in lower frame) is, overall, quite stable. Size of the darkened circles denotes relative population fragment size (modified from Théron et al. 1992 and A. Théron, personal communication).

proposed (Combes 1995) using the following, each of which has the advantage of being as parallel as possible with those terms currently used to describe the structure of free-living animal populations: *infrapopulation:* the ensemble of individuals of the same species inhabiting a host individual; *xenopopulation:* the ensemble of individuals of the same species inhabiting a population of hosts of a particular host species; *population:* the ensemble of individuals of the same species inhabiting sympatric populations of two or more host species; and *metapopulation:* the ensemble of all interconnected populations of a species within all host species, allowing for local extinction and repopulation.

In these terms the traditional sense of both "population" and "infrapopulation" are retained; the word "xenopopulation" replaces "metapopulation" in the parasite literature; and parasite populations interconnected by weak genetic exchanges form, as occurs in all species, a "metapopulation." The term "suprapopulation," previously synonymous with "population," is abandoned for the reasons discussed above.

The term "xenopopulation" (the Greek root *xeno-* can mean "host" as well as "foreign") designates a level of population fragmentation (fig. 3.2) that it is important to recognize in parasites for at least three reasons.

1. The xenopopulation groups together that fraction of the population subjected to comparable selective pressures because the individuals that compose it are all within the same host species and the same local environment.

2. Different xenopopulations often exhibit demographic parameters different from those of the more global parasite population (Pulliam 1988). That is, certain xenopopulations whose habitats (host species) are particularly favorable function as "sources" from which the parasite populations of other hosts are fed. These other populations, whose habitats (host species) may be less favorable, then function as "sinks" that contribute little or nothing to the formation of the following generation of parasites. When such a habitat is so unfavorable that it does not allow the reproduction of the parasite, these hosts act as an "absolute sink," more commonly designated a "parasitic dead end."

3. The progressive reproductive isolation between two xenopopulations is probably one of the most important mechanisms of *speciation* for parasites (see chapter 5).

Populations Associated in Communities

Almost without exception, living organisms constitute environments for numerous parasite species. These parasites come together in one host individual following very different pathways or routes. Thus, after their

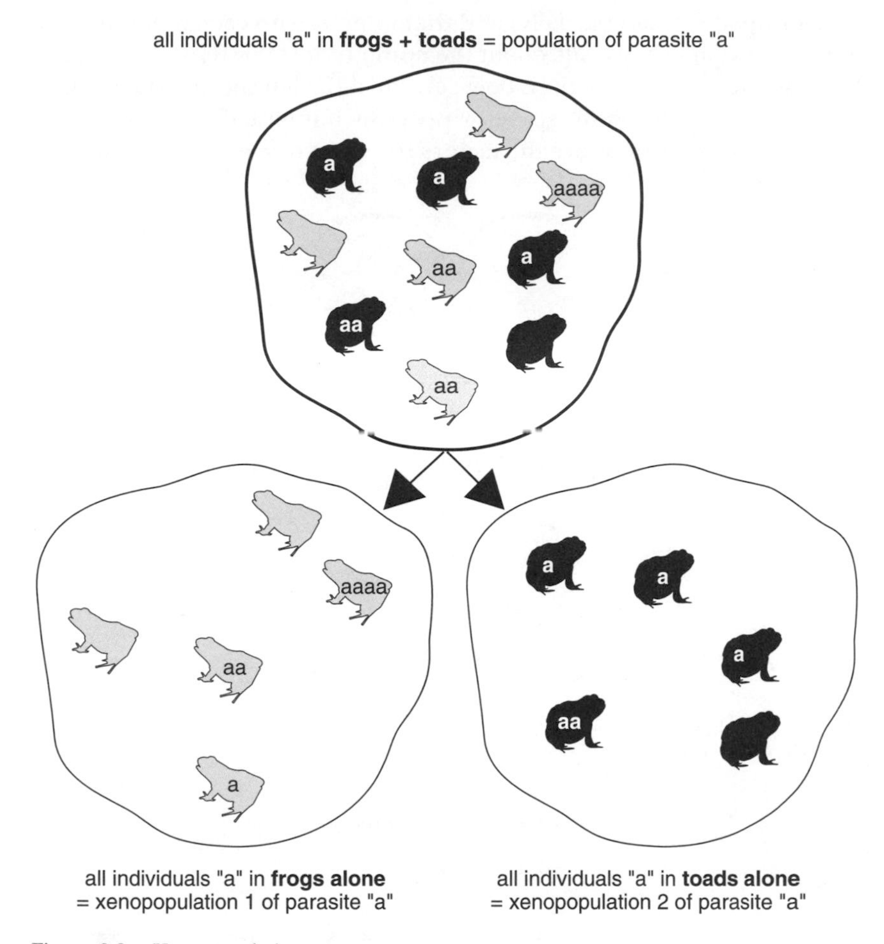

Figure 3.2. Xenopopulations.

dispersion in the ecosystem there is assembly in a common host.[2] Figure 3.3 schematizes the life cycles of a few common parasites of the dog (top), a gull (*Larus ridibundus,* center), and a mollusk (*Biomphalaria glabrata,* bottom) and shows how their different parasites converge within each.

In general ecology the term "community" designates an ensemble of free-living populations of different species that occupy the same

2. This is true not only for parasites: "An animal is never by itself in its milieu; it always maintains relationships with others either because it eats them, or because it may run the risk to be eaten by them, or because it may compete with them to acquire the same goods" (Blondel 1986).

environment (at least partially) and that interact with each other in some way. If, for example, we talk about the community of birds in a forest, it is because there are interactions between the different species, such as predation or competition for space or prey. Such interactions must be emphasized, because these are the factors that structure each community—

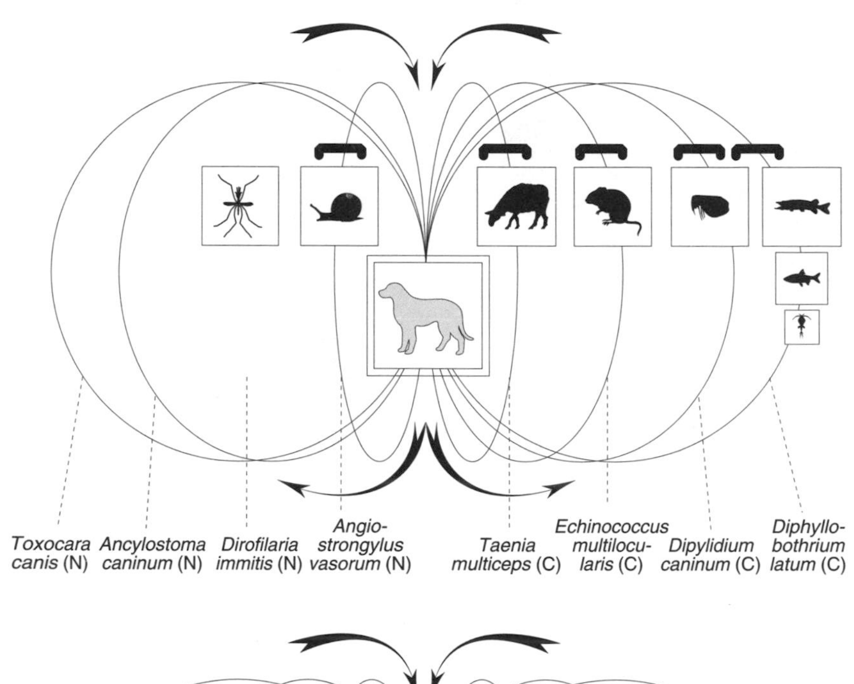

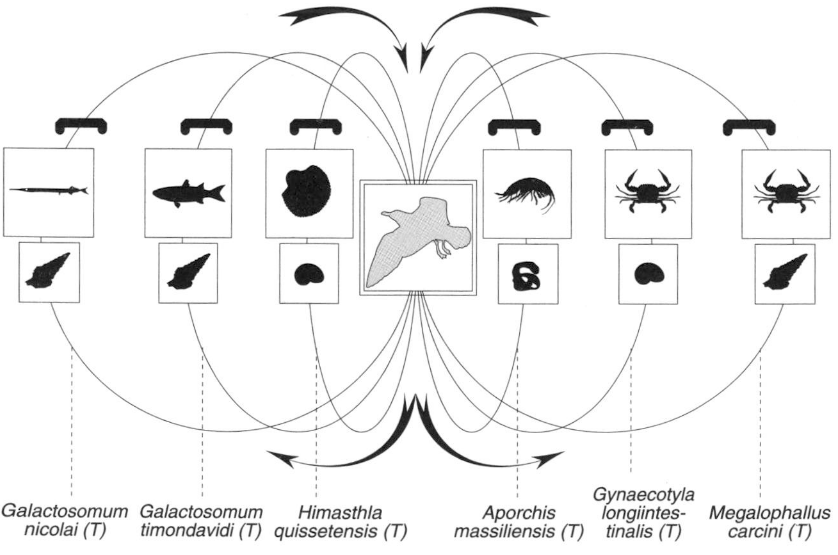

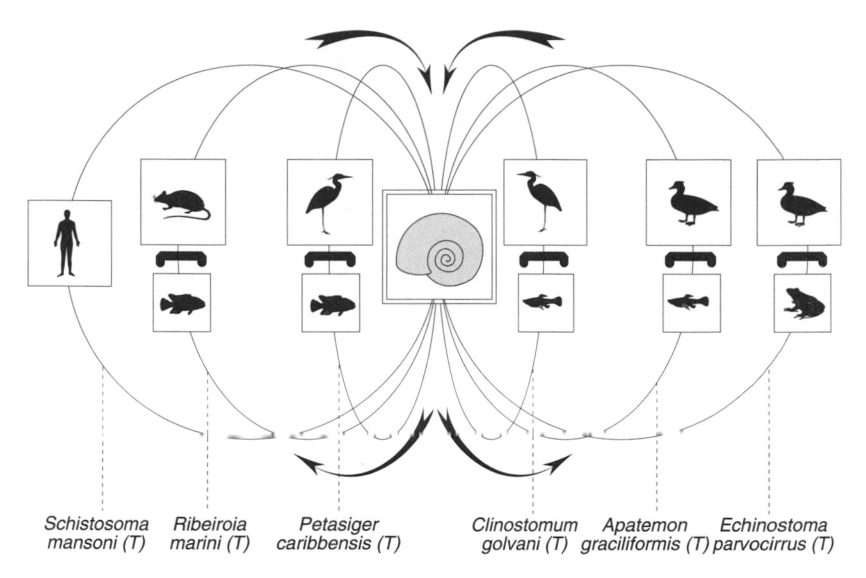

Figure 3.3. Some parasite life cycles converging toward three types of hosts (terrestrial mammal, marine bird, and freshwater mollusk). N = nematode; C = cestode; T = trematode.

that give each one its biodemographic characteristics. Similarly, interactions between the populations of parasites of different species also form parasite communities. In order for us to discuss such communities, however, it is necessary that the species involved interact either directly or indirectly and that we understand how they do so. Further, it is also necessary to understand that at the level of the community parasite populations are also fragmented in space, by host species, and by host individual, just as populations are. Additionally, because of the unique nature of the hierarchy of the interactions, the study of parasite communities also requires the use of particular terms. This has led me (Combes 1995) to propose adopting a hierarchy of terms parallel to those proposed for populations. These terms are coherent with those used in general ecology but differ from the earlier terminology used by other parasite ecologists (see Holmes and Price 1986; Sousa 1994). The proposed terms follow, with those usually used by English-language authors given in parentheses.

The *infracommunity:* the ensemble of parasites of all species infecting an individual host (for example, all those within a frog).[3]

---

3. Certain authors restrict "infracommunity" to a specific organ or location (for example, the digestive tract), which is justified in that interactions between parasites are in principle more important when they occupy the same microhabitat.

The *xenocommunity* ("component community"): an ensemble of parasites of all species infecting a defined population of a particular host species (for example, all the parasites infecting a population of frogs).

The *community* ("compound community"): the ensemble of parasites of all species infecting sympatric populations of several host species so that these parasites interact to various degrees (for example, all the parasites of a community of amphibians or all the parasites at different stages of development if life cycles are heteroxenous).

Figure 3.4 summarizes these major ideas relative to populations and communities at their different hierarchical scales and indicates the traditional terms and those proposed here.

To illustrate the use of these new terms and as a change from frogs and toads, let us use two species of mammals, the dog and cat, and three species of helminths, the cestode *Dipylidium caninum* (which parasitizes both the dog and the cat despite its name), the nematode *Toxocara canis* (which parasitizes only the dog), and the nematode *Toxocara cati* (which parasitizes only the cat):

- All the *D. caninum* within an individual dog constitute an infrapopulation, while those *D. caninum* within an individual cat constitute another such infrapopulation.
- The *D. caninum* of the dogs of one village constitute a xenopopulation, those of the cats of the same village constitute another.
- The *D. caninum* of both the cats and the dogs of the village constitute a population.
- The *D. caninum* and *T. canis* of each dog constitute an infracommunity, with the *D. caninum* and *T. cati* of each cat constituting another.
- The *D. caninum* and *T. canis* of all dogs constitute a xenocommunity; the *D. caninum* and *T. cati* of all cats constitute another such xenocommunity.
- The *D. caninum, T.canis,* and *T. cati* of all dogs and cats of a village constitute a community.

The xenocommunity (component community), as well as the xenopopulation, is a scale of analysis particularly important in parasitology because its composition is determined in part by the host's phylogeny and in part by its ecology. Based principally on the communities of fish, bird, and mammal parasites, Poulin (1995a) gives particular importance to the phylogenetic component of these associations: "Because of coevolution of host and parasites, host phylogeny is likely to be at least as important

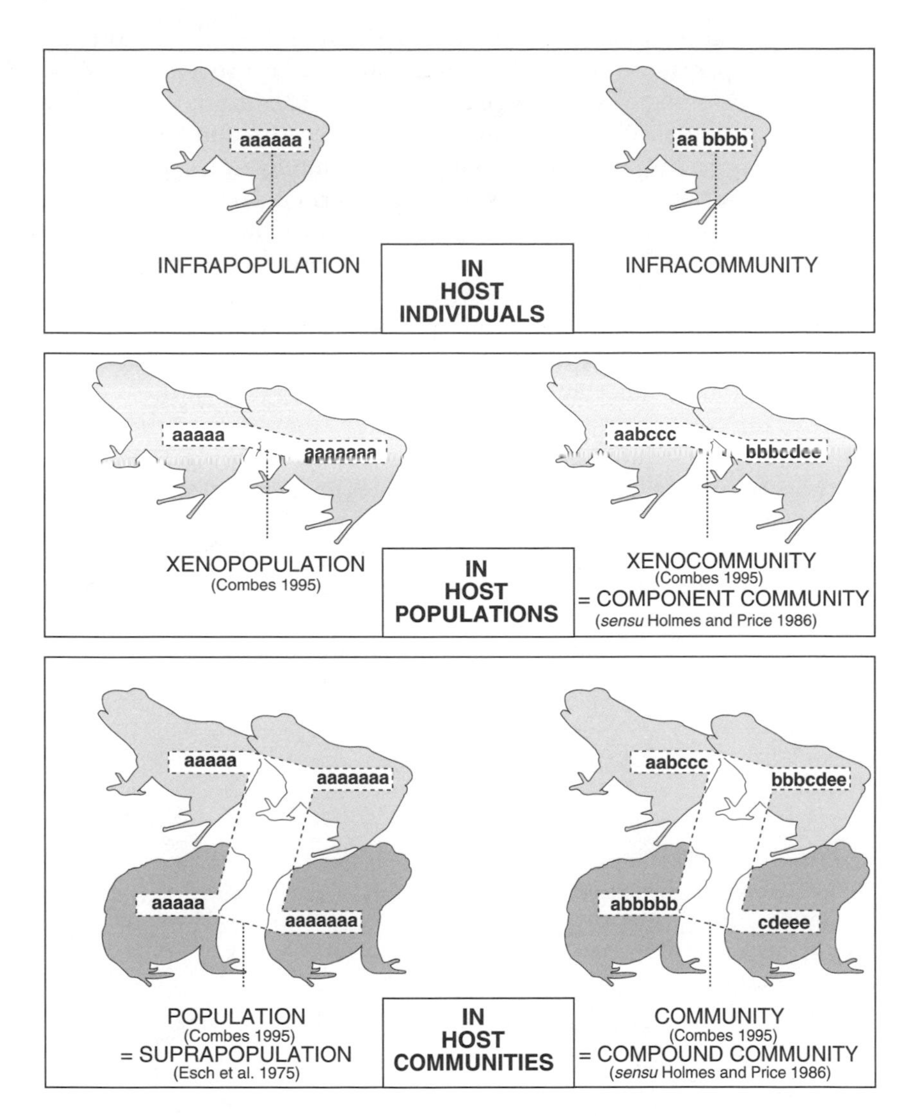

Figure 3.4. Different scales of parasite populations and communities: (1) Parasites of the individual host (top), the host population (center), and the host community (bottom); (2) monospecific assemblages (infrapopulations, xenopopulations, and populations; see diagrams on left) and plurispecific assemblages (infracommunities, xenocommunities, and communities; see diagrams on right).

as host ecology in determining the composition of the parasite community . . . the parasite component community of a host species could be more similar to that of a closely related species with different ecological characteristics than that of an unrelated species with similar ecology." Poulin supports his position by noting that immunity in particular is an important part of the host's traits that are inherited from ancestors and that this necessarily affects the composition of the communities. In fact, however, it is more likely that the comparative contributions of phylogeny and host ecology differ considerably according to host species.

One may wonder at the numerous questions that arise concerning the relationships between the different scales in parasite communities. Of these, probably the most important is the following: What is the relation between the composition of the infracommunity and that of the higher levels of xenocommunity and community? To address this question one must first recognize that parasite communities form because of a series of nested sampling events (fig. 3.5). That is, a host individual attains its *individual parasite richness* (its infracommunity) by "sampling" parasites from the generalist or specialist parasite species that surround it. The host individual's resulting species richness is then the ensemble of parasites present in it. For example, a European common frog *(Rana temporaria)*

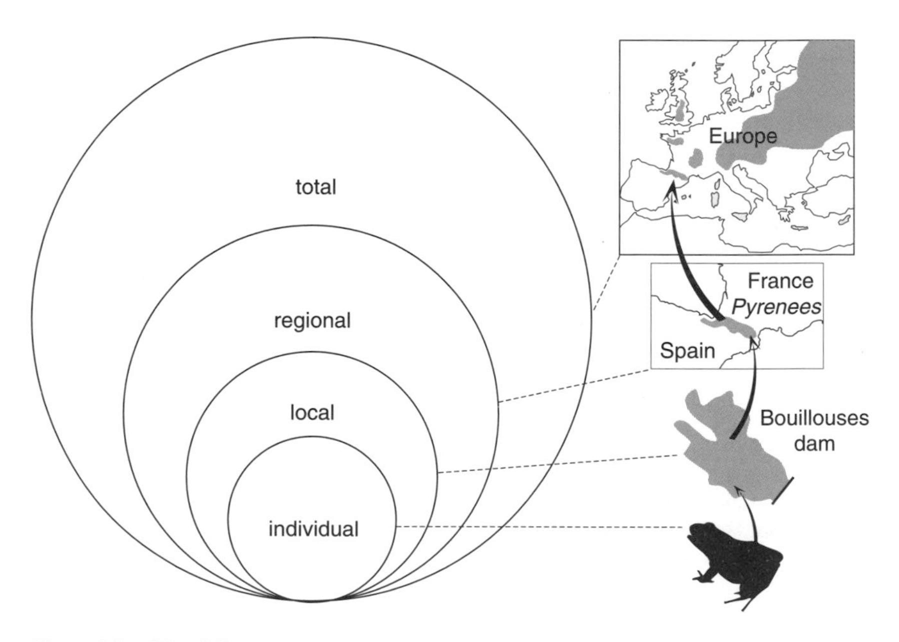

Figure 3.5. The different scales of parasite richness.

within a particular lake in the Pyrenees (Combes 1968) has a particular set of species drawn from the parasites of all such common frogs in the lake.

An infracommunity, then, is a subset of the *local parasite richness* (xenocommunity). The xenocommunity includes specialist parasites that are specific to the host species being considered; generalist parasites that also exploit other host species; and occasional "spillover" parasites whose presence is linked to the periodic occurrence of their normal host species in the habitat. As the result of this sampling the infracommunity of a particular host individual almost never includes all the parasites in theory available to it in the xenocommunity.

The xenocommunity is itself a subsample of a more *regional parasite richness* that includes the parasite species that exploits the host species in the area where its population is. For example, the ensemble of all parasites present in all populations of *R. temporaria* in the Pyrenees. Such a regional scale makes sense when it is possible to attribute to it geographic, climatic, or ecologic borders, which is sometimes not the case. Nevertheless, it is particularly interesting to analyze and explain the differences between the local richness (xenocommunity) and regional richness (see Kennedy and Guégan 1994).

Last, the regional richness is also obviously a subset of the *total parasite richness* of the host species. It is, for example, the ensemble of all parasites of all *R. temporaria* over its entire distribution. The total richness is of no help to the study of interspecific interactions, because some of the species reported in the total parasite richness may live in entirely different regions within the area occupied by the host species and never meet (see Simberloff 1990). Because of this, parasite richness expressed as a list of species is an imperfect descriptor of the actual parasite community. This is why numerous indexes of diversity, all familiar to ecologists, are used to take into account the relative infrapopulation sizes of the different parasite species. These indexes, however, are not themselves exempt from difficulty (see Simberloff and Moore 1997).

A further question concerning the interaction between infracommunities and xenocommunities is, Does infracommunity formation from the xenocommunity happen by chance or according to a particular rule or sets of rules?

The use of computerized Monte Carlo simulations allows one to make imaginary infracommunities by successively and randomly drawing species from the xenocommunity. In these simulations it is possible either to have the probability of being drawn from the xenocommunity be the same for each species or to have it determined by its abundance (see Janovy, Clopton, and Percival 1992 and fig. 3.4). The composition of the observed infracommunity may then be compared with that of simulated infrapopulations using statistical analyses (see Worthen and Rohde 1966; Hugueny and Guégan 1997).

In this way it has been shown that "poor" infracommunities may be subsets of much richer communities and that species from the xenocommunity are added to the infracommunity not by chance alone but according to a predictable order. This is, for example, the case of monogenean parasites on gills of the tropical cyprinid *Labeo coubie* (Guégan and Hugueny 1994). A possible explanation for such "nested subset patterns"[4] is simply that the infracommunities that are the richest in species are characteristic of the largest fish and that the colonization of an individual happens with defined species according to the host's size as it grows. For each parasite species there is a minimum size of host below which its presence is unlikely or nonexistent (see Simberloff and Moore 1997).

Janovy, Clopton, and Percival (1992) proposed a relatively simple model to link the composition of the xenocommunity to that of the infracommunity: if the xenocommunity has, for example, four species (a, b, c, and d), with each occurring at a different level of abundance, each infracommunity is then formed by a series of random draws from the possible infective stages present in the environment (fig. 3.6).

In reality, however, various processes can provoke departures from the predictions of this model owing to quantitative differences after recruitment, such as when competition inside the host modifies the proportion of the parasite species present. Further departures may also result from qualitative differences in hosts. The logic here is that if the xenocommunity contains, for example, ten species of parasites, then we would occasionally expect to find at least some hosts with infracommunities that harbor representatives of each of the ten species (the only condition is that the sample of examined hosts must be large enough). As Kennedy and Guégan (1996) showed, this is not always the case. Even if one captures more than one hundred eels in various locations in the British Isles, each individual host harbors a maximum of only three parasite species (exceptionally four), although the xenocommunities have up to nine members. The authors concluded that there is a saturation limit to the richness of the infracommunities and that they cannot attain the full parasitic richness available within the xenocommunity. These saturated infracommunities are therefore predictable in their low richness but not in their composition. The mechanism causing this phenomenon is not yet known.

4. The notion of a nested subset pattern was initially proposed to explain the composition of mammal communities on islands (Patterson and Atmar 1986).

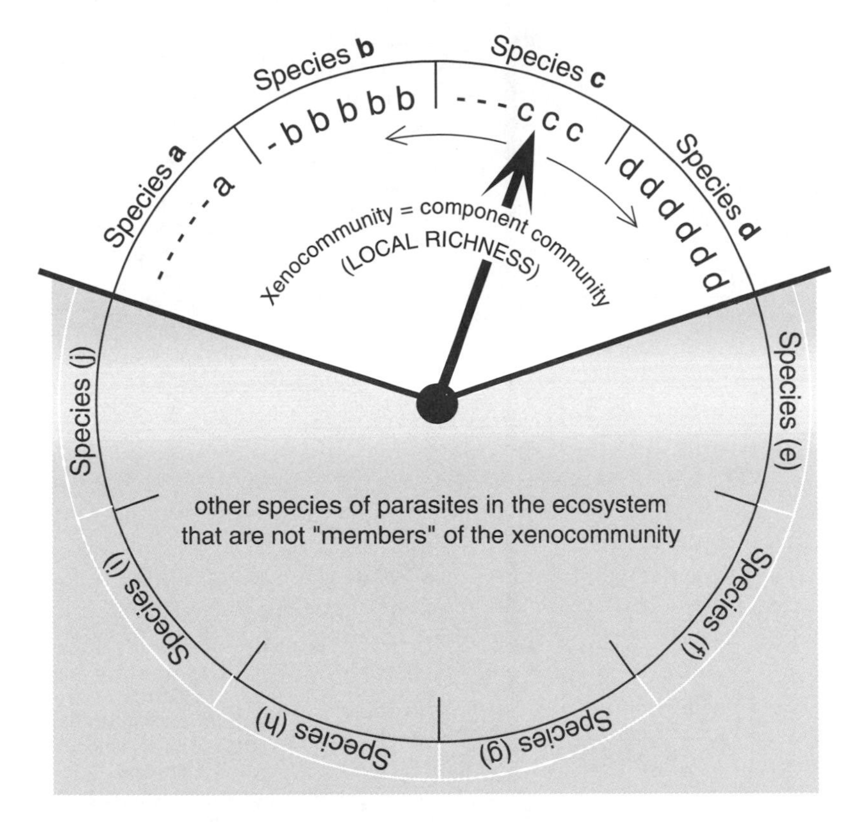

Figure 3.6. The lottery of Janovy, Clopton, and Percival (1992): each infracommunity is formed by a series of draws from the xenocommunity. In the diagram species d is supposedly the most abundant and species a the least.

Fragmentation and Life Cycles

The usual approach to the study of life cycles is principally a qualitative one where the parasitologist seeks to know what species of host a parasite is able to colonize either *longitudinally* (the sequence of hosts) or *laterally* (the spectrum of hosts). However, the quantitative dimension of life cycles is only rarely taken (see Bush, Heard, and Overstreet 1993). Let's imagine any heteroxenous life cycle that involves one member of an intermediate host species being consumed by members of one definitive host species (fig. 3.7). If we look at the level of the infrapopulation, nothing indicates that all members of a larval infrapopulation will survive or "make it" to form the adult infrapopulation when an intermediate host is eaten. For instance, if a mammal ingests a fish that harbors one hundred

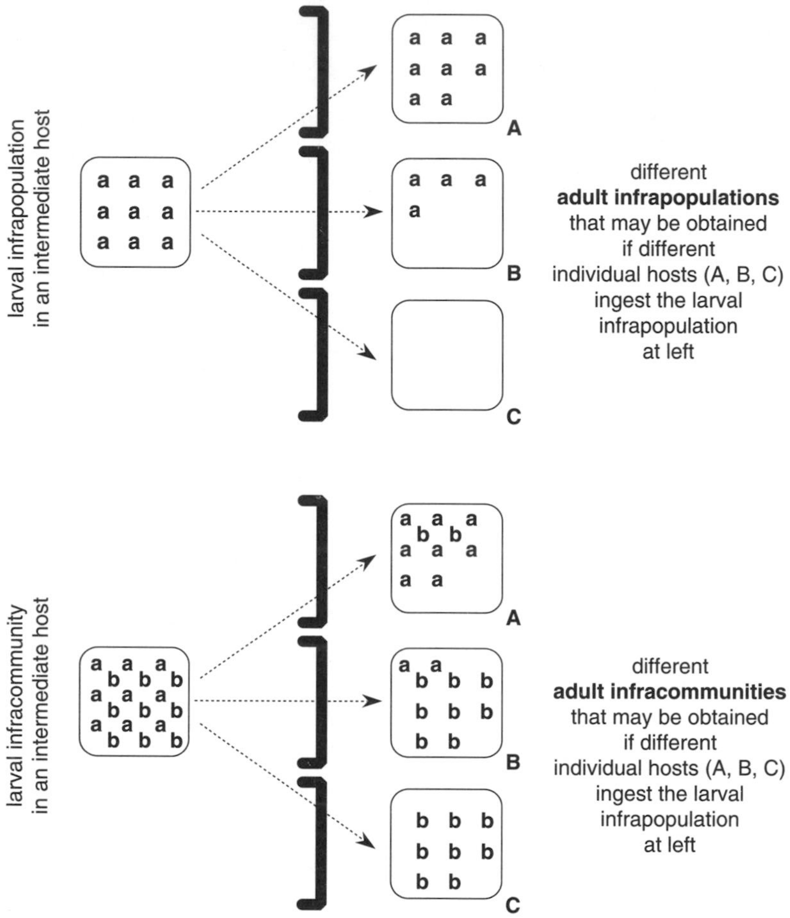

Figure 3.7. Imaginary scenarios of changes occurring between larval infrapopulations and adult infrapopulations (top) and between larval infracommunities and adult infracommunities (bottom) in heteroxenous life cycles. Changes are not identical because the intermediate host is eaten by predators of different species (designated by A, B, C); a and b represent individuals of different parasite species.

metacercariae of *Clonorchis sinensis* (these hundred metacercariae constitute a larval infrapopulation), we cannot conclude that all one hundred will contribute to the adult infrapopulation in the fish-eating mammal. This is because survival will depend in particular on both the defense mechanisms of the host and the presence of other already established parasites. Very few qualitative data relative to such establishment and survival are available, probably because such an approach would also have to take into account the number of infected intermediate hosts eaten by nonsusceptible hosts—a quantity that is most likely not negligible and that results from the lack of coincidence between encounter and compatibility filters.

If we set ourselves at the level of infracommunities, attaining quantitative data becomes even more complex because the particular infracommunity in any one intermediate host has even less chance of ending up intact in a particular definitive host than does any particular infrapopulation it might contain. To use the previous example, if the fish harbors metacercariae (or other larval stages) of, let's say, ten parasite species (the larval infracommunity), a "sorting" will occur in the predator mammal because of differences between the encounter and compatibility filters. It is even possible that out of the ten parasites present only *C. sinensis* may be able to settle in the host. Lotz, Bush, and Font (1995) showed that in certain cases the acquisition of different parasites "in packages" coming from individual intermediate hosts results in a positive association that can be observed in definitive hosts (we talk about positive association when two species are found together in the same individual host more frequently than would be predicted by chance alone).

Although it may happen that a larval community transforms en masse into an adult community in the definitive host, more often the latter only partially reflects the former in relative abundance.

4 Parasite Distribution

Are wormy people parasite prone or just unlucky?
Guyatt and Bundy 1990

An unoccupied ecologic niche, an unexploited opportunity for living, is a challenge.
Dobzhansky 1973

Distribution in the Host Species

Parasite Richness

If we consider an ecosystem as if it were composed of two categories of species, some free-living and some parasitic, we can then ask the question, How do the different parasite species distribute themselves within the space of the different host species? Figure 4.1 symbolizes an ensemble of macroparasite-host systems. It may represent, for example, a "nematode/mammal" or "monogenean/fish" association or, if we want more precision, a "monogenean/Mediterranean fish" system, and so on.

To address this question one can conceive of an imaginary sea in which there are seven species of fish (represented in the figure by their silhouettes) and six species of monogeneans (represented by the letters a to f). P is the ensemble of the six species of monogeneans and H the ensemble of the seven species of fish. The bottom part of the figure represents an imaginary distribution of the parasite species in the host species (that is, how many host species harbor zero parasite species, one parasite species, two parasite species, etc.). In short, some fish species will not harbor monogeneans (the subensemble H– in the figure), while other fish species will (forming the subensemble H+).

Moving from the imaginary to reality, that is, to data about real associations, we can use the example of fish populations in the Mediterranean. Caro, Combes, and Euzet (1997) have shown that there are about twice as many monogenean species as there are fish species and that the number of parasite species per host species is highly variable in this system. Some species of fish (and some entire families, including the Gobiidae, Syngnattidae, and Blenniidae) have no monogenean parasites. On the other hand, other species are very rich in parasites, the record being held by the spardeans *Diplodus sargus* and *D. vulgaris,* which may harbor up to eleven

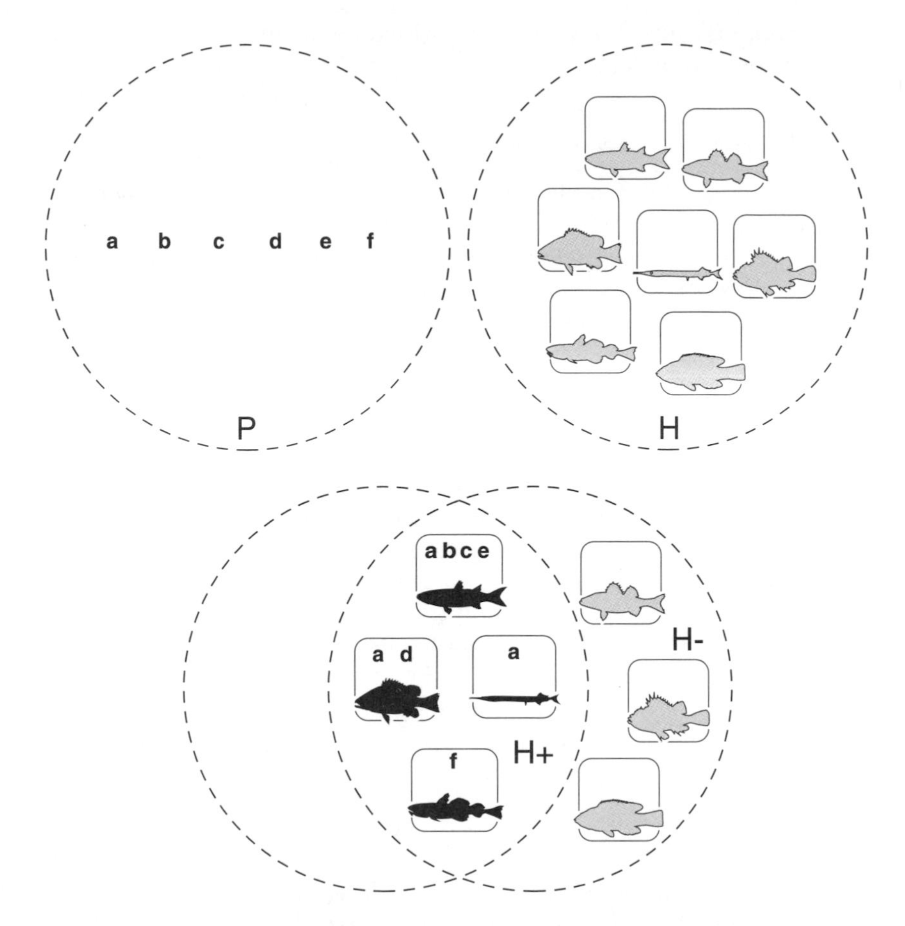

Figure 4.1. Imaginary distribution of an ensemble (P) of parasite species (a, b, c, . . .) in an ensemble of host species to form parasite-host systems. H = potential hosts species; H+ = subset of infected H; H– = subset of H that is uninfected.

and thirteen monogenean species, respectively.[1] It can be demonstrated that this distribution does not fit a random model because certain host species harbor numerous parasite species, others very few, and yet others none at all. This variable, which denotes the number of species per host, is referred to as *parasite richness*.

At the level of individual hosts, and not at the level of the host species,

---

1. These numbers may only partially reflect reality, since only 150 of 650 fish species living in the Mediterranean have been the object of thorough parasitic research.

this richness is called *parasite intensity,* which will be the subject of the second part of this chapter.

A Few Hypotheses concerning Richness

Previously we established a list of hypotheses in an attempt to explain why certain parasites exploit many host species and others very few or even only one. Let us now try to formulate hypotheses to explain why parasite richness differs according to host species. As in the case of specificity, these hypotheses are in part based on comparative analyses.

Is it possible that parasite species are distributed randomly in the space of the available host species? This hypothesis, which in effect stipulates that there is no influence of host phylogeny or ecology on a parasite's choice of a host, may be tested by determining if the observed distribution fits a calculated Poisson distribution. In fact, it appears very improbable that the distribution of parasite species within the available host species fits such a random model. Intuitively, the richness in parasitic species of the smallest mammal known, the pygmy shrew *Sonchus etruscus* (weighing in at a whopping one gram), and that of the elephant (weighing in at about seven tons) should not depend solely on randomness simply because of the vast difference in their sizes.

In the large majority of the cases parasite richness is not due to randomness but rather results from multiple factors. Some of the influences, listed below, are supported by solid evidence, while others are more speculative in nature. Moreover, yet other variables not listed probably also intervene.

Hypothesis 1: There is a correlation between parasite richness and the age of the phylum the species belongs to

This hypothesis is intuitively likely but remains to be demonstrated. One may think that the most ancient groups were the most abundantly colonized as the result of numerous lateral transfers (see chapter 5), but it is possible that in an ancient phylum parasite extinctions have been more numerous than acquisitions. Bush, Aho, and Kennedy (1990) have shown that present-day fish are parasitized less than are present-day birds but note that it is difficult to consider teleosteans as more ancient than birds! As a result it is very difficult to draw conclusions on the validity of this hypothesis.

Hypothesis 2: There is a correlation between parasite richness and the diversity of the microhabitats offered by the general body plan of the phylum

Since the host is a mosaic of sites and since parasites are more or less specialized for some of these sites, one may expect that highly complex organisms (those with complex organs providing a diversity of sites) may be richer in parasites than simpler ones. This hypothesis is true when we compare very different hosts, and it has been supported by comparing the number of described helminths in the digestive tracts of fish with those in the digestive tracts of birds (Kennedy, Bush, and Aho 1986). In this study the authors interpreted the greater parasite richness they found in birds as due largely to the greater complexity of birds' digestive tracts relative to those of fish. One may object, however, that the very different diets of fish and birds could be just as important as the diversity of the niches in their digestive tracts.

Hypothesis 3: There is a correlation between parasite richness and how ancient the species is

All species differentiate from preexisting species, either by anagenesis (the ancestral species transforms) or by cladogenesis (the ancestral species divides).[2] Given this, it is not obvious that the age of a species as determined by taxonomists influences its parasite richness. When a host species differentiates, it does not mean it does so without parasites, and nothing indicates that the frequency of parasite transfers or of their extinction follows any particularly defined patterns over the life span of a species. Further, in some regions related host species may be characterized by parasite species that differ strikingly from those in another area, as is true for eels. In this case the American species *Anguilla rostrata* (studied in Nova Scotia) and the European species *A. anguilla* (studied in Great Britain) have both been found to harbor relatively poor helminth communities that are in fact very similar. In both species the parasites specific to each are almost identical, and only a few generalist species, borrowed from other fish, differentiate them. However, the same is not true (see chapter 13) for the Australian eel, *A. reinhardtii*, which harbors a very rich helminth community. This situation has led Marcogliese and Cone (1998a) to conclude that the eels from North America and Europe are at an early phase of colonization by parasites while the communities of parasites of the Australian eel have had more time to form.

Hypothesis 4: There is a correlation between parasite richness and whether the host species belongs to a phylum that is poor or rich in species number

A logical prediction is that a high number of closely related host species within a phylum leads to increased parasite richness. In fact a phylum rich

2. Note that "anagenesis" refers to the evolution of a single species and not to the production of new species (see Brooks and McLennan 1993).

in species may be suspected to allow some parasite speciations that, once differentiated, would then allow lateral transfers that increase the parasite richness of the host species. This hypothesis was tested without significant result for insects and their parasitoids (Hawkins and Lawton 1987). On the contrary, however, both Caro, Combes, and Euzet (1997) and Raibaut, Combes, and Benoit (1998) have shown that the presence in the same environment of closely related species of the same genus, or of genera within the same family, is positively correlated with high parasite richness for both the monogeneans and the copepods of marine fish.

Hypothesis 5: There is a correlation between parasite richness and the size of the area the species occupies

This hypothesis has been the object of several thorough studies based on data in the literature. These include the works of Price and Clancy (1983), who studied thirty-nine helminth species of freshwater fish from the British Isles (fig. 4.2); that of Ranta (1992), who studied the parasites of Canadian freshwater fish; that of Gregory (1991a), who worked on the parasites of thirty-eight species of waterfowl of various origins; and that of Feliu et al. (1997), who studied the trematodes of Spanish rodents. All these investigations support this hypothesis, at least in part.

These results may be interpreted in the context of MacArthur and Wilson's (1967) theory of oceanic island biogeography, which stipulates,

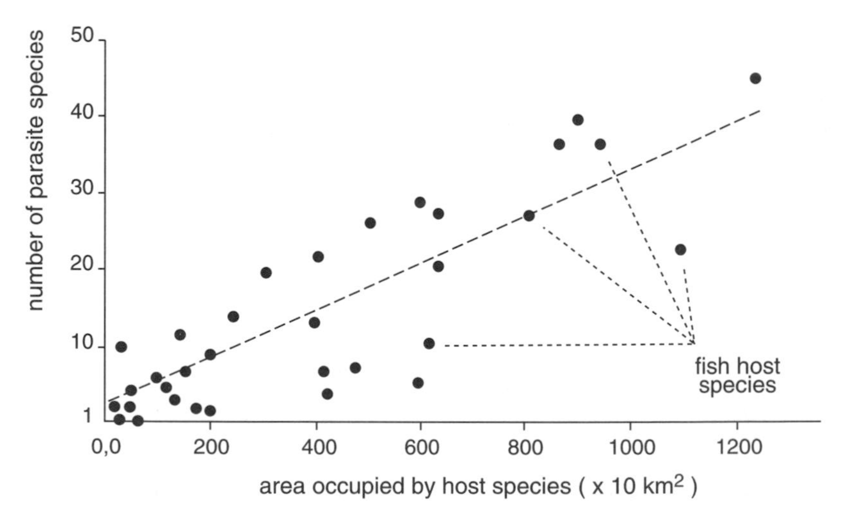

Figure 4.2. Relationships between the number of helminth species and the area occupied by hosts in freshwater fish from Great Britain (modified from Price and Clancy 1983).

among other things, that there is a positive correlation between the richness of living species on an island and the size of the island. If the correlations above reflect reality, it means that the theory of island biogeography could be applied to parasite-host systems, with the area occupied by a host species (or the size of the host, see hypothesis 9) being considered "equivalent" to the area (or size) of an island.

However, Guégan and Kennedy (1993), using Price and Clancy's data, have contested the interpretation of this idea. They have shown that seven recently introduced fish species (out of the thirty-two taken into consideration) were responsible for the previously observed correlation because they have both a limited area and a poor parasitofauna simply as the result of their recent introduction (less that a thousand years ago). It is also likely that these fish lost their parasites when they were introduced and that the transfer of indigenous parasites to these hosts may not have had time to occur.

Hypothesis 6: There is a correlation between parasite richness and a host species' diet (whether it is a carnivore, an herbivore, or an omnivore)

This hypothesis makes sense only within a given group of hosts. In the case of fish, for example, one may ask, Is parasite richness correlated with the type of diet? This problem, although apparently easy to test, has not been the object of much research, and one may expect the response to be different depending on the group of parasites studied. It is difficult to see, for example, why fish diet would affect the monogenean fauna (a holoxenous parasite) present on a host, yet it seems logical in the case of trematodes (a heteroxenous group that is often transmitted via food). Similarly, in terrestrial vertebrates it appears that being an herbivore favors the occurrence of very rich intestinal nematode faunas, and several hundred species can be counted from the digestive tracts of zebras (Lotfi S. Khalil, personal communication).

Hypothesis 7: There is a correlation between parasite richness and the diversity of the host's diet

It seems logical to predict that the more diversified the diet of the host species, the larger the probability of its ingesting various parasite vectors and, therefore, the greater its parasite richness. There is, however, no synthesis of work on this problem. Nevertheless, several studies seem to support this prediction. For example, Guégan and Kennedy (1993) have shown that omnivorous fish harbor the largest helminth communities, and Galaktionov (1996) showed that the same is true for marine birds. After summarizing his personal research and the works of various authors, Galaktionov showed that herring gulls *(Larus argentatus)*, which are euryphagous, harbor from twelve to eighteen species of platyhelminths depending on locality. This number drops to eight in the kittiwake *(Rissa tridactyla)*, which is a specialized fish eater, while the little auk *(Plautus alle)*, which is planktonophagous, is almost free of platyhelminths. However, Chabaud and Durette-Desset (1978) demonstrated that for nematodes just the opposite situation occurs. In this case the stable environment of the monophagous host probably allows the establishment of numerous parasite species, while the unstable environment of the polyphagous hosts allows the establishment of only a smaller number. These authors based their work on comparing termites, terrestrial turtles, elephants, horses, bovids, and anteaters (all relatively monophagous organisms with parasite communities that are both rich and stable) with roaches, marsupials, and squirrels (all rather polyphagous with parasite communities that are unstable and of low diversity).

Overall, what seems to count most is probably the presence in the diet of "good" parasite vectors rather than the variety of vectors. This is what Bell and Burt (1991) illustrate by writing that for freshwater fish "benthic crustaceans and fish seem to be hazardous things to eat."

Hypothesis 8: There is a correlation between parasite richness and the degree of movement (migration) of the host species

If a host species frequents a succession of different environments over time, it can be exposed to parasites that are commonly found in each of these locations. Available observations, such as those reported by Dogiel (1964), do not support this idea and instead suggest that there is a replacement of parasite species, rather than an accumulation, when hosts move to new areas. However, this hypothesis deserves to be revisited with an eye toward introducing different hierarchical scales to the analysis.

Hypothesis 9: There is a correlation between parasite richness and the maximum size reached by members of the host species

This hypothesis, considered obvious by Price (1977), has been the object of deep reflection by various authors, some basing their analysis on existing databases and others founding theirs on investigations specially conceived to address the question. Most such studies concern the richness of helminths in the vertebrate digestive tract. For instance, in Canadian freshwater fish, Bell and Burt (1991) showed that host size has a different influence depending on the parasite group. In this case the size of the host partly explains the diversity of cestodes, nematodes, and acanthocephalans but not at all that of trematodes.

In West Africa, Guégan et al. (1992) showed (fig. 4.3) that the size of cyprinid fish "explains" 77% of the variation in the number of species of

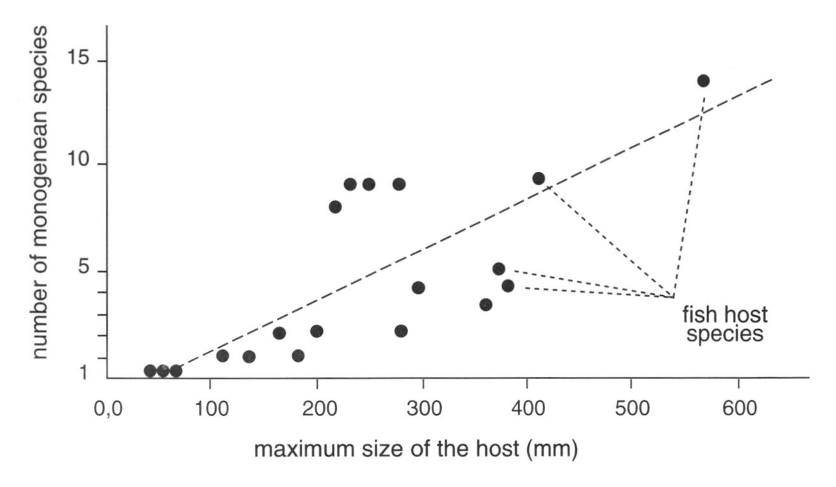

Figure 4.3. Relationships between the number of gill monogenean species and the maximum size of hosts in freshwater fish from West Africa (modified from Guégan et al. 1992).

dactylogyrid gill monogeneans on nineteen host species. As the figure shows, small species of fish have no monogeneans, the largest harbors thirteen species, and the others all fall in between.

In birds the size of the host also seems to be a determining factor in parasite richness. This has been shown by Gregory, Keymer, and Harvey (1991), who based their conclusion on data previously published by a Russian researcher (S. M. Vaidova) on the helminths of 158 species of birds from Azerbaijan. On the other hand, a study by Feliu et al. (1997) of rodents from the Iberian Peninsula involving sixteen species of hosts, seventy-seven species of cestodes, and a variety of nematodes and trematodes did not show any correlation between host size and parasite richness.

Many parasitologists intuitively believe that host size is the most important factor in determining parasite richness. If this is true it may be that parasites in certain phyla can become an obstacle to further increases in host size. It is interesting to think that dinosaurs may have disappeared because they became too rich in parasite species.

Hypothesis 10: There is a correlation between parasite richness and where the host species is found on the planet

It is well known that the diversity of living species increases as we go from the poles toward the equator. This is particularly true for both plants and insects, and it is therefore most likely that parasites obey the same rule. However, this does not imply that parasite richness necessarily increases,

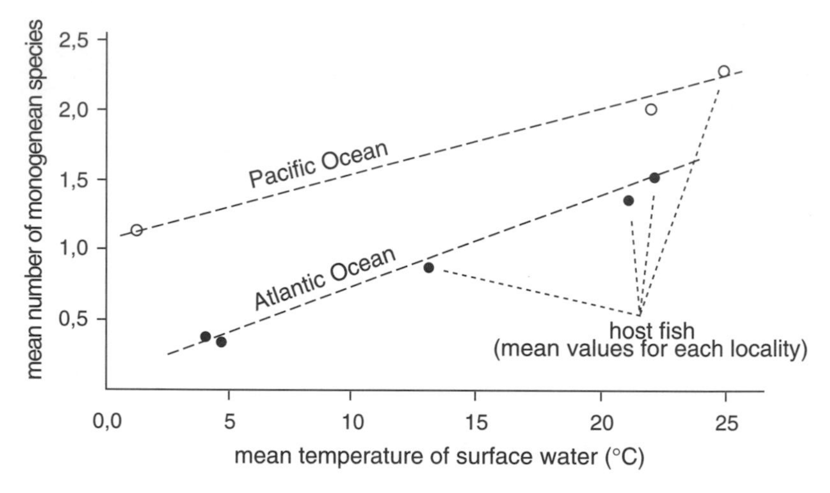

Figure 4.4. Relation between the number of species per fish and the temperature of water, after accounting for latitude (note that parasite richness is greater in the Pacific Ocean than in the Atlantic (modified from Rohde 1984).

since there may be more species of parasites but not more species of parasites per host. Thus it is very interesting to check, within a group of hosts, whether the tropical species harbor more parasites that those from temperate regions.

Some works partially address this question. For example, Rohde (1984, 1993, 1999) and Rohde and Heap (1998) showed that the richness of monogeneans in marine fish increases as one comes closer to the equator (fig. 4.4). Another estimation, communicated by Boris Lebedev, shows that the number of monogeneans increases more rapidly than the number of fish species as one approaches the equator. It also appears that small mammals of tropical regions may be richer in parasites than small mammals of temperate regions (Betterton 1979).[3]

The correlation of biological diversity with latitude has itself been the subject of numerous tentative explanations. Rohde (1992) believes that the difference in the energy received according to latitude may provoke shorter generation times, higher rates of mutations, and faster selection—from which an increase in biodiversity would then arise. One should also not forget that the surface area of tropical regions is larger than that of temperate or polar regions because of the shape of the earth.

3. We will look in chapter 20 at the problem of tropical richness in parasites relative to the emergence of hominids in warm regions.

Hypothesis 11: There is a correlation between parasite richness and the gregariousness or solitary behavior of the host species

One may expect that gregariousness makes parasitism "easier." Life in groups may even have been limited throughout evolution by the explosive transmission of certain pathogenic agents (Freeland 1976). This point has been clarified by Dobson (1988), who writes: "Where species have a complex social organization and live in very large groups, they are highly susceptible to the accidental introduction of novel pathogens."

Various analyses relative to this point have been done using data from the literature. Ranta (1992) concludes that among the freshwater fish of Canada (about sixty species) those that are gregarious are richer in parasites than those that are solitary. An analysis of parasitism of Mediterranean fish (Caro, Combes, and Euzet 1997) shows that parasitism by monogeneans is about twice as high in gregarious fish as in solitary ones. However, Poulin and Rohde (1997) do not detect this relationship between parasite richness and gregariousness.

Hypothesis 12: There is a correlation between parasite richness and the accessibility of host species in their environment

This hypothesis is certainly applicable within certain groups, and one may expect that the pressure generated by parasitism may select for hosts with "antiencounter behaviors." Seeking refuge in inaccessible places is one of these behaviors (see chapter 10), and a comparison of parasite richness in related hosts shows that those species that are difficult to reach also partially escape parasitism. For instance, Hawkins and Lawton (1987) show that insects that eat leaves (and are therefore well exposed) have a larger assemblage of parasitoids than those living in galls (protected by the gall wall), in trunks (protected by the thickness of the wood), and in roots (protected by soil).

Hochberg and Hawkins (1992) extend the notion of such refuges much further than that of a simple physical barrier and show, for example, that a species' mode of distribution in space may also influence its parasite richness. Figure 4.5 shows that the less accessible the plant environment an insect larva inhabits (bark, roots, etc.) the fewer parasitoid wasps that attack it. However, the left part of the figure shows an interesting anomaly. "External feeders" and "roller/webbers" do not have the expected parasite assemblage (dotted line) if one follows the previous reasoning to its obvious conclusion. We may deduce from this that these species have developed other adaptations (for example, immune defenses) that limit the success of the parasitoids.

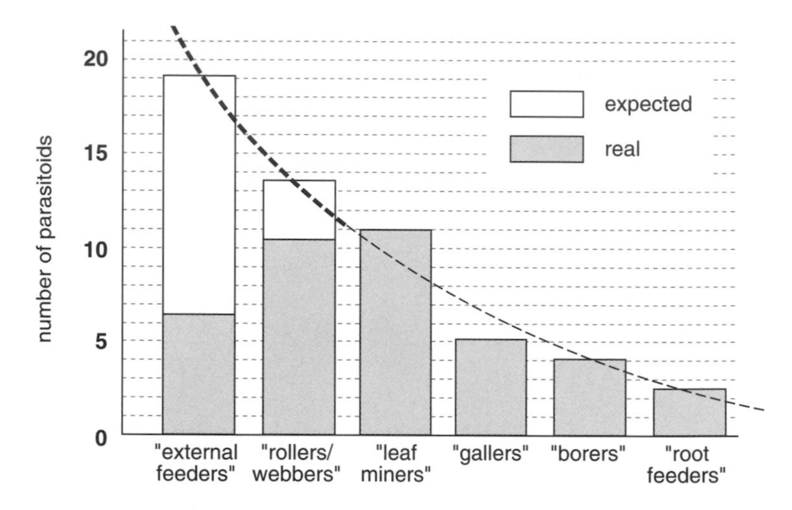

Figure 4.5. Influence of "shelter" on parasite richness: the case of insects attacked by parasitoid wasps. For both the "external feeders" and the "roller/webbers," richness is lower than logically expected, and other defense mechanisms most likely have been selected (inspired by Hochberg and Hawkins 1992).

Hypothesis 13: There is a correlation between parasite richness and the type of ecosystem the host species belongs to

This comparison can be made only if related hosts are found in different ecosystems. Simply put, the question is, Are rodents from deserts and rodents from equatorial forests characterized by parasitofauna of different richness? If we limit ourselves to vertebrates, we find rather contradictory answers to this question. Bush, Aho, and Kennedy (1990) postulate that animals living in the water or going to the water have richer fauna than do strictly terrestrial animals, while Gregory, Keymer, and Harvey (1991) note that in birds this water-dependent richness is true only for trematodes. From a comparative analysis of 488 species of vertebrates, Gregory, Keymer, and Harvey (1996) conclude that "the apparent tendency for aquatic hosts to harbor more parasite species than terrestrial ones appears to arise from covariation with confounding variables."

Hypothesis 14: There is a correlation between parasite richness and how long the ecosystem has been occupied

When organisms or groups of organisms change environments during the course of evolution, it is logical to think they may lose part or even all of their ancestral parasites and that they then may then acquire new parasite "faunas." One may predict that, depending on the case, there may be either a decrease or an increase in the number of parasites.

For example, what do we know about the helminth communities of marine cetaceans, which descended from terrestrial mammals? Two investigations show that at least some have poor helminth communities. One concerns the pilot whale, *Globicephala melas,* in the Faroe Islands (Balbuena and Raga 1993) and the other the dolphin de La Plata, *Pontoporia blainvillei,* off the coast of Argentina (Aznar, Balbuena, and Raga 1994). These works seem to indicate that the extinction of ancestral parasites has not been compensated for by more recent acquisitions in this group.

Hypothesis 15: There is a correlation between parasite richness and the degree of ploidy of the host species

Guégan and Morand (1996) have shown that the tetraploid cyprinids of West Africa are on average richer in monogenean species than the diploid ones, thus qualifying them as "strange attractors." This result was unexpected, and the authors speculate that there may not be a direct correlation between the degree of polyploidy and parasite richness, but rather a correlation with a third common variable that has yet to be identified.

Hypothesis 16: There is a correlation between parasite richness and the genetic diversity of the host species

Pariselle (1996) proposed that parasite richness is higher in host species with more genetic diversity than in those with less. The idea here is that if a host species loses alleles during a bottleneck event, it can at the same time lose some parasite species. There must therefore be a correlation between these two processes of loss (fig. 4.6A). Pariselle supports his hypothesis by a comparative analysis of genetic diversity and parasite richness in twenty- six freshwater cichlid fish of Africa in which seventy-four species of gill monogeneans were found. In short, this author shows that those species with the greatest enzymatic polymorphism also have the richest gill monogenean fauna (fig. 4.6B).[4]

Which hypotheses are to be retained? As the hypotheses listed above show, parasite richness may result from multiple causes, making it difficult to pinpoint specific sources of variation. In the end, then, the list shows with certainty only that parasite richness is linked to *host phylogeny, ecology,* or both.

4. Other hypotheses may also explain the correlation between parasite richness and host genetic diversity in this system.

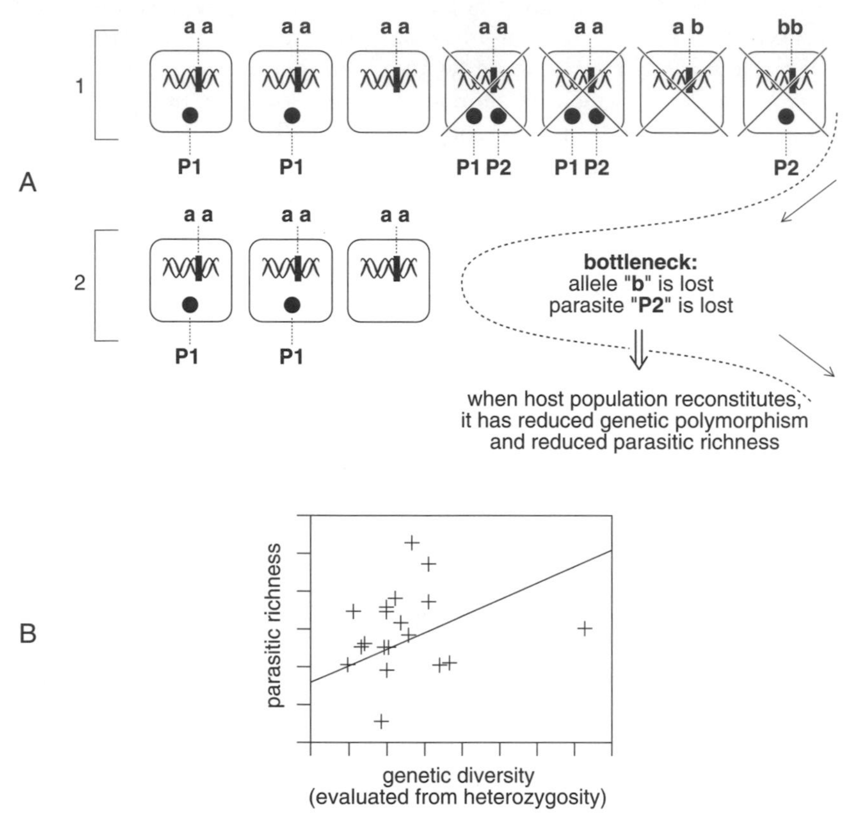

Figure 4.6. (A) Pariselle's hypothesis: if the hypothesis is correct, host species with high polymorphism should have the highest parasite richness and vice versa (inspired by Pariselle 1996). 1 = imaginary host population at stage 1 (before bottleneck); 2 = imaginary host population at stage 2 (during bottleneck); a, b: alleles at a particular locus; P1, P2: two parasite species (no link between alleles and parasites); X denotes lost genotypes during bottleneck. (B) A demonstration of the relation between polymorphism and parasite richness based on the study of monogeneans of African fish (heterozygosity and richness were corrected for various confounding variables).

Distribution in Individual Hosts

Parasite Prevalence, Intensity, and Abundance

Host species are composed of individuals that are genetically and phenotypically variable and thus do not offer identical colonization opportunities for parasites. Figure 4.7 is similar in principle to figure 4.1 except that the objects depicted are individuals and *not species*. This figure illustrates an imaginary population (P) of the infective stages of a parasite species

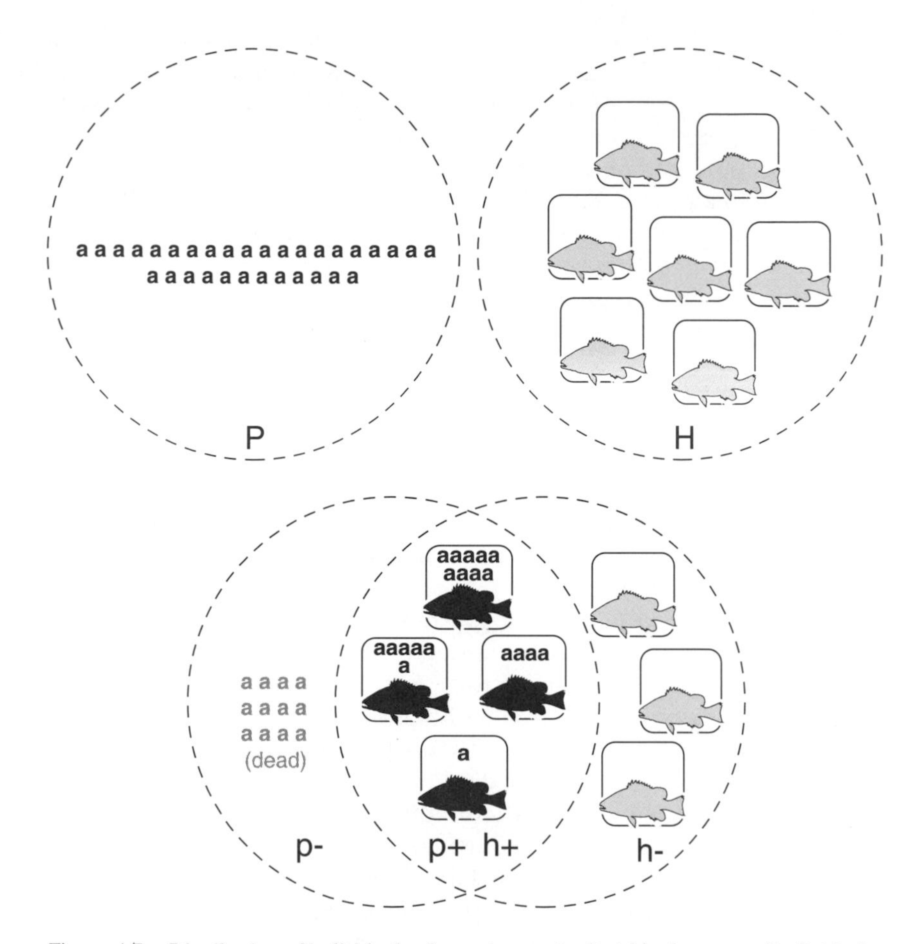

Figure 4.7. Distribution of individuals of parasite species P within the space of individuals of host species H in the form of individual parasite-host systems (compare with fig. 4.1). H+ = subset of hosts infected; H– = subset of hosts uninfected; P– = parasites that did not find hosts and that thus did not survive; P+ = parasites that infected hosts.

and the way it is distributed in an imaginary population (H) of hosts. For example, one can imagine that it is the distribution of individuals of the trematode *Fasciola hepatica* in a population of wild ungulates of a specific species. Like figure 4.1, figure 4.7 (in an imaginary way) shows the associations that may occur and the series of subensembles that can result: h– is the subensemble of potentially healthy hosts (those that have escaped infection); h+ is the subensemble of hosts infected by the parasite; p+ is the subensemble of parasites that have infected a host; and p– is the subensemble of parasites that have not infected a host.

These subensembles allow one to define the classic epidemiologic parameters of parasitism (Margolis et al. 1982; Bush et al. 1997). These are *prevalence,* or the level of parasitism, calculated as the ratio of h+ to H (h+/H); *incidence,* or the proportion of h– that become infected (h+) during a determined period; *individual intensity,* or the number of parasites found in a distinct individual of h+ (this parameter has a different meaning depending on whether the parasites multiply or do not multiply in the host); *mean intensity,* or the average number of parasites per host, obtained by averaging the individual intensities or by calculating the ratio of p+ to h+ (p+/h+); and last, *abundance,* or the ratio of p+ to H (p+/H).

In the population of infective stages (P) an important part, p–, is not associated with a host and, as a result, dies. Symmetrically, an important portion of the potential hosts, h–, are not infected, and their fitness is not decreased by parasitism.

Whatever the practical importance of these parameters, none holds as much interest as another one: the distribution of parasites in the host population. It is this distribution that allows one to fully measure the inequality of host individuals in the face of parasitism.

The term "distribution" can be applied to all living organisms, not only parasites. In a forest, for example, the trees of a particular species may be distributed in space in three fundamentally different ways. They may be perfectly equidistant from each other (the usual case when they are planted by foresters). This type of distribution is termed a regular or even distribution. They may be randomly distributed (as if they had been thrown into place with eyes closed from the top of a mountain). This is a random or "Poisson" distribution. They may form groups that are more or less separate from each other. This is an aggregated or clumped distribution, which is often referred to as "overdispersed" in the parasite literature.

To analyze the distribution of these trees in a precise manner and give it a mathematical expression, one can divide the forest into squares with sides of *x* meters, count the trees in each square, and determine the frequencies of the squares in which there are no trees, one tree, two trees, and so on. Most free-living organisms are distributed in an aggregated manner in nature because of the heterogeneous nature of habitats, with concentrations of individuals in some "hot spots" and a few others scattered about elsewhere.

In parasites the unit of analysis of the distribution has not been so arbitrarily defined: it is, quite naturally, the host individual: "Hosts are the quadrats for sampling parasites" (Shaw and Dobson 1995).

Aggregated Distributions

Let's take the example of adult cestodes of *Dipylidium caninum,* which at any time can live in the dogs of a village and are distributed among the individual dogs. Like the trees in the forest, the population of this parasite may be distributed in three ways: If the distribution is regular, then all the dogs harbor the same (or almost the same) number of *D. caninum.* If the distribution is random, it means it follows a Poisson distribution and is the result of chance alone. If the distribution is aggregated, it means some individual dogs contain much higher "concentrations" of *D. caninum* than would be predicted by chance alone and that many others are not at all or only very slightly parasitized.

There is a simple test to know if the distribution of parasites in a host population follows Poisson's law or if it is regular or aggregated. This test is to calculate the ratio of the variance to the mean: if the ratio is close to one the distribution is random; if it is less than one it is regular; if it is greater than one it is aggregated. In the last case, the higher the variance to mean ratio the more aggregated the distribution is.

Since the 1970s, numerous epidemiologic investigations have shown that random distributions of parasites in host populations (distributions that follow Poisson's law) are the exception rather than the rule. In most cases, whether in plants, animals, or humans, parasites are distributed in an aggregated manner. Thus certain hosts harbor more parasites than would be predicted by chance alone, while others obviously harbor fewer.

When the distribution is aggregated it can be represented by various mathematical models; the most commonly used being the negative binomial. This model has a parameter *(k)* that measures the degree of "nonaggregation" (the smaller *k* is, the stronger the aggregation; the higher *k* is, the less the aggregation; if *k* approaches infinity the distribution is not aggregated but random and is in fact a Poisson distribution). The first investigation showing the fit of the distribution of parasites to the negative binomial model is more than thirty years old and was done for the bovid dipteran *Hypoderma bovis* (Breyev 1968). Later work by Crofton (1971) went so far as to consider the negative binomial distribution a necessary part of the definition of parasitism. Figure 4.8 shows an example of an aggregated distribution that does not fit Poisson's law (Combes and Knoepffler 1977).

For human parasites evaluations of "distribution" parameters are indirect. They can be done for intestinal helminths by counting the number of eggs in stools and estimating the average egg production of each individual worm. This type of investigation shows not only that parasites are

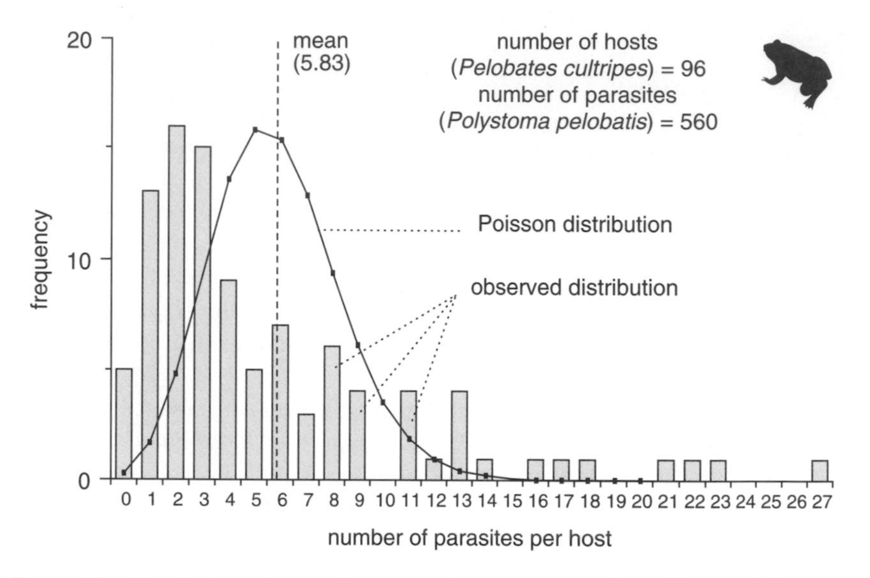

Figure 4.8. Distribution of young *Polystoma pelobatis* in a population of the amphibian *Pelobates cultripes* at the end of metamorphosis. The distribution is aggregated and does not fit a Poisson distribution (modified from Combes and Knoepffler 1977).

aggregated but that the aggregation has a comparable amplitude for the different age-classes of humans (see, for example, Croll et al. 1982).

In natural populations aggregation (overdispersion) can be self-limiting when its consequences generate underdispersion. In this case, when parasites are highly aggregated there is a density-dependent regulation (which I'll talk about in chapter 12) that, depending on the case, may induce the death of the most heavily parasitized hosts and consequently the death of a large number of parasites. These two processes therefore tend to limit the extent of the overdispersion (Anderson and Gordon 1982).

Although in general aggregated distributions often exemplify parasite populations, some regular distributions have also been recorded. In the extreme case each individual host harbors only one parasite. For example, on the Great Barrier Reef in Australia the fish *Chromis nitida* harbors a maximum of one *Anilocra pomacentri* (an isopod crustacean) that is attached on either the right or the left side just behind the eye (Adlard and Lester 1994). Why only one? This may be because there is space for only one parasite behind the eye, and if parasites were to attach on both sides the fish would be substantially debilitated and eaten by predators. Another example is *Leposphilus labrei* (a copepod crustacean), which is almost always by itself on the lateral line of the fish *Crenilabrus melops:* according to an investigation by Donnelly and Reynolds (1994), out of 1,924 parasitized fish, 1,922 harbored only one copepod, either on the right or on the left—curiously, more often on the left.

The preceding material concerns the distribution of parasite individuals at the level of host individuals. One may also wonder about the distribution of parasite individuals at other scales of groups of host individuals. Krause and Godin (1996) and Krause, Ruxton, and Godin (1999) studied the distribution of metacercariae of the trematode *Crassiphiala bulboglossa* within and between the shoals of three fish species from Lake Morice in Canada (the definitive host of the parasite is the belted kingfisher, *Megaceryle alcyon*). The lake is of modest dimensions (140 hectares), so we can consider it a relatively homogeneous environment. Metacercariae induce changes in the fish's behavior (surfacing and jerking): "[The parasites] severely influence the condition of a fish . . . and affect its quality as a shoal mate in terms of providing anti-predator benefits for other shoal members" (Krause, Ruxton, and Godin 1999). Thus, when an individual fish joins a shoal it is to its advantage to choose a group that contains fish whose behavior is modified by parasites, since these fish will attract birds to themselves and away from the others.[5]

As one might expect, the distribution of metacercariae is highly aggregated within shoals of the two species of fish that are the most parasitized by *C. bulboglossa:* the golden shiner *(Notemigonus crysoleucas)* and the killifish *(Fundulus diaphanus)*. The most interesting results, however, are from the "between shoals" analysis, for which each fish species is considered separately: the distribution is similarly not random between shoals, supporting the hypothesis of shoal choice according to the presence of parasites and showing that parasitism can play a role in determining a shoal's composition (fig. 4.9).

When possible, it is interesting to characterize the spatial distribution of parasites at different scales in the host population. For instance, Boulinier, Ives, and Danchin (1996) and McCoy et al. (1999) have analyzed the distribution of the tick *Ixodes uriae* on chicks of its seabird hosts *(Rissa tridactyla, Alca torda, Uria aalge,* and *Fratercula arctica)* in various locations in France, the British Isles, and Canada and found evidence of aggregation at the among-cliff and among-nest scales but not in chicks within nests. The authors believe such features might have implications regarding tick dispersal and host specificity.

5. Dugatkin, FitzGerald, and Lavoie (1994) showed that fish are able to distinguish conspecifics with normal swimming patterns from those whose movements are modified by a parasite.

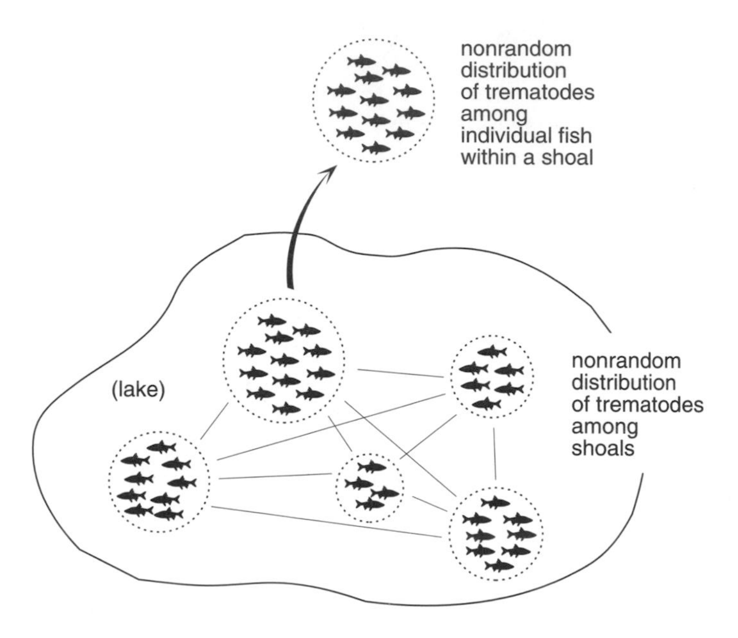

Figure 4.9. The two possible levels of aggregation, intrashoal and intershoal, in fish from a Canadian lake (inspired by Krause, Ruxton, and Godin 1999).

Why Aggregated Distributions?

The most common response to this question is, of course, that the host population shows heterogeneity relative to the parasite population (Shaw and Dobson 1995). Poulin (1998a) explored in detail the causes and consequences of such an aggregation.

A First Category of "Causes" Dependent on Parasites

The distribution of the infective stage of parasites, itself aggregated, is a cause for aggregation in the definitive hosts, an extreme example being the acquisition of the parasites as a "package." This happens each time there is an asexual reproduction that is not followed by the usual dispersion of stages to the outside, such as in the trematode of the ray *Ptychogonimus megastomus.* In this system, sporocysts containing numerous cercariae leave the mollusk and are ingested by crabs that are in turn eaten by the rays. Other examples include a trematode of birds, *Leucochloridium paradoxum* (whose complete life cycle is described in chapter 7), and two cestodes of mammals, *Multiceps multiceps* and *Echinococcus granulosus* (each of which has "budding" cysticerci). The result of such "package" acquisition by definitive hosts has been qualified by Bush,

Heard, and Overstreet (1993) as the attainment of an "instant infrapopulation."

The internal multiplication of parasites after recruitment can also yield an aggregated distribution. This is the case in viruses, bacteria, protozoans, the larval stages of trematodes in mollusks, and some rare helminths in definitive hosts (atractid nematodes and certain polystomatid monogeneans).

A Second Category of "Causes" Dependent on the Hosts

If some hosts have more parasites than would be predicted by chance alone, this could be interpreted to mean that they offer the parasites a better-quality habitat in the broadest sense. Such a high-quality habitat is a host that the parasite may encounter easily (its behaviors favor encounter) and within which compatibility is possible (its resources may be exploited and its immune defenses avoided).

Even when a host population appears superficially homogeneous, nuances in behavior and in the efficacy of the individual host's immune system create subtle differences in the quality of habitat, which then tends to aggregate populations of parasites into particular host individuals. This explains why overdispersion is commonly observed in natural populations. Moreover, such aggregation may be increased if the two filters, encounter and compatibility, are open at the same time in some individuals even though they are independent of one another. As such this process is capable of generating very pronounced aggregated distributions in host populations.

Significantly, experimental evidence shows that overdispersion occurs even where the heterogeneity of the host population is as low as possible. Roepstorff et al. (1997) infected three groups of fifty pigs with doses of one hundred (group A), one thousand (group B), and ten thousand (group C) infective eggs of the nematode *Ascaris suum*. The results showed that within each group a number of larvae, equivalent to about 50% of the given dose, was found in the intestine after seven to ten days. Between the fourteenth day and the twenty-first day there was then a considerable elimination of parasites, and only a small number of adult ascarids remained. The distribution of the parasites obtained in this way was highly aggregated in each of the three groups, even though most of the factors that generated the aggregation were accounted for (i.e., there was no alteration in pig behavior, a uniform infective dose was given to each group, no other parasites were present, and the hosts were closely related genetically).

In practice it is never easy to distinguish the role played by the encounter filter (e.g., washing or not washing hands before meals) from that played by the compatibility filter (e.g., having a genetic constitution that allows or does not allow survival of the parasite).

Variables Associated with Encounter That Influence Inequality in the Face of Parasitism

Variable 1: The Regional Environment

The life cycles of most parasites are tightly linked to the distribution of their hosts, particularly to that of their vectors when the cycles are heteroxenous.

In numerous examples of human parasites, the geographic distribution of the parasite is determined by the area occupied by its vectors. As an example, the existence of the molluskan genus *Biomphalaria* allows the life cycle of *Schistosoma mansoni* (an intestinal schistosome found in South America), while the absence of the molluskan genus *Bulinus* prevents the life cycle of *S. haematobium* (urinary schistosomiasis) in the same area. Similarly, the occurrence or lack of vectors also plays a role in other parasitic diseases such as malaria. The decrease in rain in the African Sahel between 1965 and the 1990s caused serious agricultural problems, but it also led to the quasi disappearance of the major vector for malaria, *Anopheles funestus,* in several regions. This was particularly true in Senegal and Niger, where both the prevalence and the incidence of malaria were considerably reduced (Mouchet et al. 1996).

The parasitic fauna of virtually every organism also depends on the presence of the parasites in taxonomically closely related hosts, and the more or less frequent occurrence of such reservoirs in an area is also a factor leading to inequality in the face of infection.

Variable 2: The Local Environment

At a finer scale than that of geographical regions, the quality of the milieu that directly surrounds a host individual is a further major cause of inequality leading to overdispersion in the face of parasitism. The notion of "site of transmission" used for many diseases implies that the distance between the habitat of a particular individual and a transmission site plays an important role in the distribution of parasites within hosts.

This is notably true for the waterborne parasitic diseases, that is, all those whose transmission has something to do with water because their infective stages are in water (schistosomiasis, anguillosis); in damp soil (ancylostomiasis); in a vector that lives in the water (dracunculosis); in a vector that was born in the water (malaria, onchocerciasis); or in a vector that frequents an environment influenced by water (sleeping sickness).

In a more general way the local environment also plays a role each time the existence or abundance of one of the players in the life cycle is dependent on a particular aspect of environmental conditions. Lo, Morand, and Galzin (1999) showed that the prevalence and intensity of trematodes of the territorial fish *Stegaster nigricans* are significantly higher in populations on the Great Barrier Reef than in populations from the fringing reefs surrounding Moorea Island. There is a correlation between parasitism and the invertebrate populations (mollusks, crustaceans) in these two environments, although they are separated by only a few hundred meters.

Leishmaniasis is a disease caused by protozoans of the genus *Leishmania*. The parasites are transmitted by phlebotomine flies (biting dipterans) whose larvae live in rodent burrows or in the cavities of old walls. The protozoans penetrate and multiply in vertebrate macrophages after transmission by the bite of infected flies, and the disease exists in the Old World in two forms, cutaneous (benign) and visceral (serious). It also exists in the New World in a serious mucocutaneous form.

The occurrence of leishmaniasis depends tightly not only on the presence or absence of the phlebotomine vector but also on the presence or absence of mammals that serve as reservoir hosts, themselves dependent on the local environment in a broader sense.

The research of Rioux et al. (1986) has shown that the transmission of leishmaniasis requires very well defined environmental conditions. Foci of cutaneous leishmaniasis occur in the arid regions of North Africa on the edge of the northern Sahara owing to the transmission of *Leishmania major* by the bite of *Phlebotomus papatasi*. Here humans are concentrated in villages in oases along temporary streams that are foci for *P. papatasi*, and these sites are also foci for the reservoir host of *L. major*, the gerbil *Meriones shawi* (a rodent that because of its need for free standing water remains confined to oases and that is also more or less anthropophilic). Garbage dumps in these areas are particularly favorable to the explosive occurrence of both the gerbils, which dig burrows, and the phlebotomes, whose larvae live in these burrows. Therefore the parasite circulates from gerbil to gerbil and from gerbil to human because of the phlebotomine vector. Temporal modifications of the environment also influence transmission and lead to strong epidemic "peaks" of parasitosis, with humid periods raising the numbers of gerbils (because of a major increase in plant resources) and, concomitantly, the levels of *L. major* infection in both rodents and humans.

In the Cévennes (an arid mountain region in the south of France), the epidemiological risk associated with *Leishmania* has been precisely correlated with plant strata, which itself describes the principal climatic characteristics of the region. *Phlebotomus ariasi* transmits *L. infantum* in this region, with the maximum risk for infection occurring in the zone containing both *Quercus ilex* (holm oak) and *Q. pubescens* (downy oak).

In South America human intervention has altered the transmission of leishmaniasis. For example, *Leishmania guyanensis* uses as its vector phlebotomines of the genus *Lutzomyia* in humid areas, which are themselves continually decreasing because of deforestation. At the same time *L. infantum* uses the xerophilic and anthropophilic vector *Lutzomyia longipalpis,* increasing transmission of this form to humans in the arid zones.

Knowing that the phlebotomine vectors have a relatively modest flying distance, the example of leishmaniasis demonstrates that for an individual human the choice of habitat, and therefore the choice of the immediate surrounding ecological conditions, is a major site-associated risk factor for infection.

Variable 3: Behavior

In the face of different environmental conditions it is common for members of the same species to have different behaviors. For example, Santiago Mas-Coma and his collaborators (personal communication) have compared the hosts of *Fasciola hepatica* on the European continent with those on the island of Corsica. On the Continent rats play a negligible to nonexistent role in the life cycle. In Corsica, however, rats have acquired original behaviors that lead them to eat the aquatic plants to which the metacercariae adhere, and thus they commonly become heavily infected (see Fons and Mas-Coma 1990).

Sex may also play a role in aggregation, because, as the result of their different sex-related behaviors, males and females of the same species are not necessarily exposed to certain parasites to the same degree. Silan and Maillard (1990) showed that this is the case for the sea bass *(Dicentrarchus labrax)*, which is infected by the monogenean *Diplectanum aequans* in the Mediterranean. Sea bass are, generally speaking, very mobile fish, which is probably not an ideal behavior for the parasite because it must infect its host using a swimming oncomiracidium. However, during breeding females become temporarily sedentary in rocky fractures. This immobility is enough for the parasite to increase the success of transmission (the eggs hatch in a few days), significantly increasing the intensity of parasitism in females relative to males of this host.

In birds it is sometimes just the opposite, and here the male is the victim of its parental devotion. Oppliger, Christe, and Richner (1996, 1997) showed by experimental research on the great tit that if one increases the size of the clutch the males are the ones that provide supplementary parental care by significantly increasing their feeding. Therefore the males and not the females adjust to the greater need for obtaining prey.

What is interesting to the parasitologist is that the males that work overtime to feed their young are about twice as parasitized by *Plasmodium* as are control males (prevalence of 76% versus 32%). It seems that the disease is not without consequences for the tits, because in the experiment seventeen noninfected males out of thirty in 1990 came back to the nesting site in 1993, while only two of fourteen parasitized males returned (the status of parasitized/nonparasitized being determined by a blood test). The high prevalence of parasitism when the clutch was increased may have two explanations: either the increased activity of the males exposed them to more bites by the mosquito vectors of *Plasmodium* (implicating the encounter filter), or their immune defense was not as efficient because they used more energy foraging (implying a role for the compatibility filter). Comparable results were obtained by Norris, Anwar, and Read (1995), who also worked on clutches of the great tit.

In humans it is also not rare that investigations show significant differences in infection depending on gender, with such differences linked to activities (fishing, hunting, house cleaning, etc.) that lead the different sexes to visit the transmission sites of a variety of parasites at different levels.

Although exceptional, it sometimes happens that a visible characteristic may be correlated with genes coding for behaviors that influence transmission. Such correlations have been shown in the vectors of protostrongylosis, a pulmonary disease of cattle caused by nematodes belonging to several genera (*Protostrongylus, Neostrongylus, Muellerius,* etc.). The life cycle of these parasites passes through a terrestrial mollusk that is actively infected by the parasites' larvae. The ungulate becomes infected by involuntarily eating either the mollusk or infective larvae that have left the mollusk. Experimental and field studies of snail populations have shown a correlation between decoration of the shell and polymorphism in the snails' attraction to ungulate feces, resulting in different probabilities of encountering the L1 larvae released with the feces. For example, individuals of *Cernuella virgata* with striped shells were significantly more attracted to feces than were individuals without such ornamentation (Cabaret and Vendrous 1986).

Variable 4: Social Status

Social status may influence the encounter of dominant hosts with parasites in either a favorable or an unfavorable way.

A first hypothesis, a priori the most logical, is: "The higher the status, the less abundant the parasite." This is supported in the following case. In birds, social status is important (as observed in the barnyard) and plays a particularly important role in the choice of nest site and territory

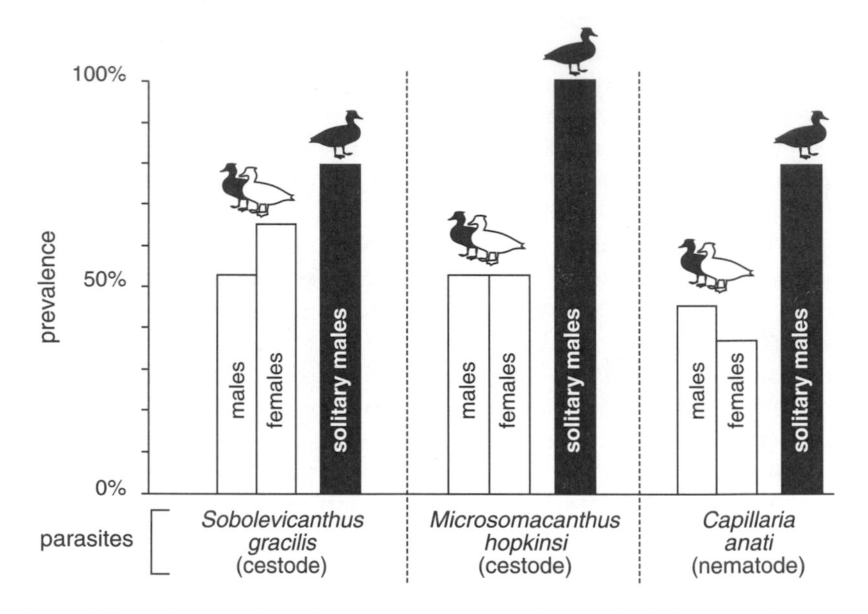

Figure 4.10. Influence of social status on parasitism. Solitary male ducks, ejected from the best habitats, are significantly more parasitized by cestodes than males or females in couples (data from Gray, Gray, and Pence 1989).

defense. The influence of social status on the distribution of the infrapopulations of various parasites was studied in late winter in the mallard, *Anas platyrhynchos,* by Gray, Gray, and Pence (1989). In this species couples and solitary males occupy different territories, with couples having a dominant status and taking more appropriate habitats than solitary males. The investigation shows that several species of helminths have their highest prevalence in solitary males (fig. 4.10). For example, the cestode *Sobolevicanthus gracilis* infects 53% of the paired males, 67% of the paired females, and 80% of the solitary males. For another cestode, *Microsomacanthus hopkinsi,* these numbers are 53%, 53%, and 100%, respectively. Similarly, these numbers for the nematode *Capillaria anati* are 40%, 33%, and 80%, respectively. The most likely explanation is that the habitats in which solitary males gather have an abundance of helminth intermediate hosts.

A second hypothesis is the exact opposite of the previous one: "The higher the status, the more abundant the parasite." This explanation is supported in the two following cases.

The reindeer, *Rangifer tarandus,* is parasitized by the protostrongylid nematode *Elaphostrongylus rangiferi,* whose intermediate hosts are terrestrial gastropods. In these hosts social status is positively correlated with size. Knowing that infection essentially occurs in the fall, Halvorsen (1985) did two investigations on marked young individuals of the same herd before and after this period of transmission. Results showed a positive correlation between weight and parasitism, with prevalence *highest in the heaviest animals* (i.e., those with dominant social status): in males, five out of ten animals weighing more than 45.8 kg were infected, while only one out of twenty-three animals weighing less harbored the nematode (infection was diagnosed from larvae in the feces). The explanation proposed was that during the winter reindeer dig through the snow to find the lichens they feed on and that the best sites (where access to lichens is the easiest) are occupied by individuals with higher social status. Because the gastropods in which the protostrongylid larvae are found are particularly abundant in lichens, those individuals that occupy the best sites subsequently become more infected than others.

Another spectacular case along these lines was discovered by Bartoli et al. (2000) in the fish *Symphodus ocellatus*. In this species the population may be divided into females, dominant nest-building males, and satellite males that do not build nests. The nesting males are larger than both satellite males and females, and *Genitocotyle mediterranea,* a trematode that parasitizes the digestive tract, is surprisingly *distributed according to this hierarchy:* females are rarely parasitized (one case out of seventy-five), while prevalence reaches 21% in satellite males and 68% in dominant males (fig. 4.11). The probable explanation here is that the males' larger size and higher social status allows them to eat prey that carry the trematode's metacercariae.

Obviously, social status does not obligatorily influence parasitism. For example, in troops of olive baboons *(Papio cynocephalus anubis)* from Tanzania (within which a clear social hierarchy is exhibited), no significant correlation has been observed between parasite richness and helminth egg output, on the one hand, and social rank on the other (Müller-Graf, Collins, and Woolhouse 1996).

Many aggregated distributions of human parasites are considered to be the consequence of individual differences in hygiene and feeding habits and are therefore linked to social status. How far one lives from transmission "hot spots" and individual behaviors related to habitat, hygiene, nutrition, protection, cultural differences, and such are so important that social hierarchy and the level of education are enough to create strong interindividual differences in infection rates.

Cultural diversity often exhibits a "collective" aspect and creates differences between "communities," in the social sense, that share the same environment (see Nelson 1990). Within the same human population sev-

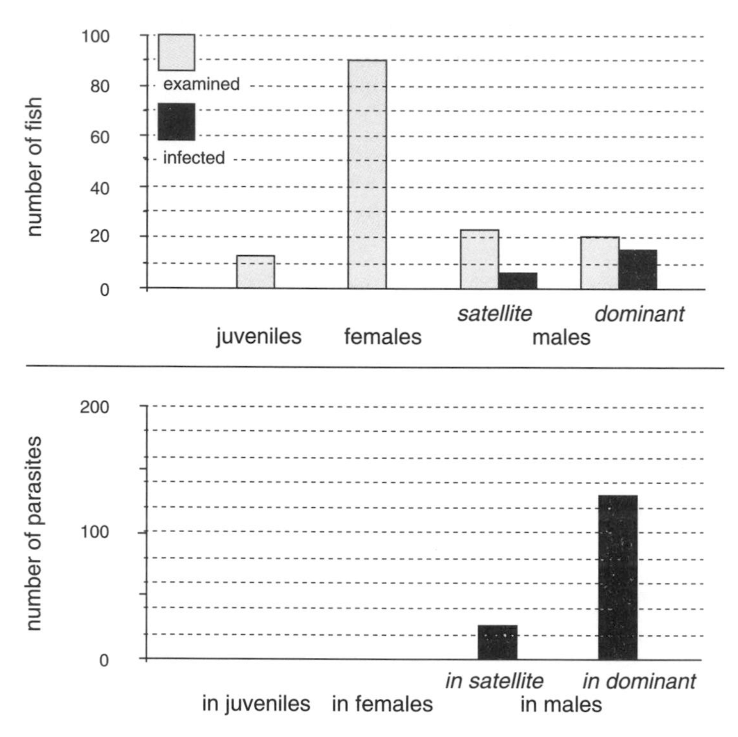

Figure 4.11. Only dominant nesting male fish, because of their large size, are exposed to parasitism by the trematode *Genitocotyle mediterranea.* Top: number of fish; bottom: total number of parasites recovered from different classes of fish (from data communicated by Bartoli et al. in press, modified).

eral such sympatric "cultural populations" may live together. Picot and Benoist (1975) have shown, for example, that on Réunion Island, infection with nematodes of the genera *Trichocephalus, Ascaris, Ancylostoma,* and *Anguillulina* is strongly influenced by the care people take of their immediate household surroundings. Even if it is difficult, particularly in this type of investigation, to isolate both specific behavioral and immunological factors, it nevertheless appears that the different "communities" of the island, characterized by different salaries, housing habits ("concern or neglect of their surroundings"), educational rules, and rituals ("ritual washing of hands"), more or less broadly open the encounter filters for these parasites. Santiago Mas-Coma (personal communication) also reported a role for malnutrition in the transmission of *Fasciola hepatica* on the high plateaus of Bolivia. Here the poorest Indian populations eat plants harvested in humid zones. These populations pay a very heavy cost (including mortality) to the parasites, with prevalence reaching as high as 90% in children. Such circumstances are certainly not isolated occurrences on the planet.

Even the structure of the habitat is responsible of the presence of certain vectors of parasitic diseases. For instance the insect *Triatoma infestans,* a vector of Chagas' disease (caused by *Trypanosoma cruzi*) in South America, lives more often in some habitats than others. In the shanty towns around cities investigations have shown that 40% of houses with cob and clay walls harbor triatomes versus only 9% of houses with concrete walls (see Mott et al. 1991).

In terms of human parasitism one could write, "Tell me what parasites you have, and I'll tell you who you are" (whether you are ignorant or educated, careless or careful, and above all, whether you are rich or poor).

Variables Associated with Compatibility That Influence Inequality in the Face of Parasitism

Variable 1: Immunity

The influence of genetic characteristics on parasitic diseases has been widely demonstrated in laboratory models, in humans themselves, and in both definitive and intermediate hosts. This is the case for infections with *Trypanosoma, Giardia, Eimeria, Leishmania, Plasmodium, Schistosoma, Ascaris, Ancylostoma,* and *Trichinella* (see Bradley 1980; Wakelin 1978, 1985). The examples cited here are limited to pathogens other than viruses and bacteria.

Inequalities linked to genes coding for characteristics of the immune system are the most important, and some individuals may acquire a fairly complete resistance to infection as the result of their genotypes. For example, sensitivity to *Plasmodium falciparum* is influenced by the presence of alleles of the HLA system (see Hill et al. 1991, 1994; Hill 1992). Two alleles, common in West Africans but rare in other groups of humans, give noticeable protection against the most serious forms of malaria. The difference in protection may result either from a higher capacity to attack the parasite or from a different amplitude of pathological reaction against it (see Carter, Schofield, and Mendis 1992). Hill et al. (1994) indicate that the high frequency of these alleles is clearly the result of selection, and it appears that this is an excellent illustration of the tendency for a host population to close the compatibility filter, although this has yet to be confirmed.

Schistosomiasis is another example of genetic diversity and immunity in humans against an endemic parasite. For example, in a village in northeastern Brazil (Caatinga do Moura), schistosomiasis caused by *Schis-*

tosoma mansoni is endemic (Dessein 1988; Abel et al. 1991; Marquet et al. 1996). Here twenty large families with a total of 269 individuals living within two hundred meters of the infectious river were studied. The number of eggs per gram of stool was determined, corrected for age and how often each individual visited the water, and compared with a model based on the hypothesis that resistance was determined by a major gene with two alleles and genotypes distributed according to Hardy-Weinberg equilibrium. The authors noted that age and the frequency of visiting water explained only 20% to 25% of the variance in infection rates; that the distribution of the most parasitized individuals was not random but, on the contrary, was aggregated in certain families; and that reinfection after treatment was much more rapid in some families than in others.

These observations strongly suggested the presence of a polymorphism in resistance within the population, and segregation analysis indicates that a gene controlling susceptibility to *S. mansoni* does in fact exist. This gene has one resistant allele (R) and one susceptible allele (S), with codominance between R and S such that heterozygotes RS have a parasite load intermediate between that of resistant RR homozygotes and susceptible SS homozygotes.

Figure 4.12 summarizes the pedigrees of three distinct families in this study, each including the founding couple and their children. Each individual was characterized by (a) the number of eggs per gram of stool (representative of parasite intensity); (b) a parasitism index calculated after correcting for age, sex, and frequency of visiting water; and (c) the proposed genotypes: RR for the individuals with nonexistent or quasi-nonexistent parasite load; SS for the individuals with very high parasite load; SR for individuals with intermediate parasite load.

Analysis of the data for family 1 indicates that two SR parents match with the existence of two susceptible homozygous SS children whose parasite loads are over three thousand eggs per gram of stool as well as the presence of SR children in the family (they could also have had RR children). In family 2 the data indicate an RR father and an SR mother, which matches with the presence of either RR or SR children. In family 3 parents and children are all interpreted as being RR, since their parasite loads were very small if not zero. The gene that controls infection levels was called SM1, while another gene, called SM2, seemed to control the severity of the disease for a given parasite load (Dessein et al. 1999).

Analyses of linkage disequilibrium using genetic markers then allowed the localization of the SM1 gene on chromosome 5 (Marquet et al. 1996), a region of the human genome that is tightly linked to the production of cytokines, molecules that play a key role in antiparasite immune defenses. Immunological analyses of antibody and cellular responses in resistant

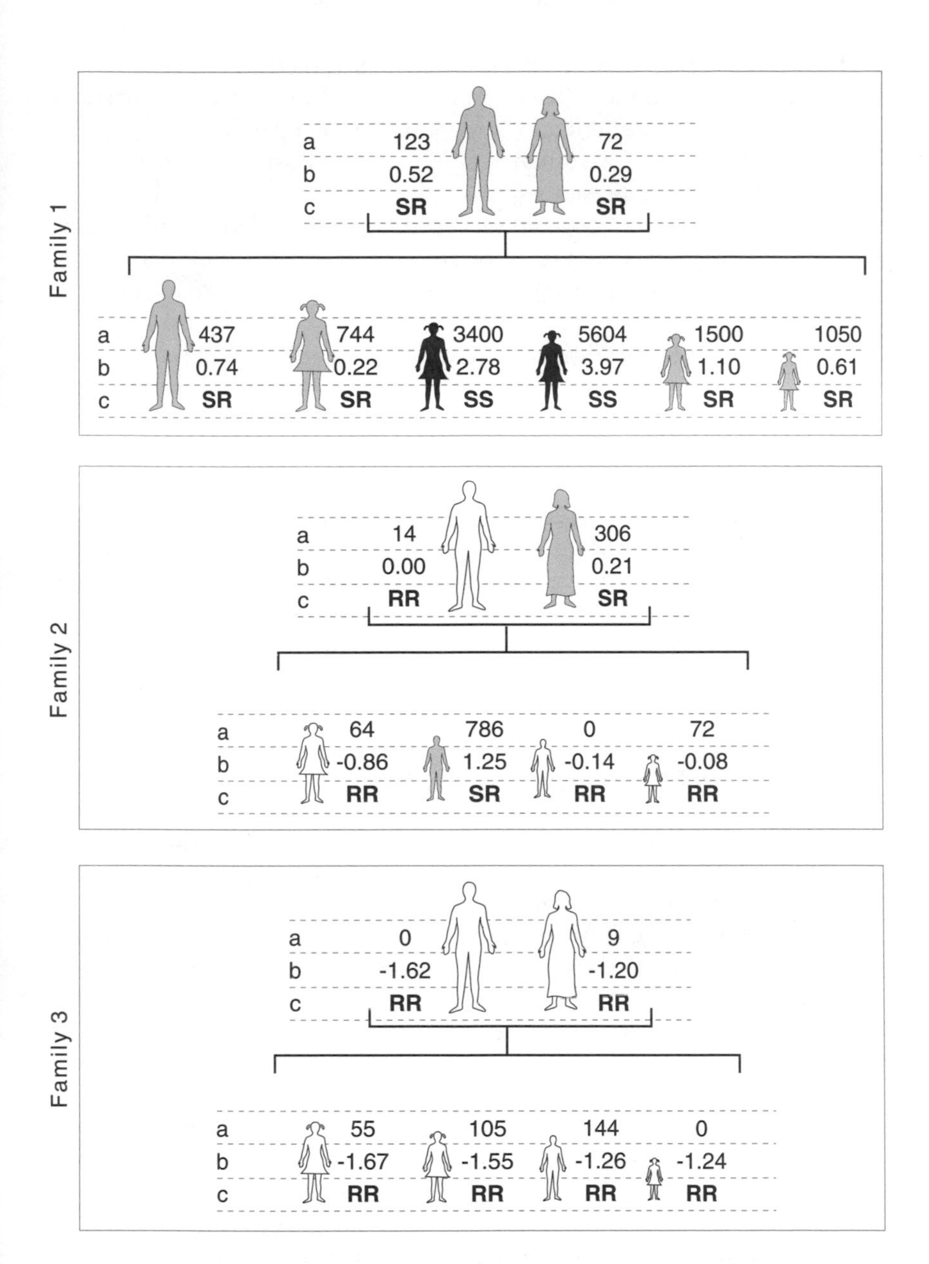

Figure 4.12. Analysis of the pedigree of three families in a village from northeastern Brazil where schistosomiasis caused by *Schistosoma mansoni* is endemic. Silhouettes indicate sex and approximate age of the individuals; see text for explanation.

and susceptible individuals from Caatinga have linked susceptibility to infection to changes in the production of some of these cytokines, particularly interleukin-4, interleukin-6, and gamma interferon.

It is possible that the abundance of intestinal parasites themselves may be influenced by genetic factors, and several works suggest a relation between certain HLA antigens and susceptibility to such infections. This is the case, for example, in the study by Holland et al. (1992) on patterns of infection and reinfection by the nematode *Ascaris lumbricoides* in Nigerian children.

Leprosy is also tightly linked to genetic characteristics, of which some control susceptibility to the disease and others control how serious it is (Abel et al. 1989). As DeVries (1991) noted, "The pathogenicity of leprosy is almost totally attributable to the immune response of the hosts towards *Mycobacterium leprae,* a virtually non-toxic intra-cellular parasite."

These effects of heterogeneity in definitive host populations are also found in intermediate hosts. Uniting parasites (schistosomes) and vectors (gastropods) in the same sentence, Southgate (1979) writes: "It is probable that most natural populations are heterogeneous, consisting of individuals varying in infectivity and susceptibility." LoVerde (1979) also wrote: "Today we recognize that within a population of *Bulinus* or *Biomphalaria,* there are individual snails each showing various degrees of susceptibility." Although most examples of unequal immunity discussed here were taken from human parasitic diseases, it is important to understand that such diversity exists in any population of living organisms and that it is the main factor responsible for aggregated distributions in nature. In host individuals whose reproduction is partially or totally clonal, the diversity in resistance is observed between clones, not individuals (Little and Ebert 1999). Although it is rare for polymorphisms in compatibility to be correlated with a polymorphism in morphology, it happens in certain mollusks. For example, in *Euparypha pisana* individuals having no colored bands are more susceptible to nematode larval stages than the rest of the population (Cabaret 1982).

Since the 1990s a particular focus has been placed on the concept of fluctuating asymmetry (the deviation from perfect symmetry of a normally bilaterally symmetrical character). In short, parasitism is expected to be associated with host developmental instability because such hosts would be more exposed or more susceptible to infection than developmentally stable ones (see Møller 1996, 1997; Thomas, Ward, and Poulin 1998). However, it is not always easy to distinguish whether developmental instability is the cause or a consequence of the intensity of parasitism. One convincing example that certain parasites choose supposedly less resistant asymmetrical individuals was provided by McLachlan, Ladle, and

Bleay (1999): the wing length of the chironomid *Chironomus plumosus* is a good predictor of infection by the mite *Unionicola ypsiliphora*. In this association the parasite infects only the adult host and thus cannot affect developmental stability (for more on the biology of *U. ypsiliphora* see chapter 19).

Variable 2: Unsuitability

The inequality of hosts in the face of parasitism may have origins that have nothing to do with immunity while still being dependent on genetic factors. This situation illustrates the differences between being unsuitable and being resistant.

A simple illustration of this type of inequality is linked to sex. Differences in parasitism between sexes have been described in higher vertebrates and are often explained by the likely influence of hormones (see Solomon 1969; Alexander and Stimson 1988; Møller and Saino 1994). One frequently reported effect is referred to as the "immunocompetence handicap of males." This handicap exists because testosterone decreases the efficacy of the tissues and organs involved in immune defense, and a number of investigations and experimental studies have shown that males are more susceptible to parasitism than females (Folstad and Karter 1992; Møller and Saino 1994). In some cases, however, it is difficult to distinguish differences between behavioral traits (those involving the encounter filter) and hormonal traits (those involving the compatibility filter). Moreover, the traits may not be independent. For example, after observing that males of the lizard *Sceloporus occidentalis* are more parasitized by *Plasmodium mexicanum* than are females, Schall and Marghoob (1995) hesitate between the hypothesis of a hormonal handicap and the fact that males are found in different locations than females at night to explain the difference.

Another striking illustration of the inequality of hosts in the face of parasitism is the role of certain genetic characteristics, including genetic diseases, in parasitism. The causative agent of malaria can penetrate blood cells only when parasites can recognize and use protein receptors on the membrane. The parasite, at this stage a merozoite, adheres to the membrane of the blood cell and induces "internalization." One often-cited protein involved in this process is the "Duffy antigen," which is necessary for the penetration of *Plasmodium vivax* into red blood cells. The absence of this antigen in black Africans seems to provide protection against infection by this particular species but not against the most pathogenic species, *P. falciparum* (see Miller 1994).

Receptivity to malaria may also be influenced by genetic anomalies of the hemoglobin of blood cells, from which the well-known link between malaria and sickle-cell anemia arises. In adults hemoglobin is formed by four protein chains: two α chains and two β chains, each attached to a heme group. These chains have 141 and 146 amino acids, respectively. In people with sickle-cell anemia a substitution of a single nucleotide on DNA induces the replacement of one amino acid by another close to the extremity of the β chains. The replacement results in the formation of hemoglobin S instead of normal hemoglobin H. The homozygous occurrence of the mutant allele S in the genotype is normally deadly, but heterozygotes (HS) have only a limited handicap and survive. The malarial agent is far from indifferent to the quality of human hemoglobin, and heterozygote carriers of the S allele for hemoglobin are protected against malaria (see Allison 1954; Friedman 1978; Friedman et al. 1979). Under these conditions, mortality in the carriers (HS) is far less than in the rest of the population. This can be summarized by saying that the homozygotes (SS) have severe sickle-cell disease, that the homozygotes (HH) have malaria, and that the heterozygotes (HS) escape both. This explains the high frequency of genetic diseases like sickle-cell anemia in countries where malaria is endemic, with the polymorphism being maintained by opposing selective pressures. Other genetic diseases of the same type, such as thalassemia, have similar consequences, although it must be noted that the conclusions stated above have sometimes been questioned. Some experiments on β-negative thalassemic mice infected by *Plasmodium berghei, P. chabaudi,* and *P. yoelii* do not show protection against infection (Clarebout et al. 1996).

The sickle-cell anemia/malaria "couple" is not an isolated case where a genetic anomaly confers protection against a pathogenic agent. Using malaria again as an example, yet another anomaly involving a membrane protein, "band 3," also confers resistance to malaria on heterozygous carriers. The mutation of the protein in this case is a deletion of twenty-seven pairs of bases (thus, nine amino acids) and is common in Papua New Guinea, where malaria is heavily endemic. As in the previous example, it apparently is maintained in the population by the pressure of the disease,

Similarly, the epidemiology of tuberculosis provides evidence that genes responsible for certain disorders of lipid metabolism (Tay-Sachs, Niemann-Pick, and Gaucher's diseases) also increase the heterozygotes' resistance to Koch bacillus, the cause of tuberculosis. At least this is the most likely explanation for the occurrence of localized high frequencies of heterozygotes. For example, Tay-Sachs heterozygote frequency reaches 11% in eastern Europe (see Rotter and Diamond 1987), where rates of tuberculosis have historically been very high.

A process analogous to that of sickle-cell anemia/malaria is unexpectedly observed in some nematode parasites and sheep. A study of polymorphism at the locus coding for the enzyme adenosine deaminase (involved in purine metabolism) has shown that heterozygous host sheep for this locus are more resistant to intestinal nematodes and that their mortality is lower than that of homozygotes. As a result, parasites contribute to maintaining this polymorphism in the population (Gulland et al. 1993).

Variable 3: Sex (Gender)

Numerous investigations in either wild or domestic animals and in human populations have shown that there are significant differences in parasitism within the same host species depending on gender. These differences were confirmed experimentally. Zuk and McKean (1996) thoroughly reviewed the data available on this subject and reminded us that a major problem is "distinguishing between differential exposure and differential susceptibility," that is, explaining the differences observed by the degree of opening of either the encounter filter or the compatibility filter.

Except for a few exceptions, males are usually more affected by parasites than females. In "higher" vertebrates it is often considered that testosterone has a tendency to open the compatibility filter in males, because the immunosuppressive effect of androgens is well known and because a positive correlation was experimentally shown between the level of testosterone and parasite success. It may also be that in some cases androgens also have a direct effect on parasite development. Zuk and McKean (1996) noted that in numerous species males compete for females and that this competition has led to the selection of traits (for example, antlers on deer) that are testosterone dependent. High levels of androgens thus give the males weapons or attractive traits for competition but also make them more vulnerable to parasites.

Variable 4: Personal History

The phenotype, and particularly the capacity to resist parasites, is influenced by each individual's past or present condition in life: "Anything that adversely affects the immune system, such as malnutrition, stress, environmental toxins or even some parasitic infections, can markedly increase the probability of diseases" (Holmes 1996).

Age is a classic variable in this context principally because it affects size and, notably, behaviors related to the acquisition of food, which often differ depending on the individual's size. Numerous works have shown that parasite communities are qualitatively and quantitatively modified throughout a host individual's life. For example, in the Mediterranean sparid fish *Pagellus erythrinus*, the trematode *Lepocreadium pegorchis* infects individuals less than fifteen centimeters long, while another trematode, *Diphterostomum israelense,* infects only individuals of the same species longer than that (Saad-Fares and Combes 1992; other examples in Dogiel 1964). Scott (1982) showed in the Gulf of Saint Lawrence in Canada that parasitism by trematodes varies according to age in certain species of flatfish but not neighboring species, which can be explained only in that some species modify their diet as they grow and others do not.

In general, the largest individuals of a certain species harbor more diversified (more species) and denser infracommunities of parasites than do their smaller compatriots. This may be explained not just by age (the older the host, the more opportunity it has had to acquire parasites) but also by size (the larger the host, the more parasites it can harbor). As noted earlier, these two hypotheses are not mutually exclusive, and their relative importance has been particularly debated concerning fish (see Poliansky and Bychowsky 1963; Dogiel 1964; and Lo, Morand, and Galzin 1998).

Independent of age and size, anything that tends to weaken the host may favor the success of parasitic infection and even influence the course of the disease. Stress or unfavorable conditions may also increase the infrapopulation size. For instance, Gelnar (1987) experimentally showed that the monogenean ectoparasite *Gyrodactylus gobiensis* multiplies much faster on freshwater fish *(Gobio gobio)* when the level of dissolved oxygen in the water is low or food is insufficient. In this case, fish infected with two or three parasites at the beginning of the experiment and maintained at reduced levels of oxygen (50% saturation) had close to five hundred *G. gobiensis* after forty-five days while controls maintained in oxygen-saturated water had fewer than twenty.

Because individual nutrition modulates the immune system, a properly balanced diet is probably one of the characteristics that most strongly influences the receptivity of a host to parasites. In poor countries malnutrition in some host individuals considerably lowers their defense threshold, and significantly (see chapter 12), the ability to regulate parasite populations is often "host mediated" as the result of host defense mechanisms.[6]

Last, parasites of some species (let alone viruses like HIV) induce an immunosuppression that may, up to a certain point, encourage invasion by still other species of parasites, beginning a devastating cascade of events as the hosts' defense system weakens with each additional infection.

---

6. See the special issue of *Parasitology* 107, suppl. 1993, on the relation between nutrition and parasitism.

5 Diversity in Time

The history of evolution and biodiversity is fundamentally a history of the evolution of species interactions.
Thompson 1999

The origin of species remains one of the least well understood and most important questions in evolutionary biology.
Tregenza and Bridle 1997

Every environment changes over time: lakes, forests, and caves do not stay the same indefinitely. On a global scale the planet has evolved, undergoing changes in the chemical composition of both the atmosphere and the oceans so that today neither is the same as it was 3.5 billion years ago when life appeared. Significantly, important climatic cycles occur over long stretches of time, such as those that caused glaciations in the Quaternary period, and all living species must adapt to these changes or perish. Entire phyla have diversified themselves, new species have appeared, and others (sometimes entire groups) have died out. In short, the free-living organisms alive today are the result of constant evolution since life first appeared, and the same is true for parasites.

New Actors

The number of species in the biosphere tends to increase because of speciation by cladogenesis and to decrease as the result of extinction. Therefore new actors in parasite-host associations may be either host species, parasite species, or both. When a host species divides into two, the parasite species either may go on parasitizing both hosts indifferently or may itself divide into two species. In addition, the parasite species may explode into two or more species outside any specific host speciation. The role of gene flow (the displacement of gametes or individuals from one population to another) is central to understanding such speciations because these flows oppose the separation of species. In general, a particular population can adapt to its environment and diverge from other populations over time only if the latter do not constantly export to the former individuals (and thus genes) adapted to other circumstances. Even genetic drift (the random fixation of certain genes in small populations in the absence of selective processes) cannot happen when such gene flow occurs (Futuyama 1998).[1]

Relative to parasite speciation we can consider the proposals of Euzet and Combes (1980), who distinguish three major types of speciation for parasites: allopatric, alloxenic, and synxenic.

Allopatric Speciation

Allopatric speciation is the best-known process of speciation and occurs when two populations of the same species are geographically separated from each other for some time and, as a result, genetically diverge either because of selection or because of genetic drift. When the separation lasts long enough the divergence may become large enough so that the two gene pools remain reproductively isolated when reunited—the most frequently used measure of the separation of species (see Brooks 1979, 1980).

Let's imagine a system composed of a host and parasite species, and let's suppose that because of climatic or topographic changes the initial population of hosts happens to be divided into several distinct populations without gene flow between them. Also, let's suppose that tens of thousands of years later (or hundreds of thousands, or millions), these populations are reunited. Three outcomes are possible.

1. Speciation may have happened at comparable speeds in both the host and parasite so that the number of host species and the number of parasite species have increased in proportion (fig. 5.1A).

2. Speciation may have happened in the host but not the parasite (fig. 5.1B). In this case if the host species are later found side by side in a common area the parasites will be one single species with an increased host spectrum.

3. Speciation may have happened in the parasite and not in the host populations, as shown in figure 5.1C. In this case if the host populations are later reunified several parasite species will be found in a single host species.

Geneticists like to think that parasite populations should diverge faster than those of their hosts because they have shorter generations and life spans, and thus case 3 should be the most frequent. This may in fact be true, since many host species harbor numerous closely related parasite species (species flocks) all thought to have descended from a common ancestor.

Petter (1966) has, for example, used hypothesis 3 to explain species

---

1. Divergence can still occur, if selection is strong enough, if the rate of gene flow is low enough (Futuyama 1998).

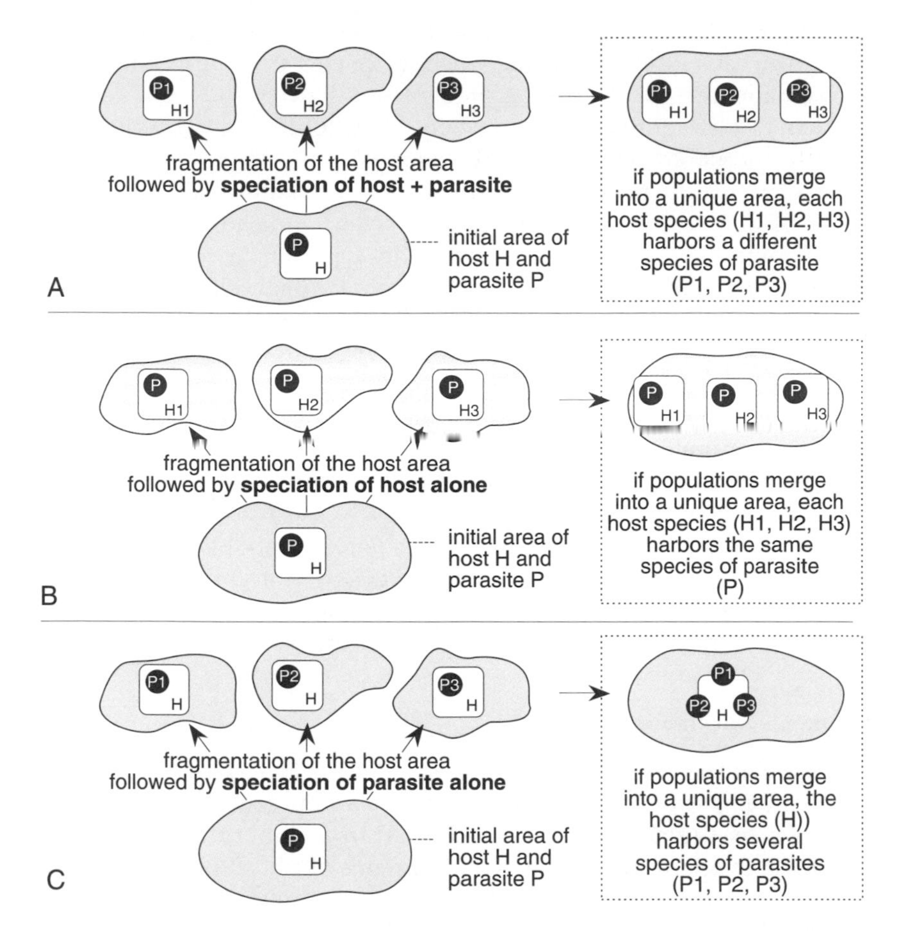

Figure 5.1. Allopatric speciation: the several alternatives when speciation concerns both host and parasite (A); only the host (B); or only the parasite (C). The bottom diagram illustrates one of the possible explanations for the species flocks that are often encountered in parasitology (from Euzet and Combes 1980). H = host; P = parasite.

flocks within terrestrial turtles. These turtles harbor huge populations of nematodes (sometimes several million per turtle) that live on the contents of the colon. Petter showed that *Testuda graeca* in the palearctic region harbors fourteen species of the nematode genus *Tachygonetria* as well as related genera, all extremely close morphologically but each also reproductively isolated. The same host individual may harbor up to eleven of these species, each occupying slightly different niches in the digestive tract. This diversity was explained by the multiple isolations of turtle populations that induced speciation of the nematodes but not the turtles.

When the isolated turtle populations fused again, their parasites did not recognize each other and formed distinct species, each differentiated by its habitat in the gut.

Genetic differentiation between geographically separated populations may be measured by the index F_{ST} (see Wright 1969). Nadler (1995) points out that the measured value of F_{ST} for populations of parasites that are far from each other is sometimes very low, indicating weakly structured populations and large gene flow. This is particularly true if the host species (or one of the host species in a heteroxenous cycle) is very mobile and disseminates the parasite over large areas, which happens in marine fish and birds. For example, McManus (1985) showed that the populations of the bird cestode *Ligula intestinalis* are not very structured genetically, although its geographic distribution is very large. In some cases however, a more defined genetic structure has been reported, such as for the populations of *Ascaris suum* (Nadler, Lindquist, and Near 1995; Anderson and Jaenike 1997).

A weak genetic differentiation between populations that are geographically distinct may go along with strong genetic variability within each of the populations, as in the cattle nematode *Ostertagia ostertagi,* studied by Blouin et al. (1992) and by Dame, Blouin, and Courtney (1993).

Alloxenic Speciation

This type of speciation, first distinguished by Euzet and Combes (1980), is a possibility with considerable theoretical and practical importance in parasitology (fig. 5.2A). It is characterized by an emergence of parasite species in sympatry and is due to the existence of a host spectrum that includes two or more species. That is, it is due to fragmentation in xenopopulations. Note that the existence of xenopopulations does not imply, a priori, any form of isolation, because xenopopulations remain entities with limited outreach while each generation of propagules issued from them mixes. The situation is different if there is a restriction of gene flow between the parasites of the two host species. Isolation may then depend on either the encounter filter or the compatibility filter.

Decrease of Gene Flow via the Encounter Filter

Phytophagous insects that are more or less specific to particular host plants allow one to understand how different mechanisms of gene flow may affect alloxenic speciation.

Suppose that an insect develops on two species of plants, X and Y, and that this insect has a genetically determined polymorphism of habitat preference. This means that within the population some individuals carry

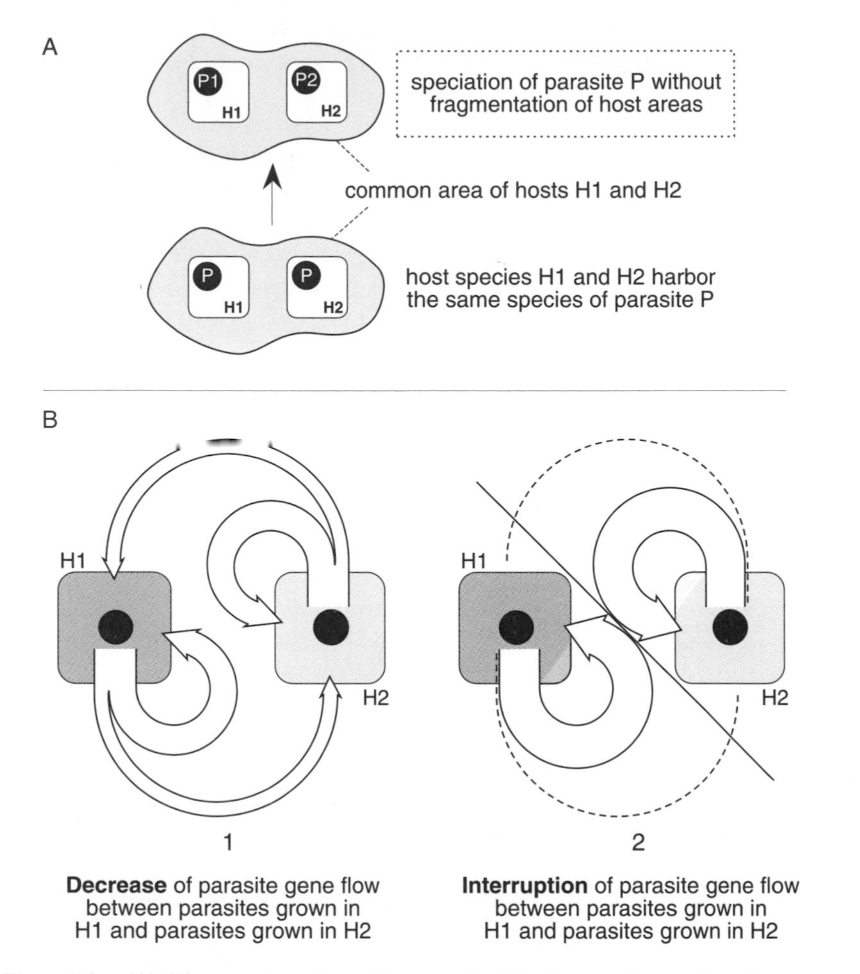

Figure 5.2. (A) Alloxenic speciation of the parasite. (B) diagram showing how the restriction of gene flow may be achieved between parasites of the same species parasitizing two host species (H1 + H2). This restriction (1) may lead to alloxenic speciation (2) induced by the fact that the same parasite species exploits two host species (from Euzet and Combes 1980). Open arrows denote degree of host use.

genes that will make them choose plant X more often while others carry genes that will make them prefer plant Y. Also, suppose that the adult insects are not very mobile and that mating occurs on the host plant. Males carrying "preference X" genes have more chance to mate with females that carry "preference X" genes than with those that carry "preference Y" genes and vice versa. The result of this process, called *assortative mating*, is a reduction or an interruption of gene flow. Although it is a sympatric situation, conditions are such that in the end two species are differentiated. Bush (1975) gave the first convincing examples of this type of isolation for parasites.[2]

The process that occurs in phytophagous insects is also applicable to parasites sensu stricto. Figure 5.2B illustrates how this process can be conceived. Let's imagine a parasite species with two different definitive hosts, H1 and H2. Let's also suppose that parasite larval stages from H1 and H2 mix and have similar probabilities of infecting H1 and H2 because there is no polymorphism orienting some parasites toward one host or the other. It is clear that in this case nothing calls for predicting the emergence of a species specific to each of the two hosts (fig. 5.3A).

If, however, some larval stages preferentially infect H1 and others H2, we are back to the case of a polymorphism of habitat preference illustrated for insects, although here it is a *polymorphism of host preference.* Parasites are the ideal candidates for assortative mating of this type, since their motility is in general nonexistent and, except for a few ectoparasites, once a parasite has acquired sexual maturity it does not leave one host individual for another. Théron and Combes (1995) showed that such a polymorphism exists in the daily pattern of cercarial shedding in trematodes (termed chronobiology) and that it may be a powerful factor in the isolation of sympatric species if a parasite infects two host species with different activity rhythms (fig. 5.3B). In schistosomes, for example, infective stages appear in the water at times of the day that are genetically determined (see chapter 7). If two principally terrestrial host species frequent the water at different hours, they can separately select parasites with different rhythms of cercarial emergence. A host with morning activity selects for cercariae with early emergence, and a host active at dusk selects for cercariae that are emitted late in the day. Morning parasites and evening parasites then have less and less chance of finding each other as adults in the same host. This type of speciation may also be said to be sympatry in space and *allochrony* in time. Selection in this case involves the encounter filter without detectable influence of the compatibility filter (Combes and Imbert-Establet 1980).

Polymorphism in preference for transport hosts in some phoretic acarids is yet another illustration of assortative mating. Here the same species of acarids may have "lineages" that are specialized for a particular host species. Since each host has a biology and habitat that is somewhat

2. Depending on the conditions under which selection occurs, only one habitat may be exploited among the other available ones, and all others may be abandoned. In this latter case, the preferred habitat that is retained may be different for different populations (see de Meeus et al. 1993).

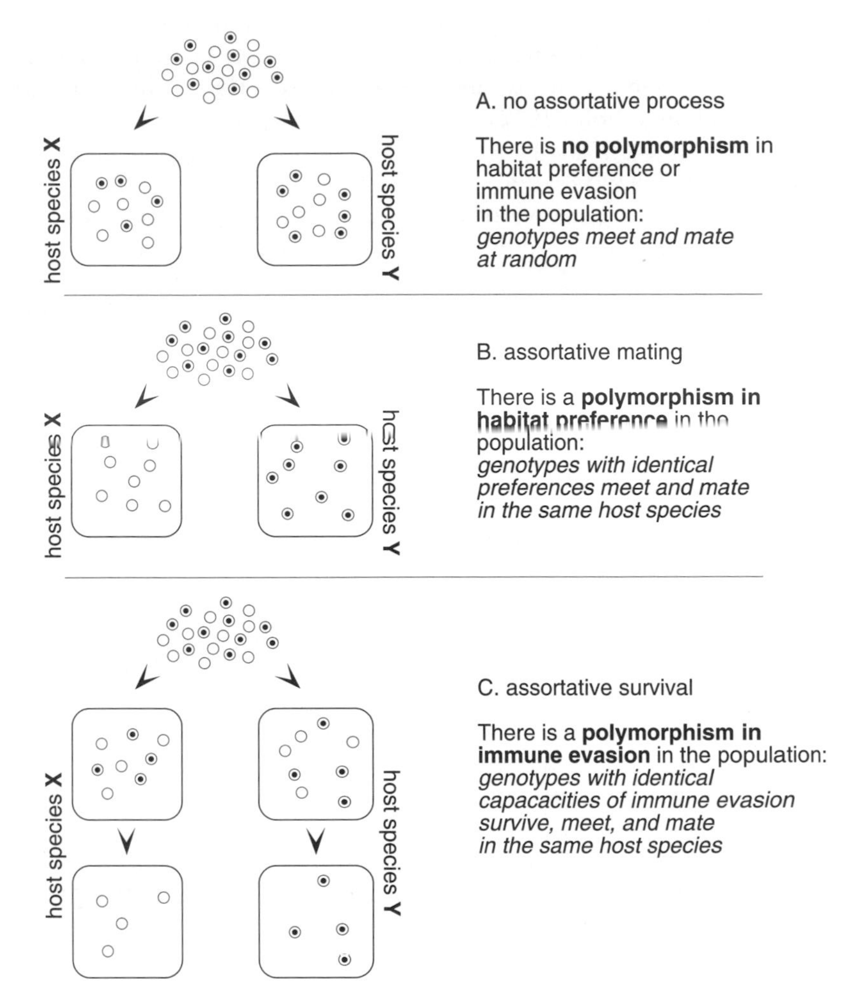

Figure 5.3. The possible mechanisms of alloxenic speciation. (A) Situation in which sharing two host species (X and Y) by the same parasite species (open circles and dotted circles) does not lead to divergence and therefore no speciation occurs. (B) Assortative mating via encounter; the population fraction that develops in X (open circles) is isolated from the one that develops in Y (dotted circles), owing to polymorphism for attraction toward the hosts. (C) Assortative survival via compatibility; a different fraction of the population is eliminated in each host species owing to polymorphism for host defense system escape.

different, and since the phoretic acarids are by definition not very mobile themselves, nothing else is needed for these lineages to be reproductively isolated. They then become sibling species (Wilson 1982; Athias-Binche, Schwarz, and Meierhofer 1992).

Decrease of Gene Flow via the Compatibility Filter

If we again take the example of phytophagous insects, isolation due to polymorphism in habitat preference may occur by a mechanism that is independent of behavior. For this to occur it is sufficient for certain individuals to have a greater fitness on plant X and others on plant Y. In this case the individuals that find themselves on the "bad" habitat are at a disadvantage, and preferential mating occurs between insects that are the most fit.

In parasitism an essential factor in survival is immune escape, which allows a parasite to resist host attack. Immune escape is a mechanism for reducing gene flow at the level of the compatibility filter because its result is assortative survival (Combes and Théron 2000) as opposed to assortative mating.

Let's take again the example of the parasite species that infects both host H1 and H2, but now let's suppose there is in the parasite a polymorphism in immune escape instead of habitat preference. That is, within the parasite population there are individuals whose genes allow them to avoid attack by the immune system of host H1 and others whose genotype allows them to avoid the immune system of host H2. It is easy to see that only parasites that have the genes appropriate for escaping the immune system of one host species will survive and therefore reproduce (fig. 5.3C) in that host. It is thus not important if sorting has not been done at the time of parasite penetration into the host, since the immune system will a posteriori proceed with sorting individuals that have the appropriate genotypes.

One must note that the concept of sympatric speciation in organisms that are highly specialized for their habitats (as parasites are) has gained considerable ground in recent years. Wilson (1992) attributes to this concept a fundamental role that "might easily have created vast numbers of insects . . . known to be specialized" and that "could also have underwritten the proliferation of parasite species."

Synxenic Speciation

Euzet and Combes (1980) further defined and outlined the concept of synxenic speciation, which is reproductive isolation occurring in a population of hosts belonging to one species without fragmentation of the

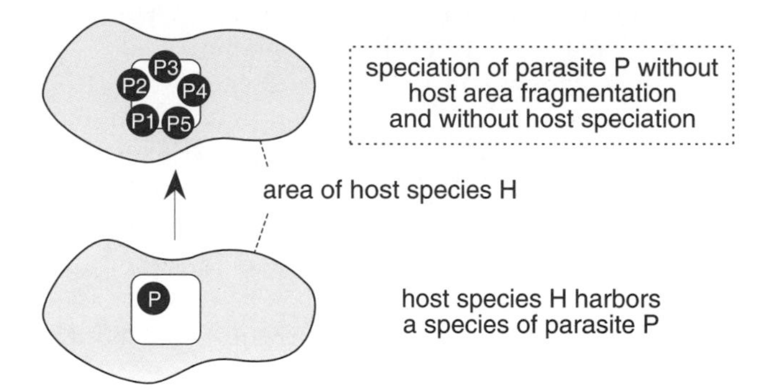

Figure 5.4. Parasite synxenic speciation: there is neither geographical separation nor exploitation of two host species (from Euzet and Combes 1980).

geographic area (fig. 5.4). Initially the hypothesis was proposed to explain the following case.

Brycinus nurse is a fish (family Characidae) that lives in several fluvial basins in West Africa. In Lake Chad, Birgi and Euzet (in Euzet and Combes 1980) identified eight species of the monogenean *Annulotrema* on the gills of the fish. Surprisingly, these species have an anatomy and morphology that is almost identical except in the copulatory apparatus; although *Annulotrema* are hermaphroditic, they copulate. Each individual thus has both a vagina and a penis, and analysis of these relatively extravagant apparatuses shows that an individual's penis can penetrate the vagina only of others of the same species. It is a "lock and key" system for which there is no master key. This led to the idea that these species could have arisen on site by mutation of a unique regulatory gene that affects both the male and the female reproductive organs at the same time. In this case there could be an immediate reproductive isolation, and therefore speciation, since only the carriers of the same mutation could interbreed. Such an ensemble of congeneric species may be characterized by having very short genetic distances even though reproductive isolation is complete.

The problem of *Annulotrema* is not different from that presented by the previously discussed turtle nematodes (Petter 1966). However, the interpretation is different and is due to the peculiarities of the copulatory apparatus. Both examples illustrate well the various mechanisms of parasite speciation. The presence in the same host population of several species of parasite of the same genus (one talks about doublet, triplet, quadruplet species, etc.), as in the cases of *Tachygonetria* and *Annulotrema,* is far from the exception, and numerous such occurrences have been reported. The best known are the fish monogeneans (it seems mostly freshwater) and the nematodes of herbivorous mammals (e.g., kangaroos, horses, zebras, and elephants). Kennedy and Bush (1992) studied this question using the data available and showed that "multiple congeners" exist in all groups of parasites, in all groups of hosts, in all environments, and in all regions of the world. Even more interesting, such multiple congeners exist even in very ancient parasites, such as the tetraphyllidean cestodes of sharks, suggesting that this is not at all a transitory phenomenon (Caira 1992).

Could it be that synxenic speciation also results from a polymorphism in habitat preference within the same host species, with the preference for habitat being for different sites? It is rather as if in a population of phytophagous insects certain individuals would prefer the leaves and others the flowers—isolation might then gradually follow. Such a scenario in parasites could easily be imagined for the polystomatid monogeneans of freshwater turtles. Most monogeneans are ectoparasites on the skin or gills of fish, and all have a swimming larva called an oncomiracidium. Polystomatids are the exception and are mesoparasites found in the bladders of amphibians and chelonians, in the pharynx of the same chelonians, or under the eyelids of hippopotamuses. Their transmission also involves an oncomiracidium. In the genus *Polystomoides* certain species live in the pharynx of turtles and others in the bladder. It is therefore easy to imagine a common ancestor for these species that could penetrate with indifference either the anterior opening of the digestive tract (ending up in the pharynx) or the posterior opening (ending up in the bladder). The occurrence of a polymorphism in the parasite population, with some individuals attracted to the anterior opening and others the posterior opening, would then have been sufficient to produce sympatric speciation via site preference in the ancestors of the present-day *Polystomoides.* The carriers of the genes for "anterior opening preference" thus would have encountered each other in the pharynx while the carriers for the gene "posterior opening preference" would have met in the bladder. We shall see below that this form of speciation is probably very ancient.

Species Isolation

In general, speciation mechanisms and, most notably, the reasons certain genomes become incompatible after a certain degree of genetic differentiation are far from well understood: "Since [Darwin] there has been much more progress in understanding the causes of adaptive change than in the mechanisms whereby new species are generated" (Coyne and Barton

1988). Nevertheless, it is possible to observe the way sibling species separate their ecological niches so that gene flow is interrupted. This does not explain either why or how species are separated, but it does allow us to propose some hypotheses. What follows refers to the work of McCarthy (1990).

Based on a preliminary investigation in a Sussex pond (England), two unrelated mollusks, the prosobranch *Valvata piscinalis* and the pulmonate *Lymnaea peregra,* were both found to be intermediate hosts of the trematode *Echinoparyphium recurvatum,* whereas other mollusks in the pond were never infected. The cercariae that emerged from each of the two species of mollusks showed the morphological characteristics attributed to *E. recurvatum,* with not one morphoanatomical difference observed between those from *V. piscinalis* and those from *L. peregra.* After leaving the mollusks, the cercariae penetrated other mollusks to form metacercariae, which then became adults in various aquatic birds after ingestion. McCarthy infected mallards *(Anas platyrhynchos)* with two batches of metacercariae, one coming from *V. piscinalis* and the other from *L. peregra.* With the eggs obtained from the adults, McCarthy then recovered miracidia that were used in an attempt to infect both healthy *V. piscinalis* and healthy *L. peregra.*

Figure 5.5 summarizes McCarthy's results. Parasites that come from *V. piscinalis* are labeled a, and those that come from *L. peregra* are labeled b.

The results obtained are the following: miracidia a infect *V. piscinalis* but not *L. peregra;* miracidia b infect *L. peregra* but not *V. piscinalis;* adults a occupy only the posterior region of the intestine of the ducks; adults b occupy only the anterior region of the ducks' intestine. adults a and b are, like cercariae, morphoanatomically indistinguishable.

In conclusion, there is not one *E. recurvatum* but two sibling species, a and b, confounded under the name *E. recurvatum.*

We may propose at least two scenarios to explain how these sibling species arose.

In the first the initial population of the common ancestors was divided into two populations that allopatrically exploited different mollusks. These two species were then reunited after separating their ecological niches in the form of the two microhabitats in the bird.

In the second scenario, the initial population of the common ancestor presented a polymorphism of preference that led to synxenic speciation within the bird. After this the two species then separated their ecological niches in the form of the two species of mollusks.

Currently it is impossible to determine which of these two scenarios may have happened. In either case it is remarkable that, whatever the scenario, the two morphologically indistinguishable species have separated their ecological niches throughout their entire cycles.

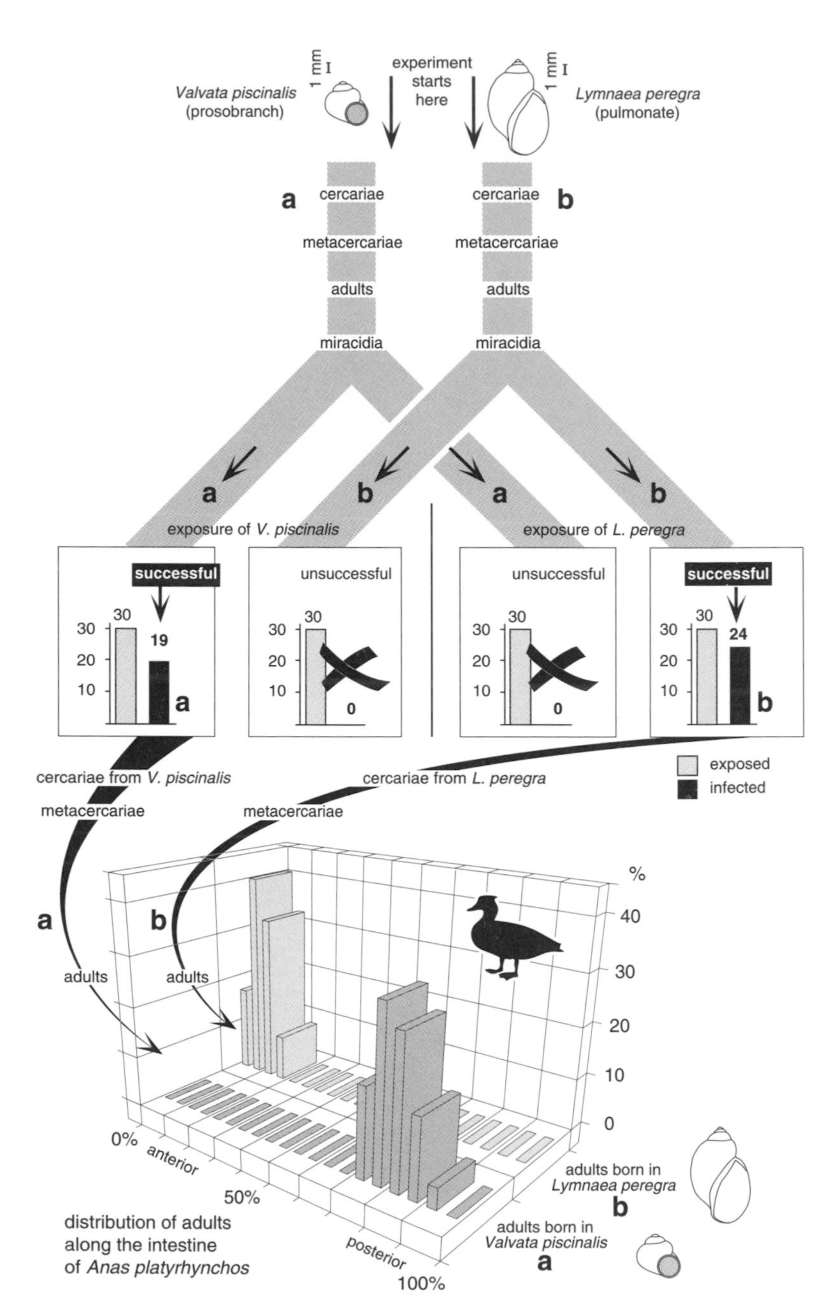

Figure 5.5. McCarthy's experiments showing that *Echinoparyphium recurvatum* is composed of two sibling species that are morphologically similar but have separated their ecological niche throughout their life cycles.

New Cycles

Because of all of the changes that could happen in the environment (most notably in the quality and quantity of species), the process of transmission that best suits a parasite today may not have been the best one in the past. Thus the life cycles observed today are likely to have become either more complicated or simpler.

Although classical parasitology textbooks depict life cycles as fixed entities, the reality in terms of evolution is very different. Life cycles, as well as morphologies, anatomies, and behaviors, appear fixed only when lit by the flash of a human life span. If we could see the cycles in an accelerated time-lapse fashion covering millions of years in just a few minutes, they would look astonishingly fluid. At present it is possible to observe species or groups of species that differ from closely related species or from other members of their phylum by possessing an original sequence of hosts whose genesis can be reconstituted. Here are two examples that illustrate first a lengthening of the cycle and then a shortening.

Selection for a Longer Life Cycle

The copepod parasites of fish are generally holoxenous, and their cycle is very simple. In short, swimming larvae hatch from eggs laid by parasitic females. After a number of molts, these larvae attach to fish and become adults. However, there is a family of parasitic copepods, the Pennellidae, that are heteroxenous.

What follows here is the cycle of *Lernaeocera lusci* as it occurs in the Mediterranean (Tirard and Raibaut 1989): females of *L. lusci* are found deep in the branchial arches of two fish, the European hake *(Merluccius merluccius)* and the poor cod *(Trisopterus capelanus)*. Females lay eggs from which swimming larvae hatch (the nauplius and then copepodid stages). Copepodids are infective, but their target is neither the European hake nor the poor cod but rather the common sole *(Solea vulgaris)*. These stages attach to the sole's gills, molt again several times, and become either male or female. Males detach to seek females and mate with them while the females are still on the sole.

What happens to males after mating? They die. What happens to females? They leave the common sole and swim. Where do the females go? To the European hake or the poor cod, and the cycle starts again (fig. 5.6).

Tirard and Raibaut (1989) called the common sole the "fertilization host" and the European hake or poor cod the "maturation host," noting also that females are heteroxenous but males are not.

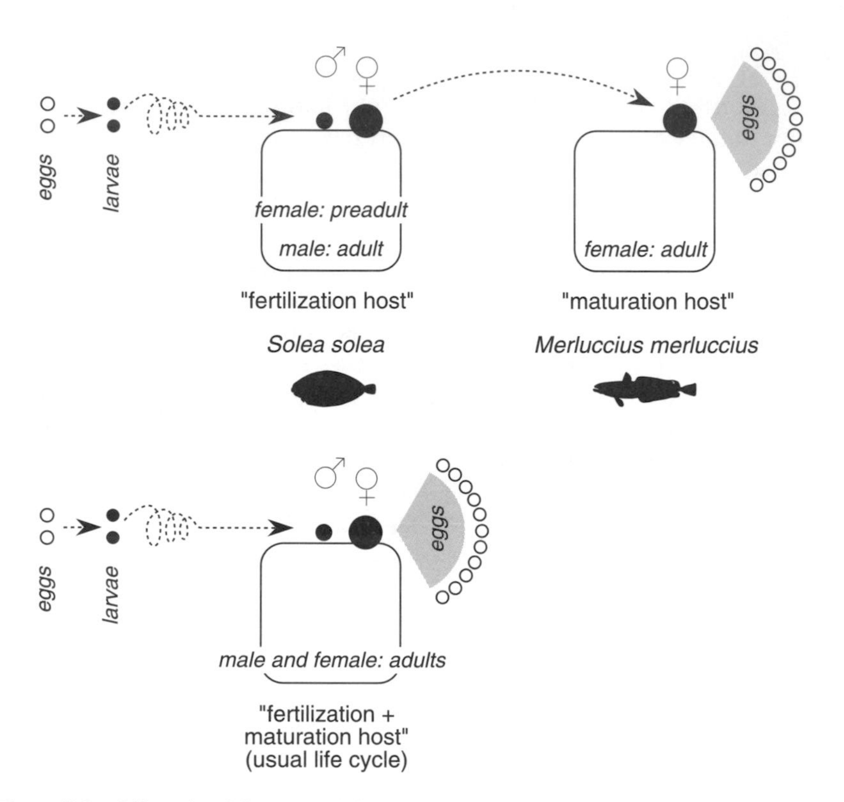

Figure 5.6. Life cycle of the pennellid copepods showing the addition of a "maturation host" relative to a normal copepod life cycle.

It seems reasonable to reconstitute the evolutionary history of the cycle of this group of copepods as follows: the ancestral situation is the holoxenous cycle on fish; the Pennellidae evolved to successively exploit two host types, the fertilization host and the maturation host; if we add that the fertilization host of certain Pennellidae (for example, members of the genera *Pennella* and *Cardiodectes*) is not a fish but a pelagic mollusk, we may hypothesize that the secondarily acquired host is the fertilization host—in the case of *L. lusci,* the common sole.

However, André Raibaud (personal communication) believes that the alternative hypothesis (according to which the maturation host would be the added host) can also be argued for because adult copepods are at times observed on the sole whereas larvae have never been found on the hake or the poor cod.

It is possible to use cycles with paratenic hosts to better understand what advantage increased complexity in a life cycle would bring. *Bothrio-*

cephalus gregarius, a parasite of turbot, and *B. barbatus,* a parasite of brill, are two taxonomically close cestodes. Both turbot and brill are flatfish, and they live together in the same environments.

Two investigations done in the Mediterranean and the Atlantic allow the examination of quantitative variables related to the transmission of *B. gregarius* (fig. 5.7). In this system, and based on three variables, all the numbers give the clear impression that the parasite of the turbot, *B. gregarius,* is more successful in transmission than is the parasite of the brill, *B. barbatus.* However, there is nothing to indicate that adult fecundity, larval or adult survival time, or differential immune reaction in the host can explain the differences. The solution was found when the cycles of both species were examined (Robert et al. 1988; Robert, Boy, and Gabrion 1990):

Bothriocephalus barbatus has the classic life cycle of a bothriocephalid, with a swimming larva (coracidium) hatching from parasite eggs. The larvae are swallowed by a copepod and transform to procercoids that then develop into infective plerocercoids. When a fish swallows infected copepods, the plerocercoids become adult parasites;

Bothriocephalus gregarius, on the other hand, is capable of selecting a "detour." If the copepod is swallowed by a small fish, such as a goby, the plerocercoid survives and the turbot then becomes infected by eating infected gobies.

It is easy to guess why the "detour" through the goby was selected. In this case the detour greatly increased the probability of transmission because turbot eat very few copepods, but gobies eat lots of copepods, and turbots eat lots of gobies.

Biochemical study of the plerocercoids found in gobies showed that they all belong to the bothriocephalids of the turbot, indicating that for the bothriocephalids of brill there was no selection for passage through the goby.

Several important conclusions can be drawn by comparing these two species of cestodes: It shows that advantageous changes in cycles may be selected for, with the result being that less advantageous changes may be discarded. It shows that the addition of an extra host in the cycle, far from decreasing the probability of completing the cycle, actually increases the probability of completion, that is, that complex cycles are adaptations. And last, it shows that parasites do not systematically "saturate" the habitat offered by hosts since, in the case of brill, the acquisition of a paratenic host would notably increase the infrapopulation of these cestodes.

Using a mathematical model of the life cycles, Morand, Robert, and Connors (1995) confirmed the following: selection of the goby paratenic host strongly increased the probability of infection of the turbot, and

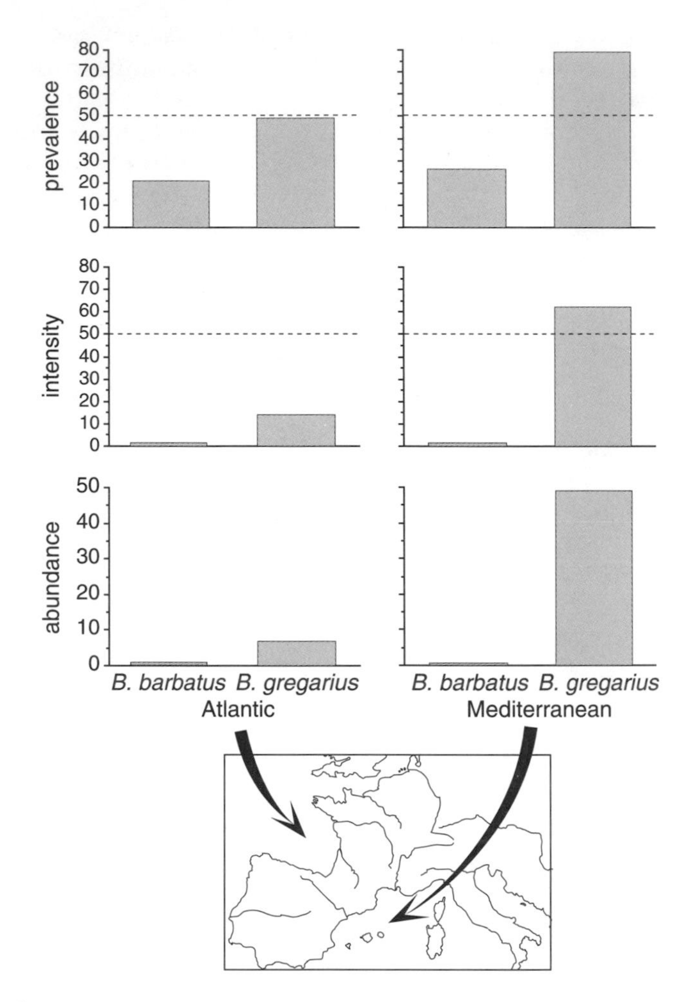

Figure 5.7. Parameters of parasitism by two closely related cestodes, *Bothriocephalus barbatus,* parasite of the brill, and *B. gregarius,* parasite of the turbot. Parasitism is much greater in the turbot and happens both in the Atlantic (left on diagram) and in the Mediterranean (right). The greater success of *B. gregarius* appears to be correlated with the intervention of an extra host (a paratenic host) in its life cycle (modified from Robert et al. 1988).

maturation of the parasite in the goby would not be selected for because of the goby's short life span. That is, becoming an adult in the goby would, in fact, decrease the parasite's life span.

In the life cycle of the Pennellidae (see above), André Raibaud (personal communication) believes that the advantage given by a two-host life cycle relative to a regular one-host cycle may be to diminish the pathogenic effect by distributing the energetic toll paid to the parasite between the two successive hosts.

Although advantages may come with increased complexity, such complexity nevertheless still has a cost. This cost is linked to the necessity of accommodating not just one but two or more hostile environments as well as to the risk that the different necessary hosts might not continue to occur in the same ecosystem. According to Morand (1996a), about a third of all parasitic protists and about two-thirds of all parasitic nematodes nevertheless have such complex life cycles, as do all known trematodes, cestodes, and acanthocephalans.

Selection for a Shorter Life Cycle

Schistosomes provide a good example of the probable shortening of a life cycle during evolution. "Classical" trematodes have a cycle with three hosts (see Shoop 1988; Steele 1985), but schistosomes, true trematodes in all other respects, have a life cycle with only two hosts. Schistosome cercariae emerge from their molluskan intermediate host and penetrate the definitive host transcutaneously, becoming adults in the circulatory system. Because the definitive hosts are birds and mammals, both large groups that appeared late in geological time, it is not unreasonable to state the following hypotheses (Combes 1991c).

1. Schistosome ancestors were classic trematodes with three hosts.

2. Since transmission with three hosts implies that the host the cercariae penetrate must later be consumed by the definitive host, the cycle could not occur normally if the cercariae penetrated a top consumer. The cycle would occur only if a cercaria became an adult in this animal via neoteny, that is, an early maturation of the genitalia (fig. 5.8). Thus one may think that selection among cercariae that penetrated top consumers conserved those that had mutated and were capable of neoteny. The first definitive hosts of schistosomes may thus have been carnivores or other animals protected from cercarial penetration (by their carapace, spines, speed, etc.).

3. Currently there are some trematodes in the family Spirorchidae that are found in freshwater turtles and that resemble schistosomes in having a two-host life cycle as well as in various other characteristics. Therefore the acquisition of a life cycle with two hosts may have occurred a truly long time ago during the Mesozoic. At that time, certain large reptiles from the Jurassic and Cretaceous periods would probably have been the terminal consumers responsible for selecting neotenic reproduction. Even if dinosaurs have not systematically been shown to live in aquatic environ-

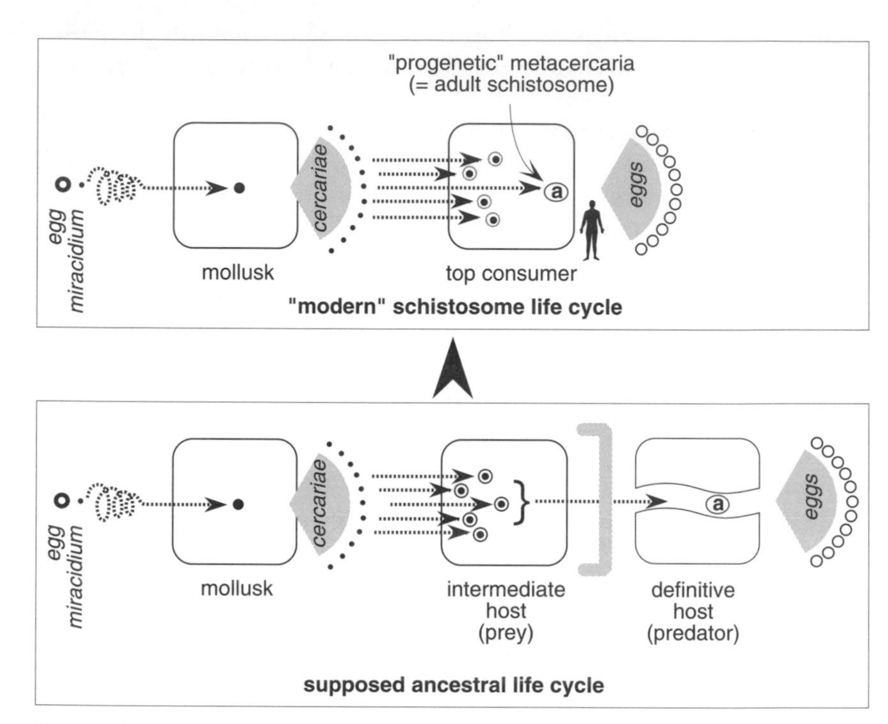

Figure 5.8. Hypothesis of the origin of the two-host schistosome life cycle (top) from a typical three-host trematode life cycle: if cercariae of a trematode penetrate a top consumer, the trematode's genes are transmitted only if the cercariae could ultimately produce adults in this host (modified from Combes 1991c).

ments, certain species probably often frequented such environments. As such, and later passing by way of birds and mammals, trematodes with such shortened life cycles could have given rise to the schistosomes of today. Spirorchidae/turtle associations of today would, then, be relict systems from the great age of dinosaurs as previously proposed by Stunkard (1923). Platt and Brooks (1997) have focused our attention on the constraints that such Spirorchidae-like parasites must have faced in becoming parasites of terrestrial vertebrates. Although the eggs that are dispersed in the circulatory system of the host would be freed when the host dies, this way of dispersing would have been much more uncertain when the hosts were only episodically aquatic.

The acquisition of separate sexes by schistosomes from surely hermaphroditic ancestors provides another particular problem. "The most puzzling aspect of schistosome biology is the evolution of the dioecious condition from hermaphroditic ancestors," say Platt and Brooks (1997).

The return to gonochorism by schistosomes after they acquire hermaphroditism appears to be the only known example of such a switch in animals (Combes 1991c). Després and Maurice (1995) suggested that a division of labor between sexes (males ensure the movement of the couple to the smallest capillaries, and females deposit the eggs there) may have allowed greater success in expelling eggs to the outside. This idea was later reaffirmed by Morand and Müller-Graf (2000). Interestingly, Platt et al. (1991) have found an authentic schistosomatid in an Australian crocodile, suggesting that such dioecy preceded the exploitation of homeotherms.

As always in reconstituting evolutionary scenarios, interpretations may diverge, and in the schistosomes we can also see trematodes that descend directly from more primitive forms before the acquisition of a cycle with three hosts. However, it must be emphasized that in another group of trematodes the analysis of phylogeny through cladistics has confirmed that the two-host cycle is derived from the three-host cycle and not vice versa. This was for the genus *Alloglossidium,* whose ancestral species had three successive hosts (mollusk-arthropod-fish), whereas certain recent species have only two hosts (mollusk-crustacean or even the unexpected cycle of mollusk-leech; see Carney and Brooks 1991).

The trematode *Asymphylodora tincae* goes even further in shortening its life cycle. This worm belongs to a family with a classic cycle that includes metacercariae forming in the molluskan intermediate host after the encystment of cercariae produced by rediae. Adults in this family live in a freshwater fish, the tench. This cycle is already shortened in the sense that the mollusks in which the cercariae are produced also harbor the metacercariae (the parasite thus saves itself the need for another dispersion into the outside environment). However, the cycle of *A. tincae* also has an ontogenic shortening in that it "skips" the metacercarial stage altogether. Here cercariae do not leave the rediae, do not encyst, and are no less infective to the tench. Once the cercariae, which are ingested along with the mollusks, reach the stomach of the fish, they develop directly into adults. In the gut the problem for the cercariae is then not unlike that of the schistosomes in that they arrived in the adequate host faster than planned. The parasite in this case takes advantage of the opportunity offered by the definitive host by cutting out an ontogenic stage (Nasincova and Scholz 1994).

Such accelerations are not reserved for trematodes. In nematodes, for instance, there are normally five successive stages, separated by molting. Whether the cycle is holoxenous or heteroxenous, it is normally the third stage (L3 larvae) that is infective to the definitive host. In some heteroxenous cycles, however, the parasite may develop in the intermediate host until the fifth or subadult stage. This shortening of the cycle guarantees rapid maturation to the adult and therefore an equally rapid dispersion of eggs. For example, the nematode *Thabdochona rotondicaudatum* uses the aquatic larvae of ephemeropterans as intermediate hosts and the fish *Notropis cornutus* as definitive hosts. Fish become infected when they arrive at the spawning site, which they frequent only briefly, and they disperse the eggs of the parasites before leaving. The eggs are ingested by the ephemeropterans, where the nematodes reach the subadult stage the next year just at the time the fish arrive again for spawning (Byrne 1992).

Selection for New Microhabitats

It is likely that numerous microhabitats have been modified over time under the influence of a variety of selective pressures. These pressures may have at least four origins: resource quality, host defense, other parasites, and predators. Each is discussed below.

1. *Resource quality:* Resource quality is not the same inside the host and on its surface. Thus it is probable that in certain groups of parasites the reproductive success of individuals that settled in a richer site could lead to the selection of their genes as opposed to others and therefore to new preferences. This is probably the way simple phoresy or ectoparasitism may have evolved toward meso- or endoparasitism. For example, some acarids originally attached to the hair of mammals for transport may have progressively passed to the "inside" of the hair follicles or even under the skin (Athias-Binche and Morand 1993).

2. *Host defense:* Hosts may obviously exercise pressure through their defense mechanisms. That birds and mammals can get rid of their ectoparasites using their beaks or teeth and sometimes their feet or hands drives these parasites to find shelter in less accessible places. In the case of meso- or endoparasites, which are particularly exposed to these types of defense systems, shelter may be found in tissues or organs where such defenses are less active.

3. *Other parasites:* Other parasites may compete for the same site or for a portion of the same site, which may then lead the less dominant species to change its niche in order to limit the effects of competition.

4. *Predators:* Predators may exercise pressures on ectoparasites that make them pass to the meso- or endoparasitic stage. The existence of these pressures is shown by the occurrence of "cleaning" animals that actively remove ectoparasites from certain fish and mammals. It is particularly interesting that some ectoparasites respond to these pressures by becoming camouflaged, just as do free-living animals subject to predation.

Whittington (1996) notes that certain monogeneans on the skin of fish are extremely difficult to see, despite their large size (up to twenty millimeters), because they are transparent, have spots for camouflage, or can find refuge in locations that are very difficult to reach. Monogeneans of elasmobranchs (sharks and rays) show a tendency to leave their natural habitats on the skin or gills to settle far away, with some species even living in the rectum or oviducts (Euzet and Combes 1998).[3] Anecdotally (although we can also take a lesson from it), it is interesting that the cleaner fish *Crenilabrus melops,* used en masse in fish farms to help salmon get rid of tegumental copepods, has its own copepod parasite. However, this copepod, *Leposphilus labrei* (mentioned in chapter 4 when I talked about regular distributions), lives inside the lateral line, where it is inaccessible to either auto- or allogrooming (Donnelly and Reynolds 1994).

New Systems

Changes in the composition of ecosystems may give the opportunity to form new host-parasite systems through a process termed "lateral transfer" (or sometimes "host switching"). The same result may be obtained if a species changes behavior over the course of evolution. It is very likely, for example, that a good number of human parasitoses did not exist in our primate ancestors and were acquired only after important and marked behavioral changes in hominid evolution.

Lateral transfer has two successive stages: the transfer itself and post-transfer events.

Parasite Transfers

Transfer Mechanisms

Lateral transfer means that at a certain point in evolution a parasite exploits a host species that it did not exploit before. This innovation may be the result of three types of events, all ecological in nature. Poulin (1992b) notes: "Ecological and physiological traits . . . lead to two unrelated hosts both being suitable for a given parasite." Either there is immigration of a species, absent up to that time, into the ecosystem where the parasite is and the parasite finds itself confronted with new "potential" niches to exploit, or a species that is present in the ecosystem changes its behavior during its own evolution and this behavioral change brings it in contact with the infective stages of the parasite, or the disappearance of one

3. In this case predatory pressure may have occurred along with the seeking of richer resources.

parasite species frees a niche for another parasite in or on a host species already present in the ecosystem.

In each case the possibility of transfer depends on the opening of both the encounter and compatibility filters. If the compatibility filter is open when the encounter filter opens (for example, after immigration or a change in behavior), the transfer may happen quickly and there is immediate uptake (fig. 5.9A). If the encounter filter opens when the compatibility filter is not open, however, considerable selective pressure is now put on the parasite to open the compatibility filter, and any compatible mutant is strongly selected. In this case there is transfer by selection (fig. 5.9B).

Among the three types of events that may induce such a transfer, two—species immigration and behavioral change—do not present any particular conceptual difficulty. For example, any peculiar or new dietary behavior carries the risk of disease transmission. The most spectacular example of this may be prions, which are pathogenic proteins that in some cases behave as infective agents. From sporadic cases around 1990 the prion-related disease called kuru spread in New Guinea as the result of eating human brains. It is by a similar process that mad cow disease (bovine spongiform encephalopathy) became epidemic in Europe in the 1990s: from sporadic cases without real incidence there arose an epidemic when contaminated animal by-products began to be used as supplements in cattle feed.

The third type of event, the opening of a niche, may not be as straightforward as the previous two, however, because it involves interspecific competition, itself continuously questioned by ecologists. The hypothesis of freeing or opening up a niche supposes that a parasite species becomes extinct either in part of the host area or completely across it and that the host species now "freed" then becomes colonized by the transfer of another parasite into the newly available niche in the host.

This hypothesis was used to explain the following situation (fig. 5.10). In the Atlantic Ocean, as in the Mediterranean, the brill *(Scophthalmus rhombus)* is parasitized by the ectoparasitic copepod *Lepeophtheirus europaensis*. In the Atlantic Ocean the flounder *(Platichthys flesus)* is parasitized by two other copepods *(Acanthochondria depressa* and *Lepeophtheirus pectoralis)*. In the Mediterranean, the flounder is never parasitized by *A. depressa* or by *L. pectoralis* (these two species being totally absent there). However, it is parasitized by *L. europaensis,* the parasite of the brill, which never happens in the Atlantic. The explanation?

The flounder is a fish of Nordic origin that entered the Mediterranean as the result of Quaternary glaciations. During this migration it apparently lost its Atlantic copepods. The flounder niche was then freed,

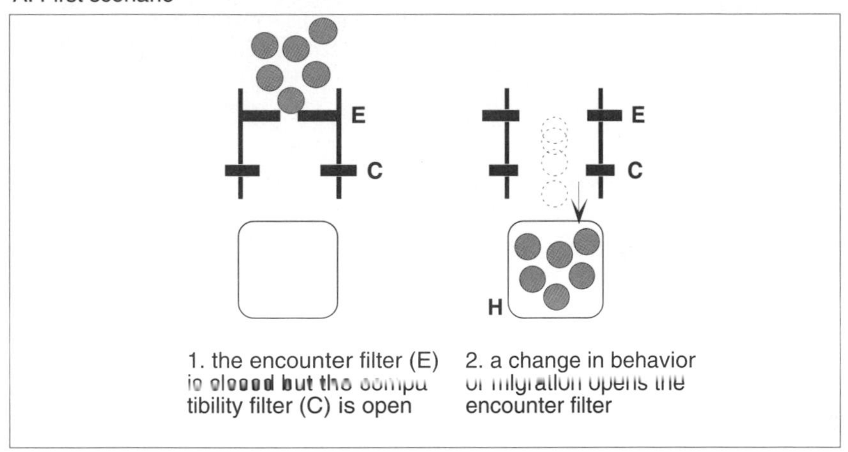

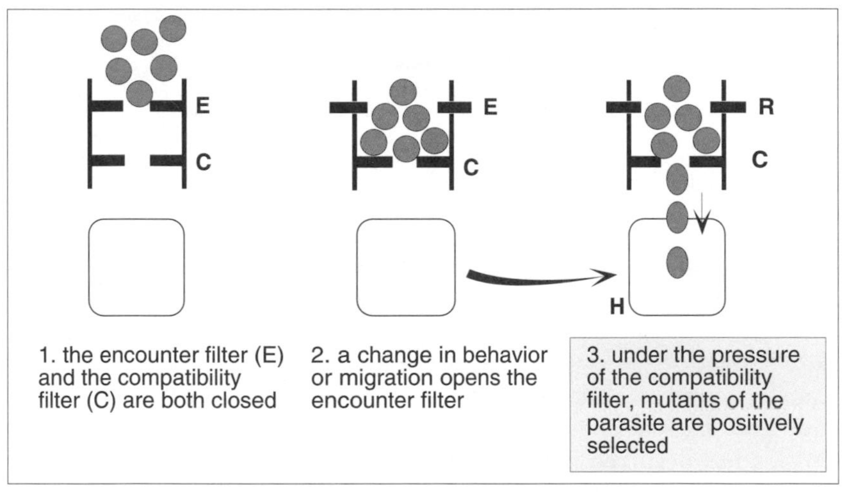

Figure 5.9. Lateral transfers (filters are represented in profile). (A) First scenario; immediate transfer (the compatibility filter is wide open when the encounter filter opens). (B) Second scenario; transfer by selection (the compatibility filter is closed when the encounter filter opens). E = encounter filter; C = compatibility filter.

allowing the parasite of brill in the Mediterranean to increase its host spectrum. In the Atlantic, the presence of two "normal" copepods on the flounder prevents this transfer (de Meeus, Renaud, and Gabrion 1990). If this interpretation is correct, it validates the hypothesis that the opening of a niche provides a situation favorable for lateral transfer. Such a lib-

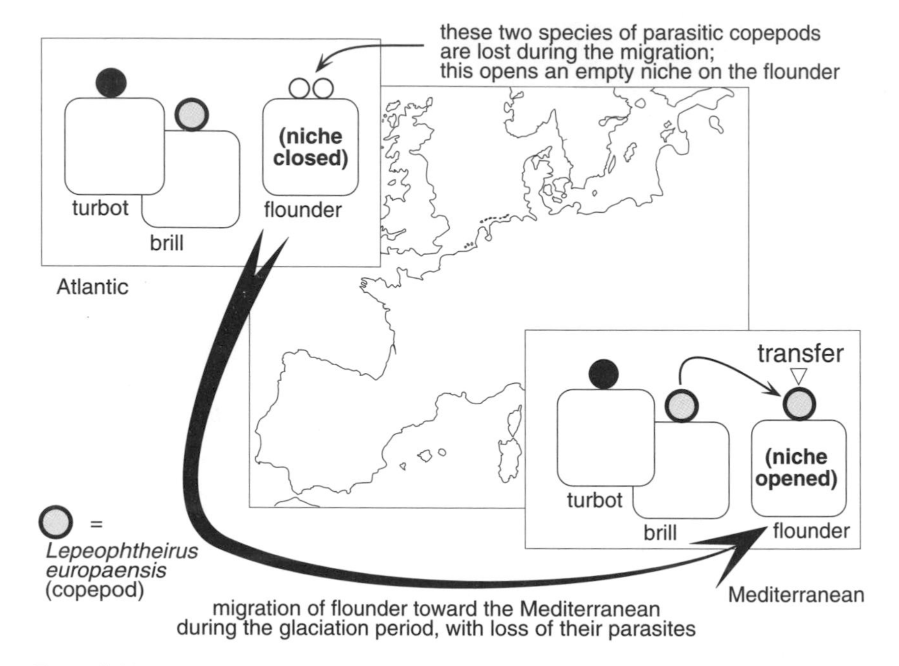

Figure 5.10. The freeing-up of a niche followed by transfer: the case of a fish copepod (from the data of de Meeus, Renaud, and Gabrion 1990).

eration of a niche has the consequence of removing what was earlier referred to as "an immunity coded for by the parasites' genes" (chapter 2).

Posttransfer Events

Transfer enriches both the host spectrum (the parasite has one more host) and parasite richness (the host has one more parasite). Once the transfer has occurred, however, there may be different consequences. The new host is inserted into the cycle without any further effect other than the numerical increase of the host spectrum. The new host is inserted into the cycle and modifies the dynamics of the cycle (for example, by allowing a demographic explosion of the parasite). The new host induces a genetic divergence between parasite populations that leads to alloxenic speciation. And last, the new host takes the place of the host of origin.

That is, the word "transfer" applies to host acquisitions that always have a similar initial process but whose long-term consequences might be very different.

Demonstration of Transfer

It is not always easy to demonstrate transfers, and such demonstrations are essentially the result of comparing host and parasite phylogenies, which can be done in four ways.

The Taxonomic Argument

First there are simple cases where it is easy to see, without intensive research, that there is no logical relation between the evolution of the parasite and that of the host. This is the case of the unexpected parasite *Oculotrema hippopotami,* found under the eyelid of the hippopotamus. This parasite belongs to the Polystomatidae, a family of monogeneans whose other members are, according to the data collected by Gueorgui Batchvarov (personal communication), parasites of both amphibians (seventy species in twelve genera parasitize the bladder) and freshwater turtles (forty species in two genera parasitize the bladder and esophagus). The presence of a unique species of Polystomatidae in the hippopotamus can logically be explained only by a lateral transfer via capture. But a conclusion should never be made too quickly. There is another polystomatid, *Concinnocotyla australensis,* that is in neither amphibians, turtles, nor the hippopotamus. Rather, it is in the Australian dipneust *Neoceratodus forsteri.* Obviously we can propose the hypothesis that it is also a capture event comparable to that of the hippopotamus. However, because dipneusts are related to the ancestors of amphibians, another hypothesis may make it the ancestor of all other polystomatids (fig. 5.11). Although the species in the dipneust is a typical polystomatid, it shows a very unusual anatomy and biology (mating occurs in the digestive tract), which in any case excludes the possibility of a recent capture (Pichelin, Whittington, and Pearson 1991).

A comparable situation is provided by nematodes of the family Cosmocercidae, all parasites of amphibians with the remarkable exception of *Nemhelix,* which parasitizes snails (Morand and Petter 1986). In certain other cases, although the transfer is obvious, it is still not fully explained because not all the actors are known. The cestode *Taenia solium* is one of two such flatworms in humans. It has hooks and infects people when they eat undercooked infected pork (a second tapeworm, *Taenia saginata,* has no hooks, and people get it by eating undercooked infected beef). The adult stage of *T. solium* is specific for humans, from which the question arises, Who was it parasitizing before humans? Our tree dwelling ancestors? This is not very probable because they were not likely to eat pigs, wild boars, or warthogs. It was probably only when they adapted to the savanna

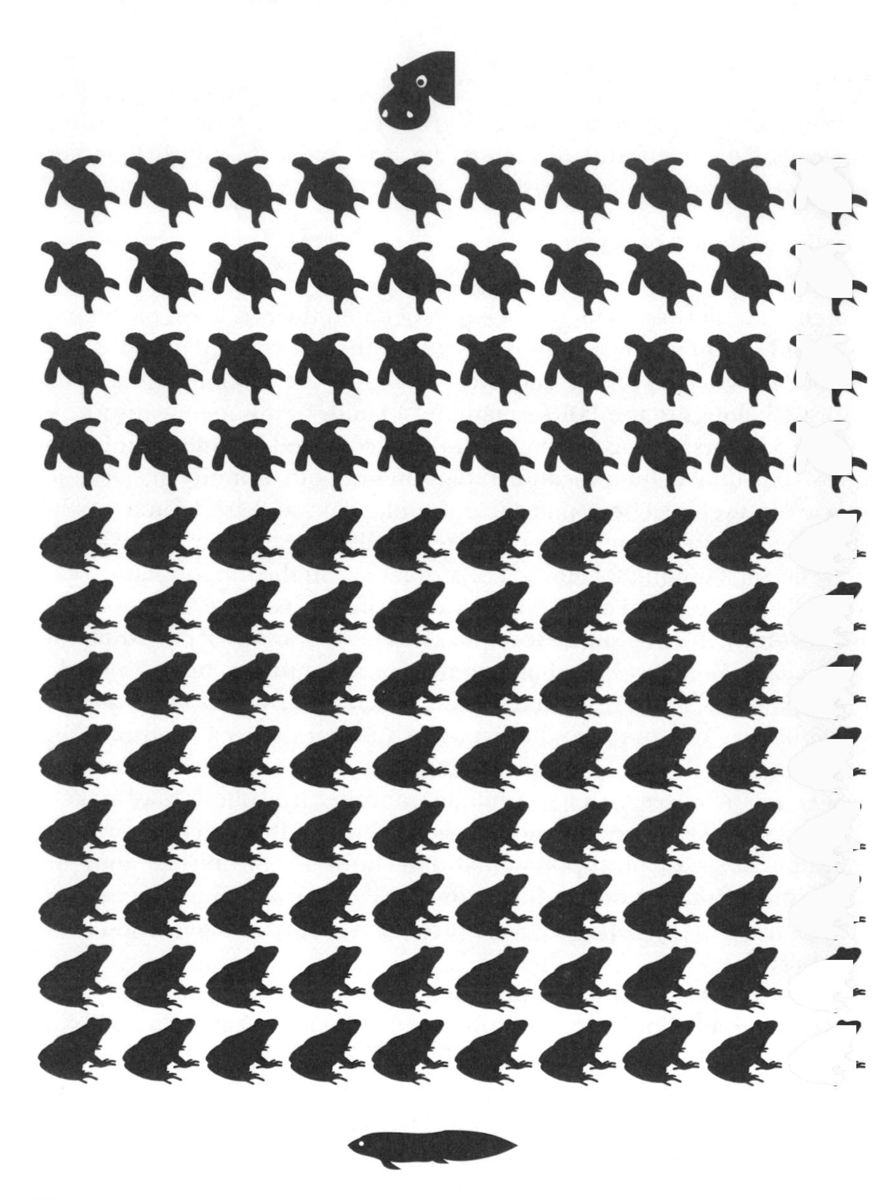

Figure 5.11. Polystomatid hosts (from G. Batchavarov's data). Each silhouette represents the host type (dipneust, frog, turtle, or hippopotamus) of a polystomatid species. The hippopotamus (top) obviously results from a lateral transfer. The dipneust (bottom) may be interpreted either as the result of a lateral transfer or as an ancestral host.

that primates started to eat such prey, perhaps by scavenging carcasses left by larger predators. Would the previous host have been a carnivore? Probably, but which one? If so, why haven't we discovered it? Could *T. solium* be the parasite of a carnivore whose extinction may have been caused by humans although they "saved" its parasite? If so, who could say humans weren't helping endangered species?

The Ecological Argument

Ecological evidence for transfer may be found in studies that show clear similarities between unrelated hosts. For instance, Quentin (1989) has shown that African rodents harbor parasites that circulate among them only occasionally. All indications are, for example, that the intestinal nematode *Parastrongyloides chrysochloris* is a parasite of the golden mole *Chrysochloris leucorhina* but that it is also occasionally found in *Lophuromys sikapusi,* a foraging rat that uses the mole's biotopes, and that it develops well there. Higher up, a tree-dwelling rodent, *Thamnomys rutilans,* borrows the nematode *Pithecostrongylus vogelianus* from the primate *Porrodictitus potto,* with whom it shares an "ecological intimacy," as noted by Quentin. In this regard, and in an even stranger occurrence, Théron (1975) has shown that the trematode *Parabascus lepidotus,* an usual parasite of bats, easily develops in the rodent *Apodemus sylvaticus* simply as the result of a convergence of diet. It is important to recognize that specificity and variability in parasites are sometimes wide enough to allow for the exploitation of an occasional host, which is always good if it ensures dissemination of an extra number of the parasite's descendants.

In fact one may argue with the use of the term "transfer" to describe the three cases just mentioned, because each is a simple localized expansion of the host spectrum. Chabaud (1965) proposed the term "spillover" (in French, *transfuge*) for such situations. The word "capture," also proposed by Chabaud (1965), applies only where speciation occurs. Xie, Bain, and Williams (1994) note: "According to Chabaud, a captured parasite is defined as a parasite which, after becoming isolated in a new host, undergoes speciation and becomes morphologically distinct from the original species."

The Molecular Argument

Modern molecular biological techniques are a precious tool in revealing transfers. Anderson and Jaenike (1997) studied the long-standing problem set forth by two nematodes, *Ascaris suum,* a pig parasite, and *A. lumbricoides,* a parasite of humans. These two species are morphologically so close that for years it seemed doubtful that they were really separate species. After studying variability in several introns from nuclear genes, single or only slightly repeated sequences, internal transcribed spacers (ITS) of the ribosomal transcription unit, and a 630 base pair fraction of mitochondrial DNA, these authors reached the following conclusions.

Based on analyses done in Guatemala in a community where both pig and human ascarids are present, it appears that multiple genetic markers characterize each. Therefore cross-infections (pig parasite in human or vice versa) and eventual genetic exchanges are extremely rare.

Analysis of samples coming from different regions of the world shows that the populations of human ascarids form a monophyletic ensemble, suggesting that all the parasites of humans, wherever they are, come from a unique transfer event.

That genetic diversity is greater in pig parasites than in human parasites implies that the pig is the ancestral host and that transfer occurred from pig to human and not vice versa.

The most ancient archaeological site where the remains of humans and pigs have been found together is dated 16,000 B.P. Since pigs were probably domesticated between 9,000 and 7,000 years ago, Anderson and Jaenike propose that the transfer must have happened between 7000 and 20,000 B.P.

These authors do not identify where the transfer may have happened, but other archaeological evidence indicates that the best candidates are the Middle East and China.

The Phyletic Argument

Various methods, founded on either morphological or biochemical characteristics as well as on DNA analyses, allow the building of phylogenetic trees.[4] Such studies, carried out in parallel on a group of parasites and on their hosts, allow one to "see" lateral transfers. These trees may be done at the level of the species as well as at the level of genera or families (see, for example, Euzet 1956). In the case of recent transfers, simply comparing the spectrum of actual hosts with the phylogeny of the hosts allows one to

---

4. There are several approaches to building phylogenetic trees: (a) the intuitive approach (also referred to as the "classical" or "evolutionary" approach) gives a certain hierarchical value to each characteristic and builds the tree as a function of this hierarchy; (b) the phenetic approach gives the same value to all characters and builds a tree based on similarity; and (c) the cladistic approach, which makes hypotheses about the ancestral (pleisomorphic) or modern (apomorphic) nature of the shared characteristics and looks for "sister groups" that share the modern characteristics.

Although molecular data are the most precise in building such trees, they also have their problems. For example, if the proportions of the bases A, C, G, and T are different, sequences of similar composition are grouped without necessarily implying a phylogenetic relationship (see Galtier and Gouy 1995).

suspect that transfer has occurred. This is the well-known case of the P element in *Drosophila.* The P element is a transposon (see chapter 1) and the best known of the mobile elements in the *Drosophila* genome. It provides an example of a parasite that probably went from one living species to another only recently (in the twentieth century). A fruit-fly genome contains a limited number of complete P elements (about fifteen) to which other "defective" (incomplete) copies are added. The extremities of the P elements have the characteristic inverted repeated sequences of transposons, and the genome of a complete P element has 2,907 base pairs, which mainly code for transposase, an enzyme that identifies sequences close to the extremities of the element and allows transposition.

There is evidence that the P element spread through populations of *Drosophila melanogaster* between 1950 and 1980.[5] Those strains that today harbor the P element are called "P strains" and the others "M strains." The distribution of the P element among the species of fruit flies does not fit with the phylogeny of the group (fig. 5.12). Within the large group of drosophilids there are four related subgroups—*melanogaster, obscura, willistoni,* and *saltans*—each one including various species. The species from the last three groups are all infected with P elements. However, in the "*melanogaster*" group only some populations of *D. melanogaster* are infected. Species related to *D. melanogaster,* such as *D. simulans,* are never found to be infected with P elements. This distribution is a paradox: when the phylogeny of the fruit flies is examined it seems that the P element was acquired by an ancestor common to the four groups about 50 million years ago but that it was not conserved in the "*melanogaster*" group. If this scenario is correct, *D. melanogaster* should not have any P elements. However, when examining the strains of *D. melanogaster* that were kept frozen in laboratories and collected at various periods, researchers became certain that the acquisition of the P element in this group was recent. *D. melanogaster* was introduced to the Americas between the sixteenth and nineteenth centuries, and the expansion of the P element in the different populations of *D. melanogaster* occurred in the 1950s. Thus these fruit flies came to harbor P elements only recently. From the American focus in the 1950s, expansion touched first Europe, then Africa and Asia, and finally

5. This invasion reminds one of HIV in human populations. However, the two invasions differ in the exact nature of the pathogenic agent (a retrovirus in one case and transposon in the other); in the fact that the P element has invaded the genome even while the virus has an extracellular phase; in the replication mode, which involves reverse transcriptase in one case and not using RNA in the other; and in the pathogenic effect, serious in one case, benign in the other. In both invasions, however, the foreign sequence behaves like a parasite of the genome, with a sudden explosion occurring in the population from an initial small focus.

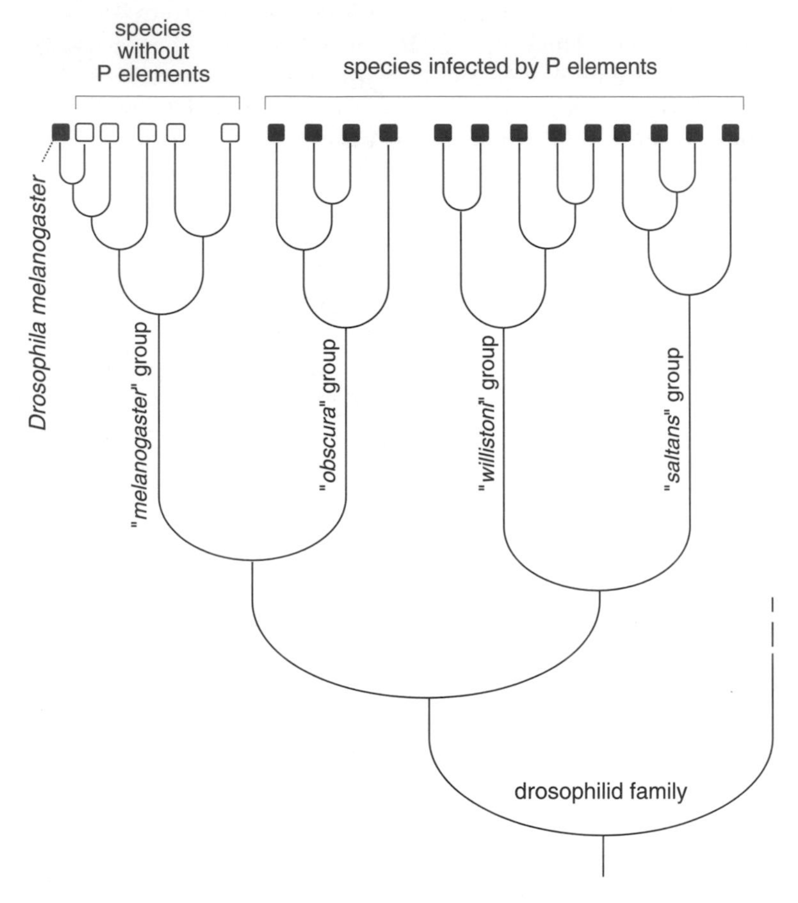

Figure 5.12. *Drosophila* groups and the P element: presence of the P element in *Drosophila melanogaster* not matching the phylogenetic tree of the *Drosophila* supports the idea that it results from a lateral transfer that may be only a few dozen years old.

the whole world. This expansion may be linked to the mixture of fruit fly populations owing to increased accidental transport. Why was *D. melanogaster,* whose ancestors seem to have lost the P element over the course of evolution, able to reacquire it recently? The most likely hypothesis is that there was lateral transfer from a species of a closely related group (probably *willistoni*). This hypothesis is supported by molecular biology, although the precise mode of such a transfer remains unknown.

With the more ancient transfers, it is the lack of congruence between phylogenetic trees of parasites and their hosts that makes one suspect that transfer has occurred. Several studies have shown that phylogenetic trees of parasites and hosts often show similarities, or congruences (parasite speciations seem to coincide with host speciations), and the points where similarities do not exist are explained by lateral transfers.[6] The first detailed study of this type was done by Hafner and Nadler (1988) on the mallophagan ectoparasites (family Trichodectidae) of rodents (family Geomyidae) and was based on the analysis of thirty-one and fourteen allozyme loci, respectively. This study showed four quasi-synchronous cospeciations and two probable lateral transfers. Guégan and Agnèse (1990) observed similar results for the dactylogyrid monogeneans of cyprinid fish in West Africa, where the presence of at least two of the parasite species was interpreted as the result of transfer. Verneau, Catzeflis, and Renaud (1997) showed, using DNA/DNA hybridization, that bothriocephalids (cestodes) of marine fish recently differentiated in an "explosive" way without visible connection to their host's speciation and with numerous lateral transfers.

It is important to note that congruence between the phylogenetic tree of hosts and that of parasites is not absolute proof of cospeciation if there is no precise dating of events. We can well imagine that a group of hosts first diversified itself and later was parasitized by a group of parasites that in turn diversified itself. In this case the phylogenetic trees could have the same shape but not coincide in time. This is the "empty barrel" hypothesis, according to which a host clade is first established and then colonized stepwise by a clade of parasites in the form of "sequential transfers . . . mediated by, for instance, . . . designs or chemicals strictly concordant with [host] phylogeny" (Lachaise 1994).

The Paleontologic Argument

An original approach to studying transfers consists of reconstructing the phylogeny of a group of parasites and comparing it with host evolution as deduced from fossils. This approach, principally applicable to vertebrate parasites, may permit dating of transfer events because of the well-established geologic chronology.

Durette-Desset (1985) analyzed the present fauna of nematodes from the trichostrongyloid group, which live in the digestive tract of mammals, by defining evolutionary lineages of the parasites using morphological criteria and comparing them with the paleontologic data of the hosts. The results show that the periods of radiation of these parasites coincided either

6. That there is cospeciation does not imply that speciation is of the alloxenic sympatric type, because the separation of the host species into two daughter species may be the consequence of geographic isolation. In this case speciation of the parasite, as well as of its host, would be *allopatric*.

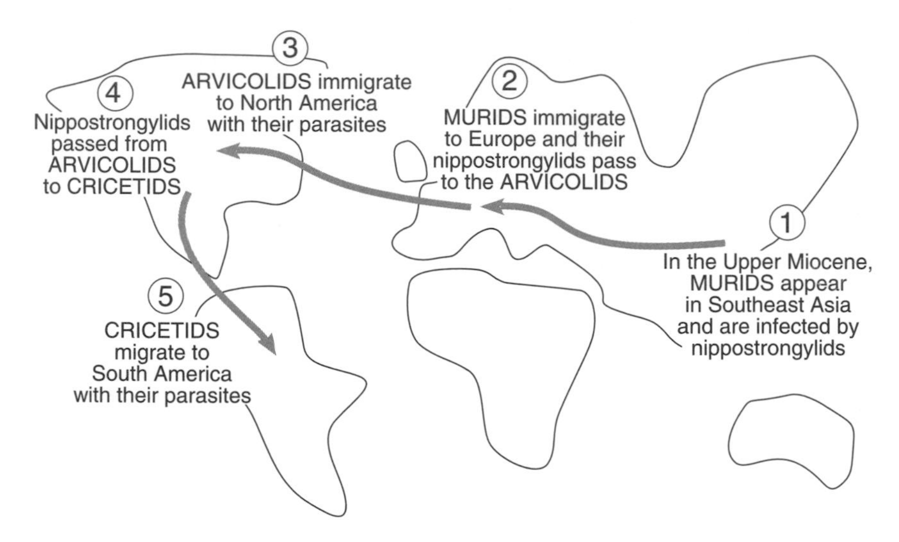

Figure 5.13. Transfers and migrations of trichostrongyloid nematodes during the Tertiary (the shapes of the continents changed during this period and thus are approximate here; from the data of Durette-Desset 1985).

with the explosive appearance of new host species or with the immigration of new groups of potential hosts offering ecological opportunities for conquest. For example, Durette-Desset demonstrates that the factor tying together nematodes belonging to the subfamily Molineinae within their numerous mammals is related not to the mammals but rather to the appearance of all the hosts at approximately the beginning of the Tertiary period. Several serial transfers were described by Durette-Desset; for example (fig. 5.13), the murids (rats and relatives) appeared during the lower Miocene in Southeast Asia, where they acquired nematodes of the subfamily Nippostrongylinae. Later, the murids emigrated toward Europe during the lower Pliocene, and the Nippostrongylinae then contaminated the myomorphs of the family Arvicolidae; during the late Pliocene, the arvicolids moved to North America with their nippostrongylids and infected the cricetids; and last, during the late Pliocene the cricetids migrated to South America with their nippostrongylids.

Similarly, the history of parasites in Australian marsupials has begun to be known through studies using morphological characteristics of parasites in concert with paleontologic data. According to Beveridge and Spratt (1996), the helminths of living marsupials have three distinct origins.

1. *Helminths acquired before the separation from Antarctica.* Marsupials probably have an American origin, and at the end of the Cretaceous period

South America, Antarctica, and Australia formed one large continent that fragmented at the beginning of the Tertiary. As a result, some marsupial species became isolated on the Australian fragment, and certain lineages of helminths alive today seem to be descendants of species that lived in the marsupials at that time. These include the nematodes, cestodes, and perhaps the trematodes.

2. *Helminths acquired by transfer during the isolation of Australia.* Australia has been completely isolated from other continents for at least 20 million years. During this period transfers occurred into the marsupials from other vertebrates. In this way marsupials have acquired nematodes and cestodes from monotreme mammals, birds, and more rarely reptiles. Marsupials, which were initially carnivores, evolved omnivorous and herbivorous forms, leading to greater complexity of the digestive tract and thus opening new niches for gut parasites.

3. *Helminths acquired after Australia and Southeast Asia came closer together.* For several million years Australia has been close enough to the continent of Asia for parasite-carrying rodents and bats to introduce themselves. Thus Australian marsupials have acquired a more recent batch of cestodes and nematodes by transfer from these hosts.

Once settled in their marsupial hosts, the "new" helminths continued evolving, either in parallel with the marsupials' own evolution or by transfer within the phylum itself. Moreover, humans have recently introduced into Australia cattle and various other animals along with their helminths, so that ubiquitous species such as *Fasciola hepatica* and *Echinococcus granulosus* have now secondarily invaded several marsupial species.

We know very little about what it takes for a transfer to happen. Bell and Burt (1991) noted that in freshwater fish there is a correlation between parasitism and diet when the analysis is done at the family level, but it is not seen at the species level. Based on these data, they deduced that there is a "substantial time-lag" between change in diet and the realization of transfers.

Faithfulness to the Site or Faithfulness to the Host?

Parasites are both faithful, since there are long periods of evolution in parallel with the host and long periods of conservation at the same site, and also unfaithful, since there are numerous lateral transfers from host to host and site to site. We may therefore ask the following questions.

Are Parasites More Faithful to the Site or to the Host?

The possibility of synxenic speciation with the separation of microhabitats implies that in this case they are more faithful to the host: the host stays the same for the lineage of parasites while the site varies. However, the opposite is possible if after adaptation to two different sites the two resulting lineages evolve by further adapting to new hosts either by cospeciating or through lateral transfer. Littlewood, Rohde, and Clough (1997) demonstrated this second possibility using DNA analysis of the monogenean family Polystomatidae. The genus *Polystomoides* occurs on all five continents and has about thirty species, all parasites of freshwater chelonians. These worms may be found in the turtles' oral cavity or bladder (each species of *Polystomoides* is present only in one of these sites). Sometimes one species may be found in the oral cavity and another in the bladder of the same host. Two hypotheses may be put forth to explain this distribution: that species occupying a particular site (for instance, the oral cavity) have given rise to species infecting the same site in different host species; and that species occupying a particular site (for instance the oral cavity) have given rise to species infecting different sites (for instance, the oral cavity and the bladder) in the same host species.

Analysis of two DNA sequences (the 28S ribosomal nuclear DNA fragment and mitochondrial cytochrome c oxydase subunit I) was done for four species of *Polystomoides* distributed as follows: *P. malayi* in the bladder of the turtle *Cuora amboinensis; P. asiaticus* in the oral cavity of the same host, *C. amboinensis; P. renschi* in the oral cavity of *Siebenrockiella crassicollis;* and *P. australiensis* in the bladder of *Emydura krefftii.*

DNA analysis has unambiguously shown that *P. malayi* and *P. australiensis* (the species found in the bladder) form one clade, while *P. asiaticus* and *P. renschi* (the species found in the mouth) form another (fig. 5.14). Therefore the two *Polystomoides* of *Cuora amboinensis* do not come from sympatric speciation within this host. Littlewood et al. (1997) concluded: "Species of *Polystomoides* infecting the same microhabitat of different host species are more closely related with each other . . . than with species infecting different microhabitats in the same host species."

Separation between the "bladder microhabitat" and "oral cavity microhabitat" lineages is therefore more ancient than is colonization of the different hosts.

Do the Most Ancient Hosts Have the Most Ancient Parasites?

Tetraphyllideans, which are cestodes of elasmobranchs (sharks and rays), are an important group of parasites because the great majority of species are oioxenous (strictly associated to a single host species). This generalized narrow specificity may be linked to the extreme antiquity of the parasite-host system. Indeed, sharks and rays are among the oldest currently living vertebrates in the world, and species looking much like those occurring today are found in sediments dating from the Paleozoic. Relative

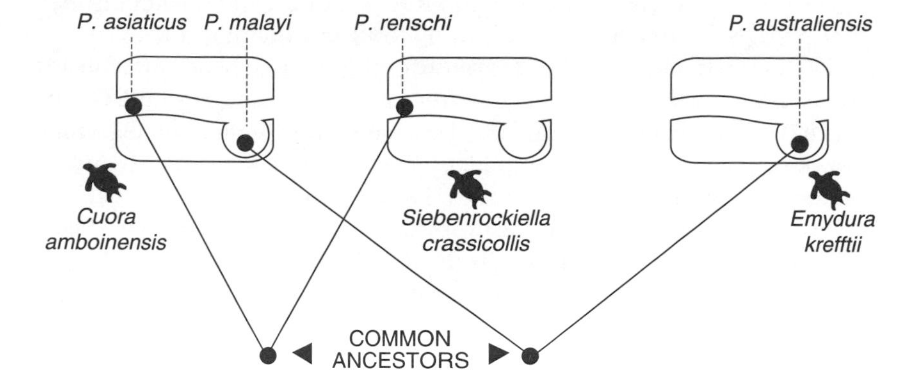

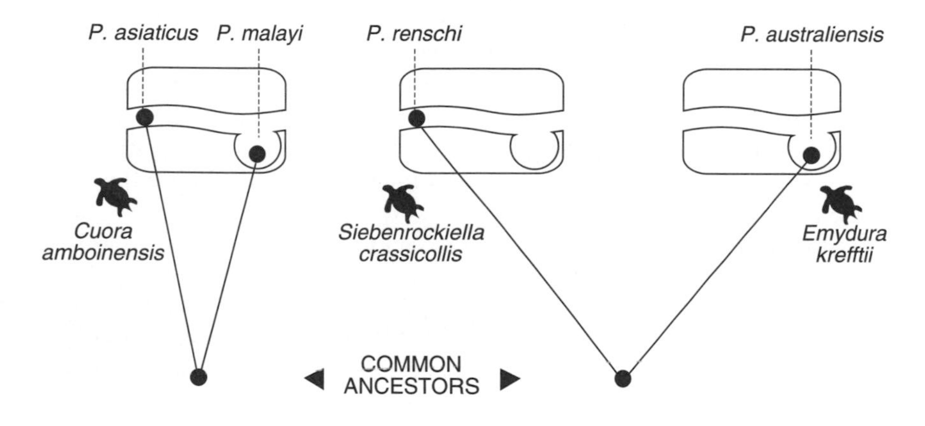

Figure 5.14. The two possible hypotheses explaining the distribution of four species of *Polystomoides* in their hosts and sites. A phylogeny based on comparison of DNA sequences has demonstrated that the first hypothesis was the correct one (from the results of Littlewood, Rohde, and Clough 1997).

to cestodes, their peculiarities compared with other Plathyhelminthes (particularly the free-living ones) testify to their own antiquity.

The current state of research on tetraphyllids shows that most species are found in only one host species (see Euzet 1956) although some rare taxa have been reported from several. However, biochemical analysis may show that these exceptions are in fact cryptic species, and it may very well be that all tetraphyllidean cestodes (about two hundred species) are oioxenous. If one admits that elasmobranchs and tetraphyllids have been associated for tens or even hundreds of million of years, then the very high specificity of their relationship may be considered an indicator that parasite-host systems grow more and more specific with time.

Another characteristic of tetraphyllids is that most of them have scolices (attachment devices) that are as extravagant and complex as they are diverse. One may wonder if in this particular case specificity is linked to the diversity offered by the elasmobranch intestine, which is itself very particular in that it contains a complex structure called the spiral valve. The narrow specificity may then result from the difficulty the parasite faces investing in a multiuse organ that would allow it to attach to various supports with the same efficiency.

Very narrow specificity occurring over an entire phylum is a powerful argument for alloxenic speciation. In this case, each time a species host exploded into two or more species (or each time a lateral transfer occurred) the process must have been followed by a parasite speciation. In other words, at least after some period, the interruption of gene flow between hosts would interrupt gene flow between parasites. This is a good argument to support the idea that the real units of the living world are in fact the supergenomes of the parasite-host systems.

Up to What Point Do Parasites Resist Their Hosts' Ecological Changes?

Some textbooks emphasize the "difficulties" involved in parasite transmission. However, what seems difficult to us may not be so for the parasites, who likely have selected for compensating adaptations. The following example demonstrates this idea: most monogeneans are ectoparasites of fish and have a holoxenous life cycle involving transmission from host to host via a swimming larval form, the oncomiracidium. As noted earlier, polystomatid monogeneans are an exception for this group, with most of them living in the bladders of amphibians. To parasitize amphibians is itself a "problem" for monogeneans, since amphibians, by definition, are only partially aquatic. Here is a series of examples by which one can see the increasing reduction, to the extreme, of the importance of aquatic transmission: *Protopolystoma xenopodis* parasitizes *Xenopus laevis,* a toad that spends most its life in permanent ponds in equatorial Africa; *Polystoma integerrimum* parasitizes *Rana temporaria,* the European common frog, which lives in the water only a few weeks each year during mating; *Eupolystoma alluaudi* parasitizes *Bufo regularis,* a toad of temporary ponds in Africa that can be found in water only for very brief periods during the rainy season; and *Pseudodiplorchis americanus* lives in *Scaphiopus couchii,* a foraging toad of the Arizona desert that has only twenty-four hours a year to lay its eggs in ephemeral ponds resulting from occasionally violent rainstorms. Figure 5.15 shows the following:

The cycle of *Protopolystoma xenopodis* is not different from that of a fish monogenean except that the swimming larva penetrates into the kidney and then the bladder rather than fixing itself on the skin or the gills.

The life cycle of *Polystoma integerrimum* involves attachment of the oncomiracidium to the gills of tadpoles and not their bladders: if the tadpole is young the parasite develops and lays eggs. If the tadpole is old, however, the parasite remains on the gills without maturing and migrates toward the bladder by crawling onto the belly during metamorphosis.

The life cycle of *Eupolystoma alluaudi* has an internal reproductive phase in the bladder during drought periods. Interestingly, there are two kinds of larvae, nonciliated ones that ensure reproduction on site and ciliated ones that leave en masse when the toad enters the water, ultimately reaching the cloacal cavity of nearby toads.

The life cycle of *Pseudodiplorchis americanus* adds the novelty of infecting the host via the nasal passages and following a pathway that involves the lungs, digestive tract, and finally the bladder!

Thus, polystomatids have found a way to counter the decreasingly aquatic nature of their amphibian hosts, and two lessons are to be drawn from the ecology of these worms. The first is that both *Protopolystoma xenopodis,* which may infect its host 365 days a year, and *Pseudodiplorchis americanus,* which can infect its host only one or two days out of the year, are successful. Prevalences for the two parasites are similar (about 40%), and *Pseudodiplorchis americanus* from the desert do as well as *Protopolystoma xenopodis* from perennial ponds. R. Tinsley, who is the author of remarkably detailed works on these polystomatids, has pointed out the extraordinary nature of this paradox, which makes the word "difficulty" seem ironic in terms of parasitic life cycles.

The second lesson is that polystomatids have had, if one may use the expression, lots of adaptive imagination in infecting hosts that are less and less aquatic. Nevertheless the swimming larva is an unavoidable phyletic constraint that forbids them from parasitizing completely terrestrial hosts.

If the ecology of the host were to undergo a drastic change, there might come a time when the parasites could no longer "follow." This

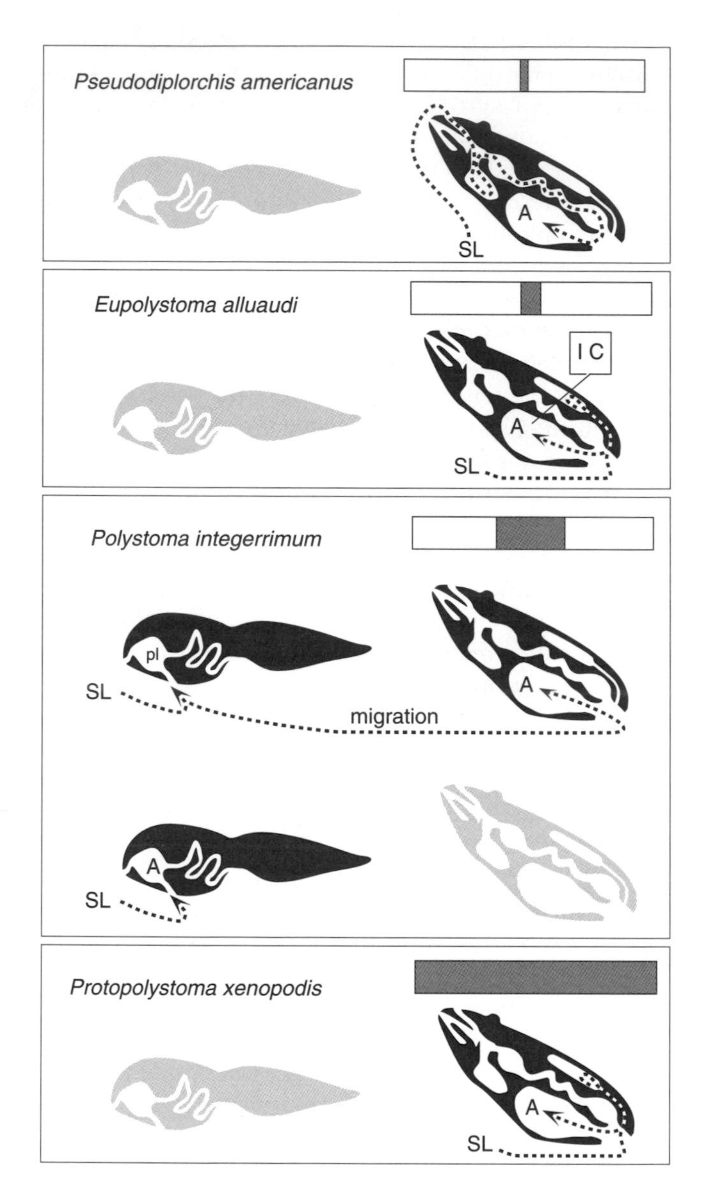

Figure 5.15. Infection of four amphibian species by monogeneans strongly differing in their ecology from the most "aquatic" (bottom) to the most "xeric" (top). Rectangles at top right in each example symbolize the duration of humid periods. SL = swimming larva (oncomiracidium); pl = postlarva; A = adult (on tadpole of *Polystoma integerrimum* called "neotenic"); IC = internal cycle.

apparently happened for the parasites of those mammals that went back to an aquatic lifestyle and gave rise to the pinnipeds (seals, sea lions, elephant seals) and cetaceans (dolphins and whales). In this case the ectoparasites of pinnipeds and those of cetaceans are totally different. The seals and other pinnipeds have lice comparable to those of terrestrial mammals, whereas whales and other cetaceans have lost their lice of terrestrial origin and instead have "whale lice"—amphipod crustaceans borrowed from fish (itself an example of a transfer).

Such a change in lifestyle (passage from a terrestrial to a marine environment) has thus eliminated many of the parasites of origin only after the stay in water became permanent, and one must note with some awe that lice have remained faithful to the seals, even surviving regular and sometimes very deep dives into the oceans. On the other hand, the parasites from fish could transfer to the mammals only once the mammals became totally aquatic.

Reflections

Fossil Parasites

Fossil parasites are extremely rare, and often the question, When during evolution did a phylum have its first parasite? remains unanswered. But it would not be fair to ignore what paleontologists have been able to conclude about parasites based on the fossil record (see Conway 1981; Boucot 1990; and Upeniece 1999). This record includes:

- some trematodes in mollusks from the Miocene (10 mya [million years ago])
- some ichneumonid, braconid, and chalcidian parasitoids in the Oligocene (30 mya)
- a tick embedded in amber from Central America dated 30 mya
- mermithids parasitizing insects in Eocene amber from the Baltic (50 mya)
- parasitic mites from dipterans in Cretaceous amber from Canada (75 mya)
- caligid copepods parasitic of marine fish from the Cretaceous (130 mya)
- bopyrid decapod crustaceans parasitic of shrimp from the Jurassic (159 mya)
- insect-induced galls from the Permian (280 mya)
- intestinal helminths from a shark at the end of the Carboniferous (300 mya)
- myzostomidians (annelid relatives) parasitic of echinoderms from the Carboniferous (300 mya)
- some type of monogenean with a sixteen-hook haptor in placoderms from the Devonian (400 mya)
- some traces evoking parasites on trilobites from the Cambrian (570 mya)

One could add that pearls, unknown in the Paleozoic, appeared for the first time at the beginning of the Mesozoic. Since present-day mussels and oysters build pearls mainly around the larvae of cestodes and trematodes, this information may provide evidence of the earliest complex life cycles in these Platyhelminthes.

More recently, and contemporaneous with early hominid evolution, dicrocoelid trematode eggs have been identified from bear coproliths found in the south of France (Jouy-Avantin et al. 1999), as has a fly *(Cobboldia russanovi)* whose larvae developed in the stomachs of mammoths 25,000 years ago (see Boucot 1990).

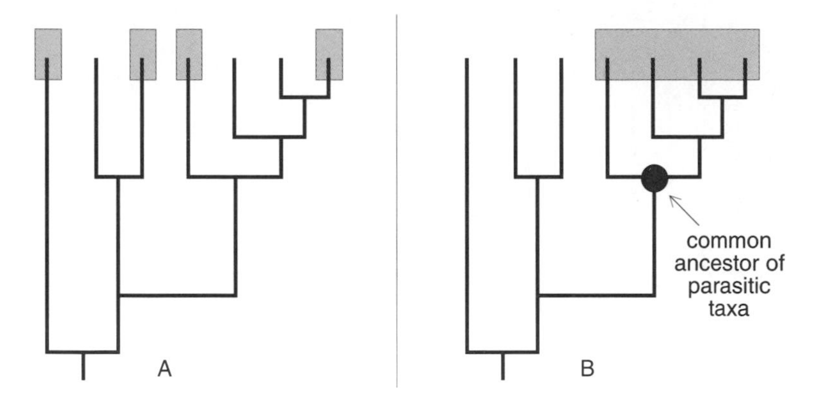

Figure R1.1. Within a given phylum, did parasitism appear independently several times (A) or only once (B)? The diagrams are imaginative. The arrow indicates the supposed common ancestor that acquired the parasitic mode of life. Gray boxes indicate the parasitic mode of life in current species.

Transfer to Parasitism

Several groups of organisms have both free-living and parasitic members, the classic example being nematodes. Concerning such a group one may ask (fig. R1.1), Was parasitism acquired only once (in which case all present parasitic species in the group would have a common ancestor), or did it happen several times (in which case parasitism would be a homoplasic character)?

The use of molecular techniques in systematics has brought forth some important pieces of information relative to this question. For instance, Blaxter et al. (1998) have shown that the phylum Nematoda includes five major clades, each with parasitic species. This conclusion was based on the sequencing of the small ribosomal subunit from fifty-three taxa. The resulting phylogeny shows that parasitism by nematodes appeared at least four times in vertebrates (see black asterisks in fig. R1.2) and at least three times in plants (white asterisks). Note, though, that the question, How many parasitic organisms became free-living? is not asked simply because no such occurrence is known with certainty to have occurred.

Do Parasiteless Organisms Exist?

When one steps back and views all living organisms, those commonly referred to as "lower" are more often found to be parasites than hosts, while those thought of as "higher" are more often found to be hosts than parasites—a fact that may simply reflect an increase in size throughout the course of evolution. That is, since hosts become the parasites' habitats, a small organism is more often fated to become parasitic of a larger organism than vice versa. As a result, by increasing complexity in higher organisms, the course of evolution has not stopped offering new microhabitats for parasites to conquer.

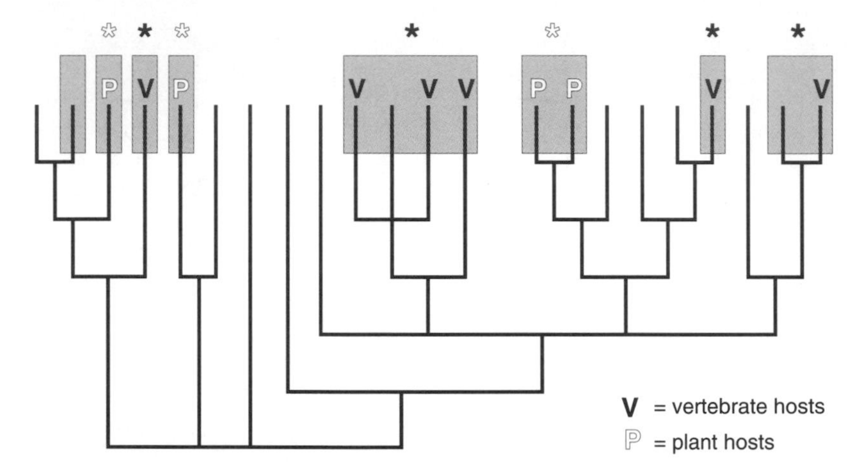

Figure R1.2. A molecular evolutionary framework allows one to propose that parasitism (gray boxes) appeared several times among the nematodes: there are at least four independent origins of parasitism for vertebrate parasites (V) and at least three for plant parasites (P). Invertebrate parasites are not shown (modified from Blaxter et al. 1998).

It is not known with certainty if there are species alive today that do not harbor parasites (although all data seem to demonstrate that every observed species harbors parasites of one sort or another). Nevertheless, we do know there are no higher-level groups that escape parasitism, which is true for parasites as well, since there are parasites that parasitize other parasites. Indeed, the living world can be pictured as a system of Russian dolls where each organism is parasitized by smaller and smaller organisms.

One may be tempted to object that such a system of smaller and smaller infectious agents must end with the tiniest occurring organisms, but it is important to remember that even bacteria can be parasitized by viruses (bacteriophages) that themselves may contain parasitic DNA.[1] Bacteria also harbor plasmids, which are small circular dsDNA molecules that exhibit all the characteristics of those types of parasites that provide advantages to their hosts (most notably the ability to confer resistance to antibiotics). Further, viruses also have their parasites, such as viroidlike satellite RNAs that occur in the capsules of true viruses, borrowing the machinery and tools required for their own replication.

Why Make It Simple When You Can Make It Complex?

The circulation of parasites within an ecosystem often appears to be a complex puzzle whose adaptive value is not always obvious. This apparently useless complexity is often the first thing to strike a biologist just discovering the parasitic

1. Phages are in effect parasitoids because they too end up killing their hosts.

world. However, remember that, whether complex or not, the life cycles of parasites are nothing other than the accumulation of simple adaptations to ensure reproduction and that these adaptations and cycles have been shaped by selective pressures over tens, if not hundreds, of millions of years. This is of course also the reason a life cycle occurring today should be interpreted not as the best possible adaptation to current conditions but rather as the best possible solution to the ensemble of selective pressures the parasite faced in the past. A good illustration of this is given by the trematode *Halipegus ovocaudatus,* a parasite found under the tongue of green frogs in Europe, whose life cycle was briefly described in chapter 1. Kechemir (1978) showed, both experimentally and in the field, that transmission requires four obligatory hosts: a mollusk in which cercariae are produced, a copepod within which cercariae penetrate and produce mesocercariae, a dragonfly larvae that ingests the copepod and within which metacercariae develop in the digestive tract, and, last the frog, which eats the dragonfly and then harbors the adult worms. Here is what is instructive: *H. ovocaudatus* belongs to a trematode family, the Hemiuridae, whose other members parasitize fish and which each have only a three-host life cycle—a mollusk, a copepod, and the fish. It therefore seems logical that cercariae would directly infect the dragonfly larvae, since they live in the same environment as the copepod, and that this would reduce the level of larval loss during host transfer. However, the parasite was obviously incapable of giving up its ancestral attraction to the copepod, and from this there arose the unexpected but efficient solution of using the dragonfly as an extra intermediate host. Thus the link between the required copepod and the new host, the frog, is easily made.

Overall, the development of complex life cycles as the result of the accumulation of more simple adaptations has allowed parasites to conquer the entire living world. This is because those parasites with aquatic ancestors have been able to obtain "vectors," in the fullest sense of the term, capable of transporting them into the realm of terrestrial ecosystems. Significantly, those free-living organisms with partly aquatic life cycles were perfectly suited for this purpose. Both insects with aquatic larvae and amphibians (see Combes 1991a) are the usual intermediate hosts of choice for those freshwater parasites that exploit terrestrial definitive hosts, and these vectors are literally, at least for these parasites, true ecosystem exchangers.

The common idea that parasites "confront enormous difficulties" in transmittal from host to host results from a purely anthropomorphic view that takes into account only the difficulties faced by a *single individual.* In reality, because of multiple adaptations the survival and transmittal of parasites is no more difficult than that of their free-living counterparts. If it were otherwise, the continued extinction of parasites would lead to, at best, the occurrence today of only those organisms that have recently become parasitic.

This said, one does not have to conclude that life cycles, even heteroxenous ones, always follow difficult or otherwise surprising routes. For example, the cestode *Dipylidium caninum,* has a life cycle with two hosts—the dog as definitive host and the dog's fleas as its intermediate hosts. Adult *D. caninum* live in the intestines of dogs, and eggs exit with the dogs' feces. Those eggs that get stuck on the fur are eaten by flea larvae, within which they become stages infective to the dog (cysticercoids). Once the fleas are eaten by the dog while grooming the cycle continues, and it is difficult to think of a simpler system of infection.

Why Take a Straight Line When You Can Wander Around?

The migration routes of parasites inside hosts are sometimes as indirect, complex, and unexpected as they are in ecosystems. For example, the larvae of ascarids hatch in the gut of humans just after they are ingested; they then perforate the intestine, enter the blood vessels, pass through the liver and heart, and arrive in the lungs, where they pierce the bronchioles, go up the trachea, pass into the esophagus, and travel back down to the gut along with saliva or food, just to become gut-dwelling adults at the same place they started from! Members of the genus *Strongyloides* follow a similarly less than straightforward route after cutaneous penetration by their larvae. Even better, when a female raccoon *(Procyon lotor)* ingests the infective stage of the trematode *Pharyngostomoides marcianae,* the parasite leaves the digestive system, passes to the lungs by way of the blood, leaves the lungs to go to the mammary glands, passes into the milk, and becomes adult in the digestive tract of the suckling raccoon (Miller 1981).

If you think such complex internal migrations are simply the reflection of uninterpretable past adaptations, you may wonder what adaptive value such migrations might have today. If *Ascaris* and *Strongyloides* undergo migrations that only bring them back to where they started, it may be because the tissues they pass through on their migration are less aggressive than the intestine in terms of chemical damage and expulsion. Read and Skorping (1995) base this proposition on a comparative analysis showing that migrating species are bigger as adults (and thus more fecund) than nonmigrant species. In other words, passing through various tissues in the host before settling in the adult microhabitat allows them to grow and become larger.

Are Parasites Similar to Cave Fish?

When the cheetah chases a gazelle its parasites also chase the gazelle (though the cheetah does the work), and when the cheetah then digests the gazelle and builds tissues its parasites, either directly or indirectly, also consume and use the gazelle. Thus, as soon as an organism enters the "parasitic mode," a whole series of functions coded for by the parasite's genome now duplicate those of the host. As long as a host can live without the parasite, such duplicated genes will likely be suppressed in the parasite's genome, which can then lose those functions ensured by the host. It is only when the parasite has a better version of a gene than that available in the host that the loss may occur in the host, which then becomes dependent on the parasite. However, it is usually the parasite's genes that are no longer necessary, as the result either of the economics of energy allocation or simply of the pressure exerted by mutation. It is this disappearance of genes that initiates what biologists in general have for some time wrongly called the "degradation" of parasites.

What happens to cave-dwelling fish that have been isolated from the light for a long time? They lose their eyes. Why? Simply because each time a deleterious mutation affected eye function, instead of being counterselected (as it would in daylight) it would persist in the population because it would not affect reproductive success. Thus bit by bit the eye would lose its function, and ultimately the entire structure would disappear. That is, the pressure of mutation in these fish would be greater than the pressure of selection.

Let's imagine that after some period a parasite that initially confronted several hosts now confronts only one. We can predict that as the result of mutation it will lose the genes that allowed it to avoid the defense systems of the no longer used hosts, since these genes are now as useless as the eyes of cave-dwelling fish. If we apply this reasoning to human schistosomes, for example, it means that *Schistosoma haematobium,* a parasite of humans in Africa since the beginning of humanity, is now more specific to humans as the result of humanity's long isolation from other possible hosts in this area. On the contrary, *S. japonicum,* an Asian parasite, would not have lost its genes, allowing it to avoid multiple immune systems because it has been in contact with humans for much less time (Asia having been populated by humans much more recently than Africa) and because it has not been isolated from other mammalian hosts (see chapter 20). However, I must add that at this time it is impossible to verify whether generalist parasites such as *S. japonicum* differ from more specialist parasites *(S. haematobium)* by having a larger repertoire of genes involved in immune evasion.

Selfish DNA

In 1980 Orgel and Crick published an article titled "Selfish DNA: The Ultimate Parasite," in which they proposed that it is not legitimate to apply an adaptive value to all sequences of DNA—that some such sequences have value only to themselves. However, applying the appellation "selfish" to noncoding sequences attributes to them the status of independent living beings. If a living being is considered selfish because it has no other goal than transmitting its own genes, then any object that has only the goal of transmitting its own gene(s) must also be considered a living being.

Would any sequence that fails to bring anything positive to a host genome be a parasite? (If nothing else, it would require some minimal amount of energy to be copied during each cell division.) Before giving an answer, I must express two initial reservations. First, the living world is not short of useless or deformed structures that take a long time to be eliminated. If genomes contain leftovers of past structures or even leftovers of failed attempts at new structures or pathways, it might mean either that it is more costly to eliminate the structure or pathway than to keep it or that such elimination occurs only very slowly. Second, that a specific sequence does not have a role in making a product does not mean it does not have another function: noncoding DNA may be maintained in genomes for mechanical or biophysical purposes, such as aiding in the curvature of chromosomes or the attachment of proteins, or because it increases the probability of crossing over.

Given these reservations and our present knowledge of the structure of genomes, it is clear that only mobile elements may be considered, for sure, to be entities independent of the host organism. That some of them may carry useful genes, which is the case for some bacterial transposons, does not change this conclusion. The principal arguments for considering that transposons and retrotransposons are indeed parasites follow.

Specificity

The specificity of mobile elements is expressed either at the level of the organism or at the level of only certain cells (for example, germline cells in the case of the P element) or for specific sites of insertion characterized by specific nucleotide sequences (specificity itself being, of course, a classic characteristic of parasites).

The Mode of Infection

To insert itself, a mobile element usually cuts the host's DNA, leaving exposed 5′ ends to which the mobile element attaches. The host then reconstitutes the missing parts, which in this case creates a short repetition of nucleotides on each side of the insertion (direct repeats). For example, each insertion of the P element of *Drosophila* induces a repetition of eight base pairs on each side of the insertion. In effect this process does not differ, except in degree, from that of the penetration and repair mechanisms occurring either when *Plasmodium* enters a red blood cell or when a schistosome cercaria penetrates the tegument of a host.

Distribution

Each type of mobile element is usually present in numerous copies in the same genome, but the number varies from one genome to another within the same species—similar to the distribution of virtually all parasites within their host populations.

Regulation

The way mobile elements regulate their abundance is comparable to the regulation of any parasite population. In the P elements, for example, an ensemble of processes controlling the activities of transposases prevents the invasion of the host genome from taking an explosive form (Rio 1991).

The Pathogenic Effect

The insertion of mobile elements has consequences for the host genome just as does a parasite that induces changes in its host's physiology. For instance, the gene in which the element inserts itself may stop being expressed, there may be a change in the expression in neighboring genes under the influence of the gene invaded (some transposons carry powerful promoters), or there may be genomic rearrangements (duplications, deletions, etc.). Certain transposons induce hybrid dysgenesis—the production of hybrids with a very high level of mutation. Some well-known mutations, such as those that affect the color of the eyes in *Drosophila,* are the result of such transpositions, and in some cases the increase in genetic diversity caused by mobile elements may be advantageous for the host population.

Transfers

Transposable elements can colonize new host populations or species, and these transfers can be reproduced experimentally. For instance, Ladeveze et al. (1998) described the experimental invasion of a strain of *Drosophila melanogaster* originally devoid of the element *hobo* by following its abundance after transfer for more than a hundred generations.

Overall, then, the terms, descriptors, and consequences just employed to describe mobile elements and their effects (specificity, infection, distribution, regulation, pathogenic effect, transfers, etc.) belong to the everyday vocabulary parasitologists use to describe parasites.

Phylogeny versus Ecology

It has always been difficult in biology to identify with certainty a cause-and-effect relationship between living beings. For example, in identifying what factors determine parasite richness, one is confronted with one variable (parasite richness) and numerous potential explanatory factors (host taxonomic position, habitat, diet, behavior, geographic area, etc.). The method of comparative analysis consists of gathering data from numerous host species and seeing whether the same factor (for example, aquatic habitat, large size, long life span, or omnivorousness) can account for parasite richness in each of the host species.

But such a comparative method confronts two problems. The first is the variable size of the samples used in the analysis. In effect, a comparative analysis is made using a database built from investigations made by different authors on different samples, which likely are of vastly different sizes. Since parasite richness is highly influenced by sample size, this variation may end up being very significant. Sampling effort, then, is a confounding factor that, unless accounted for appropriately, drastically influences the search for correlation and therefore the value of the comparative approach (Walther et al. 1995). Thus the more individuals of a particular species and the more locations examined, the closer one is to measuring *true parasite richness*. Species that are undersampled will have their parasite richness inherently underestimated.[2]

There are two ways analysts can get around the problem of "sampling effort." They can either ignore species whose sample size is below a certain threshold or correct for the sample size by calculating residuals from a linear regression between the known size of samples and their parasite richness. In this case the residuals represent the variation in parasite richness independent of sample size, and these residuals are used to pursue the comparative analysis.

A second difficulty with the comparative approach is that the potential explanatory variables are themselves linked. That is, they are not independent variables. In

2. However, Guégan and Kennedy (1996) have noted that "much of the contribution to parasite richness per host species made by sampling effort is . . . a reflection of host range," and these authors add that "host range and sampling effort correlate so closely that it is only one source variable that is functionally useful."

brief, the variable "parasite richness" is determined only in part by the current living conditions of the host species, and only if all parasite species of a particular host were to have come from lateral transfers under present living conditions would they be the sole determinants of richness. Obviously, at least some of the parasite species found in a host will have long records of coevolution with that host, and their present status in that host species is more influenced by the constraints resulting from this coevolved association than by the host's current living conditions. For instance, if two species that derive from a common ancestor have essentially the same parasite richness, it may be because they have inherited it from their common ancestor and not because they now live in a particular environment: "Closely related species are similar morphologically, behaviourally, developmentally, and ecologically" (Harvey, Read, and Nee 1995). In other words, the values of parasite richness that we have for parent species are not independent and cannot be treated as such. In fact, this is an illustration of the more general fact that current species cannot be considered independent of each other in an ecological study. Thus the effects of phylogeny must be accounted for, which is of course possible only if the phylogeny of the host species is itself known—something that is not always the case.[3]

Several methods have been proposed to carry out such a correction for phylogeny (see the fundamental work of Harvey and Pagel 1991 on this subject). The most commonly used correction is that of independent contrasts (Felsenstein 1985): the principle is to carry out the analysis not on current species but on different sister groups by estimating parasite richness values for the common ancestors of these groups.

Some authors, such as Poulin (1995a) and Morand (1997), embrace without reservation the need to correct for phylogenetic effects, while others, such as Westoby, Leishman, and Lord (1995) believe the dominant mode of evolution generates traits that are at the same time correlated with phylogeny and maintained by current selective pressures. For authors such as Westoby and colleagues a phyletically constrained quantitative trait maintained over a long period of evolution indicates that the trait considered has implausibly resisted selection over tens of thousands, if not millions, of years. According to this idea there is then no reason to give automatic priority to correcting for phylogenetic constraints, because current forces continue to favor the same traits established by past selective forces. Moreover, correcting for phylogenetic constraints also ends up treating phylogeny and ecology as two exclusive explanations, each occurring at the expense of the other.[4] Further, in all instances correction for phylogenetic effects induces a loss of power in statistical analysis. As Morand and Poulin (1998) noted, however, the value of a factor as either a "determinant" or "predictor" of richness should not be confused in such analyses. For example, Morand and Poulin showed, after

3. The effects of phylogeny are defined as similarities between species that are due to their common ancestry. This should not to be confounded with phylogenetic constraints, which are particularities of a lineage that oppose the appearance of a characteristic (see Morand 1997).

4. This debate led several people to take positions in the *Journal of Ecology* 83 (1995): 535–536, 727–734.

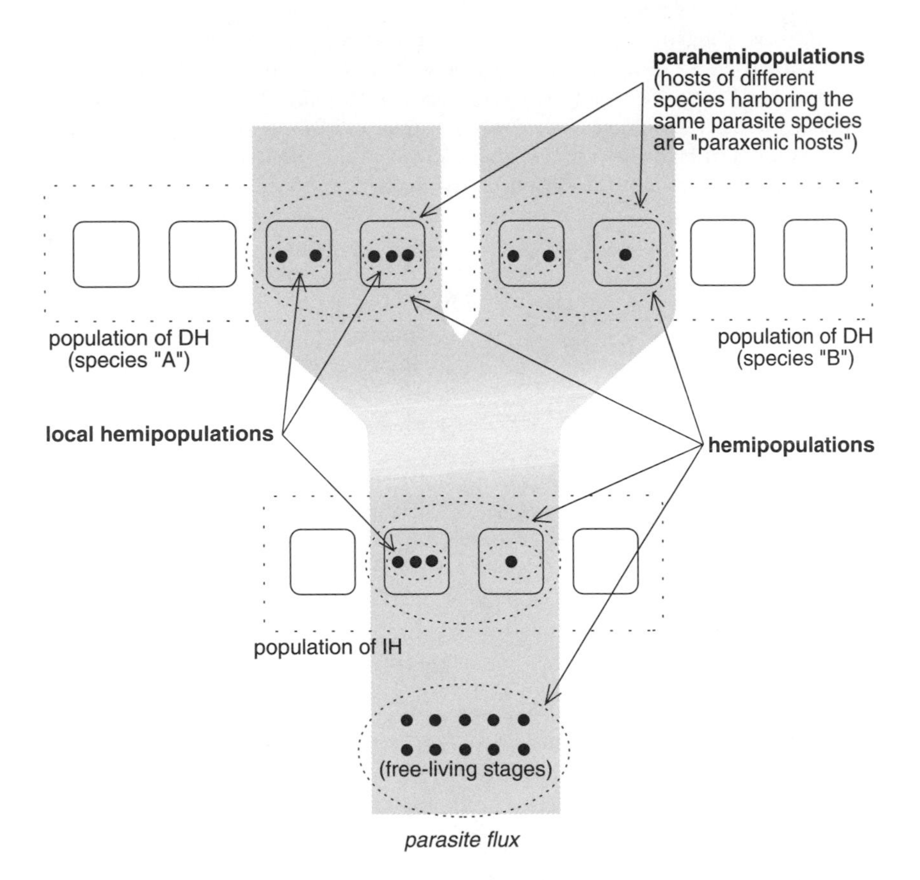

Figure R1.3. Some of the terms Russian authors use in ecological parasitology. The figure represents an imaginary parasite with a two-host life cycle and one free-living stage. IH = intermediate host; DH = definitive host.

correction for phylogeny, that mammalian body mass is not a determinant of parasite species richness, but that this does not mean size is not a *good predictor* of richness (the larger the body mass, the higher the probability of finding high richness).

Last, one cannot forget that any derived correlation is established by adding together investigations that artificially assemble genetically distinct host and parasite populations. As such, the derived databases might include studies that are not just of unequal amplitude but also of unequal quality (see Walther et al. 1995; Guégan and Kennedy 1996).

A Question of Semantics

Semantics is the science of terms and their significance. I have previously reviewed the terms used most often in Western countries to describe the different "groups" or "subgroups" of both the populations and communities of parasites and have proposed new terms to describe some of these groupings (the most important one being "xenopopulation"—used to designate the ensemble of individuals of a parasite species living in a population of a particular host species).

In parallel with the evolution of such terms in the West, an entirely different system arose in Russia, primarily under the influence of V. N. Beklemishev. These propositions were all published in Russian (and therefore reached few readers outside the old Soviet bloc) and as such have been virtually ignored in the West. Granovitch (1999) has recently summarized Beklemishev's nomenclature and has added his own suggestions and insights. Figure R1.3 summarizes a small part of this Russian vocabulary for parasite populations, and from this we can see that the Russian terms "hemipopulation" and "local hemipopulation" are, respectively, synonyms of "infrapopulation" and "xenopopulation." Granovitch calls "paraxenic" the different species of hosts that harbor a given parasite at a given stage and "parahemipopulations" the corresponding hemipopulations (for instance, the adults of *Fasciola hepatica* in rabbits, sheep, and cattle in the same area constitute three parahemipoplations).

Even if certain Russian terms contradict the vocabulary now in use in the West (for instance, "metapopulation" is used in a sense different from that used by population geneticists), the "Russian" system deserves to be considered for at least three reasons: it is coherent in that it applies to both free-living and parasitic members of populations and communities and as such has a general value; it is the result of a long-standing and excellent tradition of solid fieldwork and thought in ecology; and it has been used and is still used in a number of potentially important Russian papers that should perhaps not remain ignored by Western scientists.

Part 2

Genes in Durable Interactions

6 Parasite-Host Coevolution

It is selectively advantageous for a prey or a host to decrease its probability of being eaten or parasitized. It is often selectively advantageous for a predator or parasite species, and much more often for a predator or parasite individual, to increase its expected rate of capture of food. Every species does the best it can in the face of these pressures. . . . Biotic forces provide the basis for a self-driving perpetual motion of the environment and so of the evolution of the species affected by it.
Van Valen 1973

Evolution: It's the Others

Selection occurs because individuals with characteristics that are the best adapted to their environments have a greater probability of reproductive success (fitness) than others, thus increasing the frequency of their genes in a population. This is, in essence, the mechanism of natural selection. However, it is difficult to explain the evolution of life solely by adaptation to the physical environment alone. Once the population is adapted to its physical milieu no further selection should occur, because all further changes in DNA would then be unfavorable or, at best, only neutral and therefore useless.

That evolution has kept going for at least 3.5 billion years and has led to the appearance of more and more complex organisms has a plausible explanation if we associate the concept of the adaptive landscape of Sewall Wright (1932) to the work of Leigh Van Valen (1973).

The "Adaptive Landscape" of Sewall Wright

An adaptive landscape may be envisioned as a bumpy surface on which peaks ("adaptive peaks") correspond to gene combinations that confer on an organism the greatest reproductive success. Figure 6.1 represents such an imaginary landscape, with the points along the x and y axes representing various gene combinations and those along the z axis representing fitness. Note that the diagram gives the impression that there are only two genetic dimensions, x and y, whereas in reality there are a large number. In this model a population along the axis is represented by a

The opening subheading is a parody of the line "L'enfer, c'est les autres" (literally, "Hell is other people") in *Huis clos (No Exit)*, by Jean-Paul Sartre.

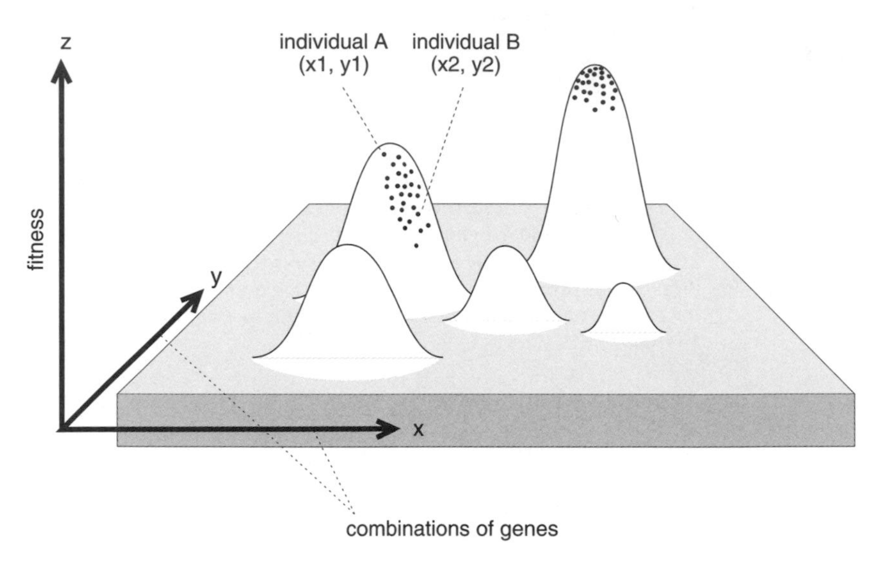

Figure 6.1. Adaptive landscape (after Wright 1932).

cloud of points (each point being an individual), some closer than others to a peak (they are more fit). For example, in the figure the combination of genes x1, y1 is "better" than the combination x2, y2 because the individual represented by this particular combination is "higher" up the adaptive peak. In a variable population half the height of a peak (left in the figure), those individuals closest to the peak have the highest reproductive success and are thus better able to transmit their genes to future generations. Over time the population progressively "slides" toward the tip of the peak as the "evolutionary lag" (sensu Stenseth and Maynard Smith 1984) between the momentarily adapted and the best possible adaptation decreases; in the end, the population occupies the peak itself (as on the right in the figure). Wright also supposes that because of genetic drift individuals can explore the valleys between peaks and in that way go from one peak to another and thus end up occupying the highest peaks in the landscape.

The Red Queen Hypothesis of Leigh Van Valen

In the fundamental work where Leigh Van Valen (1973) proposed the Red Queen hypothesis, the most important component of a species' environment is thought to be represented by the other species it interacts with in its ecosystem. Van Valen deduced that any change in one of the interacting species modifies the environment of all the other species it interacts with. In short, the closest adaptive peak is displaced, forcing the species to face new selective pressures. In this model resources are in limited quantity, and whatever is taken by one species is not available for another. That is, the species play a game "with sums that equal zero."

Van Valen started out by observing that the extinction rates of taxa (species, genera, and families) as obtained through paleontologic data are fairly consistent in a given group (for example, in the brachiopods, ammonites, teleosteans, rodents, etc.) and independent of the species' life spans. In the perpetual interaction proposed by Van Valen, the "losers" are regularly eliminated, producing a consistent rate of extinction.

Fundamentally, Van Valen proposes that even if the physical-chemical environment does not change, living species should evolve in order to respond to the continued changes of the other species forming the ecosystem they live in. Each species is under the pressure of numerous other species. The "Red Queen" title of this hypothesis comes from Lewis Carroll's story *Through the Looking-Glass*. At one point Alice runs alongside the Red Queen and notices with surprise that the surrounding landscape does not change. The Queen then explains that they must run just to stay in the same place: "Now, *here*, you see, it takes all the running *you* can do to keep in the same place." In short, each species must change just to stay in the race and maintain the same adaptive quality. Obviously a given species may exert unequal selective pressures on the surrounding species, and it is when two particular species exert selective pressures on each other over a long period that biologists speak of coevolution.

The Orchid and the Butterfly

I can perhaps best illustrate the concept of coevolution with one example taken from outside the world of parasites, which has the particular advantages of being extremely demonstrative and of having been understood by Darwin himself. This is the association formed by some Madagascar orchids and their butterfly pollinators (see Nilsson 1988). These orchids have nectaries—straight cylindrical tubes that secrete a sweet liquid at their bottoms that butterflies drink. In order for the orchid to reproduce (to transmit its genes) small masses containing the pollen, called pollinia, must adhere to the butterfly's head, be transported to another plant, and contact the female part of the flower. As such, taking up pollen is an "all or nothing" event: either the butterfly takes away some pollinia or it takes none. However, in order to take up pollinia the butterfly's head must strongly strike a column "strategically" placed above the nectary. If access to the nectar is too easy, the butterfly feeds and leaves without pollinia.

Such a mechanism was selected in plants with long nectaries, which oblige the insect to forcefully enter the flower and push on the column to reach the "treat" at the bottom. If the plant has short nectaries its genes are not transmitted, since the butterfly does not take up the pollinia and the trait "short nectaries" is counterselected. In parallel, the lengthening of the nectaries has selected for individual butterflies with longer proboscises, which allow access to longer nectaries. Butterflies with short proboscises cannot reach the nectar and therefore are not fed and cannot reproduce—the trait "proboscis too short" is selected against.

A never-ending series of selective events may occur, with each partner's evolution being determined by the selective pressures exerted on it by the other. Experimental evidence indicates that if the nectaries are shortened using a constriction, the insect absorbs the nectar with no problem, and neither physical contact with the male parts of the flower nor pollination occurs. It has been demonstrated that this system functions well only if the ratio of the length of the nectariferous tube to the length of the proboscis is more than one, ensuring that the butterfly must press hard into the nectary to obtain food.

What, then, is the result of this double evolution? Orchids with extremely long nectaries and butterflies with extremely long proboscises. For example, the orchid *Angraecum sesquipedale* has nectaries twenty-eight to thirty-two centimeters long, and its pollinator, the butterfly *Xanthopan morgani*, has a proboscis twenty-five centimeters long!

Such a series of reciprocal selective pressures, perfectly conforming to the hypothesis of the Red Queen, is called coevolution (see Thompson 1994 for a detailed work on mechanisms of coevolution as well as numerous other examples). To be accurate, I must note that Wasserthal (1997) proposed a different explanation for the example above: butterflies with long proboscises could have been selected for by the pressure of butterfly predators waiting on the flowers (in this case, a long proboscis would allow them to drink the nectar without getting too close). Secondarily, selection would then elongate nectaries so that the function "pollen carriers" continued to be ensured. The plant therefore would not have exerted any selective pressure on the butterfly. However, Nilsson (1998) retains a preference for Darwin's original explanation.[1]

---

1. Darwin knew only the orchid *Angraecum sesquispedale,* but he correctly predicted the existence of the butterfly. As noted by Nilsson (1998) "[Darwin] predicted the existence of a giant pollinator hawk moth in Madagascar that was, tongue-wise, capable of handling the orchid's extraordinary nectar position. . . . The pollinator candidate hawk moth . . . was later recovered and described in 1903."

Significantly, selective pressures on their own are insufficient to induce coevolution; there must be genetic variation on which pressure can be exerted. In the case of the orchids and butterflies, nothing would happen if variation in the length of the nectary and proboscis were not continually renewed through recombination and mutation. Further, note that coevolution involving free-living species does not necessarily result in the evolution of the entire genome of the organisms involved, as the example of the orchids and their butterflies shows well. Nothing indicates that the leaves or the roots of the plants, or for that matter the wings or the legs of the butterfly, are important in the process of nectary and proboscis lengthening.

When pressures that are exerted on one species come from a series of ecologically closely related species, it is difficult to distinguish the pressures exerted by each, and we then speak of a coevolution guild. Last, one must not fall victim to the idea that a given species coevolves everywhere and always with the same partner species. There is increasing evidence that such species interactions change with time and place, sometimes rapidly, and Thompson (1999) speaks of a geographic mosaic of evolving interactions to describe the situation in nature.

Selective Pressures in Parasite-Host Systems

The Notion of Virulence and the Cost of Virulence

In parasite-host associations coevolution is transformed into an adaptive cascade that, in general, concerns not just a particular organ of each partner but numerous characteristics of the phenotype.

Parasites and hosts have neither the same possibilities nor the same constraints. The unique characteristics of parasite-host associations allow one to define three fundamental concepts important in understanding the costs to each partner that are not found in free-living associations: virulence, the cost of virulence, and optimum virulence.

To understand these concepts we must first understand that without the parasite the host has maximum fitness (determined by prevailing abiotic and biotic conditions), while the parasite without the host has a fitness equal to zero (fig. 6.2, top). The consequence of this asymmetry is that comparing parasite fitness with and without the host makes no sense, while comparing host fitness with and without the parasite allows one to define *virulence*—one of the most important parameters in the study of parasite-host systems. The fitness of a parasitized host is determined by various factors, such as the availability of spatial and energetic resources as well as predation and both intra- and interspecific competition. The

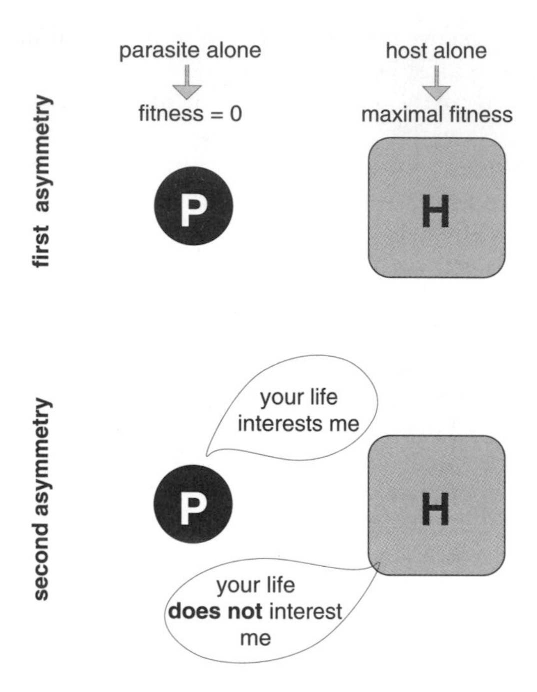

Figure 6.2. The two asymmetries between host and parasite.

fitness of an infected host is also limited by competition with the parasite, which may hijack part of the available resources or otherwise damage the host. The resulting *parasite-induced loss of fitness* is virulence (Ebert and Hamilton 1996).

In this sense, for the host the parasite is only a competitor, while for the parasite the host is both a competitor and a resource. For the parasite, the host is the object of the competition itself. Therefore the life of the parasite is not of interest to the host, whereas that of the host is of interest to the parasite even when the host ends up dying (fig. 6.2, bottom). This second asymmetry leads to the concept of the *cost of virulence.*

Parasite fitness is determined by the success of its transmission to other hosts and by nothing else. Poulin (1998a) summarizes the situation perfectly when he says: "The host is simply an ephemeral resource that the parasite uses to maximize its fitness. . . . From the parasite's perspective . . . what happens to the host as a consequence of its exploitation by the parasite may be of no importance." However, this does not mean the parasite should not "care" at all about the life of its host. The host is not only the base of resources for the parasite, but also its habitat and its vehicle. It is therefore to the parasite's advantage for the host to live as long as it is

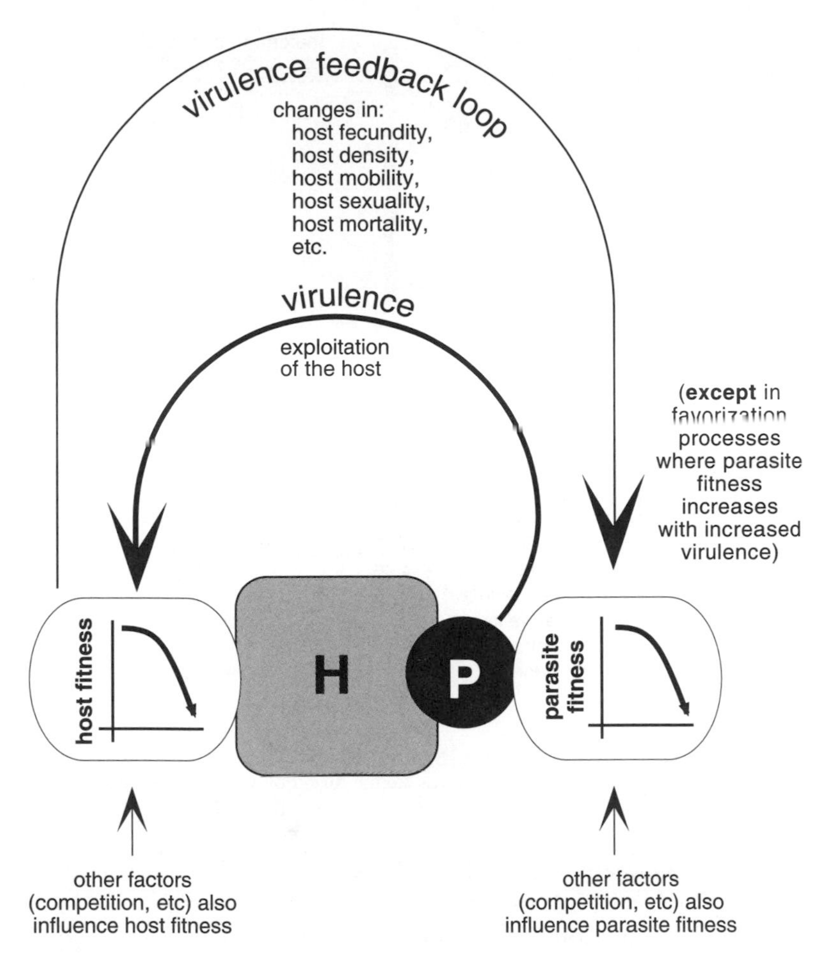

Figure 6.3. Virulence negative feedback loop: by modifying certain components of host fitness, virulence limits itself (note that other factors may also influence both host and parasite fitness).

needed to ensure dissemination. Once the parasite no longer "needs" the host as a resource, its fate is of no interest.

From this conflict arises the cost of virulence: since the parasite needs its host (or its host's descendants) to live, an exaggerated virulence may impair the parasite's fitness as well. The decrease in host fitness induces a concomitant decrease in parasite fitness (fig. 6.3). As a result there is pressure for the self-limitation of virulence. Virulence self-limitation then leads to the idea of an *optimal virulence* (fig. 6.4), which is the best possible

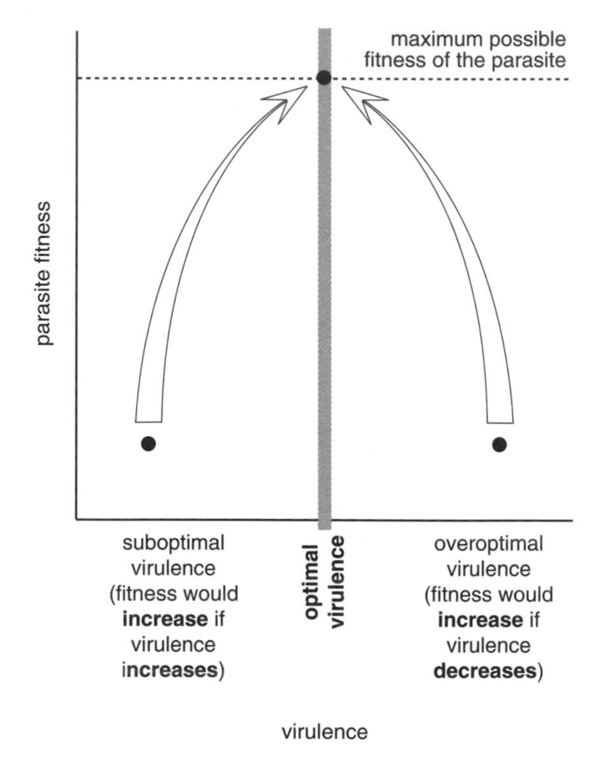

Figure 6.4. Optimal virulence. Arrows indicate the direction virulence will evolve in if at the time the association occurs virulence is either above or below the optimal level.

compromise between the benefits and costs associated with virulence.[2] As Ewald (1995) notes: "It is this tradeoff that is at the heart of current theory about the evolution of virulence." Optimal virulence is an equilibrium where "a decrease or increase in virulence is associated with a loss of fitness" (Ebert and Herre 1996). Poulin (1998a) explains this concept while emphasizing the aspect of host death: "The optimal level of virulence . . . is the level below which improvements of the transmission rate are still possible without being annulled by the death of the host, and above which any increase in transmission rate is offset by high host mortality."

The notion of an optimal virulence implies that two traditionally held views are both incorrect: that the parasite should exploit the host as much as possible, and thus be as virulent as possible in order to ensure

2. Of course, for the host the only optimum virulence is zero—the absence of the parasite.

abundant resources and a high rate of multiplication; and that the parasite should become less virulent over evolutionary time because the host and its descendants are necessary to maintain the parasite population.

Ebert and Herre (1996) illustrate the notion of optimal virulence with a striking example: the mite *Dicrocheles phalaenodectes* is a parasite of the ear of certain noctuid moths but is found on only one ear. The explanation is that if no ear is parasitized the parasite's fitness is zero, and if both ears are parasitized the moth cannot detect predatory bats and the parasite's fitness is also zero since the hosts are more readily eaten. The only "good" virulence is therefore infection of just one ear.

Several important points must be made concerning virulence.

1. Virulence is different from pathogenicity. The latter concerns particular aspects of the host individual, such as its morphology, anatomy, metabolism, behavior, and reproductive activity, whereas virulence applies only to the consequences of the parasite on the transmission of the host's genes. Although the pathogenic effect is important on the human scale of suffering, it is virulence that is important in terms of gene frequencies and evolution. The decrease in the host's reproductive success may be significant (up to the point of castration) or almost undetectable. In either case it is the result of the hijacking of resources to benefit the parasite as well as of the toxic effects and damage caused, for example, by the waste products of the parasite in the host organism.

2. Virulence is a parameter used to characterize the parasite, but it is measured in terms of host fitness. Poulin (1998a) writes: "The level of harm done to the host is often the main criterion used to categorize symbioses in which the host does not benefit." The origin of this counterintuitive notion may be sought in the traditional image of the parasite as an organism perceived only through its exploitation of the host.[3]

3. The term "virulence" is far from satisfactory: while it has the advantage of avoiding the lengthy expression "parasite-mediated loss of fitness in the host" it has the serious disadvantage of evoking only images of viruses when, in fact, it is applicable to any parasite (virus, bacteria, protistans, and metazoans). Also, it is used with different meanings by agronomists, for whom it is the capability to infect a plant ("this strain of fungus is virulent to this wheat variety") and by physicians, for whom it is the pathogenic effect observed in patients ("the flu virus is very virulent this year").

4. Last, virulence is difficult to measure (see Poulin 1998a; Poulin and Combes 1999), so it is sometimes defined in more accessible ways, such as "parasite-mediated morbidity and mortality" (Levin 1996) or "parasite-

3. This concept might be termed "exploitativeness."

induced host mortality" (Ebert and Mangin 1997). The difficulty of direct measurement therefore makes virulence a tempting object for indirect measurement. One of these indirect approaches, popular in the 1990s and discussed earlier, consists of measuring the degree of fluctuating asymmetry in hosts, a factor considered to be an important indication of host quality and thus fitness (see Møller 1996; Agnew and Koella 1997).[4] Optimal virulence may also be deduced from mathematical models, which most often analyze the relation between the level of transmission and parasite-induced mortality (see Poulin 1998a for a brief overview of this approach).

In this chapter "virulence" will always designate the parasite-induced loss of fitness of the host.

The Notion of Resistance and the Cost of Resistance

Strictly speaking, resistance against a parasite consists either of closing the encounter filter by selecting new behaviors or closing the compatibility filter by selecting new defense mechanisms. However, the term "resistance" is in practice often restricted to the defense mechanisms intervening *after infection,* that is, to the closing of the compatibility filter. The focus in this chapter is on this restricted definition of resistance.

Host resistance is nothing other than virulence against the parasite, since it induces in the parasite a "host-mediated loss of fitness" and even host-induced parasite mortality. The more efficient its resistance against a parasite, the greater the fitness of the host. Further, the cost of resistance for the host depends on the metabolic expense involved (fig. 6.5).

While it is true that such expenses may be huge, any change selected to fight against the parasite implies that resources allocated to resistance would be taken from other important functions: that is, the host organism must make a trade-off to further defend itself (Connors and Nickol 1991). Such active mechanisms of resistance would consume part of the energy normally allocated to movement, reproduction, and so on, and are compensated for by decreasing the energy invested in these processes.[5] For example, there may be competition for substances such as amino

4. "Fluctuating asymmetry, which represents small random deviations from otherwise bilateral symmetry, is a measure of the phenotypic quality of individuals" (Møller 1995).

5. It must be added that the cost of resistance against a parasite is not fundamentally different from either the cost of resistance of insects against a toxic product, such as a pesticide, or that of an herbivore against its predator. Insects select resistance mechanisms against substances used to destroy them, and the reproductive cost of these mechanisms may be enormous. For instance, within a few dozen years mosquitoes have developed throughout the world that are resistant to insecticides because of selection for the massive

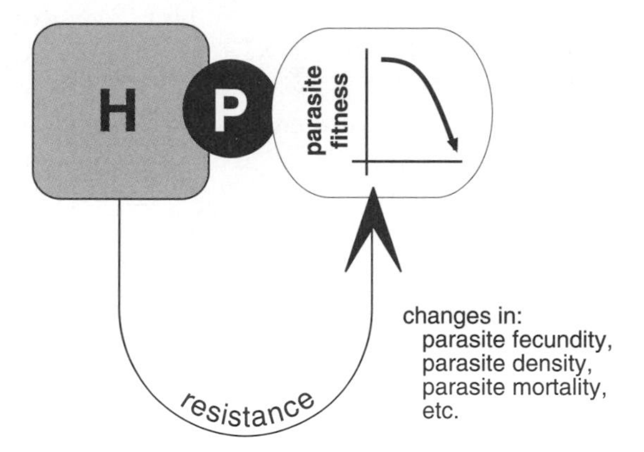

Figure 6.5. Host resistance, which decreases parasite fitness, is similar to host "virulence," although there is no feedback loop as in parasite virulence.

acids between mechanisms of resistance and those linked to reproduction in the host (Ferdig et al. 1993).

Just as there is an optimal virulence for the parasite, there is an optimal level of resistance for the host (fig. 6.6). Optimal resistance is the best possible trade-off between the advantages of resistance and its cost.[6] Various observations or experiments delineate the cost of resistance to infection in parasite-host systems. One such example includes populations of the bacteria *Escherichia coli* that are resistant to the bacteriophage T2. In this case resistant populations have a lower growth rate than do susceptible populations (Levin and Lenski 1985; see Keymer and Read 1991). The most likely explanation is that to enter the cell the phages must first recognize surface molecules whose normal function is capturing nutrients. Those bacteria that became resistant did so by selecting particular molecular variants that are less efficient at capturing nutrients and are not recognized by the phages. A cost of resistance has also been demonstrated in the relation between pathogenic fungi and host plants, as seen,

---

production of a group of enzymes, called esterases, that may represent as much as 10% of their total protein. Significantly, however, such resistant mosquitoes are quickly counterselected in the absence of insecticide. In terms of predators, resistance may induce the expenditure of energy that is translated into such defense mechanisms as increased carapace thickness (the armadillo), the flourishing of spikes (the hedgehog), venomous teguments (the toad), and so on. Such protective mechanisms make sense only if enemies actually exist; otherwise they would truly be wasted investments.

6. The costs of resistance cannot be ignored in making certain vaccines, since vaccination would make no sense if the cost induced by the vaccine would be higher than the cost of parasitism (see Behnke, Barnard, and Wakelin 1992; Gemmill and Read 1998).

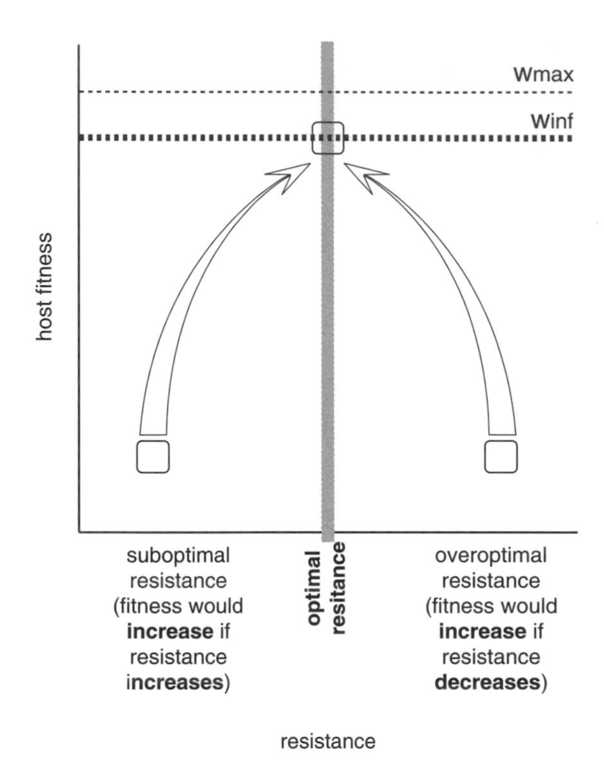

Figure 6.6. Optimal resistance. W max: best possible fitness of a host when parasite is absent; W inf: best possible fitness of an infected host. Note: W max – W inf = virulence.

for example, between different strains of the fungus *Puccinia* and various wheat varieties. In this case fungal strains that have "useless" virulence genes (ones that are not adapted to the particular host variety encountered in a particular place) have reduced reproductive success (see van der Plank 1968) relative to those that have more useful adaptations. Finally, Kraaijeveld and Godfray (1997) have selected strains of *Drosophila melanogaster* that are able to encapsulate the eggs of the parasitoid wasp *Asobara tabida*. These authors have shown that the resistant fruit flies were, however, less competitive when feeding conditions were not good.

Resistance is at the same time both frequency dependent and pathology dependent: in order for resistance to be selected, the parasite prevalence must be high and the pathogenic effect strong enough to alter host reproductive success (Minchella 1985). In terms of the adaptive landscape, the parasite would reduce the height of the peak but would not move it (there would be no better peak for the host than the one it is on at present, despite the cost of the parasite). Therefore, little to no response from the host to parasite-induced selective pressures is possible when the cost of eliminating the parasite is higher than the cost of managing it: "Although host defence can prevent the buildup of large parasite loads, energetic constraints will probably prohibit the evolution of perfect defence against parasites" (Clayton 1991a).

Along these lines Møller and Erritzoe (1998) have shown that there is a relation between parasite pressure and what is invested in immune reactions. Starting with the idea that migratory birds are exposed to at least two parasite faunas during their annual movements, they showed, through a comparative analysis involving the bursa of Fabricius and the spleen, that these primary immunodefense organs were better developed in migratory birds than in sedentary ones.

On the scale of the individual, Poulin, Brodeur, and Moore (1994) considered that the cost of resistance might vary during the host's life span. In short, resisting infection appears to be a better deal for younger individuals than for older ones, since the young ones have a longer reproductive life in front of them if they win. That is, even if the cost is high, the benefit in terms of total reproductive success for younger individuals may be higher than the cost expended in defense over the long term.

The importance of balancing costs and benefits in the fight against pathogens, whatever they may be, might best be understood by the following example: What town would ruin itself to get rid of every last mouse? Mice are obviously a nuisance, but we end up forgetting about them if they are not obviously damaging. An all-out antimouse fight would make sense only if mouse damage increased to the point where such a war was worth it; otherwise a few traps in the attic (a few antibodies here and there) are sufficient. As such, the notion of a "bearable cost" of parasitism is very different for the ecologist and the physician. To the ecologist the expression does not have the same sense of urgency as the presentation of a parasitized and suffering human being to a medical doctor.

The study of virulence has shown us that the cost of virulence involves a feedback loop: beyond a certain degree of lost fitness in the host because of the parasite's influence, the parasite itself may pay the consequences. Is there something similar for resistance? That is, does the host pay a price to decrease the parasite's fitness?

The logical a priori response is no: the more the resistance increases, the more the parasite is weakened or eliminated, and thus the more the host's fitness increases. Figure 6.5 is built based on this "a priori." However, the mafia strategies of some parasites (see chapter 10) are at least one demonstration that this a priori idea is not totally correct.

The Arms Races of Parasite-Host Systems

The previous example of orchids and their pollinating butterflies concerns an association in which each partner finds an advantage, with coevolution resulting from the *selfishness* of each partner: the butterfly's interest is feeding itself, not the transport of the orchid's pollen, and the orchid's interest is to reproduce, not to feed the butterfly. (After all, why spend energy feeding someone else?) The evolution of the orchid thus is not directed "toward" the butterfly, and the evolution of the butterfly is not directed "toward" the orchid—in both cases it is for themselves.

In other associations, however, the partners have totally divergent needs. For instance, the only interest of a parasite is to exploit the host, and the interest of the host is to avoid being overexploited. In this case coevolution takes the form of an "arms race," which is nothing other than an escalation of adaptive inventions or weaponry on each side. In fact, two distinct arms races are involved in a parasite-host association, with the battlefront of the first being the encounter filter and that of the second the compatibility filter (fig. 6.7).

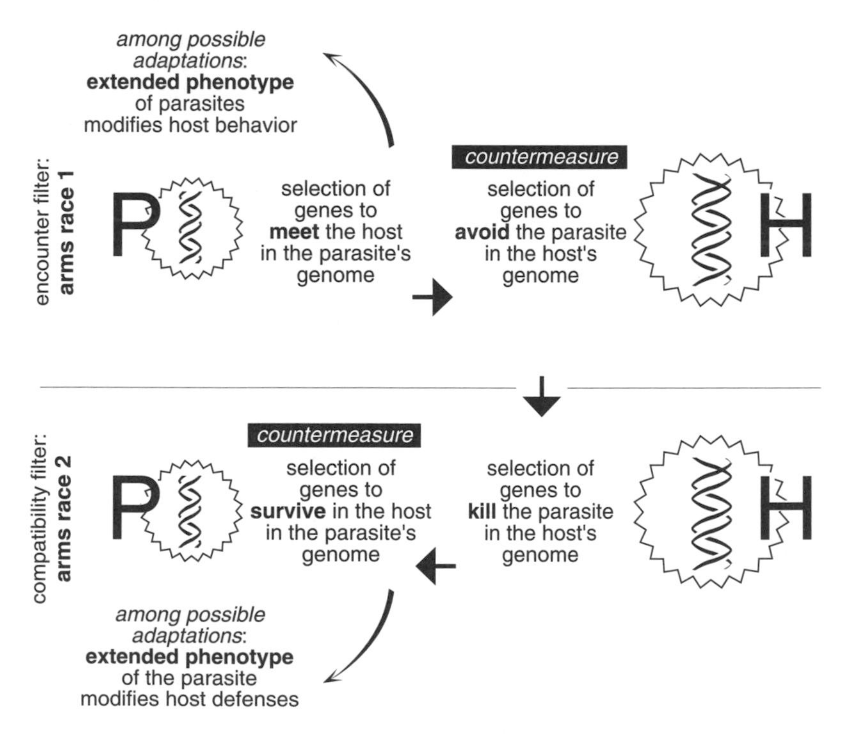

Figure 6.7. The two arms races in parasite-host systems. Note the advantage of the extended phenotype among the adaptations selected in parasites.

At the level of the encounter filter, the "encounter" arms race has the following results.

In parasites, there is selection for genes (or combinations of genes) that give the best possible chance of encountering compatible hosts. These are the "*encounter genes,*" and they may determine, for example, behaviors that orient infective stages (larvae or adults) toward the environment where potential hosts may be found and also structures associated with behaviors that attract potential hosts.

In hosts, there is selection for genes (or combinations of genes) that determine behaviors to avoid the parasites. These are "*avoidance genes,*" and they may be of various natures. For example, they may determine behaviors that deflect potential hosts from encountering infective stages when these stages may be directly detected; and behaviors or structures that makes access to the potential host more difficult.

Poulin, Brodeur, and Moore (1994) emphasized two important points concerning the parasite-host relationship at the level of encounter: all parasites need to interact with hosts, whereas not all hosts encounter parasites; and hosts are involved in a large number of arms races and parasites in only a few (see also fig. 9.2 in this book).

At the level of the compatibility filter, the results of the "compatibility" arms race are as follows.

In the hosts there is selection for genes (or combinations of genes) that provide the parasite with the most unfavorable or hostile environment possible as the result of defensive mechanisms. These are the "*killer genes,*" and they may differ depending on whether the organism is ectoparasitic or not: ectoparasites may be eliminated by structures or specific destructive behaviors, with or without the intervention of auxiliary organisms; and all parasites, except in very rare exceptions, are attacked by internal defense systems that may themselves differ depending on the type of host.

In parasites there is selection for genes (or combinations of genes) that provide the best potential to adapt to the hostile environment, particularly to escape defense reactions. These are the "*survival genes.*" These genes are various in nature and may determine, for example, a tendency to remain invisible to the hosts' recognition molecules or a renewing of surface molecules that forces the host to continuously modify its defensive activity, or the production of molecules to counter or nullify host killing mechanisms.

Each filter is, then, a "*crossed phenotype*" coded for by genes that belong to the genomes of the two organisms whose interests are divergent. Poulin, Brodeur, and Moore (1994) note that the selection of innate mechanisms of resistance in hosts is expensive and thus of little advantage if encounter with the parasite is rare, whereas the cost of acquired

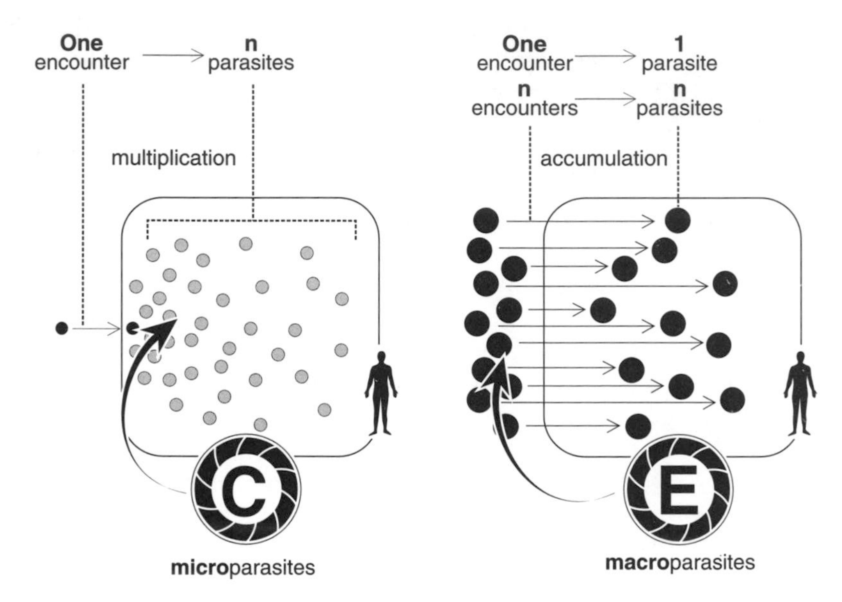

Figure 6.8. Predictions: the compatibility arms race (C) "dominates" in parasitic diseases involving multiplication in the host, whereas the number of encounters does not count. The encounter arms race (E) "dominates" in parasitic diseases where postinvasive multiplication does not exist and parasites accumulate in the host according to the number of encounters.

resistance has the advantage of being paid by the hosts only if they are infected.

Figure 6.7 emphasizes the frequent role of each organism's extended phenotype in the two arms races. In the encounter arms race the extended phenotype of the parasites' encounter genes may, for example, modify the behavior of the upstream host in a way that favors transmission to the downstream host, while in the compatibility arms race the extended phenotype of the parasites' survival genes may modulate the efficacy of the hosts' internal defense system.

Last, depending on the type of parasitosis, both races do not have the same importance, and it therefore seems reasonable to make two predictions (fig. 6.8).

In diseases caused by the multiplication of microparasites, one encounter may be enough to induce infection. Here the seriousness of the disease comes from the proliferation of the parasite inside the host, and the prediction would be that in such systems selection happens principally on compatibility.

In diseases caused by the accumulation of macroparasites, the seriousness of the disease generally comes only from the number of encounters with the infective stages, since there is usually no further multiplication within the host. The prediction in this case is that selection will principally occur on encounter.

Note, however, that such predictions cannot be absolute, since the two arms races are not independent: if changes in behavior are more expensive than internal defense reactions, a host may respond to an increased encounter frequency by increasing its immune capabilities.

7 Genes to Encounter

What does the starling do in the face of prey that are both available and parasitized?
Moore 1983a

The snail might as well be flashing neon signs on its head that read, "Eat Here."
Rennie 1992

Produce More or Produce Better?

For the vast majority of parasites, if not all, the transmission of offspring into the next generation is a time of tremendous loss. When a parasite population is in balance its gains and losses are comparable. However, the real extent of such gains and losses is rarely known. As Rea and Irwin (1994) note, "Although parasites have enormous fecundity in practice, precise estimates are very difficult to obtain and frequently relate to only one life-history stage."

Let's follow, for example, the losses and gains of an imaginary cohort of schistosomes (fig. 7.1). According to various measures, both direct and indirect, it is thought that a miracidium that has penetrated a mollusk may give, over a period of several weeks, from 100,000 to 200,000 cercariae, and that, depending on the species, an adult couple can lay between a hundred and a thousand eggs a day—an amount corresponding to from several hundred thousand to several million eggs over the duration of its life (three to four years). Using an estimate of a million eggs, each containing an embryo that ultimately gives rise to 200,000 cercariae, just one adult couple could theoretically have 200 billion descendants! However, if the schistosome population remained stable over time the couple would actually have not 200 billion descendants but instead just two.

Such numbers, however theoretical, help us understand why in general there has been selection for highly prolific reproductive systems in parasites. It is important to remember, though, that the costs of reproduction (linked, for example, to reduced host survival or weakening), and therefore the trade-offs between fecundity and other functions, remain to be studied in parasites.

Still other numbers provide us with an idea of parasite fecundity. For example, in *Plasmodium* (the causative agent of malaria) there are several asexual reproductive events, each referred to as schizogony. As the result

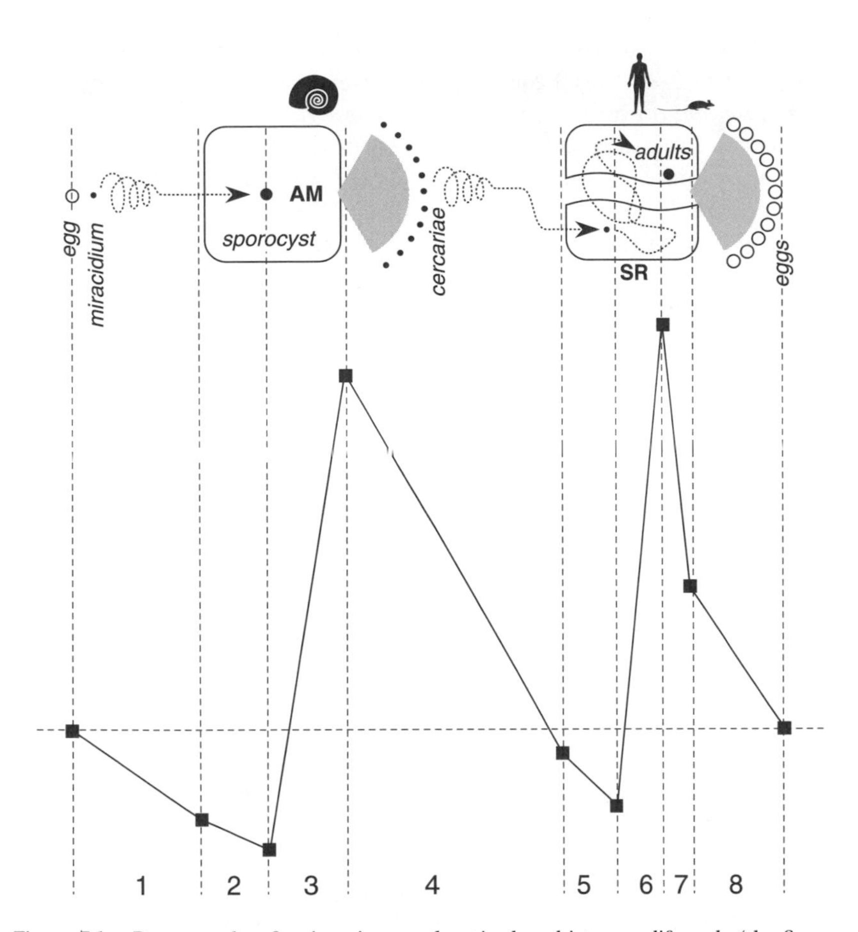

Figure 7.1. Demography of an imaginary cohort in the schistosome life cycle (the figure represents the supposed fluctuations of a cohort, not of a population; the curve is arbitrary and does not correspond to any concrete mathematical model or data). (1) Loss: some miracidia do not encounter a mollusk host. (2) Loss: some miracidia are destroyed by the defense system of the mollusks. (3) Gain: asexual reproduction gives rise to several generations of sporocysts and then cercariae. (4) Loss: some cercariae do not encounter a vertebrate host. (5) Loss: some schistosomules and adult worms are destroyed by the vertebrates' immune systems. (6) Gain: sexual reproduction gives rise to eggs. (7) Loss: part of the eggs are retained in the vertebrate body. (8) Loss: eggs are deposited out of reach of water. AM = asexual multiplication; SR = sexual reproduction.

of these multiple fissions, up to 10,000 cells will be produced in the mosquito, another 30,000 in human hepatic cells, and from eight to thirty-two every two to three days—depending on the species—by schizogony in each red blood cell. Similarly, a human infected with the dysenteric amoeba *Entamoeba histolytica* can evacuate up to 50 million infective cysts a day, while one infected with *Ascaris lumbricoides* can pass up to 64 million eggs a year (1,700 times the ascarid's own weight). The human tapeworm *(Taenia saginata)* produces about 10 billion eggs in its average eighteen years of life. One can only imagine the number of eggs produced by pseudophyllidean cestodes of the genus *Polygonoporus,* parasites of cetaceans and real monsters measuring almost forty meters each (making them the longest known animals on the planet). How about cestodes of the genus *Tetragonoporus,* which have 45,000 proglottids, each containing male and female reproductive systems? Given this, it is easy to understand the reflection of Schmidt and Roberts (1977): "There are few whales and the ocean is large."

It therefore does not seem a stretch to conclude that, even if some free-living animals may produce large quantities of eggs or other propagules when their lifestyle imposes it (as in the case of many pelagic or fixed animals), parasites are probably unbeatable when it comes to fecundity. Since numerous parasite functions such as mobility, prey capture, and sometimes even digestion are ensured by hosts, this is not surprising, and a considerable part of the parasites' energy can therefore be assigned to reproduction. This idea is echoed by Jennings and Calow (1975): "In terms of fitness, it is more advantageous to form gametes than to convert excess food material into long-term reserves." This is reflected in reality by how large genitalia are in most metazoan parasite bodies and is well illustrated by the extraordinary drawing found in Baer (1952) showing the insect nematode *Sphaerularia bombi* as nothing more than an enormous uterus packed with eggs and a tiny attached lateral appendix harboring all its other organs! However, it is not certain that such massive fecundity is positively correlated with the complexity of transmission. Morand (1996a, 1996b) demonstrates, for instance, that there is no correlation between having a monoxenous or heteroxenous life cycle and the total number of offspring a female nematode produces.

MacArthur and Wilson (1967), on the one hand, and Pianka (1970) on the other developed the idea that free-living organisms may be *r* or *K* selected depending on environmental conditions, which raises the question whether parasites can similarly be considered *r* or *K* strategists.[1]

The concept of *r* and *K* selection is not often applied to parasitism (Rea and Irwin 1994; Esch, Hazen, and Aho 1977) simply because parasites'

---

1. When the environment of a species is unsaturated (weak competition with no density-dependent constraint on population growth), selection favors a large production of potential descendants and limited survival, which is *r* selection. When the environment is saturated (heavy competition with density-dependent constraints), selection favors producing fewer potential descendants better armed for competition and having a longer survival time than the adults, which is *K* selection.

heavy fecundity would normally classify them as *r* strategists, a position that is not deniable since parasite transmission entails such heavy mortality. However, parasites are not "*r* strategists" in the classical sense for at least two reasons: some adult parasites may survive a very long time, a characteristic more often found in "*K* strategists"; and parasite infective stages, even if small and numerous, don't often leave their survival to chance alone, itself another characteristic of "*K* strategists."[2] This alternative to increasing fecundity in parasites has been termed *favorization,* a concept that implies selecting life history traits that favor increased survival and transmission of infective stages to hosts (Combes 1991d).

To fully understand this concept one must first recognize that the process of selection in parasitic life cycles occurs at two levels. The first level pertains to the sequence and spectrum of hosts. At this level, selection occurs according to both the physical environment of the parasite and the diversity of potential hosts. It is not surprising, for example, that benthic fish get their cestodes and trematodes by eating prey that live on the bottom of the sea whereas pelagic fish get theirs by eating prey that live in the water column. Here selection would tend to not keep a sequence of hosts that successively included a fish and a lizard if nothing linked the fish and the lizard (they had no contact, did not belong to the same food web, etc.).

The second level of selection occurs in the form of "fine-tuning," which retains characteristics that optimize the success of the transfer from host to host(it keeps those traits that increase the probability that individual infective stages will survive and encounter appropriate hosts). In other words, selection at this level would tend to keep genes or suites of genes that code for successful strategies or behaviors in the larvae that increase survival and transmission. That is, selection at this level would be for processes of favorization.

In short, an increased probability of transmission—the further opening of the encounter filter—may thus be obtained either by increasing the production of infective stages ("producing more offspring") or by increasing the probability that the infective stages will survive as the result of favorization ("producing better offspring").

Favorization

Favorization consists of an ensemble of adaptive characteristics selected for in parasite populations by the successful encountering of hosts (see Combes 1980). The novelty of the concept is that it allows us to recognize

---

2. Infective stages of parasites are most often larval forms, active or not. But in parasitoid insects adults, equipped with elaborated sensorial equipment, are the stages that search for the hosts.

that phenotypes can be selected and sorted: either among the *phenotypes of parasites themselves* or among the *extended phenotypes of the parasites in their hosts.*

In the first case characteristics of the parasites are selected for (shapes, colors, behaviors, etc.), whereas in the second case characteristics of the hosts that can be "manipulated" by the parasites are selected for (shapes, colors, behaviors, etc.). Such manipulation occurs because of the nature of the durable interaction between the parasite and its host, which allows the parasite to extend its phenotype into the host in the form of either physical or behavioral attributes.

The recognition that parasite phenotypes favoring transmission are extended into the domain of the host yields an extraordinary expansion of the range of theoretically possible responses that natural selection can act on—something still not fully understood or appreciated by many parasitologists, let alone evolutionary biologists and ecologists.

In terms of behavior, the increased probability of encountering a host results from responses to stimuli that include those coming from the downstream host or its environment (fig. 7.2), where the response occurs either in the free-living infective stage or in the form of a manipulated upstream host, and those coming from the free-living infective stage or manipulated upstream host (fig. 7.3), where the response is in the downstream host.[3]

The behavioral approach will serve us as a guide, and it is in the context of the manipulation of hosts that the concept of favorization through the extended phenotype takes on all its meaning.[4] In the gene pool of parasites there is selection for alleles that alter the behavior of the upstream host to favor transmission. The extended phenotype of the parasite in the upstream host is nothing other than a pathogenic effect, and favorization is nothing other than a pathogenic effect that is selected for because of its value in completing the cycle. That effect is not in itself the result of natural selection but is selected for as soon as it increases the reproductive success of the parasite. If in a parasite population there is genetic diversity relative to such "pathogenicity," and if certain of these pathogenic effects increase the probability of encounter with the host, the genes that induce these effects are then selected for.

---

3. Remember that during a transmission event the upstream host is where the parasite comes from and the downstream host is where the parasite goes to.

4. To give more homogeneity to the examples illustrating favorization, most of those discussed here will involve trematodes, which have been particularly well studied and demonstrate all types of selective processes well. In parallel with the models given by trematodes, other significant examples will be provided from other parasite groups.

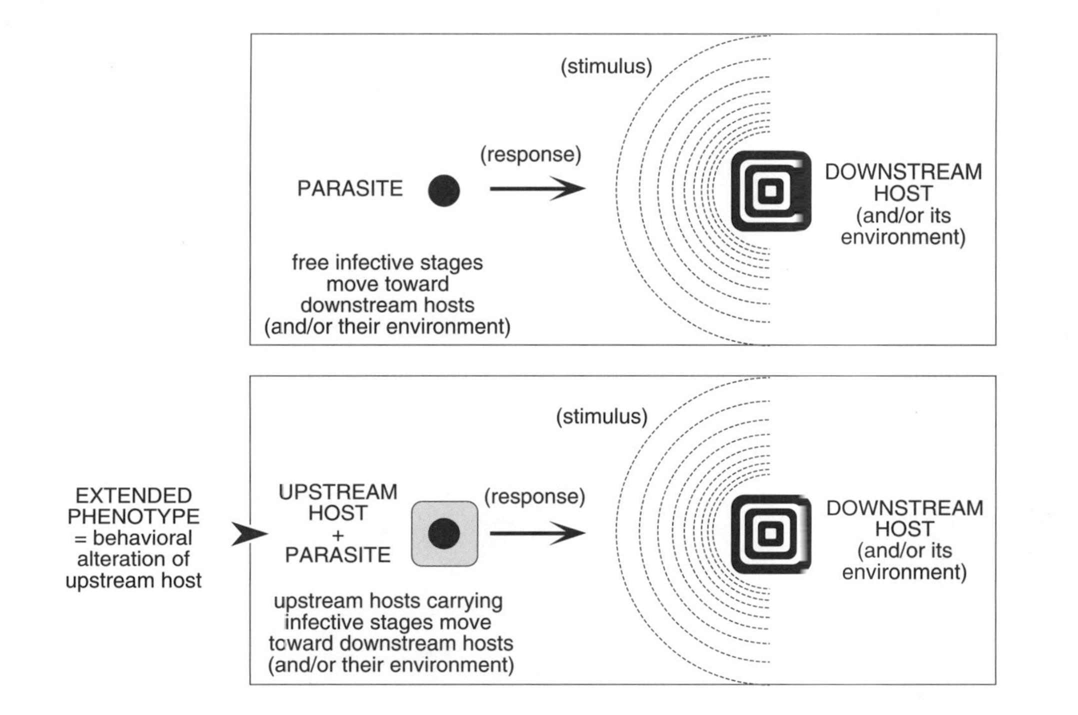

Figure 7.2. Favorization. Stimuli coming from the downstream host or its environment. Top: response of the free-living stage of the parasite. Bottom: response of the manipulated upstream host.

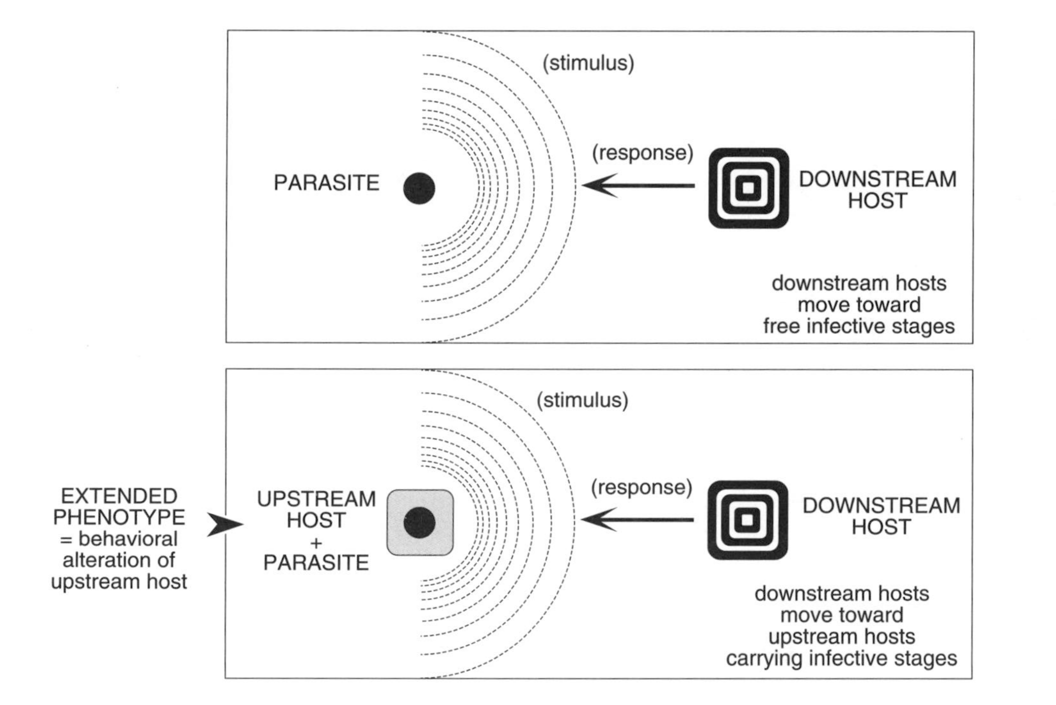

Figure 7.3. Favorization. Stimuli coming from the free-living infective stage of the parasite or from the manipulated upstream host. Top: response of the downstream host to stimuli coming from the parasite. Bottom: response of the downstream host to stimuli coming from the upstream host.

The pathogenic effect in question may result from the parasite's physical presence being more or less annoying according to the organ concerned (a parasite on the skin and one in the nervous system will not have the same effects on behavior). Additionally, it may also result from the secretion/excretion of bioactive molecules; or to the hijacking of part of the metabolites belonging to (or destined for) the host (see Holmes and Zohar 1990).

A question that then arises is, Does a trade-off exist between fecundity and favorization? We may suppose that there is a negative correlation between investment in fecundity and investment in favorization, although this still needs to be demonstrated by research on the reciprocal costs of reproduction and favorization.

Stimulus: Downstream Host or Its Environment. Response: Free-Living Infective Stage

A parasite's seeking a downstream host because it has a free-living and active infective stage is similar to one person's needing to meet another. To do so, he or she must go to the *right place* (the place of the rendezvous) at the *right time* (the time of the rendezvous).

The Right Place

The stimuli a parasite's free-living stage must respond to by moving to a more or less specific place may have three origins (fig. 7.4): (1) the habitat frequented by the downstream host—for example, the swimming larva of a trematode may direct itself toward the water surface or an adult parasitoid may be attracted by the plant used by its caterpillar target; (2) the habitat modified by the presence of the downstream host—for example, the swimming larva of a trematode may orient itself using a gradient of host secretions or an adult parasitoid may be attracted by the odor of the caterpillar-damaged plant; or (3) the downstream host itself when there is contact with it—for example, trematode cercariae may recognize the host tegument by its lipid composition or temperature, and a parasitoid may recognize a caterpillar because of its cuticular hydrocarbons.

Depending on the life cycle, these different stimuli may act in isolation, in synergy, or in succession, and distinguishing the different sources of stimuli is often very difficult.

Parasite Responses to Stimuli Coming from the Downstream Host

Studies by Chernin (1970), Wright (1971) and Roberts, Ward, and Chernin (1979) on schistosomes (Trematoda) all reference parasite aquatic

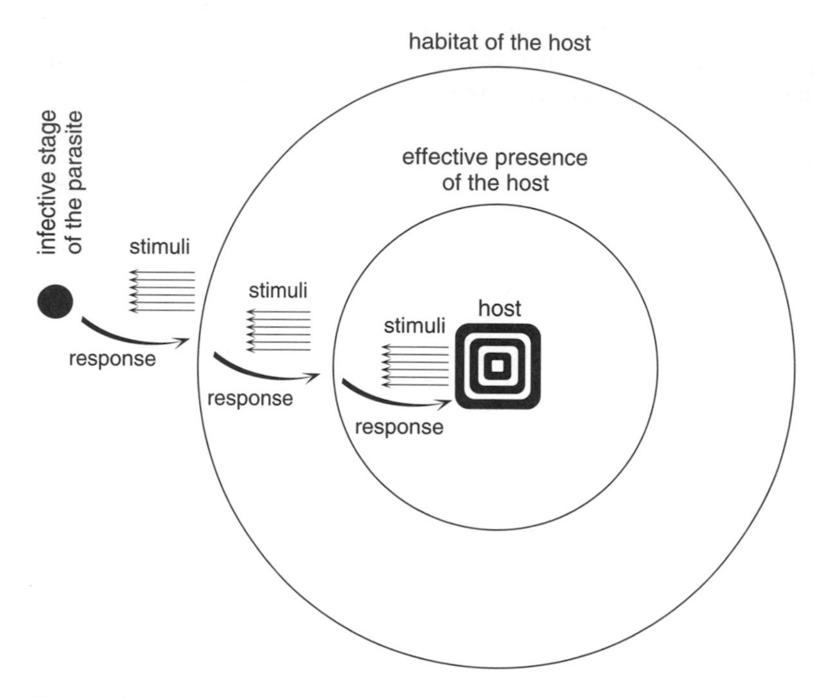

Figure 7.4. The three main origins of stimuli that may be perceived by parasite infective stages.

stages and their downstream molluskan hosts. Their results show that miracidia modify their swimming behavior when they approach a zone where there is a concentration gradient of substances emitted by the mollusk. As the result of this gradient the pathway followed becomes more sinuous, and in some cases larvae swim faster. These changes increase the chance of contact with a host (see Wright 1971). Sometimes, however, the effect is more direct: Ulmer (1971) showed that miracidia of *Megalodiscus temperatus* (a trematode parasite of amphibians that develops in the pulmonate mollusk *Helisoma trivolvis*), are directly attracted to the mollusk if the mollusk is set in the experimental tank in advance.

The searching ability of the free-living stages varies over their life span even though it may be very short. For example, Fournier et al. (1993) have shown that trematode larvae at first disperse in the environment rather than seek immediate encounters with their hosts. The study of parasite attraction to invertebrates is an active area of research today, and much remains to be discovered about the substances emitted by hosts and the receptors on parasites that allow them to identify these substances. Moreover, although trematode miracidia can detect signals coming directly from hosts, this does not seem to be true for all free-living aquatic stages. For example, the cercariae of the same trematodes discussed above appear to invest much more effort in perceiving environmental characteristics than in detecting the hosts themselves (Combes et al. 1994).

Some of the most remarkable examples of host localization by parasites are found among the arthropods, whose advanced sensorial receptors let them use very sensitive search strategies. This is the case of the copepodids (copepod infective stages) that target fish. Fraile, Escouffier, and Raibaut (1993) showed, using a rigorous experimental and mathematical analysis of the swimming patterns of the copepodids of *Caligus minimus,* that these stages are very sensitive to the mucus of their host, the sea bass *Dicentrarchus labrax.*

In the terrestrial environment tick larvae are also able to detect the warm-blooded hosts they must attach to, using very elaborate sensorial receptors. In general ticks wait for their hosts on plants, and on the tarsi of their first pair of legs they have a complex sense organ (the organ of Haller) that has receptors sensitive to contact, temperature, humidity, and a variety of chemical substances.[5]

Yet another amazing adaptation is that of the phoretic mite *Histiostoma laboratorium,* which may jump as high as five centimeters (Binns 1982) to attach to flying fruit flies (the equivalent of five hundred meters for a human, according to Françoise Athias-Binche, personal communication). The mites appear to detect fruit flies because of the noise of their wings and respond rapidly because of their extraordinary reflexes.

Parasite Response to Stimuli Coming from the Environment of the Downstream Host

For some parasites stimuli coming from the environment (light, temperature, gravity, pressure, oxygenation, etc.) appear to be the only factors involved in orienting the movements of the free-living stages. Such movements are independent of the downstream host itself and happen even when the host is absent. For example, many trematodes live in lagoons, and their cercariae are "emitted" at the bottom of the lagoons from benthic mollusks such as prosobranchs or bivalves (fig. 7.5). Bartoli and

5. Ticks are particularly important to discuss, since they are both parasites and vectors of parasites such as viruses, bacteria, protozoans, and filarial worms (Aeschlimann 1991). One of the best-known parasites vectored by ticks is the spirochete *Borrelia burgdorferei,* which causes Lyme disease and affects numerous mammals besides humans. The ticks' ability to detect their hosts is selected in their genome and therefore is favorable to the transmission of the parasites they vector.

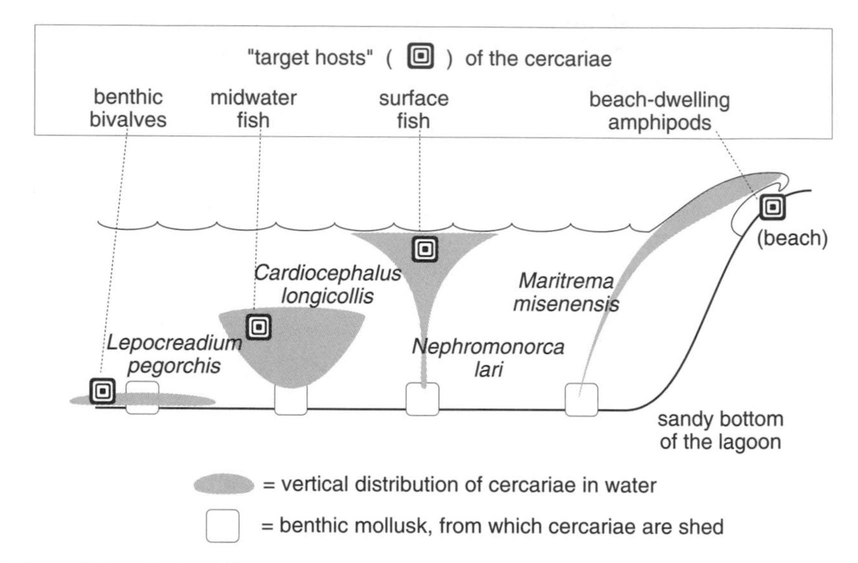

Figure 7.5. Modes of dispersion of several species of trematodes in a Mediterranean littoral lagoon. All cycles have three hosts: the definitive host is a fish in the case of *Lepocreadium pegorchis* and a bird for the three other species (modified from Combes et al. 1994).

Combes (1986) have shown that the behavior of the cercariae after they emerge depends on the type of host they must penetrate. For instance, cercariae of *Lepocreadium pegorchis* are weak swimmers and have a positive geotaxis, which favors their contact with the sediment-dwelling bivalves they must penetrate and encyst within (cercariae are siphoned in by the bivalve's respiratory current). Cercariae of *Cardiocephalus longicollis,* on the other hand, actively swim and have a negative geotaxis so that they swim directly toward the surface, stop briefly, then slowly descend while scanning the middle stratum of the lagoon for the sparid and belonid fish in which they must encyst. Last, cercariae of *Nephromonorca lari* actively swim up to the water/air interface and, as in the previous example, take brief rests as they move. However, in this case the scanning is done closer to the surface, which favors encounter with surface fish, such as atherines, that are their next hosts.

The most spectacular behavior is that of the cercariae of *Maritrema misenensis,* which has been described by Bartoli and Prévot (1978). These cercariae must infect small amphipods (principally *Orchestia montagui*) that stay above water level among dry leaves of neptune grass. To reach their well-hidden hosts the aquatic cercariae use two suckers to attach to the inner side of the water/air interface (in the process forming a bubble) and wait to be projected onto the amphipods by small waves and spray.

Movement patterns comparable to those observed in cercariae occur in the terrestrial environment in certain nematodes. For instance, larvae of *Trichostrongylus tenuis,* a nematode parasite of the Scottish red grouse, must be ingested by the bird to become adults in the digestive cecum. The larvae, after a free-living period in the soil, crawl onto the base of heather and settle on the tips of budding branches, where they can wait as long as several months before being ingested by birds that eat the buds.

Some parasites obtain an advantage by being farther away from an environment rather than closer to it. For example, cestode segments (proglottids) packed with eggs are usually evacuated with their host's feces. Staying on site is all right if the intermediate host that ingests the eggs is coprophagous, but not if it has finer tastes. In these species proglottids leave the feces on their own and move a short distance away. A still more original mode of transport is used by the nematode *Dictyocaulus viviparus.* Adults parasitize the lungs (bronchi) of domestic cattle and have been the cause of severe losses on farms and ranches (Robinson, Poynter, and Terry 1962). The eggs of the parasite are evacuated with the feces and rapidly give rise to free-living larvae. The herbivorous host becomes infected as it eats the larvae while grazing. To reach the surrounding grass, the nematode larvae actively climb onto the sporangiophores of a mushroom of the genus *Pilobolus* (a phycomycete) that grows in the cow pie. *Pilobolus* have explosive sporangia (owing to osmotic pressure) that ensure spore dispersion, and the nematode larvae group themselves (up to fifty at a time) on the outer surface of the sporangia, taking advantage of the explosion to disperse up to three meters away.

Most such spatial strategies of encounter are characteristic of free-living stages, but there are also parasitic stages that can move toward their target. This is the case with certain nematode microfilariae that must be taken up by biting arthropods and that "know" where the vector preferentially bites. Microfilariae of *Onchocerca tarsicola,* a parasite of cervids, are born in the leg articulations (where the adults are) but migrate to the ear skin, where vector dipterans take them and transmit them to other cervids (Schulz-Key and Wenk 1981). In general, all such microfilariae that require ingestion by biting arthropod vectors locate in host sites compatible with this mode of transmission.

Response to Stimuli Coming from both the Environment and the Downstream Host

The most elaborate processes of searching for hosts are found in the insect parasitoids, which lay eggs or larvae on or in other arthropod eggs or larvae. Although the parasitoid-host relationship must be considered a classical parasite-host relationship during the time the host survives, adult female parasitoids detect signals (chemical, visual, sounds, vibrations) absolutely analogous to those predators use to detect their prey. Convincing demonstrations of this have shown a genetic-based diversity of behaviors in natural populations on which natural selection can act (see Wajnberg, Rosi, and Colazza 1999).

The use of chemical signals is the most frequent, and parasitoids have apparently selected these signals for their efficiency in locating targets without the need to detect the precise source of the signal, which may come from the target insect itself, from its environment, or from both. The interaction is said to be "tritrophic" because it involves three sources: the plant host, the insect host, and the parasitoid. Signals coming from the plant that are used to find hosts are called synomones.

In general, all such signals are specific (see Vinson 1976; Cortesero, Monge, and Huignard 1993), and the parasitoid is sensitive only to those coming from both the part of the plant (for example, seeds) where the target phytophagous insect is and the stage of the host (in general, the larval stage) within which the eggs will be laid.

Here are examples that illustrate the variety of the stimuli used by adult parasitoids in target location.

1. *Acrolepiopsis assectella* is a lepidopteran (butterfly) whose caterpillars are phytophagous, specializing on leek leaves. *Diadromus pulchellus* is a parasitoid hymenopteran that lays eggs in the butterfly's chrysalis. The female parasitoid is strongly stimulated by odors released from leek leaves when they are attacked by caterpillars and by the insects' feces. In both cases it recognizes volatile sulfur compounds (disulfides) and uses them to successively localize first the host habitat and then the host itself. Bacteria in the digestive tract of the host insect produce the volatile sulfur compounds from precursors found in the host plant, and the search for the host is completed when the parasitoid identifies host-produced nonvolatile substances. It is interesting that male *D. pulchellus* are also sensitive to stimuli coming from host plants, which allows them to find females (Lecomte and Thibout 1986; Thibout 1988; Thibout, Guillot, and Auger 1993).

2. *Eupelmus vuilleti* is a parasitoid of the Bruchidae (Coleoptera) that develops in legume seeds. This species illustrates what may be called a learning process—using individual experience to modify behavior. Such training may be done during the adult life of this parasitoid, but it may also occur earlier, either immediately after emergence from the host or even before emergence. These parasitoids are generalist that can exploit

several host species, which themselves can develop on several plant hosts. However, they learn at a particular stage of their life cycle to search for a particular host-plant complex. The most amazing aspect of this system is preemergence training. This occurs in those *E. vuilleti* that are sensitive both to the odors of the insects and to the odors of the seeds in which the insects spend their larval stage: after emergence, sensitized adult females search for the same insect-seed complex to lay their eggs (Cortesero and Monge 1994; Cortesero, Monge, and Huignard 1995; Monge and Cortesero 1996).

3. *Mamestra brassicae* is a nocturnal lepidopteran whose caterpillars devour cabbage leaves as well as a variety of other cultivated plants. Females emit signals in the form of volatile substances (pheromones) whose role is to attract males. These pheromones are adsorbed by the inner face of the leaves where females lay their eggs. *Trichogramma evanescens* is a tiny hymenopteran (less than half a millimeter long) that lays its eggs inside the eggs of *M. brassicae*. Noldus (1989) has shown that pheromones of female *M. brassicae* are recognized not only by the males of the species but also by females of the parasitoid, whose activity is diurnal. Parasitoid females then identify the leaves within which female *M. brassicae* oviposited during the previous night, go to the inner side of one of these leaves, and oviposit in the lepidopteran's eggs. The pheromones (an intraspecific chemical signal) are thus exploited as a kairomone (an interspecific chemical signal used to the advantage of the receiver). Interestingly, sexual pheromones are not the only signals trichograms recognize; they are also sensitive to substances present on scales that have fallen from their hosts and, as in the previous example, to the synomones of the plant hosts. Further, these tiny wasps are even able to identify a substance coming from their own eggs; thus they can identify host eggs that are already parasitized and avoid laying two eggs in the same host egg.

4. *Oomyzus gallerucae* is a parasitoid that lays its eggs in the elm leaf beetle, *Xanthogaleruca luteola*. Parasitoids of eggs "face the problem that chemicals from eggs are the least detectable, but most reliable host-finding cues, whereas synomomes from the plant are the most detectable, but least reliable ones" (Meiners and Hiker 1997). Natural selection has retained in *O. gallerucae* a curious process that solves the difficult problem of detecting its host's eggs. Experiments using olfactometry (Meiners and Hiker 1997) showed that *O. gallerucae* was not attracted by either intact elm leaves or *X. luteola* eggs, nor was it attracted by a combination of odors from both the leaves and eggs. However, elm leaves damaged by adult leaf beetles on which eggs had been deposited

were attractive. The authors concluded that depositing eggs on the leaves released volatiles from the plant, which then attracted the wasp. This attraction is complemented by further parasitoid sensitivity to signals coming from the host insect's feces. Once in the eggs' neighborhood, the wasps randomly move about and identify potential host eggs thanks to contact kairomones. *O. gallerucae* has been introduced into the United States several times in an attempt to control elm leaf beetles.

5. Parasitoids of root-feeding larvae also must detect extremely concealed hosts. *Trybliographa rapae* is a solitary larval endoparasitoid of *Delia radicum,* a fly that spends its larval development stage inside crucifer roots. Bioassays using a linear and a four-arm olfactometer (Anne-Marie Cortesero, personal communication) have shown that host-derived cues play a minor role in host location—frass produced by feeding larvae only increased the locomotor activity of the females. However, plant-derived cues played a major role in locating hosts. The odors from *Brassica* plants that were infested by root-feeding larvae were attractive for the female parasitoid. The production of parasitoid-attracting volatiles not only was restricted to infested roots but also occurred systemically throughout the whole plant and particularly the leaves. Females were not attracted to uninfested or artificially damaged plants. However, applying crushed larvae to artificially damaged plant tissue increased the response of *T. rapae.* This is therefore the first evidence that herbivore-induced synomones exist in plants damaged by concealed root-feeding larvae.

6. *Ichneumon eumerus* lays its eggs in the caterpillar of the lepidopteran *Maculinea rebeli* and can detect hosts even in apparently inaccessible habitats. Caterpillars of *M. rebeli* live in the nests of the ant *Myrmica schencki.* In this case *I. eumerus* is able to deposit eggs on *M. rebeli* caterpillars at the bottom of the ants' nests. Thomas and Elmes (1993) have shown that the parasitoids recognize the nests of the appropriate species by odor and that they likely identify the nests where host caterpillars are by sound (caterpillars emit a characteristic stridulation). Last, when the wasps penetrate the ants' nest they secrete a substance that triggers a fight among the ants, allowing them to quietly lay their eggs on the caterpillar.

Here we must be reminded that in parasitoid insects adults search for the host when ontogenesis is complete. In other parasitic metazoans, such as helminths or crustaceans, the search is done by larvae at the very beginning of ontogenesis using sensorial equipment that cannot be easily compared with that of adult arthropods.

The Right Time

A successful encounter strategy calls for the reciprocal location of parasite and host not only in space but also in time, and parasites select responses that are synchronized to a particular moment that favors host encounter. The patterns used as a selective frame of reference are circadian rhythms, that is, they are linked to the regular alternation of day and night.

The Cercarial Clock

The emergence of schistosome cercariae has been studied in detail by Théron and his collaborators (Théron 1984, 1985a, 1985b; Chassé and Théron 1988; Théron and Combes 1988; Théron 1989). *Schistosoma mansoni,* the causative agent of human intestinal schistosomiasis, has larvae that develop in small freshwater pulmonate mollusks of the genus *Biomphalaria.* A month or so after the snail is infected with miracidia of *S. mansoni,* clouds of swimming larvae, the cercariae, emerge. Each cercaria has two parts: a body, which will ultimately become the adult parasite, and a bifurcate tail that aids in propulsion. The total length of these cercariae is about one-third of a millimeter.

The cercariae have no way to feed in the outside environment and have limited energy resources. Although they may live up to twenty-four hours, their activity strongly decreases a few hours after emergence, and during this short time they must find their definitive host—normally humans, but also black rats, *Rattus rattus.* When cercariae contact the host they lose their tails, and their bodies penetrate the skin within a few minutes. Each cercaria's body ultimately gives rise to an adult if it is not destroyed by host defense reactions.

The chronobiology of the cercariae's emergence from the mollusk can be easily studied using an automated chronocercariometer. This simple system sets side by side twenty-four containers filled with water. Each hour an infected mollusk is moved from one container to the next. The water in each container is then filtered, and the cercariae on the filter are counted, yielding an hour-by-hour "profile" of cercarial emergence over twenty four hours.

Studies by Théron and his colleagues have shown that there is a precise rhythm of emergence that can be reproduced in different mollusks infected with the same species of schistosome; that this rhythm is characterized by the start of emergence in the morning, a peak in the middle of the day, and a stop in the evening; that this rhythm is synchronized by the alternation of day and night and that it is possible to shift or reverse it using an artificial photoperiod; that the rhythm is not linked to molluskan activity; that the rhythm is independent of the effective presence of humans or rodents; that the rhythm is genetically determined by the parasites' genes; and that parasite populations sampled from different transmission sites show emergence rhythms with different maximal peaks between 11:00 A.M. and 5:00 P.M.

The two last aspects of this favorization process are particularly interesting. First of all, the demonstration of genetic determinism of emergence rhythms indicates that it is possible to discuss the process in terms of both selection and adaptation. Second, the demonstration of diversity in the rhythm of emergence by different isolates supports an adaptive value for this process relative to the success of transmission. Théron (1984, 1985a, 1985b) has shown that these rhythms of emergence are linked to the roles played by the definitive hosts—rats and humans (fig. 7.6). Depending on the sites of transmission of *S. mansoni* in Guadeloupe, the human and black rat play roles of different importance as definitive hosts. The profiles of cercarial emergence are correlated with these sites and may be characterized as "early" for some and "late" for others. In more urbanized zones, humans are the most infected and the selected patterns of cercarial emergence in schistosome populations are of the early type. In forested zones away from houses, rats play the main role and the selected rhythms are of the late type. In the palustrian mangrove zones where humans and rats are both infected, the rhythms observed are of the intermediate type.

This means that, within a species, there is genetic diversity in cercarial chronobiology and that different alleles are selected for in different parasite populations because of different environmental and transmission conditions. The result of this selection is the optimal opening of the encounter filter for each population. Further, this means that the different *S. mansoni* populations are relatively isolated from one another, or at least that selection for different chronobiological phenotypes occurs "faster" than any tendency to homogenize them through gene flow.

This is, using the expression of Dobson and Merenlender (1991), an example of a "fragmented arms race." By this these authors mean that if the adversary is not the same at different transmission foci, then the parasites will develop different weapons.

It is very likely that each time there is a "definitive host polymorphism" relative to transmission sites or foci, particularly depending on who the dominant hosts are (humans or other vertebrates), such a "chronobiologic polymorphism" in cercarial emergence will be observed. This is in fact true for the other human schistosomes as well. For example, *Schistosoma mansoni, S. haematobium,* and *S. intercalatum* also show

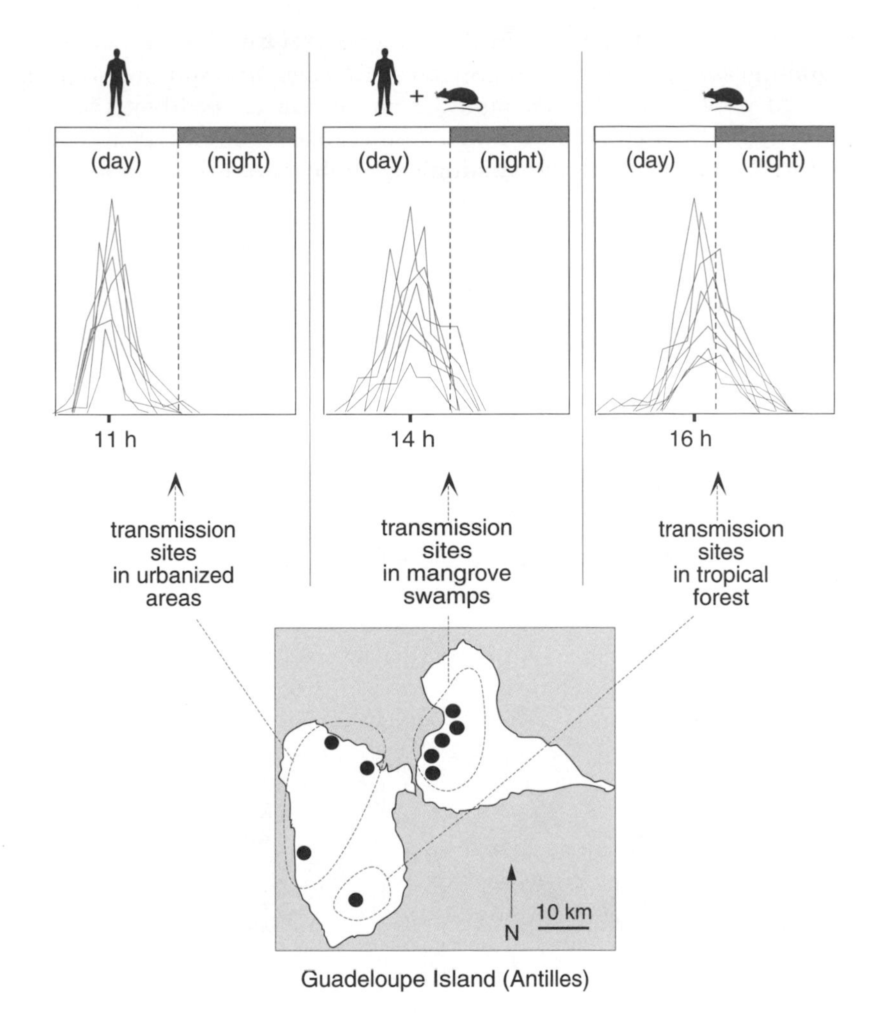

Figure 7.6. Polymorphism in the chronobiology of cercarial shedding for *Schistosoma mansoni* in Guadeloupe (adapted from A. Théron's original diagram).

characteristic "profiles of emergence" (Pagès and Théron 1990a) with each having diurnal emergence (fig 7.7A, B, C). Animal schistosomes, however, clearly differ in their profiles: on the one hand *S. bovis* and *S. curassoni,* both ungulate parasites, emerge very early in the morning (Mouahid et al. 1991) (fig. 7.7D, E), whereas *S. margrebowiei,* also a parasite of ungulates, has two emergence peaks, one at dawn, the other at dusk (Raymond and Probert 1992) (fig. 7.7F). *S. rodhaini* (a specific parasite of rodents) emerges late in the evening (André Théron, personal communication, fig. 7.7G). One species of *Plagiorchis* is shown in figure 7.7H to provide another example of late cercarial shedding (André Théron, personal communication). The regularity of emergence profiles has been useful in epidemiological studies where specific schisto-

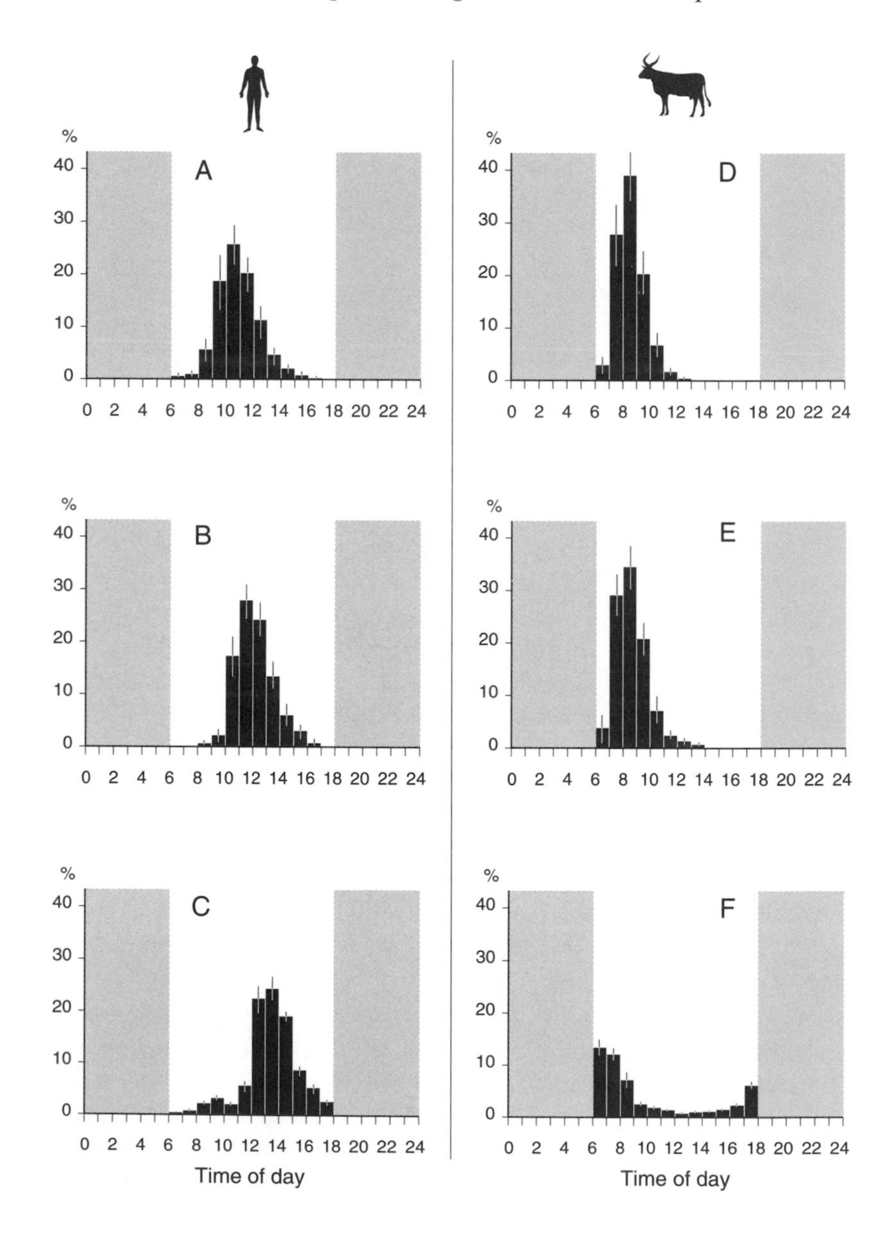

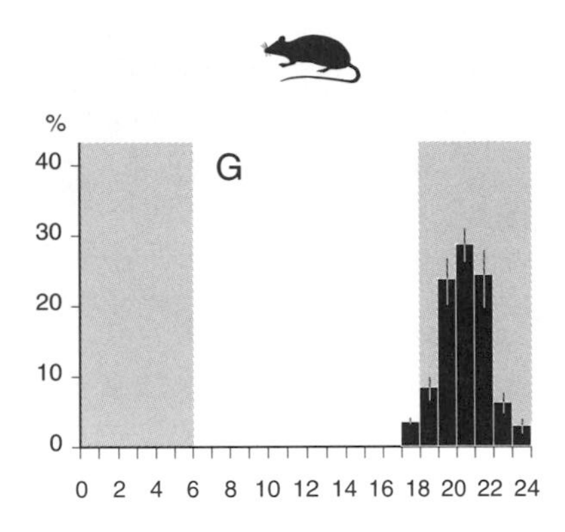

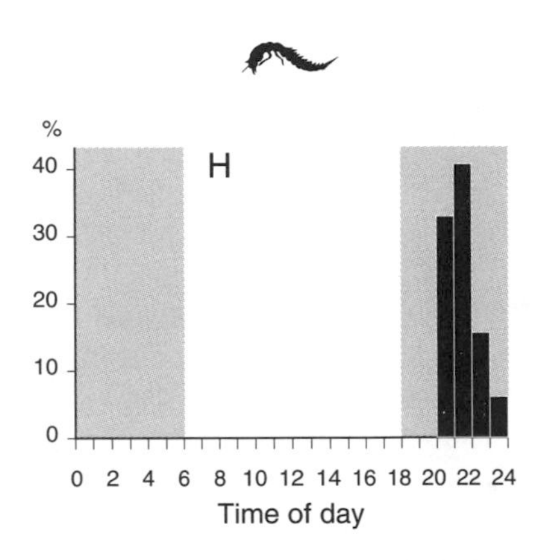

Figure 7.7. "Average" cercarial emergence patterns. A, B, C: from *Schistosoma* whose cercariae penetrate humans—they emerge mainly toward the middle of the day. D, E, F: from *Schistosoma* whose cercariae penetrate ungulates-they emerge in the morning, but note the particular case F *(S. margrebowiei)* that shows two daily peaks. G: from *S. rodhaini,* whose cercariae penetrate rodents-they emerge mainly at the start of the night. H: from *Plagiorchis neomidis,* whose cercariae penetrate insect larvae—they emerge at the start of the night as in *S. rodhaini,* although the definitive host is different (data from A. Théron and various other authors).

some cercariae having similar morphoanatomical characteristics are otherwise difficult to distinguish (Mouchet et al. 1992).

Figure 7.8 synthesizes the cercarial emergence patterns currently known for schistosomes. Note that there appear to be four favored periods during the day: a morning period (between about 6:00 and 9:00 A.M.) only

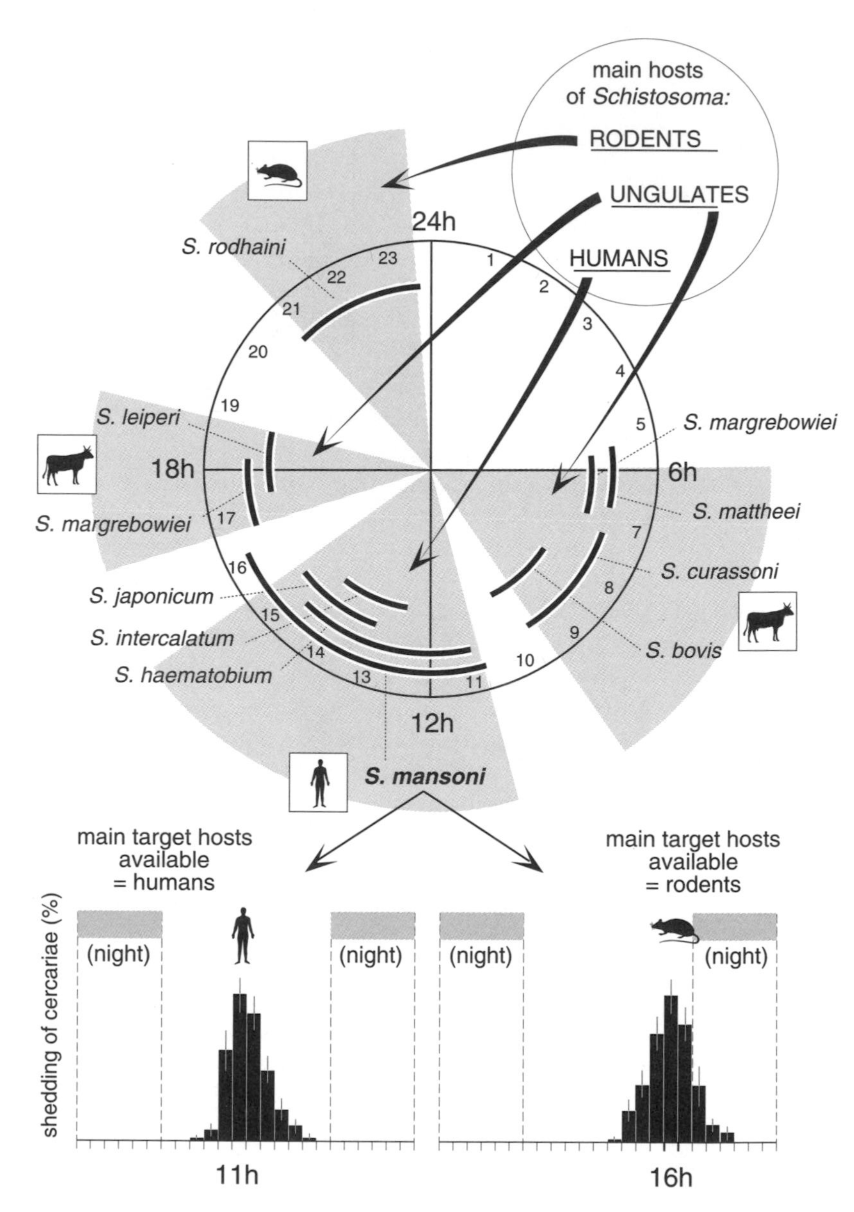

Figure 7.8. Synthesis of the cercarial shedding periods as they are currently known for species of *Schistosoma*. The periods indicated correspond to the most current patterns and show local variations only for *S. mansoni*.

for ungulate schistosomes; a midday period (between 11:00 A.M. and 3:00 P.M.) where all human schistosomes are grouped; a late afternoon period (between 5:00 and 7:00 P.M.), also for ungulate schistosomes; and a period at the beginning of the night that characterizes the rodent schistosomes.[6]

It appears that *S. japonicum* emerges in the middle of the day (Théron and Xia 1986) as observed in other human schistosomes. However, since this species infects numerous vertebrates, we may expect that the rhythms might not be the same for strains from different regions. Several studies have shown that *S. japonicum* from the two main foci in mainland China (Anhui and Sichuan) and those of Taiwan and the Philippines do not use the same definitive host spectra. Thus it is unlikely that the same "chronobiologic profiles" will be selected for. Disappointingly, research on this interesting system has thus far been done on characteristics such as allozymes without an ecological perspective.

Emergence profiles of *S. haematobium* have indicated that variations also occur in this species (Kechemir and Théron 1997; N'Goran et al. 1997). Humans seem to be its only host in nature, so the explanation for these variations cannot be those used for *S. mansoni*. According to N'Goran, Brémond, and Sellin (1997), in Ivory Coast the differences correspond to light levels in either the savanna or the forest.

The rhythm of cercarial emergence is an obvious encounter strategy selected for by the best possible coincidence between cercarial emergence and the periods when hosts most often visit aquatic environments. It is clear, for example, that humans frequent water mostly in the middle of the day (for agriculture and domestic work, swimming, and children's games). Further, it is clear that rodents are more active at dusk and that ungulates often visit ponds at the beginning or the end of the day.

The following points must be added concerning cercarial strategies for encountering their hosts:

1. the patterns of emergence of trematode cercariae that are taxonomically different or have different life cycles may be similar if the rhythms of activity of their target hosts are comparable. As such, the emission of *Plagiorchis neomidis* cercariae occurs in a manner almost identical to that of *Schistosoma rodhaini,* although all the protagonists are different. The cycle of *P. neomidis* has, successively, the mollusk *Radix limosa,* the aquatic insect larva *Sialis lutaria,* and the water shrew *Neomys fodiens*. Cercariae emerge at the start of the night, which coincides with the maximum activity of *Sialis* larvae (Théron 1976) (fig. 7.7H).

6. There are other *Schistosoma* species parasitizing rodents that are not included here, such as the enigmatic *S. sinensium* of China and Thailand, but their cercarial emergence rhythms are not known.

2. Even if the pattern of emergence follows a circadian periodicity, it is not certain that light acts directly on the parasite in all cases. *Bucephalus polymorphus,* a parasite of freshwater fish, emerges from the bivalve host mainly at the end of the night. The cercariae then rest on the bottom, and their long floating filaments allow them to stick to fish as they swim by. This timing of emergence is probably linked to the high activity of fish at dawn, when they are most likely to contact cercarial filaments (Wallet, Théron, and Lambert 1985). It is thought that the cercariae simply accumulate over the course of the day without any particular pattern and passively emerge when the mollusk filters the most.

3. It is striking that when the target does not have a discernible activity pattern, as in the cycle of *Fasciola hepatica* (whose cercariae encyst on aquatic plants), no emergence rhythm is observed (Bouix-Busson, Rondelaud, and Combes 1985).

4. Several authors have noted that the infectiveness of cercariae is not always greatest when they emerge from the mollusk. Only 20% of the cercariae of *Plagiorchis elegans* are infective when they exit *Stagnicola elodes.* This number rises to 75% four to six hours later, leading Lowenberger and Rau (1994) to think it favors cercarial dispersion in the aquatic environment, thus decreasing the risk of massive infection of insect second intermediate hosts.

The Filarial Clock

Filarial worms offer another type of favorization using circadian rhythms. These are long, thin nematodes that as adults live in the blood and other tissues of mammals, birds, reptiles, and amphibians. The worms are transmitted when biting vectors feed, and the parasite larvae obtained this way (microfilariae) must develop for a time within the vector before infecting a new host.

The classic example is that of Bancroftian filariasis, which is caused by *Wuchereria bancrofti,* a human parasite transmitted when blood-borne microfilariae are taken up by mosquitoes of the genus *Aedes* or *Culex.* Microfilaremia (the number of microfilariae per unit volume) in skin blood (the only blood accessible to mosquitoes because of the length of their biting apparatus) follows a circadian rhythm (Hawking 1967). As in schistosomes, the rhythms differ depending on the exact conditions of the life cycle: in African *W. bancrofti* microfilaremia is highest in the evening, when mosquitoes bite the most, and very low at the end of the night and during the day. The rhythm is clearly marked, with 99% of the microfilariae "disappearing" from peripheral blood during the diurnal phase. In *W. bancrofti* from Polynesia (the "Pacific" strain), microfilariae are in the peripheral blood during the day. This correlates directly with the biology of the local vector, *Aedes pseudoscutellaris,* which bites during the day. However, the rhythm is less marked than in the previous case, and there is only a 60% decrease in nocturnal microfilaremia compared with maximum levels.

Other microfilarial rhythms are well known, such as for the human parasite *Loa loa* (present in peripheral blood during the day and linked to the activity of its horsefly vector, *Chrysops*), and the parasite of the dog *Dirofilaria immitis* (a nocturnal rhythm linked to the activity of its mosquito vectors in the genus *Aedes*). There is, then, a superposition of two favorizations in these filarial worms: an encounter in space (concentration in cutaneous vessels) and an encounter in time (concentration at the "right time").

For some time it was thought that microfilarial rhythms were analogous to the previously discussed cercarial rhythms, that is, that there was a massive release of microfilariae into the bloodstream at specific hours. This does not happen, however, and the rhythm is the result of migration rather than time of emergence. In fact, microfilariae live a fairly long time (several weeks) and migrate daily between peripheral and deep blood vessels, mainly those in the lungs (Hawking 1962). If humans change their rhythm (that is, if they sleep during the day and are active at night), the parasites also change after some delay, which indicates that whatever synchronizes the rhythm it is not directly linked with day and night but rather depends on physiologic parameters in the host that themselves have a rhythm. For instance, night watchmen have microfilarial rhythms that are the reverse of more common patterns, and they probably do not participate much in transmission. As with cercariae, there are arhythmic species of filariae, also probably linked to particular transmission conditions.

Significantly, the chronobiological adaptations of microfilariae and cercariae, although often lumped together in parasitology textbooks, have several important differences.

In both cases the parasite "sets its clock" by the downstream host. However, in human disease the emergence of schistosome cercariae is adapted to human activity whereas the migration of microfilariae is adapted to vector activity.

Cercarial rhythm is *directly* synchronized by a factor from the outside environment (photoperiod) that is independent of the upstream host (the mollusk), whereas microfilarial rhythm is *indirectly* synchronized by photoperiod; the microfilariae are sensitive to internal rhythms of the upstream hosts (vertebrates) that are themselves synchronized by the photoperiod.

The short survival of cercariae does not allow them a second chance if they miss the rendezvous, whereas microfilariae, which live longer, have repeated chances to be transmitted (as often as mosquitoes bite their hosts).

Other examples of encounters in time analogous to that of cercariae and filariae are continually being described among parasites, with rhythms found to exist each time the downstream host itself has a rhythmic activity. If such emergence patterns are to be selected, genetic diversity must allow for such selection, and the downstream host must be accessible during a predictable and limited time.

For instance, rhythms whose adaptive significance are probably linked to host encounter are also known to occur for parasitic protists (e.g., the production of gametocytes by *Plasmodium* or the excretion of coccidia oocysts in the intestinal lumen). This has led Hawking (1975) to write that in *Plasmodium* "the biological purpose of the synchronous accurately timed asexual cycle is to make the gametocytes match the mosquitoes." *Plasmodium* chronobiology allows "chronotherapeutic" applications where the efficacy of medications is optimized by timing their intake relative to parasite emergence (Landau et al. 1991).

Numerous "chronobiological hatchings" have been described for helminth eggs that give birth to short-lived aquatic larvae (in this way resembling cercariae). Hatching happens at the best time for the larvae to encounter hosts. This is especially well known in monogeneans such as *Entobdella soleae,* a parasite of sole (Kearn 1973); *Diplozoon homoion,* a parasite of barbels (MacDonald and Jones 1978); and *Polystoma integerrimum,* a parasite of frogs (MacDonald and Combes 1978). It is also well known in cestodes with aquatic life cycles, such as the bothriocephalids *Bothriocephalus barbatus* and *B. gregarius,* whose larvae (coracidia) must be ingested by copepods (Berrada Rkhami and Gabrion 1986).

Adult parasitoids, whose ability to localize their hosts was described above, also do not neglect to select for temporal coincidence with their hosts' activity, and their exploration of the environment occurs at very precise times over the course of the day. As in cercariae and filariae, depending on local conditions there is selection for different chronobiological characteristics in parasitoids (Fleury et al. 1991). Although in certain cases selective pressures may not come from host activity (notably in parasitoids such as *Trichogramma* that lay their eggs in their victim's eggs), the selective value of parasitoid activity rhythms is still probably linked to host encounter.

Stimulus: Downstream Host or Its Environment. Response: Manipulated Upstream Host

We shall see further on that the alteration of certain traits (color and behavior) in the upstream host frequently advances favorization by "signaling" the upstream host to the downstream predator. However, manipulations of upstream hosts that make them sensitive to stimuli coming from downstream hosts or from their environment are exceptions. Here are two examples.

The trematode *Gynaecotyla adunca* was studied by Curtis (1987, 1990, 1993). Adult *G. adunca* live in the digestive tract of larids (gulls and their allies) that eat marine fish. The miracidia hatch from eggs evacuated with the bird hosts' feces and infect the marine mollusk *Ilyanassa obsoleta,* which normally lives a few meters deep in the water. Cercariae emerge from the snail but have no tails. The next stage, the metacercaria, encysts in various species of amphipod crustaceans in the family Talitridae (including *Talorchestia longicornis* and *T. megalopthalmia*) and in a small crab *(Uca pugilator),* all of which live on the beach at the extreme limit of the water. Birds become infected by eating crustaceans harboring the metacercariae.

One might guess that the transmission of *G. adunca* between the mollusk and the crustacean is a problem because the mollusks live in deep water at some distance from the beach while the crustaceans live on the beach. Moreover, cercariae, which must go from the mollusk to the crustacean, have no tails, so they are poor swimmers and can move only short distances.

Curtis (1987) has shown that the mollusks' vertical distribution is modified by parasitism. Infected individuals go up toward beaches and release cercariae close to amphipods and crabs (fig. 7.9). A precise analysis has shown that parasitized mollusks repeatedly go toward beaches and that these excursions are programmed so the mollusks remain out of the water on sandbars only when low tide happens at night. This means that mollusks go up only during a high tide that precedes a nocturnal low tide, not one that precedes a diurnal low tide.

The mollusk's manipulated behavior is not due to its being parasitized by just any trematode, since *I. obsoleta* also harbors other trematodes that do not induce such excursions. Moreover, if a mollusk harbors *G. adunca* and any of several other trematodes at the same time, this behavior still occurs. Therefore it is *G. adunca,* and only *G. adunca,* that manipulates its upstream host's behavior by having it deposit cercariae within reach of the downstream host. According to Curtis, that infected mollusks move so they are on the beach during nocturnal and not diurnal low tides has value because night provides some protection against cercarial desiccation and also corresponds to the crustaceans' maximum activity period (Curtis 1990, 1993). Significantly, nothing is known about the physiological mechanism(s) by which *G. adunca* manipulates its host's behavior, although it is likely the parasite induces a modification of the mollusk's responses to environmental stimuli (light, gravity, or water agitation).

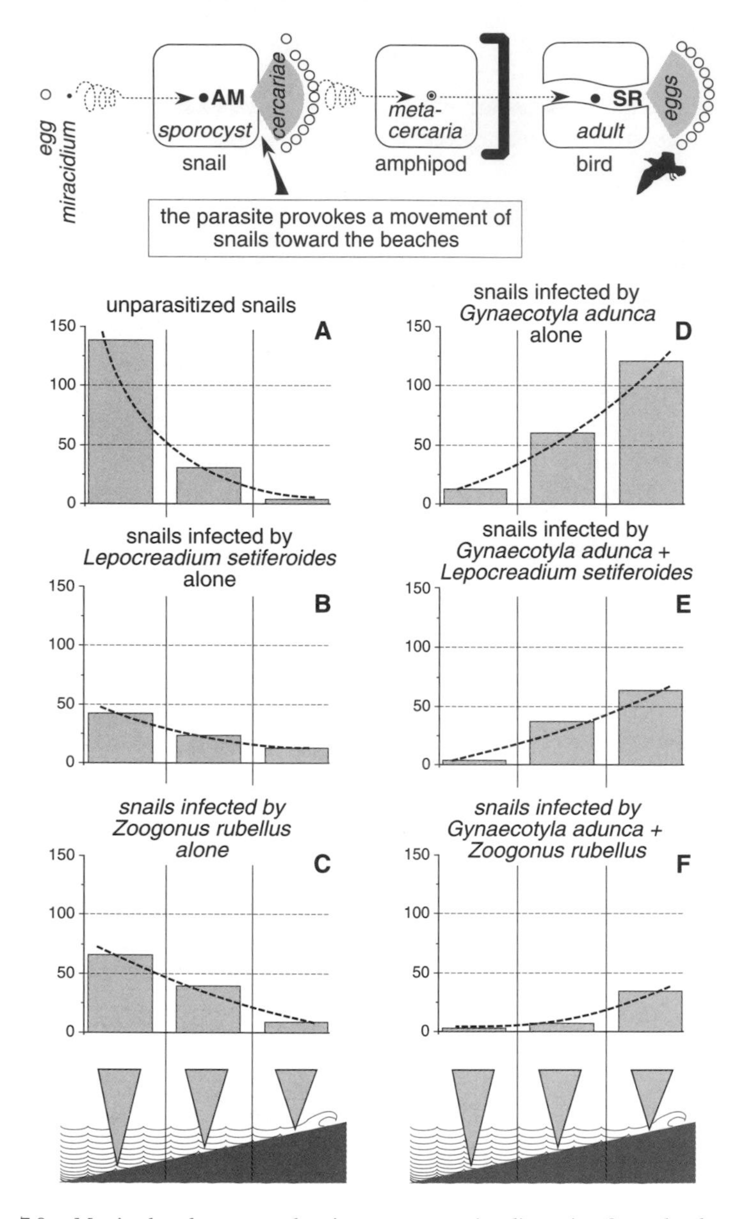

Figure 7.9. Manipulated upstream host's response to stimuli coming from the downstream host's habitat. Top: mode of transmission of the trematode *Gynaecotyla adunca.* Bottom: location of the mollusks *Ilyanassa obsoleta* on the shore relative to their being infected by the trematodes *Gynaecotyla adunca* (A and B), *Lepocreadium setiferoides* (C and D), *Zoogonus rubellus* (E), and *G. adunca* and *Z. rubellus* (F). Each time a mollusk is found to be parasitized by *G. adunca,* it tends to move toward the beaches that are frequented by the crustacean's downstream host (from data of Curtis 1990). AM = asexual multiplication; SR = sexual reproduction.

A potentially interesting manipulation, as original as the previous one but not yet confirmed scientifically, is suggested by the occasional migration of dragonflies in Europe. These migrations may be triggered by the presence of metacercariae from trematodes of the genus *Prosthogonimus*. Although not well documented, the hypothesis is interesting: there are sometimes spectacular migrations of several million dragonflies (particularly *Libellula quadrimaculata*) in Europe, with one instance darkening the sky over Montpellier, France, for half a day at the end the nineteenth century (see Aguilar, Dommanget, and Préchac 1985). Significantly, metacercariae of *Prosthogonimus* in extreme abundance are thought to trigger synchronous emergence of the dragonflies, followed by massive flights along rivers that would then attract birds (definitive hosts), thus favoring transmission (Dumont and Hinnekint 1973).

There are still other curious or weird life cycles in which the best chance of transmission is obtained not by affecting a movement of the upstream host toward the downstream host but by decreasing the movements of the upstream host. This is observed on two scales exemplified by the isopod crustacean *Anilocra pomacentri,* which decreases movement of its parasitized hosts, and by the trematode *Parasymphylodora markewitschi,* which paralyzes part of its host. In both cases the parasites mediate inhibition of movement and thereby favor transmission.

In brief, *A. pomacentri* parasitizes the reef fish *Chromis nitida*. According to Adlard and Lester (1994), infection of juvenile fish from parasitized adults happens in the following way: Most *C. nitida* live on the outer slope of the reef, but some also live in the openings and depressions of coral heads, where they exhibit strong territorial behavior. During the winter one-year-old fish normally go to meet those on the outer slope of the reef, but those infected with the isopod do not. Thus, when a new generation of juveniles arrives to settle in the coral heads, the parasite larvae are present to infect them. The authors suggest that the parasite inhibits gonadal development in the fish, which in turn modifies hormonal secretions so the fish fail to migrate toward the outer slope of the reef.

P. markewitschi, on the other hand, has the following life cycle: The molluskan first intermediate host is the prosobranch *Bythinia tentaculata,* the second intermediate host another aquatic mollusk (most often *Lymnaea limosa*), and the definitive host the fish *Leuciscus cephalus*. Therefore two mollusks succeed each other in the life cycle, one producing cercariae and the other harboring metacercariae. Curiously, the transfer from mollusk to mollusk happens this way: The cercariae, which have no swimming appendages, emerge from *B. tentaculata* and migrate to the mollusk's tentacles, where they form a living coat. Importantly, *B. tentaculata* move with their tentacles pointing forward. If an individual that harbors cercariae

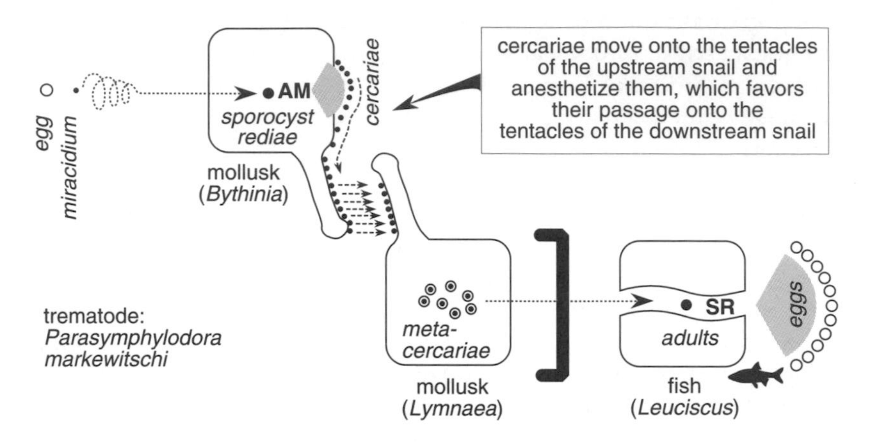

Figure 7.10. Favorization in the transmission of the trematode *Parasymphylodora markewitschi.* Anesthesia of the tentacles of the molluskan first intermediate host favors the passage of cercariae to an obligatory second intermediate molluskan host (from data of Lambert 1976). AM = asexual multiplication; SR = sexual reproduction.

knocks into the other mollusk, the parasites immediately attach to the new host using their ventral suckers and penetrate the body, where they encyst and form metacercariae (fig. 7.10). Favorization consists of anesthetizing the *B. tentaculata* tentacles when cercariae are attached—normally any knock into an obstacle would provoke the mollusk to retract the tentacle, but that does not happen when the tentacle carries parasites. This insensitivity to obstacles gives the cercariae a chance to detach and pass into the second mollusk (Lambert 1976). In this exceptional and little-known life cycle there is therefore a moment of "no response" on contact with the downstream host.

Let me end this discussion of manipulation of the movement of upstream hosts with the following idea: In the life cycle of human schistosomes, where the mollusk is aquatic and the definitive host terrestrial, the return of the parasites to water in the form of eggs is not guaranteed. In the case of urinary schistosomiasis it would therefore be of great advantage for the schistosome to manipulate infected people to pee directly into the water, thus favoring transmission. Although the parasite has not succeeded in this tour de force, there is an example that shows such a manipulation is not unthinkable. This is the case of *Gordius,* a bizarre parasite (actually a nematomorph) (see Schmidt-Rhaesa 1999) whose larvae live in nonaquatic insects (for example, bees) and whose adults live in the water. Returning to the water is obviously not easy when one is in a bee. However, when an infected bee flies over a pond it commits suicide by voluntarily diving into the water, where the parasite larvae are then immediately released (cited by Dawkins 1990)!

Stimulus: Free-Living Infective Stage. Response: Downstream Host

In some parasite life cycles the infective stages remain free-living until eaten by a potential host. An example involves the trematode *Gorgodera euzeti,* which starts its development in various species of small freshwater mussels of the genus *Pisidium.* The cercariae must encyst as metacercariae in larvae of an aquatic insect (the mayfly *Sialis*), which ultimately metamorphose into winged insects (imagoes) eaten by the red frog *Rana temporaria* (where adults live in the urinary bladder). It is between *Pisidium* and larval *Sialis* that favorization occurs (Combes 1971b). Here cercariae do not swim but stay attached to or near the mussel. The huge tails of the cercariae (compared with their body size) resemble worms crawling in the mud, and the *Sialis* larvae actively eat them. These fake worms are digested, while the cercariae's bodies penetrate the hemocel and encyst. Transmission is thus ensured as metacercariae pass during metamorphosis into the emerged imagoes, which are then eaten by the amphibians (fig. 7.11, top).

Leucochloridium paradoxum is another trematode with an exceptional life cycle that includes only two hosts, a snail and a bird. The mollusk is a succinean *(Succinea putris)* that lives along the edges of small rivers and streams. The cercariae, however, are not dispersed outside the mollusk, and the metacercariae remain prisoners inside the sporocysts in which they were produced. In this case the sporocysts develop in the snails' tentacles, greatly distending them. The consequence is one of the best-known examples of favorization—whereas in most trematodes sporocysts are colorless and transparent, in *L. paradoxum* they look like insect caterpillars: they are long, ornamented with transverse colored bands, and undulate continuously. Attracted by these signals, birds eat the infected snails and at the same time the hundreds of infective stages of *L. paradoxum* in the sporocysts (fig. 7.11, bottom). Thus, although the parasite stays inside the upstream host, it nevertheless attracts the downstream host via its phenotype.

It is curious that similar adaptations are selected in totally different environments. Baer (1952) reported the case of *Ptychogonimus megastomus,* a marine trematode parasite of, successively, a scaphopod mollusk *(Dentalium alternans),* decapod crustaceans (crabs), and various sharks and rays. In this cycle sporocysts packed with cercariae exit the mollusk. Although they do not mimic caterpillars as do those of *L. paradoxum,* their shape

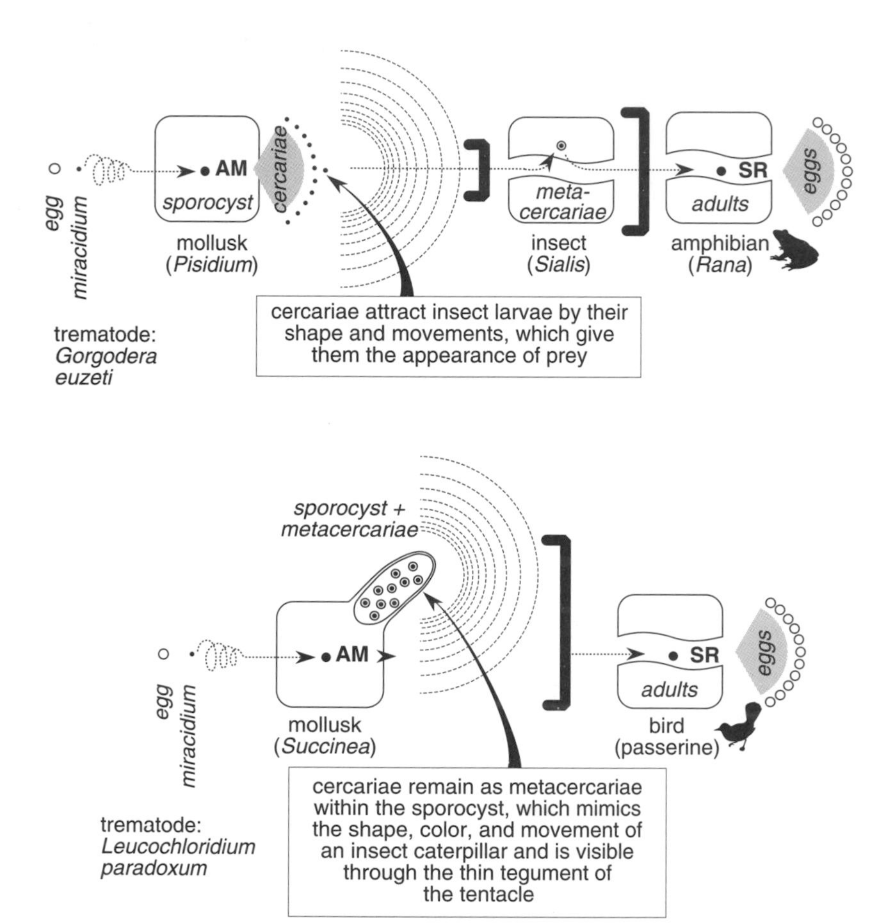

Figure 7.11. Responses of the downstream hosts to stimuli created by the parasite. Top panel: transmission of *Gorgodera euzeti* (from Combes 1968). Bottom panel: transmission of *Leucochloridium paradoxum* (from various authors). AM = asexual multiplication; SR = sexual reproduction.

and movement nevertheless make them look like appetizing prey to the crabs, which actively eat them.

Prey mimesis in unionid mussels is even more curious, because here it is not the larvae that mimic prey, but rather the adult that produces them. Adult unionids are large clams that live in the mud at the bottom of rivers and ponds, and their larvae (glochidia) are parasites on fish gills. In unionid species of the genus *Lampsilis* the "mother" clam accumulates hundreds of thousands of glochidia between its valves and hangs out a piece of its mantle that looks like a brightly colored and motile small fish.

When a big fish approaches, attracted by the pseudoprey, the clam "spits in its face," and a cloud of infective larvae ends up being taken in by the fish's respiratory current, with many attaching to the gill filaments.

These examples are spectacular because they concern free-living stages that are able to move in some way. When a nonmotile stage (eggs or larvae encysted in the outside environment) "must" be eaten, processes to attract the hosts' attention are generally not as elaborate but still exist. For example, some cestodes whose intermediate hosts are aquatic invertebrates have large eggs with walls rich in nutrients. According to Holmes (1976), this provides a "nutritional reward" for hosts that eat them.

Finally, we must include the magical fungus that makes flowers appear where they are not. The rust *Monilinia vaccinii* parasitizes myrtle bushes and at one point in its cycle depends on various pollinator insects to transport spores from infected leaves to healthy plants. The insects must therefore be attracted to both the leaves and the flowers of a bush. It resolves this problem by manipulating the tissues of the leaves so that a pseudoflower appears, packed with rust spores. The pseudoflower attracts insects both by its ultraviolet-reflective "color" and by nectar production: "The *Monilia*-infected, discolored leaves evidently mimic the flowers of their host in yielding sugary rewards to the pollinator-vectors and, to some extent, in providing them with ultraviolet-reflective patterns analogous to nectar guides" (Batra and Batra 1985).[7]

Stimulus: Manipulated Upstream Host. Response: Downstream Host

Bringing the Upstream and Downstream Hosts Closer Together

Modifying the behavior of a parasitized host may favor transmission in holoxenous life cycles. This is in general the classic situation of "contagion," where the probability that a pathogenic agent will pass from an infected host to a healthy one is increased if the "disease" provokes individuals to gather around the "patient."

Scott (1985) studied the mode of transmission of the monogenean *Gy-*

7. Some modern marketing techniques used to attract our attention are nothing other than applications of the favorization techniques used by parasites. In products that resemble other products to attract consumers, economists distinguish counterfeiting, which is ostensibly the exact reproduction of an original product (such as fake Lacoste shirts); imitation, which appropriates only certain characteristics of a known product; and the *parasite,* a particular form of imitation that tries to recreate "the atmosphere" of a product in order to benefit from its image (as by using colors, forms, or graphics that trigger a buying reflex without legally amounting to counterfeiting or imitation). For economists to use the term "parasite" tells one a lot about the "blind" behavior of both hosts and consumers.

rodactylus bullatarudis, an ectoparasite living on the skin of the guppy *Poecilia reticulata*. Like all *Gyrodactylus*, this oviparous species does not have swimming larvae. Instead, juveniles attach to the skin of their mother's host, and transfer from host to host is made by physical contact between fish. The atypical swimming of parasitized fish attracts the attention of healthy guppies and makes them come closer, increasing the level of transmission. Most likely the same thing happens for all related species.

Transmission of the rabies virus is similar. In this case the upstream host (the fox) is the one that seeks contacts with the downstream host (a healthy fox). Rabies from foxes *(Vulpes vulpes)* has spread in Europe during the past thirty years, moving an average of thirty kilometers west each year. The virus, abundant in saliva, is transmitted by biting. Studies using radio-tracking have compared the movements of infected and uninfected foxes (Artois, Aubert, and Stahl 1990) and suggest that infected foxes travel farther in twenty-four hours than uninfected ones. Moreover, infected individuals tend to station themselves close to the limits of their territory, something rarely done by uninfected foxes (Artois et al. 1991). This aberrant behavior obviously increases the probability of contact between the foxes of neighboring territories. Since the virus makes infected foxes more aggressive, bites are inevitable, and the abundance of the virus in the saliva almost ensures that it will be transmitted with just one bite (Artois, Aubert, and Stahl 1990).

Having the Upstream Host Eaten by the Downstream Host

Healthy and parasitized individuals often behave differently. Such changes in behavior may have no effect on transmission but may cause the parasitized upstream host to be eaten by the downstream host, thus favoring passage.

In all cases when an intermediate host must be eaten by the downstream host, a conflict of interest is inevitable between the parasite and the host. As Milinski (1990) noted, "The host and the parasite compete for the host's behaviour." Thus any change in the intermediate host's behavior that increases the probability of its being eaten increases the parasite's fitness and decreases the intermediate host's fitness. The mechanisms that allow the host to be eaten may range from the simple mechanical deterioration of an organ to a true physiological manipulation of the entire host.

Moving the Host Closer to the Predator

Increasing the probability of the intermediate host's being consumed may require manipulating its movements as *Gynaecotyla adunca* does in

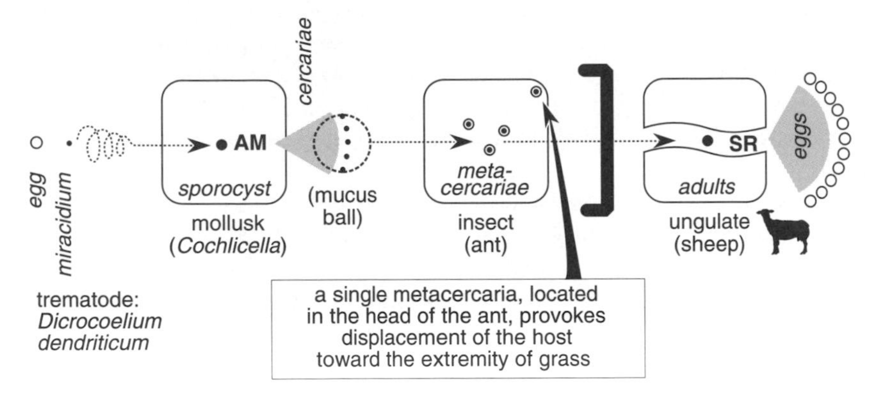

Figure 7.12. Responses of manipulated upstream hosts to the downstream host's habitat: transmission of *Dicrocoelium dendriticum.* AM = asexual multiplication; SR = sexual reproduction.

mollusks. However, in this case, the manipulation makes the intermediate host move closer to the predator within which the life cycle must continue. A well-known example is the trematode *Dicrocoelium dendriticum,* also called the small liver fluke. The cycle is not exactly orthodox: a terrestrial mollusk (from which cercariae exit), an ant (where metacercariae are), and a sheep (with adults in the bile ducts). This host sequence raises two questions: How do cercariae infect ants in the absence of water? And how do sheep become infected if they eat grass and not ants? In brief, two favorization events occur.

In the first favorization, the cercariae "swim" in small balls of mucus that are launched, like spitballs, from the mollusk's pulmonary cavity. These sweet treats attract ants that eat them, and as soon as cercariae are swallowed the second favorization occurs: one of them (and only one!) migrates to the ant's brain and induces a modification of its behavior. The parasitized ant climbs to the tip of a blade of grass, attaches using its mandibles, and waits for a sheep to ingest it involuntarily while grazing (fig. 7.12). Interestingly, the ant stays at the tip of the grass only during the cool hours of the day. If it is not eaten, it goes down when it gets hot, feeds with its fellows, then climbs back up the next night to again await a sheep (see Hohorst and Lämmler 1962; Carney 1969; Spindler, Zahler, and Loos-Frank 1986).[8]

8. The metacercaria in the brain of the ant loses its capacity to develop into an adult. Its behavior therefore is the equivalent of suicide. Selection of such a behavior explains itself by kin selection (all ingested cercariae are clones having their origin in the same miracidium and therefore have the same genes).

There is also a fungus in the genus *Entomophtora* that parasitizes ants and induces a behavior similar to that of *D. dentriticum*. In this case, however, the infected ant's nocturnal climb onto a grass tip ends with its dying during the night, then it quickly dries out (Loos-Frank and Zimmermann 1976), most likely a manipulation that favors fungal spore dissemination.

Less spectacular, but still as efficient, is the behavior of mosquito larvae (*Aedes*) infected by metacercariae of the trematode *Plagiorchis noblei*. In this case infected larvae spend more time than healthy larvae at the surface of the water, which certainly makes them more accessible to definitive hosts (birds and rodents). The peculiarity, however, is that the aberrant behavior occurs only when three or more metacercariae are present. As long as the larvae carry only one or two metacercariae, they stay at or near the bottom, close to the mollusks that emitted the cercariae (Webber, Rau, and Lewis 1987). In this case it seems that lightly infected larvae are programmed to orient themselves toward the source of infection whereas heavily infected larvae orient toward the definitive host.

Making the Host More Vulnerable to the Predator

Another good way to have a prey vector eaten is to weaken its capacity to flee so it becomes a designated victim for the predator. Metacercariae of *Cainocreadium labracis* are carried mainly by gobies *(Gobius niger)* and adults by sea bass *(Dicentrarchus labrax)*. Maillard (1976) has shown that some metacercariae locate at the base of the muscles of the goby's fins, thus decreasing its agility, while others encyst mainly around the ventral fins, which in the goby are fused into the shape of a sucker. Infected gobies thus have decreased agility and cannot attach and hide well under rocks, so the most heavily infected gobies are most likely to be eaten by the predatory sea bass (fig. 7.13, top).

In the cycle of *Diplostomum spathaceum* the metacercariae are in freshwater fish and adult parasites are in piscivorous fish (Crowden and Broom 1980; Chappell 1995). As in the previous case, the metacercariae locate in a favorite place on the fish's body, but this time it is the lens of the eye. No matter where the parasites penetrate a fish, they gather in the lens, without exception and through more or less complicated pathways. Their accumulation in this organ greatly decreases the fish's visual acuity and therefore its ability to detect predators. Hence fish that are most heavily parasitized are most likely to be eaten (fig. 7.13, bottom). Other species of *Diplostomum* have metacercariae in the retina, the cerebellum, the optical centers of the brain, or other nervous system regions, with comparable consequences.

Multiceps multiceps is a cestode with a two-host life cycle. Here sheep be-

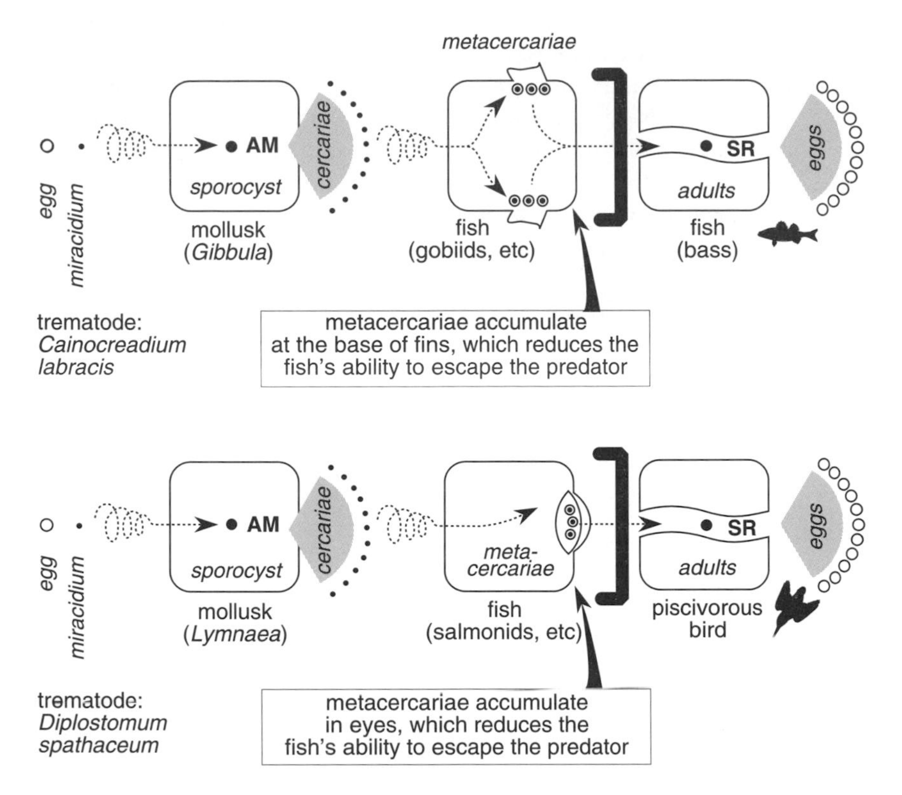

Figure 7.13. Responses of downstream hosts to stimuli created by a handicap of the parasitized upstream host: transmission of *Cainocreadium labracis* and *Diplostomum spathaceum*. AM = asexual multiplication; SR = sexual reproduction.

come infected by ingesting parasite eggs, after which they harbor budding larvac (cysticerci) that usually localize in the brain. A carnivore (wolf, coyote, fox, dog, etc.) becomes infected by eating infected sheep, after which it carries adult worms in its gut. Sheep that have cysticerci in their brains suffer from a serious motor problem known as "gid," which leaves them easy prey for the carnivore definitive hosts. Still other cestodes (*Echinococcus multilocularis* and *E. granulosus,* for example) also weaken their hosts via budding cysticerci in various places. It is highly probable that such asexual reproduction of cestodes in their intermediate hosts was selected because of the resulting pathogenic effects (thus the favorization) and not necessarily for the demographic consequences. Often a parallel is made between asexual reproduction of larval trematodes in mollusks and larval cestodes in rodents or herbivorous mammals. However, while it is likely that trematode asexual reproduction was selected because it amplifies larval populations, this is not so for cestodes such as *Multiceps* and *Echinococcus.* Cestodes have no need to produce multiple scolices at the larval stage, since adult worms can multiply their proglottids virtually infinitely. Further, adult cestodes with larval budding are usually very small, and nothing indicates that total egg production is increased by obtaining numerous adults rather than a single large cestode. On the contrary, it is clear that larval budding has a very high pathogenic effect that favors the capture of the intermediate host by the carnivore definitive host.

Larvae of the nematode *Baylisascaris procyonis* lodge in the brain of North American mice and squirrels and induce a defect in motor coordination so they are more easily caught by raccoons, the definitive hosts. The mechanism of this manipulation is not identical to that of the cestodes in sheep brains, since there is no larval multiplication, but its influence on host behavior and transmission is the same (see Sheppard and Kazacos 1997).

Nervous or sensorial organs are not the only "sensitive" organs in which parasites may be located. For example, the estuarine fish *Cyprinodon variegatus* harbors metacercariae of *Ascocotyle pachycystis* in Florida, and adult worms are found in predatory fish. In this case hundreds of metacercariae are gathered in the arterial bulb, where they decrease blood flow, which in turn impairs swimming performance and increases vulnerability (Coleman 1993).

Making the Host More Accessible to the Predator

Another way to attract the downstream host is to make the parasitized intermediate host easier to eat. The trematode *Meiogymnophallus fossarum,* when adult, is a parasite of the oystercatcher, a shorebird that eats bivalve mollusks (not just oysters). In the western Mediterranean, Bartoli (1973a) showed that metacercariae of *M. fossarum* occur in bivalves, including cockles *(Cardium)* and the clam *Tapes.* The parasite is thus on the right track to "reach" the bird, but why not ease the bird's task? In *Cardium* the metacercariae are localized on the edge of the mantle and induce irregularities in the growth of the shell, leading to gaps where the valves close (fig. 7.14, top). In *Tapes,* metacercariae induce the animal to flip over so the valves open toward the top and not, as usual, toward the bottom (Bartoli 1973b). In both cases access to infected prey is easier for the predator bird, which uses its specially shaped bill to open the bivalve and eat the soft parts.

In New Zealand the trematode *Curtuteria australis* has a life cycle similar to that of *M. fossarum.* Thomas and Poulin (1998) have shown that the metacercariae locate in the foot of the bivalve mollusk *Austrovenus stutch-*

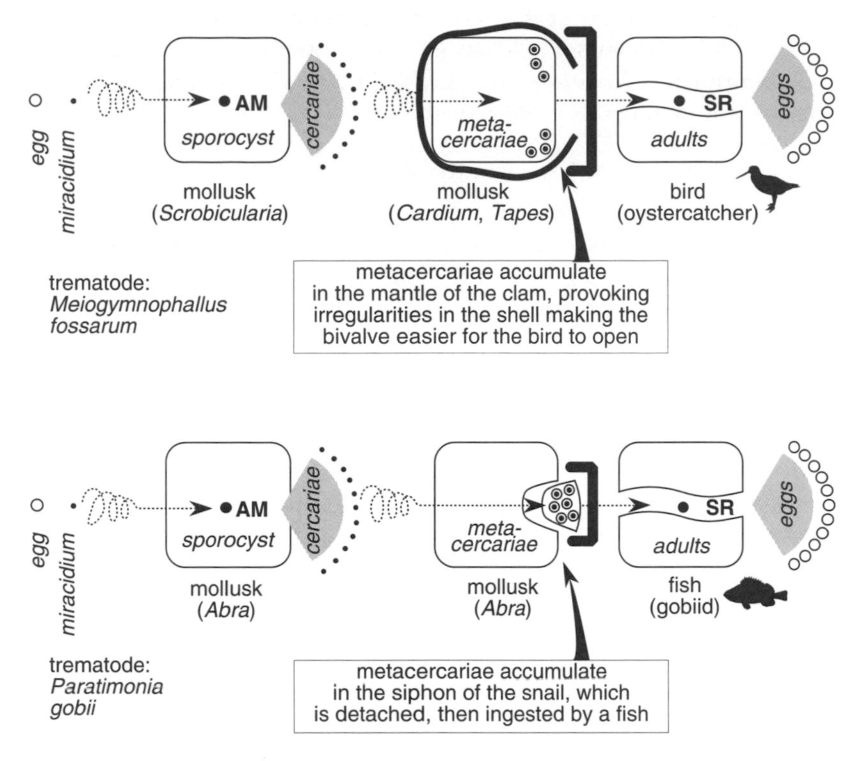

Figure 7.14. Responses of downstream hosts to stimuli created by specific pathologies in parasitized upstream hosts: transmission of *Meiogymnophallus fossarum* and *Paratimonia gobii*. AM = asexual multiplication; SR = sexual reproduction.

buryi and make it shorten, which in turn prevents the bivalve from burying itself correctly in the sediment. Observations on predation show that the parasitized bivalves are thus more accessible to their bird definitive host.

Paratimonia gobii, another trematode, develops in bivalve mollusks, this time in the genus *Abra*. Curiously, cercariae are emitted into the water column but encyst in other *Abra* as they are vacuumed in via the inhalant current of the respiratory siphon. Adult *P. gobii* live in the intestine of gobies, small fish of shoreline bottoms. In this case the parasite has not "chosen" a simple solution to favor transmission to the hosts. Gobies cannot ingest the mollusks because they are almost as big as the fish themselves. The selected solution is the following (fig. 7.14, bottom): metacercariae accumulate in the siphon of the mollusk, which then causes autotomia (a rupturing and severing of the siphon). Autotomia can occur in the absence of parasitism, but it is accelerated by the presence of *P. gobii* metacercariae. Once detached, the siphon moves about on the bottom like a small benthic invertebrate, attracting gobies that then ingest these small parasite stuffed morsels (Maillard 1976).

Making the Host More Visible to the Predator

The last means used to get an infected host eaten, which may be the most common strategy of all, consists of making the host more conspicuous. Here manipulation can reach very elaborate levels. For example, *Polymorphus paradoxus* is an acanthocephalan whose infective stage, the cystacanth, develops in gammarids (amphipods) and whose adults mature in ducks that eat the infected gammarids (fig. 7.15, top). Cystacanths are usually at no particular location within the gammarids but are most often found in the hemocoel.

Holmes and Bethel (1972) carried out a study in Canada on this system that has since become a standard reference and has also played a major role in convincing mainstream ecologists of the reality of favorization. In this classic study, the authors showed that parasitized gammarids exhibit a positive phototactic response that makes them swim up to the surface and stay in well-lighted zones instead of down to the bottom where they are less conspicuous. They also turn in circles on the surface of the water if they are disturbed ("skimming" behavior) and have a strong tendency to attach to floating substrates or to the feathers of the ducks themselves ("clinging" behavior). These aberrant behaviors of parasitized gammarids significantly increase the probability of their being eaten, particularly those that hang on the ducks' feathers, which may then be swallowed during preening.

Microphallus papillorobustus is a trematode whose life cycle and favorization processes were studied by Helluy (1981, 1982, 1983, 1984) in lagoons along the Mediterranean. Adults locate in the digestive tract of gulls, and the eggs they lay are passed with the feces. The miracidia penetrate a gastropod mollusk, *Hydrobia ventrosa,* and cercariae exit the mollusk to penetrate small amphipods of the genus *Gammarus* (most often, *G. insensibilis*), where they then encyst as metacercariae. After infected gammarids are eaten by a gull, the metacercariae develop into adults (fig. 7.15, center).

Helluy first noticed that the metacercariae are found in a particular site in the gammarids. Whether gammarids are captured in the natural environment or are experimentally infected, the metacercariae locate almost exclusively in the head at the level of the ganglia, which form the amphipod's "brain." Experiments have shown that it is only after the brain is parasitized that other cercariae will settle in other regions of the body, particularly in the thoracic segments. When healthy gammarids are first put in contact with mollusks that emit cercariae, most of the metacer-

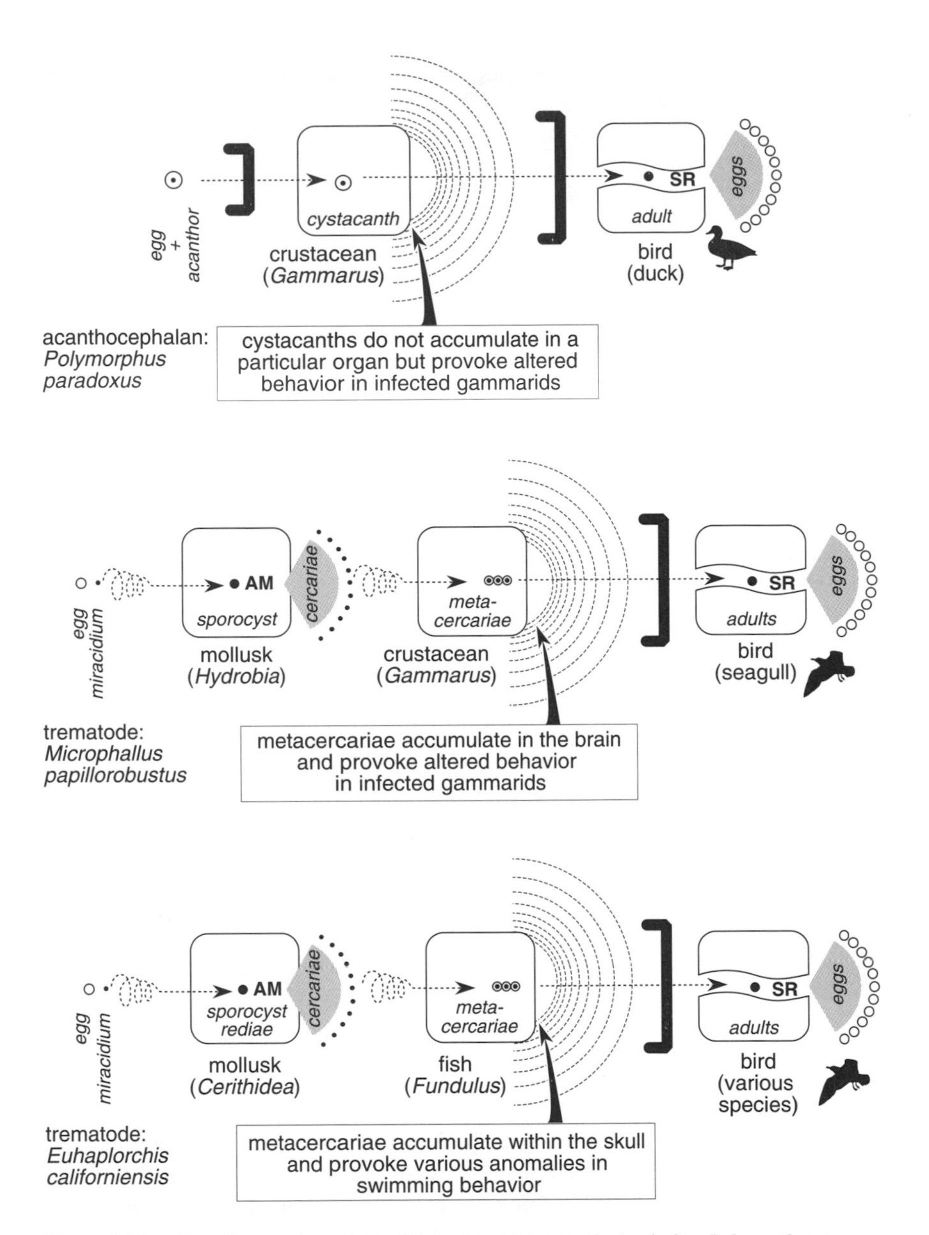

Figure 7.15. Top: favorization in the life cycle of the acanthocephalan *Polymorphus paradoxus*. Center: favorization in the life cycle of the trematode *Microphallus papillorobustus*. Bottom: favorization in the life cycle of the trematode *Euhaplorchis californiensis*. AM = asexual multiplication; SR = sexual reproduction.

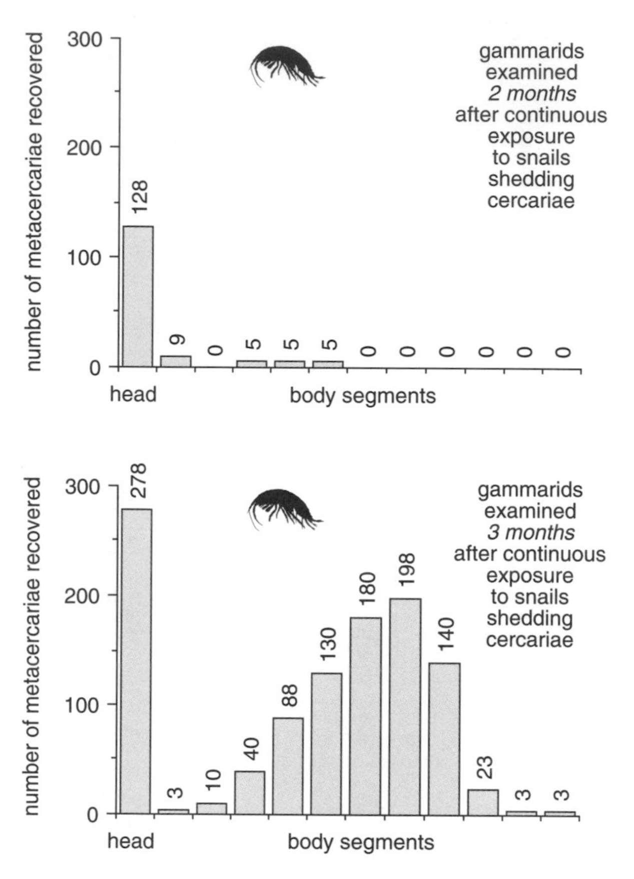

Figure 7.16. Distribution of the metacercariae of *Microphallus papillorobustus* in infected amphipod bodies. At the start of the experiment (top) parasites are almost all in the head, where they manipulate the amphipod's behavior, making it "crazy." It is only later that parasites are also recovered from the amphipod's abdomen (bottom, from Helluy 1982).

cariae localize in the brain (fig. 7.16). If the experiment is extended, metacercariae are found in both the brain and the rest of the body.

The presence of the parasite in the brain is correlated with behavioral changes in the gammarids. "Normal" gammarids—those not carrying parasites—show a whole series of behavioral responses that serve to partially camouflage them from birds: they live at the bottom of the lagoon rather than at the surface, and at the smallest warning they immobilize themselves in aquatic vegetation. Gammarids with parasitized brains have at least three "crazy" behaviors: they are attracted by light instead of fleeing from it; they frequent the surface of the water rather than the bottom of the lagoon; and they react to danger instead of immobilizing themselves.

Further, Helluy experimentally demonstrated that "crazy" gammarids had more chances of being eaten than did "normal" gammarids. To do so she installed gulls in an aviary with a large tank at its center to mimic a lagoon. In the tank were placed equal quantities of "crazy" gammarids that harbored metacercariae and "normal" gammarids that did not. After twenty-four hours all gammarids left in the tank were recovered. Analysis of recovered individuals conclusively showed that the "crazy" gammarids were about four times as likely to be eaten as "normal" ones.

Euhaplorchis californiensis is a common Californian trematode with a life cycle similar to that of *Microphallus papillorobustus* but with the Pacific killifish, *Fundulus parvipennis,* as its second intermediate host. What is surprising in this case is that, as in the example of *M. papillorobustus,* natural selection led to localization of the metacercariae in the vicinity of the brain (fig. 7.15, bottom), where more than 1,500 metacercariae may be found in one fish. Lafferty and Morris (1996) have shown that parasitism induces abnormal behavior in the fish ("flashing," "surfacing," "contorting," "shimmying," "jerking"), and that these behaviors tremendously increase predation by birds. Based on outdoor experiments, Lafferty and Morris estimated that after twenty days about 80% of the fish that were infected by more than 1,400 metacercariae were eaten, versus approximately 20% of those infected by fewer and only 2% of those that were uninfected. As in the case of *M. papillorobustus,* one effect of the large number of metacercariae was to induce the hosts to swim at the surface, increasing predation by their avian hosts.

In rats parasitized by *Toxoplasma gondii,* a protist that lives in the brain, the capacity to respond to something new decreases ("neophobia") and nocturnal activity increases. Rats then are more vulnerable to predation by cats, within which the parasites must continue their development. Among several parasites commonly found in rats, including *Leptospira* sp., *Cryptosporidium parvum, Coxiella burnietti, Hymenolepis nana,* and *Syphacia muris,* only *T. gondii* requires in its cycle that the rat be eaten by a predator—it is also the only one to induce an increase in nocturnal activity. Moreover, the specific rat behaviors affected by *T. gondii* are only those that would truly increase the probability of predation by cats (Webster, Brunton, and MacDonald 1994).

Numerous other processes of favorization operate much like the previous systems. These include, among others, acanthocephalan/amphipod systems such as *Polymorphus marilis/Gammarus lacustris* and *Corynosoma constrictum/Hyalella azteca* (Bethel and Holmes 1973), for which the definitive hosts are waterfowl; acanthocephalan/insect systems, such as *Moniliformis moniliformis/Periplaneta americana* (Moore 1983b), for which the definitive hosts are rodents; nematode/isopod systems such as *Dispharynx nasuta/Armadillidium vulgare* (Moore and Lasswell 1986), for which the definitive hosts are birds; and protist/vertebrate systems, such as the coccidian *Sarcocystis rauschorum* in the lemming *Dicrostonyx richardsoni* (Quinn, Brooks, and Cawthorn 1987), for which the definitive host is the snowy owl, *Nyctea scandiaca.*

The host's phenotype may also be manipulated to make it more visible to a predator for reasons other than movement. Such manipulation has previously been termed "demimetization" (Combes 1983a), and an example involves *Clinostomum marginatum,* a trematode parasitic in piscivorous birds (mainly small herons). Here the metacercariae are in fish (mainly of the genera *Lebistes* and *Gambusia*) in tropical ponds, and the birds become infected when they eat parasitized fish. Normally the fish have dark backs and thus are not very visible from the top, since the darker color mimics the bottom. This protection is a disadvantage for the parasite, however, since metacercariae will transmit their genes only if ingested by the birds. The trick the parasite performs (obviously the result of natural selection) is to form large white spots on the fish's back that then attract the attention of predators. When approaching the pond it is easy to see the white spots on the fish that carry metacercariae, whereas the healthy fish are much less visible (fig. 7.17).

In the preceding examples the upstream host carrying an infective parasite "makes itself visible" to the predator definitive host thanks to a variety of abnormal behaviors or physical changes. When the "predator" is a biting arthropod, however, it is likely more advantageous to have the upstream host (the one that must be bitten by the vector) stay quiet rather than be more visible. This is the reason *Plasmodium* (the causative agent of malaria) renders vertebrate hosts rather lethargic, thus altering nor-

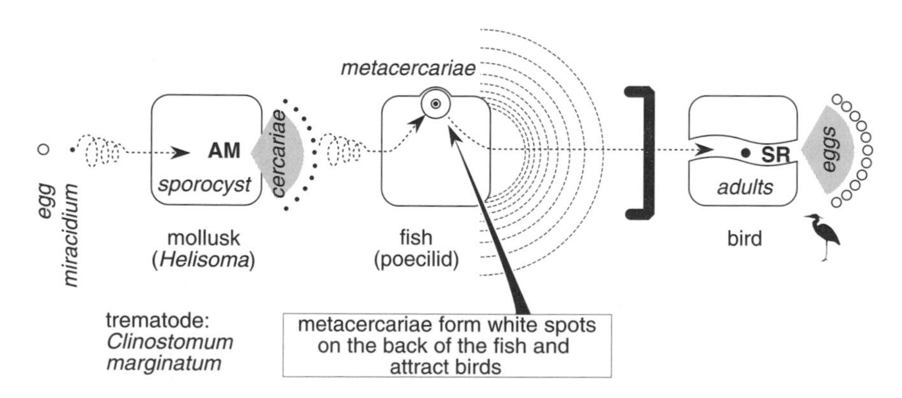

Figure 7.17. Responses of the downstream hosts to the "demimitization" of the parasitized upstream hosts: transmission of *Clinostomum marginatum.* AM = asexual multiplication; SR = sexual reproduction.

mal "antimosquito" behavior (Kingsolver 1987). *Plasmodium berghei, P. chabaudi,* and *P. yoelii,* for example, all make their infected rodent hosts consenting victims in this way. Significantly, this abnormal behavior coincides with the most infective period of the day for the mosquito host (Day and Edman 1983).

Manipulation

Not all processes of favorization use host manipulation by extending the parasites' phenotype (see fig. 7.3). However, this type of manipulation is frequent, if not omnipresent, when transmission depends on the consumption of an upstream host by a downstream predator.

The Origin of Manipulation

What can be said about the phyletic origin of manipulation? In other words, is the character "manipulation of the host" inherited by several species from a common ancestor, or is it independently acquired by each species?

Poulin (1994b) writes: "No doubt several living parasite species share an ability to manipulate host behaviour that was inherited, perhaps without modification, from a common ancestor. On the other hand, it is also likely that host manipulation has evolved independently in different, more distant, parasite taxa." Situating himself in the midst of a controversy (Is it more advantageous to manipulate the host or simply to be in a milieu where immunity is not very active?) to explain the localization of the metacercariae of *D. spathaceum* in the eye of its fish intermediate host (see fig. 7.13, bottom), Poulin postulates that changes in host behavior may sometimes be a fortuitous payoff arising from other adaptations.

The unorthodox idea that phyletic constraints may explain certain such manipulations was originally proposed by Moore and Gotelli (1990). By this they mean there would be a phylogenetic "memory" of favorization that "may have been adaptive in the ancestral association" and would have today lost its significance in some parasite-host systems.

The Mechanisms of Manipulation

When I defined durable interactions at the start of this book, I noted that one characteristic of such associations is that one of the "partners" can send signals of some sort to the other. In certain cited examples favorization may have been a modification pertaining to a morphological or an anatomical characteristic (for instance, the weakening of the host's fins by *Cainocreadium labracis*). In many other cases, however, the manipulation is done partially or totally at the molecular scale, and it is possible that the two processes complement each other. For example, *Diplostomum spathaceum* and *Microphallus papillorobustus* compression of nerve centers by metacercariae may be happening along with secretion of active molecules by the parasites.

In general we must keep in mind that the presence of a foreign genome (that of the parasite) in another living organism cannot not interfere with the other organism's phenotype. Direct products of the parasite's genome (neurotransmitters, hormones, growth factors, enzymes, various signals, etc.) or metabolic products (wastes, secretions, etc.) must necessarily influence at least some of the host's vital processes. The resulting alterations may even be expected to alter the "normal" responses to "normal" signals in the host. For instance, cockroaches *(Periplaneta americana)* respond differently to pheromones from members of their own species depending on whether they are infected with the acanthocephalan *Moniliformis moniliformis* (Wilson and Edwards 1986; Carmichael, Moore, and Bjostad 1993).

It is not surprising that such mechanisms have been retained and amplified by selection when they increase the probability of transmission of the parasite and thus its fitness. Few studies offer precise information about these mechanisms, but those of Helluy (1988) and Helluy and Holmes (1990) should be cited. These authors have analyzed the *Polymorphus paradoxus/ Gammarus lacustris* relationship, starting with the hypothesis that the gammarids' behavior was being manipulated because the gammarids' neurons recognized a neurotransmitter secreted by the parasites.

These researchers have studied the influence of injections of a series of substances (serotonin, octopamine, dopamine, noradrenaline, L-enkephalin, cinanserine, and GABA) known to be involved in neurotransmission in various organisms. The results with serotonin were the most interesting. If parasitized gammarids are put in an aquarium that is partly lighted, they have a strong tendency to gather in the lighted part, whereas most uninfected gammarids stay in the dark part. The experiment consisted of injecting saline containing serotonin into healthy gammarids and observing their behavior. The hypothesis was that serotonin-treated gammarids would move toward the lighted part of the aquarium. The assay was done in parallel with a control assay containing gammarids injected with saline alone to ensure that the saline injection did not have similar results.

The results of one of these experiments are shown in figure 7.18. The injection of saline itself does not induce any significant change, but saline plus serotonin induced increased visitation of the lighted part of the aquarium within about one hour.

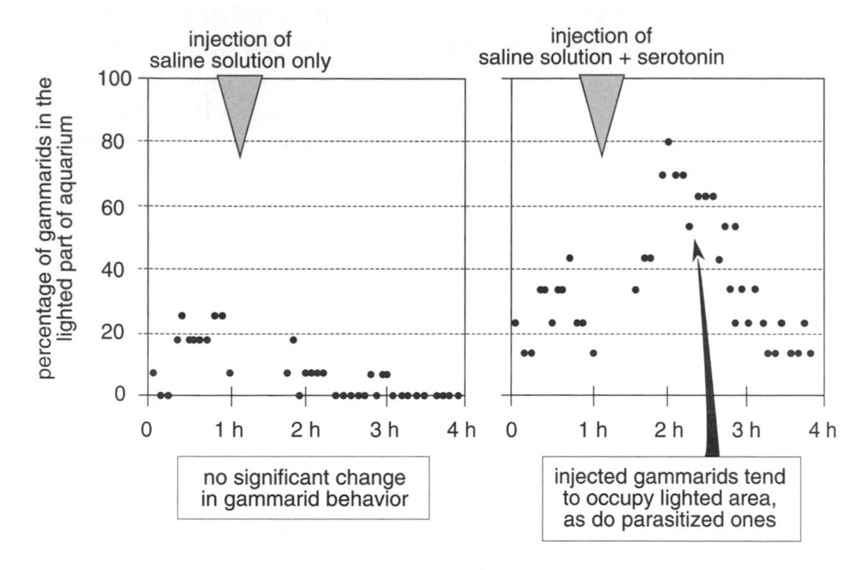

Figure 7.18. Simulation of reactions of gammarids parasitized by the acanthocephalan *Polymorphus paradoxus* after serotonin injection (from Helluy 1988).

A logical conclusion is that injecting serotonin modifies the gammarids' behavior the same way parasites do, which supports the hypothesis that parasites may affect serotonin-related circuits in the host. It may be that the parasite releases its own serotonin; releases an unidentified substance that induces the release of serotonin by the host; releases another substance that mimics serotonin; or induces a slowing of the gammarid's serotonin degradation.

Is the hypothesis of the release of serotonin by the acanthocephalan defensible? In favor of this it must be noted that serotonin has been localized in the nervous system (cerebral ganglion) of some acanthocephalans. In *Macracanthorhynchus hirudinaceus* eleven neurons of the eighty-six identified in the cerebral ganglion use serotonin as a neurotransmitter. The cerebral ganglion of *Polymorphus paradoxus* has seventy-eight neurons, and some of them are certainly serotoninergic. Thus, in theory anyway, serotonin may be produced by the larval cystacanth and diffuse through its tegument. There are also neurons with serotonin receptors in crustaceans (to which gammarids belong), and small traces of serotonin are found in their hemolymph. Therefore nothing prevents the parasite's serotonin from being recognized by the host's neurons, thus inducing a change in its behavior.

On the other hand, we may object that the doses Helluy used (ten milligrams per gammarid) were very high. In brief, although there is some serotonin in the crustaceans' hemolymph, the doses in these experiments were 10,000 times higher than those detected. Thus the state of this research leaves many questions, and Helluy and Holmes (1990) state that comparative studies (using for example, immunohistochemistry) could confirm eventual changes at the level of serotonin in the gammarids' brains when infected. Could we also attempt to "cure" the crazy gammarids by injecting them with 5,7-dihydroxytriptamine, which induces the degeneration of serotonin neurons?

The Benefits of Manipulation

Even though favorization has been a popular subject of study by parasitologists over the past several years, it is not always as firmly demonstrated as one might think. The reason is that it is difficult, if not impossible, to design experiments or collect quantitative data pertinent to these processes in nature. Nevertheless, it is still absolutely necessary to empirically demonstrate the adaptive value of the manipulation, and only based on such proofs (or at least more solidly supported supposition) can any such change in host physiology or behavior rightly be considered advantageous for the parasite (Poulin 1995b).

Therefore, when there is a change in the behavior of a parasitized host other hypotheses must also be tested before we conclude that such a change is the result of manipulation by the parasite to benefit its transmission.[9] These hypotheses are that there is a benefit for the parasite but that it is not linked to its transmission; that there is no benefit for either the parasite or the host; and that the benefit is for the host. It is also possible that there is a benefit for both the parasite and the host.

Benefit for the Parasite but Not Linked to Its Transmission

Evolution is not lacking in adaptations that, initially selected for one purpose, have been "used" and sometimes perfected in the frame of reference of another. For instance, birds' wings were certainly not initially selected because they allowed flight, and it may be the same for many parasite adaptations, particularly those affecting manipulation.

For example, in the life cycle of *Diplostomum* the presence of metacercariae in the eye is usually interpreted as being selected for because it gives the parasite an advantage by making the fish blind and thus easy prey for the next host. However, another interpretation is that these

9. See a detailed discussion on relationships between parasitism and hosts' behavioral changes in Keymer and Read (1991).

trematodes may occur in the eye or the nervous system simply because these tissues are poor in immune reactions (Szidat 1969). In short, there might have been selection for metacercariae settling in sites that are less and less aggressive, ultimately leading them to the eye. This hypothesis is logical, but in my opinion the process occurs at the same time as favorization and not instead of it. Clearly, two such distinct pressures may work together in selecting the same trait.

No Benefit for either the Parasite or the Host

Earlier I emphasized that any pathogenic effect caused by a parasite on its intermediate host favors transmission even outside any selective process on the parasite's genome. I noted that a clue to the occurrence of favorization would be the localization of the parasite in a particularly "sensitive" organ. Comparing two serious parasitoses of sheep will show that distinguishing between selection and nonselection is sometimes difficult.

One major parasitosis is caused by *Multiceps multiceps*. As previously discussed, the life cycle of this cestode has two hosts: a sheep that becomes infected by eating eggs and that harbors the parasite (generally in the brain) in the form of budding larvae (cysticerci), and a carnivore (wolf, dog, etc.) that becomes infected by eating the sheep and that then harbors adult parasites in its intestine. Sheep that are infected by the cysticerci exhibit "gid," a "whirling" disease that makes them easy prey for carnivores. It is justifiable to consider the localization of the parasite in the brain of the sheep and the resulting serious pathogenicity a process of manipulation.

The second major parasitosis is caused by *Oestrus ovis*. This parasite is a dipteran that has free-living adults; its parasitic larvae occur in the nasal sinuses of sheep and other herbivores. Females of *O. ovis* dive-bomb the sheep's nostrils and deposit their larvae (not eggs) in a fraction of second. Almost immediately the larvae reach the sinuses, where they grow and induce a serious pathogenicity, notably characterized by heavy and difficult breathing. Later the larvae leave the nostrils and pupate on the ground. Since they are present close to the sheep's brain, the larvae sometimes induce whirling. Even in the absence of this symptom, however, the sheep are handicapped and have a hard time breathing. Hence, if a predator shows up these infected sheep are probably as easy prey as those infected by *M. multiceps*.

In both cases, therefore, the parasite induces selective predation by the carnivore, although there is an essential difference: in *M. multiceps* predation by the canid is favorable to the parasite; in *O. ovis* such predation leads to the death of both the parasite and the sheep.

Thus, here are two parasites that by their precise localization in the host equally decrease their host's fitness, but with diametrically opposite consequences for themselves. It is therefore certain that the pathogenic effect induced by *O. ovis* exists for reasons that have nothing to do with parasite transmission. The simplest explanation is that because of its location and its need to grow, *O. ovis* cannot be less pathogenic without jeopardizing its own reproductive success.

In some cases the benefit given by the manipulation seems obvious but leaves some doubt about its effect on transmission. This is the case for the pseudophyllidian cestode *Spirometra mansonoides,* whose life cycle has the following steps: from the egg a free swimming coracidium hatches, which is then ingested by a copepod, where it develops into a procercoid. If the copepod is eaten by a reptile (a snake) or rodent (mouse), the procercoid develops into a plerocercoid in the body cavity. If the snake or the mouse is eaten by a carnivore (a cat), the plerocercoid becomes an adult cestode that then lays eggs.

Let's look at one of the steps: the transfer of the parasite from the mouse. If juvenile mice are hypophysectomized they stop growing because the secretion of growth hormone (GH) is inhibited. If these mice are then infected with *Spirometra* plerocercoids, they immediately restart their growth. Infecting normal mice accelerates their growth, and infected animals rapidly become huge (Mueller 1974). *Spirometra* larvae are known to produce a molecule (*Spirometra* growth factor [SGF], also called plerocercoid growth factor [PGF]) that has the same effect as the growth hormone without having the same chemical composition. SGF is a protein that is first expressed as a precursor in the plerocercoids' membranes and then is cut into a functional state and released. It is a cysteine-proteinase of 27.5 kilodaltons showing about 50% homology at the nucleotide level with a mammalian cysteine-proteinase but none at all with growth hormone, even though it interacts with GH receptors (Phares 1996). Besides its effect on growth, SGF seems to play a role in the digestion of host tissues, and it may also have an immunosuppressive effect on host defense mechanisms by splitting molecules involved in host immunity (Phares 1987).

The interpretation that comes to mind in terms of favorization is that the gene that codes for SGF is the result of selection: cats certainly prefer fat mice to slimmer ones and probably catch them more easily, since they run slower. This appears to be an ideal favorization scenario, since the chances of the parasites' ending up in the cat's intestine are then considerably increased. Some parasitologists consider this interpretation obvious; others question it because the parasite acts on mice but not, as far as we know, on the snakes that in nature seem to be the usual intermediate host. Therefore a certain amount of reserve is necessary, and objections to such "adaptationism at any cost" must always be kept in mind.

Benefit for the Host

Changes in host behavior linked to parasitism may happen without a role that favors transmission (Poulin 1994b). However, these changes may have value for the host and therefore remain an adaptive response. Often it is not easy to make a choice between these two alternatives. In this regard, few cases have been discussed as thoroughly as the behavioral change in the bumblebee, *Bombus lucorum,* when parasitized by the larvae of the conopid dipterans *Sicus ferrugineum* and *Physocephalus rufipes* (Schmid-Hempel and Müller 1991; Müller and Schmid-Hempel 1992). Bumblebees live in colonies with a queen and workers, and it is the workers that collect pollen and nectar from flowers for the colony. With rare exceptions, only the queen produces eggs and thus transmits her genes. Therefore the labor of the workers (all are the queen's daughters) indirectly aids in the transmission of their genes through the queen, but only indirectly.

Conopids attack bumblebee workers while they are flying near flowers and lay one egg in a bee's abdomen. The egg rapidly hatches, and the larva feeds at the host's expense. The host is seriously affected by the parasitoid and dies twelve days after infection. Infected bumblebee workers change their behavior significantly. Schmid-Hempel and Müller (1991) have shown that if workers are captured as they enter or leave their hive, parasite prevalence is low (12%), whereas it is high when workers are captured in the fields (76%). Thus we may deduce that parasitized *Bombus* do not often visit the hive, that they spend much more time outside, and that they probably stop working for the collective. This behavioral change occurs when the conopid larva is still very small and apparently causing little damage to the host.

These data can be explained by several hypotheses. First of all, nothing prevents the behavioral change from being just a simple pathogenic effect of the parasitoid with no particular value for either the parasitoid or the host. A second hypothesis is that the modified behavior confers some advantage for the parasitoid. With this altered behavior the infected bumblebee worker might then better ensure spatial distribution of the parasitoid or allow the parasitoid to emerge and continue its development outdoors and not in the humid confines of a bumblebee hive.

A different view (Poulin 1992a) would be to consider the change favorable to the host, since when the bumblebee worker is away from the hive it is not using up "banked energetic reserves." It might also attract new conopids to itself, alleviating attacks on its sisters (this is kin selection: the behavior increases the reproductive success not of the individual itself but rather of the individuals that share, at least in part, the same genes). Or it might attract various predators (birds, spiders) and by this adaptive suicide benefit the colony.

Another study shows how cautious one must be when interpreting changes in behavior. Moore and Gotelli (1992) have shown that roaches (*Supella longipalpa*) seek darker shelters when they are infected with larval stages of the acanthocephalan *Moniliformis moniliformis* (whose adults are in rodents). This contradicts all research done on similar life cycles that shows that infected intermediate hosts have a tendency to go to better-lighted areas than do uninfected hosts. Here the observed change in behavior may decrease the probability of encountering the definitive host. Therefore some pathologies may have nothing to do with "encounter genes."

Benefit for both the Parasite and the Host

There is at least one life cycle for which we can reasonably formulate the hypothesis that the manipulation benefits both the parasite and the host. This is the life cycle of the trematode *Podocotyloides stenometra,* found in the Hawaiian Islands. Cercariae of this worm are produced in mollusks and encyst in two species of coral, *Porites compressa* and *P. lobata.* Infected polyps are white to bright pink and lose their ability to retract into their cups. The definitive hosts are coral-feeding butterflyfish of the genus *Chaetodon.* The fish preferentially graze infected polyps, since they cannot retract when the predator approaches. This is, then, a case of favorization halfway between "more accessible" and "more visible."

Aeby (1991) carried out a unique experiment comparing the growth of infected versus uninfected corals and of parasitized corals protected from fish by a cage versus parasitized unprotected corals. After sixty days the following conclusions were drawn (fig. 7.19).

In the comparison of infected and uninfected coral (experiment A), the growth of coral that harbored metacercariae was strongly affected, since weight gain was half that observed in healthy corals (although metacercariae are apparently inactive in polyps, their presence induces a decrease of the number of zooxanthellae and, as a result, a perturbation in both calcium metabolism and coral growth).

Relative to the effect of protection (experiment B) parasitized corals that were exposed to predation had, at the end of the experiment, three to four times fewer metacercariae than protected corals, showing the effect of predation by the fish. Further, although heavier damage could have been expected, since predation is added to the effect of parasitism in the case of parasitized corals exposed to predation, the weight gain of parasitized exposed corals was similar to that of parasitized but protected

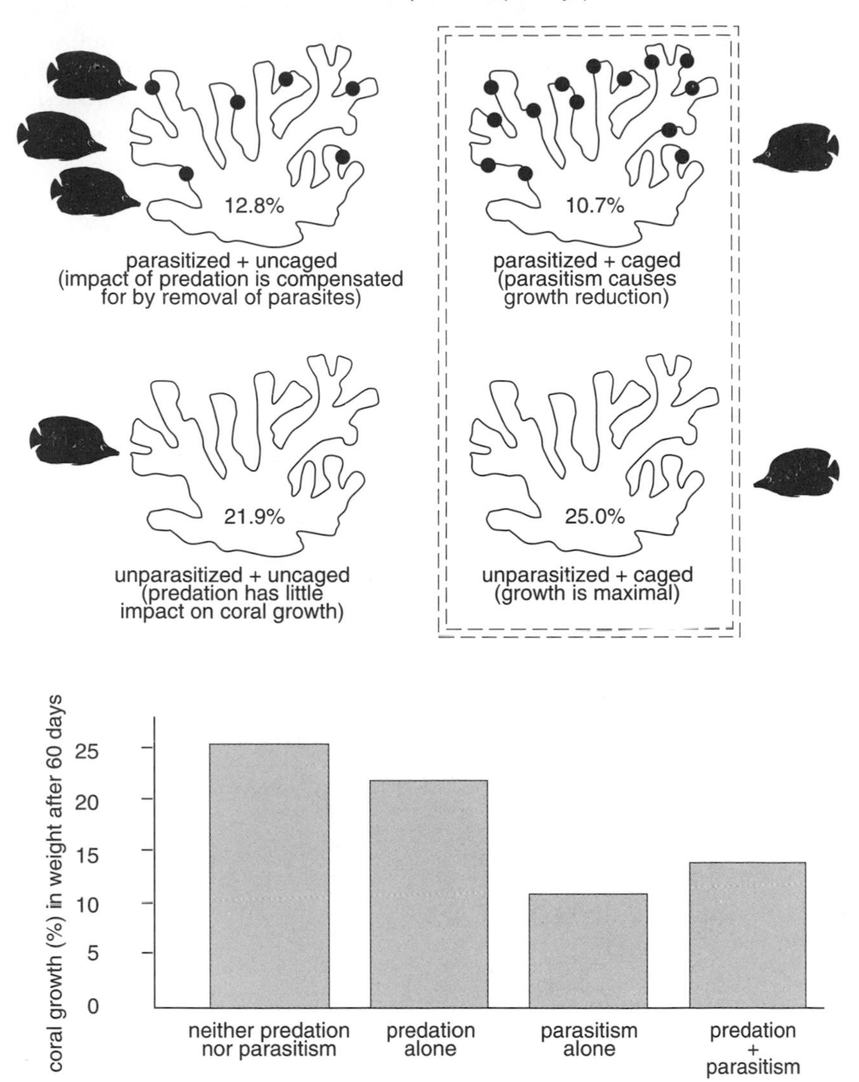

Figure 7.19. Experiment of Aeby (1991). Effect of parasitism on coral growth in the presence or absence of predation. See text for details.

corals. The probable explanation is that "the negative effect of predation may have been offset by the positive effect of parasite removal" Aeby (1991). This shows clearly that favorization, by inducing the consumption of parasitized polyps, was advantageous for the host while ensuring the transmission of the parasite. It can even be imagined that in some cases predation by fish might have a positive effect on the coral: this would obviously be possible only if predation does not affect the entire coral.

The Cost of Manipulation

That prey manipulation is costly to the parasite is not disputable. For example, secretion of an active molecule by *Polymorphus paradoxus,* the migration of *Diplostomum spathaceum* cercariae in the fish's body to reach the lens, and other manipulations all expend energy.

Poulin (1994b) showed that natural selection must lead to an *optimal manipulative effort* that takes into account manipulation costs and benefits for the parasite, and he reminded us that the more numerous the number of infective stages, the less advantageous manipulation is. "Logically, what a particular species invests in fecundity need not be invested in favorization, and vice versa" (Combes 1991d).

Poulin suggests that the optimal manipulative effort should be *inversely proportional* not only to fecundity but also to the following.

The usual size of parasite infrapopulations: when infrapopulations are large the individual effort of manipulation is reduced by sharing the costs among individuals. If, on the other hand, parasites are likely to find themselves alone in the host, the full cost of such manipulations must be borne by each parasite.

The longevity of both the parasite and host: if the parasite survives for a long time in the host and if the host also survives a long time during infection, then it may be advantageous for the parasite to be "patient" rather than manipulative.

The intensity of predation by the definitive host: if the definitive host usually eats large quantities of prey susceptible to carrying the infective stage, it is less and less advantageous to invest in such manipulations.

On the other hand, Poulin also suggests that the optimal manipulative effort should be *proportional* to the usual prevalence of the parasite. If prevalence is high, transmission between prey and predator may be easily accomplished without manipulation, and the benefits of manipulation drop to zero. However, as Poulin also notes, this situation may trigger competition between conspecific parasites to have the prey that harbor them be eaten. This competition could then lead to either an increased manipulation or a polymorphism in the manipulation itself.

An original way to avoid the energetic costs of favorization is to pirate the favorization process of a manipulative species. An example is the trematode *Maritrema subdolum.* Transmission of this parasite was studied by Helluy (1982) in parallel with that of *Microphallus papillorobustus.* The life cycles of the two trematodes are very similar and include the same mollusks, the same gammarids, the same gulls, and the same ecosystems. The diagram in figure 7.15 (center) represents the life cycle of both *Ma. subdolum* and *Mi. papillorobustus* except for one detail: the metacercariae of *Ma. subdolum* do not go to the gammarid's head but instead settle anywhere in the body. Experimentation (fig. 7.20) has shown that if healthy gammarids are placed with mollusks that emit cercariae of *Ma. subdolum,*

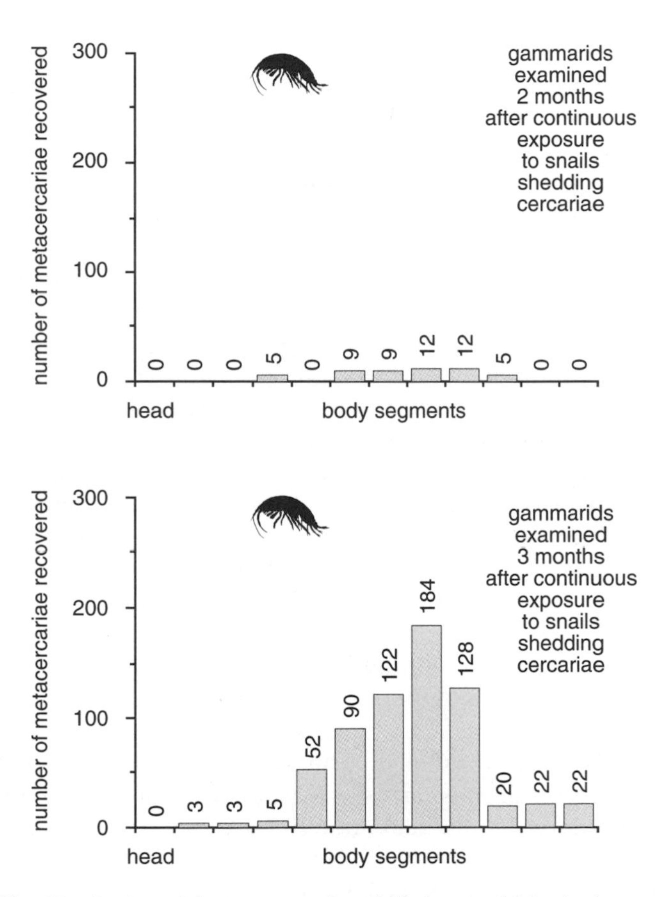

Figure 7.20. Distribution of the metacercariae of *Maritrema subdolum* in the gammarids' bodies: contrary to the case in *Microphallus papillorobustus* (fig. 7.15), there is no attraction for the cephalic region even at the beginning of the experiment (from Helluy 1982).

metacercariae in the gammarids are found in the abdominal region and not in the head after three weeks (compare with the results presented in fig. 7.16).

At this stage of her work, Helluy had two hypotheses to work with: that this species had not invested in favorization but only in increased fecundity or that the metacercariae acted at a distance, as do the previously discussed cystacanths of *Polymorphus paradoxus.*

Further research (Nathalie Grande, unpublished data) has shown that neither of these hypotheses was in fact correct. Instead, *Ma. subdolum* "pirates" the favorization process by preferentially penetrating gammarids that are already parasitized by *Mi. papillorobustus* and hence are already "crazy." An investigation done in a natural population of gammarids in Camargue (France) has shown the following: out of 464 gammarids, 377 were parasitized by *Mi. papillorobustus* and 202 by *Ma. subdolum.* A simple calculation shows that *Ma. subdolum* associated with *Mi. papillorobustus* could have been expected in $(377 \times 202) \div (464 \times 464)$ cases, or about 35% of the time if the association was due to chance alone. However 195 out of the 202 gammarids infected by *Ma. subdolum* were also infected by *Mi. papillorobustus,* which instead is almost 97% of the cases! Nathalie Grande (unpublished) proposed three hypotheses to explain the preferential infection of "crazy" gammarids by *Ma. subdolum:* that the cercariae of *Ma. subdolum* may be capable of recognizing a gammarid parasitized by *Mi. papillorobustus;* that the cercariae of *Ma. subdolum* may migrate toward the water surface and thus to the zone where parasitized gammarids with *Mi. papillorobustus* already are; and that mollusks parasitized by *Ma. subdolum* may come close to the surface to emit their cercariae.

Whichever hypothesis is correct, *Ma. subdolum* clearly takes advantage of the favorization provoked by *Mi. papillorobustus,* and in any case the life cycle of *Ma. subdolum* offers a nice demonstration of parasitism's spirit of "genetic economy." Parasites always appear ready to have others ensure their vital functions (most of the time the host, but in this case another parasite). This is, then, an example of Richard Dawkins's extended phenotype within the parasite world itself: *Ma. subdolum* uses to its advantage the extended phenotype of *Mi. papillorobustus* and is thus a parasite of a parasitic process.

The case of *Maritrema subdolum,* which chooses hosts with behaviors already modified by another parasite, is probably not unique. One may wonder, for example, if the trematode *Renicola buchanani,* which is in the liver of all fish parasitized by *Euhaplorchis californiensis* (see above and fig. 7.15), also takes advantage of the favorization induced by the latter. *R. buchanani* has the same definitive hosts (piscivorous fish) as *E. californiensis* and in this way may take advantage of behavioral problems induced by the presence of *E. californiensis* in the host's skull (Lafferty and Morris 1996). Later we will see more instances of such "pirating" among parasites, such as an insect parasite of plants that lays its eggs in a gall on the plant, itself induced by another parasitic insect (chapter 19).

The process of transmission of *M. subdolum* should not be confused with that in which the parasites randomly take advantage of the favorization obtained by other species. For instance, transmission of various species of cestodes of mammals implies a budding cysticercus, usually in a vital organ, that weakens the carrier (an herbivore) and favors predation by the definitive host (a carnivore). This is again the case of *Multiceps multiceps,* whose cysticercus is voluminous and located in the sheep's brain. However, sheep that harbor *M. multiceps* and that are devoured by predators as a result often transmit other taxonomically independent parasites that have similar life cycles—for example, the larvae of another cestode *(Taenia ovis),* a pentastomid *(Linguatula serrata),* and a coccidian *(Sarcocystis ovicanis),* none of which seem to preferentially infect *M. multiceps*–infected sheep themselves.

The Timing of Manipulation

Manipulation is advantageous only if the change in the host happens while the parasite is infective, generally after maturation. Several studies demonstrate this, and modification of the intermediate host's behavior effectively starts only once the parasite can survive in the definitive host. For example, Galaktionov (1993) carried out research on four species of microphallid trematodes, *Microphallus pygmaeus, M. piriformis, M. pseudopygmaeus,* and *M. triangulatus.* Their life cycles are as follows: the first host is a mollusk of the genus *Littorina* and the definitive host is the eider duck, *Somateria mollissima* (whose feathers are used to make quilts). Between mollusk and duck there is no intermediate host because cercariae, produced in sporocysts, remain on site in the mollusks. Therefore it is by eating mollusks that carry infective metacercariae that eiders become infected.

Galaktionov (1993) showed that infected mollusks locate at distinctive places on the shore depending on the stage of development of the sporocysts (young, not yet packed with metacercariae and thus not yet infective, or older and carrying infective metacercariae).

For example, when one divides the thallus of the alga *Fucus* into approximately three zones, the littorinids infected by young sporocysts are found at the base, where they are difficult for the birds to reach, while those infected with older sporocysts locate mainly at the top, where they are much easier to capture (fig. 7.21). Galaktianov's numbers are demon-

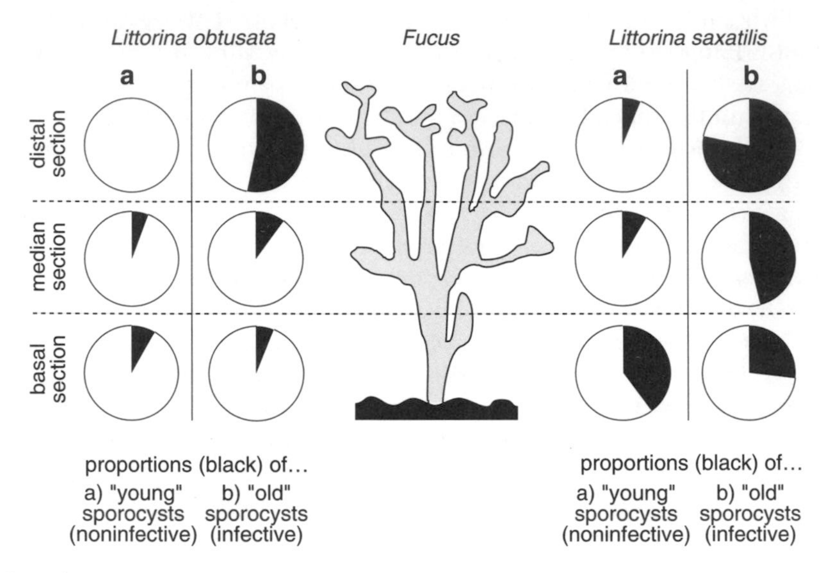

Figure 7.21. Favorization in the life cycle of *Microphallus pygmaeus:* distribution of two species of littorinids depending on whether they harbor infective sporocysts. In the case of *Lymnaea obtusata* (left) as well as *L. saxatilis* (right), carriers of noninfective (young) sporocysts are in the majority on the basal part of the fucus, while carriers of infective (older) sporocysts are found mostly on the terminal portion of the blades where they are more accessible to potential bird hosts (adapted from Galaktianov 1996).

strative, and there is no reasonable explanation other than favorization to explain why the manipulation of the host's behavior occurs only when the parasite is infective, thus preventing birds from eating parasites that have not yet reached the stage of development required for successful transmission.

Still other studies have shown that healthy littorinids at low tide normally shelter either in the empty carapaces of barnacles (Granovitch and Sergievskii 1989) or in algal cavities (Mikhailova, Granovitch, and Sergievskii 1988), whereas those that harbor mature sporocysts of *Microphallus pygmaeus* stay outside where they are easier prey for the birds. That is, everything happens as if the parasite inhibits the mollusks' capacity to respond normally to the retreating sea and as if the parasite manipulates the mollusk host to make it go (or stay) where it has the highest probability of being eaten.

A similar timing can be observed with the pseudophyllidean cestode *Schistocephalus solidus*. The parasite (in the form of a procercoid) is found in copepods of the genus *Cyclops*. Infected copepods show increased activity that may have a favorable influence on predation by small fish

(Wedekind and Milinski 1997). However, there is an increase in activity only when the procercoids are mature and, more important, when there are two (Urdal, Tierney, and Jakobsen 1995). The parasites are next found in the form of a plerocercoid in small freshwater fish such as the stickleback *Gasterosteus aculeatus*. Several studies have shown that the behavior of *G. aculeatus* is modified by the presence of *S. solidus*. In particular, warnings by other fish induce an increase in activity in parasitized sticklebacks, making them easier for predators to capture (Giles 1987). However, other studies have shown that plerocercoids are infective only if they have had time to grow in the stickleback (their weight must be over fifty milligrams) and that host behavior is modified only after they have passed this critical threshold (Tierney and Crompton 1992). Moreover, it appears that the fleeing behaviors of sticklebacks that carry immature parasites may even be more efficient than those of healthy sticklebacks!

Another example of timed manipulation of the host phenotype exists in the life cycle of the cestode *Hymenolepis diminuta*, whose intermediate host is the mealworm *Tenebrio molitor* and whose definitive host is the rat. When mealworms are infected by cestodes their defensive glands (which give *Tenebrio* its bad taste) respond less to stimulation and contain fewer toxins (toluquinone and m-cresol) than those of healthy mealworms (Blankespoor, Pappas, and Eisner 1997). This increases the probability of their being eaten by rats. It seems that the parasite hijacks metabolites that are normally used by *T. molitor* in the functioning of the defense system, and the inhibitory effect on the glands occurs nine days after infection, just at the time cysticercoids are infective.[10]

Responses to Manipulation

When a parasite is in an intermediate host that must be eaten, we have the sharpest possible conflict between a parasite and its host. The parasite will transmit its genes only if the intermediate host is eaten, from which arises the selective pressure for the manipulation, and the intermediate host will transmit its genes only if it is not eaten, from which arises the selective pressure to oppose the manipulation.

It is therefore very probable that there is selection for "antimanipulation" genes in populations of manipulated hosts. However, this is an area of research yet to be explored, leading Poulin, Brodeur, and Moore

10. Chemical defenses of certain species of insects are more perfected than those of *Tenebrio* and thus are more expensive to inhibit. This is probably why arthropods with strong chemical protection are only rarely used as hosts in predation-dependent parasite life cycles.

(1994) to note that "we are unaware of any published example of host opposition to behavioural manipulation."

Favorization in Natural Populations

There is no reason to systematically doubt the deductions made from morphoanatomical or experimental observations in the laboratory. However, any demonstration of favorization, and particularly of host manipulation, in natural environments would be of great value. I cite here three ways to demonstrate, or at least strongly suggest, that favorization occurs in natural populations.

Favorization Deduced from Epidemiology

Pseudoterranova decipiens is a nematode that lives in the stomach of gray seals *(Halichoerus grypus)*, where it lays its eggs. Eggs fall to the bottom of the sea and give birth to larvae that are ingested by benthic crustaceans. When these crustaceans are eaten by various fish, including cod *(Gadus morhua)*, the larvae localize in the fish's muscles. When fish are captured by seals, the parasites become adults and the life cycle continues (fig. 7.22). Therefore there is a succession of free-living phases (eggs and larvae) and three parasitic stages (in the crustacean, fish, and seal). What is interesting to us here is the transmission from fish to seal.

Investigations in the field allows us to know the average number of infective stages per cod, to estimate the number of fish caught by one seal in a given time frame, and to estimate the life span of an adult nematode. These three variables then let us calculate what the average parasite load in a seal "should be." Because the real load may be known from field in-

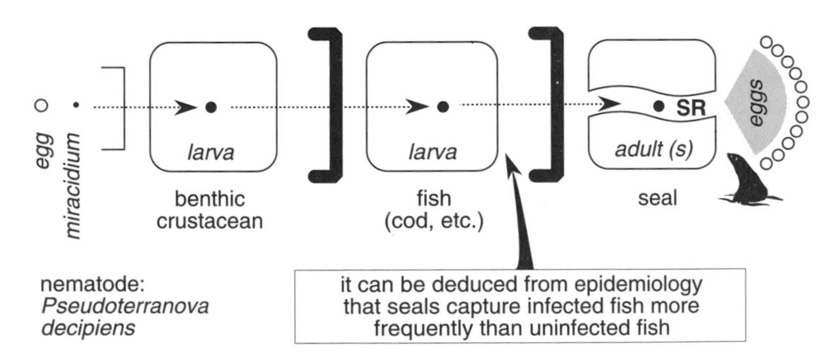

Figure 7.22. Favorization deduced from epidemiology: the life cycle of the nematode *Pseudoterranova decipiens*. SR = sexual reproduction.

vestigations, it is then possible to compare the expected value with the observed. The study of Des Clers and Wootten (1990) shows that seals have more parasites than expected based on the calculations. The only logical explanation is that favorization is occurring and that seals preferentially catch those cod that are the most parasitized, probably because they do not react as fast or as well to attacks. The hypothesis necessitates only that the distribution of larvae in the cod be strongly aggregated, which it is.

Studies also show that favorization, at least in certain cases, is far from "specific." Such processes favor the ingestion of the parasite not only by species in the host spectrum but also by other species in the ecosystem that have similar ecologies and behaviors. In the case of *P. decipiens,* when cod are weakened they are victims not just of seals but also of all other predators, including other fish. There is, then, a "wide range of encounter" that explains the possibilities of lateral transfer at the evolutionary level.

Favorization Deduced from Field Experiments

Moore (1983a) is to my knowledge the only researcher to have truly verified the efficacy of favorization under totally natural conditions. *Plagiorhynchus cylindraceus* is an acanthocephalan (fig. 7.23) with cystacanths that develop in pill bugs *(Armadillidium vulgare)* and that as adults live in the digestive tract of starlings *(Sturnus vulgaris)* and other birds. Infected pill bugs show three changes in behavior: less marked hygrotactism, less attraction for shelter, and a preference for well-lighted areas. Thus the

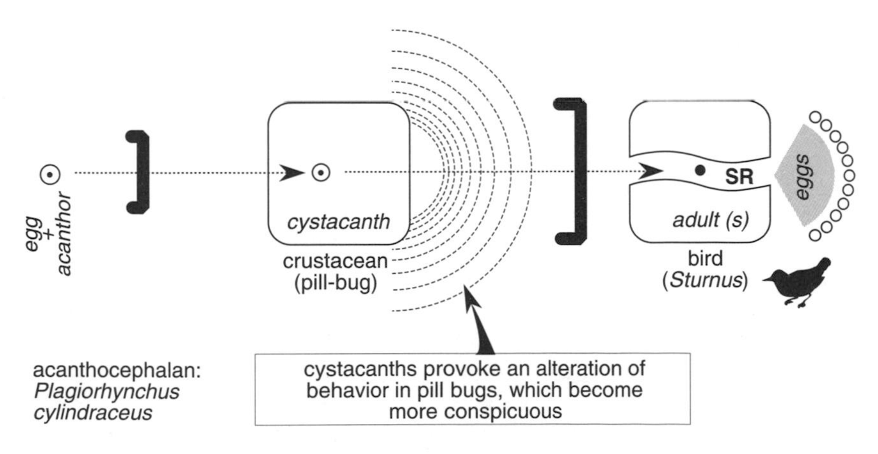

Figure 7.23. Favorization in the life cycle of the acanthocephalan *Plagiorhynchus cylindraceus.* SR = sexual reproduction.

parasitized upstream host behaves so that its level of predation by the downstream host should logically be increased.

First, Moore installed starling nests on the campus of the University of New Mexico at Albuquerque. During the reproductive season chicks were ligated by placing strings around their necks. The chicks were not killed or injured, but they could not swallow prey brought back by their parents. Moore observed that chicks were fed about one pill bug every ten hours. By analyzing the prevalence of *P. cylindraceus* in the pill bugs on campus (0.4%) and in the mouths of eighteen-day-old starlings (32%), a simple calculation allowed her to see that if parent starlings were taking pill bugs randomly, the prevalence in chicks should have been half the number actually found. Therefore there was an increased predation on infected pill bugs versus uninfected pill bugs, and this difference was correlated with the observed manipulation of the pill bugs' behavior by the acanthocephalan.

Favorization Deduced from Mathematical Models

Since experimentation similar to that carried out by Moore is impossible in most cases, mathematical modeling is yet another useful approach to confirm the role of favorization in the natural environment. However, let me first state that favorization is not always quantifiable. For example, attempting to quantify the advantage trematode cercariae receive from being able to swim in an aquatic milieu containing their living targets does not make much sense and would be like trying to quantify the advantage a fox has in visiting the henhouse.

A modeling approach essentially focuses on manipulation and its outcome: when transmission of a parasite involves a predator-prey system, modeling this system allows one to quantify the effect of manipulation of the intermediate host by the parasite as well as the effect of capturing infected prey relative to uninfected ones.

Using modeling, Dobson (1988) showed that changes in behavior that make the carriers of infective stages more susceptible to predation both increase the parasite's reproductive success and decrease the minimum number (threshold number) of definitive hosts required to maintain the parasite in nature. Among other consequences, parasite-induced behavioral changes in hosts may be especially favored in fragmented or very small habitats; habitats not frequently visited by predators; or habitats with intact populations within which the parasite is newly introduced.

Dobson also insists that favorization cannot go "too far" without the risk of destabilizing the parasite-host system.

Genes Not to Encounter

In each of the preceding examples there is in the parasites' genome a selection for what I have called "encounter genes." These refer, of course, to those genes that function to increase encounters with hosts in which parasites can continue their development. That is, they refer to genes that open the encounter filter.

One may also ask, however, Are there some situations in which parasites choose to close the encounter filter? The answer is yes, and this happens in the following situations.

First, it must be understood that new adaptations to encounter hosts may be selected only if they increase the parasites' fitness. We have seen (chapter 6) that there is an optimal virulence above which any increase in damage inflicted on the host is reflected in a decrease in the parasites' fitness. In this situation no characteristics to further open the encounter filter may be selected.

Second, we must also keep in mind that encounter is advantageous only if it happens with adequate hosts—those that are neither unsuitable nor resistant. This is certainly one of the true problems parasites face: a mechanism of favorization is advantageous only if it leads to the "right" host and not the "wrong" one. It would be disadvantageous, for example, if weakening a prey animal by manipulation made it more consumable by nonhost predators than by host predators.

Third, the free-living stages of many parasites, particularly those that live in plankton, are prey for many organisms. We may wonder, for instance, if all cercarial emergence patterns were selected for by pressures linked to host encounter because of this. An alternative hypothesis is that in some cases such rhythmic patterns were selected to avoid predators: hence the rhythm of emergence may not always overlap the hosts' rhythm of activity but rather may be opposed to a predator's rhythm (Shostak and Esch 1990). There may thus be a cost to having an emergence pattern, and if this is so, selective pressures may be opposed to one another, with selection retaining the compromise that gives the best cost-benefit ratio.

Fourth, once in the host, parasites have the advantage of not being eaten by predators if predation is not necessary for the continuation of the life cycle. Host manipulations toward sheltering are mainly observed in parasitoids.[11] For example, diapausal larvae of the parasitoid *Aphidius nigriceps* induce their host aphids, *Macrosiphum euphorbiae,* to move toward

11. For a general overview of parasitoid strategies, see Vinson (1975).

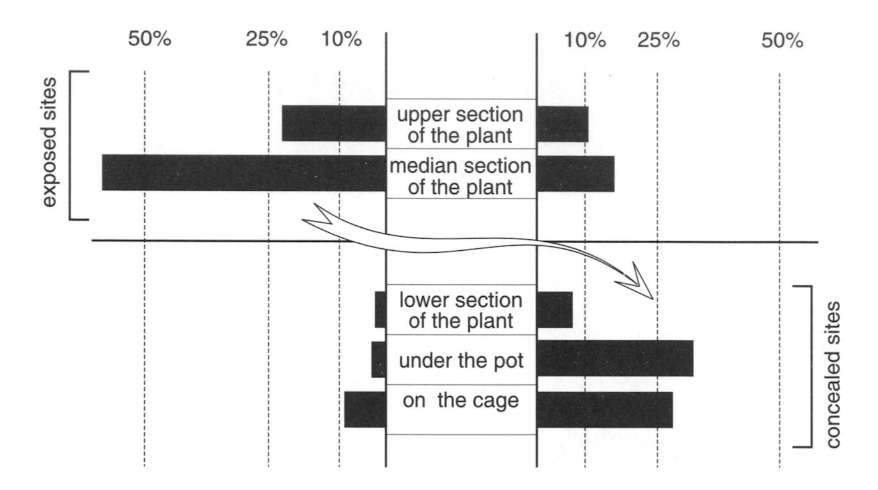

Figure 7.24. Distribution of *Macrosiphum euphorbiae* mummies containing nondiapausing (left) and diapausing (right) *Aphidius nigripes* on tomato plants: diapausing individuals tend to select sheltered overwintering sites (symbolized by the arrow) over the entire height of the plant (adapted from data of Brodeur and McNeil 1989).

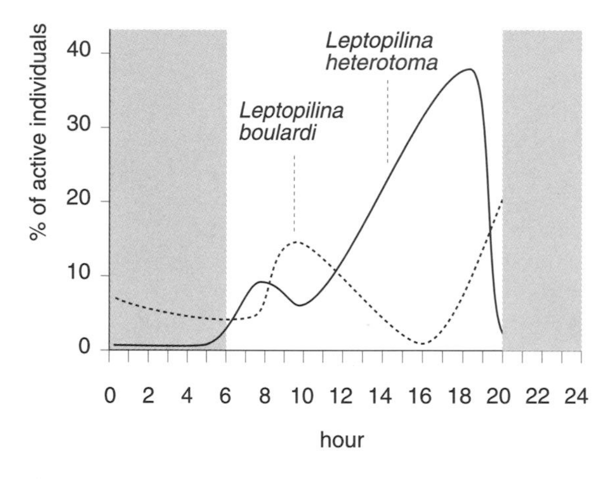

Figure 7.25. Differences in activity period between two species of parasitoids exploiting the same species (modified from Fleury 1993). Shaded areas = night

habitats where the risks of predation by various invertebrates and hyperparasitism by other parasitoids are less than in the aphids' normal habitat (Brodeur and McNeil 1989, 1992; fig. 7.24). Nondiapausing parasitoids also modify the aphids' behavior, but in a different way: in this case hosts leave the inner side of the basal leaves they are on and mummify on the outer surface of apical leaves where there is less parasitoid pressure

(Brodeur and McNeil 1992).[12] Fritz (1982) notes that a manipulation that decreases the level of host predation is exactly the opposite of one that increases the level of predation when parasite transmission occurs from prey to predator, but that the positive result on the parasite's fitness is the same.

Last, when there is competition between parasites, it may be advantageous for some genes to be selected that avoid encounter "with others."

This avoidance of competitors seems to happen between species of adult parasitoids that attack the same host species. Logically, since the host species is the same, the selected rhythms should be similar. However, Fleury (1993) showed that infection happens as if different parasitoids species were trying to avoid each other (fig. 7.25). In his study, all three wasps, *Asobara tabida, Leptopilina heterotoma,* and *L. boulardi,* attack fruit flies and are present in the same locality. The first species is active in the morning, the second in the afternoon, and the third in the evening. If the origin of this shifting is really a tendency to avoid each other, it shows, as in the case of predation affecting cercarial emergence patterns, that contradictory pressures may interfere at the level of the encounter filter and oppose the selection of genes ideally suited for host encounter.

Once parasites are settled in the host, the encounter with newcomers is avoided by "closing the door in their faces," that is, by helping the host close the compatibility filter. We shall see later that some parasites can do so by "immunizing" their hosts against their own species as well as others.

---

12. To my knowledge, this is the only experimental demonstration of a parasite's ability to differentially manipulate a host depending on its physiological state.

8 Genes to Avoid

Mate choice in experimentally parasitized rock doves:
lousy males lose.
Clayton 1990

Attack and defense invariably favor diversification.
Frank 2000

We have just discovered the many types of behavioral adaptations that through their actions allow parasites to encounter their hosts. There are, however, far fewer behavioral adaptations that allow hosts to avoid parasites. This is most likely due to three reasons.

First, in the arms race of encounter the outcome of the confrontation is more vital for the parasite than for the host. In a sense the parasite is fighting for its life—if it doesn't find a host, it dies. A host, on the other hand, usually runs the risk only of decreased health or increased exposure to predation. Things are different in predator-prey systems where, as Dawkins and Krebs (1979) and then Poulin, Brodeur, and Moore (1994) remind us, the prey (the equivalent of the host) is running for its life while the predator (the equivalent of the parasite) is running only for its meal.

The second reason is that in the arms race for compatibility the host has a second chance to get rid of the parasite (compliments of its immune system), whereas a prey item caught by a predator has little or no chance to escape (fig. 8.1).

The third and final reason is that selecting behavioral traits that allow a host to avoid a parasite is costly, and such behaviors often interfere with all sorts of other activities. Here we will look at some host adaptations to avoid encountering the infective stages of parasites.

Optimizing Appearance

Prey species' use of shape and color to hide themselves is a favorite subject of numerous textbooks. Notably, in predator-prey systems the selective pressures due to predators often maintain appearances that camouflage an animal in its habitat (see Pasteur 1981). In this context zebras have left the specialists perplexed. Even stretching our imagination, it is

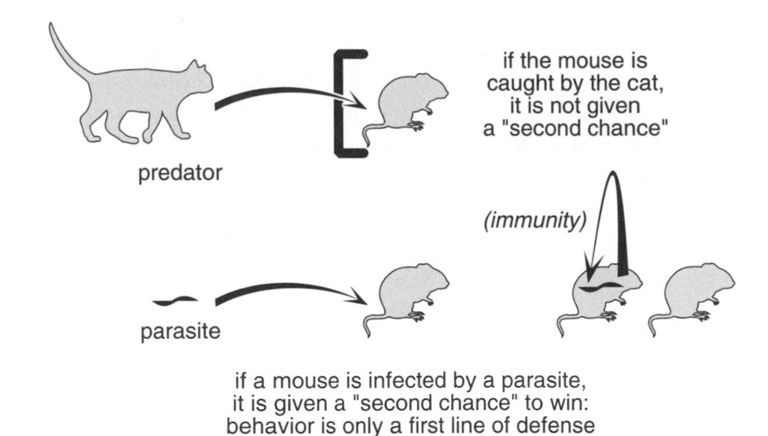

Figure 8.1. Prey have only one line of defense; hosts have two.

difficult to affirm that their black-and-white stripes were selected to camouflage them from their enemies, since the stripes do not do a very good job of imitating the grass of the savanna and zebras are the only inventors of this mode of camouflage. It has been suggested that although zebras are highly visible close up, they are not so from far away and that lions are nearsighted—an idea that is not very convincing. One might say zebras invented the bar code, which proclaims their identity to the image analyzer, the predator.

This is why Waage's (1981) explanation of the zebra's stripes in terms of parasitism is particularly attractive. Several studies have shown that biting arthropods are attracted by large unit surfaces, a response that was probably selected because it favors the detection of large animals. Specialists have used this fact in controlling some parasitic diseases by developing very efficient traps. For example, "Challier-Laveissière" traps are large panels of black or blue fabric, based on the fact that tsetse flies are attracted by contiguously colored surfaces. These traps are extremely efficient in villages in riverine forest and gallery areas, where their attractive surface is treated with an insecticide that kills thousands of tsetse flies every day (Challier and Laveissière 1973). Thus one may suppose that zebras have selected their stripes as a means of camouflage from the tsetse fly and thus avoid both the annoyance of a bite and the infection it might transmit. Two serious arguments support this hypothesis: first, zebras are in fact rarely bitten by tsetse flies; and second, zebras living in areas where the dipterans are most abundant have the most marked stripes. In terms of evolution this may be interpreted as showing that the pressure exerted by a parasite may be greater than that exerted by predators—a good lesson for ecologists, who still often forget parasites.

If this hypothesis is correct, then other African mammals such as impalas that have a uniform coat (a nice target for tsetse flies) should perhaps protect themselves by some other adaptation. In fact, a continuous vibration of the skin in these animals makes it difficult for biting arthropods to obtain their blood meal and therefore to transmit trypanosomes. In the end, zebras and other ungulates "tell us" that they prefer being eaten by lions to getting sleeping sickness. This makes sense, since not being alert because of trypanosomes in the circulatory system is certainly a good way to ensure being eaten by lions.

Optimizing Behavior

Detecting Parasites

Antiencounter behaviors would most likely be more common if the potential hosts could see the infective stage of the parasite coming—just as potential prey can often see their predators. This ability is rather exceptional, and in general infective larval stages are small. A young African entering a body of water may see a crocodile but not schistosome cercariae or anguillulid larvae; otherwise the response would simply be to avoid the transmission site. However, in some cases antiencounter behaviors may have been selected because the potential host does in fact detect the infective stages. Ungulates, for example, limit their contamination by larval ticks in this way.

Sutherst et al. (1986) studied cattle behavior experimentally relative to the tick *Boophilus microplus*. Different numbers of tick larvae (none, 1,000, 3,000, 9,000, 27,000, and 81,000) were set in circles one meter in diameter on a grass pasture. Two groups of bulls were used during two successive days of testing, over which time the lateral displacement of the larvae was found to be negligible. Each group was left in the paddock for twenty-four hours, and the bulls' behavior was observed for three hours each day. Observers did not know which circles contained the ticks, and the response of the bulls as they arrived on a circle was noted as "obvious avoidance" or "no response." After forty-eight hours, the larvae remaining on the grass were recovered using a vacuum cleaner. Results showed that the proportion of avoidance reactions was strongly density dependent (see the black dots and solid lines on fig. 8.2). The percentage of attached larvae became smaller and smaller as the number of larvae set in the circles increased (open squares and dotted lines on the same figure). Bulls could see larvae because as the hosts approached the dark brown ticks moved onto the upper surface of grass blades, leading the authors to conclude

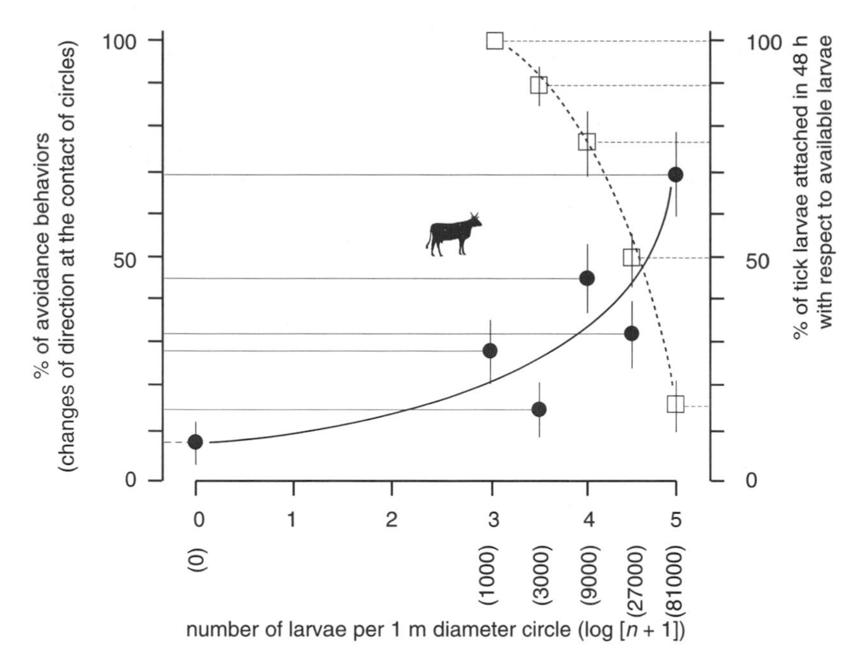

Figure 8.2. Experiments of Sutherst and colleagues: avoidance of tick larvae by cattle (inspired by Sutherst et al. 1986).

that "whilst cattle appear unable to detect the larval progeny of a single female *B. microplus* (ca. 2,000 larvae), they are capable of detecting heavily contaminated foci of tick larvae."

Cattle are also capable of limiting infection by larvae of the nematode *Dictyocaulus viviparus,* but in this case there is no real detection of the parasites. Instead, they avoid eating grass growing close to bovine feces (Michel 1955).

Along these lines, females of the bird *Centrocercus urophasianus* seem able to see some ectoparasites, or at least the hematomas they have induced, on the air sacs of males. These sacs swell during strutting, and females choose healthy males over parasitized ones. Females even refuse males carrying artificial hematomas glued to their sacs, and as a precaution they shake themselves actively and preen for several minutes after each copulation (Spurrier, Manly, and Boyce 1990).

Sheep also seem able to detect a parasitic dipteran, *Oestrus bovis.* As noted earlier, adults of *O. bovis* drop their larvae in the nostrils of a sheep, whose protective attitude is to walk with its nose as close as possible to the ground or in the wool of the sheep in front of it. An aptitude for detecting the flies seems to be selected for according to the pressure the parasite exerts, because in countries where these parasites do not exist the sheep are not as gregarious (Philippe Dorchies, personal communication).

Hart (1994) cited several other cases where hosts detect parasites before infection, and we shall see later that mutualists may contribute both to insects in their fight against parasitoids and to plants in their fight against phytophagous insects.

Hiding

The pressure of parasitoids is particularly strong on their hosts, since these hosts end up dead. Thus when any gene of a parasitoid host determines a behavior that closes the encounter filter (prevents attack by the parasitoid), it must be vigorously selected (Perez-Maluf et al. 1998). Accessibility of the host is important for successful oviposition by any parasitoid, and individuals within the same host species whose genetically determined spatial position makes them more accessible are more often parasitized.

For example, in the fruit fly *Drosophila melanogaster* there is a gene (denoted *for,* which stands for foraging) that has two alleles, *forR* and *forS.* The allele *forR* determines the characteristic "rover," which codes for rapid displacement of the larvae in the food source. The allele *forS* determines a "sitter," which codes for slow displacement of the larvae. In some parts of the world *D. melanogaster* larvae are attacked by two species of hymenopteran parasitoids, *Leptopilina boulardi* and *Ganaspis xanthopoda.* The first one seeks *Drosophila* larvae randomly by probing its ovipositor here and there in the substrate, whereas the second seeks larvae by "sound"—it periodically stops to listen for vibrations the larvae generate as they move within the substrate. Because of these different modes of seeking larvae, *L. boulardi* lays its eggs mostly in the "sitter" larvae (little mobility) and *G. xanthopoda* does so in the "rover" larvae (more mobility and greater "noise").

Carton and David (1985) and Carton and Sokolowski (1992) have shown that, depending on the dominant parasitoid in the local environment, there is selection for either the *forR* gene or the *forS* gene. That is, in the presence of numerous *L. boulardi* the *forS* gene is counterselected, since this parasite lays its eggs more often in the "sitter," whereas the opposite is true when *G. xanthopoda* dominates. Thus the *Drosophila* population modifies its gene frequency in a way that closes the encounter filter. These results are principally experimental, but many clues lead us to believe that this selection also occurs in natural populations. In at least one oasis (Nasrrallah) in Tunisia, for example, the rarity of the phenotype "sitter" is correlated with the large abundance of *Leptopilina.* Similarly, such a correlation was shown to exist between the burying behavior of the larva (Sokolowski and Carton 1989; Rodriguez et al. 1991) and the presence of parasitoids attacking the resultant buried pupa (Carton and Sokolowski 1994).

A comparable situation is illustrated by larvae of the butterfly *Aonidiella aurantii,* which live in citrus fruit, and their parasitoids. Some larvae stay close to the surface and others live deep within the fruit: the former are about thirty times more often infected by two species of parasitoids than those that live deep in the fruit (Murdoch et al. 1989). One would logically guess that if there is genetic variability for the preference to exploit either the surface or the inside of the fruit, then in this case selection would widely favor the "deep" host larvae.

Hochberg and Hawkins (1992) have shown that for parasitoid hosts in general the notion of such refugia must be understood in a broad sense: it can mean both spatial heterogeneity in the host's distribution (which decreases the efficacy of the parasite's search, as illustrated above) and occupying environments that are difficult to reach (roots, wood, galls, etc.; see chapter 4).

Forming Groups

Various species of horseflies in the genus *Hybomitra* attack heifers. When the horseflies attack, the heifers, which are normally dispersed while grazing, form grazing lines, with the dominant animals well protected at the center. The grazing front continuously moves, and the animals continue feeding (fig. 8.3). If the horsefly attack becomes worse, a more defensive configuration appears: the heifers form bunches and completely stop grazing (Ralley, Galloway, and Crow 1993). Although horseflies are themselves not parasites sensu stricto, like all biting arthropods they are potential vectors for parasites.

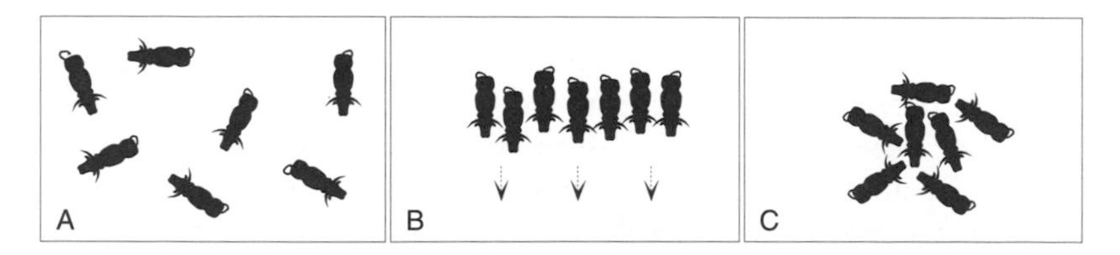

Figure 8.3. Heifers grazing (A) in the absence of horseflies, (B) in the presence of horseflies, and (C) in the presence of massive attacks of horseflies (modified from Ralley, Galloway, and Crow 1993. The position of the heifers was interpreted from photographs taken at ground level by the authors and redrawn as if seen from above).

Changing Habitat

Animals that make seasonal migrations are not rare, and the value of these movements is generally explained by enhanced access to high-quality food or by avoidance of predators. Several works, however, also invoke a probable role for parasites in migration. For example, Folstad et al. (1991) looked to see if there was a correlation between the amplitude of the migration of the reindeer *Rangifer tarandus* and the abundance of one of its parasites, the dipteran *Hypoderma tarandi,* in mountainous northern Norway. This fly lays its eggs in the fur of the reindeer about July, and the larvae penetrate and move beneath the skin. After three to four months the larvae form cysts from which they escape during the next spring and drop to the ground to pupate. Adults then leave the pupae, and the cycle starts again. Reindeer form herds of a few dozen to several hundred head, and the calving grounds and summer pastures of some groups overlap, qualifying them as "nonmigratory." In other groups, however, the calving grounds and summer pastures are separated by tens or even hundreds of kilometers, and the herds therefore may be classified as "migratory." A study carried out on 964 males showed that the animals that belong to migratory herds are significantly less severely attacked by the dipterans than those that belong to the nonmigratory herds. The most likely explanation is that the abundance of flies is determined by pupal density and that migratory herds drop fewer larvae within their summer pastures than nonmigratory ones. The authors ruled out the possibility that the correlation was caused by other variables such as local climate conditions, herd density, or a genetic or nutritionally based susceptibility. Therefore one may conclude that reindeer with migratory behavior have a selective advantage over sedentary ones.

Should one perhaps think of pressure from parasites whenever an animal migration is unexplained? For instance, 99% of the fish species that live in the lagoons of Pacific atolls have larvae that develop for long periods outside the lagoon. The larvae leave the lagoon right after they are born, spend several weeks out at sea, then come back to complete their growth within the confines of the lagoon (Dufour and Galzin 1993). This migration may be an adaptation to flee either predation or parasitism. Given the numerous consumers of plankton in the sea, however, predation may be just as important there as it is in the lagoon, whereas parasitism, notably by those species that use benthic mollusks as intermediate hosts in their life cycles, must not be as important in the open sea. Interestingly, human populations also emigrate outside such endemic zones, and we have all heard of African villages in the Sahel abandoned as their inhabitants flee the river blindness caused by *Onchocercus.*

Knowing that parasites are more abundant in an aquatic milieu than in the terrestrial environment and that most parasites originate in the water, nothing prevents one from imagining that parasite pressure may have played a role (among other selective pressures) in forcing vertebrates to leave the water. That parasites were able to "catch up" with terrestrial animals later on is then a simple question of coevolution.

Changing Nests

Bird ectoparasites (particularly dipterans and hemipterans) are strongly pathogenic and induce noticeable costs, principally in the young. These pressures result in numerous antiencounter behaviors. A bird's ability to choose its nest based on perceived numbers of parasites was experimentally analyzed by Christe, Oppliger, and Richner (1994) in great tits, *Parus major*, which were offered the following choices of overwintering nest sites: between empty sites and used sites containing nests devoid of parasites (experience a), and between empty sites and sites containing nests infected by the hematophagous flea *Ceratophyllus gallinae* (experience b).

In each case the sites formed a pair and were separated by less than a meter. The results were as follows: in experience a, seventeen birds chose the empty site while eighteen chose the site with the "healthy" nest; in experience b, twelve birds chose the empty site and only one chose the site with the infected nest.

It is therefore clear that it is not the existence of an already built nest that influences the tits' choice but the presence of the parasite *C. gallinae*. This behavior clearly prevents encounter.

The cliff swallow *(Hirundo pyrrhonota)* has also been studied in this regard, and the behavior of couples that build nests is strongly adapted to avoid ectoparasitism (the most pathogenic ectoparasite being the hemipteran *Oeciacus vicarius*). When the swallows return to nesting areas from their migration, they are able to recognize nests that are the most heavily infected by *O. vicarius* larvae, and they then select the cleanest nests. Further, the probability of building entirely new nests is higher in larger colonies (the most infected) than in smaller ones (Brown and Brown 1986).

Various passerines commonly construct "clean" nests rather than reusing old ones, based on the same strategy. The European starling has a slightly different behavior but achieves the same antiencounter effect: it garnishes its nest with plants that repel parasites (Clark 1991). The plants chosen are particularly effective against the development of acarids.

Similarly, some birds that nest in colonies may alternate sites or totally desert sites where nests are subjected to high levels of parasite pressure. This is the case for both pelicans (King et al. 1977) and cliff swallows (Loye and Carroll 1991). This strategy of fleeing parasitism is justified in that the abundance of ectoparasites increases year after year if nests are constantly occupied. On islands where precise nest site information was recorded over many years for gulls of the species *Rissa tridactyla,* the percentage of chicks as well as the percentage of nests that harbor ticks *(Ixodes uriae)* increases regularly with time (Danchin 1992). In the cliff swallow, the dispersion of the young far from where they were born is more marked when there is a high level of ectoparasitism, and it has been suggested that, in general, dispersion of the young in animals that have a colonial tendency may be a response to the pressure of ectoparasitism (Brown and Brown 1992). In response to these host strategies, ectoparasites have selected the ability to wait longer and longer to neutralize the effect of the temporary abandonment of the nest.

Choosing One's Partner

Clayton (1990) designed an experiment to show that the choice of a sexual partner may be influenced by parasite load. It used the pigeon *Columba livia* and two species of phthiropterous insects, *Columbicola columbae* and *Campanulotes bidentatus,* both commonly harbored in the bird's feathers. These parasites transfer from one bird to another during body contact, particularly during mating, and they exclusively eat feather barbules, which they can digest because of symbiotic bacteria.

During the experiment, pairs of males, one lousy and one clean, were presented to female pigeons. Pairs were formed with birds as similar as possible in terms of age, weight, overall color, and sexual experience. In sixteen cases females chose clean males, in five cases they chose the lousy males, and in eight cases they refused both. The difference between the choice of lousy and clean males was statistically significant (fig. 8.4) and the second step of the study consisted of determining how females could identify the lousy males. Contrary to the case of *Centrocercus urophasianus* and the hematomas cited earlier, *C. livia* could not see either the parasites themselves or their traces. Three hypotheses were then equally possible. Either a lousy male preened itself more often than a clean male; the plumage of a lousy male was less bright because of parasite damage; or a lousy male did not have as good a courtship display as a clean male and was therefore not as persuasive.

Intuitively, the first hypothesis might seem the best. In fact, however, observations showed that preening was exactly the same in parasitized

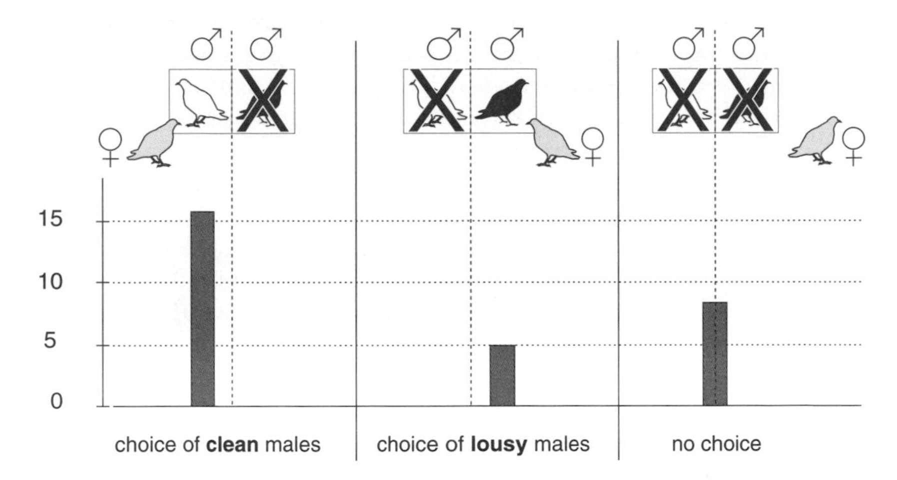

Figure 8.4. Experiment of Clayton (1990): an example of transmission avoidance. Clean males are represented in white and lousy ones in black.

and healthy birds. The parasites that were studied do not induce itching, since they eat only the barbules and never go to the skin. Preening is thus independent, at least in this case, of infection.

Since the first hypothesis was not supported, the second hypothesis—that females noticed duller plumage—was looked into. The researchers concluded that this also was not a good explanation because the phthiropterans devour only the barbules in the basal and median regions of the feathers and never the distal region, which is the only area visible from the outside. In short, lousy and clean males are identically colorful and bright. Since the experimenters could not distinguish the pigeons on this criterion, they concluded that the females probably couldn't either.

Only the third hypothesis of a less successful display by the lousy males remained. The observations showed that the lousy males spent significantly less time than clean males in parading. Therefore this hypothesis is most likely the good one. But there was one difficulty: How did being parasitized by insects that eat the median and basal barbules of feathers influence the length of the male's parade? The explanation is that the parasites impair thermoregulation—parasitized pigeons get colder faster and spend more energy maintaining their body temperature, so they have less energy available for the display. The damage is so real that even the survival of the lousy pigeon is affected during the winter.

In short, then, sexual selection by females is linked to the level of parasitism in the males, and it is likely that the importance females give to the males' display was selected for to prevent infection by lousy males. Any female that becomes infected probably also has a diminished fecundity because of the impairment of thermoregulation, leading Clayton to speak of a "transmission-avoidance model." This is, then a striking example of closing the encounter filter by selecting an appropriate behavior.

Educating Humans

In humans the cost of behavioral modification to avoid parasitism is both cultural and material. Public health personnel know how difficult it is to modify human habits in this regard, and an example of this difficulty is dracunculosis, a disease caused by the filarial worm *Dracunculus medinensis.* Adult parasites, both males and females, live in the subcutaneous tissues of humans, affecting several million people in arid tropical regions. The females induce ulcerations (usually on the legs) that allow them to evacuate their larvae when people enter water. Once in the water the larvae are ingested by small copepod crustaceans of the genus *Cyclops,* and after several molts they become infective. If humans swallow the crustaceans as they drink water, the larvae then cross the intestinal wall to become adults. Unlike other filariae, *D. medinensis* is not transmitted by a biting arthropod, so control of the disease seems easy: water can be filtered with any available fabric capable of retaining the copepods. The disease has regressed a lot in some areas (India, the Middle East, and Uzbekistan) but is, paradoxically, still present in numerous African countries where social habits, especially concerning water, are deeply ingrained and often hard to change. After seeing what little effect advice given seventy-five years ago had on an African village, Chippaux (1993) wrote: "It is out of the question to critique our predecessors or to doubt their good faith. The mistake comes from the fact that the importance of social habits was underestimated, and they are particularly strong when it comes to water."[1] Despite this pessimistic observation, education and information have been shown to improve many habits important to the transmission of parasites to humans. These include checking food for parasites before it is eaten, cooking food, disinfecting water, avoiding contaminated sites, and protecting against arthropod vectors.

The efficacy of such education, however, is often limited because of its real cost in material terms (money). The best example is Chagas' disease in South America, a parasitosis caused by *Trypanosoma cruzi.* The vector of this parasite is the kissing bug, which tends to be found in precariously

1. This example gives me the opportunity to note that there is a difference between parasite avoidance by animal hosts and by humans. In humans, avoidance is transmitted not by genes but by education and information. The process is therefore "Lamarckian" (transmission of acquired characteristics) and not "Darwinian."

built cob houses. The kissing bugs are not found in more permanent structures, so a simple improvement in the quality of life avoids the vectors and therefore their parasites.

A very interesting effect on primates other than humans has been reported by McGrew, Tutin, and File (1989) and McGrew et al. (1989). These authors studied parasites of green monkeys *(Cercopithecus sabaeus)*, patas monkeys *(Erythrocebus patas)*, and Guinea baboons *(Papio papio)* in Senegal and found that neither green monkeys nor patas monkeys were infected by *Schistosoma mansoni* whereas Guinea baboons were. These workers suggest that the reason *S. mansoni* is absent from *C. sabaeus* and *E. patas* "may be that both prefer to drink from flowing or temporary water-sources, which are less likely to have the molluscan intermediate hosts," whereas *P. papio* "is infected presumably because it uses stagnant water-sources" (McGrew et al. 1989). "Culture" is generally supposed to characterize only humans. But if we suppose that the drinking behavior of young green and patas monkeys is learned from their mothers (which is of course not demonstrated), then these observations might reflect a first cultural attempt at limiting infection by a parasite in the primate lineage. Other observers (Galat-Luong and Galat, in press) report an even more "cultural-like" behavior: in Senegal, baboons and chimpanzees were seen drinking from holes they dug by hand in the banks of watercourses, which efficiently filtered the water. The research work demonstrated that river water was polluted by various bacteria, whereas water from the "holes" was entirely safe, meaning the behavior protected the primates from infection.

Optimizing Life Histories

"Life history" means those biodemographic parameters that are subject to natural selection, such as size at birth, age at sexual maturity, number, size and sex ratio of descendants, frequency and duration of reproductive periods, and length of life. As Stearns (1992) noted, "the complexity and interest of life history evolution arises because organisms have evolved so many ways of combining these traits to affect fitness."

Such life history traits are in competition with one another for the energy resources of an individual, and the energy allocated to one is not available for another. Therefore life history traits evolve under the influence of selection to optimize these allocations.

The Theory

Ecologists who work on free-living animals admit that the pressure of predation is one of the forces that molds the life history traits of prey species.

It is also starting to be recognized that such traits may also have been molded in part by natural selection as a means for limiting encounters with parasites. In particular, the growth rate (more or less rapid) and the reproductive age (more or less precocious) "are crucial for maximization of fitness" (Koslowski 1992). As a result it may therefore be advantageous for hosts to modify life history traits if doing so can reduce the pressure exerted by parasitism.

The argument for this can be made in the following way. Let's consider, for instance, the duration of the prereproductive period and the other life history traits associated with it. If the pressure exerted is strong enough, parasites might be capable of modifying this duration for one of four reasons (Hochberg, Michalakis, and de Meeus 1992).

1. Fundamentally, the presence of a virulent parasite could lead to the selection of shorter prereproductive periods in the host. This is because the host individual that reproduces when young, before the pathogenicity of the parasite is fully developed, has a selective advantage over the individual host that reproduces when older.

2. The evolution of longer prereproductive periods in a phylum could occur if it was paired with the evolution of mechanisms for resistance, which would then reduce parasite pathogenicity (that organisms with longer prereproductive periods, such as the "higher" vertebrates are overall also those that have the most elaborate defense mechanisms supports this idea).

3. If a certain number of other life history traits are correlated with the duration of the prereproductive period, such as the duration of life and the maximum size of the species, then parasitism could play a role in the duration of the prereproductive period if it affected the evolution of the ensemble of these characteristics. For example, trematodes could have contributed to the evolution of a shorter life span with only one reproductive period in their molluskan vectors because of the very strong selective pressure exerted by the parasite larvae (Bayne and Loker 1987). In this case, to reproduce quickly would then be an adaptive response by the mollusks to the reduction (and sometimes the complete arrest) of fecundity induced by the presence of sporocysts and rediae and to the high probability of being infected by miracidia throughout their life.

4. Parasitism could induce some shifting in the time of reproduction (Forbes 1993). For example, when the parasite develops quickly and then disappears, the host could decrease its reproductive effort at the time it is parasitized and increase it when the infection is over. If the parasite develops slowly at first, the host could increase its reproductive effort during the initial phase of infection and thus compensate for any later parasite-induced decrease.

These effects would not be the simple consequence of parasitism but rather would constitute adaptive responses in the hosts and thus be classified as true antiencounter mechanisms.

The Facts

There are few concrete data capable of supporting the previous theoretical predictions, but listing them does indicate that there is a broad domain to explore. Of the works done in this area, some compare populations in space and time whereas others are experimental studies.

The selection of rapid maturation under the pressure of parasites has been demonstrated by Lafferty (1993) in a particularly original work involving populations of the mollusk *Cerithidea californica*. These mollusks are vectors of numerous species of trematodes whose adults are principally found in shorebirds. The study compared parasitic prevalence with the size of mollusks reaching sexual maturity in a series of isolated locations. The prediction was that if life history traits evolved under parasitic pressure, then *C. californica* should be reproducing at a younger age where prevalence was the highest.

The prediction was verified despite the difficulty of neutralizing the effect of other factors acting on growth (resource availability, latitude, etc.). The results showed that mollusks are small at maturity when parasite prevalence is high (fig. 8.5), but it had to be verified that the change in the life history trait "age at maturity" was due to selection and not to an undetected environmental factor or factors. To address this, the author exchanged mollusks between two populations with different characteristics. The results here were also positive in that descendants of the moved mollusks reached maturity as in their original population. These results led Lafferty to conclude that the mollusks had limited the risk of parasite-induced castration by reaching sexual maturity faster. Statistically speaking, this adaptation allowed the mollusks to have a longer "reproductive life" than they would otherwise have had.

The only alternative hypothesis is that the mollusks may be able to "detect" the risk of parasitism (for example, from the density of miracidia) and respond to a strong risk by accelerated maturation. This hypothesis, which supposes phenotypic plasticity rather than genetic variation, seems less well supported by the results obtained relative to the first hypothesis. Note also that this work was made possible because of the relative isolation of the *C. californica* populations. Had gene exchange between populations been occurring, such local adaptation would not have appeared. Also, research like that on *C. californica,* which compares populations in space, does not indicate how long it takes for such changes in life history traits to occur.

The evolution of a life history trait in time may be shown, however, if we

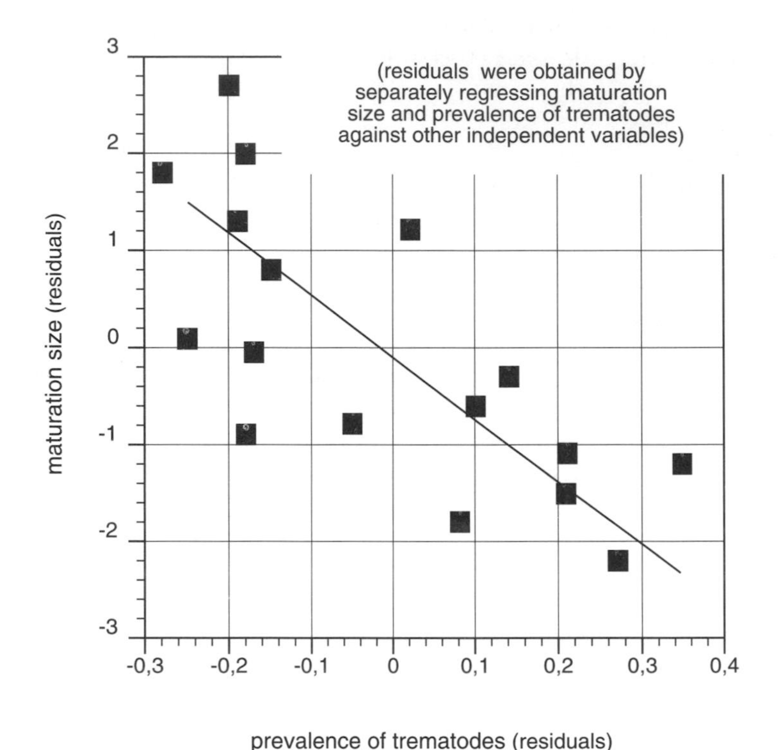

Figure 8.5. The relation between the strength of "parasitism" and size at maturity in *Cerithidea californica* (after Lafferty 1993; line fitted by eye).

access paleontologic data. *Transennella tantilla* and *T. confusa* are two small bivalves from the west coast of North America that are heavily parasitized by trematodes whose adults live in fish (Ruiz 1991). These trematodes induce partial or total castration of the mollusks, and as in the previous case, infection is age dependent (prevalence increases in the bivalves with age).

One of the trematodes *(Parvatrema borealis)* forms metacercariae in the mollusks, and these metacercariae leave traces that are perfectly identifiable in fossil shells.

Paleontologic data indicate that the parasites existed 2 million years ago and that within this time the average size of the shells decreased from 10.4 millimeters to 6.6 millimeters.

Although other factors may be responsible for the reduction in bivalve size, Ruiz concluded that strong parasitic pressure may have played an important role.

In birds, producing several clutches containing a few young is not the same as producing only one clutch of many. As previously discussed, then, the seasonal increase in nest ectoparasite populations may favor selection for fewer clutches (de Lope and Møller 1993). Richner and Heeb (1995), however, think the effect may be different depending on the type of parasite involved. By analyzing the effects of fleas and mites on various life history traits in the great tit, these authors showed that, depending on the parasite, it may be in the best interest of the bird to select smaller clutches (the number of parasites increases less quickly if there are fewer chicks) or, on the contrary, larger clutches (where there is "dilution" of the parasites available when chicks are more numerous).

According to Christe, Møller, and de Lope (1998), asynchrony in the development of the young of one clutch may have an adaptive value partially linked to ectoparasite avoidance. The idea rests on three observations: often a hatching asynchrony occurs between eggs of the same clutch, leading to a hierarchy of weight within the nest; ectoparasites have an aggregated distribution among chicks; and there is a positive relation between nutritional status and immunocompetence. Within a nest the least resistant individuals (the last to hatch) suffer more parasitism than others and are referred to by Christe, Møller, and de Lope (1998) as the "tasty chicks." Although other explanations are possible, asynchrony of nestling development, while sacrificing part of the clutch, may give the "eldest" a better chance of survival (fig. 8.6). Christe, Møller and de Lope

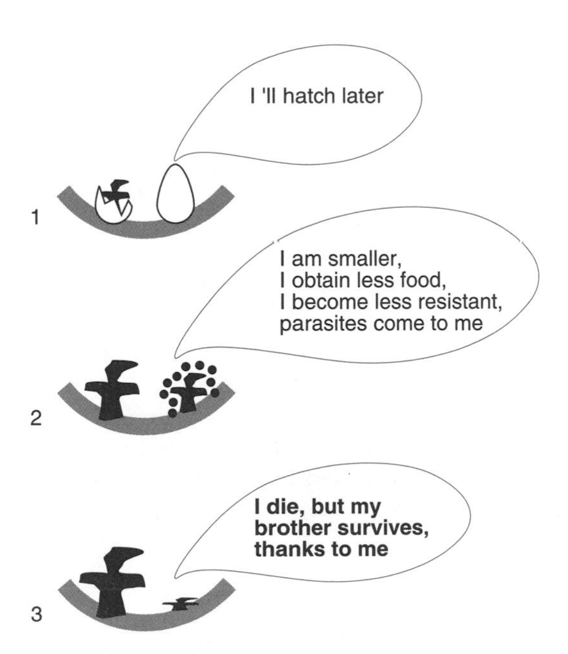

Figure 8.6. The "tasty chick" (inspired by Christe, Møller, and de Lope 1998).

remind us that some birds systematically lay two eggs but that the first chick to hatch always kills its sibling. These authors suggest that "adding a nestling that may attract parasites in favour of its larger sib, at least during the critical first half of the nestling period, may be of selective advantage." Genes or combinations of genes coding for the life history trait "two eggs laid" would then have the same value as an "avoidance gene."

Dying for One's Genes

In terms of evolution what counts is saving one's genes, not one's skin. If dying allows one to save one's genes, then dying becomes one way for hosts to fight parasites. This obviously requires further explanation.

One way for a host to kill its parasites is to die, since it will take the parasites with it in the "sacrifice." However, since dying drops the individual host's reproductive success to zero, there is apparently no chance for the sacrifice to be selected as a response to parasitism because the genes leading to it are never transmitted. The reasoning takes a different turn however, if we consider inclusive fitness instead of individual fitness.

Inclusive fitness is the sum of the fitness of an individual and that of all its close relatives, which share a part of the genome (see Hamilton 1976 and Maynard Smith 1989a concerning theories of social behavior, and Wickler 1976 concerning their application to parasitism). As soon as we take into account inclusive fitness, it becomes perfectly possible for the sacrifice to be selected. This would happen if the individual's sacrifice protects other individuals that carry a significant proportion of the same genes. In other words, an individual can increase its inclusive fitness by diminishing the risk of infection for its relatives.[2] To decrease the risk of infection of its kin (kin selection), the parasitized individual may make itself more visible to predators by moving more or by increasing its exposure. Its parasites are then eaten at the same time as it is.

Shapiro (1976) described a very convincing example of sacrifice in this vein, and the title of his article, "Beau Geste," is in French extremely evocative. Larvae of the butterfly *Chlosine harrissii* (a North American species) normally live on the undersides of leaves. When parasitized by a braconid hymenopteran, however, they make themselves as visible as possible on the tips of tall grasses, where they are then devoured by predatory birds. Obviously the more abundant and spatially close the kin group, the better the process functions. In other words, the process is more actively

2. Note that the idea of inclusive fitness leads to the image of a living world governed only by DNA. Wickler (1976) writes: "The genes run the individual in their own interest . . . the organism's behaviour is only the DNA's way of making more DNA."

selected under such conditions, and it would be useless to protect unrelated individuals by one's own sacrifice, since inclusive fitness would not increase. In the case of *C. harrissii* these conditions are met because the caterpillar's populations are strongly interbred, often being founded by one female who has been fertilized by only one male. Shapiro also notes that parasitized caterpillars develop more slowly so that their sacrifice occurs after healthy caterpillars pupate. Therefore the sacrifice does not lead to the development of a search image in predators that would then put the healthy caterpillars at risk.

Another example of such suicidal behavior was reported by McAllister and Roitberg (1987) based on a comparison between two areas with different climates. In the regions of British Columbia where the climate is dry, pea aphids *(Acyrtosiphon pisum)* are parasitized by the wasp *Aphidius ervi*. When parasitized, the larvae have a tendency to drop to the ground and desiccate, taking their parasites with them. This behavior does not exist in regions with a humid climate, where larvae that fall to the ground do not dry up.

Intuitively, we may feel that even if there is a benefit there is also a cost associated with such a sacrifice. This is true, since the individual that sacrifices itself evidently stops the transmission of its own genes. However, certain parasites appear to make an amazingly boneheaded move: they castrate their hosts (as do many trematodes in mollusks and crustaceans in other crustaceans) or they end up killing the host (most parasitoid insects). In this case the reproductive success of the host is nil, and the cost of the sacrifice drops to zero, since fitness cannot drop lower. If fitness as the result of the sacrifice is zero, there is only the question of what benefit is left.

It is almost certain that such "sacrifices" are more frequent in nature than they seem, simply because not enough research has yet been done on this fascinating and deserving subject.

9 Genes to Kill

Host-finding and settlement by parasites can be analyzed using well-developed ecological and behavioral models for habitat choice and patch settlement. The difference, however, is that in parasite settlement, the habitat is able to fight back.
Hanley et al. 1996

One should expect the immune response of a host to be optimized to the extent that low intensity infections will be tolerated if the costs of sterilizing immunity (complete removal of a parasite) outweigh the benefits.
Sheldon and Verhulst 1996

A host may identify a parasite at two times: before and after infection. Let's return for a minute to what I have called "the second chance for the host." If an antelope does not detect a cheetah nearby (by sight, sound, or odor), it will most likely pay the ultimate price for this failure—it will get no second chance. However, if the same antelope does not identify schistosome cercariae while wetting its feet on the side of a pond, it will become infected. But in this case the antelope has a second chance: although a prey animal never kills its predator, the antelope can kill the parasite thanks to its immune system.

In short, infective parasite stages are not detectable most of the time because of their very small size. In the case just cited it is impossible for the antelope to identify the schistosome cercariae by sight, smell, or any other clues any more than a fox could detect a cestode cysticercus in the vole it is chasing or a restaurant customer amoeba cysts on the plate. We saw previously that host avoidance genes only exceptionally involve the visual detection of infective stages. Without extending the analogy too far, one might say the prey "knows" it is going to be eaten whereas the host does not "know" it is threatened.

Identification of parasites by hosts is thus almost always "postinvasive." If antiencounter mechanisms have not been selected or have not functioned, the parasite has won the first part of the battle. The second part, however, consists of the host's selecting ways to get rid of the parasite and the problems it may cause—so long as doing so justifies the expense.

Some of these ways call for behavioral changes while others call for immunity.

Efficient Arms

Killing from the Outside

Antelope incisors are separated by a space and look something like a wide-toothed comb. This characteristic attracted the attention of Georges Cuvier himself, who thought it might be an adaptation to remove ectoparasites, particularly ticks, from the fur. This scraping or cleaning the coat is called grooming in mammals and preening in birds, and both hold important places in animals' daily lives. Numerous observations since Cuvier's time have confirmed that antelopes do remove ectoparasites from their fur with their teeth. Depending on the species of antelope, the number of grooming periods may vary from six hundred to two thousand in twelve hours (Hart et al. 1992).

The efficacy of this process in mammals has been demonstrated in an ingenious experiment by McKenzie (1990) (see also McKenzie and Weber 1993): twelve impalas were captured on a reserve, and the spaces between the teeth on one side of their mouths were filled with dental cement before the animals were released. When the same animals were recaptured one month later, an exhaustive count of the ectoparasites found on their fur showed that the side that the impalas could not groom (because of the cement) carried up to ten times as many parasites as the other side (fig. 9.1).

Whenever possible the first step in getting rid of intruders is apparently mechanical removal. A dog does this when it eats its fleas, and the grouper does the same when it allows a cleaner fish to remove its parasitic crustaceans and monogeneans. It seems obvious, then, that structures and behaviors that allow grooming were selected under parasitic pressure. This is supported by the work of Cotgreave and Clayton (1992), who showed that bird species harboring the most ectoparasitic species spend more time grooming (up to 20% of their time each day) than those that harbor only a few.

The invasion of the mite *Varroa jacobsoni* into European and American beehives gives indirect support to selection for such cleaning behaviors. *Varroa jacobsoni* is not very pathogenic in the Asian bee, *Apis cerana,* but is devastating in the hives of the European bee, *A. mellifera,* into which it was introduced.[1]

1. This mite invaded Europe, North Africa, and North and South America in the 1970s and 1980s. The acarids feed on the hemolymph of their hosts as larval and adult stages (Ritter 1981).

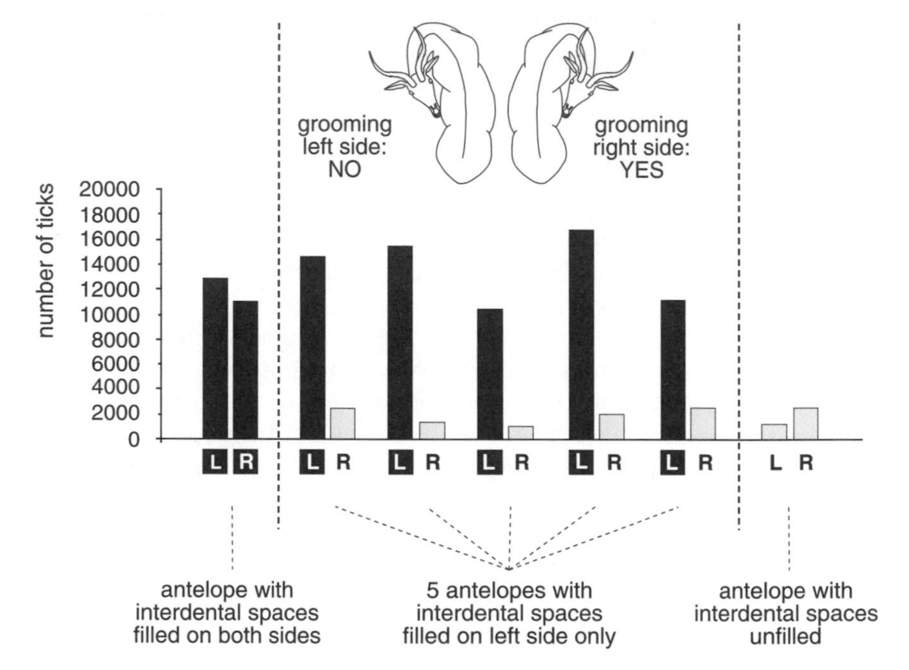

Figure 9.1. Results of the experiment of McKenzie and Weber (1993), which involved plugging the spaces between the incisors on one side of impalas' mouths with dental cement and measuring the effect on ectoparasitism. The side where the incisors were plugged (black bars: left side of individuals 1 to 5), had almost ten times as many ticks after one month as the side where the incisors were not cemented (after original data provided by A. A. McKenzie).

The weak pathogenicity of *V. jacobsoni* for *A. cerana* is in part due to the bee's secretion of antiacarid chemical substances. However, another mechanism, true grooming, also plays an important role in this species, and populations of *V. jacobsoni* never reach high densities in this bee's nests. Better still: if this autocleaning is not enough, a parasitized worker performs a cleaning dance involving lateral movements of the abdomen that alerts other workers, which then come and bite the acarids until they detach (allogrooming).

V. jacobsoni is highly pathogenic for *A. mellifera* because the workers of this species do not practice allogrooming. However, a strain of *A. mellifera* has been selected that knows how to get rid of *V. jacobsoni,* a good indication that grooming derives from the selective pressure exerted by parasites and that one day European bees might naturally select genes for recognizing the mite and eliminating it through active cleaning (Ruttner and Hanel 1992; Moritz 1994).

Cleaning behaviors, then, are one of the host genome's first responses to parasite attack. That some parasites can use such grooming to favor their transmission should not be too surprising. For example, *Heligmosoides polygyrus* is a nematode of mice and other rodents with a direct life cycle, and its free-living infective stages are passed with the hosts' feces. Hernandez and Sukhdeo (1995) have shown that *H. polygyrus* are able to direct themselves toward the hosts' habitats; that they attach themselves to the rodents' coats; and that they are ingested during grooming. Since mice spend 20% to 50% of their waking time cleaning themselves, their auto- as well as allogrooming behaviors only favor the parasite's propagation.

Collaboration with others may be very important in order to kill from the outside. In vertebrates allogrooming between conspecifics is frequent, and collaborative efforts to remove ectoparasites may also occur between different and even distantly related taxa. For instance, fish such as the cleaning labrid *Labroides dimidiatus* eat the ectoparasites from the skin and even fins of other fish (see Arnal, Morand, and Kulbicki 1999), and collaborations between tickbirds or oxpeckers and large ungulates are also well known. Dickman (1992) indicates in his review work that the red-billed oxpecker, *Buphagus erythrorhynchus,* can reduce tick numbers by up to 99%. This author also recounts the case of the fossorial blind snake *Leptophlops dulcis,* which eats insect larvae and thus reduces ectoparasitism in the nests of screech owls, *Otus asio*—the owls carry the snakes alive to their nests.

Killing from the Inside

Although we may have the impression that hosts have not selected any spectacular or extremely ingenious ways to avoid parasites in the arms race involving encounter, this is not so in the arms race involving compatibility. In this case hosts have fully blossomed in terms of closing the door on parasites. The mechanisms in place are so complex in "higher" vertebrates that, despite the remarkable development of both general and parasitological immunology over the past twenty years, discoveries and surprises continue to accumulate at unprecedented rates. The immune system of vertebrates is the most efficient way to fight pathogens after they have invaded the hosts, and it is a killing machine that shows no mercy to either internal or external parasites. Ectoparasites (such as ticks) are not sheltered from it, and perhaps only the parasites of bird feathers are not subject to immune system attack, simply because of where they live. When any foreign organism penetrates a vertebrate, the immune system triggers a cascade of reactions that fight the invading organism as long as its surface molecules are identified as "nonself." A general characteristic of this type of defense system is that it adjusts the strength of its reactions to its opponents (that is, the immune response is generally proportional to the aggression). According to the degree of "perfection" of the system, the parasite may simply be recognized as nonself, or the identification may be extremely specific. In either case immune recognition is entirely molecular, and parasite molecules are identified by receptor molecules on the membranes of specialized cells or free in the blood or plasma.

Molecular defense systems exist in all living organisms. Bacteria, for example, confront bacteriophages (viruses with parasitoidlike behavior) thanks to recognition mechanisms for nonself coded for either by the bacterial chromosome itself or by plasmids; restriction enzymes (also coded for by the bacterial chromosome or by plasmids) that cut the DNA of invasive phages at specific sequences; and mechanisms of self-protection such as the methylation of base sequences that could result in their being targeted by enzymes. Thus all the fundamental characteristics of the vertebrate immune system are already present in the defense system of bacteria.

In mollusks, which have been studied in detail because of their role in transmitting schistosomes and other trematodes of humans and domestic animals, there is a defensive reaction that has, as in vertebrates, both a cellular component (hemocytes) and a humoral component. In these organisms the cooperation between these systems allows the encapsulation or destruction of parasites (Bayne and Yoshino 1989; Van der Knaap and Loker 1990; Bayne 1990, 1991, review in Adema and Loker 1997). Here the infective stages of the parasites can trigger a defensive reaction as soon as they penetrate the mollusks. During the infection the number of circulating hemocytes changes, and these cells use phagocytosis (against bacteria) or encapsulation (against helminths) to clear the infection. As in the defensive systems of more complex organisms, pathogens are destroyed by lysosomal enzymes, proteolytic enzymes, and reactive oxygen intermediates, and the evidence indicates that these responses are mediated by an active and complex cell-to-cell communication system involving cytokines (Connors et al. 1998).

In insects, also very well studied because of their roles as vectors of numerous human parasites (viruses, protists, filarial worms, etc.), the defensive reaction occurs because of phagocytosis, encapsulation by hemocytes, and melanization (melanin deposition) around the parasites. Depending on the case, melanization may happen with or without hemocytes (see Christensen and Severson 1993). In these hosts the cooperation of hemocytes when they fight a large foreign body is spectacular: they flatten and form a capsule around the intruder, producing desmosomes to guarantee its solidarity, and they secrete toxic molecules toward the inside of the capsule.

In general, most vertebrate immunologists consider the invertebrate defense system "inferior" only because it lacks genes coding for immunoglobulins (decreasing response specificity) and because there is no solid evidence of memory in invertebrates (speeding up future responses to the same pathogen).

The immune systems of vertebrates, including humans, appears to be an arsenal as diversified as it is redoubtable, with mechanisms for recognizing nonself elements that are nonspecific and others that are rigorously specific, cooperation between numerous cell types, memory, and chemical signaling substances (interferons, etc.) able to inhibit pathogen multiplication or regulate other parts of the system itself. In fact immunologists have given the name "natural killer cells" to some of the vertebrate lymphocytes involved in the response, which tells much about the offensive nature of this "defensive" system. The complexity of the vertebrate immune system and its efficiency are further enhanced by immune memory, which triggers rapid and specific responses against a pathogen that has previously been encountered. This memory is more or less faithful and more or less durable depending on the parasites concerned. The genes that code for this aspect of "immunity" have evolved to such a point that the least expected arms can be anticipated. Vertebrate immune systems can recognize billions (if not trillions!) of different antigens in their size range, some of which they have never encountered and may never encounter. This diversity in the proteins that recognize antigens and trigger their destruction is vastly greater than the number of genes that can code directly for them in the genome (there are only about 50,000 genes in the human genome), and these proteins can be built because of special rearrangements of the DNA within the cells that produce them—a process that is itself quasi-unique among living organisms. These genes code for T lymphocyte receptors and for the antibodies (immunoglobulins) that are produced by B lymphocytes. In addition to this special recombination, there are also hypervariable regions in the genes (Reynaud et al. 1995) as well as the unexpected incorporation into normal genes of mutated regions of pseudogenes (Reynaud et al. 1985). Once an antigen is recognized by the immunoglobulin products of these recombination events, there is a clonal expansion of the specific lymphocytes that have receptors corresponding to that antigen, causing the explosive expansion of only those cells capable of responding to that specific antigen. Yet other molecules on the membrane of cells form the major histocompatibility complex (MHC), which is also associated with the recognition of pathogens. Here there may be as many as one hundred alleles at each loci that code for these molecules in humans.

Overall, this exceptional and highly polymorphic arsenal of molecules produced by the vertebrate immunodefense system can be explained by the selective pressures exerted by having to deal with a wide array of parasites (Potts and Wakeland 1990). The vertebrate immune system has coevolved with an enormous ensemble of pathogens that are constantly modifying themselves over time. In mammals, for example, it is likely that at any one point in time the number of predator species that confront a prey species, the number of prey species consumed by a predator species, and the number of host species exploited by a parasite species are limited. However, the number of parasite species a host must defend itself against is not (fig. 9.2).

One question must then be asked: Why do "microparasites" (viruses,

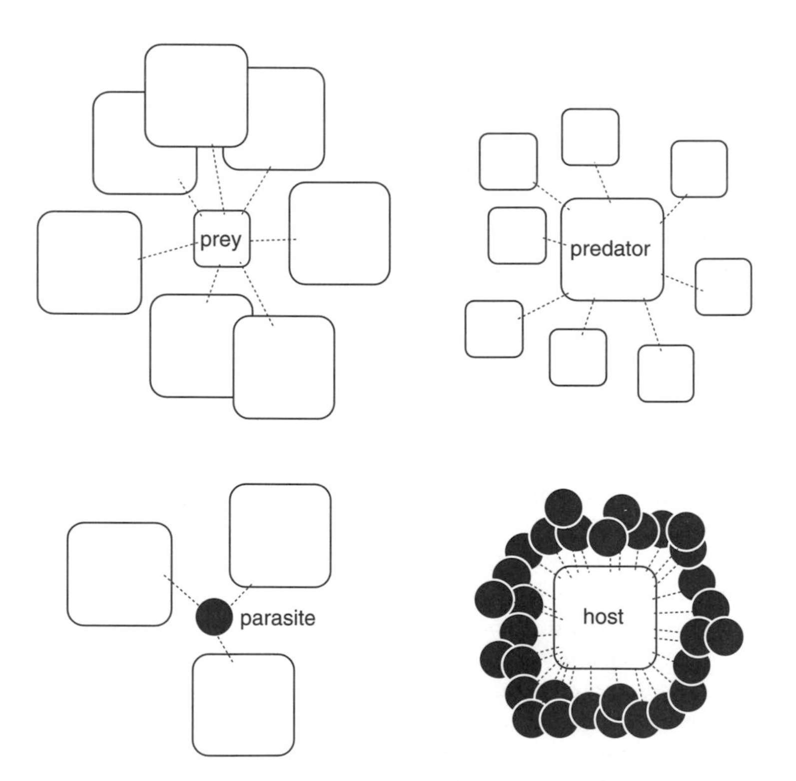

Figure 9.2. Hosts (for example, mammals) confront more parasites than parasites confront hosts, prey confront predators, or predators confront prey (each symbol represents a species).

bacteria, protists) induce more efficient immune responses than "macroparasites"? (This factor makes it particularly difficult to produce anthelmintic vaccines.) The explanation is probably that given by Anderson (1979), speaking about microparasites: "The development of effective immunological protection to such parasites is undoubtedly linked, in evolutionary terms, with the parasites' ability to display direct population growth within the host. If unconstrained, such a growth would inevitably lead to host death."

During the past several years there has been increasing interest in defense mechanisms, which appear to be present in all living organisms, including plants. These include the antibacterial and cytolytic peptides, which constitute a defense system that is not specific but is both rapid and efficient. These peptides are fairly small (a few dozen amino acids) and attack the membranes of target cells by attaching to phospholipids. For example, when confronted with bacteria, insects produce peptides (cecropins, attacins, diptericins, apidaecins, defensins, etc.) whose actions complement the effect of lysozyme, which is also always present.

We are not forbidden to be imaginative about killer genes. For example, Profet (1993) suggested that menstruation in mammals could have been selected to fight bacteria and other sexually transmitted pathogens. Here the blood flow of menstruation would have a particular effect on numerous cells of the immune system in the uterus. If this hypothesis is correct, there should be a certain correlation between the importance of hemorrhage and parasite richness in mammals, although this has yet to be tested.

An Example of Killer Genes

As I noted above, insects defend themselves against parasites that are too large to be phagocytosed by a process of encapsulation involving their hemocytes. Fruit flies *(Drosophila)* are attacked by various species of parasitoid wasps, and within the same species some strains can encapsulate the wasps' eggs but others cannot. Carton, Frey, and Nappi (1992) and Carton and Nappi (1997) have shown the existence of a major gene involved in host resistance in this system, which would act very early in the defensive response—at the level of recognition. In this research two lineages of *Drosophila melanogaster* were isolated from natural populations, one resistant (having the gene[s] to kill the parasite) and the other susceptible to a specific strain of the parasitoid, *Leptopilina boulardi*. Experiments involving crosses (reciprocal F_1 hybrids, reciprocal backcrosses, and reciprocal F_2 hybrids) between these two lineages have shown that the model that best fits the results obtained involves a single major segregating locus with two alleles, resistant and susceptible; a complete dominance of the resistant allele over the susceptible one; and a limited influence of minor modifying genes acting on the major locus. The locus coding for resistance of *D. melanogaster* against *L. boulardi* was called *Rlb* and two alleles, *Rlb+* (resistant) and *Rlb–* (susceptible), were denoted, with *Rlb+* being dominant over *Rlb–*. Genetic manipulations showed that transferring just chromosome 2 from the resistant to the susceptible strain provided resistance to the susceptible strain (and vice versa), indicating that the gene *Rlb* is therefore on this chromosome (Carton and Nappi 1997). Various other experiments involving recombination allowed the localization of the gene in a precise region (the 55D–55F interval) of the right arm of chromosome 2 (Hita et al. 1999; Poirié et al. 2000). Significantly, knowing the precise location of the gene should then allow it to be cloned in the near future (Hita et al. 1999).

Vass, Nappi, and Carton (1993) asked, Does the absence of a gene to kill correspond to a global malfunctioning of the immune system or to failure to recognize the specific parasite *L. boulardi?* In other words, is the killer gene in this case a generalist or a specialist? These authors then went on to show that the strains susceptible to *L. boulardi* reacted against another parasitoid, *Asobara tabida,* thus indicating that there was no immune deficiency in the strain susceptible to *L. boulardi* but rather specificity to this parasite. Further research confirmed that there were in fact two distinct loci, one *(Rlb)* coding for resistance to *L. boulardi* and another one *(Rat)* coding for the resistance to *A. tabida,* with the locus coding for resistance against *A. tabida* also having two alleles, *Rat+* and *Rat–* (Benassi, Frey, and Carton 1998). The allele frequencies vary in natural populations, but it has been difficult for the authors to find a relationship with parasite pressure. In short, several parasite species are present in the same habitat (Carton et al. 1991), and the host genome is thus involved in a diffuse coevolutionary process. Other experiments suggest that these specific killer genes are, above anything else, genes for nonself recognition, since numerous other genes are involved (particularly those concerned with melanin synthesis) in building the immune reaction associated with encapsulation in these hosts (Carton and Nappi 1997).

Limited Arms

There are observations showing that parasite elimination by the host organism may be total, as with numerous viral or bacterial infections. This is more rarely the case for macroparasites, against which the efficacy of the killer genes is in general limited. Nevertheless a total elimination occurred during the experiment done by Jackson and Tinsley (1994) on the dynamics of infection of the amphibian *Xenopus laevis* by a unique monogenean, *Gyrdicotylus gallieni,* that lives in the mouth cavity of its host. *Gyrdicotylus gallieni* does not have a swimming larval form, and host-to-host transmission occurs by direct contact between the two hosts, with the parasite crawling from one to the other. If healthy xenopids are infected with even one parasite, an infrapopulation of several dozen to several hundred parasites forms rapidly following reproduction on the site. However, after a maximum of five months, and often sooner, the infection totally disappears even though the parasites can be maintained for several more months if changed to a different host. Similar infrapopulation dynamics were observed for *Gyrodactylus,* another viviparous monogenean (see, for example, Harris 1988). The logical interpretation of these experiments is that after some time host resistance can totally eliminate the parasite infrapopulations through self-cure.

In most cases, however, killer genes simply limit the macroparasite load to a "tolerable" level. That is, the death of an infrapopulation is not always selected for, and a form of obliged coevolution can then install itself. McKenzie's experiment on impala ectoparasites showed us that the animal does not get rid of all its ticks. To fight against all the ectoparasites, to the very last one, would probably have more negative consequences for the impala than would tolerating a small residual parasite load, simply because this fight would decrease the time spent feeding and protecting itself against predators.

The nematode *Strongyloides stercoralis* has the rare distinction among parasitic metazoans of multiplying on site. This multiplication has huge potential consequences, as shown by the explosive and fatal hyperinfection this worm induces in immunosuppressed patients. However, in the patient who is not immunosuppressed the parasite multiplies slowly enough that the pathogenic effect is less. This response then allows the parasites and the host to live together for many years. The host does not kill this parasite but contains the infection within tolerable limits (see Hawdon and Schad 1991).

Better yet, certain hosts seem to take care of their parasites. For example, certain lizards harbor their ectoparasites in small depressions in their bodies called "acarinaria" or "mite pockets." There are acarinaria on iguanas, geckos, lizards, chameleons, and skinks, and there is a positive correlation between the existence of acarinaria in certain reptiles and their harboring mites, even though some geckos of the genus *Rhacodactylus* have mite pockets and no mites while other species have mites that often settle outside the mite pockets (Bauer, Russel, and Dollakon 1990).

What is the adaptive significance of the acarinaria? They exist at birth in individuals that have never been infected by mites and are therefore coded for by the lizards' genome. In this sense they are not an extension of the phenotype of the mites, even though they favor the parasites by protecting them from being easily detached. Moreover, nothing indicates that lizards have any type of advantage in having the mites. According to Arnold (1986), mite pockets could be a defense mechanism. Mites that live in the reptiles' acarinaria belong principally to the family Trombiculidae, aggressive parasites that actively eat their host's epithelium and cause considerable damage. The mite pockets' "reason for being" would be to concentrate the mites where they cause less damage. Significantly, these areas contain numerous lymphoid cells, and one could term this the "damage limitation hypothesis" (Arnold 1986). Instead of promoting true killer genes, selection in this case would have kept genes that maintain a sustainable level of damage.

10 Genes to Survive

The extraordinary brood-mimicking vocalizations of the common cuckoo chick are the key stimulus that leads to adequate host provisioning.
Davies, Kilner, and Noble 1998

Some parasites have converted the phagosome from a slaughter house into a sanctuary.
Moulder 1979

The Eggs' Story

The Others' Nests

The European and Asian cuckoo *Cuculus canorus* does not build a nest.[1] During the reproductive season the eggs of *C. canorus* may be found in the nests of various other bird species. Taking advantage of the momentary absence of an incubating female passerine, the female cuckoo lays her eggs in the following manner: she sits on the nest, swallows an egg, and replaces it with one of her own—all in about ten seconds. She then does this again in another nest, and then again and again. From this time on the female cuckoo has no interest in her own eggs, which are dispersed in nests with adoptive parents that incubate and feed the cuckoo chicks as their own. Since the young cuckoo hatches first and eliminates the eggs of its host (the term "false friend" has never been more appropriate), and because the cuckoo is a fairly big bird, the "parasitized" adoptive parents often end up feeding birds bigger than themselves.[2]

To explain the attack and survival mechanisms of this parasite-host sys-

1. Reported here are predominately the results of Brooke and Davies (1988), who worked in England, where eggs of *Cuculus canorus* are found mainly in the nests of the meadow pipit *(Anthus pratensis),* reed warbler *(Acrocephalus scirpaceus),* dunnock *(Prunella modularis),* and less often the robin *(Erithacus rubecula)* and white wagtail *(Motacilla alba)*. A second species of European cuckoo, the great spotted cuckoo, *Clamator glandarius,* has a different biology that will be mentioned briefly. There are, particularly in Africa and Australia, many other species of cuckoos and parasitic birds other than cuckoos (such as cowbirds in the United States), as well as cuckoos that are not parasitic.

2. Winfree (1999) notes that cuckoo nestling behavior was reported by Aristotle 2,350 years ago and that Edward Jenner, better remembered for preventing smallpox by inoculating humans with cowpox, first described it in detail in 1788.

tem I will use somewhat unusual but easily understandable terms, such as "self eggs" (to designate the eggs of the host bird) and "nonself eggs" (to designate the eggs of the parasitic bird).

What Weapons Do the Hosts Have?

Certain birds, such as warblers, reject objects introduced into their nests by pushing them over the side, as when one of their eggs is replaced by an artificial egg that is not identical to theirs. When the bird cannot expel the intruder, rejection sometimes takes another form—simply abandoning the nest. Birds that act this way are called *rejecters,* though even they may be cheated if the introduced object imitates their eggs very well. Other birds, such as the dunnock, are termed *acceptors* and do not show this behavior. Instead, these birds continue to incubate the foreign egg or object in their nest. While the behavior of rejecter birds is similar to the mechanism of nonself rejection in classical immunology, the acceptors are the equivalent of organisms without an immunodefense against a pathogenic agent.

Brooke and Davies (1988) made artificial eggs, some of which imitated those of passerines and some of which did not, and placed them in a variety of bird nests. Figure 10.1 synthesizes part of the results these researchers obtained. The results show that mimetic eggs are much better accepted than nonmimetic eggs (the results being striking for the reed warbler); that the quality of the rejecter is not the same for all species (that is, the reed warbler defends itself much better than the meadow pipit, and the meadow pipit much better than the pied wagtail); and that the dunnock is almost totally defenseless (even accepting eggs that are totally black or white).

That is, there is variation in the repertoire of nest defenses against brood parasites such as cuckoos, and therefore the recognition and destruction of nonself eggs could be selected for in bird hosts. At first one may get the impression that in all cases of nest parasitism it would be to the advantage of the bird host to select just such a behavior. In reality, however, things are not so simple, because rejection includes the possibility of mistakes. In short, if the threshold of detection of nonself eggs is too fine, the bird host may reject one or more of its own eggs even if there is no cuckoo egg in its nest. Davies, Brooke, and Kacelnik (1996) presented a mathematical model showing that the host species may have no advantage in developing such rejection behavior because of the associated risks, and all available evidence suggests that in fact, owing to the cost of recognition, rejection is selected against when parasitic birds are absent in an area. Another model suggests that egg rejection should in-

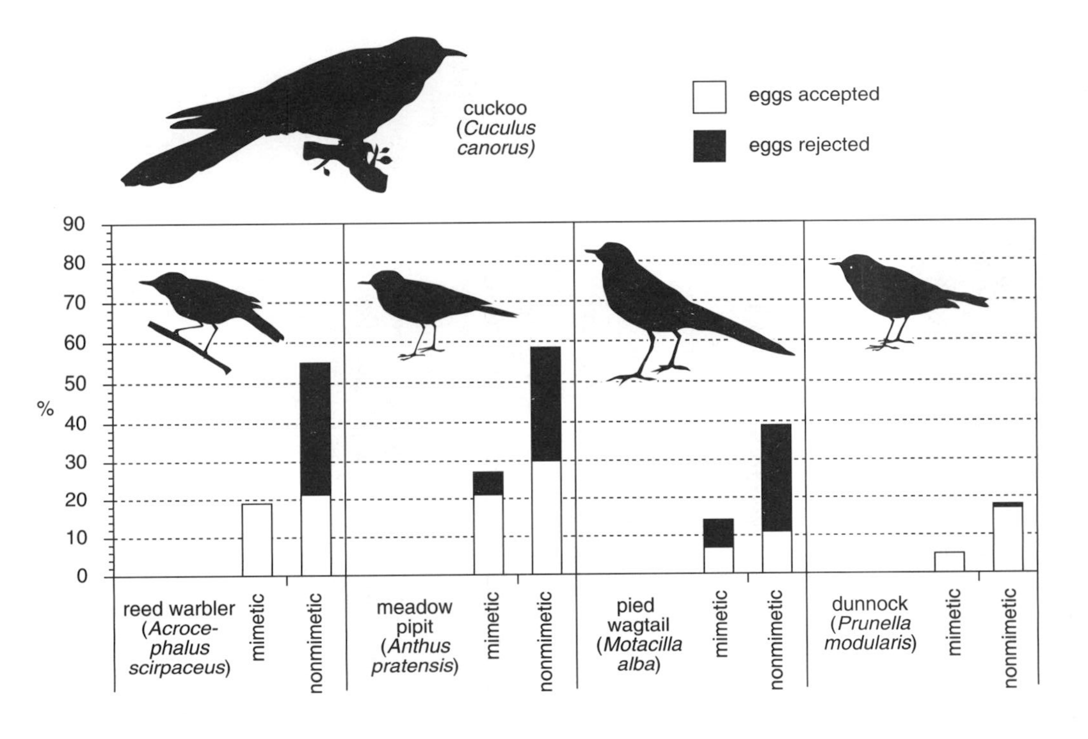

Figure 10.1. Acceptance and rejection of mimetic and nonmimetic eggs experimentally introduced into the nests of a variety of birds (modified from Brooke and Davies 1988).

crease as the rate of parasitism increases and egg mimicry deteriorates and that the erroneous rejection of their own eggs by host females should be expected at intermediate levels of egg mimicry (Rodriguez-Girones and Lotem 1999). Further, the data also indicate that rejection can spread rapidly if a population of acceptors is invaded by a brood parasite. For example, in Africa village weavers *(Ploceus cucullatus)* are parasitized by cuckoos of the genus *Chrysococcyx* and show high levels of discrimination. About 1900, village weavers from Africa were introduced to Hispaniola (the West Indies), where no brood parasite was present, and Cruz and Wiley (1989) demonstrated that in the absence of parasite pressure on Hispaniola the rejection rate had strongly declined by 1985. However, the shiny cowbird *(Molothrus bonariensis)* a brood parasite, was introduced to Hispaniola at about that time and is at present exploiting the village weavers. As might be expected, Robert and Sorci (1999) have shown that rejection behavior in the weavers has been selected again in just twenty to twenty-five years, with *P. cucullatus* currently rejecting about 90% of the cowbird model eggs placed in nests. In this case Robert and Sorci consider that genetic processes could have been complemented by learning as brood defense is reacquired.

Rejecting cuckoos' eggs as protection against brood parasitism is not the only mechanism of defense that may be selected, however. Cuckoos need a tree to stake out the nests of their prospective hosts, and when bird hosts nest on the ground they may chose a site far from trees to escape surveillance and avoid being parasitized (Alvarez 1993). *Acrocephalus scirpaceus* and *Cercotrichas galactotes,* for instance, both nest in littoral plants and show preference for sites far from trees, even to the point of competing with others for these low-risk habitats (Oien et al. 1996). Further, not all the members of a particular host species are necessarily parasitized by cuckoos even if cuckoos are present in their area. According to Lindholm (1999), that different host species most often live in distinctive habitats leads to a fragmentation of the female cuckoo population and adaptation to a particular host within the larger area. Thus there would be a metapopulation of cuckoos relative to hosts, and certain suitable patches (host populations), particularly isolated ones, may not be inhabited by cuckoos.

What Weapons Do Cuckoos Have?

One might think that cuckoos parasitize only acceptor birds, but this is not true. Therefore it is not surprising that the selective pressure exerted by the rejecters would allow selection for adaptations that result in occasional acceptance of cuckoos' eggs. If we look at the ensemble of cuckoo species (and not only *Cuculus canorus*), these adaptations include size mimicry, ornamentation mimicry, and cryptic coloration. As I mentioned earlier, parasitic cuckoos are relatively big, but their eggs are, all things being equal, fairly small. Moreover, parasitic cuckoos' eggs are smaller than those from nonparasitic cuckoos. In certain species cuckoo eggs are exactly the same size as the hosts', while for others the match is not as good. The reduction in size of the cuckoos' eggs is an adaptive response to their hosts' mechanism of rejection, since nonself eggs that are the most similar in size to self eggs have more chance of being accepted, even by rejecters.

Most passerine eggs are identifiable by the ornamentation on their shells—their background color, their reflectance, and various dots, spots, or marks. Cuckoos' eggs mimic such ornamentation on their hosts' eggs, and in some cases, such as Indian cuckoos, the perfect mimicry of ornamentation as well as size can dupe even the most accomplished ornithologist.

Certain bird species build their nests in very dark places, such as tree hollows or rock fractures, and there are cuckoo species that parasitize these birds as well. In this case the cuckoos do not imitate their hosts' eggs but simply lay black ones. In these species, the nonself eggs mimic not the eggs of the bird host but rather the dark environment—making them as inconspicuous to the host as are its own eggs.

Is the Cuckoo a Specialist or a Generalist?

The problem of parasite specificity is particularly clear in cuckoos: If the cuckoo knows how to make only one type of egg, imitating only eggs from a particular bird, it can successfully parasitize only that bird. But if the cuckoo knows how to make several types, imitating eggs from several bird species, it can then parasitize all of them.

The European cuckoo is not specific, which implies that it knows how to imitate the eggs of several bird species. Does it really do it? The answer is, of course, both yes and no. No, because a specific female cuckoo makes only one type of egg (for instance, if it imitates the marsh warbler's eggs it will lay its eggs each season only in the nests of marsh warblers) and yes because different females have different adaptations—each can imitate eggs from only one type of bird (some lay their eggs in marsh warblers' nests, others in wagtails' nests, etc.). Further, it is also known that a female cuckoo lays her eggs in the same type of nest where she was born, and in a given locality it appears that it is almost always the same bird host that is parasitized. It is thus said that there are several races or "gentes" in *Cuculus canorus,* each specializing in a different host species.

One may therefore ask, What is the origin of this specialization? According to Brooker and Brooker (1990), selective pressures having their origin in intra- and interspecific competition may have contributed to the evolution of this specificity in cuckoos. However, the need to avoid host defense mechanisms seems to be the more natural explanation. Support for this comes, for example, if we look at the nests of dunnocks *(Prunella modularis)*, a typical acceptor. Here there is no egg mimicry; it appears that members of a particular lineage of cuckoos lay eggs in dunnocks' nests and that the eggs they deposit are spotted whereas the dunnocks' eggs are uniformly blue. That is, the absence in the cuckoo of an egg that mimics that of its acceptor bird is a counterintuitive yet strong argument that the strategy "mimetic egg" has evolved mainly under the pressure of host defense mechanisms.

Specialization without Speciation?

The division into separate lineages of cuckoos as the result of host specialization brings out a difficult question: How is the characteristic "choice of the host" maintained if females mate randomly with males, and if they do not mate randomly, then how do they chose the right mate? The hypothesis that several species, each one oioxenous, were confounded under the name *Cuculus canorus* has sometimes been proposed. However, Marchetti, Nakamura, and Lisle Gibbs (1998) studied the mating patterns of *C. canorus* using microsatellite markers in a region of Japan where a new host, the azure-winged magpie, *Cyanopica cyana*, appeared about thirty years ago. In this study female cuckoos were shown to have strong host preferences, with the males mating indifferently with females specialized for different hosts, thus invalidating the hypothesis that several species may be confused under the name *C. canorus*. The results obtained instead suggest that traits linked to egg mimicry are inherited from the mother while the indifferent choice of the males maintains panmixia within the species. One hypothesis is that genes coding for egg coloration may be located on the female-specific W chromosome, while another is that lineages differ in mitochondrial DNA and not nuclear DNA (since mitochondrial DNA is transmitted only by females, they could then mate with any male without altering their adaptation to a specific host): "Such sex-linkage would allow the preservation down the matriline of favourable combinations of genes for egg mimicry, regardless of the male genotype" (Lindholm 1999). Thus cuckoos apparently resolve the paradox by specializing without speciating, although more work on these matrilineal hypotheses awaits.

What Happens after the Cuckoo Egg Is "Accepted"?

Another enigma of coevolution between the cuckoo and its hosts is what happens after the egg has been accepted. Even rejecter birds that normally can identify badly imitated eggs later tolerate, if they make a mistake, cuckoo chicks that absolutely do not resemble their own progeny. That the cuckoo's trickery can be prolonged after hatching is surprising, and one can only suggest that this is a totally different problem. Just as the young goose, famous from Konrad Lorenz's work, imprints on and follows the first living being it sees, female birds may raise any chick that hatches in their nest without trying to identify it. A variety of experiments done on reed warblers have shown that this may in fact be so, since these birds cannot distinguish their own chicks from foreign ones, whatever they may be. Further, Davies, Kilner, and Noble (1998) have shown that the begging call of a cuckoo chick parasitizing a reed warbler nest closely resembles that made by a whole reed warbler brood.

In a sense, then, cuckoos are a unique model to understand the nature of immune evasion in parasitism simply because in most cases the parasites effectively escape their host's defense mechanisms by trying not to be identified, just as do cuckoos.

Let's finish with this sentence from Dawkins (1982): "Host birds may be very good at resisting psychological manipulation, but cuckoos might become even better at manipulating."

THE BUTTERFLIES' STORY

Cuckoos do not ask their bird hosts to come and get their eggs, but such brazenness does occur in a very different parasite-host system, one involving a parasitic butterfly and an ant host. Butterflies of the genus *Maculinea* are very pretty diurnal lycaenids with blue wings that fly in Alpine meadows. Here we will follow the annual cycle of the species *Maculinea rebeli* with Elmes, Thomas, and Wardlaw (1991) and Hochberg et al. (1994).

During July and August males and females mate, and eggs are deposited on the flowers of the blue gentian, *Gentiana cruciata*. From the eggs hatch minuscule caterpillars that penetrate to the inside of the flowers, feed on the growing seeds, grow, and molt several times. At the end of August the first unexpected event happens: caterpillars (now in their fourth and last larval stage) pierce the fruit and drop outside. If they drop onto a leaf of the plant, they crawl until they reach the edge and drop again. This may happen several times until the caterpillars are on the ground, then they stop moving. Significantly, the exiting and dropping of caterpillars happens not just at any time but only at the end of the afternoon.

In the Alpine meadows where the butterflies live there also occur several species of ants, one being the species *Myrmica schencki*. Although the butterflies do not seek out plants near the ants' nests to deposit their eggs on, there nevertheless is a correlation between the butterflies' egg distribution and *M. schencki*'s habitat.

Following the ants' behavior over the course of a summer day will be instructive. During the hot hours the ants stay inside their nest of underground galleries and occupy themselves in various domestic chores, primarily feeding the queen and larvae and cleaning the chambers. After about 6:00 P.M., with a peak around 7:00 P.M., the workers leave the nest to forage.

If an ant knocks haphazardly into a caterpillar, it palpates it with its antenna and almost immediately (between one and four seconds after the encounter) picks it up and delicately carries it toward the nest. This behavior is well known in other circumstances in the ants' life: when a nest has been disturbed and larvae have been dispersed outside, the workers will take lost larvae and bring them back to the nest in the same way. Significantly, the caterpillars have a morphology, principally in the anterior region of the body, similar to that of an older ant larva, and their cuticle contains chemicals that imitate those of the ants' larvae. As Elmes, Thomas, and Wardlaw (1991) note, the caterpillar looks and smells so much like the ant's larva that the instinctive response of any *Myrmica* is to immediately take it back to the safety of the nest. Given the foraging and search habits of ants, the caterpillars that fall from gentians never wait long before being adopted—on average about twenty minutes.

The adoption of caterpillars by ants *(M. schencki* or, with less success, *M. sabuleti)* is indispensable to their survival, since if they are not adopted caterpillars survive a maximum of only one or two days after they fall to the ground. Once the caterpillar is in the nest, however, it either shares the larval dormitory or has its own chamber. In either case it is permanently surrounded by four or five workers that feed it solids and liquids, clean it, and if necessary even transport it. These workers enjoy several caterpillar secretions, and in this way the caterpillar pays its room and board. The caterpillars grow during the winter, increasing their weight from two to sixty-five milligrams. This is then followed by a hibernation period when they lose more than a third of their weight. In the spring, when temperatures reach 14°C, the caterpillars are fed again. and their weight ultimately reaches one hundred milligrams. By then it is the beginning of the summer, and the caterpillars pupate and metamorphose into adults. When the adult butterfly finally emerges it runs rapidly out of the nest, takes its first flight, and searches for mating partners and gentians.[3]

Survival Strategies

Strategies of Discretion

Wasson (1993) sees in any multicellular organism a community of cellular populations in equilibrium. The introduction of a parasite into this community corresponds to the arrival of an exotic population within the balanced system, and a new equilibrium must then be reached. The parasite is in competition with the other elements of this community, and if not recognized as nonself could discreetly enter and develop without trouble. When a parasite penetrates a host organism, the host's immune system "sees" the surface of the parasite. That is, it is exposed first and foremost to the molecules the parasite releases on entry or that it exposes on its surface. More precisely, characteristic regions of these molecules, or epitopes, are recognized, and the most efficient way for the parasite to survive in this hostile environment, like both the cuckoo and the butterfly, *is to not be recognized as different.*

The selection of surface molecules that are as little immunogenic as possible, that is, that are the least visible to the immune system of their host, may be the ultimate weapon of choice for most parasites. Nematodes are very efficient in this domain, and various studies have shown that their cuticle is only slightly immunogenic, if at all, thus allowing them to avoid encapsulation in invertebrate intermediate hosts as well as to escape destruction by antibodies in definitive hosts. The nematodes' cuticle (more exactly the epicuticle) is principally made of a particular protein, cuticuline, which when associated with lipids is immunologically inert (Betschart, Marti, and Glaser 1990; Rudin 1990). What is most amazing is that free-living nematodes living in fresh water, sea water, soil, and plants also have such a nonimmunogenic epicuticle, thus allowing nematodes to become parasitic with only a few new adaptations. Add to this that their general shape need not change much to become parasitic, and we have organisms exquisitely preadapted to enter hosts and get around their immune defenses.

A different survival strategy, and one no less discreet than a total lack of immunogenicity, is selection for surface molecular communities that mimic that of the host (fig. 10.2A). If some surface molecules of the par-

3. Although the biology of *Maculinea* may at first appear as an exceptional curiosity, there are more than 18,000 species of Lycaenidae known to be associated with ants in some way (J. A. Thomas, personal communication).

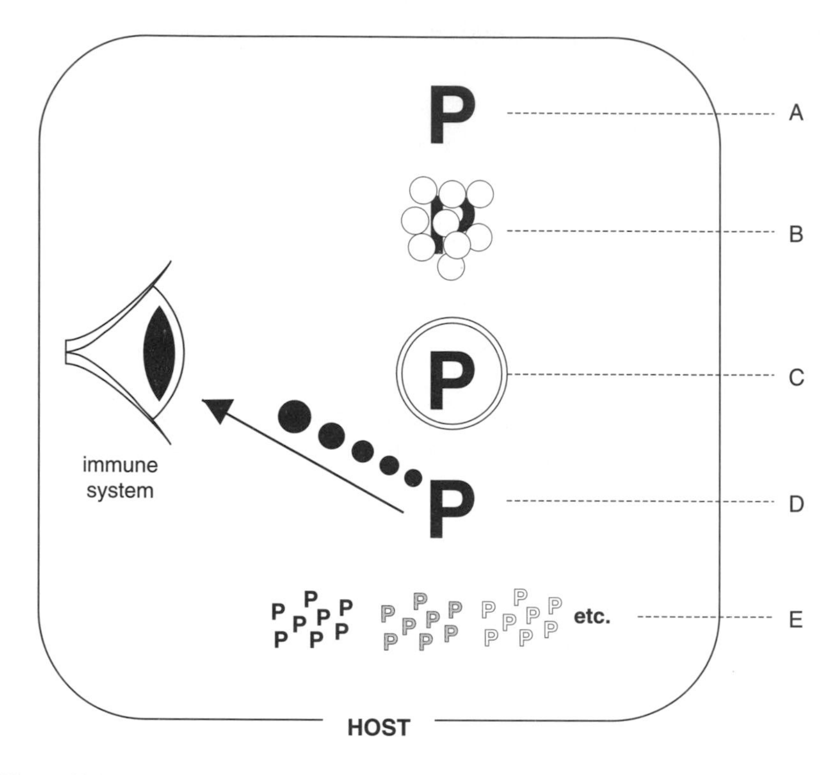

Figure 10.2. Main survival strategies of parasites in their hosts. (A) antigenic communities; (B) antigenic camouflage; (C) seclusion; (D) immunosuppression; (E) antigenic variability.

asite (principally proteins, but also lipids, carbohydrates, or complex associations of each), show shapes similar to those of host molecules, it is difficult, if not impossible, for the host's immune system to detect them. Such nonantigenic surface communities result from selective pressures exerted by immune systems, and any mutation that induces a change in the conformation of a parasite's molecule to make it resemble a host molecule has a selective advantage. The same result may be obtained by transferring genes from the host genome into the parasite's genome, which would shortcut the number of generations necessary for such a community to evolve by successive mutations. Gene transfers may be done by such things as mobile DNA elements, but this and other processes of transfer are still not well understood.

This pressure from the immune system may also lead to selection of surface structures on the parasite that can adsorb host molecules to accomplish the same end. This constitutes an antigenic camouflage that renders the parasite "invisible" to the immune system, since it is hidden under a coat of "self" (fig. 10.2B). In this case it is not necessary for highly specific receptors to be present on the parasite's surface. Rather, it is enough that the surface shows a structure that passively adsorbs host molecules.

Parasites that live in the hemolymph or blood of their hosts must remain particularly inconspicuous or discreet because they are continuously surrounded by defense cells and surveillance mechanisms. Schistosomes, for example, have antigenic communities for both mollusks and vertebrates, the two host groups they successively confront in their life cycle (see Damian 1964; Yoshino and Bayne 1983). Dissous et al. (1990) and Dissous and Capron (1995) have shown that the tropomyosin of *Schistosoma mansoni* is 65% homologous with that of one of its molluskan intermediate hosts, *Biomphalaria glabrata*, far greater than with that of *Drosophila* or humans. This homologous protein is present in the tegument of the larval stage (sporocyst), which lives in contact with the mollusk's hemolymph. Dissous, Grzych, and Capron (1986) have also shown that both the cercaria and miracidium abundantly express on their surfaces an oligosaccharide structure also found in the molluskan host. Even so, it is in their definitive hosts that the schistosome's ability to camouflage itself has been best demonstrated. Here the borrowed host molecules have been found to be glycolipids and glycoproteins that are part of the hosts' major histocompatibility complex. Such antigenic camouflage is associated with the development of an immunity against "newcomer" schistosomula that are not yet protected by such a disguise. In this case parasitologists speak of "concomitant" immunity in the vertebrate host, which protects against new infection by the same species without eliminating the parasites that are already there.

However, according to Sorensen (1999), worms of the challenge infection eventually replace those of the primary infection. This is true, for instance, when pigs are exposed to successive infections by cercariae of *Schistosoma japonicum* originating from two strains distinguishable by analysis of a particular mitochondrial marker, indicating that the idea of concomitant immunity should probably be revisited.

On their side, larval schistosomes seem able to borrow molecules from their molluskan host that mask their surface. In this case specific sugars from the mollusk's plasma have been detected on the sporocyst's tegument. Even if these sugars may be metabolites used by the parasite for its growth, their presence can only decrease the efficacy of the snail's defense mechanisms (Yoshino, Cheng, and Renwrantz 1977).

Strategies of Protection

In a variation of a strategy of inconspicuousness, the parasite is not camouflaged as previously described but rather seeks refuge from the host's defense system (fig. 10.2C). The word *seclusion* is often used to designate some parasites' ability to locate in the least hostile of possible habitats within their hosts. For instance, the very act of adhering to its host's tissues seems to protect a parasite, at least partially, from the cells of the host's defensive system. Monconduit and Prévost (1994) noted that the eggs of the parasitoid *Asobara tabida* strongly adhere to the tissues of *Drosophila melanogaster* in the hours following oviposition, whereas they adhere much less strongly to those of *Drosophila simulans*. Further, the parasitoid is rarely encapsulted in *D. melanogaster*, but it is always destroyed in the larvae of *D. simulans*. From this the authors deduced that attachment to host tissue protected the parasites from the host's circulating hemocytes, and electron microscopy confirmed that only very limited areas of egg chorion remain exposed to hemocytes when adhesion to the tissues of *D. melanogaster* is complete (Eslin et al. 1996). In this vein it is logical to think that the intracellular life of some parasites has been selected as a means to shelter them from host defense reactions, which are in general highly efficient in the blood, lymph, or interstitial fluids outside the cell. Viruses are usually intracellular, as are numerous bacteria such as *Listeria monocytogenes* (the causative agent of listeriosis), which penetrates macrophages as well as other cells. Similarly *Mycobacterium leprae*, the bacteria that causes leprosy, penetrates several cell types, and intracellular protists are also common (see Mauël 1996). *Bonemia ostreae*, for example, penetrates oyster hemocytes, *Leishmania*, human macrophages, *Plasmodium*, hepatic cells or erythrocytes (depending on the case considered), *Babesia*, lymphocytes and then erythrocytes, *Theileria*, lymphocytes, and *Trypanosoma cruzi*, macrophages and certain muscle cells. Among multicellular parasites larvae of *Trichinella* (a nematode) penetrate muscle cells, and the trematode *Apatemon graciliformis* has cercariae that penetrate their fish hosts' oocytes (Combes and Nassi 1977).

It is remarkable that in many such cases the colonized cells are those of the host immune system itself: to live in macrophages or lymphocytes not only is hiding oneself, it is also "cuckolding" the immune system. Continuing to reason in an evolutionary approach, we can expect that intracellular parasites, under the pressure of the host's immune defense, would decrease the time spent outside the cell as much as possible. In fact, as soon as they exit a cell most such parasites rapidly adhere to other cells and penetrate. Penetration of the host cell is often remarkably rapid (fifteen seconds for *Toxoplasma gondii*), and in *Plasmodium falciparum* infection the presence of particular proteins on the erythrocyte surface may induce a pileup of cells in blood vessels—sometimes interpreted as an adaptation favoring passage from one erythrocyte to another. Still other parasites induce healthy cells of the same type to adhere to the cell they are in so as to reduce exposure. For instance, HIV induces membrane changes on the host lymphocyte so that other lymphocytes adhere to the infected cell. When adhesion is accomplished, the infective stages of the virus then pass to the adjoining cell without having to go through the plasma.

The transfer of parasites from one cell to another may also take advantage of particular properties of the cells: *Listeria monocytogenes,* after diving into a macrophage, induces the formation of a thin and rigid pseudopod shaped from actin filaments. When an infected macrophage touches a healthy macrophage, the latter then phagocytoses the bacteria-containing pseudopod and itself becomes infected (see Donelson and Fulton 1989).

Intracellular life does not protect parasites from all dangers, however, and parasites are often not as invisible from the outside as we may think. Transformations induced in erythrocytes by *Plasmodium* to ensure the energy needs of the parasite lead to surface changes in red blood cells that cannot fail to be perceived by immune cells. Moreover, and even if the stay in plasma is brief, parasites may be recognized by antibodies that attach to them and that, when dragged into the host cell, may continue to act (this is the case, for example, when *Plasmodium* sporozoites penetrate hepatic cells). Moreover, cells that are infected by an intracellular pathogen may themselves be destroyed. Cytokines secreted by Th1 T cells stimulate the activity of NK (natural killer) cells, cytotoxic T cells, and macrophages that act together to eliminate infected cells, while those secreted by Th2 cells stimulate the production of antibodies and are more involved in defense against extracellular parasites such as helminths.

Once in the host cell, the parasite confronts intracellular defenses that involve the fusion of the vacuole the parasite is in with lysosomes packed with destructive enzymes. These new pressures in turn lead to other parasite adaptations (see Ojcius and Dautry-Varsat 1995; Channon and Kasper 1995). *Toxoplasma gondii,* for instance, prevents fusion of lysosomes with its vacuole, and *Trypanosoma cruzi* leaves its vacuole to directly contact the cytoplasm. *Leishmania,* on the other hand, is simply resistant to the intracellular enzymes poured on it, and many intracellular parasites have sophisticated antioxidant systems to neutralize or inhibit host cell oxidative defense mechanisms.

Cysts often result from a combination of parasite secretions and host reactions and are another form of protection from the ambient milieu, which can either be outside a host (in this case, protection mainly against desiccation) or inside (in this case, protection against attacks by defensive systems). Sometimes, despite the presence of a cyst wall, there may be exchange of nutrients or molecules that are used to manipulate the host. In general, though, it seems that encysted stages in animal intermediate hosts are not very specific, and we should recall here that narrow specificity is a result of the difficulty of avoiding defense systems, all differing from each other. It is therefore not surprising that encysted stages, which are protected from such systems, exhibit little specificity. This is the case, for example, in trematodes. Here encysted stages (metacercariae) are much less specific than the nonencysted stages of these parasites (see Combes et al. 1994).

Seureau and Quentin (1986) have shown that during the evolution of certain parasite lineages such as insect nematodes, a strategy of seclusion has been increasingly successful. Basically, the more ancient the parasite-host system, the more the parasite localizes in protected places where the host's attacks are less and less efficient. Still other parasites practice antigrooming or antipreening strategies by settling in zones inaccessible to the host's cleaning tools. Clayton (1990), for example, has shown that phthiropterans attached to the barbs of bird feathers react to preening: "at rest" they are hooked to feathers with their bodies oriented in haphazard directions. As soon as they are alerted by a pull on the feather, however, they become rigorously parallel and therefore are protected by the barbules and inaccessible to the bird's searches.

Strategies of Manipulation

Parasites use host manipulation to limit or prevent attacks from defense systems (fig. 10.2D). Such manipulations probably owe a lot to the fact that proteins have conserved amino acid sequences throughout evolution because selective pressures (induced by their function) have eliminated most nucleotide substitutions or other changes that do not work. We know today that numerous such vertebrate genes are already present in the form of related sequences in invertebrates and even protists (Cooper et al. 1995). This conservation of sequences therefore allows parasites to act on the host phenotype with a minimal adaptive cost.

A demonstration of this conservation of sequences in parasites can be distilled from the work of Duvaux-Miret et al. (1992), who showed the presence of peptides present in humans (alpha-MSH [melanocyte-stimulating hormone]; ACTH [adrenocorticotropic hormone]; and beta-endorphin) in all stages during the development of *Schistosoma mansoni*. These three hormones are produced in humans by the adenohypophysis and derive from a common precursor, proopiomelanocortine.

The interest of these molecules is that in a complex manner they suppress immune system function. For example, alpha-MSH and ACTH (via its conversion in alpha-MSH) have anti-inflammatory properties because they inhibit certain categories of leucocytes, while beta-endorphin inhibits antibody production.

The dipterans *Hypoderma bovis* and *H. lineatum* use a similar manipulation. These flies lay their eggs on bovids' legs, and their larvae migrate to various organs. The larvae secrete two enzymes, hypodermines A and B, that depress the immune response, particularly the C3 fraction of complement—a factor that plays a central role in the induction response (Boulard 1988, 1992).

In invertebrates, hemocytes seem to be the main targets of immunosuppressive processes, and manipulation by the parasite has been demonstrated in vitro by altering hemocyte function. Excretory-secretory products of trematode larvae recovered from culture medium modify molluskan hemocyte activity by inhibiting enzyme synthesis, decreasing superoxide and hydrogen peroxide H_2O_2 production, modifying phagocytic activity, and so forth (see Yoshino and Lodes 1988; Connors and Yoshino 1990; Connors, Lodes, and Yoshino 1990; Loker, Cimino, and Hertel 1992). While some of these processes are suppressed by the direct action of parasite products on hemocytes (e.g., the production of a superoxide scavenging molecule by the parasite), others appear to be due to the parasite's ability to modulate the initiation of the host's defense system by altering levels of a host immunoreactive cytokine-like molecule (SnaIL-1), the vertebrate analogue of which has neuroimmunomodulatory activity in mollusks (Granath, Connors, and Tarleton 1994; Connors, Buron, and Granath 1995; Connors et al. 1998; Clatworthy 1998).

Strategies of Confrontation

Trypanosomes are flagellated protists that parasitize various organisms, including plants. In humans they cause sleeping sickness in Africa and Chagas' disease in South America (the two diseases are very different both in the biologies of the parasites and in their pathogenic effects). What follows concerns the causative agent of sleeping sickness, *Trypanosoma brucei*, known to occur in Africa under three "forms": two that affect humans (*T. brucei gambiense* in the west and *T. brucei rhodesiense* in the east) and one (*T. brucei brucei*) that affects only animals (see Godfrey et al. 1990). *T. brucei* has a heteroxenous life cycle that involves humans (or other mammals) and tsetse flies (*Glossina palpalis* and *G. morsitans*). These parasites have the peculiarity of reproducing only asexually (if sexual reproduction does occur it is extremely rare) and thus are genetically distinct clones.

The first essential characteristic of trypanosomes is that one individual (that is, one cell) exposes on its membrane only a few proteins, of which one is highly dominant. This protein is formed from two chains of about five hundred amino acids each and has some glycosylated parts. On average, there are about 10 million molecules of this dominant glycoprotein on one trypanosome.

In the same species not all individuals, including those of the same clone, express the same dominant glycoprotein on their surface. There are numerous possible glycoproteins, all differing in the amino acid sequence in a variable region that represents about four-fifths of the molecule. This diversity of potential types gives those molecules the name variable surface glycoproteins, or VSGs (see Pays and Steinert 1988). Within the membrane the constant region is inserted toward the inside of the cell and the variable region toward the outside. All other surface proteins (non-VSGs) are one hundred to ten thousand times less abundant and are "hidden" among the VSGs on the surface (Overath et al. 1994).

There are about one thousand genes capable of coding for VSGs, although only one gene at a time is expressed in an individual cell. As a result, a given membrane shows only one type of VSG. Moreover, there are only a few locations on the chromosomal regions where VSG genes may be expressed, and all these genes are close to one telomere on one of the largest chromosomes, very close to the 3′ extremity of its coding strand.

Expression of a VSG gene (transcription into messenger RNA and then translation to protein) requires two successive conditions: that the gene to be expressed is duplicated, with the copy inserting itself into one of the expression regions near the telomere, and that the VSG genes situated in other telomeric regions remain silent.

This is why a given individual of *T. brucei* can express only one type of glycoprotein on its surface among all that can potentially be made. Further, when a gene takes the place of another in the expression site, these genes can exchange sequences so that an expressed gene is a patchwork of both "old" and "new" genes.

In humans the parasite develops mainly in the blood, where it multiplies very rapidly via binary fission. The resulting dynamics of the parasitemia are particular: there are waves of high parasitemia separated by periods when the parasites all but disappear. "Explosions" and "drops" in number happen every seven to ten days, with "drops" corresponding to the destruction of about 99.9% of the parasites by the immune system. All trypanosomes of the same wave express the same VSG, and trypanosomes from two successive waves have different VSGs. For each infection by a given clone of *T. brucei* the order of successive VSG waves is generally the same. On the contrary, the trypanosomes forming in the tsetse fly form an infrapopulation in which each individual cell may express a different VSG.

The description above allows one to understand the evasion strategy of *T. brucei:* when the tsetse fly injects trypanosomes, one of the types, characterized by the expression of a particular VSG (call it A), multiplies faster than the others and forms the first "wave." Immediately afterward the immune system produces antibodies against A and eliminates the vast majority of these trypanosomes. However, among the ensemble of trypanosomes present a few (maybe one in one million) escape immune destruction either because they have changed the expressed VSG gene or because they come directly from the stock of trypanosomes containing a different VSG originally injected by the tsetse fly. These trypanosomes, which are all covered by a different glycoprotein (call it B), multiply to form the next wave. Anti-A antibodies are ineffective against B, so the immune system must produce anti-B antibodies, and so on (fig. 10.2E).

This strategy of immune evasion, termed *antigenic variation,* results in a never-ending fight where the parasite is continually decimated and continually revives. Thus this strategy is different from that of cuckoos, schistosomes, and the endocellular parasites cited previously, which all essentially hide, in one way or another, from the host defenses.

Other protists, such as *Plasmodium,* also use antigenic variation. For example, during infection of the monkey *Macaca sinica* by *Plasmodium fragile* there are several peaks of parasitemia characterized by the expression of different antigens on the membrane of the parasitized red blood cells, which also appear (as for trypanosomes) in a predictable sequence. When *Plasmodium* expressing a particular antigen are experimentally transferred to monkeys with antibodies directed against this antigen, the parasites immediately express a new antigen (Handunnetti, Mendis, and David 1987).

Variation induced by immune system pressure may exist even in parasite clones. For example, merozoites (the stage involved in the passage from red blood cell to red blood cell) of *Plasmodium knowlesi,* a parasite of the rhesus monkey, express on their surface a 140 kilodalton protein (David et al. 1985). When, about the thirtieth day after infection, monkeys are immunized by injection of the purified protein, the parasites (which come from one clone and are thus expected to be genetically identical) are no longer expressing the 140 kilodalton protein (David et al. 1985). The hypothesis is that individuals who have lost the gene have been immunoselected (Hudson, Wellems, and Miller 1988). This example shows how difficult it is to find a "target" for the development of a truly protective vaccine for this organism. Variation of *Plasmodium falciparum* clones has also been shown to occur even in vitro (and thus in the absence of immune system pressure) at a rapid pace (Roberts et al. 1992), and in *Plasmodium yoelii,* a parasite of African rodents, there are even dozens of genes encoding variable members of a particular protein family of which only a single gene is transcribed in each nucleus produced during schizogony; each individual merozoite derived from one parent schizont then expresses a different form of the same protein (see Snounou, Jarra, and Preiser 2000).

Strategies of Collaboration

We know (chapter 2) that a decrease in the defense response caused by parasite X may also profit parasite Y and allow it to infect the same host. It is therefore not surprising that certain parasites may have used this possibility as a weapon. Insect hosts of parasitoids defend themselves because their hemocytes encapsulate and melanize invading organisms. The mechanisms that allow parasitoids to exploit their hosts and prevent this destruction are varied. In numerous species of hymenopteran parasitoids the female's ovaries have been shown to contain polydnaviruses, or PDVs, so named because their genome is fragmented and divided into fractions, all of which must be present in order for the virus to replicate (see Fleming 1992). In at least some species, such as *Campoletis sonorensis,* the viral DNA is integrated into the host DNA. The PDVs' genome codes for between ten and twenty-five proteins, and DNA analysis has shown that PDVs from different species of parasitoids belong to different "species."

In the parasitoid the viruses replicate abundantly in the cells of the female reproductive system and then release their viral particles into the lumen of the ovaries (Buron and Beckage 1992). When female parasitoids inject their eggs into their caterpillar hosts, they also inject the infective PDV particles. Experiments show that the PDVs prevent parasitoid eggs from being encapsulated by the hosts' hemocytes, and it is thought that proteins coded for by the virus genome could saturate lectins on the surface of the hosts' hemocytes and thus prevent the parasitoids from being recognized. Dupas et al. (1996) have shown that viruslike particles present in a gland connected to the reproductive system of *Leptopilina boulardi* show differences depending on whether the parasite is able to survive *Drosophila melanogaster*'s defenses: viral particles of "immunosuppressive" strains were round and had several vesicles, whereas "nonimmunosuppressive" strains were elongated with many fewer vesicles. This is an indirect demonstration that survival genes are probably present in the viral particles, their DNA being integrated into the parasitoid genome. The presence of these survival genes was definitely clarified in a study of the genetic determinism of virulence by Dupas, Frey, and Carton (1998), who demonstrated that transmission was Mendelian, with the number of genetic factors involved being close to one.

Such collaboration between two genomes to avoid host defenses occurs in other parasite-host systems as well. This is the case for bacteria *(Xenorhabdus* and *Photorhabdus)* that are associated with certain nematode parasites of insects *(Steinernema* and *Heterorhabditis)*. These nematodes require the presence of the bacteria to metabolize insect host tissues and transport them in their digestive tract to be injected into the insect. However, bacteria may be released only after the nematodes have destroyed antibacterial peptides that the insects normally direct against invading bacteria. Further, the bacteria also secrete a factor that inhibits activation of prophenoloxydase, a key molecule in the insects' cellular defense system (see Akurst 1983; Brehélin et al. 1989; Boemare et al. 1997).

Relative to trypanosomes, which must also avoid the tsetse fly's mechanisms of defense, some rickettsia-like organisms have been found to be associated with some tsetse fly populations. These organisms give an advantage to their host, since adults emerge more successfully when they are present. The most surprising thing, however, is that these rickettsia-like organisms make the flies receptive to trypanosomes and thus are responsible for their being vectors of sleeping sickness (see Hide 1994).

Strategies of Recuperation: Turning Negatives into Positives

To use the host's own defenses to one's advantage is certainly the most elegant of solutions for a parasite. *Leishmania,* for example, is a protist that lives in human and dog macrophages and uses its hosts' own immune systems to penetrate these cells. In general, when a foreign organism is detected in the blood an ensemble of protective proteins in the serum (termed complement) is activated. One of the activated complement components is called the C3 fraction, and the macrophages have receptors for this molecule. *Leishmania*'s entry into the cells happens in the following way (see Mosser and Brittingham 1997): Promastigotes (the stage that penetrates macrophages) are recognized as nonself by the organism, which triggers the activation of complement, causing C3 molecules to coat the promastigote. Then, thanks to this blanket of C3 molecules on their surface, the promastigotes are recognized by the surface receptors of macrophages, which phagocytose them and become infected. Once inside the cell, the parasite uses a variety of mechanisms to survive.

Other parasites induce real antibody conflicts, as do schistosomes. Here protection against the parasites depends on immune mechanisms involving IgE and IgM, two types of immunoglobulin (antibody) molecules produced by the host. There is a particularly good correlation between high levels of IgE specificity directed against schistosomes and a low level of infection and reinfection in people (Capron et al. 1989; Capron, Dombrowicz, and Capron 1999). This immunity builds slowly in humans starting in early adulthood, so children are most often heavily infected. It seems that an efficient immune response is slow to occur in children because of the preferential production of another category of antibodies, IgG4, that is also directed against the parasites and their eggs but has the effect of blocking the action of IgE (Butterworth et al. 1987).

Schistosomes even use the host's own defense system to ease their reproduction. Among the molecules produced by T cells in the immune system is TNF‡, which triggers the formation of granulomas around eggs trapped in tissues—the parasite eggs use these host-produced granulomas as vehicles to deliver themselves to the lumen of the intestine for excretion, and schistosomes even use TNF-α as a factor to stimulate reproduction. In the absence of TNF-α, female schistosomes lay fewer eggs, and fewer eggs are excreted to the outside (Amiri et al. 1992; Sher 1992).

Interleukin-7 (IL-7) could also be another molecule originally selected to regulate the parasite-host relationship that over the course of the arms race between the parasite and the host became beneficial and perhaps necessary for infection. This idea derives from the observations that an injection of recombinant IL-7 in the skin before exposure to cercariae has a positive effect on the recovery rate of adult *S. mansoni* (Wolowczuk et al. 1997); that production of IL-7 is stimulated in human skin grafted onto SCID (severe combined immunodeficient) mice exposed to cercariae (Roye et al. 1998); and that in mice not producing IL-7 schistosomes do not develop normally, remain as dwarfs, and show signs of sexual immaturity (Wolowczuk, Roye, et al. 1999). Although no study on the changes in parasite fitness has yet been carried out, these experiments lead one to conclude that the penetration of the schistosomula through the skin triggers the expression of the IL-7 gene and that transmission of *S. mansoni* is at least in part dependent on the presence of IL-7 (Wolowczuk, Nutten, et al. 1999). In the experiment where recombinant IL-7 was directly injected in the skin (Wolowczuk et al. 1997), the authors observed an increase in the liver pathology (faster growth of granulomas). Since the formation of granulomas in the intestine favors the passage of eggs through the intestinal wall (Doenhoff 1997, 1998), an aggravation of pathology owing to IL-7 might then increase egg output and thus favor transmission of *S. mansoni* to the vector mollusk.

Yet another use of the host's defense mechanisms for the benefit of the parasite has been suggested by Coustau et al. (1996), who showed that when a parasitoid *(Leptopilina boulardi)* develops in a sensitive strain of

Drosophila melanogaster (see chapter 9), there is a strong activation of genes that code for the antibacterial peptides cecropine and diptericine. However, expression of these genes does not slow down the parasitoid's development and therefore cannot be considered an efficient defense mechanism for the fruit fly larva. One hypothesis is that this response may be a diffuse reaction of the defense system in the presence of the parasite, since the activation of such antibacterial genes is not specific. Another hypothesis (Yves Carton, personal communication) is that this gene activation results from a manipulation by the parasite, which would have an advantage in "cleansing" its host of any annoying bacterial development, most notably bacteria that may have been introduced by the parasitoid during oviposition.

Mafialike Strategies

A particularly original trait, selected by parasites to oppose host resistance, may be termed "mafia behavior." Soler, Møller, and Soler (1998) designate with this expression a strategy that, for some parasites, consists of imposing extra costs on hosts that resist too well—"imposing costs in the absence of compliance." In this way parasites can then "teach" their hosts that it is better to pay than to increase their defense.

Cuckoos seem to have mastered this survival weapon well. Zahavi (1979) was the first to suggest that a parasitic bird could force its bird host to tolerate nonself eggs by making the consequences of rejection more damaging than acceptance. This is possible only if the bird host can raise at least part of its own young along with those of the cuckoo. Thus in some cuckoos there would be selection for countermeasures opposing rejecter behavior. Simply put, cuckoos would "punish" rejecters by returning to the nest and stamping on or devouring their eggs (Soler et al. 1995), which would make their reproductive success drop to zero and prevent the diffusion of "rejecting genes" in the host population. This conclusion was based on study of the *Clamator glandarius/Pica pica* system by Soler, Møller, and Soler (1998), and figure 10.3 summarizes the two alternatives offered to the bird host when confronted with the cuckoo's "mafia behavior."

Soler, Møller, and Soler (1998) suggest that such mafia-style strategies are evolutionarily stable "due to the maintenance of the relationship between mafioso and host at equilibrium by the mafioso, who could change its pressure on the host depending on host defences." The authors also suggest that these strategies could be much more common than currently thought.

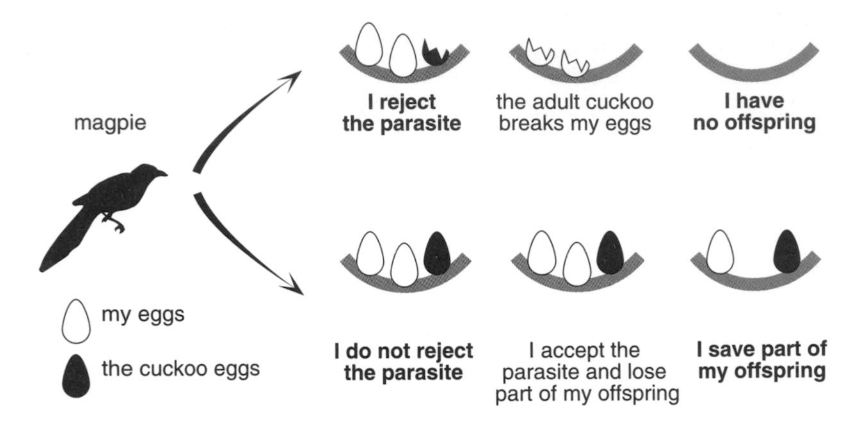

Figure 10.3. "Mafia behavior" in cuckoos (inspired by Soler, Møller, and Soler 1998).

Strategy: Flowers for the Hosts

The last strategy might also be termed "seduction." This approach consists of bringing innovative genes to the association, which then leads to mutualism. To decrease a host species's defenses it is enough for the parasite to give the host a benefit larger than the cost it imposes. In this case any resistant individual that avoids or rejects invaders is counterselected, and susceptibility genes are spread and become more common.[4]

Parasitic birds allow one to comprehend how such a seductive parasite could make its host's defenses drop. The species that exemplifies this strategy best is not the European cuckoo and its hosts but other birds that form a similar parasite-host system in Central America. The parasitic bird *Scaphidura oryzivora* lays its eggs in the nests of several species of oropendolas in Panama. Although young *Scaphidura* are raised by their bird hosts, one difference from the European cuckoo is that they do not destroy their hosts' eggs as do the cuckoos. In this case certain populations of oropendolas reject *Scaphidura*'s eggs while others do not. Smith (1968) has shown that the "acceptor" characteristic coincides with the invasion of the nests by dipterans of the genus *Philornis,* whose larvae cause serious and sometimes lethal damage to the chicks. However, parasitic *S. oryzivora* chicks eat the dipteran larvae and enjoy them so much that they even clean their fellow nestlings (the host's chicks).

4. This situation is similar to the following: You invite a guest to your table and make him pay more for the meal than it costs you. You have no advantage in defending yourself against the presence of this guest, since his presence is beneficial overall (it is on this simple principle that restaurants are founded).

It is clear that where *Philornis* is present the bird host has an advantage in tolerating the parasitic bird. In this case the benefit given by the parasite is larger than the cost it incurs, and genes are positively selected from any oropendola that does not reject the eggs of parasitic birds, since this bird has a better chance than others of having abundant descendants. In short, tolerance and intolerance toward the parasitic birds effectively coincide with the presence and absence of the dipterans. Where the dipterans attack young oropendolas, the host populations are "acceptors" (they do not reject the eggs of the parasitic birds). Where the dipterans are not present, however, populations of the same bird hosts are "rejecters" and eliminate, at least in part, the parasites' eggs.

It is remarkable that selection of the characteristics "acceptor" and "rejecter" is apparently very quick, since it concerns particular populations of the same host species. Moreover, the *Scaphidura*/oropendola story shows the point where discussions about limits between parasitism and mutualism are useless. How would one qualify *Scaphidura?* As a "parasite" where there are no dipterans and a "mutualist" where there are?

As this example has shown, a good way to get oneself accepted is to bring to the association genes whose expression benefits the partner. The useful genes a parasite brings are flowers that it offers to its host. These flowers allow it to be accepted, to be invited, or even to be indispensable. This has the double effect of avoiding an attack by the host and exchanging the name "parasite" for that less derogatory term "mutualist."

11 How Do Coevolutions Evolve?

Evolutionary biologists have much to gain from a closer association with parasitologists.
Harvey 1989

Natural selection does not necessarily favor peaceful coexistence.
Ewald 1994a

The Evolution of Ideas

Until the beginning of the 1980s it was generally thought that a parasite-host system that had had little time to evolve would be characterized by strong virulence. That is, the "new" parasite would strongly reduce its host's reproductive success. It was even deduced that the characteristic "strong virulence" was primitive and was replaced, with time, by the evolved characteristic "weak virulence." Many authors even went so far as to consider, at least implicitly, that mutualistic associations ("symbiosis" in the European sense) were the end product of this evolutionary process. Levin (1996) summarized the ideas of this time in the following way: "A fully evolved parasite would not harm the host it needs for its survival, proliferation, and transmission." The idea that pathogenic agents become less and less pathogenic or even beneficial to their hosts over time may have come from the following imprudent sentence of a great parasitologist of the nineteenth century, Pierre J. van Beneden: "The parasite practices the precept—not to kill the fowl in order to get the eggs."

Historically, the first author to voice doubts about the evolution of a decreasing virulence in parasite-host systems was Gordon H. Ball, who as early as 1941 wrote in an article ignored by numerous workers, "A parasite may choose the course of manifest destiny and find aggressiveness more attractive and more valuable than an existence of peace." However, it was only in the 1980s that a profound theoretical reflection on the evolution of virulence emerged, and nowadays it is accepted that virulence is just another adaptive characteristic that increases or decreases over time in response to host or environmental pressures. In short, it is recognized today that certain parasite-host associations may evolve toward a more peaceful coexistence whereas others may evolve toward stronger virulence or even pass through high and low virulence phases.[1]

The symbolism of encounter and compatibility filters allows one to easily understand that the evolution of virulence has different outcomes relative to the initial conditions of the system (those that prevailed at the time the parasite-host system was formed). When a novel parasite-host system is first formed (for example, after a migration or change in behavior of potential hosts), virulence may be "below or above the level that presumably would result in the highest fitness" (Ebert and Herre 1996). Both filters may be only slightly open, either because encounters are not frequent or because the host's immunity is naturally efficient. In this case (fig. 11.1, top) selective pressures exert themselves principally on the parasite population and not the host population. If genetic diversity allows, there is then selection in the parasites for genes that improve the frequency of encounters or defense system evasion or both such that the abundance of the parasite increases generation after generation. Thus virulence has a tendency to increase with time in this situation.

On the other hand, both filters may be relatively wide open, either because encounters are frequent or because the host's defenses are not very effective. In this case (fig. 11.1, bottom) pressures are exerted both on the host and on parasite populations. As before, and if genetic diversity is great enough, there is selection in the host population for characteristics with a tendency to avoid encounter, strengthen defense systems, or both. Further, there is at the same time selection in the parasite population for lowered virulence in order to limit the negative feedback of host death on parasite fitness, and in this case virulence would be expected to decrease with time. Last, there may even be counterresponses relative to encounter and compatibility, with, for example, selection in the parasite for genes to encounter, leading to the selection of genes to kill in the host, and so forth.

If one takes a quantitative approach to the idea of filters, one sees that virulence is related to parasite abundance (fig. 11.2), which results either from multiplication in the host (microparasites) or recruitment of

1. The evolution of a "novel" infectious or parasitic disease may be compared to the introduction of a free-living species into a new environment: if the introduced species has a very high fitness in this milieu, it acts as an invader that first dominates all indigenous species. The latter, being submitted to strong selective pressure, progressively make a comeback (if they survive), and there is a decrease in the perturbation of the system over time. Alternatively, if the introduced species is partly adapted to the milieu, at best it can increase its fitness with time by selecting new characteristics, and in this case there is an increase in the perturbation of the system over time.

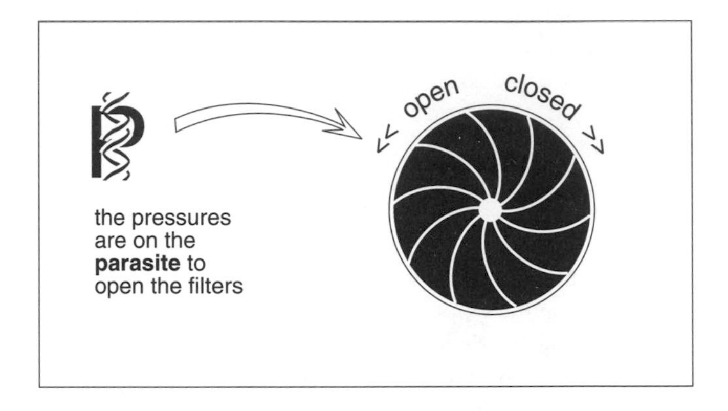

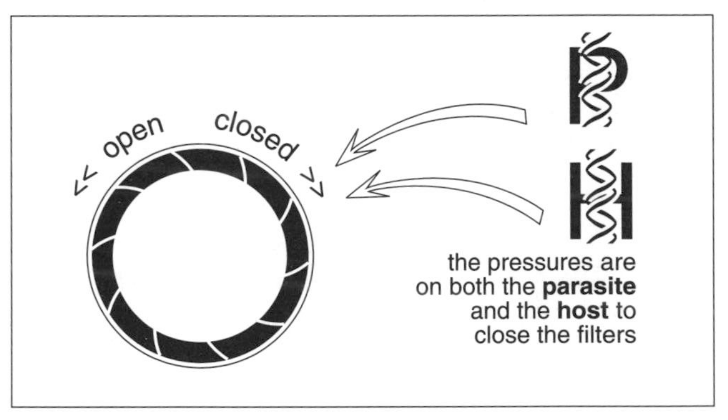

Figure 11.1. If the filters are almost closed when the parasite-host interaction begins, selective pressures are mostly on the parasite, and virulence increases with the selection of genes that tend to open the filters (top). If the filters are wide open (bottom), the opposite happens. Encounter and compatibility filters are depicted in a single filter on the figure. P = parasite; H = host.

individuals from the outside (macroparasites). In numerous parasitic diseases the problem is not the isolated pathogenic effect induced by an individual parasite but rather the effect induced by the parasite infrapopulation as a whole. This explains why the vast majority of studies and models focused on the evolution of virulence concern microparasites, whose processes of internal multiplication after infection are much easier to deal with than are the process of continued macroparasite recruitment, which must take into account many kinds of difficult events.

Parasite virulence is extremely variable depending on the species, and Ewald (1995) wonders: "What are the factors that drive some parasites toward intense exploitation of hosts and others toward mild coexistence?"

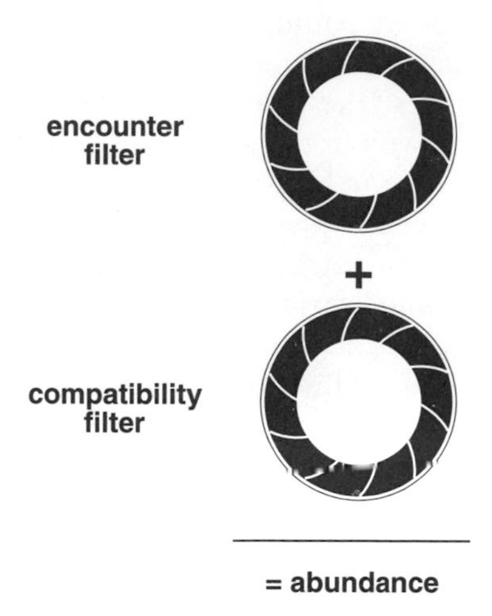

Figure 11.2. Quantitative use of the symbols for the encounter and compatibility filters: the degree of opening of the filters determines the parasites' abundance, which is the essential variable virulence depends on, particularly for microparasites.

In what follows we will examine the factors that supposedly first orient selection toward strong virulence and then those that act in the opposite way. Finally, we will see that the virulence observed in many populations is not always what was first expected.

Factors Driving Parasites toward the Intense Exploitation of Hosts

Programmed Host Death

The pathogenic effect of a parasite makes no sense for the parasite and is not itself selected for. Rather, it appears instead to be a by-product of the interaction between the parasite and host (Ebert and Hamilton 1996), and in most cases mathematical models reveal that the host's death from infection either is of no consequence or is in fact unfavorable to the parasite.

An important exception to this scheme exists, however, in those biological life cycles where host death is necessary for transmission (this exception is noted in fig. 6.3 by the terminology "except in favorization processes"). This parasite-programmed host destruction happens in two cases.

1. In numerous life cycles parasites must pass from prey to predator, and a marked pathogenic effect thus greatly increases the probability that

the prey will be eaten. Being pathogenic thus also increases the probability of the parasite's transmitting its genes, and selection would then favor very pathogenic parasites (the "illness for transmission hypothesis" of Ewald 1995). However, even in this type of life cycle there is an optimal level of virulence that cannot be passed without reducing the parasite's fitness, and an excessive pathogenic effect may induce death before the prey is caught by the predator, which would then lead to the death of the parasite instead of allowing its transmission.

2. In certain microparasitic life cycles transmission occurs when the parasite literally bursts out of the host to the outside. The host's death is thus necessary as soon as its exploitation by the pathogenic agent no longer allows further multiplication. In this regard Ewald (1995) mentions the example of the dinoflagellate *Pfiesteria piscimorte,* which infects estuarine fish and kills them before being dispersed into the environment.

Interparasite Competition

Most reflections on, and models of, the evolution of virulence reduce the problem to simple situations involving only two "partner" species (the pathogen and its host), and only rarely have researchers attempted to link virulence to intra- and interspecific competition. A process of selection may occur by competition among conspecific parasites within the same host individual, a hypothesis defended by Levin (1996), who stipulates that in microparasites that multiply inside the host a selection of the most virulent parasites may occur within the host itself. This selection would be the result of pressures exerted by the host's defense/immune system, and parasites that best resist host defenses are the best competitors and therefore multiply the fastest. For example, in the case of HIV this increase in virulence may explain the progression from being seropositive to the state of disease. Such a process of competition/selection inside the host can only induce a continuously renewed increase in virulence at each infection of a new host individual and is opposed to the logic of natural selection at the level of the population. This opposition derives from the fact that within a few months or years in an individual host a pathogenic agent such as a virus may have more generations (thus more opportunities to mutate) than a metazoan has over hundreds of thousands of years! Thus Levin (1996) talks about the "short-sighted evolution in infected hosts," adding that "myopia is a fundamental premise of the theory of evolution by natural selection." When the time between generations is very short, myopia is such that the parasite that multiplies within a host does not see further than the host individual in which it temporarily resides.

Attempts to model the relationship between virulence and interspecific competition have also been made. For instance, May and Nowak (1995) considered opposing situations in which two or more parasites coming from different species or strains of the same species exploit the same host individual. In both cases one of the parasites was supposed to infect the host either after the first parasite did so or after all the others had settled, which may happen when mutants appear in infected hosts. In the first case, which May and Nowak called "superinfection," the most virulent parasites took over from the less virulent because, for example, they multiplied more rapidly. In the second situation, called "coinfection," there was no competition between the parasites of the different species or strains. May and Nowak showed that these two situations are similar to "contest" competitions (in the case of superinfection) and "scramble" competitions (in the case of coinfection).[2] In short, in superinfection one of the species or strains eliminated all the others. In coinfection, on the other hand, there is also a tendency for the predominance of the most virulent species or strains, but virulence is much more alike between the different surviving parasites.

Van Baalen and Sabelis (1995) approached the same question using a different model, which considered several quantitative aspects: "Whether high or low virulence pays off . . . depends on abundance and virulence of the other pathogens in the population. . . . Reducing virulence will only pay when there are not too many virulent pathogens. . . . Increasing virulence allows pathogens to steal hosts from other pathogens, but this approach only pays when they are sufficiently abundant." Van Baalen and Sabelis reached the conclusion that at the beginning of an epidemic, since multiple infections are few, virulence may be limited. When a pathogen becomes endemic, however, different strains' sharing the same hosts can lead to a conflict of interest that could only increase virulence, even in hosts that are not coinfected.

These models support Levin's idea that evolution is shortsighted and suggests that interparasite competition occurs to the detriment of hosts. For instance, Taylor, Mackinnon, and Read (1998) have demonstrated that for the parasite *Plasmodium chabaudi,* which causes malaria in rodents, infections with several clones are more virulent than single-clone infections. Using identical inoculation densities, these authors showed that mice lost more weight and had lower blood counts in mixed-clone than in single-clone infections. However, they believed this increase in

2. In "contest" competition the winners obtain the resources necessary for their reproduction while the losers have no access to resources. In "scramble" competition the resources are shared between all the individuals.

virulence would be explained not by a more rapid replication of parasites, as supposed by the mathematical models above, but rather by the fact that mounting an immune response against more than one genotype is more costly for the hosts, which have to divert more energy to resistance.

Factors Driving Parasites toward Mild Coexistence

Host Resistance

Numerous host traits exert selective pressures on the parasite. To pierce teguments, inactivate enzymes, resist blood flow, and such, the parasite must spend energy, and this need has helped mold parasite adaptations throughout evolution. Nevertheless, host resistance has imposed the highest costs on parasites, and once again the relationship between parasitic cuckoos and their bird hosts serves as a good example of just how a mechanism of resistance by the host is set up and how the cost is imposed on the parasite.

Soler and Møller (1990) studied nest parasitism in the magpie, *Pica pica,* by the great spotted cuckoo, *Clamator glandarius.* These authors have shown that (fig. 11.3): where the bird host has never been in contact with the parasitic bird the magpies tolerate foreign eggs even if they are not mimetic; where the parasitic bird and the bird host have "known each

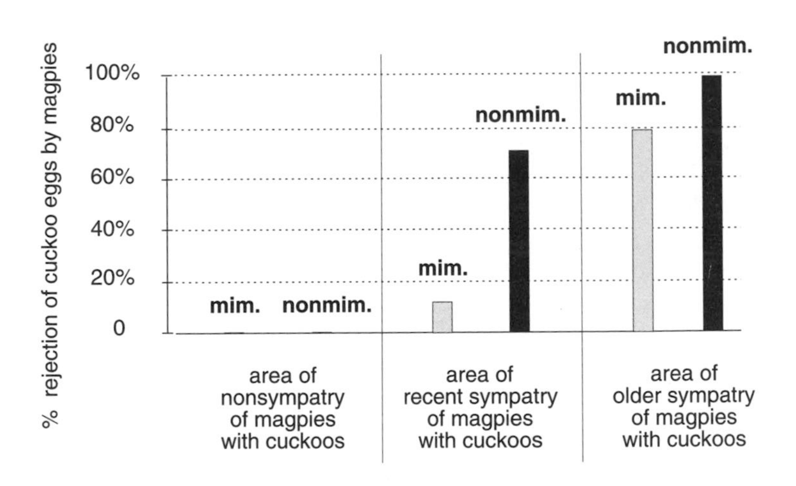

Figure 11.3. Selection for resistance (rejection of cuckoos' eggs) by magpies (after Soler and Møller 1990).

other" for some time the magpie begins to reject foreign eggs, particularly if they are not mimetic; and where the parasitic bird and the bird host have lived together for a long time in sympatry, the magpies reject all nonmimetic foreign eggs and a large portion of the mimetic ones as well.

The authors conclude that magpies have selected the capacity to recognize and reject nonself eggs with an efficiency about proportional to the duration of exposure to the cuckoos, with the resulting efficacy of rejection by magpies and the destruction of cuckoos' eggs giving a measure of the decrease in reproductive success experienced by the parasite.[3]

Logically, so long as genetic diversity allows it there should always be in hosts a selection for either genes to avoid or genes to kill, or both. This is not always the case, though, because of the cost of resistance (see chapter 6). That is, "an uncostly resistance would spread very rapidly in the host population and the underlying genes should go to fixation within a few generations" (Sorci, Møller, and Boulinier 1997).

An example showing how expensive selection for an avoidance behavior may be for the host involves the life cycle of the trematode *Meiogymnophallus fossarum* (see chapter 7). When an oystercatcher selects seashells that are already half open because of infection by the parasite, it follows the general tendency of spending the least energy possible to get the most from its meal. But the bird also becomes infected, which can reduce its life span or its fecundity and thus its fitness. Therefore, if within the oystercatcher population certain individuals (because of a different genetic constitution) were able to chose healthy seashells (with closed valves), they would escape parasitism, and their genes would be the ones most likely to be transmitted and spread (fig. 11.4). In this case one would expect that selection would favor the appearance in oystercatchers of rejection behavior against infected seashells, since they are identifiable

---

3. Note, however, that the conclusions of Soler and Møller (1990) are the subject of some controversy. In particular, Zuniga and Redondo (1992) doubt there is a difference in sympatry between two of the localities in Spain, which are only sixty kilometers apart. These authors have also suggested that differences in the magpies' capacity to respond may be due to the cuckoos' density and not to the duration of their association, and Lotem and Rothstein (1995) have summarized the state of this controversy. Nevertheless, when mechanisms of rejection are not selected by bird hosts, parasitic birds may be serious threats. In North America, cowbirds *(Molothrus ater)*, which sometimes form flocks of several million individuals, lay their eggs in the nests of more than two hundred species of birds, and only a few host species have developed rejection behavior against the eggs of the parasite. This shows that selection of mechanisms of rejection is not automatic and that it can occur only if genetic diversity provides an opportunity. The problem set by *M. ater* is huge, and conferences are organized around it. Campaigns have been organized to destroy tens of thousands of these birds, though without much success, and several host species are threatened and probably without much hope (see Rothstein, Robinson, and Robinson 1994).

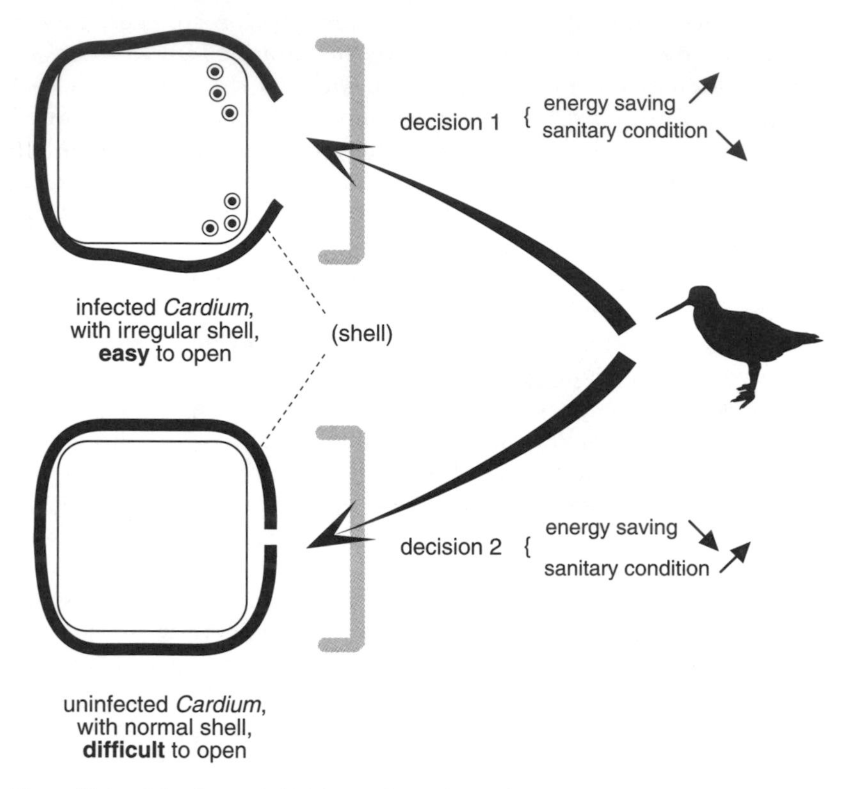

Figure 11.4. Selection and decision making: the trade-off between ease of capture and parasitism in oystercatchers.

from the outside. This has not happened, or at least it does not seem to have happened, apparently for the following reason: If the parasite costs the bird less in terms of reproductive success than does the energy-intensive consumption of closed seashells, there is no reason or possibility for the selection of genes to modify the bird's trophic behavior. In this case the parasite renders the optimal foraging of the host less optimal than one would expect, but not enough to make it a disadvantage. Lafferty (1992) summarizes this type of situation: "An individual predator can benefit from parasites if costs of parasites are moderate and prey are sufficiently modified by parasites. . . . The parasite provides a delivery service for hard-to-get-prey."

It is not easy to find examples in which the diet choice of a potential host is modified to avoid parasitism. Lozano (1991) writes, "I propose that animals could change their optimal diet such that it . . . takes into account the possible harm done by parasites" but does not find convincing examples in the literature and concludes, like Lafferty, that "the effects of parasites on diet choice would . . . only be observed if, in terms of fitness, there is a high cost to being parasitized."

Thus selection has a tendency to maintain the parasites' cost to the host at moderate levels, below the threshold where selection of an efficient defense mechanism would be triggered. If this threshold is high, the pathogenic effect may be marked: for example, Wedekind and Milinski (1997) have shown that copepods infected with larvae of the cestode *Schistocephalus solidus* were easier for fish to catch than healthy ones; yet though *S. solidus* is strongly pathogenic for the stickleback host, *Gasterosteus aculeatus,* there is no selection in the fish for any type of avoidance behavior.

If the virulence of a newly introduced parasite triggers strong defense mechanisms (avoidance or immunity) in the host population, selection should tend only to decrease virulence. Unpublished data of P. M. V. Martin and C. Combes suggest, for example, that the spirochete causing syphilis is less virulent in humans today than it was a few centuries ago. In the fifteenth and sixteenth centuries syphilis tended to produce repugnant cutaneous lesions on infected individuals. Thus sexual intercourse was less likely to occur with a person having such lesions than with someone with more discrete clinical signs. As early as 1497, Professor Widman of Tübingen recommended avoiding sexual contact with women with pustules. The decrease in the occurrence of overt lesions within a few centuries could be the result of a selective process making the bacteria more discreet, less virulent, and better transmitted. At this point one should also recall Clayton's experiments on the ectoparasites of bird feathers, which are transmitted from one bird to another during mating. In this case the female pigeon avoids copulating with parasitized males, which are identified by their shorter display. This sexual selection means the parasite must limit its effects on infected pigeons in order to ensure transmission. We can thus predict that above a certain threshold in the duration of the parade transmission would become more and more problematic, and "any directly transmitted parasite would suffer from such cue-reading by individuals of the choosy sex, resulting in selection for a reduction in virulence among such parasites" (Møller and Saino 1994).

Autolimitation of Virulence (the Virulence Feedback Loop)

Decrease in Host Fecundity

Like all living beings, parasites have the natural tendency to reproduce as much as possible. However, the fact that their habitat is alive sets some limits on its exploitation. Virulence is, then, a double-edged sword:

"Parasite reproduction is traded against the negative effects of parasite proliferation on host fitness . . . the parasite faces a dilemma: should it speed up its own reproduction, with the potential damage to host survival, or manage host survival to increase its transmission?" (Sorci, Møller, and Boulinier 1997).

For the parasite not to be able to depress its host's fecundity below a certain point seems simply to be common sense. But Anderson and May (1982) have shown that if parasites easily find new individuals to infect, the decrease in the host's fecundity does not decrease the parasite's reproductive success, and there is no reason for its virulence to decrease over time. Selection acts on transmission of parasite genes to the next generation, not on the long-term view of its future. Host castration, however, has often been perceived to have an adaptive value for the parasite, since once the host's reproduction stops resources are released that can be used by the parasite (Obrebski 1975). As Poulin (1998a) comments, "Host castration may be the ideal strategy of host exploitation: by attacking non-vital organs, castrators do not reduce the host life span, and they can obtain a high transmission rate without trading off longevity."

Limited virulence will manifest itself only in the special case where the reproductive success of an individual parasite and its individual host are linked. This happens when transmission is vertical, that is, when pathogens are systematically transmitted from generation n to generation $n + 1$. Herre (1985, 1993, 1995) studied infection by nematodes of the genus *Parasitodiplogaster* in the hymenopteran pollinators of fig trees. In this case gravid female wasps penetrate receptive figs, pollinate some of the female flowers, then lay their eggs in others and die there. Thus it is possible, at maturity, to count both the number of females that entered the fig (dried bodies are found) and their descendants (young wasps emerging from seeds). Specificity in this association is narrow, and each hymenopteran species pollinates one particular species of fig tree. The nematodes' life cycle follows that of their hymenopteran hosts: adult nematodes emerge from the females' dead bodies and lay their eggs in the fig. The larvae hatch just when the young wasps emerge, and they immediately infect them.

Herre has analyzed dozens of associations differing by the number of females that penetrate each fig. In certain species there is almost always only one founder wasp, whereas in others the rule is "multifoundation." When there is only one founder, nematodes can parasitize only the founder's descendants, and thus transmission is vertical. When there are several founders, however, the nematodes may parasitize descendants of several wasps; hence transmission is at least in part horizontal. In the first case (only one founder), the reproductive success of the nematode is

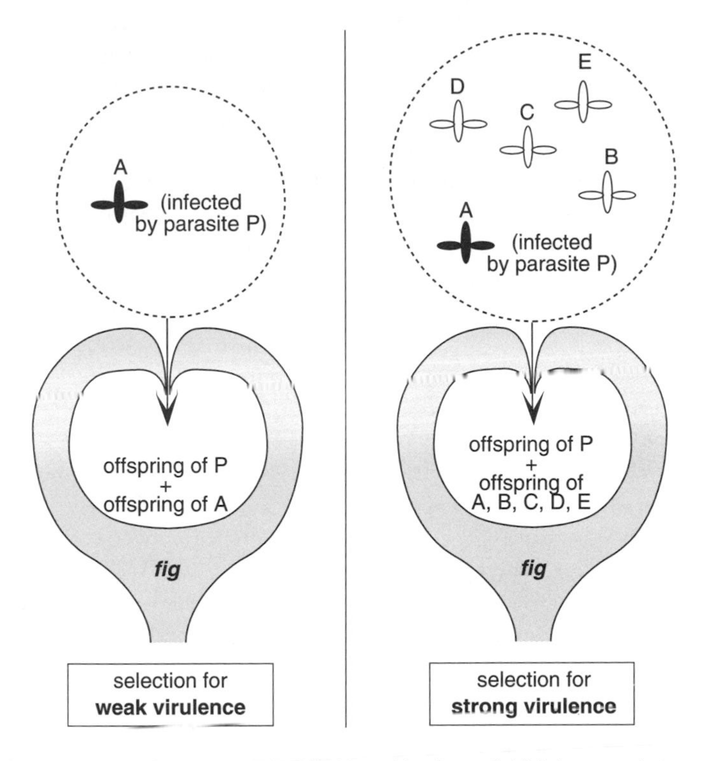

Figure 11.5. Relation between vertical (left) and horizontal (right) transmission and virulence (from the work of Herre 1993).

totally dependent on the reproductive success of the wasp: if the adult nematode is too pathogenic, the host individual will not lay as many eggs, and nematode descendants will have fewer wasps to infect (fig. 11.5). Thus we may logically expect the nematodes in such "monofoundation" systems to be less pathogenic than those in multifoundation systems. In fact, in multifoundation systems the reproductive success of a particular nematode does not depend only on the reproductive success of its own host, and Herre's data completely support this hypothesis. Nematodes from multifoundation systems are in fact more virulent (they alter the host's reproductive success to a greater extent) than are nematodes from monofoundation systems. Poulin (1998a) notes: "Demonstrating that highly virulent nematode species are more fecund, for instance, than less virulent nematode species would complete nicely the story."

Clayton and Tompkins (1994, 1995) carried out a similar demonstration with two groups of feather mites, both ectoparasites of the pigeon *Columba livia*. In this case the mites that are mainly transmitted horizontally are so virulent that host fitness may drop to zero (they make adult birds so agitated that they cannot care for their eggs properly), while mites transmitted vertically (phthiropterans) are less so.

Yet another series of studies shows another aspect of the link between host fecundity and parasite fitness. Minchella and LoVerde (1981) have shown that if mollusks *(Biomphalaria glabrata)* are infected with miracidia of *Schistosoma mansoni,* the fecundity of the snails increases (more eggs are laid) before they become castrated by the parasite (fig. 11.6A). Thornhill, Jones, and Kusel (1986) found that increased egg laying occurred not only in infected snails but also in those that had been exposed to miracidia without being infected, leading them to suggest that the increased fecundity benefited the parasite by contributing to the pool of future susceptible hosts. Last, Théron, Rognon, and Pagès (1998) demonstrated that when miracidia had a choice between snails of various ages, they preferred to penetrate subadults, where such increased egg laying could occur. More miracidia penetrated each snail (fig. 11.6B), and there was also a higher prevalence in this subadult group when several age-classes of mollusks were exposed together (fig. 11.6C). If miracidia penetrate a very young mollusk, castration is total and the snails genes are not transmitted, and older individuals are either unsuitable or resistant to parasite development. Théron, Rognon, and Pagès (1998) concluded: "The parasite has been selected for specific locating and recognition mechanisms increasing the infection rate of sub-adult snails . . . one can consider sub-adult individuals among the snail host population as having the most favorable 'space-time-energy budget' for parasite development and reproduction, and as constituting the best compromise for parasite fitness." Even if transmission is not strictly vertical in this system, parasites still infect their hosts under the most favorable conditions to maintain fecundity. The explanation may be that transmission is only partially vertical, since mollusks are not very mobile and the trematodes, even after their transfer in the definitive host, have more chance of infecting descendants of the snail in which they developed than others. Moreover, one cannot forget that descendants of infected snails carry the genes of susceptibility for the parasite and that the parasite transmits its genes only if it encounters such susceptible individuals. Studies by Herre (1993) and Théron, Rognon, and Pagès (1998) show that there is no absolute opposition between horizontal and vertical transmission; rather, there is a continuum. The closer to "vertical" transmission is, the stronger the pressure to limit virulence.

Møller and Erritzoe (1996) elegantly demonstrated the link between the mode of transmission and virulence, beginning with the hypothesis that if virulence is high, hosts must then develop particularly efficient

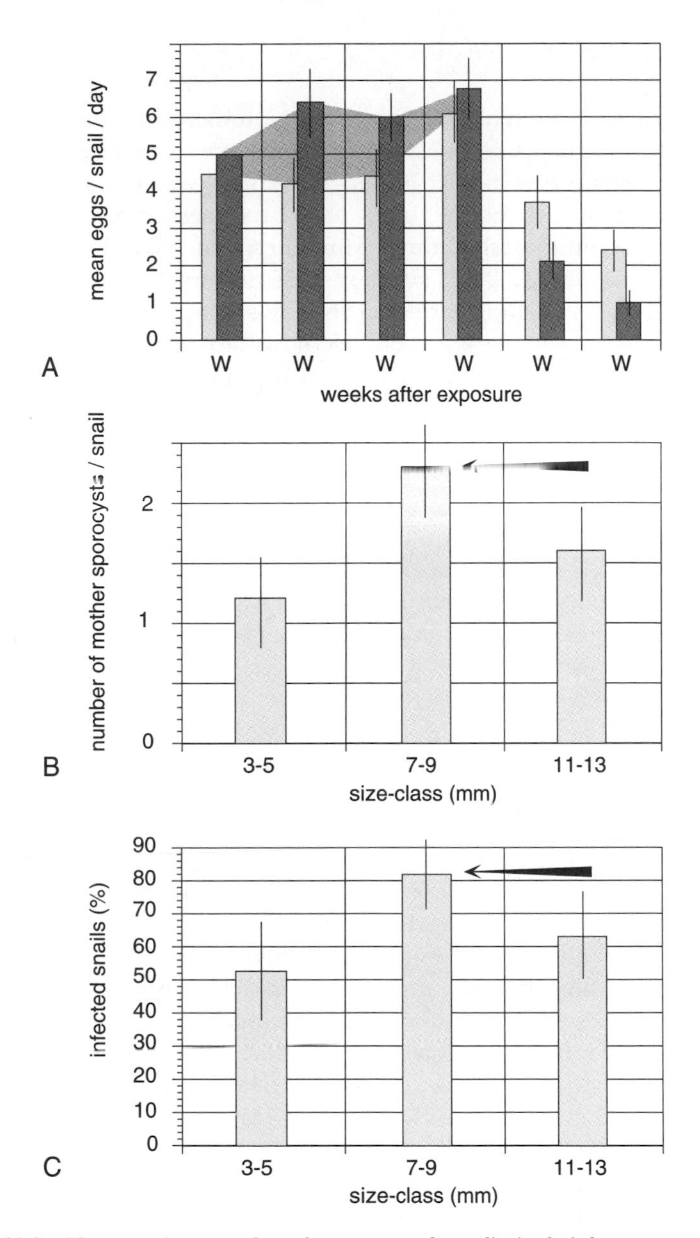

Figure 11.6. How parasites are selected to preserve fecundity in their hosts even when transmission is not strictly vertical. (A) Increased oviposition (shaded area) in *Biomphalaria glabrata* exposed to *Schistosoma mansoni* miracidia. Black bars: exposed (infected plus uninfected); open bars: controls (modified from Minchella and LoVerde 1981). (B) Number of mother sporocysts per exposed snail determined for three size-classes of *B. glabrata* exposed in groups to *S. mansoni* miracidia (after Théron, Rognon, and Pagès 1998). (C) Prevalence of infection for three size-classes of *B. glabrata* exposed in groups to *S. mansoni* miracidia (after Théron, Rognon, and Pagès 1998). Arrows indicate the size-class with the best success of infection.

defense mechanisms. They compared the volume of the bursa of Fabricius and the spleen of birds in hole nesters versus open nesters and in colonial nesters versus solitary nesters. The bursa of Fabricius has a role in immune defense in juvenile birds, and the spleen serves that function in adults. The results of this comparison are spectacular: the bursa of Fabricius was, on average, 249% larger in hole nesters than in open nesters and 319% larger in colonial breeders than in solitary breeders. The numbers for the spleen are 38% and 65%. This was explained by the fact that hole nesters and colonial nesters frequently reuse already built nests. According to Møller and Erritzoe, this pattern would favor horizontal transmission and an increase in virulence that in turn induces selection for larger defense organs.

Vertical transmission is frequent in mutualistic systems, and by definition virulence is null in these interactions. In "cellular endosymbiosis" vertical transmission of the mutualist is via the egg. This is the case, for example, of algae in green hydra, and vertical transmission is "the first step on the road to becoming a single unit of evolution" (Maynard Smith 1989b).

Host Survival

Several authors have emphasized that in hosts with long life spans it may be advantageous for the parasite to decrease its virulence in order to take full advantage of the longevity of its habitat (Lenski and May 1994). In a more general way this virulence limitation may happen whenever the transmission of a parasite depends on continuation of its host's life. This is the case for biting arthropods: any parasite that killed its vector host would substantially decrease its own transmission rate. The most evident affirmation of this is that certain parasites go so far as to commit suicide by apoptosis in cases where other regulatory mechanisms cannot limit the growth of their infrapopulation. Programmed cell death, or apoptosis, is a genetically regulated process of cellular suicide that plays an important role in metazoan life. It occurs when a cell has been damaged or when it is simply no longer useful to the organism. The "decision" for each cell to live or die (not in a passive manner, but actively) is influenced by signals coming from other cells in the organism. Altruism has sometimes been credited with playing a role in programmed cell death, with the decision to die being beneficial for other cells having the same genome as the dying cell. Apoptosis plays a role in virulence limitation in parasite populations, particularly if the infrapopulations involved have a clonal genetic structure. For instance, in the life cycle of the protist *Trypanosoma cruzi* there is an alternation of dividing forms between the insect and vertebrate hosts, with nondividing forms produced in readiness to settle in the new environment after transmission from one host to the next. The forms that divide in vectors are called epimastigotes, and these have been experimentally shown to undergo apoptosis (Ameisen et al. 1995). This programmed cell death limits the number of parasitic cells in the insects, which in turn contributes to the survival of the vector-parasite couple. Thus in this case programmed cell death limits virulence in the vector and ensures transmission.

Decrease in Host Density

Recall (Anderson and May 1982) that the major problem for the parasite is encountering new hosts to infect. If, because of infection, new host individuals become rare, then a weaker virulence will be selected. For instance, "chestnut blight" is caused by the fungus *Cryphonectria parasitica,* which is transmitted from chestnut tree to chestnut tree by phytophagous insects. There are virulent and hypovirulent strains of the parasite, and the genes for hypovirulence are carried by a double-stranded RNA molecule that is itself a parasite of the fungus (see Taylor, Mackinnon, and Read 1998). Hypovirulence thus is due to the weakening of the fungus by its hyperparasite. If the two strains are in competition on the same tree the virulent strain, which develops faster, is more competitive and eliminates the hypovirulent strain so that the host tree dies quickly. At the scale of the chestnut tree population there is a slow die-off, as happened in both North American and European forests during the first part of the twentieth century. However, as chestnut trees became too rare the least virulent fungi progressively gained the advantage, since fungi that do not kill their hosts too rapidly have more chance to be transmitted by insects than those that kill too quickly (Newhouse 1990). When host density is not great enough, parasites face the same problem in transmission whether of not their transmission is vertical.

Very close to the evolution of hypovirulence in chestnut blight is the evolution of reduced virulence in the myxoma virus of rabbits. This virus, originally from America, causes only a benign cutaneous fibrosis in its hosts of origin *(Sylvilagus brasiliensis* and *S. bachmani).* The virus is transmitted by insect bites (fleas or mosquitoes depending on the foci). It was introduced to Australia and Europe in the 1950s to control rabbit populations, and it killed up to 99% of the rabbits present (Kerr and Best 1998). However, rabbit populations have since become progressively less and less affected by the virus, even though the disease still kills numerous individuals.

Where was this change in virulence coming from? It is now accepted that the virus has reduced its virulence in a relatively few years, more so than the rabbits have increased their resistance. Why? As summarized by

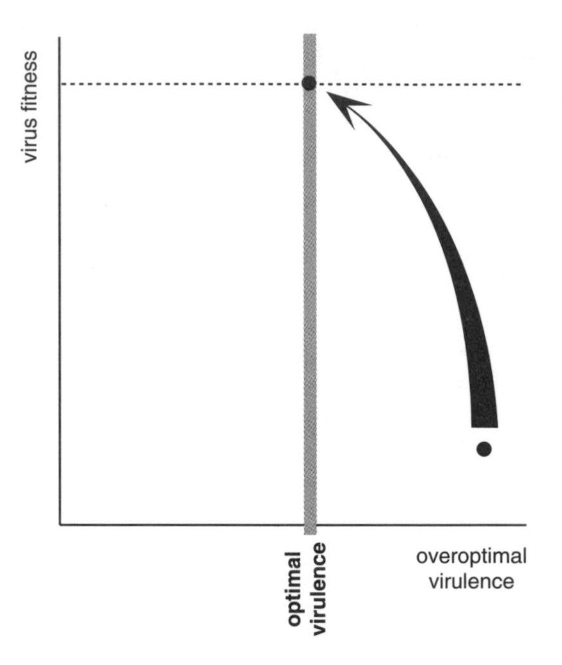

Figure 11.7. Supposed change in the virulence of the myxoma virus in European rabbits within forty years.

Levin (1996), this is probably because virulence had too high a cost. The most virulent forms of the virus killed rabbits too quickly for successful transmission to occur before the hosts died. Here natural selection has stabilized the system in a compromise state where the virus is sufficient to infect the biting insect vector but does not kill the rabbit too rapidly (fig. 11.7).

Finally, certain parasites (which may be more numerous than is currently thought) have cyclical changes in virulence linked to changes in the conditions of their transmission. Dye and Davies (1990) report the work of Strelkova, from the Institute for Tropical Medicine in Moscow, which shows that *Leishmania major* shows seasonal variations in virulence. *L. major* is a parasite of the great gerbil, *Rhombomys opimus,* and its vector is the sandfly *Phlebotomus papatasi.* The relative abundance of virulent and less virulent strains of *L. major* collected from gerbils changes dramatically with the season: during the summer strongly virulent strains dominate, and during winter less virulent strains do so (virulence in this case being determined in the laboratory from the rate of development of cutaneous ulcers). The precise reason for the change is not known, but it may be linked to changes in host density with the seasons.

Decrease in Host Mobility

In addition to requiring that their hosts' fecundity and density be ensured, parasites may also require assurance of their mobility. Ewald (1995, 1996) pointed out the importance of recognizing host mobility in virulence evolution: If in the environment there is a very mobile vector able to transport a parasite from one host to another, then the parasite may select virulent strains that decrease host mobility. If transmission implies that the hosts themselves remain mobile, then parasites must select less pathogenic strains that affect mobility only slightly. Ewald believes that pathogenic strains of the intestinal amoeba *Entamoeba histolytica* are selected for in regions where transmission occurs mainly via polluted waters (water being the mobile vector), because the hosts do not have to move. On the other hand, weakly pathogenic strains are selected where transmission occurs from human to human. Ewald deduced that supplying potable water may therefore reduce not only the frequency of infection in people but also the amoebas' virulence.

In general, Read and Harvey (1993), and Ewald (1995) believe that holoxenous parasites are less virulent than parasites transmitted by biting vectors. For holoxenous parasites transmission requires that the host have some minimum level of mobility, whereas biting vectors are favored by immobility.

Counterselection of the Useless

Useless Resistance

Not only is resistance costly, but it may not benefit the host. This "useless" defense by the host leads to expense without efficacy in the fight against the parasite. Although it gives the impression of the occurrence of strong virulence, it does not benefit the parasite either.

Useless resistance is illustrated in the study of Sakanari and Moser (1990), who looked at the consequences of the introduction of the striped bass, *Morone saxatilis,* on the West Coast of the United States about one hundred years ago (fig. 11.8). In this region the introduced fish was added to the host spectrum of the larval cestode *Lacistorhynchus dollfusi,* which does not infect *M. saxatilis* on the East Coast. Sakanari and Moser have shown that those *M. saxatilis* that currently populate the East Coast (their origin) develop much more marked pathology to the cestode's larvae than do the descendants of the introduced fish on the West Coast. It is not likely that the parasite has modified its virulence on the West Coast, because the introduced fish is only one among many hosts. More likely the striped bass on the West Coast has, by selection, increased in a few

West Coast:
Lacistorhynchus dollfusi
is present and uses various species of intermediate hosts

East Coast:
Lacistorhynchus dollfusi
is absent

Scenario:

1. The fish are introduced on the Pacific Coast (early twentieth century)
2. The cestode switches to this new host species and is highly virulent
3. One hundred years later the virulence has been reduced.

Figure 11.8. Process of virulence reduction occurring between the beginning of the twentieth century and the present after the fish *Morone saxatilis* was introduced on the West Coast of the United States (after data of Sakanari and Moser 1990).

dozen generations its ability to limit the pathogenic effect, most notably by better control of the inflammatory process.

Useless Virulence

Just as useless resistance may be counterselected in hosts, any gene or combination of genes in the parasite that are responsible for "useless virulence" are also likely to be counterselected. Useless virulence is defined as virulence that "can only reduce the fitness of both the parasite and the host" (Combes 1997). A logical prediction is that this useless virulence must be eliminated by natural selection and that "this would result in a fitness increase for the host, but also for the parasite" (Ebert 1995).

Useless virulence may be illustrated by the nematode *Anguillicola crassus* in its recently acquired host, the European eel *Anguilla anguilla.* Adult parasites live in the swim bladder and lay eggs that are evacuated outside when they contain L2 larvae. Larvae hatch while in the water and are eaten by copepods, within which they molt to the L3 larvae. When infected copepods are ingested by eels, the L3 larvae migrate toward the swim bladder by breaking through the intestinal wall, molt to the L4 larvae in the swim bladder wall, and finally develop to adults. According to Haenen et al. (1989), L3 larvae follow abnormal pathways throughout their migration to the swim bladder, most likely because they do not know how to interpret signals (certain substance concentrations in tissues) present in the newly acquired European eel hosts. These abnormal pathways cause serious lesions, and this useless pathology should logically be counterselected over time. However, it is also probable that when useless virulence is not counterselected it is because gene flow linked to a broad host spectrum opposes such selection.

What Causes the Mismatch of Current Virulence with Optimal Virulence?

Fluctuations in Space

Various parasite and host populations are genetically distinct because in different geographical locations the populations are not subjected to exactly the same selective pressures. Adaptation to the local partner (often called local adaptation; see Kaltz and Shykoff 1998) may be better for the parasite than for the host or vice versa because of such difference. If the parasite is better adapted to exploit the host in one locale, its success is likely to be better in sympatric rather than allopatric host populations. Conversely, if the host is locally better adapted to resist the parasite, the parasite's success will probably be better in allopatric host populations than in sympatric ones. Ebert and Herre (1996) note that local adaptation is nothing other than *one of the dimensions of parasitic specificity* expressed "on a micro-evolutionary scale." Such local adaptation may be studied by observing natural confrontations, by setting up experimental confrontations between parasites and hosts from different geographic regions, or by modeling.

Natural Confrontations

The introduction of "exotic" pathogens is often responsible for the spread of serious infectious or parasitic diseases in animals. For example, in East Asia the nematode *Anguillicola crassus* is not known for its strong pathogenicity.[4] However, as soon as it was imported into Western Europe (first Germany in 1982, then Hungary, Holland, France, etc.) and North America it became a redoubtable pathogen of indigenous eels. The most serious damage was recorded in Balaton Lake in Hungary (Moravec 1992; Molnar, Szekely, and Baska 1991), where heavy pathogenicity of *A. crassus* in its new hosts was partly due to useless virulence, as discussed above, but also to an excessive opening of the encounter and compatibility filters. The widening of the filters' opening in this case may be explained by several hypotheses: that there may be, in the nematode population of origin, a polymorphism in pathology and the introduced

4. Other species of *Anguillicola* that have coevolved for long periods with their hosts cause limited damage. This is the case, for example, of *Anguillicola australiensis* and the eel *Anguilla reinhardtii* in Australia, which have been studied in detail by Kennedy (1994).

population may have been, by chance, heavily pathogenic (the founder effect); that the new host has not selected defense mechanisms to limit parasite load or, on the contrary, that its inflammatory reactions are exaggerated; that the parasite develops better in European copepod intermediate hosts, for which the parasite is as brand new as for the eels, resulting in the demographic expansion of larval stages; or that transmission may be amplified in Europe by the presence of numerous fish paratenic hosts for which the parasite is also new. Kennedy (1994) prefers this last hypothesis: "The use of such hosts by *A. crassus* in Europe is a reflection in some way of the parasite being in an alien ecosystem."

In short, it seems that in this recently created allopatric combination there is at the same time both a "bad" local adaptation on the part of the eel and too much of a good thing in terms of transmission success for the parasite.

In the domain of human diseases the incidents of heavy mortality provoked by the introduction of a "new" pathogen are too numerous to be counted. Black (1992) states that 50 million Native Americans died because of the viruses and bacteria brought by Europeans, and Martin and Combes (1997) estimated that 80% of the population of the Marquesas Islands and 90% of that of Rapa Island in the Austral Islands have disappeared because of pathogens imported with European immigrants. In Polynesia, history retains the names of the boats, such as the *Britannia* and *La Magicienne,* whose passage was followed by catastrophic epidemics of flu, dysentery, smallpox, or typhoid fever. On the other hand, pathogens present in assaulted populations have sometimes opposed the invaders. For example, in 1895 the French expedition to Madagascar counted 5,756 dead—25 soldiers killed by fighting and 5,731 killed by malaria and other "local" diseases (Aubry 1979). All these examples clearly show the local adaptation of host populations to the parasites they have confronted for a long time.

Experimental Confrontations

Ebert (1995) has experimentally studied the relationship between the freshwater planktonic crustacean *Daphnia magna* and the microsporidian *Glugoides* (= *Pleistophora*) *intestinalis.* The life cycle is holoxenous: parasites live in the intestine and form spores that are evacuated to the outside; new hosts become infected by ingesting these spores. The author experimentally infected crustaceans from various geographic regions (England, Germany, and Russia) using microsporidians from England. Such geographic distance is supposed to correlate to genetic distance and serves as "an estimate of how novel the host is for the parasite." Ebert then comparatively analyzed parasite load and host growth, survival, and fecundity in the different combinations. The experiments showed that it was in sympatric combinations that, on average spore loads were largest, the capacity for infection was greatest, and the growth, survival, and fecundity of *Daphnia* were most affected.

Therefore there is a positive correlation between the success of the parasite and the geographic proximity of the populations used in these experiments: the parasite was more pathogenic in *Daphnia* "it knew" than in those it had never been in contact with.

Ebert's research may be interpreted as follows: Virulence was increasing as the association became older. In this case the compatibility filter was opened by selection of parasite genes that allowed a more efficient exploitation of the local host population. Ebert notes that there also is a local adaptation in the host for its parasite. With an equal parasite load, *Daphnia* in sympatric combinations have a less marked reduction in fecundity. On the other hand, Hochberg (1998) points out that since superinfection can occur in microsporidians, there may be a dose effect such that "the result involving virulence may uniquely reflect variation in some measure of transmissibility," and Ebert (1998) agrees that "the effects of differential infectivity and differential within-host growth rate cannot easily be disentangled."

In the case of schistosomes confronted with their mollusk vectors, experimental work has shown that molluskan strains of local origin are often more vulnerable to parasites of the same region than to parasites of foreign origin. In other words, sympatric combinations are more compatible than allopatric combinations. Such situations allow one to suppose that the parasite has increased its fitness over time by adapting to the local host population.

Xia, Jourdane, and Combes (1998) used two distinct experimental approaches (exposure to miracidia and microsurgical transplant of sporocysts) to demonstrate that there is such local adaptation of *Schistosoma japonicum* for its vector mollusk, *Oncomelania hupensis*. This study used strains of the parasite and host coming from two foci of mainland China (the provinces of Anhui and Sichuan) twelve hundred kilometers apart. In all experiments developmental success was significantly higher in sympatric than in allopatric combinations (fig. 11.9). Further, the weak compatibility obtained in some allopatric combinations was the result of a rejection process that occurred at several stages of development, sometimes as long as several weeks after initial parasite-host contact. Because of the techniques used (sporocyst transplant), this study is one of the few that can distinguish between compatibility and encounter without ambiguity.

Manning, Woolhouse, and Ndamba (1995) obtained similar results for

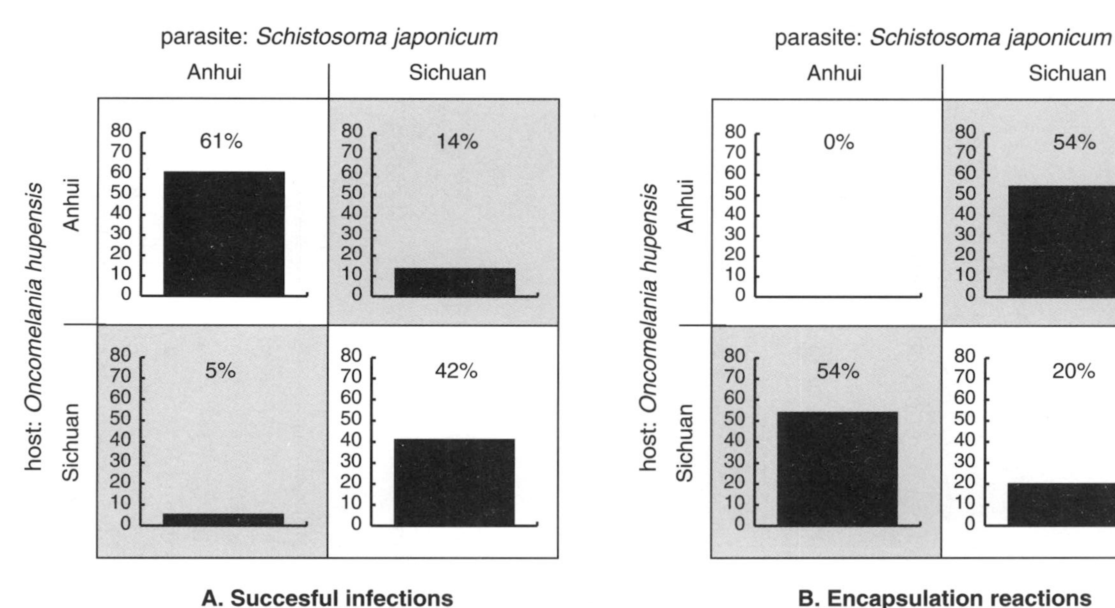

Figure 11.9. The experiments of Xia, Jourdane, and Combes (1998). The success of infections (determined by cercarial shedding) is larger in sympatric combinations (in white). The percentage of rejections (determined by the percentage of encapsulation after microsurgical transplantation of the parasite) is larger in allopatric combinations (in gray) (after the data of Xia, Jourdane, and Combes 1998).

the *Schistosoma haematobium/Bulinus globosus* system when they exposed miracidia from parasite populations separated by even shorter distances—about fifteen kilometers.

Models

It is logical to predict that the partner with the higher genetic variability will improve its fitness faster, at the expense of the other. This variability is influenced by both mutation rate and the time between generations. Hence parasites should most often have the advantage and thus evolve faster than their hosts. As a result, parasites should also be better adapted to sympatric hosts than to allopatric hosts. The available genetic diversity of the parasite may also come as the result of population movement, and Gandon, Michalakis, and Ebert (1996) suggest, based on a mathematical model, that if parasites migrate from one population to another more often than hosts do, there is a good chance that parasites may encounter here or there host populations with little resistance. In this case parasite populations would be, on average, better adapted to sympatric host populations. They also propose, in contrast, that if hosts migrate from one population to another more often than parasites do, there is a good chance for the hosts to encounter parasite populations that they resist well. In this case parasite populations would be, on average, more adapted to allopatric host populations.

Also based on mathematical modeling, Morand, Manning, and Woolhouse (1996) suggest that there may be cyclical fluctuations in the frequencies of virulence and resistance alleles in schistosome-host associations. Here cases where the worms infect allopatric hosts better than sympatric ones would be explained by a temporal shift in the fluctuations of the allele frequencies in the parasites relative to those in the host.

It is currently difficult to determine the most common direction in which virulence evolves when a "novel" parasite-host association occurs: Does it increase or decrease? The conventional wisdom dictating that obligatory parasite-host associations evolve from strong virulence toward peaceful coexistence is certainly to be rejected (Toft 1991a). Supporters of an increase in virulence note that although in natural confrontations we often have the feeling that virulence decreases over time, this is because of the involuntary importance given to striking examples that are exceptional. This point of view seems to be supported experimentally, but such support concerns only invertebrates.

In fact, a number of human diseases that give the impression of being serious when they emerge (or reemerge) and then decrease in seriousness over time are not isolated cases. Levin (1996), although defending the idea that virulence more often increases over time, notes that "many observations are consistent with conventional wisdom about parasite-host coevolution." Thus the question about how the evolution of virulence most frequently proceeds is still open.

Fluctuations in Time

Parasite-host associations may be characterized temporally by a virulence that is far from its optimum simply because genetic or environmental events may provoke sudden changes in the system.

Transitory Maladaptations Owing to Genetic Events

During the past several years numerous studies have been carried out on the sudden increases in virulence seen in several diseases, particular those important to human health. Mutations sometimes allow pathogenic agents to suddenly increase their "success" in hosts whose populations are then strongly affected. In this case everything happens as if coevolution occurred in steps, with the pathogens temporarily taking a substantial advantage. This temporary advantage, termed an evolutionary lag, is defined by Stenseth and Maynard Smith (1984) as the time, for a given species, between the momentarily adapted state and the best possible adaptation. In a host-pathogen association, if the evolutionary lag increases to the pathogen's benefit we may say the host "loses contact" with the pathogen. In this case there is a "transitory maladaptation" of the host population, which may be qualified, depending on the case, as a simple epidemic (e.g., the plague epidemic in the Middle Ages or the Spanish influenza of 1918–19 with its 20 million deaths); as an emergent disease (hemorrhagic colitis in North America because of a particular clone of *Escherichia coli* in the 1980s and 1990s); or as a reemergent disease (tuberculosis due to *Mycobacterium tuberculosis,* with 3 million deaths per year in the 1990s).

The concept of an evolutionary lag illustrates the contrast that exists between the rapid renewal of genetic diversity in many pathogenic agents and the slowness of natural selection in hosts. This rapid renewal of diversity in pathogens is due to the nature of genomes themselves, with noneukaryotic pathogenic organisms generally having higher rates of mutation than eukaryotes. The result is a large genetic diversity such that mutants, which are relatively rare in a given environment, may rapidly invade a host population if a favorable change occurs in that environment. It is interesting that the virulence selected in natural environments and that selected in culture media may be very different. In viruses, for example, mutation rates are remarkably high, and they also have the property of recombining within different "species" in infected hosts. It is thought that the influenza virus comes back in some years in Western countries in particularly virulent "forms" because it undergoes complex exchanges with similar viruses of various wild and domestic animals, such as ducks and pigs, that live in close contact with humans in some other parts of the world. During each epidemic this viral recombination produces a virus that is unknown to the human immune system (Yasuda et al. 1991). In HIV infection viruses belonging to different groups (within HIV type 1 there are, for instance, three main groups called M, N, and O) can recombine and give rise to new and highly replicative variants (Peeters et al. 1999).

In bacteria there are even mutator alleles that suddenly increase the rate of mutation (by a factor of one hundred to one thousand) by altering the efficacy of the DNA repair enzymes (Haraguchi and Sasaki 1996; Taddei, Matic, et al. 1997, Taddei, Radman, et al. 1997). For example, a mutation in one locus coding for a particular protein, such as adhesin, is sufficient to modify the success of the pathogenic agent in its host. A mutation of one locus involved in virulence is often linked to particular repetitions of nucleotides in sequences coding for such molecules, and the number of repetitions may strongly influence the expression of the gene. This is known, for example, in *Neisseria, Yersinia,* and *Haemophilus,* and "the high frequency of mutators associated with pathogenic bacteria suggests the evolution of increased mutation rates is relevant to the evolution of parasites" (Taddei, Radman, et al. 1997).

The genomes of "higher" organisms (with their longer generation times) are not always competitive against such brutal and sudden processes and can really respond only by the slower selection of new adaptations. Even the fabulous diversity of bird and mammal antibodies is not enough to protect them from some new or renewed pathogens capable of such rapid recombination.

Transitory Maladaptations Owing to Environmental Events

The encounter between parasites and hosts may be deeply influenced by the survival of infective stages in the outer environment. Among microparasites, enteropathogenic bacteria such as *Salmonella* are dispersed to the outside and then ingested with food, such as shellfish. The risk of *Salmonella* contamination therefore depends on their survival first in "brown" water and then in the waters of the lagoon where the shellfish are raised. Virulence decreases with exposure to sunlight and salinity because of the alteration of genes involved in virulence (Caro 1998).

Among macroparasites a life cycle such as that of the liver fluke *Fasciola hepatica* is also strongly dependent on outside conditions. Fascioliasis is typically a disease of accumulation, becoming more serious with greater exposure to infective stages. In humid years the snail intermediate hosts *(Lymnea trunculata)* reproduce abundantly in flooded pastures and infective metacercariae live longer. All these conditions are favorable to the parasite and further open the encounter filter, causing parasites to accumulate in cattle in unusual numbers. In this case host fitness may be decreased to the point of death for the most heavily infected individuals, and meteorological conditions are so closely linked to the realization of the life cycle that in the early 1990s it was possible to make available a software model allowing farmers to predict infection peaks linked to different climatological parameters. Predicting such "epizootics" is not simple though, since dry years may also be periods of high transmission because the snails and cattle are forced to gather in the smaller and smaller remaining humid zones.

De Lope et al. (1993) experimentally demonstrated the influence that environment may have on virulence. In the introduction to their article these authors define the problem as follows: "It is inherent in the definition of parasitism that parasites are costly to their hosts. These costs can be measured in absolute terms but may vary in relative magnitude as a consequence of variation in the access of hosts to essential resources. If parasites are costly to their hosts in terms of fitness, they should be so to a larger extent under stressful environmental conditions." The authors field tested the prediction that the same parasite load has a greater effect on the hosts' reproductive success if environmental conditions are unfavorable using as a model the house martin, *Delichon urbica,* and its hemipteran ectoparasite *Oeciacus hirundinis.* This parasite, found in the swallows' nests, sucks the chicks' blood and is sometimes lethal.

The experiment, carried out in Badajoz, Spain, on 2,700 nests of *D. urbica,* consisted of "manipulating" the level of infection by *O. hirundinis* in nests and analyzing the consequences of this manipulation on the reproductive success of the birds throughout the summer. Nests were divided into three categories: nests "without parasites" were fumigated twice with insecticide before the swallows' arrival and again between the first and second clutches; nests "with weak parasitism" were fumigated only once before the swallows' arrival and then contaminated with ten *O. hirundinis;* and nests "with heavy parasitism" were fumigated only once before the swallows' arrival and then contaminated with one hundred *O. hirundinis.* Further, the authors attempted to neutralize as much as possible any other factors, such as the physiological condition of parents, clutch size, and insecticide effect between treatments.

Figure 11.10 shows that, because of either exchange between neigh-

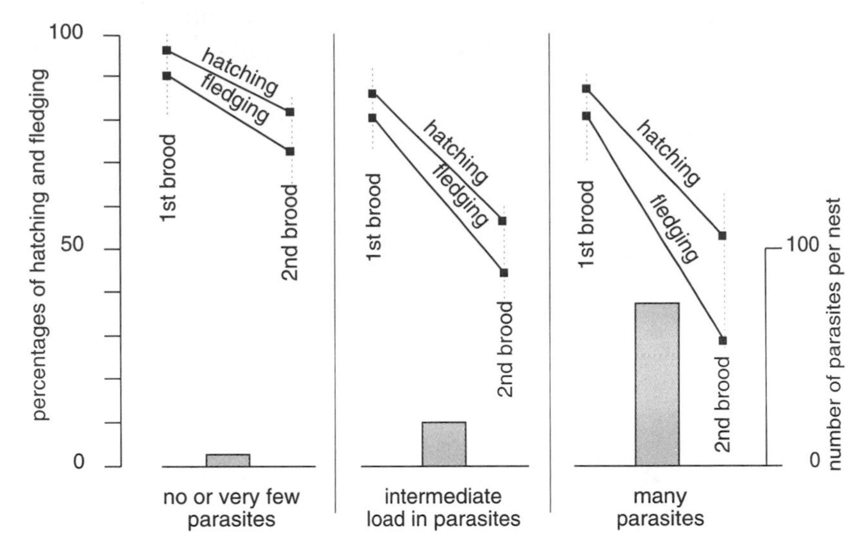

Figure 11.10. Manipulation of parasitism showing that virulence of nest parasitism differs according to environmental conditions (after the work of Lope et al. 1993). Shaded bar indicates number of parasites per nest.

boring nests or parasite mortality, the nests did not strictly remain with zero, ten, and one hundred parasites throughout the experiment. However, the differences between the three types of nests remained great enough to be considered significant.

Over the course of the experiment the authors precisely estimated various parameters associated with the swallows' reproductive success for two successive clutches during the March and July breeding periods. The average percentages of egg hatching and young fledging are presented in figure 11.10, and the results obtained are particularly demonstrative: in nests with "zero parasites" (in reality slightly contaminated by transfers from neighboring nests), the proportion of hatching and fledging are close to 100% for the first clutch and decrease to 80% for the second clutch; in nests with "a few parasites," the proportions for the first clutch remain little affected but drop to 50% for the second clutch; and in nests with "many parasites," the proportions for the first clutch remains little affected but drop drastically for the second.

This experiment demonstrates, as de Lope et al. (1993) write, that "ectoparasites were indeed more costly to house martin hosts in the second clutch." It is spectacular that the parasites have almost no effect on the reproductive success of the first clutch when spring conditions supply the birds with abundant food, whereas they deeply affect both hatching and fledging when the aridity of the Spanish summer renders prey more and more rare. The explanation is that the environmental conditions change between March and July, and the parasites become more virulent when conditions are less favorable.

This experiment is particularly instructive in that it distinguishes the part of virulence that belongs to the parasite's *genes* and the part that belongs to the genes' *expression*. It is certain that, overall, virulence as a selected strategy did not change between March and July. What did change were the conditions in which the parasites' genes were expressed, and as a consequence the parasites induced a loss of fitness in the host only at that time.

Durable Maladaptations

In certain parasite-host associations maladaptation may also be durable. In this case optimal virulence is not selected and virulence durably remains either suboptimal or superoptimal.

Durable maladaptation is not unique to parasite-host systems; it is often encountered in free-living animals. For example, Storfer and Sih (1998) have analyzed the antipredator behavior of the salamander *Ambystoma barbouri* in habitats with different selective pressures owing to the presence or absence of carnivorous fish. Such fish are present in waters that are permanently present and absent in waters that temporarily dry out, while the salamanders are present in both types of habitat. These authors have experimentally compared the behavior of salamander larvae from both habitats in the presence of the fish. Larvae coming from habitats with permanent waters were able, up to a certain point, to escape the fish by either fleeing or decreasing their activity. Larvae coming from habitats "without fish" were vulnerable because they did not have such antipredator behavior (or had it to a lesser degree). Logically we may deduce that such antipredator behaviors were selected only in environments where the fish lived.

However, as the authors note, the most interesting aspect of the work was that "larvae from populations containing fish that were more isolated from fishless populations showed stronger antipredator responses than less isolated populations." This means that the quality of antipredator adaptation depends to a large extent on population isolation: if populations from sites with fish are close to populations from fishless sites, there is gene flow between the two sites. These genetic exchanges between populations then prevent local adaptation from fully occurring by diluting "adapted" genes with "nonadapted" ones.

Similarly, Blondel et al. (1992) have shown that in southern France blue tits live mainly in habitats containing deciduous trees that provide them with an abundant resource of caterpillars about May 15 each year. Birds lay their eggs at about this time, and the nesting period matches the period of greatest food abundance. On the island of Corsica, however, blue tits live mainly in habitats dominated by evergreen trees. In this location caterpillars appear one month later than on the mainland, and the blue tits also reproduce one month later. In both France and Corsica the nestlings are fed adequately, and breeding success is high. What is interesting is that in southern France there are also some woods dominated by evergreen trees, and in these areas the blue tits are "maladapted": they lay their eggs too early (about May 15) and thus have little food and reduced breeding success. The explanation is that the timing of tit reproduction is genetically determined (Blondel, Perret, and Maistre 1990), and there is sufficient gene flow from populations breeding in the deciduous trees to populations breeding in the evergreen trees so that local adaptation of the latter is prevented by genes coming from the former (Dias and Blondel 1996).

As in salamanders and tits, similar local maladaptations occur in parasites, except that parasites are maladapted not to a predator or prey but to a particular host species (Combes 1997). This happens when a parasite has a wide host spectrum because if a parasite circulates in numerous distinct host organisms, optimal virulence cannot be selected in all the different hosts at one time. That is, in those hosts that are the least exploited the parasite may remain, maladaptively, more pathogenic for that host. For example, *Schistosoma japonicum,* one causative agent of human schistosomiasis in the Far East, parasitizes about forty mammals besides humans. In terms of sheer numbers it may be more abundant in nonhuman hosts than in humans, which may explain why it is so pathogenic in man. One might think, along with Combes (1997), that "it is preferable for a host to have its own parasites than to share them."

Cyclical Maladaptations

Arms races in which hosts and parasites alternate having the advantage have been described for a long time in gene-for-gene systems and recently in more complex associations.

Historically such gene-for-gene models have concerned particular associations of organisms in which a small number of major genes control both virulence and resistance. Such models describe particularly well certain plant/bacteria associations as well as certain plant/parasitic fungi associations (see, for example, Frank 1992, 1993). These relatively simple models can be contrasted to more complex associations that are much more difficult to model because of the complex combination of genes involved.[5]

In gene-for-gene type models (fig. 11.11) a host that expresses the character *R* (resistant) at locus *X(h)* successfully opposes infection with a parasite expressing the character *Avr* (avirulent) at locus *X(p)*. Any other combination leads to infection with the parasite.[6] Such gene-for-gene models predict coevolution between the loci of the opposing parasite and host characters, and several important issues result from this game of hide-and-seek among genes. The most interesting is the installation of cycles in which populations characterized by one or the other character alternate. In practice the number of loci that oppose each other may be high, which then leads to a high degree of polymorphism. This is the case, for example, in the flax *(Linum ultissimum)*/rust *(Melampsora lini)* system, a pathogenic fungus/plant host association involving twenty-nine such loci. Interestingly, the genes that correspond to these loci seem to constitute gene families, according to their sequences, and thus are likely derived from a single common ancestral gene. In this system it is sufficient for only one of the locus couples to determine incompatibility and thus for the combination not to be viable.

The biochemical mechanism of incompatibility in such systems involves a signal molecule produced by the parasite gene that would normally be recognized by a receptor molecule produced by the corresponding host gene, which would then trigger resistance. As noted above, the gene for host resistance is generally designated by *R* and the gene for parasite avirulence by *Avr*. The avirulence character (a protein) coded by *Avr*

---

5. "Virulence genes" are understood to be loci that directly or indirectly code for characteristics associated with the pathogenic effect and that are implicitly responsible for the host's loss of reproductive success. A gene for virulence may code for various characteristics involved, for example, in host attraction, host recognition, adhesion to the host, exploitation of the host's resources, multiplication in the host, or evasion of the host's immune defenses. In bacteria such "virulence genes" are often carried by plasmids. Poulin and Combes (1999) have suggested restricting the term "virulence genes" to genes that have been selected for their immediate virulent effect, such as in cases where harming the host can promote parasite transmission. "Resistance genes" are understood to be loci that code for characteristics associated with host defense. A gene for resistance may code for a molecule involved in either parasite recognition or parasite destruction. In most parasite-host associations it is probable that several thousand parasite and host genes are involved in virulence and resistance. These genes remain mostly unidentified, and studies of their selection belong to the field of quantitative genetics (see, for example, Maynard Smith 1989a).

6. Note that the genotype called "resistant" in the host is infected if it encounters the virulent genotype of the parasite, and that the genotype called "avirulent" of the parasite is not infective if it encounters the resistant genotype of the host.

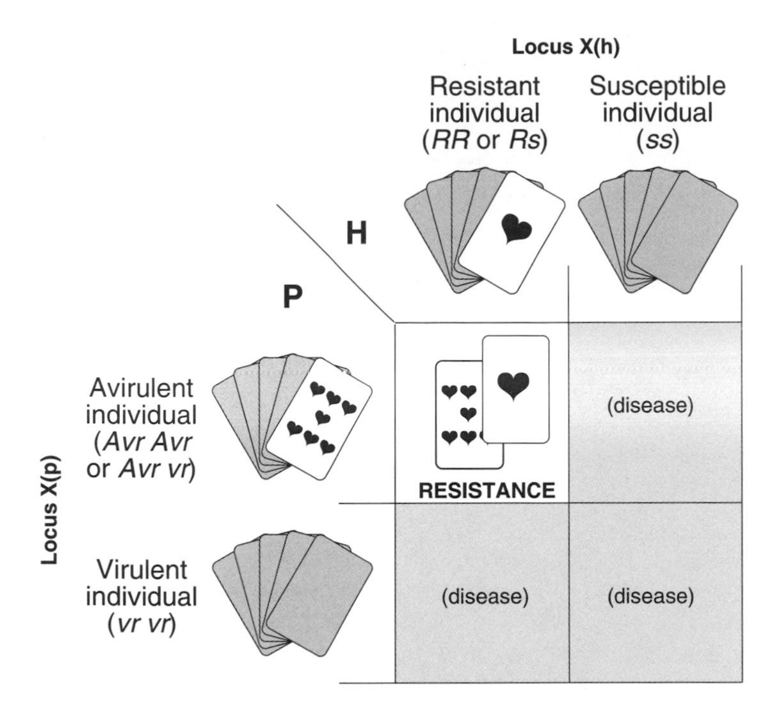

Figure 11.11. A symbolic representation of the gene-for-gene model: if the parasite P has a card (here the seven of hearts) that is recognized and dominated by a host card H (here the ace of hearts), there is host resistance. In all other cases (if the parasite does not have the seven of hearts or if the host does not have the ace of hearts), there is host susceptibility. Resistance genes and avirulence genes are dominant so that a resistant host may have the genotypes *RR* or *Rs* and an avirulent parasite may have the genotypes *Avr Avr* or *Avr vr*.

is expressed continuously, and in the presence of its corresponding gene *R* events unfold as if *Avr* were to code for an "antigen" that would be recognized by an "antibody" coded for by *R*. If *Avr* is present in the pathogen and *R* in the plant, there is *recognition* and attack of the pathogen; if *Avr* or *R* (or both) are absent, the pathogen may then develop. In general, the allele *Avr* (avirulent) is dominant over *vr* (virulent) and the allele *R* (resistant) is dominant over *s* (susceptible), the result being that heterozygous pathogens are avirulent and heterozygous plants resistant (fig. 11.11). Numerous such avirulence genes (in pathogens) and resistance genes (in plants) have been cloned (see, for example, Hammond-Kosack and Jones 1996; Van den Ackerveken and Bonas 1997; and Bonas and Van den Ackerveken 1997).

Variations of the gene-for-gene model exist and include the matching allele model, in which each host's allele confers resistance to one

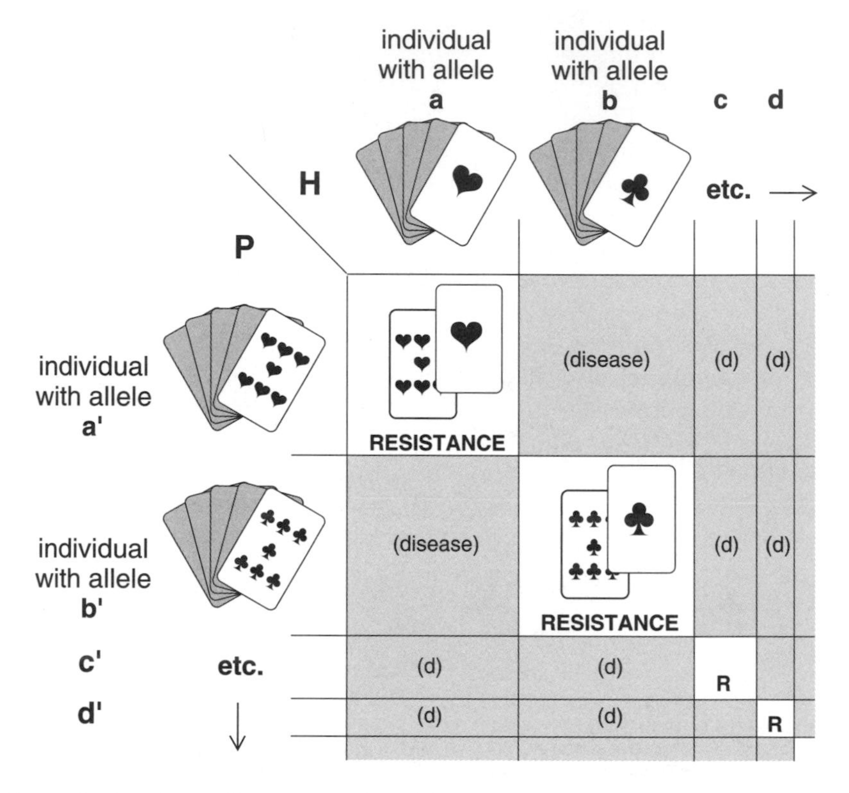

Figure 11.12. A symbolic representation of the matching allele model (heterozygotes are not considered on the figure): parasites may successfully invade a host when there is no match between parasite and host alleles. H = host; P = parasite.

"matching" parasite allele (fig. 11.12), and the "matching genotype" model, in which each host genotype confers resistance to one "matching" parasite genotype. In both cases the biochemical basis of the interactions is presumed to be similar to that described above.[7]

In animals, even if major genes may determine resistance, it is unlikely that such gene-for-gene models completely apply (see Sorci, Møller, and Boulinier 1997). For example, for associations involving an undetermined number of genes, Soler, Møller, and Soler (1998) have proposed

7. The main difference is that in the gene-for-gene model there is only one allele that confers resistance at a particular locus of the host and only one locus conferring avirulence on the corresponding parasite locus. It is sufficient for one of these alleles not to be present for the infection to be possible. In contrast, in the matching allele model there is one particular host and parasite locus with an undetermined number of alleles, and each host allele codes for a molecule recognizing one of the molecules produced by the parasite.

an "intermittent arms race hypothesis" based on their observations of cuckoos and their bird hosts in the region of Cadiz, Spain. As previously discussed, the great spotted cuckoo *Clamator glandurius* lays its eggs in the nest of the magpie *Pica pica*. According to hunting data, the parasitic bird has been in Cadiz only since the beginning of the 1960s. Magpies at that time were very vulnerable to parasitism but have quickly improved their rate of rejecting cuckoo eggs, thus closing the compatibility filter. Soler and colleagues postulate that other factors have also been involved in the decrease in parasitic pressure applied by the cuckoo, such as an increase in nest density and high breeding synchrony resulting in a "swamping effect" and the closing of the encounter filter (fig. 11.13).

In the face of such evolved defense mechanisms the cuckoos may have

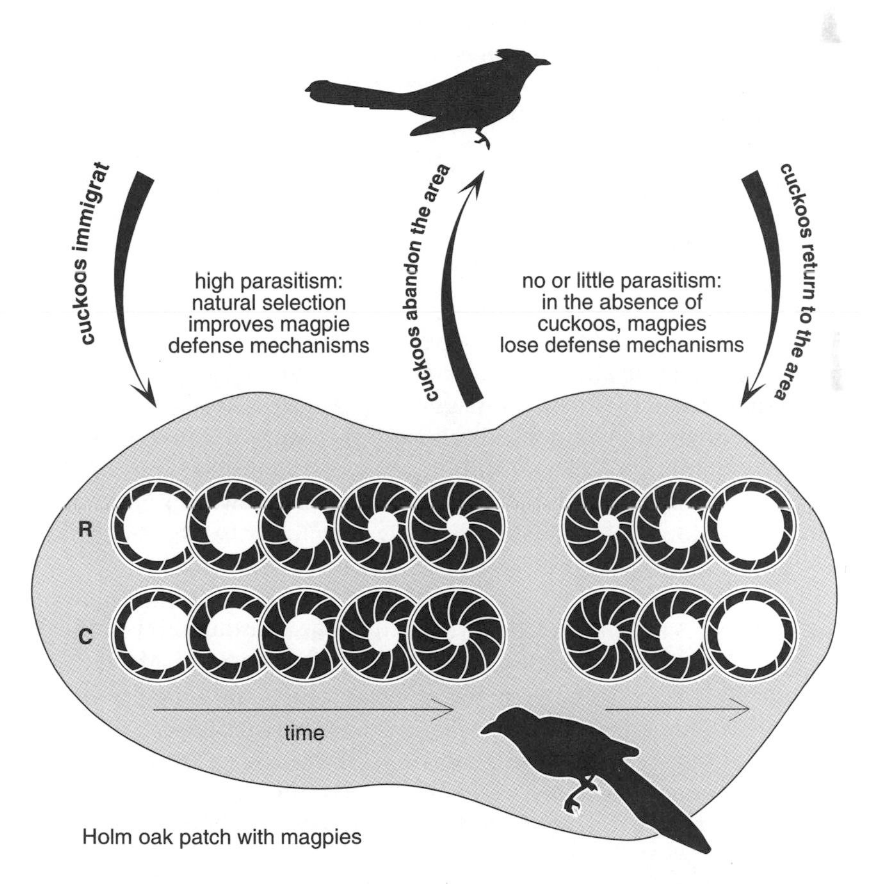

Figure 11.13. Hypothesis of the periodic migration of cuckoos to and from a locality each time the magpies' defenses become too efficient (after the data of Soler, Møller, and Soler 1998).

abandoned the field to migrate to zones where resistance was less efficient: "When half (or more) of the population have become rejecters, parasites may benefit by abandoning that host population and starting to exploit naïve host populations" (Soler, Møller, and Soler 1998). In the region studied, parasitism of magpies by cuckoos reached a maximum in 1992 and then the cycle reversed: in some patches of oaks *(Quercus rotondifolia)* separated by treeless areas, prevalence dropped from more than 40% to zero within two years. Soler and colleagues suppose that defensive mechanisms would then be lost by the magpies when the cuckoos are absent, so that after some time the cuckoos may return. Such an arms race starts "with clear advantage for cuckoos and finishes with hosts as temporary winners, but after a period during which the hosts have lost most of their defensive mechanisms the arms race may start again" (Soler, Moler, and Soler 1998). In intervals of a few dozen years the alternation of periods of high parasitism with periods of very little or no parasitism may thus be observed in a distinct locality. Such an intermittent arms race may also be occurring with European cuckoos, *Cuculus canorus,* but may involve much longer periods.

It should not go unnoted that there are distinct similarities between the cycles predicted by the gene-for-gene models and the hypothesis of Soler, Moler, and Soler (1998).

Transgenesis

Genetic engineering allows the introduction of one or another gene into a foreign genome to obtain "transgenic" organisms. The methods used are either physical or biological (the use of natural vectors, including viruses or mobile DNA elements that can insert themselves into the host genome), and the control of several major parasitic diseases endemic to human and domestic animals will likely benefit from this technology through the creation of transgenic resistant vectors.

For example, one can imagine "building" transgenic mollusks unable to transmit schistosomiasis (Richards, Knight, and Lewis 1992) as well as transgenic mosquitoes unable to transmit malaria (Crampton et al. 1990; Kidwell and Ribeiro 1992). The basis for this reasoning is that such vectors can be obtained once the genes involved in resistance are identified, methods of insertion of these genes into the genomes are found, and transgenic animals are successfully introduced into sites of transmission.

However, though the biotechnological approach of transgenesis seems fairly easy, there are at least three major difficulties from an ecological perspective that should not be forgotten (Jarne 1993). These include introducing the "good" gene or genes for resistance at the right place so that there is expression of the gene(s); obtaining competitive transgenic vectors capable of replacing the natural vectors; and avoiding the very rapid selection of parasite genes for virulence that are capable of neutralizing the genes for resistance.

We may in fact worry that the genetically modified vectors would be counterselected in the environment because of the nature of their resistance. In this case control would be possible only by "flooding release" (that is, using massive and renewed releases), which would be expensive and possibly not sufficient to significantly decrease the rate of transmission of the parasite.

Nevertheless, should controlling parasitic diseases with such genetically altered vectors become possible, we may end up with a decisive weapon that might not leave parasites the time needed to select new adaptations before they are wiped out.

Reflections

Manipulation of Humans

Can the parasites' phenotype extend into the humans' phenotype?

First example: favorization intervenes in the transmission of the pinworm *Enterobius vermicularis,* an intestinal nematode of humans, particularly children. In the case of the pinworm the upstream host and the downstream host are the same (humans), and the eggs laid around the anus by the female worm are immediately infective to children with dirty and wandering hands, who may then readily reinfect themselves or their buddies while sharing a snack. Further, the parasite "knows" all too well how to manipulate the child's behavior in order for its eggs to end up on dirty fingers: by inducing an intense itching in the anal area. The child scratches, doesn't wash, . . . and there you have it.

This is a simple but efficient behavioral manipulation, and added to it is a component of chronobiological manipulation, since the irritation is caused by female worms that lay their eggs around the anus, most often between 9:00 P.M. and midnight, a time when children often suck their thumbs while sleeping.

Second example: Have you ever wondered why you sneeze when you have a cold? To get rid of the viruses? More likely to transmit the viruses to your neighbors. In effect, sneezing and coughing are both favorization processes, just as is the diarrhea induced by most of the parasites that actively multiply in your intestines (such as the bacteria that cause cholera or the protozoan *Giardia,* which causes backpacker's disease). Although diarrhea may certainly be considered a simple manifestation of pathology, it is also very efficient at disseminating infective stages to the outside environment. As such it is difficult, if not impossible, to deny an adaptive value to the pathogenic agent that takes this route to the outside world, which then makes it highly suspicious as an agent of manipulation in humans.

However, for humans, and as for any other host, one must be cautious about affirming that a behavioral change results from manipulation on the part of a parasite and, most important, that such a manipulation would have an adaptive value for parasite transmission. Nevertheless an astonishing study by Flegr and Hrdy (1994) demonstrates that toxoplasmosis modifies both the behavior and the personality of men (but apparently not women). Fifty-six men and 34 women, all infected, were compared with 139 men and 109 women, all healthy. The analysis showed that personality factors in the males (such as being reserved or vivacious, disciplined or not disciplined, tolerant or dogmatic) and maintaining self-control were all significantly (and in some cases very significantly) altered by infection with *Toxoplasma gondii.* Although the importance of the behavioral changes is obviously not the same for humans and for other animals (notably rodents) that are normally hosts for the parasite, the basic mechanisms are nevertheless probably the same (Flegr, Zitkova, and Kodym 1996). Moreover, all the observed alterations increase the aggressiveness of the possessor host and hence would promote capture by a potential definitive host. As an aside it is interesting that of the forty-three university professors involved in the study, of whom twenty-nine were negative for toxoplasmosis and fourteen positive, all but one of the department heads were negative.

Another investigation along these lines, this time involving 191 young women, also gave original results: intelligence, guilt proneness, radicalism, and "high ergic tension" increased in subjects with latent toxoplasmosis. The difference between the two studies was explained thus: "The biologists represent a highly atypical population sample which differ in many personality traits from a general population" (Flegr and Havlicek 1999).

Manipulation of Vectors

When transmission occurs because of a biting vector, this vector is the upstream host and the bitten organism is the downstream host. As we have seen, the necessity for a parasite to get from one host to another often leads to parasite-mediated behavioral changes in hosts in order to increase the probability of the transmission.

Based on the present state of knowledge, however, it does not seem that there is much, if any, manipulation of biting vectors to make them more efficient at finding their "victim," which is the downstream host for the parasite. Such manipulations, if they occurred, would likely concern either the vector's ability to search for the downstream host's habitat (relative to humidity or temperature gradients) or its capacity to locate the downstream host itself (through visual, biochemical, or other stimuli it emits). Moore (1993a) analyzed the literature on vector manipulation and found no convincing data to support its occurrence. In short, biting arthropods, such as mosquitoes and ticks, have developed very elaborate mechanisms to encounter their victims, and both parasite and vector target the same organism. This elaborate and efficient targeting system makes it likely that the performance of the vectors cannot be much improved on by the parasite. Nevertheless, such manipulations should not be excluded from consideration a priori because of this, and though they are probably rare, there is no reason for them not to occur.

Although it is unlikely that a vector's targeting ability can be manipulated, once the vector has hit its target behavioral alterations that would favor transmission may in fact occur. Two types of such alterations, almost the opposite of each other, have been reported.

The change most often described in vectors that carry parasites (mosquitoes, phlebotomines, glossines, etc.) is a reduction in the volume of each blood meal, which is then compensated for by more frequent feeding. This is the case, for instance, in trypanosome-infected tsetse flies, where the gathering of trypanosomes close to the fly's biting apparatus disrupts the absorption of blood. The insect vector then needs to feed more often, increasing the frequency of transmission (see Molyneux and Jenni 1981; Rossignol 1988).

A second behavioral alteration at the target involves parasites that seem to in-

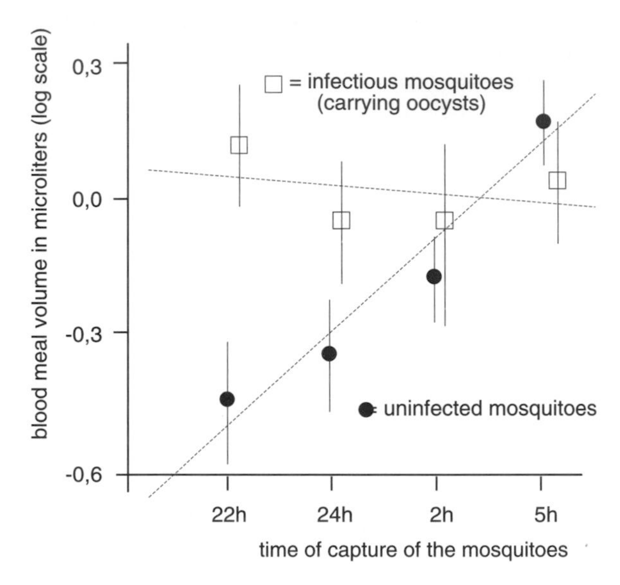

Figure R2.1. Infected mosquitoes maximize blood intake whatever the time of night (after Koella and Packer 1996).

duce "resilience" or persistence in certain vectors to maximize transmission. By capturing uninfected mosquitoes four times a night and measuring the blood they ingested during each blood meal, Koella and Packer (1996) showed that the quantity an uninfected mosquito took in during each meal increased over the course of the night, probably because later in the night the mosquitoes were less often disturbed while biting. Mosquitoes infected with *Plasmodium falciparum* and *P. vivax,* however, fed to a maximum whenever they bit during the night (fig. R2.1). Without knowing the precise mechanism, we can deduce, since the length of a meal increases the probability of the parasite's being ingested, that infected mosquitoes are more tenacious than uninfected mosquitoes in completing their blood meal, either because they induce fewer rejection reactions or because they come back to finish their meal if disturbed.

Since long life in the biting vector is favorable to transmission, we may expect that selection would eliminate any useless pathogenicity in the vector and that, in addition to increasing biting frequency, selection might also increase the vector's life span. This reasoning is logical, but it is true only in part because the presence of the parasite cannot fail to cause some minimum damage or cost to the vector. According to Rowland and Boersma (1988), mosquitoes *(Anopheles stephensi)* infected by *Plasmodium yoelii,* a rodent parasite, have their flight activity (measured acoustically with a microphone) reduced by one-third relative to that of controls. These authors conclude that "the small reduction in flight capacity observed in these experiments may be the best compromise the parasite has achieved under the circumstances."

Parasites that are transmitted by injection may also take advantage of manipulations initially selected to benefit the vectors. For instance, biting arthropods have in their saliva molecules that mimic naturally occurring host vasodilators (see Champagne 1994). This manipulation by the vector is a real bargain for the parasites they transmit, since vasodilation favors both their uptake (from the host to the vector) and their injection or transcutaneous penetration (from the vector to the host).

Favorization versus Phyletic Constraints

If evolution had resulted in whales before tetrapods, we might bet that whales would have gills instead of lungs. However, whales were built on the terrestrial plan and therefore must come to the surface to breath—even with all the problems inherent in living in the ocean. This is an example of a phyletic constraint: a restriction linked to the phylum in which the organism's ancestors originated.

Favorization can sometimes appear as a correction for some phylogenetic constraints. For example, in the case of *Gynaecotyla adunca* (discussed in chapter 7) one may first think it would be more economical for the parasite to "make" cercariae that are good swimmers or, even simpler, to choose other hosts. However, if we knew the conditions under which the life cycle of the ancestors of *G. adunca* evolved we most likely would understand why certain constraints were conserved and, instead of being eliminated, have been compensated for by the manipulation of the intermediate host: since its cercariae do not swim well enough to go toward the downstream host, the parasite manipulates the upstream host and makes it move instead.

The Cost of Immunosuppression

Immunosuppression or immunodepression resulting from a parasite's manipulation of the host's defense mechanisms is an efficient survival mechanism even though there may also be a cost to the parasite.

Let's consider HIV. It is during the period of "seropositivity" (before immunodeficiency occurs) that the virus most efficiently transmits its genes to healthy individuals via intercourse. As soon as the immunodeficiency shows up, however, opportunistic pathogens take full advantage and ultimately induce the death of the patient, thus interrupting virus transmission.[1]

---

1. The major opportunistic parasites involved in AIDS are viruses (cytomegalovirus . . .), bacteria (pneumococci . . .), fungi (*Candida albicans,* the causative agent of candidiasis, and *Cryptococcus neoformans,* the cause of some meningitis), protistans (*Toxoplasma gondii,* the cause of toxoplasmosis, and *Cryptosporidium,* responsible for intestinal problems), and organisms of uncertain taxonomic position (*Pneumocystis carinii,* a causative agent of pneumonia). Immunosuppression does not, however, affect all parasites. For instance, HIV does not increase the pathogenic effect of *Plasmodium falciparum.* Work done on a "retrovirus/mouse" model indicates that the virus might protect the host against the cerebral phase of malaria and that the higher the state of immunosuppression, the higher this pro-

This cost of immunosuppression exists for all parasites that select this type of strategy. For instance, according to Loker and Adema (1995) the existence of certain trematodes with either sporocysts or rediae in their life cycles may be linked to this cost. That is, certain species of trematodes produce cercariae in sporocysts that have no digestive system and that feed by absorbing nutrients from the host mollusk across their tegument (hemolymph), while others instead produce cercariae in rediae, which do have a digestive tract and actively ingest the mollusk's tissues. The digestive tract of the rediae is preceded by a muscular pharynx that also allows the ingestion of sporocysts from other trematode species present in the same mollusk (a property that originally made them possible candidates for schistosome control—see Lim and Heyneman 1972; Combes 1982). In short, Loker and Adema believe that those trematodes with sporocysts and those with rediae have different survival strategies in their vector: not being seen or being camouflaged for the trematodes with sporocysts and immunosuppression for those with rediae. The latter thus are exposed to increased competition, since they then facilitate the invasion of their host by other trematodes, leading Loker and Adema to suggest that the pharynx of the redia and its capacity to devour sporocysts would have been selected as a defense mechanism against opportunistic parasites and not necessarily as a way to acquire more resources.

The Individual's Sacrifice

We have seen how the sacrifice of a parasitized individual could in certain instances be a defensive mechanism against parasitism thanks to the protection it would provide to kin. However, it is very likely that, once again, parasites (or at least some of them) have found a way to counter such host sacrifice. Smith Trail (1980) believes that certain life cycles with long sequences are nothing other than a response of the parasite to host sacrifice, since any parasite capable of surviving in the predator of its host would then be strongly selected for. Along these lines it should be noted that nothing more resembles the behavior (unfavorable for the parasite) of *Chlosine harrissii* parasitized by a hymenopteran than that (favorable to the parasite) of ants parasitized by a trematode.

One characteristic of the living environment of parasites must not be forgotten—their habitat itself is edible. It has sometimes been said that the adaptive value of complex life cycles is the avoidance of competition between adult parasites and their progeny, but they might also be nothing other than an adaptation to the "edibility of the host" (that is, an adaptation to a habitat that is potential food for a predator).

Smith Trail (1980) writes two riveting sentences about this subject: she successively says that there are conditions that "make suicide a viable alternative to life" and that complex life cycles are "a parasite's way of making the best of a bad situation."

---

tection; experimental infection by *Plasmodium* of mice already infected by the virus (for a month) is not followed by manifestation of cerebral malaria. This "protection" may be due to the production of interleukin-1, a cytokine that inhibits the production of other cytokines (gamma interferon and tumor necrosis factor alpha) involved in the appearance of cerebral malaria (Eckwalanga et al. 1994).

Who Runs Faster?

In the case of the orchids and butterflies discussed earlier in the text it is justified to consider (using the terminology of the Red Queen hypothesis) that the two partners run at the same speed because neither can pass the other. In parasite-host systems, however, the hosts constantly run behind the parasites, which are always ahead because of their greater capacity to renew their genetic diversity. This is what Gemmill and Read (1998) express when they write that "large asymmetries between host and parasite generation times may leave hosts 'lagging' behind pathogens in coevolutionary arms races." If by chance the host succeeds in running faster than the parasite, the parasite is then eliminated, since it brings nothing positive to the system. That is, those associations in which the host succeeded in keeping ahead of the parasite no longer exist.

Pathogenic Effect and Virulence

It is sometimes difficult to differentiate between the pathogenic effect and virulence. Let's take an example showing that these two concepts are in fact different.

Mollusks that serve as intermediate hosts to trematodes have, depending on the species, different life spans (in general between a few months and a few years). Some trematodes are parasitic of the mollusks' gonads and others of the digestive gland or kidney. If the parasite is in the gonad it does not affect the survival of its host, since the gonad has no essential physiological function. If the parasite is in the digestive gland or in the kidney, however, it affects essential physiological functions such as digestion or excretion. Taskinen, Mäkelä, and Valtonen (1997) have hypothesized that when a trematode exploits a mollusk with a long life span the parasite has advantage in adopting a "damage avoidance tactic," that is, it is better off settling in the genital gland rather than in other organs: "Natural selection should maximize the trematode's total life-time reproductive output, or in a mollusk, the total life-time number of cercariae produced."

Taskinen and colleagues note, for example, that if a mollusk is capable of living fifteen to twenty years, as are certain species of *Anodonta* from temperate regions, then it is in the best interest of the mollusk to select for a trade-off between the number of cercariae produced during each reproductive season and the number of reproductive seasons it can have. They demonstrate the validity of this hypothesis using observations on the development of two trematodes of the genus *Rhipidocotyle* in *Anodonta piscinalis* in Finland and by a comparative analysis of the relation between the life span of mollusks in general and the type of organ parasitized. This analysis shows that in those mollusks whose life span is less than two years the trematodes usually exploit the digestive gland, whereas in those mollusks whose life span is over four years the tendency is to occupy the gonad.

By settling in the gonad the trematode exerts a minimal pathogenic effect on its host. As Poulin (1998a) notes, "Host castration may be the ideal strategy of host exploitation: by attacking non-vital organs, castrators do not reduce the host life-span, and they can obtain a high transmission rate without trading off specificity." When a parasite is in the gonad, however, its virulence is maximal; its presence

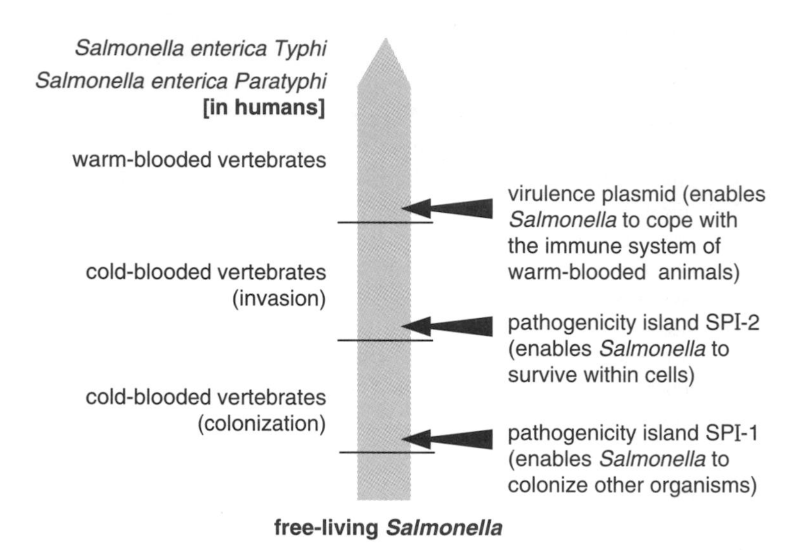

Figure R2.2. Hypotheses concerning virulence acquisition by *Salmonella* (modified from Heesemann, Schubert, and Rakin 1999).

makes the mollusk's fitness drop to virtually zero as the result of being either partly or completely castrated.

Thus the smaller the pathogenic effect, the higher the virulence and vice versa. Obviously this is not what happens in all parasite-host associations.

Pathogenicity Islands

Bacteriologists call "virulence genes" all those genes that are supposed to distinguish parasitic bacteria from their free-living bacterial kin.

For instance, those genes that allow bacteria to adhere to cells,[2] that let them penetrate and survive in cells despite their defense mechanisms, and that then allow them to multiply are all "virulence genes." These genes (there are often dozens of them) are not randomly distributed in the genome but are usually grouped into particular zones in the bacterial genome (or in plasmids). These groups are termed "pathogenicity islands," or "pais" for short (see Hacker et al. 1997). Pais are mobile because of their excision and integration abilities, and they are subject to exchange between different species of bacteria. Heesemann, Schubert, and Rakin (1999) summarize this property by saying that microorganisms "acquire pathogenicity in quantum leaps." They suppose, for example (fig. R2.2), that the evolution of path-

2. Adherence of certain disease-causing microorganisms to a cell is as important for their survival as is the attachment of a helminth larva to its potential host's tegument. For example, the capacity for adhesion of different species of *Candida* is correlated with their virulence.

ogenic bacteria in the genus *Salmonella* occurred from a free-living ancestor by the successive acquisition of two different pais (SPI1 and SPI2) followed by the acquisition of a plasmid carrier of the virulence genes necessary to cope with the highly developed immune system of warm-blooded animals (*Salmonella,* under slightly different forms, causes the typhoids and paratyphoids in humans). A comparable evolutionary model had been proposed by Heesemann, Schubert, and Rakin (1999) for *Yersinia* (*Y. pestis* causes bubonic plague, which has been responsible for devastating epidemics throughout human history). The mobility of the pais could also explain why some strains that are only slightly pathogenic sometimes appear and multiply in asymptomatic carriers. That is, periods of little pathogenicity are an advantage to the bacteria because host populations can then reconstitute themselves, with full pathogenicity being restored later by the transfer of "pathogenicity determinants." Such cyclic processes may explain the epidemic mode of certain infectious diseases, and nothing indicates that such processes are particular to bacteria alone.

Hide or Fight?

The title of this section is borrowed from Hochberg (1998), who asks: Under what conditions is it advantageous for a potential host to invest in mechanisms that prevent infection by parasites ("avoidance") as opposed to investing in mechanisms to attack the parasites after infection ("kill")? That is, when is it better to "hide" and when is it better to stand and "fight" (fig. R2.3)?

Hochberg applied this question to parasitoids and built a mathematical model in which there are evolutionary options: either there is selection in the host allowing the host to hide or there is selection for encapsulation and killing. Further,

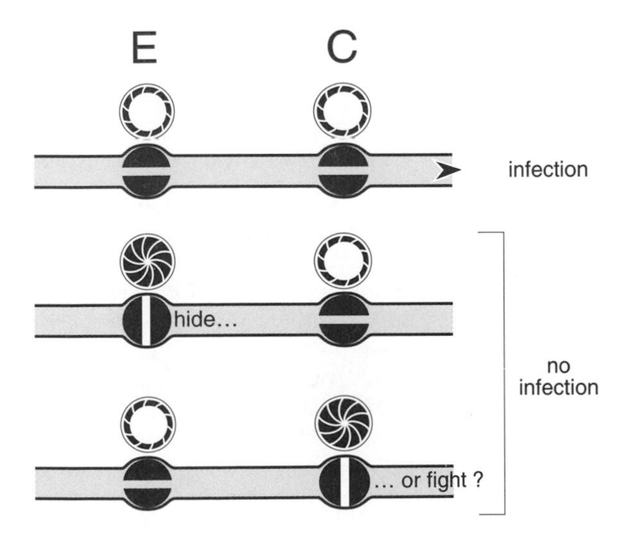

Figure R2.3. Hide or fight?

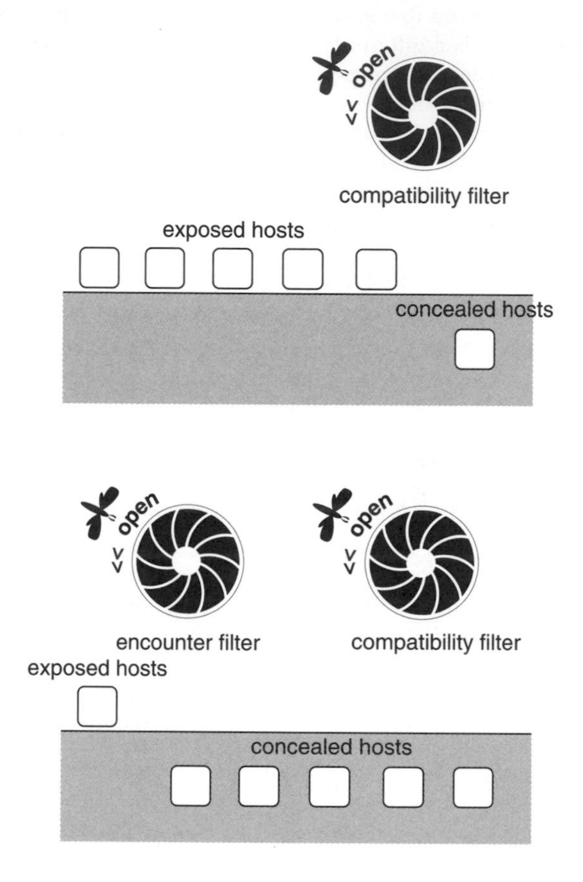

Figure R2.4. Hochberg's model.

either there is selection in the parasitoid for an antidissimulation characteristic or there is selection for the antiencapsulation characteristic.

The model indicates that if the hosts are abundant or the level of dissimulation is moderate, there is then selection in the parasitoids for antiencapsulation rather than antidissimulation because the parasitoid easily encounters nondissimulated hosts. In this case selection in the parasitoid genome tends to open only the compatibility filter (fig. R2.4, top). If on the other hand hosts are not abundant or the level of dissimulation is high, then there is selection in the parasitoid in equal proportion for both the antiencapsulation and antidissimulation characteristics, because the parasitoid must avoid competition within its offspring for the rare nondissimulated hosts while at the same time cheating the host's defenses. Selection in the parasitoid genome then tends to open both the encounter and compatibility filters (fig. R2.4, bottom).

Thus, in both situations the antiencapsulation characteristic is selected for in the parasitoid, which is easily understandable, since otherwise encapsulation means almost certain death. However, the best defense for the host is selection for dissimulation characteristics.

Although, like any model, Hochberg's work involves a great many simplifications, its results nevertheless show that selecting an "encounter" characteristic may be compensated for by selecting a "compatibility" characteristic and vice versa. Thus, for the host the option of closing the encounter filter is the most advantageous whenever it is possible. However, its best interest is probably to maintain the ability to invest in either the encounter filter or the compatibility filter, since success in the same parasite-host association likely depends on local transmission conditions.[3]

Fight or Run Away?

Dispersal is another mechanism hosts use to lower the fitness consequences of parasitism: individuals can leave areas where the risk of being parasitized is high. Here the choice is between the selection of genes to kill and the selection of genes to avoid: "A host might face the following dilemma: to leave a patch where the risk of parasitism is high and eventually pay the cost of dispersal (an individual can never be sure that elsewhere it will find a safer patch) or to stay in the risky patch and resist" (Boulinier et al. 2001). Various examples of dispersal related to parasitism exist. For instance, Van Vuren (1996) has shown that ectoparasite load is higher in dispersing individuals of the yellow-bellied marmot, *Marmota flaviventris*. Along these lines hematophagous parasites in nests have been shown to have a negative impact on reproductive success and are responsible for the abandonment of colonies, as demonstrated by the studies of Boulinier and Lemel (1997) and Danchin, Boulinier, and Massot (1998) on the kittiwake *Rissa tridactyla* and its parasite the seabird tick *Ixodes uriae*. In the common lizard *Lacerta vivipara*, Sorci, Massot, and Clobert (1994) have shown that the level of mite infestation on pregnant females was related to the proportion of dispersing offspring (males tend to disperse more and females less). Similarly, in the great tit, *Parus major*, Heeb et al. (1999) found that a greater proportion of male fledglings were recruited from flea-infested broods than from uninfested ones, leading them to conclude that this bias could result from sex-related differences in mortality after leaving the nest as well as to differences in dispersal.

3. Hochberg (1998) notes that the parasitoid's always investing in antiencapsulation appears as a "twist" to the "life-dinner principle" of Dawkins and Krebs (1979). In fact this is not really surprising, because as soon as a parasitoid is in its host, they both run for their lives.

Part 3

Durable Interactions and the Biosphere

12 Parasites against Their Conspecifics

Which particular features of the dynamics of such fecund organisms are able to create the stability necessary to generate long-term prevalence patterns?
Keymer 1982

The Debate about Regulation

A population whose size varies slightly over time or that varies according to a regular cycle is said to be stable or predictable, whereas one whose size varies irregularly is said to be unstable or unpredictable (see Pimm 1991). A population is stable because of a regulatory process and not (or only very rarely) by chance alone.

Regulation consists of maintaining a population, by density-dependent mechanisms, at a smaller size than the theoretical carrying capacity of its environment. In density-dependent mechanisms the effect of a factor is correlated with population density. Because of this relationship, when a population approaches or passes a certain threshold, density-dependent factors act as constraints whose effects become more serious with increasing population size. Mathematically speaking, the relation between regulation and density may be either linear or, more frequently, nonlinear. Density-independent mechanisms such as climatic events, on the other hand, limit population growth but do not regulate it. In short, density-dependent factors cause a reduction in a population only if population size increases, whereas density-independent factors affect a population whatever its size.

Any environment capable of sheltering living beings has "finite" dimensions and resources and cannot harbor a population larger than a certain size: a pond cannot have an infinite population of frogs or a forest an infinite population of boars. Population is regulated by feedback mechanisms that involve increased mortality, decreased fecundity, increased emigration, or limited immigration.[1]

A fundamental characteristic of the structure of parasite populations is that when regulation exists it occurs independently in each infrapopulation. This is the general rule except for parasites whose free-living stages

1. Regulation is a universal process in the dynamic of life, and it can be observed to occur from gene expression on up to ecosystem functioning passing by way of the physiology of organisms.

have a long enough life span outside the host to be subjected to pressures, such as predation, that can have a regulatory effect in that realm. The trigger of regulation in a parasite population therefore normally depends not on its global population size or density but rather on that of its infrapopulation. The result is that the degree of aggregation in the population exerts a strong influence on the threshold that triggers regulation: the more aggregated the population, the earlier regulation occurs in highly parasitized hosts and the more marked the stabilizing effect will be in these hosts (Anderson and May 1978a). In fact, such regulation may happen only in a few highly infected host individuals.

The apparent complexity of parasite life cycles may leave one with the idea that parasite populations have little chance to reach the threshold for density-dependent regulation. If this were the case, parasite populations would be unstable and nonregulated because they are never saturated.

The notion that parasite populations are nonsaturated has been defended by Price (1980), who noted that for certain parasite populations prevalence and intensity are so small that intuition suggests they are far from saturated. No regulatory process then affects the population, since the limiting mechanism is density independent. Thus the population is limited upstream from the infrapopulation because of, for example, huge losses in free-living stages.

Numerous parasites (particularly helminths and other metazoans) show high fluctuations in population densities either irregularly or seasonally, and such fluctuations are by their very nature an indication that populations are not permanently saturated. Moreover, some taxonomically and ecologically related parasite-host systems may be characterized by very different parasite intensities, also suggesting that those with low intensities are not saturated. For instance, the cestode of the brill discussed earlier (see fig. 5.7) does not exploit the whole space or all of the resources available to it relative to the cestode of the turbot, which reaches much higher intensities in a comparable habitat.

However, just as there are arguments to support instability and nonregulation, there are also arguments to support stability and regulation in parasite populations. First of all, one must be cautious before concluding that a population is unstable: it is possible for the population size to fluctuate cyclically and for regulation to happen other than at the time of observation. Further, the host under study may be accidental or accessory, or the regulation of the parasite population may occur in an undetected major host. Stability over time is far from an exception in natural populations of parasites, and as Keymer (1982) emphasized, this is the best cue for regulation because it is unlikely that fluctuating exterior factors would explain such stability. Stable populations correspond to situations in which either parasite multiplication on site or a high rate of infection is such that a density-dependent safety valve is necessary: when the population size passes a certain threshold, either survival or the capacity to multiply decreases. When infrapopulation recruitment is entirely exogenous (no multiplication on site) the flux may be small enough for regulation not to occur. This is not possible when parasites multiply on site, and a regulatory mechanism limiting multiplication is unavoidable in such cases. Further, it is difficult to reconcile the following two statements with the third: (a) parasite populations are almost always aggregated; (b) aggregation favors the appearance of regulation; and (c) parasite populations are usually not regulated. If a and b are true, then c cannot be true.

Therefore it seems reasonable to accept the idea that besides those populations that are rarely regulated, numerous other parasite populations are regulated in their hosts. Poulin (1998a) notes: "As aggregation of parasites is the rule, we may expect density-dependent regulation to be common in parasite populations." In holoxenous life cycles this regulation happens necessarily in the only host, whereas in heteroxenous life cycles it may happen in any one host or in several hosts. Nevertheless, regulation in only one host during the cycle is sufficient for the overall population to be regulated (Keymer 1981).

Regulation

Modes of Regulation

Here parasite regulation will be considered to have three mechanisms of operation: *decision-dependent* regulation, where the infective stage "knows" the state of infection of a potential host and avoids it if it is already infected; *competition-dependent* regulation, where the availability of resources or any of several active elimination processes limit the population size; and *host-death-dependent* regulation, where the most heavily infected hosts die, taking their parasites with them.

Let's imagine, for example, that the hosts are dogs and the parasites hookworms. Let's also imagine that each dog can harbor a hookworm population of only a certain size and that each dog becomes contaminated a little more each day because of the transcutaneous penetration of new larvae. Figure 12.1 illustrates the theoretically possible outcomes. First, there may be no regulation after infection if the entering flux of parasites remains low enough that the worm's infrapopulation never reaches saturation (taking into account natural mortality). In this case the dog does not die but owes its survival to random exterior factors that limit

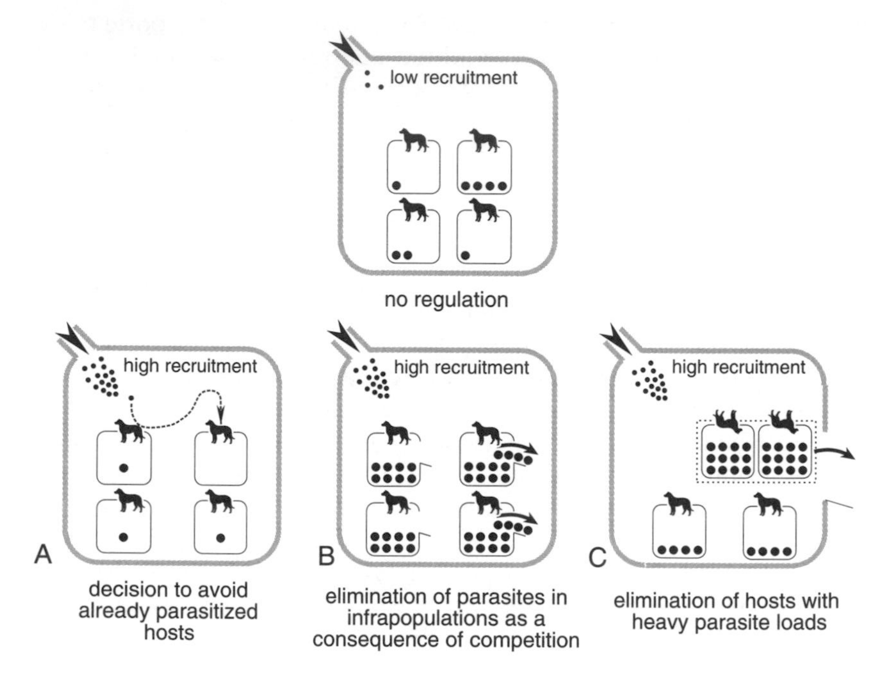

Figure 12.1. The three theoretically possible modes of regulation (A, B, and C) of parasite populations.

infection. This corresponds to a situation of instability, since no regulatory mechanism exists (fig. 12.1, top).

In terms of *decision-dependent* regulation nothing forbids one to imagine that when a hookworm larva touches the dog's tegument it detects a signal indicating that the dog's intestine is already saturated and therefore does not penetrate (fig. 12.1A).

Relative to *competition-dependent* regulation, it may also be that above a certain load of adult parasites newcomers are unable to develop because of those parasites already present or that those already present trigger their elimination. The dog may therefore continue getting infected, but its parasite load would not increase (fig. 12.1B).[2]

In the case of *host-death-dependent* regulation there is no influx of larva sufficiently low, and no regulatory process within the host sufficiently powerful, to prevent the parasite population from surpassing some

2. Parasite population turnover has been the subject of only a few studies, and the apparent stability of certain infrapopulations may hide the more or less rapid renewal of their individuals.

threshold level. The dog therefore dies (unless the veterinarian can save it), taking all its worms with it (fig. 12.1C).

Obviously the three processes are only imaginary in the case of hookworms, but we can also examine some concrete examples.

Decision-Dependent Regulation

One way to limit intraspecific competition is for the individual to "decide" whether to occupy a particular milieu depending on its current occupation by other individuals of the same or different species. Although dispersion of free-living animals (reptiles, birds, mammals, etc.) takes this constraint into account, it is exceptional for parasites to be able to make decisions about host occupation because infective stages do not "know" what inhabitants might already be there. Ancylostome larvae penetrate dogs without taking into account whether they are already infected, and our supposition is therefore inexact.

But exceptions exist! For instance, certain parasitoids "know," because of chemical signals, if the insect hosts they are going to oviposit in are already infected. In these parasitoids only one individual may develop in an individual host (solitary parasitoids), and a female getting ready to oviposit in an already parasitized host usually "changes its mind" and looks for another target.

Competition-Dependent Regulation

This form of competition may have two modes. In competition by exploitation the use of resources by some individuals decreases their availability for others, whereas in competition by interference one competitor has a direct effect on another. In general, competition by interference is considered to be selected for only if resources are particularly disputed by opponents (Schoener 1982). Here specificity influences intraspecific competition, and we may suppose that very specific parasites have a large probability of being in the same hosts whereas generalist parasites are likely to be more exposed to interspecific competition.

Competition by Exploitation

When individuals of the same species (conspecifics) are numerous, the first consequence is that some individuals are forced to occupy less favorable or more marginal situations and the niche tends to enlarge. The second consequence is that some individuals do not survive as long as others (see Miller 1967; Branch 1984).

Parasite infrapopulation growth in host individuals resembles that of a free-living population in a finite space. The growth in the number of sheep after their introduction into Tasmania and the growth of schistosome sporocysts forming an infrapopulation in a mollusk vector (Gérard, Moné, and Théron 1993) have strikingly similar curves: high growth (exponential) at the beginning followed by a slowing down and, finally, a "plateau" approximately corresponding to the carrying capacity of the environment.

The parasitology literature abounds with examples of competition by exploitation in which survival, size, or the fecundity of a given parasite in a given host species is subjected to negative density-dependent effects. Cestodes of the genus *Hymenolepis* have been the favorite model of numerous such studies on this form of competition: two months after the infection of rats with dosages ranging from one to twenty cysticercoids of *H. diminuta,* the length, weight, and egg production of each cestode is found to be inversely proportional to the dosage given (see Crompton 1991). The effect is so marked for fecundity that the highest production of eggs per infected rat is obtained with a dosage of only one cysticercoid (fig. 12.2A).

The trematode *Echinostoma togoensis,* an intestinal parasite of mice, lays on average 1,870 eggs/individual/day when by itself, 950 eggs/individual/day when there are 5 of them, 830 eggs/individual/day when there are 75, and 360 eggs/individual/day when there are 150 individuals (José Jourdane and Sim-Dozou Kulo, personal communication). Contrary to the previous example, and despite a decrease in fecundity, an abundant infrapopulation nevertheless remains more "profitable" for the species than a small one (unless, of course, there is a difference in egg quality).

Similar results are obtained for the monogenean *Polystoma integerrimum* when the data are expressed in terms of fecundity relative to biomass (fig. 12.2B; Combes 1972).

As shown by comparisons of cercarial production for both *Schistosoma mansoni* and *S. rodhaini,* each of which parasitizes mollusks of the genus *Biomphalaria,* the production of cercariae by larval trematodes in mollusks is not a function of the number of miracidia that penetrate.

For *S. mansoni:* when mollusks are infected with only one miracidium the average production is about 170 cercariae/day/mollusk. When the infective dose is two miracidia the production increases to 400 cercariae/day/mollusk, while for ten miracidia it reaches 500. Thus there is an increase in the production of cercariae with an increase in the dosage of miracidia, although this increase is not linear—a clear indication that regulation is occurring (Théron 1985a; Gérard, Moné, and Théron 1993).

For *S. rodhaini:* whatever the dose between one and ten miracidia, the

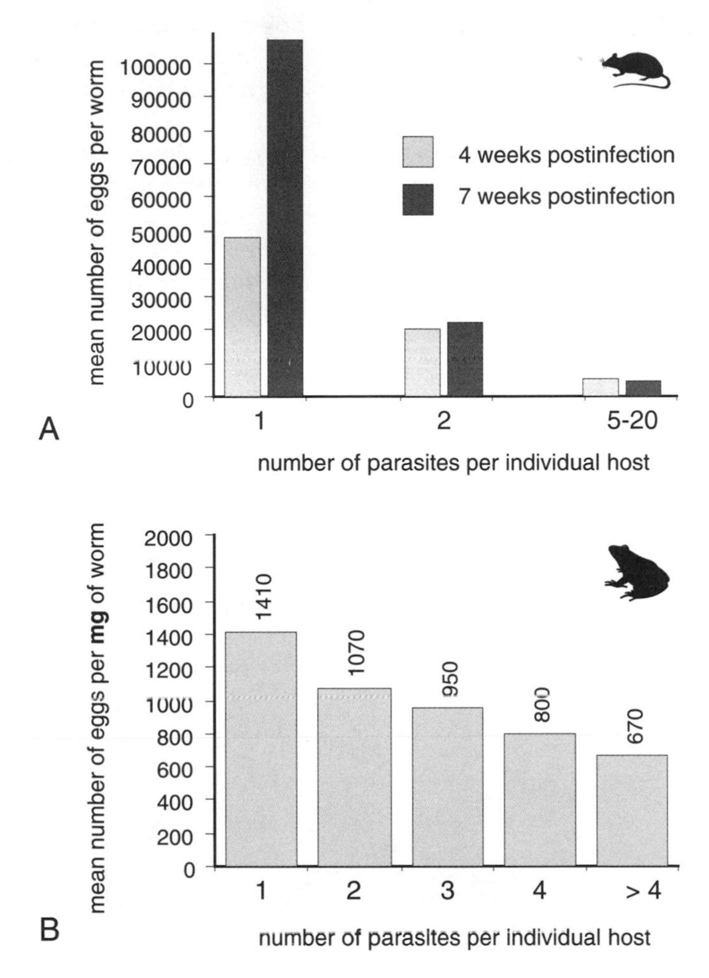

Figure 12.2. (A) Production of eggs of the cestode *Hymenolepis diminuta* according to the number of parasites per host (modified from Hesselberg and Andreassen 1975). (B) Production of eggs by the monogenean *Polystoma integerrimum* (from Combes 1972).

production of cercariae is the same (about 500 cercariae/day/mollusk), meaning that *S. rodhaini* saturates the host with as small a dose as one miracidium owing to the intense larval multiplication that occurs in this species. Thus, above one miracidium, regulation prevents any increase even if the dosage of infective larvae is greatly elevated (Touassem and Théron 1989).

When there is regulation in microparasites that multiply within the host (mobile elements, viruses, bacteria, and protists), competition also may concern exploitation. In these cases, however, immunity may also intervene to limit the population.

Competition by Interference

Competition by interference frequently plays a role in parasite population regulation. In most macroparasites infrapopulations are formed by the progressive recruitment of infective stages, and interference-dependent regulation then eliminates newcomers.

This mechanism has been well illustrated using the following example, which has the advantage of concerning natural populations. Maillard (1976) studied parasitism in a teleost, the sea bass *Dicentrarchus labrax,* by the trematode *Cainocreadium labracis.* This author noted that, whatever the season, sea bass harbored numerous young or very young *C. labracis* and relatively few adults. He therefore divided the infrapopulations into four size-classes, A, B, C, and D, and obtained the graph shown in figure 12.3. Only one explanation was possible for this repeated outcome: sea bass receive a constant and large influx of infective metacercariae with their food, and most "newcomers" are eliminated after they start growing. Elimination occurs between stages B and C, as is clearly shown by the arrow on the figure, so that the infrapopulations of adults do not exceed a certain level. Since the intestinal environment of the fish seems to pro-

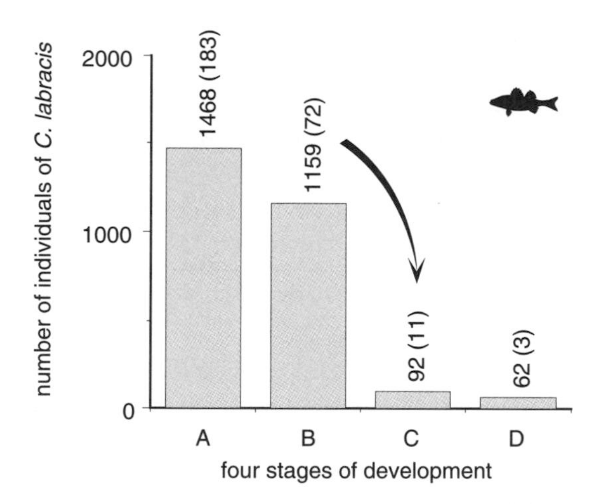

Figure 12.3. Example of regulation of one population: intensities of the marine fish trematode *Cainocreadium labracis* strongly decrease between size-classes B and C, and the number of adults (D) varies only slightly. Thus it is likely that regulation occurs between stages B and C (numbers in parentheses = mean intensities) (modified after Maillard 1976).

vide many more resources than the parasites can take, it seems logical to think that an interference-dependent regulation occurs without host mediation in this case.

Another model of intraspecific competition by interference is given by insect nematodes, most notably those of roaches. Several species of nematodes live in the digestive tract of roaches, and their life cycles are simple and direct (the insects ingest eggs containing an infective larva). Studies have not demonstrated a marked interspecific competition between the species but have, on the contrary, noted an extraordinarily efficient intraspecific competition. Most often the infrapopulation of a given species contains only a few individuals whatever the dose of ingested larvae given: one male and one female for the nematodes *Thelastoma bulhoesi* and *Leydinema appendiculatum* in the American cockroach *Periplaneta americana* (Noble 1991). Not many more occur in other cases: in *Blatella germanica,* for example, there are up to three *Blatticola blattae,* but always a maximum of only one male under both experimental and natural conditions (Morand and Rivault 1992). However, that several species of nematodes share the same niche (or at least apparently the same niche) is troubling here, since it means that the biotic capacity offered by the host's intestine is not saturated by only one species of parasite. As Adamson and Noble (1992) noted, "Each species limits its own numbers such that the carrying capacity is not reached; this in turn allows the colonization of the same niche by other species." Why, then, would a particular species limit its own numbers? The mechanism probably involves competition by exploitation between females (which are large) and interference between males (which are small), and the first adult male may release sex-specific toxins that would kill its conspecifics so there would always be at least one female and one male in the infrapopulation (Zervos 1988a, 1988b).

In parasitoids competition by interference is often ferocious, and it gives an important role to priority, with either the first or the second attacker winning, depending on the species. For example, the wasp *Venturia canescens* parasitizes the phycitid moth *Ephestia kuehniella* and is a solitary parasitoid that can decide not to oviposit in an already infected host (decision-dependent regulation). However, it can also accept "superparasitism" and oviposit in an already parasitized host. When superparasitism occurs, two larvae end up in the same host individual and a serious competition occurs, with one killing the other. It is not always easy, however, to know if the insect that emerges from its host's body corresponds to the first or second larva deposited. Sirot (1996) used two strains of *V. canescens* ("Valbonne" from France and "Oxford" from England) that could be distinguished by a genetic marker for pigmentation. This author showed

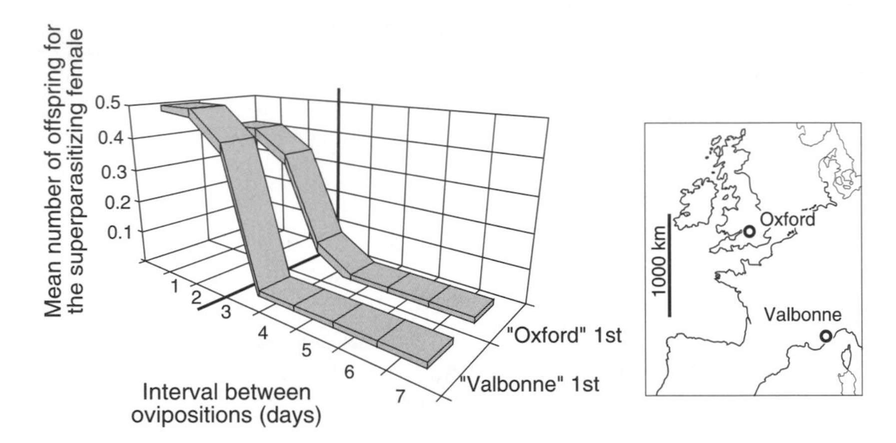

Figure 12.4. Superparasitism in the parasitoid hymenopteran *Venturia canescens*. Whatever the oviposition order, overparasitism "pays off" only if the time between ovipositions is less than three days (from the data of Sirot 1996).

that if the time between the two ovipositions is less than three days, the probability of winning is about the same for the first or second comer (with a small advantage to the Valbonne strain over the Oxford strain); but if the time between the two ovipositions is more than three days, there is a priority effect and the first occupier always wins (fig. 12.4).

Numerous parasitoid species have selected adaptive mechanisms to prevent superparasitism, and acceptance or rejection of superparasitism is influenced by both host and parasite population densities. In certain cases (for example, in species with high fecundity and short life spans) it may be advantageous to "accept" superparasitism and "fight to the death" with conspecifics (Sirot, Ploye, and Bernstein 1997).

Host-Mediated Competition

It is important to note that the host also plays a role in parasite infrapopulation regulation, since defense mechanisms are usually proportionate to the parasite load: a dog will scratch more often if it has more fleas. Host defense mechanisms may prevent sudden overloads if one recruiting event brings in too many infective stages for the capacity of an individual host, or it may prevent progressive overloading if recruitment events are too frequent or parasite multiplication is too rapid.

Sudden overloads lead to the elimination of part of the candidates for settlement in a host. For instance, when cows are infected with fewer than 1,300 metacercariae of *Fasciola hepatica,* there is proportionality between the dosage and the number of adults recovered. However, as Dawes and

Hughes (1970) wrote: "When the level of infection was increased to 2500 cysts, the number of flukes reaching the bile ducts was drastically reduced and at higher levels (5000 and 10,000 cysts) many immature flukes were trapped in the liver parenchyma." In this case the "trapping" of the young flukes in the parenchyma, just before they settle in the bile ducts (the flukes cross the liver to reach the ducts) is due to a host defense reaction that is proportional to the aggression.

Progressive overloads often lead to more and more efficient parasite elimination over the course of successive infections. When two experimental infections are carried out on the same host individual with a certain time lag, the second infection is often not as successful as the first.[3] In schistosomes, for example, this effect is observed in both the mollusk and the vertebrate. For instance, Sire, Rognon, and Théron (1998) have shown that if snails *(Biomphalaria glabrata),* are exposed to *Schistosoma mansoni* miracidia, reinfection is not possible fourteen days after exposure even if miracidial dosages are increased. In this case miracidia in the challenge infection that penetrate the mollusk stop developing after five to six days and then die. Although the authors invoke the possibility of a direct antagonism between parasites, the most likely interpretation is that the mollusk's internal defense system becomes operational about the second week after penetration of the first miracidium and successfully opposes the development of further miracidia while sporocysts of the first infection remain protected (fig. 12.5A).

In vertebrates a very similar process has been known for some time with schistosomes (Dean et al. 1978; Smithers and Gammage 1980): if laboratory mice or black rats are exposed to two batches of *Schistosoma mansoni* cercariae with a three-week interval between infections, the first exposure results in the development of many more adults than does the second (often the proportion of adults to cercariae is about 30–40% for the first infection and 5–10% for the second; fig. 12.5B). The explanation is that the surviving schistosomes from the first batch induce the appearance of antibodies active against young parasites of the second wave but not against established adults, so newcomers are the victims of immune defenses triggered by on-site parasites (Smithers and Terry 1969). Terry (1994) writes, "Schistosomes derived from early infections will, in conjunction with host immunity, restrict the levels of reinfection which might otherwise lead to the death of both hosts and parasites." As noted in chapter 10, however, Sorensen (1999) could identify with certainty *Schistosoma japonicum* coming from two successive infections and showed that worms

3. This is the principle behind vaccination in vertebrates.

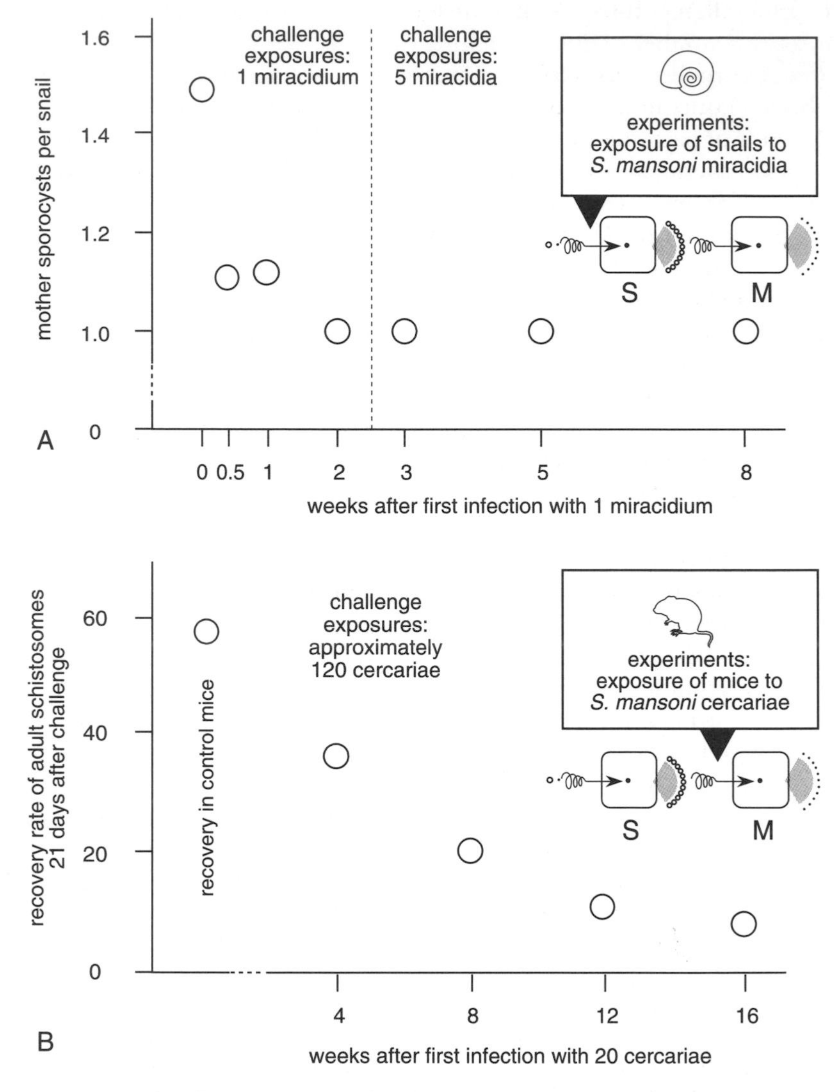

Figure 12.5. (A) Mollusks *(Biomphalaria glabrata)* susceptible to *S. mansoni* cannot be re-infected more than two weeks after the first infection with only one miracidium. Even if the snails are exposed individually to large doses of miracidia, only one sporocyst (that of the first infection) develops ($n = 20$ for each group) (after Sire, Rognon, and Théron 1998). (B) Mice that are susceptible to *S. mansoni* and infected once with cercariae of *S. mansoni* become more and more resistant to new infections in the weeks following the first infection (after Smithers and Gammage 1980). S = snails; M = mice. Confidence limits are omitted.

of the challenge infection may also replace worms of the primary infection.

In the mollusk as well as in the vertebrate this type of immunity is termed "concomitant immunity" because it occurs in parallel with the presence of parasites that have previously settled. Because it is impossible to apply the same experimental protocols to both miracidia and cercariae, the curves displayed previously in figures 12.5A and 12.5B thus result from different procedures. Nevertheless it is striking that concomitant immunity expresses itself similarly in both the intermediate and vertebrate hosts in this system.[4]

Regulation by a density-dependent progressive host response is also well illustrated by the experiments of Lysne, Hemmingsen, and Skorping (1997), who exposed cod to cercariae of the trematode *Cryptocotyle lingua* in cages in the open sea. After a certain period of exposure to cercariae, the number of metacercariae per fish reaches a plateau, and there is a decrease in the variance-to-mean ratio owing to a decrease in the variance (which means there is a progressive saturation of the resource host at slightly different levels depending on the host individuals). Observations show that the metacercarial infrapopulations of the trematode are regulated not by parasite density-dependent competition and not by the mortality of highly infected hosts, but rather by a host-mediated density-dependent response. Here the success rate of new infections tends toward zero in most parasitized cods while they continue to establish themselves in the rest of the population.

An original case of host-mediated interference is observed in some nematodes where "excess" worms are not eliminated but instead take a "wait and see" approach that wastes fewer individuals while conserving a regulatory effect. In short, numerous species of nematodes are capable of facultative diapause, which consists of developmental arrest in particular

---

4. The role of immunity in reinfection of vertebrates by cercariae has been questioned, and the outcome may be the result of an entirely different process. In brief, the settling of schistosomes induces a pathogenic condition known as leaky liver that consists of blood vessel enlargement in the liver because of, at least in part, an increase in portal vein blood pressure. The consequence of this effect is to prevent the necessary stopping of new schistosomules in the liver. This mechanism was demonstrated for *Schistosoma mansoni* in both laboratory mice (Wilson 1990) and black rats, the natural hosts of this parasite (Establet and Combes 1992). Moreover, studies of a particular strain of mice with a genetic abnormality also support this explanation: these mice have a congenital leaky liver independent of infection. First exposure of these mice to *S. mansoni* cercariae has shown that they are resistant to the parasites without otherwise having acquired any special immune capacity (Coulson and Wilson 1989). This implies that the dilation of hepatic vessels as the result of initial infection likely causes failure of subsequent infections in normal mice, and the same may be true in humans.

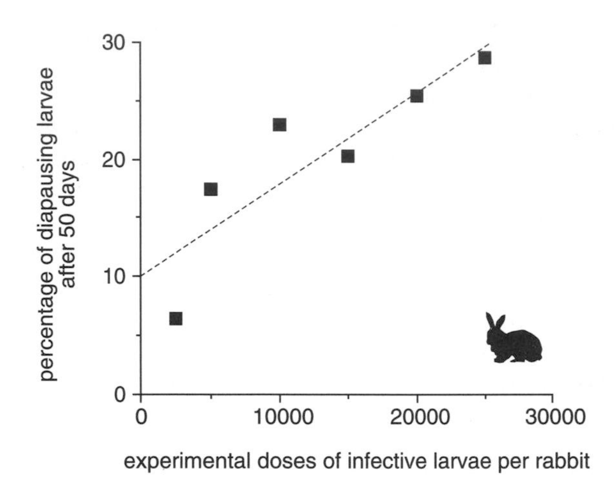

Figure 12.6. Nematode diapause: the number of larvae that enter diapause increases as competition increases (at higher "doses" of infection) (after the data of Russell, Baker, and Raizes 1996).

circumstances and may affect all or part of the stock of infective larvae. In some cases this diapause interrupts the life cycle at a time of the year when, for climatic reasons, free-living larval development in the outside environment would be difficult or impossible. In other cases, however, diapause is clearly correlated with a large parasite load in the host. It is thus a density-dependent process that limits both intraspecific competition and the pathogenic effect, since nematode larvae induce less damage than do adults. Such developmental arrest is known to occur in numerous nematodes, for example *Ostertagia ostertagi* and *Cooperia curticei* in sheep and *Obeliscoides cuniculi* in rabbits. Figure 12.6, constructed from the data of Russell, Baker, and Raizes (1966), clearly shows that, in the same experimental infection, the proportion of larvae that are in diapause fifty days after infection increases with the dosage of infective larvae used.

Nematode diapause may be viewed as an *extension of the host phenotype into the parasite.* By arresting young stages waiting for free space and thus avoiding disadvantageous competition, the parasites use to their advantage mechanisms that have been selected to destroy them. This is one of the greatest successes in parasitism, or even in the living world in general, and it reminds one of what occurs with marsupial embryos. Marsupials are not infertile when they nurse a baby, and fertilization may occur while one youngster is completing its development in the marsupial pouch. The new embryo, a brother or sister of the preceding one, interrupts its own development and is able to remain in embryonic diapause for up to two hundred days. Nematode diapause also reminds one of certain insects whose larvae also show diapausal polymorphism, with some developing into adults after a year and others after two or three, thus providing insurance against bad climatic years.

It is, of course, possible to see in nematode diapause only benefit to the host and to consider it only as a means of defense. A different and more likely perspective, however, is to see, as Schad (1977) did, an adaptation of the parasite to survive difficult periods, which could be either external climatic factors or internal overpopulation. Thus the effect of diapause is to soften otherwise notably violent fluctuations owing to seasonal peaks of transmission, and the ability to arrest development leads Hawdon and Schad (1991) to consider it a preadaptation to parasitism in nematodes.

Host-Death-Dependent Regulation

If none of the previous processes of regulation occurs, and if recruitment goes on, the pathogenic effect of the infrapopulation naturally increases. In this case the host dies, and all its parasites die with it. When parasite distribution is aggregated, the hosts carrying the most parasites are the ones most likely to disappear. That is, it is not extra parasite individuals that disappear but rather the hosts that carry the denser infrapopulations.

This does not mean, however, that in a given host population there is a threshold or "fatal number" of parasites above which all host individuals will die, since genetic diversity and the state (age, nutrition, etc.) of each host individual obviously also play a role. One individual may die with a small parasite load while another might survive with a far heavier burden. Further, parasite distribution within the host population is not static because opposing processes have a tendency to accentuate aggregation in some hosts and reduce it in others (Anderson and Gordon 1982). Any such heterogeneity tends to create overdispersion, while any density-dependent process tends to create underdispersion.

Host death is not necessarily "natural," and most living beings die as the victims of others. In predator-prey systems in particular, small differences in the ability to flee let some individuals be devoured rather than others. Often this "small difference" is nothing other than a difference in parasite load.

For instance, Hudson, Dobson, and Newborn (1992) showed that red grouse in Scotland that were killed by predators were heavily parasitized by the cecal nematode *Trichostrongylus tenuis*. Infected grouse may be more susceptible to predators because parasitized animals are weakened and so are captured more easily, because they spend more time looking for food, or because they have a different social behavior. Further, parasitized birds, particularly incubating females, are more easily detected by their odor. When healthy females incubate, dogs can detect them only within a very short distance (half a meter) because incubating females do not eat and thus do not leave feces to indicate their presence. By provoking lesions in the cecal mucosa, however, the nematodes make parasitized females easier to smell and detect, so they are more often eaten.

Finally, it cannot be forgotten that the pathogenic effect is rarely due to just one parasite species. Thus there may be elimination of hosts carrying the most abundant infracommunities (not just infrapopulations). The summation of the pathogenic effects of several species of parasites likely becomes more marked as the processes that result in aggregation "coincide." This means that a cause provoking the aggregation of parasite X may be the same as that affecting parasites Y, Z, and so on, each with similar modes of transmission or each eliciting host defense mechanisms that help this to happen. For example, an investigation done on ducks in Lithuania (Kontrimavicius 1987) showed that nine ducks out of thirty-three (27%) carried three species of acanthocephalans. These ducks harbored 77%, 82%, and 90%, respectively, of the total number of each acanthocephalan species in the sample. Here aggregation was comparable for the three species and occurred in the same host individuals, most likely because of similarities in transmission.

Two reservations must be expressed about host-death dependence. First, some aggregated distributions cannot contribute to parasite regulation but, on the contrary, contribute to parasite success. Aggregated distributions favor transmission only if host death is programmed into the life cycle. For instance, the consequences of the aggregation of infective stages in an intermediate host that must be eaten (which is the case in numerous parasitic transmissions) are exactly the opposite of a regulatory effect: an overload of parasites in the intermediate host increases the chances for successful massive transmission. Thus it is only in definitive hosts and in nonconsumable vectors that aggregation may be a regulatory process. In other cases "host death" may act as an amplifier. For example, Gordon and Rau (1982) have demonstrated the "disappearance" of sticklebacks with metacercarial loads of the bird trematode *Apatemon gracilis* that were above a few dozen per fish. It is likely that most fish that died had been caught by the bird definitive hosts, so the parasites had attained their goal.

The same amplification effect of aggregated distributions happens when dispersion of the propagules is concurrent with the death of the host. This is the case of the trematode *Aphalloides coelomicola,* which becomes an adult in the coelom of the fish *Potamoschistus microps* (a goby). Trapped in the body cavity, the parasites cannot evacuate their eggs

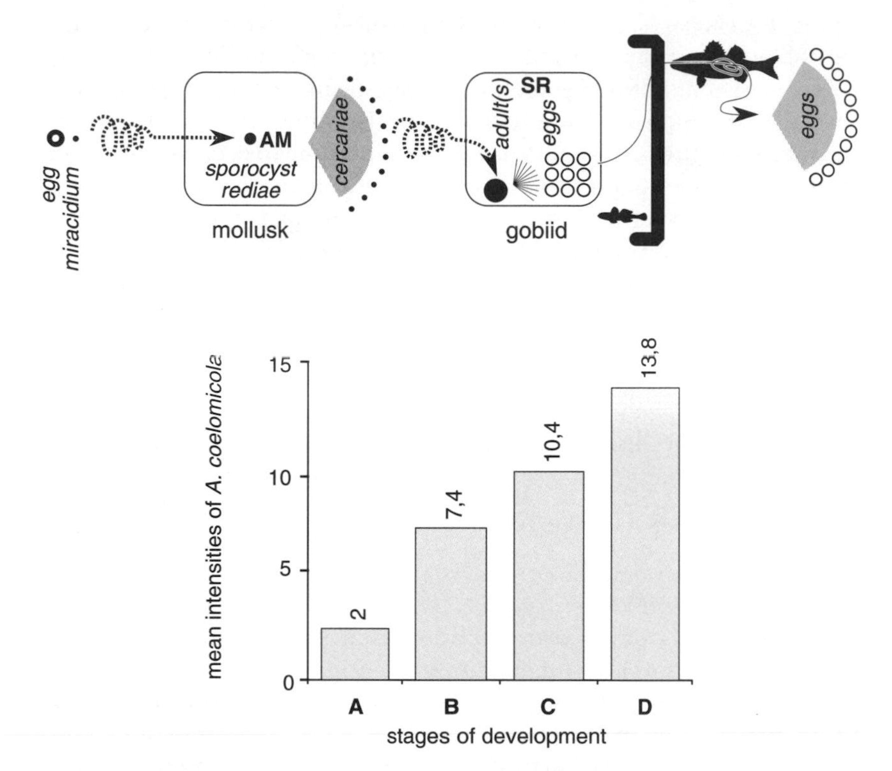

Figure 12.7. The "nonregulation" of adult trematodes in the life cycle of *Aphalloides coelomicola.* Top: biological life cycle. Bottom: progressive accumulation of metacercariae, as testified to by the increasing number of the trematode size-classes (A, B, C, and D) (compare with figure 12.3) (after the data of Maillard 1976).

through the intestine as trematodes usually do, and they are freed only when the fish dies. Thus death is necessary for transmission. Maillard (1976) has shown that adults accumulate without any density-dependent effect, and dispersion occurs only when a predator eats the goby (fig. 12.7).

The second reservation concerning host-death dependence is that the link between the degree of parasite aggregation and the efficacy of the regulatory effect can be questioned. A mathematical model (assuming parasite-induced host mortality to be the only cause of decreased fitness for both host and parasite) allowed Jaenike (1996) to visualize two opposite situations.

In one, if host fitness is deeply affected by small parasite intensities but less and less so when the number of parasites increases, then the average parasite fitness decreases when the degree of aggregation increases, whatever the mean parasite load.

In the other, if host fitness is little affected by small parasite intensities but decreased fitness becomes more and more marked when parasite load increases, then for small parasite loads the mean parasite fitness decreases when aggregation increases and there is the possibility of host-death-dependent regulation (for heavy parasite loads mean parasite fitness is larger at intermediate levels of aggregation).

We see that in the second situation if mean parasite loads are high an increase in the degree of aggregation may increase the mean fitness of the parasites and not have the expected regulatory effect. Jaenike concluded: "Aggregation can increase the impact of parasites on their host population, particularly when parasites are at low densities, and increase mean fitness of parasites when parasite densities are high."

The Distinction between the Types of Regulation

Distinguishing between the types of regulation occurring in parasite populations may be difficult.

One difficulty is that it is rarely easy to distinguish competition without host mediation from host-mediated competition. Most likely both play a role in numerous regulations, but it is also likely that the role of the host is underestimated in many associations. For example, the regulation of infrapopulations of larval trematodes (sporocysts or rediae) may arise from exploitation-dependent competition, or it may arise from an elevated molluskan defense system that is proportional to parasite density.

The same dilemma occurs for almost all parasitoses. However, "immunity" as an explanation is not always logical. It is commonly suggested that aggregated distributions (parasite densities higher than randomly expected in certain host individuals) are due to variable qualities in the host's immune mechanisms. In this case it is illogical to suggest that high parasite densities are linked to weak immune defenses, that regulation is linked to high parasite densities, and that regulation is linked to strong immunity. These three propositions are not compatible, and it seems more coherent to believe that the consequences of weakened defenses are high parasite densities and that these in turn trigger a higher level of competition for resources among the parasites themselves.

A second difficulty resides in distinguishing between competition-dependent regulation and control of the population by processes that are not density dependent. The following study illustrates this difficulty well (Granath and Esch 1983).

Bothriocephalus acheilognathi is an intestinal parasite of the mosquito fish, *Gambusia affinis.* In North Carolina, infrapopulation density regularly varies with the season, density being maximal in winter and minimal in the summer. A first possible interpretation is that the population drop in the hot season either is due to temperature, because mortality of the cestodes increases, or occurs because immune defenses become more efficient when temperature increases. In this case control of the population would be density independent (even when immune defenses are involved they intervene in response not to an increase in parasite density but rather to an increase in ambient temperature). Thus this would not be regulation as normally defined.

However, a second possible interpretation is that high winter densities are due to the presence of immature cestodes; it is in the spring, when the temperature starts rising, that the parasites mature and experience intense growth (from 0.1 to 4 centimeters). This growth may generate intraspecific competition for space or resources, eliminating some of the worms and thus reducing the density observed in the summer. In this case there is density-dependent regulation expressed in biomass but not in individuals, since the number of worms does not change.

It is likely that the second interpretation is the correct one, although the two interpretations demonstrate that a seasonal correlation between parasite density and an external environmental factor (here temperature) is not sufficient to confirm that density fluctuations are directly determined by the exterior factor.

The Benefits of Intraspecific Competition

Individuals from the same parasite species are not identical, and except for infection by the single clone of some protists there is genetic variability between individuals found in the same host individual.[5]

Neither decision-dependent regulation nor host-death-dependent regulation allows the direct confrontation of different genotypes, and it is only when regulation is competition dependent that conspecific but genetically distinct parasites confront natural selection. If competition is serious the fittest individuals survive, statistically speaking.

Cohabitation between different genotypes, however, may take unexpected forms in which competition can lead to an increase in transmission. Such a process thus has consequences opposite those of regulation, as illustrated here by the experimental works of Taylor, Walliker, and Read (1997a, 1997b) and Taylor and Read (1998).

Taylor, Walliker, and Read (1997a, 1997b) studied rodents with mixed

5. In fact, even within a clone somatic mutations may produce genetic diversity.

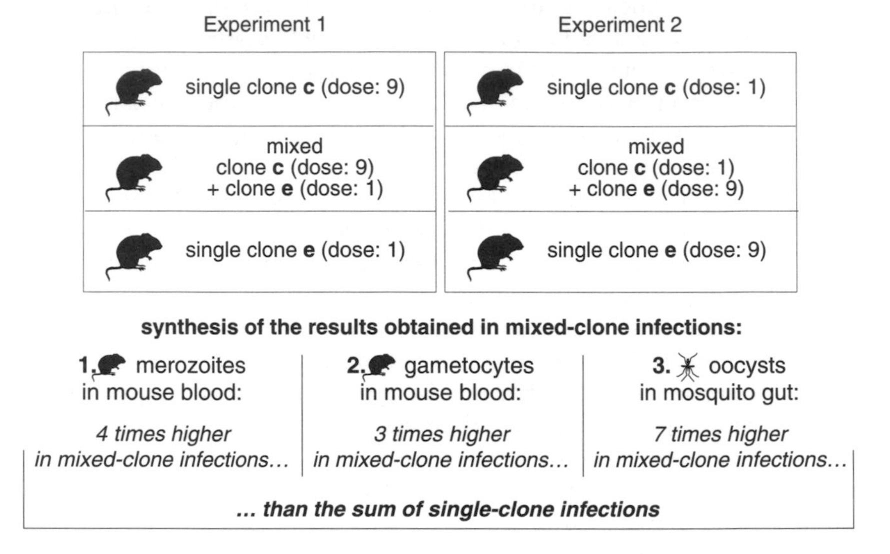

Figure 12.8. Experiments showing the issue of competitive infections between two clones, here called c and e: mixed infections lead to higher transmission rates than single ones. Note that the asexual parasites in the mixed infections are four times the sum of the controls only at the very end of the infection (during the main part of the infection, mixed infections can have parasitemias the same as or lower than those of the control group) (inspired from Taylor, Walliker, and Read 1997a, 1997b).

infections of *Plasmodium chabaudi,* the causative agent of malaria, in rodents. In a typical infection with *P. chabaudi,* merozoites (an asexual form of the parasite) divide every twenty-four hours and infect up to 40% of the red blood cells within about eight days. The authors analyzed the success of two clones of *P. chabaudi* introduced into mice as a mixed infection in 1:9 and 9:1 ratios (the clones were distinguishable by immunological and genetic techniques). Mosquitoes were fed on mice at fourteen days postinfection, the number of oocysts were recorded, and the mixed-clone infections were compared with equivalent single-clone infections (fig. 12.8). These experiments gave two important results.

First, an analysis of the dynamics of infection in mice and of transmission to mosquitoes showed that the initially rare clone was heavily suppressed at the beginning of mixed-clone infections but finally ended up contributing most of the parasites transmitted: a clone "can produce as many or more oocysts when it is the initially rare clone in a mixed infection as when it is infecting a host on its own" (Taylor, Walliker, and Read 1997b). An initial explanation would be that the originally rare clone, being dominated in the asexual phase of its development, had diverted its resources to producing many sexual forms (gametocytes). Another explanation would be that the gametocyte densities are simply proportional to the merozoite densities and that a change in the clonal composition of the asexual infection occurred before gametocyte production. Taylor and Read (1998) favor the second explanation even though it was difficult to assess and believe that the particular dynamics of the mixed clones could be the result of interactions with the immune system, with the initially rare clone eliciting a less efficient immune reaction than the dominant one. In other words, the presence of the dominant clone would tend to open the compatibility filter for the rare clone.

The second result was that in all experiments mixed-clone infections were more successful at transmission than single-clone infections, with some giving rise on average to seven times as many oocysts in mosquitoes as the sum of equivalent single-clone infections (fig. 12.8). This was due to greater gametocyte densities on the day the mosquito fed and not to increased gametocyte infectivity, indicating that transmission efficiency was increased by the competition between the two genotypes of the same parasite species in the same host individual. Taylor, Walliker, and Read (1997b) conclude: "It may be that malaria parasites have evolved to maximize transmission from mixed-genotype infections." That mixed infections lead to a higher transmission than the sum of equivalent nonmixed ones means that intraspecific competition favors parasite transmission in this association—a result that may have important implications in our understanding of transmission dynamics and genetics in human malaria.

13 Parasites against Competing Species

The study of factors determining the composition and structure of natural communities has always been at the core of ecology.
Poulin 1995a

The Debate about Interactivity

The Isolation/Interaction Distinction

The forces that are classically considered to structure free-living animal communities are predation and interspecific competition.

In parasite communities, predation apparently plays little or no role, and it is competition between species, when it exists, that appears to represent the essential structuring force.[1]

As in intraspecific competition, interspecific competition may occur either via exploitation if it results from resource limitation or by interference if it results from an active process.

Also, it is at the infrapopulation scale that intraspecific influences are exerted, but interspecific influences operate at the infracommunity scale. However, although the infracommunity is well defined in space, it is not as well defined in time simply because fluctuations at this level are only rarely known with precision. As Simberloff and Moore (1997) write, "Parasite infracommunities are temporally fuzzy entities." Moreover, they are most often described as they are observed at the time the host is being examined and are, like Schrödinger's cat, destroyed or altered when viewed.

Depending on the presence or absence of interspecific competition, both "isolationist" and "interactive" infracommunities can be distinguished (Holmes and Price 1986; also see Bush et al. 1997). An *isolationist infracommunity* is a simple assemblage in which infrapopulations are independent of each other. In these communities the parasite species ignore each other as much as if they lived in different host individuals and

1. The concept of structure covers species number, the list of species present, the quantitative hierarchy of populations, space and resource distribution among species, and the mode of energy circulation. As noted by Janovy, Clopton, and Percival (1992), "The presence of a parasite is an evolutionary phenomenon, whereas the population structure is of ecological origin." Kuris and Lafferty (1994) note that "the more a community differs from a random association of species, the more 'structured' it is." The "descriptors" necessary for the study of parasite communities are discussed in detail by Janovy, Clopton, and Percival (1992) and Simberloff and Moore (1997).

exert no pressure on one another. Further, there is no saturation of the host, which means there are vacant niches and new species may enter the host without perturbing the system. Isolationist infracommunities, much like nonregulated infrapopulations, are not in equilibrium and are therefore not predictable.

An *interactive infracommunity,* on the other hand, is characterized by stabilizing forces that are due to constraints the species exert on each other because the available environment is saturated and niches overlap. In these communities parasite species exert selective pressures on each other, which then induces the selection of traits that tend to limit competition by separating niches in one way or another. In this case there are no more vacant niches, and new species can be added only at the price of the local extinction of another. Further, it is generally thought that such competition is more active if the parasite species involved are related (Goater, Esch, and Bush 1987). Interactive infracommunities, much like infrapopulations that show density-dependent regulation, are in equilibrium and are predictable entities—their structure (number, the identity and abundance of parasites, the way space and resources are shared, etc.) is predictable because stabilizing interspecific forces tend to reproduce the same communities in the ensemble of host individuals within a given host population.

When a population of species A is in competition with a population of species B, each population may temporarily restrict its niche (the opposite of what happens in intraspecific competition) because each species remains competitive in the most favorable or preferred part of its own niche (called the "preferendum"). This allows one to distinguish what is referred to as the *fundamental niche,* which characterizes a species in the absence of other species, from the *realized niche,* which characterizes a given population subjected to the constraints exerted by other species. Recognizing and comparing fundamental and realized niches allows one to determine, in the case of parasitism, whether an infracommunity is structured by competition. Thus, demonstrating interaction between parasites is extremely important: if any type of competition is present and if different species of parasites are frequently found in the same hosts, an interparasite durable interaction is added to the parasite-host durable interaction. In this case we may expect reciprocal selective pressures between parasite species and *parasite-parasite coevolution.*

The Principle of Niche Displacement

In an interactive community the niches of parasite species may be modified in two main ways, with the outcome being either interaction, meaning that the species displace their microhabitats in the host, or exclusion, meaning that one species is eliminated at the expense of the other.[2] Interaction and exclusion, in this sense, are nothing other than the application of the concept of "character displacement" in classical ecology (see Brown and Wilson 1956), and four conditions must be met for the displacement of characteristics to occur.

First, the niches must be occupied. This occupation may result from the use of the same host or microhabitat or, more seriously, from being a member of the same guild. Note that different parasite species in the same hosts may be in competition even without sharing microhabitats or belonging to the same guild, since the first scale on the hierarchy of resources for a parasite is the host individual taken as a whole.

Second, there must be genetic diversity in the parasites relative to microhabitat preference.

Third, any change must be advantageous, since niche modification may be costly. For example, if two parasite species are in competition in the duodenum, it is clear that competition would be avoided if one species could move its habitat to the jejunum. However, if the quality of the food available in the jejunum is clearly inferior to that in the duodenum, then there are few chances for the individuals that settle in the jejunum to increase their reproductive success, and such a change would be possible only if competing with the other species was too costly.

Holmes (1987) has insisted that one final condition is also essential for selective pressures to appear: the species must encounter each other frequently and in "substantial" density. Just as intraspecific competition requires a sufficient number of actors, interspecific competition has more chance to induce the displacement of characters if the infracommunities are rich in species and the infrapopulations are dense.

The importance of "frequent encounter" led Bush and Holmes (1986) to apply to parasite communities the notion of *core* and *satellite* species, originally developed for free-living organisms by Caswell (1978) and Hanski (1982). Based on studies of species frequencies in fragmented habitats, these authors noted that the component species can be separated into two groups: those that are frequently found in patches, are generally in abundance, and are the predictable part of the community constitute the core species of a community; and those that are rarely found in

2. Although there is no inconvenience in using the terms "interaction" and "exclusion" in everyday language, we need to be aware that "interaction" designates, in a restrictive way, the separation of niches at the microhabitat scale whereas "exclusion" designates the separation of niches at the scale of the host species. In reality there are, in one case or another and at the same time, both interaction and exclusion.

patches, are generally not in abundance, and are the nonpredictable or only slightly predictable part of the community constitute the satellite species.

The notion of core and satellite species is applicable to parasite infracommunities ("patches"), and it is between the core species, and only there, that it is reasonable to search for interactive processes and to identify the evolutionary consequences of the interaction. The interest of the core/satellite distinction is that infracommunities may be studied mainly by analyzing the core species, which are the most susceptible to interacting and contributing to community structure and predictability. Satellite species, on the other hand, interact very little and are not predictable (or are only slightly predictable); therefore they play little or no role in structuring the infracommunity. Core species are often specific for the host species considered, but cases may also occur in which generalist species behave like core species. Satellite species are often not specific or are spillovers that are core species in another host. The application of the concept of core and satellite species to parasitism as such is not, however, without its critics. For example, Poulin (1998a) notes that parasites come in a variety of sizes and that infrapopulation data based only on counts may be of little value: "Clearly, a species that is a satellite in terms of numerical abundance can be a core species in terms of biomass." Moreover, it may not always be possible to distinguish core species from satellite species. An investigation by Bucknell et al. (1996) on 150 horses in Australia, for example, showed that there is a continuous distribution between small and large prevalences in a community of thirty-one nematode species.

Modes of Niche Displacement

There exist "current" (observable) and "past" (unobservable) forms of niche separation, each concerning either the microhabitat or the host species. From this it is easy to see that there are four total forms of niche partitioning if we stick only to the "spatial" aspects of the niche.[3]

1. Current interaction (current displacement in the microhabitat). Here both parasite species partition their microhabitat when they effec-

3. These are the "principal" or "main" dimensions of the niche. However, other modes of separation or partitioning exist, such as exploitation of hosts of different ages and specialization for specific resources ("guild separation"). There may even be more subtle processes of partitioning. For example, the number of host individuals where both species cohabitate decreases if distribution becomes more aggregated, which is effectively the same as setting aggregation to the level of limiting competition (Dobson 1985).

tively confront each other in the same host individual, but only in this case (Holmes called this process interactive segregation).

2. Past interaction (past displacement in the microhabitat). In this case selective pressures have modified the preferences of parasite species until the microhabitats became separated and the species currently occupy distinct microhabitats even when they are not at present confronting each other (Holmes called this process selective segregation).

3. Current exclusion (current displacement in the host species). When the two species are in competition in the same host individual, one causes the other to disappear. That is, the two species cannot be found simultaneously in the same host individual.

4. Past exclusion (past displacement in the host species). As the result of competition in the past between the two species, one has been forced to no longer exploit a specific host species: the two species cannot be found simultaneously in the same host species.

Figure 13.1 outlines the separation of the niches at the level of both the microhabitat and the host for two species of parasites, P1 (small circles) and P2 (crosses), that initially colonized the digestive tract of the same host species. The figure simplifies the concept of the niche by representing only its spatial dimension.[4] On the top portion (A) of the figure both species prefer different regions of the digestive tract so that they are not in contact and there is no competition for space. In part B, both species have the same "preferendum" and are in competition. If the two species enter into competition, this imaginary parasite-host association may evolve as follows.

In B1 the interaction leads to a displacement of the microhabitat only when infrapopulations of the two are effectively present in the same host individual.

In B2 there is no current interaction, but there have been a series of displacements of one or both species as the result of pressures the two species exerted on each other in the past.

In B3 the interaction occurs in the form of exclusion, with one of the two preventing the other from settling in the occupied host individual.

In B4 there is no current exploitation of the same host species by the two parasite species; the host is abandoned by one parasite because of selective pressures that the two species exerted on each other in the past.[5]

---

4. In most studies of infracommunities the sharing of spatial resources is the only dimension considered; studies of the sharing of energetic resources are not as advanced.

5. It is always difficult to confirm that there has been past exclusion (The "Ghost of Competition Past"; Connell 1980). That is, it is difficult to determine if a host species is no longer in the host spectrum of a parasite because it has already been excluded by a more

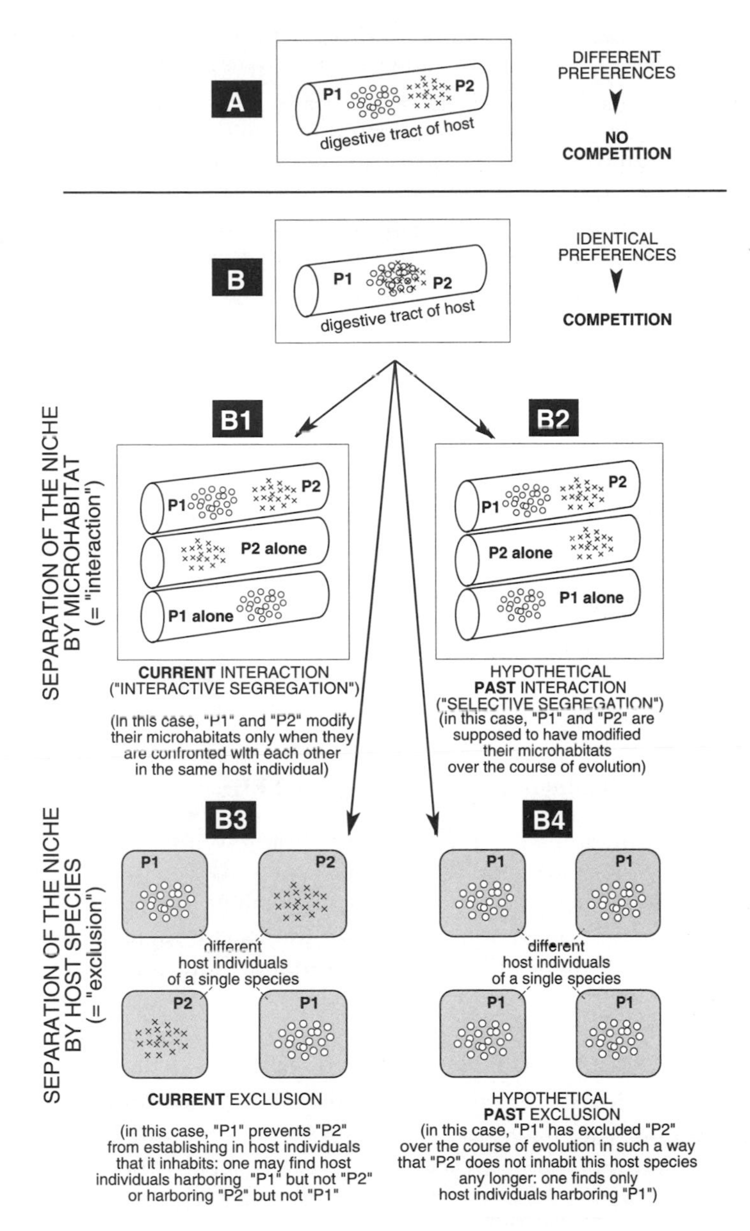

Figure 13.1. Different possible evolutions of the niche of two parasite species occupying the same organ (in this diagram the digestive tract is sketched as a section of a cylinder). See text for explanation.

It is possible that competition by exploitation leads mainly to microhabitat displacement and that competition by interference leads mainly to exclusion.[6] As previously indicated, other dimensions of the niche may also limit competition, and recognizing "exploited resources" as a dimension is important because two closely related species may continue to exploit the same host and even the same organ if the resources utilized are distinct. For example, Butterworth and Holmes (1984) have studied two trematodes, *Pharyngostomoides procyonis* and *P. adenocephala,* both parasites of the raccoon, *Procyon lotor.* When both species are in sympatry and thus effectively confront each other, they are of different sizes and consequently do not eat the same food particles. *Pharyngostomoides adenocephala* reaches a larger size in the southern United States, although, curiously, in Canada it is *P. procyonis* that is larger. When both species are geographically isolated and thus do not confront each other in *P. lotor,* this character displacement is not observed.

Note that competitive exclusion is one component of the compatibility filter, since it is not only host genes but also parasite genes that allow the parasite species to survive in a host after infection. In effect, the genes of the dominant species complement or replace the host genes in opposing the other parasite species. This possibility has already been mentioned in chapter 2, where it was described as *immunity coded by parasite genes.*

How Does One Distinguish Past from Present Effects?

In order for two or several populations in competition to survive, some mechanism must exist that prevents one of them from winning and eliminating the other. The most common mechanism postulated involves the fact that the niches of the two species are not exactly identical, allowing each to "win" in a slightly different environment. The question immediately becomes one of determining whether the niches were different before the two species entered into competition or if they became different because of competition (in this last case the current situation would then be an example of the "Ghost of Competition Past"). Outside parasitology,

---

competitive species. However, note that such an outcome is observable today. Competition by exclusion is the basis for both the extinction of species owing to introductions of foreign plants or animals and for biological control (the introduction of a plant or animal species with the objective of making an undesirable organism disappear).

6. Stock and Holmes (1987) have shown, for example, that the "large" cestode, *Dioecocestus asper* (6.7 grams average weight), induces an erosion of the intestinal mucosa in grebes that in turn leads to fewer "small" cestodes and other helminths and therefore to an exclusion process.

most opinions support the first proposition, as illustrated by these quotations: "Adaptations already possessed at the time of meeting are the principal determinants of coexistence" (Grant 1975) and "The mechanism of niche differentiation . . . seems unlikely to have commonly arisen by species having diverged by coevolution . . . it is more likely that they diverged as they coevolved separately so that, when they later came together, they coexisted because they had already become adapted to different resources" (Connell 1980).

The choice between the two propositions is, however, particularly difficult to make when it is a question of parasites. For example, in microhabitat displacement (the selective segregation of Holmes) the final result obtained is very similar to that produced by an independent settlement without competition. If situation A (fig. 13.1), in which there was never an interaction, is compared with situation B2, which resulted from a long history of selective pressure, the localization of the two parasite species would be the same in the end. Therefore choosing between isolationist and selectively segregative communities is as difficult for parasitologists as for classical ecologists, if not more so.

For Price, who tends to look at populations as not being saturated, most parasite communities are immature, with interspecific competition being as negligible as intraspecific competition (see Price 1984). Referring to the work of Connell (1980), Price (1987) writes this sharp retort: "[In the absence of experimental tests] the Ghost of Competition Past remains a figment in the theoretical ecologists' mind." For Price the restriction of the microhabitat dimension, as well as restriction of the host dimension, is unlikely: if there are not more species in a parasite community it is not because candidate species have been rejected, but rather because there are no more candidates. This position explains why the search for "vacant niches" (supposedly supporting Price's position) has been particularly active in parasitology in recent years. This aspect of parasite ecology fits itself into a more general debate in ecology, since ecologists are far from agreement on whether ecosystems are saturated.

This problem can be illustrated with a typical example, that of the aporocotylid trematodes of the marine fishes *Sebastes caurinus* and *S. maliger,* investigated by Holmes (1971). Two species of Aporocotylidae live in the circulatory system of *Sebastes.* By studying their precise distribution, Holmes showed that the niches of the two overlapped only very slightly (fig. 13.2). *Psettarium sebastodorum* lives mainly in the heart of the fish, while *Aporocotyle macfarlani* lives in the branchial arteries. The two species have different morphologies adapted to their respective habitats, and depending on whether the two parasites coexist in the same host individual, their niches are modified in an almost imperceptible manner.

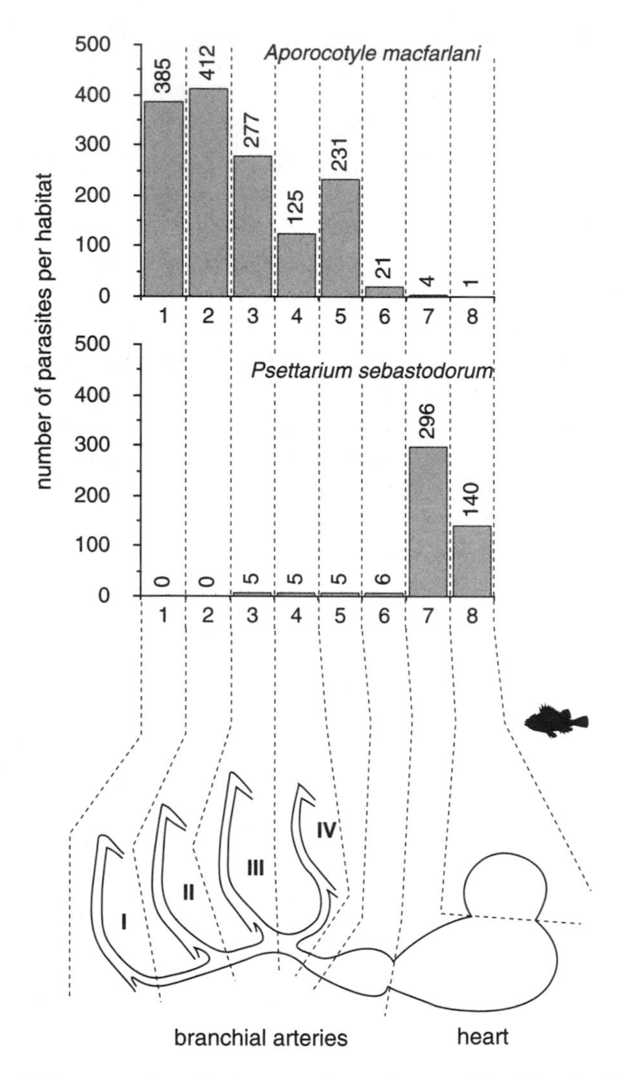

Figure 13.2. Niche separation of two trematode species parasitic of the circulatory system of a marine fish *(Sebastes)* (after the data of Holmes 1971).

In this case Holmes sees an interactive segregation that is quasi-perfect, whereas Price sees proof that the two species are not, and have never been, in competition. This debate leads one to wonder if parasites are specialists so that interspecific competition is decreased or if competition is light because they are specialists. Analysis of the phyletic relationship between the blood flukes of fish by Brooks (1985) has supported

Holmes's position: the localization of *A. macfarlani* in the afferent arteries of the gills appears to be a "derived" (modern) character of the genus *Aporocotyle,* while occupation of the heart by *P. sebastodorum* is a shared character of the *Psettarium* ensemble. This analysis thus supports the idea that *A. macfarlani* has displaced its niche toward the branchial arteries to avoid competition with *P. sebastodorum.*

If we examine all the work on infracommunities, we can arbitrarily group the results into three categories: those that detect no interaction ("isolation"); those that show the displacement of the niche of one parasite in the presence of "the other" ("interaction"), while making note that if the separation of niches is genetically determined the current interaction may no longer be observable; and those that demonstrate the partial or total disappearance of a parasite in the presence of "the other" ("exclusion"), taking into account that at the evolutionary scale one of the parasites in the competition may have abandoned a host species, so that exclusion is no longer observable.

These three levels will serve as a guide to review the state of the problem of host saturation.

Isolation

British Eels

Kennedy (1990) carried out a study on the intestinal helminths of 842 eels *(Anguilla anguilla)* from thirty-nine localities in Great Britain. Here are the results.

1. Infracommunities are poor: only 49% of the eels carried intestinal helminths; 77% of the infected eels carried only one parasite species; 20% carried two; and only 3% carried three. Even where xenocommunities are the richest, most eels carried no parasites or only one parasite species.

2. Any of the five specific or eight generalist helminths, or even other species usually specific for other hosts, may be dominant (core) species.

3. In each infracommunity there is at most one core species, the most commonly dominant being those already cited for the xenocommunities.

4. A core species in an infracommunity may be a satellite species in another and vice versa.

5. Any helminth species may coexist with any other.

6. There is a limited specificity for microhabitat that allows wide occupation, and no indication of competition is evident.

From this investigation Kennedy concluded that infracommunities are poor because of the low rate of parasite transmission and not because of competitive exclusion. That most eels carry no helminth species or only one indicates that host populations are largely unsaturated: there is a preponderance of vacant niches, and nothing indicates that regulation occurs at the level of the infrapopulations, which are not themselves in equilibrium. That is, the intestinal helminths in eels of the region studied must be considered to be unstructured or only slightly structured and to be made up of no predictable (or only slightly predictable) assemblages. Kennedy et al. (1998) obtained similar results in a study of the eels of the Tiber River in Italy.

Other Examples

Other investigations support the conclusions made from the study of eel parasite communities. For instance, Molloy, Holland, and Poole (1995) described a helminth-poor community in trout (549 *Salmo trutta* examined), with few likely interactions and domination by the acanthocephalan *Pomphorhynchus laevis* (58% prevalence). Fiorillo and Font (1996) investigated four species of *Lepomis* (Centrarchidae) and also concluded that the parasite communities of these fish were of an isolationist type. In this case eight helminth species were observed, and one trematode, *Barbulostomum cupularis,* and one acanthocephalan, *Leptorhynchoides thecatus,* were found to occupy the duodenum. Although the two were at the same time both common and abundant, they showed no signs of shifts in their respective niches. Similarly, Sasal, Niquil, and Bartoli (1999) demonstrated that there was little competition and a lack of niche saturation in trematode communities of the digestive tract of Mediterranean sparid and labrid fishes.

Four species of salamanders in the Appalachian Mountains of the United States most often carry no or only one species of helminth (79% out of 397 individuals examined) (Goater, Esch, and Bush 1987). Those found were mainly nematodes with direct life cycles with, rarely, trematodes and one cestode. Moreover, mean intensities were always small.

Likewise, Choe and Kim (1987) investigated the ectoparasite communities of birds from Alaska, which led them to conclude that parasite spatial distribution is "largely a reflection of rigid preferences for certain microhabitats rather than the outcome of competitive dynamics."

Certain populations of gulls have parasite communities that are of an isolationist type. Several investigations of these birds seem to show that although their parasite communities are, overall, rich in species, it is of little consequence at the infracommunity level because prevalences and intensities are small (Hair and Holmes 1970; Kennedy and Bakke 1989). In general, between ten and forty helminth species are identified in the gull populations, while in each host individual there are only a few species (between one and five), each represented by a modest number of individuals.

Similarly, Czaplinski (1975) showed that eight species of cestodes of the wild mute swan *(Cygnus olor)* from Poland were distributed sequentially in the bird's intestine but did not note any niche displacement linked to competition.

Last, investigations on two populations of the bat *Eptesicus fuscus* from Wisconsin and Minnesota (Lotz and Font 1985) also showed no signs of interaction between intestinal helminth species, although species numbers were high in each location (twenty-one and thirteen, respectively). In each case the different helminth species occupied well-defined parts of the digestive tracts, but prevalences and intensities were small and the niches were not influenced by interspecific competition.

INTERACTION

Lesser Scaup from Canada

Lesser scaup *(Aythia affinis)*, small North American ducks from western Canada (Alberta), have served as models in one of the most detailed studies published on interspecific competition in parasites.

Bush and Holmes (1986) analyzed the intestinal helminth infracommunities of forty-five scaup from thirteen lakes in Alberta. Fifty-two species of helminths were identified (trematodes, cestodes, nematodes, and acanthocephalans), and a total of about one million individuals were found with, on average, 22,000 parasites per scaup.

The authors recovered data on the position of the helminths along the anteroposterior gradient of the scaup's intestine and were fully aware that while doing this they were leaving aside other aspects of the parasites' ecological niche, such as their orientation relative to the intestinal mucosa or their selective use of certain resources. In short, the investigation revealed that in this case infracommunities had high degrees of organization and, consequently, predictability.

A Defined Group of Helminth Species Is Dominant

The results obtained by Bush and Holmes (1986) indicate that infracommunity structure is strongly associated with the distinction of core and satellite species in the scaup.

There was a very significant correlation ($p < 0.001$) between prevalence and intensity, with the most common parasites also showing the heaviest loads and vice versa.

Eight parasite species, representing 91.7% of all parasites, are present in a majority of birds and are defined as core species. These include six cestodes *(Hymenolepis pusilla, H. spinocirrosa, H. abortiva, H. tuvensis, Retinometra pittalugai,* and *Fimbriaria fasciolaris)* as well as one trematode *(Apatemon gracilis)* and one acanthocephalan *(Polymorphus marilis).*

Thirty six species, defined as satellites, represented in total only 1% of all parasites and were most often found as immatures.

Analysis of the cooccurrence of the core species showed that their association was not fortuitous but instead was rather predictable: when one of the eight species was found the probability of encountering any of the other seven was greater than 0.5. In contrast, nothing indicated that the presence of the satellite species was predictable.

As expected in such studies, there may be some uncertainty in classifying some of the species into the different categories. For instance, the cestode *Retinometra pittalugai* was considered a core species but could have just as easily been excluded, since it infected only thirty-two of the forty-five birds and because it occurred in modest intensities, averaging only 180 parasites per host. Similarly, a group of eight species of intermediate status between core and satellite status, which Bush and Holmes qualified as "secondary," are far closer to being satellite species than core species because the eight of them together represent only 7.3% of all parasites observed.

The Presence of a Dominant Species Is Correlated with the Presence of Other Dominant Species

When a scaup carries numerous individuals of one parasite species, other core species are also abundant and vice versa. Statistically speaking, it may then be deduced that members of core species respond together to some common factor. The most likely explanation resides in the heterogeneity of scaup populations: there are differences in diets between individuals and, more likely, individual differences in their intestinal environments, making them more or less welcoming (immunity?) to some helminths.

Most Dominant Species Are Host Specific

The studies done on the parasites of various birds from Canadian lakes has allowed the delineation of the host spectra for at least part of their helminths. As expected, some are generalists (found in about the same abundance in several species of birds), while others are specialists (found abundantly in a particular host and sometimes as accessory species in other birds). In the case of scaup, six out the eight core species are specialists. These include five of the six cestodes *(Hymenolepis pusilla, H. spinocirrosa, H. abortiva, H. tuvensis,* and *Retinometra pittalugai)* as well as the acanthocephalan *(Polymorphus marilis).* The other two core species are generalists (the cestode *Fimbriaria fasciolaris* and the trematode *Apatemon gracilis*), and the six specialists represent 85% of all recovered parasites.

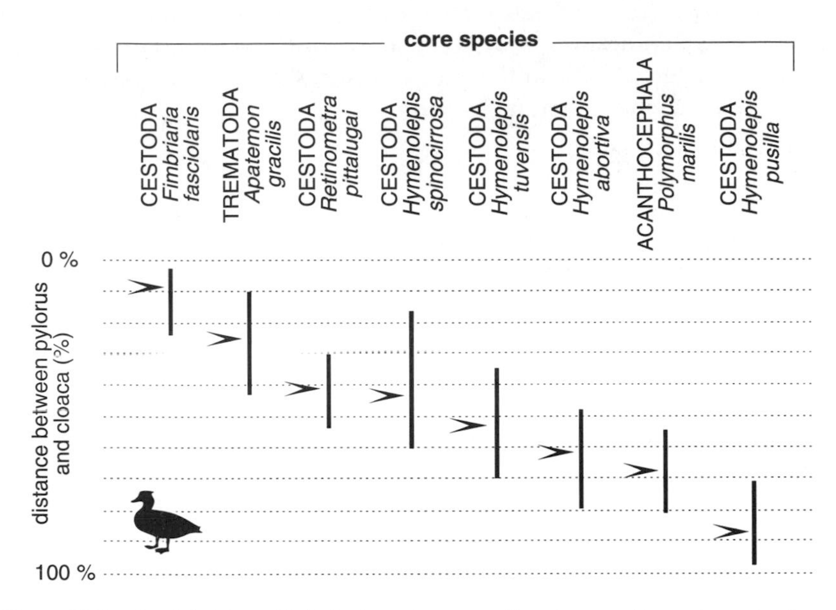

Figure 13.3. Spatial niches of helminths in the scaup digestive tract. For each species the vertical line spans the positions of the most anterior and most posterior individuals while the arrow equals the position of the median parasite (after the data of Bush and Holmes 1986).

On the other hand, all species classified as satellites are spillovers whose main hosts are something else, which explains why most of them were found in the scaup only as immatures.

Each Dominant Species Occupies a Predictable Position

Bush and Holmes (1986) noted the position in the intestine of the helminths found (fig. 13.3) and characterized the habitat of each species by the average position of both the median parasite (horizontal arrow on figure) and the average positions of the most anterior and posterior individuals (black vertical line).

Thus it appears that each core species occupies a defined position, with a range that varies very little from one bird to another. That is, each one occupies a highly predictable location in the scaup's gut.

Dominant Species Occur Successively according to a Predictable Sequence

When the sequence of positions for the eight core species was determined, it is clear that this sequence is predictable. In fact no bird exhibited a sequence of parasite spatial preference that was different from the "consensus sequence" of figure 13.3, and the distribution of the median location was significantly ($p < 0.001$) more uniform than random. Core species were "organized" in the intestine and shared the available space in the gut in an almost equitable fashion.

Realized Niches Are Smaller Than Fundamental Niches

The fundamental niche of a helminth species may be determined by superimposing the distributions of all species in all studied host individuals; the realized niche is the distribution observed in each individual host.

If the fundamental niches of various parasite species in scaup are superimposed, one sees that the adjacent species overlap strongly. However, if realized niches are superimposed (of course as many superimpositions may be done as there are hosts), adjacent species are still seen to overlap, but much less so. Therefore it is easy to presume that in this case the difference between fundamental and realized niches results from "competitive restriction" and thus that the infracommunity is interactive.

The Abundance of a Species Influences the Distribution of Adjacent Species

If interaction exists (if in fact competition occurs), then growth in the size of the infrapopulations may be expected to increase this competition and thus to influence parasite distribution along the intestine. Bush and Holmes (1986) showed that this prediction occurred for at least two species with high intensities, the most demonstrative example being that of the two adjacent cestodes *Hymenolepis spinocirrosa* and *H. abortiva*. These two cestodes belong to the same "suite" of parasites. That is, they are vehicled by the same intermediate host, the amphipod crustacean *Hyalella azteca,* and their intensities generally vary together.

Figure 13.4 shows that when cestode intensities increase *H. abortiva* colonizes more and more anterior regions, which in turn forces *H. spinocirrosa* to move forward. The correlation between the most anterior individual of *H. abortiva* and the most posterior of *H. spinocirrosa* is positive and highly significant ($p < 0.0001$; less than one chance in ten thousand for the observed positions to be due to chance alone). Moreover, *H. microskrjabini,* a secondary species, is moved from its position in the second quarter of the intestine into the first when the two other species are abundant.

Based on the information above, Bush and Holmes thus concluded that helminth infrapopulations in scaup are structured by interactions between species. In the predictable environment of the digestive tract (because of morphological or biochemical gradients) the community is organized in a predictable way. More precisely, there is a group of predictable dominant species that occurs along with less predictable or nonpredictable subordinate species. Typically, the segregation of niches is at

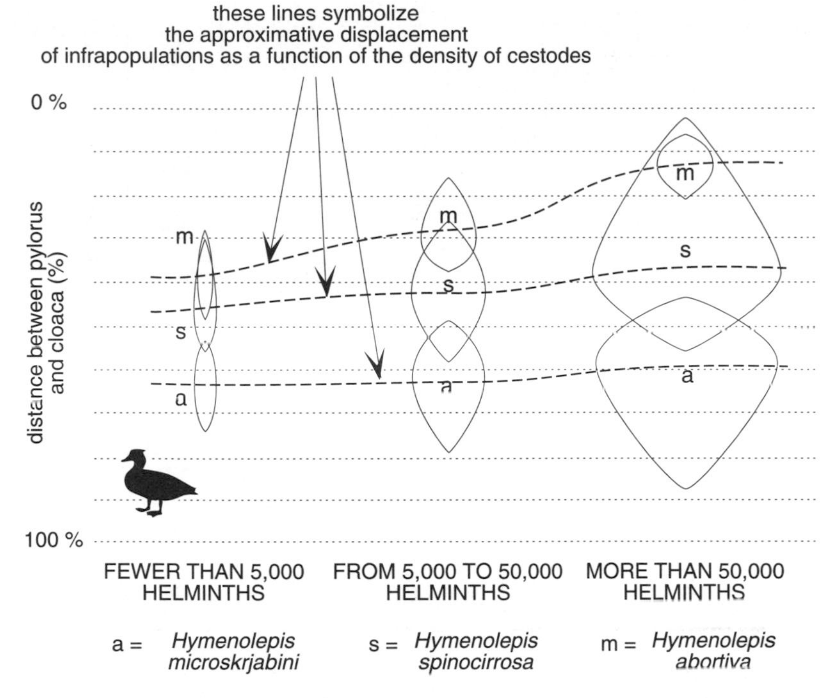

Figure 13.4. Modification of the infrapopulation position of three species of cestodes in the digestive tract of scaup according to interspecific competition (after Bush and Holmes 1986). Sizes of enclosed areas = relative sizes of populations.

the same time both interactive and selective, with *interactivity* being shown by the restriction of realized niches and the forced displacement caused by adjacent species. Moreover, the community is also *selective,* since some of the parasites' differences are genetically fixed, as shown by the invariability of the sequences of the different infrapopulations.

Coots in Poland

Results supporting microhabitat displacement were also obtained by Pojmanska (1982), who studied parasitism in the coot *Fulica atra* in central Poland. Each of three species of cestodes, *Diorchis brevis, D. inflata,* and *D. ransomi,* infects from 60% to 80% of the birds, sometimes with relatively high densities (several hundred parasites). The spreading of parasites throughout the digestive tract as the result of intraspecific competition and, more important, displacements owing to interspecific competition were also observed in this system.

Figure 13.5 was constructed from data obtained from 5 of the 173 birds

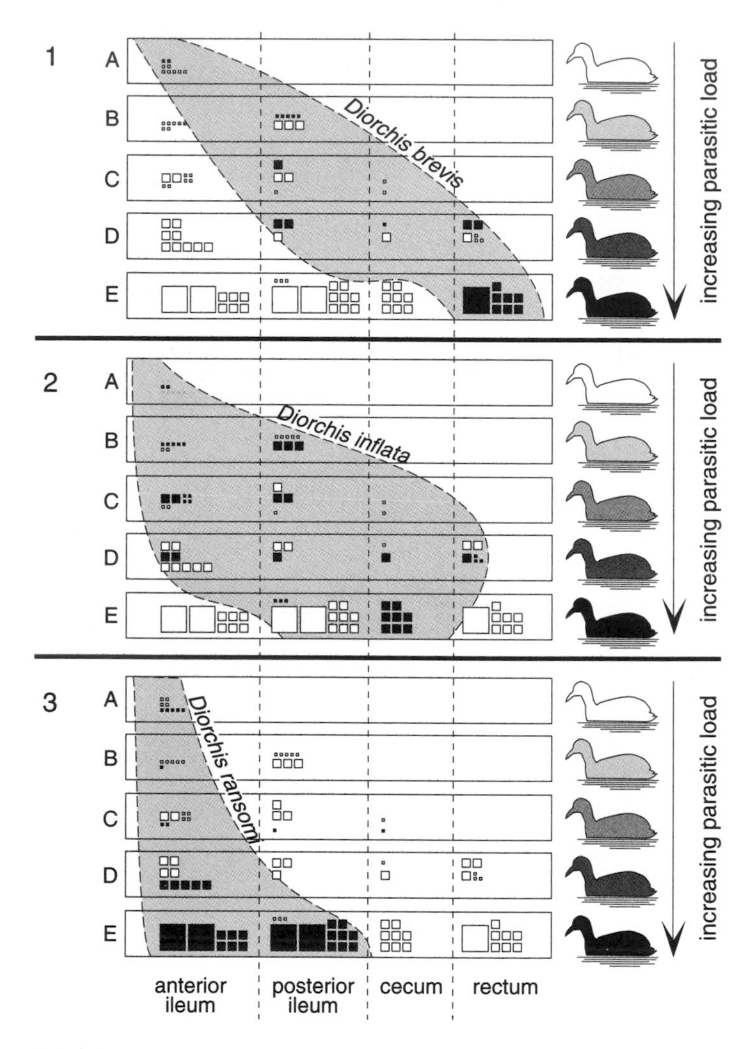

Figure 13.5. Distribution of cestodes of the genus *Diorchis* in coots for parasite density values increasing from individual A to individual B. Each small square represents one cestode, each medium square represents ten cestodes, each large square represents one hundred cestodes. The three graphs are identical except that in each the black square represents the parasite species being considered and the open squares represent the two other species (after part of the data of Pojmanska 1982).

Pojmanska examined. These birds were chosen because they were spectacularly different in their overall parasite load. The five coots are classified from A to E with increasing densities of cestodes (14 in A, 44 in B, 132 in C, 271 in D, and 997 in E), and black squares represent the three separate *Diorchis* species. It is clear that when cestode densities increase (from top to bottom on each graph), *D. brevis* is totally rejected from the anterior ileum toward the rectum; *D. inflata* is rejected toward the posterior region of the tract while still maintaining intermediate positions, notably in ceca; and *D. ransomi* appears as the winner of the competition, since it obviously keeps the ileum for itself. In this infracommunity each species occupies a larger space as density increases, but the entire distribution shifts so that infrapopulation overlap does not increase.

Sharks from Virginia

Elasmobranch cestodes are characterized by a narrow specificity and are almost the only intestinal helminths of their hosts. As such, shark tapeworms are a unique model because they are not subject to perturbation by the occurrence of generalist parasites. Moreover, elasmobranch intestines are also unusual because they are straight and most often harbor a "spiral valve" with variable numbers of turns. Thus this intestine has more gradients than a classical intestine and is ideally suited for the study of infracommunities. Cislo and Caira (1993) studied forty-nine individuals of *Mustelus canis* and showed that all the helminths present were cestodes, of which three species were frequent (prevalences above 50%). These species exhibited a characteristic distribution in the digestive tract, and it is particularly interesting that the two species belonging to the same genus (*Calliobothrium lintoni* and *C. verticillatum*) are those that have best separated their microhabitats. These results are original in that although the niches are remarkably distinct (fig. 13.6 top), the comparison of cestode positions and sizes in monospecific or plurispecific infections gives no indication that the species actually interact.

The investigation shows that the posterior region of the intestine is not occupied (fig. 13.6, bottom), a frequent case in communities whatever their hosts. In the particular case of *M. canis* we also see that the percentage of parasites in spiral turns regularly decreases from the pylorus to the rectum. This gives the impression that the more anterior habitat is better and that species that do not occupy the very anterior regions are those that have displaced their habitat to avoid the more competitive species. If this is true, the current distribution showing no detectable competition would be the result of a previous competition—there would be ghosts in the elasmobranch spiral valves.

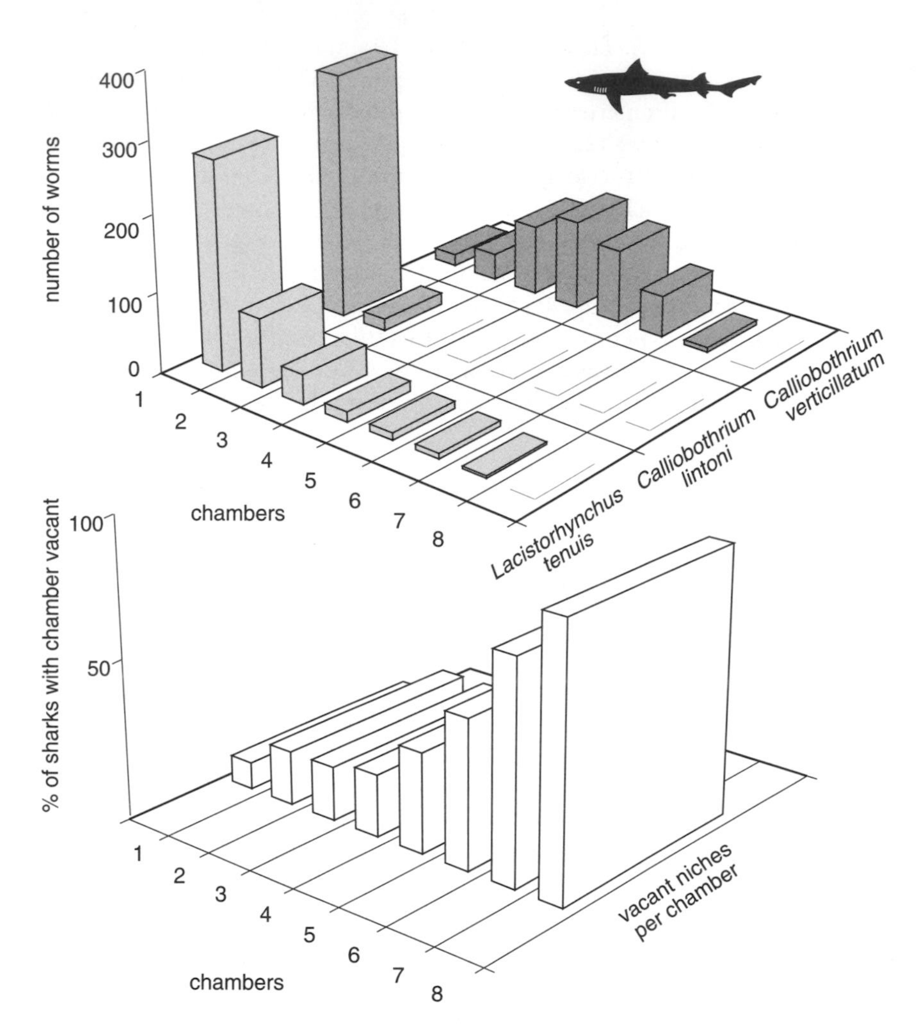

Figure 13.6. Top: distribution of the main helminth species in the spiral valve of the shark, *Mustelus canis*. Bottom: percentage of vacant niches.

Thus we are again faced the "Price-Holmes" dilemma, outlined above. Do the two *Calliobothrium* species occupy distinct microhabitats because they have distinct preferences (the "Price" interpretation) or because one or the other has displaced its niche in order to limit competition (the "Holmes" interpretation)? The second interpretation seems the most likely, since the two species concerned belong to the same genus.

The Gills of Fish

Besides the intestine, fish or elasmobranch branchial systems are another location that allow the very precise study of interspecific competition. For example, Sanfilippo (1978) has shown that two monogeneans in the genus *Ligophorus, L. vanbenedenii* and *L. szidati,* both parasites of the Mediterranean mullet *Liza aurata,* distribute themselves differently on the gill depending on whether infections of host individuals are monospecific or bispecific (fig. 13.7). When infections are monospecific, *L. vanbenedenii* and *L. szidati* have similar distributions on the four gill arches (this distribution indicates the

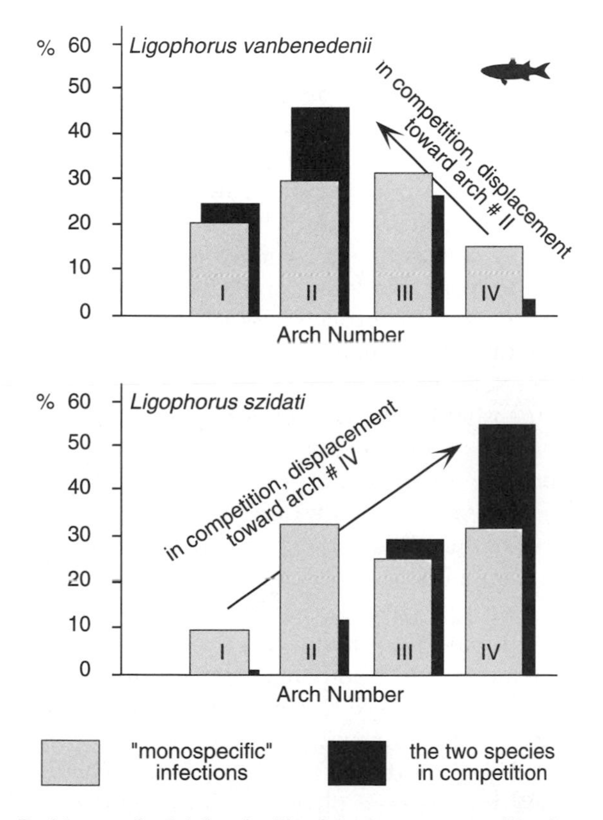

Figure 13.7. Position on the fish host's gills of the monogeneans *Ligophorus vanbenedenii* and *L. szidati,* depending on whether infections are monospecific or bispecific. In monospecific infections (gray bars) preferences of the two species do not appear clearly distinct. In bispecific infections (black bars) *L. vanbenedenii* (top) has a tendency to move toward arch II and *L. szidati* (bottom) has a tendency to move toward arch IV (after Sanfilippo 1978).

fundamental niche of each species). When infections are bispecific, however, *L. vanbenedenii* has a tendency to occupy only arches III and IV, that is, the posterior arches (this distribution indicates the realized niches).

This competition for space, which is discreet but real, is even more interesting because the niches of gill monogeneans are not saturated (Rohde 1979).

Silan et al. (1987) showed that things can be more complex and that two variables have to be taken into account when studying gill parasites: host age (branchial surface increases with age and water currents become stronger) and parasite age (young and adult parasites occupy different positions). These variables are involved, for instance, in the distribution of monogeneans on the gill in the sea bass *Dicentrarchus labrax*. Here, when juvenile *Diplectanum* become adults they migrate from the base to the tip of the gill filaments, and when water currents become stronger as the fish grows older, the parasites seek more "sheltered" zones on the gill apparatus.

The results of other studies of gill parasites are even more interesting because they lead one to question the link between phyletic relationship and competition (Patrick Silan, Charles Bilong-Bilong and Eve Le Pommelet, personal communication). For instance, the freshwater fish *Hemichromis fasciatus* is parasitized in Cameroon by five gill monogeneans belonging to two genera (three species of *Cichlidogyrus* and two of *Onchobdella*). In this case the high competition leading to niche segregation (either interactive or selective) that would be expected within each genus does not happen—whether the congeneric species are by themselves or associated on the same host individual, their situation is very similar (fig. 13.8). On the contrary, a high degree of segregation exists between the two genera, since *Cichlidogyrus* essentially occupies the distal part and *Onchobdella* the basal part of the gill filaments. This situation does not agree with the usual notions about niche separation and, most notably, with Rohde's ideas on distribution that have been referred to several times here. According to Rohde, the need to reinforce reproductive barriers and select distinct requirements should lead congeneric species to rapidly separate their niches. Similar results were reported by Dzika (1999), who showed that two congeneric monogeneans, *Pseudodactylogyrus anguillae* and *P. bini*, are similarly distributed on the gills of *Anguilla anguilla* in Poland. That there is no association between phylogeny and competition shows that "gill monogeneans" offer nice research opportunities in the future.

Squirrels from Pennsylvania

All the previous examples concern studies done on natural populations of hosts and parasites. As shown in the following study by Patrick (1991),

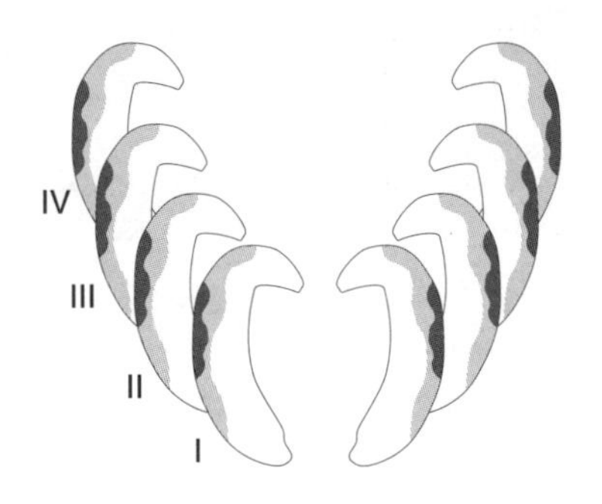

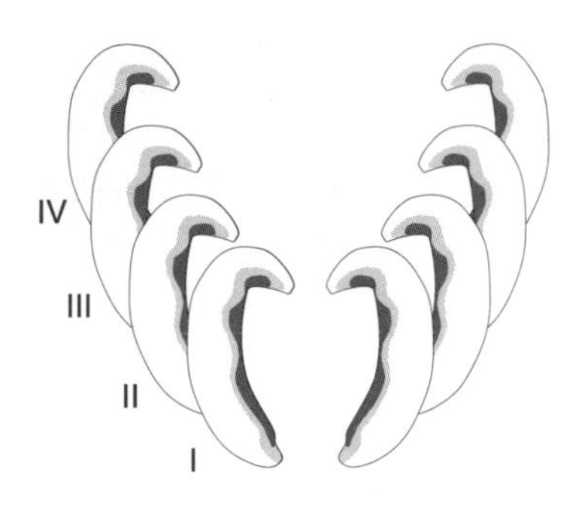

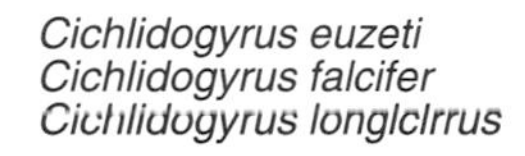

Figure 13.8. Position on the fish host's gills of three species of the genus *Cichlidogyrus* and two species of the genus *Onchobdella*. Note that the two genera occupy different zones on the gill system, while within each genus the species occupy similar locations (these locations were grossly sketched on the figure and do not show parasite densities with precision) (from unpublished results of P. Silan, C. Bilong-Bilong, and E. Le Pommelet). Gray = light densities; black = high densities.

however, it is also possible to experimentally "manipulate" a parasite community to demonstrate that it is interactive.

In Pennsylvania, the flying squirrel *Glaucomys volans* is parasitized by two core species of intestinal nematodes: *Strongyloides robustus* and *Capillaria americana*. In this case a synthetic helminthological map of the intestines of ten squirrels was generated (fig. 13.9A) to outline the location of each helminth, with position being denoted as the distance to the pylorus and expressed in terms of its percentage of total intestinal length. The distributions obtained were different for the two species: *S. robustus*, the more abundant of the two, is found very close to the pylorus, while *C. americana* was found farther down. Moreover, the two overlapped only in a very small zone.

Patrick wanted to test the hypothesis that the fundamental niche of *S. robustus* was more spread out than that observed, with its realized niche being restricted because of the presence of the second nematode, *C. americana*. To do this the author dewormed a group of ten squirrels with an anthelmintic. After making sure the squirrels were free of worms, he reinfected the animals with known dosages of infective *S. robustus* larvae. By establishing maps similar to those described above, he found that at comparable intensities *S. robustus* spread significantly farther down the intestine when it was alone than when it occurred in the presence of *C. americana* (fig. 13.9B and C). A statistical test (Mann-Whitney) showed

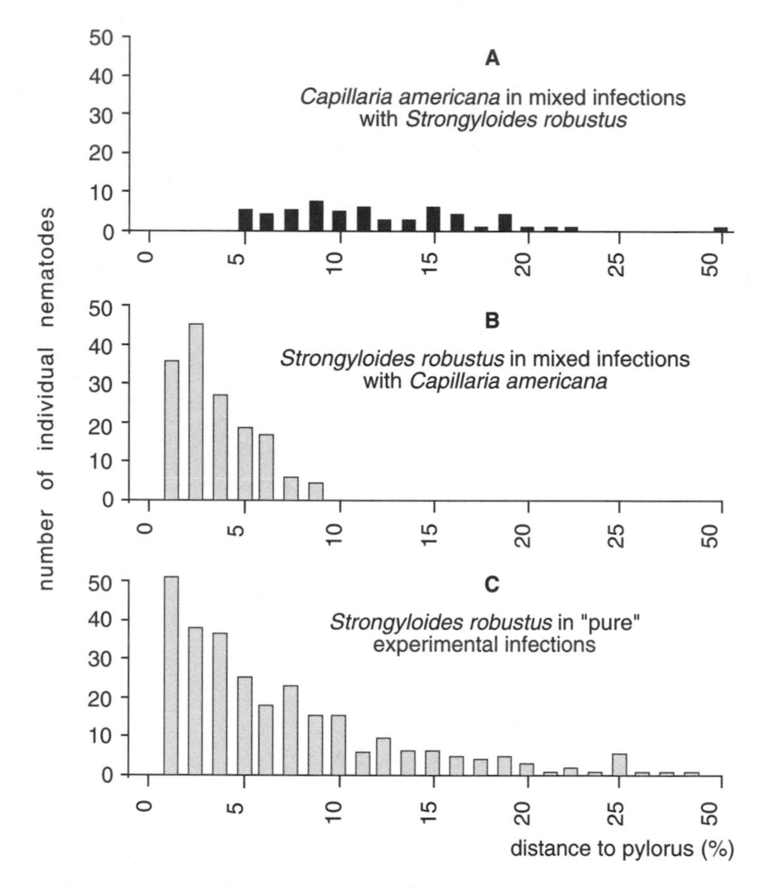

Figure 13.9. Occupation of the squirrel digestive tract by *Strongyloides robustus* in the presence or absence of *Capillaria americana* (modified after Patrick 1991).

that the difference was significant ($p < 0.05$), and we can conclude that in this case competition plays a role in the organization of the community.

Other Examples

Still other investigations or experimental studies support the ideas of Bush and Holmes (1986); here is a brief summary of some of them.

Holmes (1990a) studied the infracommunities of intestinal helminths of the rockfish *Sebastes nebulosus* from the Pacific coast of North America with a method similar to that used to study the scaup. Here the infracommunities in the fish are rich (a total of twenty-seven species of helminths were identified), interactive, and characterized by core species (from four to nine depending on the population), and opportunities for colonization remained because there were vacant niches. Further, and despite variation between individuals and seasons, infracommunity structure was largely predictable in a given location. Interestingly, the core species of *S. nebulosus* are generalists rather than specialists, which is different from the situation in scaup. Moreover, *S. nebulosus,* with its highly structured intestinal helminth communities, is not necessarily representative of other fish, which often have poorer communities. Nevertheless, other infracommunities of intestinal parasites of fish also show indications of being interactive. This is the case, for instance, for *Sciaena umbra* in the Mediterranean (five trematode species, of which two are core species; Holmes and Bartoli 1993) and *Amia calva* in the United States (five species of trematodes, two cestodes, five nematodes, and one acanthocephalan; Aho, Bush, and Wolfe 1991). Species richness is, in fact, not a prerequisite for interactivity. For instance, the cestode *Proteocephalus filicollis* and the acanthocephalan *Neoechinorhynchus rutili* are in competition and show interactive segregation in the stickleback *Gasterosteus aculeatus.* In this case, when each species is by itself its infrapopulation is dispersed over a large portion of the digestive tract. When the two species are associated, however, *P. filicollis* occupies the most anterior positions and *N. rutili* the most posterior positions in the intestine (Chappell 1969).

Even though European eels may be characterized by having "species"-poor helminth communities (see above), Kennedy (1995) has shown that, in contrast, Australian eel communities are particularly rich and that almost all their helminths are specialists. This richness, which Kennedy considers "the most diverse yet reported from any species of fish anywhere," is explained as follows: "Rich parasite communities are . . . characteristic of old host species and of species in their heartland."

In certain parasite-host associations interactions occur in an unexpected way. McInnes, Crompton, and Ewald (1994) studied the intestinal distribution of the nematode *Porrocaecum spirale* and the acanthocephalan *Centrorhynchus aluconis* in 109 tawny owls *(Strix aluco)* in Scotland. These authors showed that in concurrent infections (thirteen tawny owls), both helminth species were located more anteriorly than in single-species infections.

Communities of grebes analyzed by Stock and Holmes (1988) in the same regions as the scaup discussed earlier (Alberta, Canada) are composed of numerous helminth species, with infrapopulations always on the same order of magnitude in all host individuals. In this case realized niches are much smaller than fundamental niches; there is little overlap between adjacent species, and positions change in the presence of other species. Hence these communities are also interactive. Further, vacant niches also appear to exist, mostly in the posterior part of the intestine, probably because (as in *Mustelus*) this region is poor in resources and not because of a lack of interactive processes in the anterior part of the intestine.

Studies also carried out in Canada by Goater and Bush (1988) on the curlew *Numenius americanus* showed again that to be interactive an intestinal helminth community does not require many species or individuals. In fact, curlews carry only two core species, the cestodes *Dictymetra numenii* and *D. nymphae,* both specific to this host. Prevalences are 100%, and the average intensities are 100 to 130 *Dictymetra* per bird. Despite these parasite loads, which are far from the thousands of cestodes found in scaup, the two cestodes clearly interact: when there is a high relative intensity of *D. numenii,* which lives at the start of the intestine, *D. nymphae* is forced to move farther down.

The study of Moore and Simberloff (1990) on the helminth communities of the American quail *Colinus virginianus* in Florida is the only one to take into account both parasite density and biomass, with each parasite's biomass calculated based on volumetric and morphometric measurements. This investigation shows that in this case the occupation of space (helminths' position in the digestive tract) is linked to intraspecific competition but that a given species' biomass may be negatively correlated with another species' infrapopulation density (interspecific competition). This is the case for two species of cestodes in the quail, where one influences the other but not in a reciprocal manner: in the presence of *Railletina colinia* the average (individual) biomass of *R. cesticillus* is diminished; and in the presence of *R. cesticillus,* the average (individual) biomass of *R. colinia* is not affected.

Hence interspecific competition is occurring, and this result is particularly interesting because the influence is "asymmetrical" and because biomass and fecundity are usually correlated.

It would be wrong to believe that in vertebrates only gills and intestines can be used to study the differential localization of infrapopulations. Any organ showing gradients will allow such comparisons. For instance, Kennedy and MacKinnon (1994) have shown that two species of *Thelazia* (*T. skrjabini* and *T. gulosa*—both nematodes transmitted by flies that were briefly discussed in chapter 1) show distinct preferences in their locations within the eyes of cattle (eyelids, various glands, etc.), indicating that they also form an interactive community.

Exclusion

Cercarial Factories

Competition between larval trematode stages for the exploitation of their molluskan hosts provides one of the most striking examples of competitive exclusion observable in the parasitic realm.[7]

Not Much Space for Many

Trematodes transform their molluskan intermediate hosts into cercarial production factories. The problem is that such factories are highly sought after (see Esch and Fernandez 1993). The "microecosystem" represented by each mollusk thus is a privileged place for competition between different species of trematodes. As noted by Fernandez and Esch (1991a), "The individual snail represents a highly limited resource from an energetic and spatial point of view."

The intensity of interspecific competition within the trematode/mollusk system has two main causes. The first is that the trematodes of different species found within the same molluskan species form a true guild (the "infraguild" for Fernandez and Esch 1991a) whose members exploit almost the same resources. In short, in an individual mollusk there are very few possibilities for niche separation, and in most cases each trematode species uses the maximum available resources in the host individual. The second cause is that the number of trematode species capable of exploiting the same mollusk species is often very large, so many different parasites often want the same host.

Kuris (1974, 1990), with good reason, compared the trematode parasites of mollusks to insect parasitoids that systematically kill their hosts. Two essential characteristics make them similar: one is that, because of parasitic castration in one and host death in the other, the fitness of the parasitized individual in each case drops to zero. The second characteristic is that in both cases the pathogenic effect is not

---

7. In this book I consider multiplication of trematodes within their molluskan hosts as both larval and asexual processes. However, Teresa Pojmanska and Bojena Grabda-Kasubska, who translated this book into Polish, noted that trematode life cycles could also be interpreted as an alternation of generations, in which case the parasitic stage in mollusks would not be "larval." Moreover, multiplication of sporocysts and rediae could be either polyembryonic or parthenogenetic and would thus not be just a simple asexual process. A. Dobrovolskij (Saint Petersburg State University) and G. Ataev (Russian State Pedagogical University) consider that there is virtually no doubt the multiplication of sporocysts and rediae is parthenogenetic (personal communication, July 2000).

intensity dependent, since only one parasite, either through multiplication or through growth, is sufficient to exert a maximum pathogenic effect on the host.

The fundamental conditions required for competition (numerous competitors and limited resources) thus occur together in a host mollusk, and trematode larval stages are a priori differentially "armed" for the resulting competition. Certain species produce cercariae in sporocysts and others do so in rediae (at the start of the infection, multiple sporocysts and rediae always arise in a primary or mother sporocyst, itself issued from the miracidium). Sporocysts are absorbing larval stages that feed by nutrient transport across the tegument, and rediae are predatory larval stages with a digestive tract and muscular pharynx. Rediae are, in theory, capable of devouring sporocysts (a particularly efficient form of competition by interference, since it is in essence predation).

There are two approaches to demonstrating competition between trematodes in mollusks: experimentation (simultaneously or successively exposing the same mollusk to the miracidia of different species); and field investigation (deducing competition from the degree of cooccurrence of particular species).

The Experimental Demonstration of Exclusion

Most experiments in this arena demonstrate that possessing either sporocysts (S) or rediae (R) in the life cycle often determines dominance and that priority effects also exist. For instance, Basch, Lie, and Heynemann (1969) exposed the pulmonate snail *Biomphalaria glabrata* to miracidia of both *Cotylurus lutzi* (S) and *Schistosoma mansoni* (S) at different intervals. The simultaneous exposure to miracidia of both species led to temporary double infections in which *S. mansoni* finally eliminated *C. lutzi*. Further, initial infection by *S. mansoni* prevented secondary infection by *C. lutzi*, but primary infection by *C. lutzi* did not prevent secondary infection by *S. mansoni*, which then progressively dominated *C. lutzi*. Basch, Lie, and Heynemann (1970) also showed that *Paryphostomum segregatum* (R) may completely eliminate *Ribeiroia marini* (R) in both *Biomphalaria glabrata* and *B. straminea*, and Donges (1972) has shown that in *Lymnaea stagnalis*, *Echinoparyphium aconiatum* (R) dominates *Isthiophora melis* (R), which in turn dominates *Echinostoma revolutum* (R).

Along this line, Fernandez and Esch (1991a) took some *Helisoma anceps* that had been naturally infected with *Haematoloechus longiplexus* (S) and overinfected them with *Halipegus occidualis* (R). After about ten days of emitting both types of cercariae, *Halipegus occidualis* became the only cercariae produced. Dissection of the mollusks showed that only a few sporocysts of *Haematoloechus longiplexus* were still in the gonad while most of the available space was occupied by *Halipegus occidualis* rediae.

Ataev and Dobrovolskij (1992) showed that *Philophthalmus rhionica* (R) cannot develop successfully in the prosobranch *Melanopsis praemorsa* if it is already infected with *Metagonimus yokogawai* (R). In this case the first invader wins the conflict, and then often both the host and its trematodes die. On the other hand, if *M. praemorsa* that are already parasitized by *Mesostephanus appendiculatus* (S) are infected by *P. rhionica* (R), *P. rhionica* develops while *M. appendiculatus* diminishes in number and size. In both cases, however, domination by one of the trematodes is a progressive process, with the dominant species first occupying vacant space and only later invading the space occupied by the dominated species. After three hundred days the dominated species is still present, although in much smaller numbers, and the authors do not say whether they observed predation of sporocysts by rediae.

Such experiments tend to show that when two species in competition both have either sporocysts or rediae, then one can dominate and eliminate the other and that when one has sporocysts and the other one rediae, the rediae almost always win. Direct predation between different species of trematodes has been observed only rarely, however: Lie, Basch, and Heyneman (1968) observed rediae of *Paryphostomum segregatum* eating rediae from *Echinostoma lindoense,* and Kwo, Lie, and Owyang (1970) observed rediae of *Echinostoma audyi* eating primary sporocysts of *Fasciola gigantica.* Thus predation is most likely only a minor component of interspecific competition in snails.

Niche displacement is not frequently reported in mollusk-trematode systems, although Jourdane and Mounkassa (1986) observed it in the mollusk *Biomphalaria pfeifferi.* The primary sporocyst of *Schistosoma mansoni* is usually localized in the foot of this host. However, when the mollusks were exposed to miracidia of both *S. mansoni* (S) and *Echinostoma caproni* (R) at the same time, 90% of the sporocysts of *S. mansoni* were found in the nerve ganglia, blood sinuses, and kidney. These habitats appear to be shelters (refugia) and it is remarkable that *E. caproni* does not occupy the space freed up by *S. mansoni* (its primary sporocysts are in the heart or the aorta, and sporocysts are in the genital gland). Jourdane and Mounkassa indicate that the organs used as refugia by *S. mansoni* have, in general, little immunological activity. This suggests that the presence of *E. caproni* would induce an immunodefense response in the mollusk, which would explain why *S. mansoni* flees toward more protected microhabitats. Along these lines Kuris (1990) has postulated that when the same mollusk host produces cercariae from two or more trematode species, the trematodes may be exploiting at least partially different areas in the host.

The Dilemma of Field Investigations

Field investigations into competitive exclusion have brought about both the most data and the most debate. It is difficult to interpret abnormally rare cooccurrences, because if a heterogeneity of infection exists in the mollusk population, or if sampling is involuntarily biased away from dual infections, then double infections are necessarily rare and failure to detect them does not mean there is interspecific competition (the outcome is similar to the Wahlund effect, which leads to a reduction in the number of heterozygotes when genotypes are sampled in two populations).

There are at least three reasons for host heterogeneity:

– *Spatial:* Fernandez and Esch (1991a) showed that trematode infective stages may differ both quantitatively and qualitatively over a distance of even a few meters.
– *Temporal:* based on results reported by Goater, Esch, and Bush (1987) on communities of salamander helminths, Janovy, Clopton, and Percival (1992) demonstrated that the composition of infracommunities varied considerably depending on the season.
– *Age:* if mollusks of different age-classes are mixed in the sampling and if there is "niche partitioning" along host age (if different trematode species infect mollusks of different ages), then double infections will occur in a small number of hosts by chance alone without competition's being the cause.

These reservations aside, apparent competitive exclusion between trematodes has nevertheless been frequently reported in natural populations of mollusks. The fact that in nature the cercariae emerging from a single mollusk rarely belong to several species was observed early on by authors such as Sewell (1922) and Dubois (1929), and numerous investigations in the 1990s have brought similar results. However, the interpretation of the respective roles of heterogeneous sampling and competitive exclusion varies depending on the worker.

It's Heterogeneity, Not Competition

A two-hectare pond in North Carolina called Charlie's Pond has been used by Esch and his colleagues as a model to study trematode communities in freshwater mollusks (Fernandez and Esch 1991a, 1991b;

Williams and Esch 1991; Snyder and Esch 1993). The comparison between two pulmonate species found in the pond was particularly instructive.

Helisoma anceps is a relatively immobile snail; individuals are able to recruit trematode larval stages only within an area of a few meters. Although, 30% to 58% (depending on the season) of *H. anceps* were infected by eight species of trematodes, only seven multiple infections were encountered out of 1,485 *H. anceps* (0.5%) examined by Fernandez and Esch (1991b). *Physa gyrina,* on the other hand, is a mobile species that moves about in the pond and thus recruits trematodes over a much larger area. This snail carries a total of six trematode species, and in 467 individuals examined, Snyder and Esch (1993) detected 86 double infections, or 18.4% of those examined.

By marking and recapturing some individuals, Fernandez and Esch (1991a) were able to note several replacements of one species by another, showing that interspecific competition, and even competitive exclusion, occurred between trematodes. However, based on the comparison of results between *H. anceps and P. gyrina,* Snyder and Esch (1993) concluded that the rarity of double infections was more likely due to spatial and temporal heterogeneity than to exclusion.

Curtis (1997) questioned the role of heterogeneity in sampling and believed that too often different age-classes, locations, or both are mixed in the analysis of results. Basing his conclusion on over fifteen years of dissections of nearly 12,000 *Ilyanassa obsoleta,* he showed how difficult it is to use such data to demonstrate competitive exclusion. If this is the case, apparent competitive exclusion may reflect nothing other than the mixing of data.

It's Competition, Not Heterogeneity

Populations of *Cerithidea californica,* a prosobranch very abundant on the Pacific coast of North America, have been the subject of several detailed studies. Sousa (1990) sampled, over seven years, the *C. californica* of Bolinas Lagoon (twenty-four kilometers northwest of San Francisco) in two microhabitats separated by some 750 meters. Over this time only ninety-one multiple infections were recorded out of 4,462 infected individuals (a total of 25,859 snails were examined). This author concluded that "several lines of evidence indicate that strong antagonistic interactions between larval parasite species occur within individual snails."

Kuris (1990), based on the data of Martin (1955, in Kuris 1990) for the same mollusk, recorded 3% multiple infections for 12,995 individuals of *C. californica* examined in upper Newport Bay (California) and established a true competitive hierarchy in the trematode guild of this mollusk. To establish this hierarchy, Kuris used both direct criteria (for instance, observation of the replacement of one species by another in the same host mollusk) and indirect criteria (the different prevalences of two species in young and old mollusks). From this analysis the dominant species were found to belong to the Echinostomatidae, Philophthalmidae, and Schistosomatidae (genus *Austrobilharzia*), followed by the Heterophyidae, then the Notocotylidae, the Renicolidae, and finally the Strigeidae and Microphallidae (note that *Austrobilharzia,* which is at a high level in the hierarchy, is by exception a trematode with sporocysts).

Kuris and Lafferty (1994), by regrouping various authors' data for almost 300,000 mollusks, noted that out of the almost 25% that were infected by trematodes only 4,300 showed double infections, whereas their calculations expected over 14,000. They concluded that the small number of multiple infections usually encountered in nature is not the result of environmental heterogeneity but rather is due to species interactions. Kuris and Lafferty believe that though environmental heterogeneity is a significant factor, its usual effect is to intensify competition: the recruitment process comes first, and only later is competition possible, which is then intensified if trematode species share aggregation factors (which means they are recruited at the same location).

Lafferty, Sammond, and Kuris (1994) collected five hundred individuals of *C. californica* in five microhabitats, each covering a ten-meter stretch of tidal channel and separated from one another by about fifty meters in Carpinteria Salt Marsh (California). Mark and recapture experiments showed that infected and uninfected snails randomly mixed within a site over a period of one week, and each site could be considered to be deprived of heterogeneity. A total of nine trematode species were found, with overall prevalences varying from 42% to 83% depending on the microhabitat. Overall, only seven double infections were recorded, which led the authors to conclude that this very small number could be explained only by competitive exclusion and that "even the seven double infections that we did observe could represent competitive events in progress."

In contrast to these studies is the work in Florida of Bush, Heard, and Overstreet (1993), who reported more than 16% double infections in *Cerithidea scalariformis,* a mollusk closely related to the previously cited *C. californica.*

In general mollusks, which can be considered "cercarial production factories," are ideal models to determine if "what happens in the infracommunity" has consequences for the composition of the xenocommunity.

One could believe that with time mechanisms of competitive exclusion should lead to less diversity in parasites through the disappearance of certain species, that is, to a poorer xenocommunity in a specific molluskan host. As shown by the seven-year study by Sousa (1990) of the trematodes of the Californian mollusk *Cerithidea californica,* however, this is not what happens. Although exclusions are the rule at the infracommunity level, they have no long-term influence on the xenocommunity, which unpredictably fluctuates depending on location and year. It is likely that exclusions are indeed responsible for the extinction of some local trematode populations but that recolonizations are just as numerous thanks to input from other parts of the metapopulation. Overall, then, intertrematode exclusion in mollusks leads to situations where xenocommunities may be rich but each infracommunity is poor, if not monospecific (limited to one infrapopulation). Armand M. Kuris (personal communication) believes that subordinate species must have some specific features relative to the xenocommunity parasitizing a particular species of mollusk: in order to exist, it must have either a higher recruitment rate in the snail population or another compensatory life history feature such as a wider definitive host range or habitat refuge.

The processes of exclusion by competition in mollusks makes those trematodes with rediae, such as the echinostomes, possible auxiliaries in the fight against those with sporocysts, such as schistosomes, since both often develop in the same molluskan species (see Lim and Heyneman 1972; Combes 1982).

Frogs from the Pyrenees

When there are double infections by two parasites that are suspected of partially excluding each other, it is easy to show whether competitive exclusion occurs by comparing the number of associations actually observed with the expected number calculated from the prevalences for each of the two parasites (the null model). If, for example, parasite A is present in 20% of the hosts and B in 30%, one expects to encounter A and B in 20% × 30% of cases, or 6% of the hosts examined (six hundred cooccurrences in ten thousand). If the observed percentage is significantly smaller than the calculated one, there is competitive exclusion (in fact, the parameters used to calculate the null model's values are already affected by the interaction, so it underestimates the number of cooccurrences; see Lafferty, Sammond, and Kuris 1994).

The trematode *Gorgodera euzeti* and the monogenean *Polystoma integerrimum* share the same host, the European common frog *Rana temporaria,* in the Pyrenees (Combes 1968) as well as the same microhabitat, the frog's urinary bladder. However, they do not belong to the same guild, since *G. euzeti* seems to "graze" on bladder epithelium and *P. integerrimum* feeds on blood by piercing the blood vessels of the bladder. Further, their mode of infection is different, with *G. euzeti* cercariae developing in freshwater bivalves of the genus *Pisidium* (see life cycle in chapter 7, fig. 7.11) and metacercariae forming in insect larvae of the genus *Sialis*, which are eaten by frogs as imagoes. Frogs thus start to accumulate *G. euzeti* at about age two or three. *P. integerrimum*, on the other hand, infects amphibians when they are still tadpoles. Here the larvae first attach to the gills and then migrate to the bladder during metamorphosis, and after two or three years the adult polystomes measure more than one centimeter in length.

An investigation of common frogs from a series of lakes in the Pyrenees gave the following results: number of *Rana temporaria* examined, 1,941; number of *R. temporaria* carrying *G. euzeti*, 576; number of *R. temporaria* carrying *P. integerrimum*, 280; number of *R. temporaria* carrying both *G. euzeti* and *P. integerrimum*, 39.

Calculating the prevalence of each parasite then allowed the calculation of the probable number of frog hosts in which the two parasite species should be found in association by chance alone. This number is $(576 \times 280)/1{,}941 = 83$, which is quite different from the 39 observed. Statistical analysis (chi-square) showed that the difference between the calculated and observed numbers had less than one chance in a thousand of occurring by chance alone ($p < 0.001$), leading to the conclusion that the presence of one parasite species limits the other despite the fact that they belong to different guilds and the habitats are apparently far from saturated. In this case it can be presumed that *P. integerrimum* prevents or slows down the settlement of *G. euzeti* and not the opposite, since *P. integerrimum* is recruited by the frogs during metamorphosis and *G. euzeti* only much later. Thus it is likely that the presence of adult *P. integerrimum* diminishes settling success for juvenile *G. euzeti* (Combes 1983b). Further, this example illustrates partial exclusion by a priority effect, since there are individual hosts in which both parasite species are found side by side.

Jackson and Tinsley (1998) have shown partial exclusion between two monogeneans of the genus *Protopolystoma*, *P. fissilis* and *P. ramulosis*, which were found together less often than expected by chance alone on the African amphibian *Xenopus fraseri*. If we suppose that the first parasite that settles somehow opposes the second's settlement, there also may be a priority effect as in the previous example. Competition is demonstrated in the experimental observation that if the two species are in the same host individual, eggs from concurrent infections show reduced viability, possibly because mating between the taxonomically related parasites gives rise to nonviable eggs.

Human Schistosomes

An assortment of experimental data seems to explain the peculiar geographical distribution of certain human schistosomes through a somewhat original exclusion process involving sexuality (Tchuem-Tchuenté et al. 1993; Tchuem-Tchuenté et al. 1997). In general, except in rare circumstances, the distribution of the schistosomes follows that of their molluskan vectors. However, the distribution of one of the human schistosomes, *Schistosoma intercalatum,* has intrigued researchers for some time, and in this case the vector is the pulmonate gastropod *Bulinus forskali,* which is widely distributed over the entire African continent except for the Maghreb.[8] In this instance, though, the distribution of *S. intercalatum* is much smaller than that of the vector, with *S. intercalatum* essentially limited to one focus in Cameroon and several other minor foci in central Africa and Mali.

This peculiar distribution was at first suspected to be a problem of incompatibility between *S. intercalatum* and *B. forskalii* outside transmission areas. Experimental investigation, however, quickly showed that, on the contrary, *B. forskalii* from regions as diverse as Egypt and the Ivory Coast were perfectly receptive to *S. intercalatum* from Cameroon, for instance. As a result, the constant movement of humans within contemporary Africa should logically distribute *S. intercalatum* over almost the entire continent, which does not happen.

Two solutions to this dilemma have been proposed, each involving the two other human schistosomes of Africa, *S. haematobium* and *S. mansoni.* Although only the second hypothesis (involving *S. mansoni*) calls for exclusion, both will be summarized here to highlight the differences between them.

Hypothesis 1: Schistosoma haematobium *and* S. intercalatum *both belong to the same group of terminally spined schistosomes and thus are closely related species that likely separated only a relatively short time ago*

The two species can cross and give fertile hybrids (recall that schistosomes have separate sexes) that show various characteristics intermediate between the parents (Taylor 1970; Pagès and Théron 1990a). Field observations in the region of Loum, Cameroon, have shown that such crosses

---

8. *Schistosoma intercalatum* from Zaire is a particular problem and is not considered here. The experiments reported here concern the "Lower Guinea strain."

occur in nature and are favored by the arrival of *S. haematobium* in recently deforested zones (Southgate, Van Wijk, and Wright 1976). In this way there may be a progressive dilution of the *S. intercalatum* genome into that of *S. haematobium* outside the forested areas. Thus *S. intercalatum* would subsist only in forests, that is, where *S. haematobium* does not penetrate because its vector, *Bulinus truncatus,* is absent. As a result, it appears that hybridization between *S. intercalatum* and *S. haematobium* has led within the past thirty years to the disappearance of *S. intercalatum* in certain African foci following environmental changes. This is supported by the fact that *S. intercalatum* was the only schistosome present (and in abundance) in the region of Loum in 1969. *Schistosoma haematobium,* however, was reported in this region in 1973, and in 1996 an investigation carried out on 426 schoolchildren showed that half had eggs of *S. haematobium* in their urine but no eggs of *S. intercalatum* present in their stools (see Tchuem-Tchuenté et al. 1997).

Hypothesis 2: Schistosoma mansoni *belongs to an evolutionary lineage (the "laterally spined schistosomes"), different from the terminally spined schistosomes to which* S. haematobium *and* S. intercalatum *belong*

Although *S. haematobium* and *S. intercalatum* can hybridize, *S. mansoni* and *S. intercalatum* cannot, and no hybrid descendants between the two species have been obtained in the laboratory despite numerous attempts. Curiously, however, *S. intercalatum* and *S. mansoni* are able to form interspecific couples: if mice are experimentally infected with male *S. mansoni* cercariae and female *S. haematobium* cercariae, or vice versa, "heterologous" couples are easily obtained. Based on this, researchers have performed multiple experiments and induced a surprising and asymmetrical form of competition. Some of these experiments include the following combinations and outcomes (fig. 13.10) (Tchuem-Tchuenté et al. 1993).

1. If a mouse is infected with male *S. mansoni* and with equal quantities of both *S. mansoni* females and *S. haematobium* females, almost 100% of the couples formed are *S. mansoni/S. mansoni,* indicating that males of *S. mansoni* prefer the females of *S. mansoni.*

2. If a mouse is infected with male *S. intercalatum* and with equal quantities of both *S. intercalatum* and *S. mansoni* females, about 75% of the couples formed are *S. intercalatum/S. intercalatum* and 25% are *S. intercalatum/S. mansoni.* Male *S. intercalatum* thus recognize *S. intercalatum* females but are less specific in mate choice than are male *S. mansoni.*

3. If a mouse is infected with equal quantities of *S. mansoni* and *S. intercalatum* males and with only female *S. intercalatum, S. mansoni/S. intercalatum* pairs are as numerous as *S. intercalatum/S. intercalatum* couples.

4. If a mouse is infected with equal numbers of *S. mansoni* and *S. inter-*

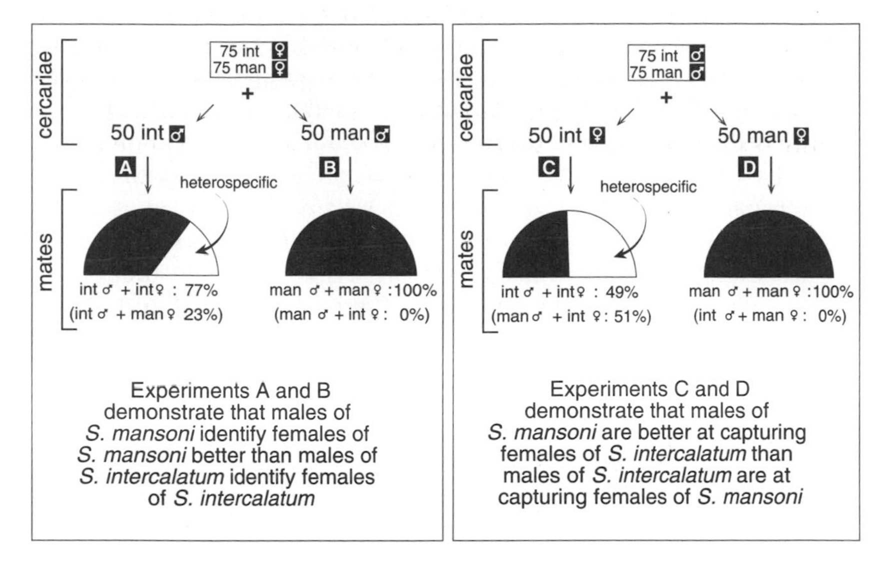

Figure 13.10. Differences in competitive values for the possession of females between two schistosome species: *Schistosoma mansoni* (man) and *S. intercalatum* (int). The figure summarizes the results of a series of experiments showing that *S. mansoni* males recognize "their" females better than *S. intercalatum* recognize "theirs" (experiments A and B) and that *S. mansoni* males compete for *S. intercalatum* females in the presence of *S. intercalatum* males but the opposite does not happen (experiments C and D). These results may explain why *S. intercalatum* cannot maintain itself in regions of Africa where *S. mansoni* also occurs.

calatum males and with *S. mansoni* females, the resulting couples are all *S. mansoni*/*S. mansoni*.

The interpretation of the first two experiments (A and B in the figure) is as follows: even if heterologous couples are possible, each species still prefers to mate with a homologous partner, although *S. mansoni* is better at doing so. Interpretation of the other two experiments (C and D in the figure) is as follows: in situations of competition for females, male *S. mansoni* are more efficient than are male *S. intercalatum*, since they can "capture" some females of the opposite species whereas *S. intercalatum* males cannot.

The overall interpretation is that the attraction between male and female *S. mansoni* is stronger than the attraction between male and female *S. intercalatum* and that male *S. mansoni* are more combative or attractive than male *S. intercalatum* and dominate them for the possession of females, whatever their identity.

This is one of the most original aspects of competition between parasites ever observed, and the ability of *S. mansoni* to appropriate *S. intercalatum* females may explain the restricted distribution of *S. intercalatum* in Africa if its fitness is strongly decreased when in sympatry with *S. mansoni.*

It is not impossible that such mate-mediated competition may be involved in the geographical distribution of other human schistosomes as well. It is troubling, for example, that replacement of *S. haematobium,* the causative agent of urinary schistosomiasis, by *S. mansoni,* the causative agent of intestinal schistosomiasis, has been reported several times in the literature (see, for instance, Colette et al. 1982).

The role of exclusion at such a geographical scale may also be exemplified by two species of acanthocephalans of freshwater fish, *Pomphorhynchus laevis* and *Acanthocephalus anguillae.* These two parasites avoid each other completely under natural conditions—not only are they not found in the same host individual but they are very rarely found in the same stream (Kennedy, Bates, and Brown 1989).[9] However, although acanthocephalans, like schistosomes, have separate sexes, it does not necessarily mean their lack of occurrence in the same host is due to heterologous pairing as in the schistosomes.

---

9. This investigation illustrates how difficult it is to conclude that exclusion has occurred, since there are other possible explanations for the two acanthocephalans' distribution (Bates and Kennedy 1991). For example, the two species have different crustacean intermediate hosts *(Gammarus pulex* for *Pomphorhynchus laevis* and *Asellus aquaticus* for *Acanthocephalus anguillae)* whose abundances may be different in different rivers. On the other hand, the two species may coinfect the same eel under experimental conditions.

14 Parasites and Host Individuals

Even classically "benign" parasites . . . can reduce host condition through the accumulation of subtle energetic costs over time.
Booth, Clayton, and Block 1993

Parasites exploit their hosts as energetic and habitat sources at the levels of the genome, the cell, the organism, and the superorganism.

Exploitation at the Level of the Genome

Transposable elements, repeated sequences, and pseudogenes all are nucleic acid molecules mixed in with functional sequences of a genome, and they all seem to provide no advantage to the host. Although geneticists hesitate to qualify such DNA sequences as "parasitic," parasitologists recognize that these noncoding sequences act as parasites that interact with their hosts' genomes. These sequences build molecules from material (phosphates, sugars, and bases) borrowed from the environment of the host genome and use enzymes coded for and produced by the host genome to assemble themselves.

As soon as DNA molecules behave as parasites of other DNA molecules, density-dependent processes of regulation similar to those described previously for more classical parasites can then be expected. For transposable elements, insertion may have at least three consequences that can result in regulation.

First, the insertion of selfish DNA may disturb normal gene expression. As a result its multiplication may be limited, as for all parasites, by its deleterious effect on the host. That is, the best chance for a transposable element to transmit itself is for the host genome to transmit its own genes; if a host individual's reproductive success is adversely affected by mutations induced by insertions, then the transposable element is itself counterselected. Transposable elements are usually more numerous in heterochromatin than in euchromatin, and the absence of selection against insertional mutations in heterochromatin (because it is genetically inert) probably explains their specific accumulation in these chromosomal regions (Charlesworth, Jarne, and Assimiacopoulos 1994).

A second consequence is that the insertion of transposable elements may disturb genetic recombination. If recombination occurs between identical copies of the same noncoding sequence on nonhomologous sites ("ectopic" recombination), it may lead to deleterious chromosomal rearrangements and therefore to elimination of the sequences involved (Charlesworth, Sniegowski, and Stephan 1994).[1]

Finally, some forms of intraspecific competition (self-regulation) may exist between noncoding sequences, lowering the level of transposition when the number of copies increases.

No mechanism of regulation is known for repeated sequences other than transposable elements, such as for the hundreds of thousands of *Alu* sequences in the human genome. Nevertheless it is almost certain that a selective process prevents such sequences from becoming even more invasive than they already are.

Exploitation at the Level of the Cell

One of the levels in the host at which parasites can take control is the cell. Among the best-studied examples are *Plasmodium,* which inhabit red blood cells; *Thelaria,* inhabitants of lymphocytes; and *Trichinella,* which live within muscle cells.

Red blood cells have no nucleus at maturity and exhibit a simplified metabolism with not much more than glucose being required from the host's plasma. *Plasmodium,* taking the form of a minuscule cell, the merozoite, penetrates the red blood cell, which then ensures its multiplication. Within forty-eight to seventy-two hours, depending on the species, many new merozoites are formed and burst from the red blood cell ready to infect new red blood cells in a continuous cycle of penetration, multiplication, and eruption.

Once invaded by a new merozoite, a red blood cell is rapidly transformed into a cell showing completely new characteristics and needs (fig. 14.1). It has new characteristics because the cell now has a genome (that of the parasite—it had lost its own during differentiation), and new needs arise because it must take from the blood substances not normally required by a healthy cell that are needed for the parasite's multiplication. That is, the red blood cell is transformed from relative inactivity into

1. However, this hypothesis is not supported by Hoogland and Biémont (1996) who demonstrate that there is no particular recombination of insertions in genomic regions with low recombination levels as would be expected if the hypothesis was correct. Curiously, the transposable element called "hobo" of *Drosophila melanogaster* is even more abundant in regions with high recombination. This could be explained either if the genome is more accessible to hobo insertions in regions of intense recombination or if this transposable element takes advantage of the recombination machinery for its insertion.

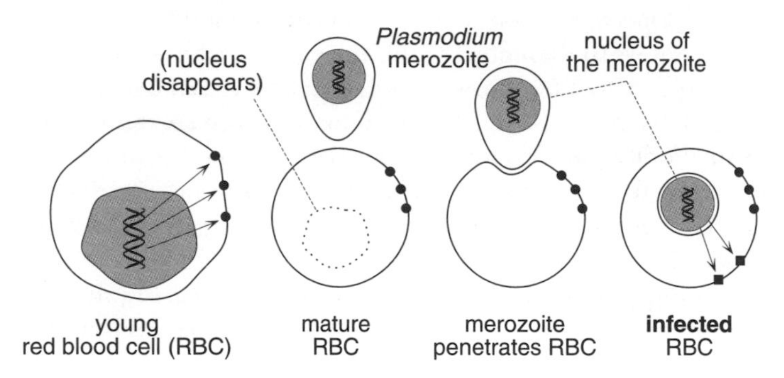

Figure 14.1. How a red blood cell loses its set of genetic information as it matures, and how it acquires another set when parasitized by *Plasmodium*. Arrows symbolize the expression of genes coding for proteins that insert themselves in the membrane of the red blood cell (RBC); solid circles = membrane proteins coded for by host cell nucleus during cell development; solid squares = membrane proteins coded for by the parasite.

a bustling factory: glucose consumption is increased by a factor of ten or more, and amino acids, lipids, and other substances become necessary to fulfill its demands, even as the digestion of hemoglobin provides some of these needed products.

The presence of *Plasmodium* in the red blood cell also induces changes in the protein composition of the host cell membrane even though the parasite is not itself in contact with the membrane, leading one to wonder how the parasite induces such changes (see Newbold and Marsh 1990; Sherman and Winograd 1990). All normal eukaryotic cells have various systems of vesicles that ensure the internal trafficking of molecules to and from the membrane, but red blood cells do not. It may be that the parasite induces the appearance of such vesicles in the cell's cytoplasm to ensure protein transport between the membrane enclosing the parasite and the host's peripheral plasma membrane. Alternatively, it may be that the parasite remains linked to the outside by a thin canal to allow molecular trafficking in both directions. Regardless of how the needed proteins get there, these membrane proteins then provide the functions necessary to supply *Plasmodium* with the appropriate metabolites. The presence of new proteins on the surface of parasitized blood cells does not, however, fail to catch the attention of the host's defenses, and selective pressure may have retained the best compromise between the parasite's risk of starving to death and its risk of being recognized, attacked, and killed by the host's immune system.

Some of the proteins inserted into the parasitized red blood cells' membrane are responsible for the seriousness of the illness caused by

Plasmodium falciparum. These proteins appear as small "knobs" (about 70 to 160 nanometers in diameter) when merozoite release is close at hand; they cover up to 5% of the blood cell's surface and make it adhere to the endothelial cells of venules (cytoadherence). In the cerebral circulation obstructions formed in this way may be fatal, and cells can also accumulate in the placenta in pregnant women. Such cytoadherence happens only in certain species of *Plasmodium* and, depending on the species, may occur in a variety of host organs (for instance, the heart, liver, or lungs). The protein responsible for cytoadherence in *P. falciparum,* termed knob associated protein (KAB), inserts itself into the plasma membrane while also linking to submembrane cytoskeletal components (spectrin, actin, and ankyrin) that give the cell its rigidity.

The adherence of parasitized blood cells to endothelium presents an interesting problem: What is its value for *P. falciparum*? The two most commonly proposed hypotheses are that cytoadherence makes it much easier for the merozoite to infect healthy blood cells because blood flow is slowed and the density of red blood cells is high at the site of obstruction and that cytoadherence keeps parasitized cells away from the spleen, where the phagocytic cells of the immune system are most efficient.

Before infecting red blood cells, *Plasmodium* first infects liver cells in the hepatic portion of its life cycle, also imposing transformation on these cells even though they are nucleated. As in red blood cells, proteins coded for by the parasite's genome insert themselves into the hepatocyte's plasma membrane, which may then become the target of immune effector mechanisms. For example, CD8 lymphocytes attack infected hepatocytes by identifying associations between major histocompatibility complex (MHC) type I molecules and parasite epitopes on the cell's surface.[2]

The modifications a parasite imposes on its host cell may occur very early in infection and likely makes use of preexisting signals or signal mechanisms. This is the case for *Trypanosoma cruzi,* a protist responsible for Chagas' disease in South America. Unlike *Plasmodium* in its blood phase, *T. cruzi* invades various nucleated cells, particularly macrophages. Invasion here includes adhesion of the parasite to the membrane; internalization of the parasite in a vacuole; rupture of the vacuole about two hours after penetration, with passage of the parasite into direct contact with the cytoplasm; and multiplication. During the process numerous lysosomes normally confined to the proximity of the nucleus of the host

---

2. The possibility that *Plasmodium* might be attacked by the immune system while in the liver raises hope that an antimalaria vaccine could act at this first stage of the disease and thus prevent the second, more serious, stage in red blood cells (Mazier 1993).

cell move toward the zone of penetration and fuse with the forming vacuole surrounding the parasite. These lysosomes move along cytoplasmic microtubules and, significantly, penetration is inhibited if the microtubule network is damaged or if the lysosomes are otherwise prevented from fusing with the parasite vacuole by a biochemical process. On the other hand, cells become more susceptible to infection if the movement of perinuclear lysosomes toward the periphery is eased or induced (Tardieux et al. 1992). That is, it seems that lysosome movement and fusion are necessary for *T. cruzi* to invade macrophages. Since the parasite penetrates without phagocytosis (without pseudopod formation), the lysosomes are likely to be the source of membrane material for building the vacuole instead of the macrophage's plasma membrane. Thus the parasite in this case would be taking advantage of preexisting signal mechanisms (normally responsible for the recruitment of lysosomes, their fusion with phagocytic vesicles, and the destruction of internalized objects) to ensure cell invasion.

The intracellular parasites that induce visible modifications of their host cells are numerous. For instance, microsporidia of the genera *Glugea* and *Nosema* invade and induce hypertrophy in certain cells in their fish hosts with or without the division of their nucleus (see Weidner 1985). In this case the invaded host cells are completely reorganized to the benefit of the protist and are called *xenomes*, indicating that the cells are under the control of a foreign *(xeno-)* genome.

Another remarkable example of exploitation associated with the manipulation of a host cell is given by *Theileria. Theileria parva* is the causative agent of a serious and often fatal cattle disease, theileriosis, mostly known from central and eastern Africa. Theileriosis is transmitted to cattle by the tick *Rhipicephalus appendiculatus,* which belongs to the family Ixodidae. Ticks pick up parasites from an infected animal, and after they develop in the salivary glands they are injected into another animal with the tick's saliva during a blood meal. As in *Plasmodium,* the parasites are injected in the form of sporozoites; as soon as they enter the bloodstream they penetrate T lymphocytes, where asexual reproduction (schizogony) occurs. This schizogony is similar to what occurs in *Plasmodium,* but it takes place in lymphocytes instead of erythrocytes. A further difference relates to the extension of the phenotype of *Theileria* into its host cells; to examine this one must recall two of the normal activities of healthy lymphocytes.

The first involves immunity: T cells are one of the key elements of the immune response against any foreign agent that penetrates an organism. Antigens (all or part of the invasive agent) are presented to some T cells by macrophages and are recognized by surface receptors in the T cell's membrane. Because millions of possible antigens exist, natural selection in vertebrates has "created" millions of possible receptors, each capable of recognizing a specific antigen. Each T cell expresses on its surface only one kind of the millions of possible receptors, and the efficacy of the immune system is due in large part to the specificity of this system. When a particular antigen penetrates an organism, it is recognized only by lymphocytes carrying the receptor that corresponds to that antigen. The specific lymphocytes carrying this specific receptor for that antigen start proliferating in the form of clones to induce a larger specific response against the aggressor exhibiting the antigen. This clonal proliferation is induced by a cascade of signals, with antigen recognition by the T cell ultimately leading to release of interleukin-2 (IL-2). IL-2 released by the T cell in turn binds to IL-2 receptors (IL-2R), which then triggers a series of events leading to cellular division and the activation of other cell types to join in the fight against the invasive objects carrying the antigen originally presented by the macrophage.

The second normal lymphocyte process that is important to remember relative to *Theileria* is the way the lymphocytes multiply, although this in itself is not original. In short, lymphocytes multiply via binary fission (mitosis), where the mother cell replicates its chromosomes, a mitotic spindle is formed, and replicated chromosomes are equally shared between the two resulting daughter cells.

The two previous processes (clonal proliferation and mitosis) are modified by *Theileria* as follows.

1. When T cells are infected by *Theileria,* they no longer require stimulation by an antigen to undergo proliferation. Instead, an internal "signal" coming from the parasite activates the IL-2/IL-2R system (multiplication will stop immediately after infected lymphocytes are treated with a substance derived from naphthoquinone that is toxic for *Theileria*).

2. The divisions of the parasite and lymphocyte become synchronous, with the parasite nuclei incorporating lymphocyte mitotic spindles in order to be pulled toward the mitotic poles and distributed between the two resulting daughter lymphocytes.

Taking control of the multiplication processes of the host cell in this way thus ensures the parasite the automatic multiplication of its habitat while preventing its contact with host serum and thus with host antibodies (Dobbelaere et al. 1988; Heussler et al. 1992). Similarly, when *Theileria* are in their vector ticks, they penetrate and also manipulate the phenotype of the tick's salivary gland cells, inducing in particular a steep increase in glycogen reserves and hypertrophy.

Along the lines of host cell manipulation, note that bacteria also use their host cell proteins to travel inside cells. *Shigella* (the causative agent of shigellosis, a serious dysentery in children from areas with poor sanitation) moves in intestinal cells by inducing the polymerization of actin G into "actin motors" (Prévost et al. 1992). Additionally, one can also cite HIV, which by adhering to CD4 molecules and lymphocyte coreceptors triggers several cascades of events, including kinase and transcription factor activation, that in turn trigger apoptosis (programmed cell death) in the host cell.

Protists and viruses are not the only organisms capable of entering the host cells and reorganizing them to their advantage, and some metazoans even go so far as to reshape the cellular environment. *Trichinella spiralis,* although a metazoan with differentiated organs, penetrates only the host's muscle cells and no others, causing serious problems if the number of larvae is high. As soon as a *Trichinella* larva penetrates a muscle cell, it completely reorganizes the cellular environment to its own advantage without killing the cell (see Despommier 1993). This transformation, which occurs within about twenty days, totally eliminates the actin and myosin filaments of the muscle cell. At the same time, it causes an increase in the size of the nucleus (suggesting intervention with the host genome) and the differentiation around the cell of a capillary network that will carry nutrients to the parasite.

The example of *Trichinella* shows that the extended phenotype may rise above the level of the parasite's immediate surroundings and brings up the fascinating problem of how a parasite can command host morphogenetic mechanisms such as angiogenesis (blood vessel proliferation). Plant galls are yet another example of parasites reorganizing the environment to their own benefit. In this case the gall develops around particular dipteran (cecidomyiid) or hymenopteran (cynipid) larvae, and it is interesting that in these cases the extended phenotype crosses the boundary separating the plant and animal kingdoms without difficulty.

Exploitation at the Level of the Organism

Host Health

Paradoxically, the parasite's interest is to exploit its host as much as possible while at the same time keeping it in as good a health as needed. However, if there is a ceiling above which virulence cannot climb, there is also a floor below which it cannot go. Just because a parasite does not appear to hurt its host does not mean no harm occurs; as small as it might be, there is always a minimum cost to parasitism.[3]

3. To make the concept more concrete, note that an ant weighing one milligram that embarks in a Boeing 747 to fly from Paris across the Atlantic costs the airline ten thousand molecules of fuel.

These three studies demonstrate that we need to temper the widespread notion that some parasites are totally nonpathogenic.

1. Connors and Nickol (1991) studied the effects of the acanthocephalan *Plagiorhynchus cylindraceus* in the starling, *Sturnus vulgaris*. Their results showed that this intestinal parasite, which up to that time was considered only slightly pathogenic, in fact greatly modifies the flow of energy in its host. After experimentally infecting each host with fifteen cystacanths, the authors detected a significant perturbation in basal metabolic and thermal regulatory capacities. This disruption was accompanied by a significant weight loss in infected individuals even though they ate more food, most likely because the parasite affected digestion or intestinal absorption. Further, parasitized birds were not as resistant as controls to the stress induced by colder temperature (cold-stressed infected birds lost substantially more weight than did controls), and the only birds to die over the course of the study were infected cold-stressed females.

2. Booth, Clayton, and Block (1993) compared the effects of "heavy" and "light" loads of feather lice in birds. In this case parasite loads in the birds were experimentally manipulated by preventing preening. A small metal device inserted between the mandibles, anchored to the nostrils, left a space of one to three millimeters between the mandibles, preventing the birds from completely closing their beaks to preen. In the experiment, all birds captured in the field in May in Illinois had their beaks modified, and some of them ("light load" birds) were fumigated to destroy parasites. Other birds were not treated and were considered "heavy load" birds. The treatments were repeated in July, August, and October, and in February of the next year twenty-four birds were recaptured. Overall their average feather lice loads at the time of recapture were 450 on "heavy load" birds and 104 on "light load" birds. Birds with heavy loads showed reductions in feather mass and also weighed less than birds with light loads. Further, the metabolic rate of birds with heavy loads was 8% higher than those with light loads even though their body temperature was the same. Such changes indicate that they raised their metabolic rate to compensate for the parasites' effect on insulation. From this we may deduce that winter survival might not be as good in birds harboring heavy lice loads because of decreased thermoregulatory abilities. The authors concluded that although the effect of parasites on wild hosts' fitness is not negligible, it is nevertheless very difficult to measure.

3. Ratti, Dufva, and Alatalo (1993) demonstrated an unsuspected cost of parasitism in flycatchers *(Ficedula hypoleuca)*, migratory birds whose spring arrival in Europe occurs over several days. Males parasitized by an apparently nonpathogenic trypanosome arrive, on average, two days af-

ter healthy birds. This delay reduced the infected birds' reproductive success, because males that arrived first had more opportunity to become polygamous and thus to transmit their genes.[4]

It is very likely that numerous parasites normally thought to be "nonpathogenic" render infected animals more vulnerable to predation because they either are less able to flee or spend more time foraging. It is also very likely that many parasites render hosts less competitive reproductively because of subtle yet significant demands or shifts in energy use. It is thus imperative not to confuse reduced pathology with the absence of pathology, since a variety of subtle and not readily apparent effects can occur. That is, assessments of localized pathologies and gross measures of body mass do not suffice to gauge the effects of infection on hosts, and it is dangerous not to address these hidden impacts if we are to make meaningful conclusions about host ecology and behavior (Connors and Nickol 1991).

For human health the notion of considering parasites to be of little or no pathogenicity is as dangerous as it is in ecology, and parasites that seem more or less well tolerated by the body may in fact have totally underestimated consequences. Among these is the intellectual underdevelopment of children, especially when numerous pathogenic agents can accumulate in individuals, as occurs in many underdeveloped countries. Nematodes particularly seem to cause such effects (see Nokes and Bundy 1994) because they may have both a direct effect (the weakening of cognitive functions) and an indirect effect (school absenteeism).

The opposite of cases where pathogenicity is scarcely apparent are cases where the entire organism is under the control of the pathogenic agent. This is an extreme form of the durable interaction, and two examples where parasites deeply modify the anatomy, physiology, and behavior of their hosts in order to maximally exploit them are easily highlighted: schistosomes in their molluskan intermediate hosts and endoparasite crustaceans such as *Lernaeodiscus* and *Sacculina*.[5]

Schistosomes and Mollusks

The development of trematodes in mollusks represents one of the most completely elaborated examples of control in the parasite world. In general, the mollusk-trematode association has three characteristics: first, in

4. However, the hypothesis that infected birds may have been weakened by something else, leading to infection by the trypanosome, cannot be totally discarded.

5. A fuller review of exploitation by parasites at the scale of the host organism can be found in Hurd (1990).

terms of biomass the parasite-host relationship is exceptional: Gérard (1993) has noted that the parasite volume/host volume ratio is about 0.25 for schistosomes in their molluskan intermediate host, whereas it is only 0.000005 for the same parasite in its human host. Second, the parasites multiply on site, which implies both a high metabolic demand and the necessity for a regulatory process. Finally, as soon as it is infected a mollusk becomes an entirely different organism controlled by a double genome comprising that of the trematode, which becomes the director, and its own genes—with only those functions serving the parasite being conserved.

To understand this exploitation it is important to remember that, most often, infection of a mollusk by a trematode has two phases. In the first phase (the prepatent period), penetrating miracidia produce either sporocysts or rediae in the mollusk's visceral mass (the later stages have a digestive tract but the former do not). In schistosomes there is no redial stage, and the prepatent period lasts about four weeks, followed by the patent period. In this second phase the sporocysts (or rediae in other trematodes) ultimately produce cercariae, which actively exit the mollusk. In schistosomes the patent period usually lasts several months or until the host dies.

Théron and his students studied how resources are affected in the pulmonate snail *Biomphalaria glabrata* when parasitized by *Schistosoma mansoni.* Their research produced several interesting results, including demonstrating that a parasitized individual does not show an increase in trophic activity. Thus the relatively large quantity of energy the developing parasite requires is taken from what is normally available to the snail because of altered allocations that favor the parasite. Second, the part of the host energy budget the parasite garners energy from differs between mollusks that are parasitized during their growth period and those parasitized at maturity (Théron, Gérard, and Moné 1992b; Théron and Gérard 1994). When the mollusk is parasitized during its growth phase (when miracidia penetrate juvenile mollusks), sporocyst production during the prepatent period commandeers energy normally allocated to the development of the snail's reproductive system, which fails to form (fig. 14.2), whereas cercarial production during the patent period reroutes part of the energy normally used for growth, which is slowed.

Alternatively, if the snail is parasitized when mature (when miracidia penetrate an adult mollusk), sporocyst production (the prepatent period) mobilizes the snail's energy reserves (glycogen, etc.), which are then used up (fig. 14.3A). Cercarial production during the patent period, however, reroutes energy normally allocated for snail reproduction to the parasites so that those hosts, originally with normal reproductive systems, stop laying eggs (fig. 14.3B)

Last, the hijacking and rerouting of the host's energy budget alters the

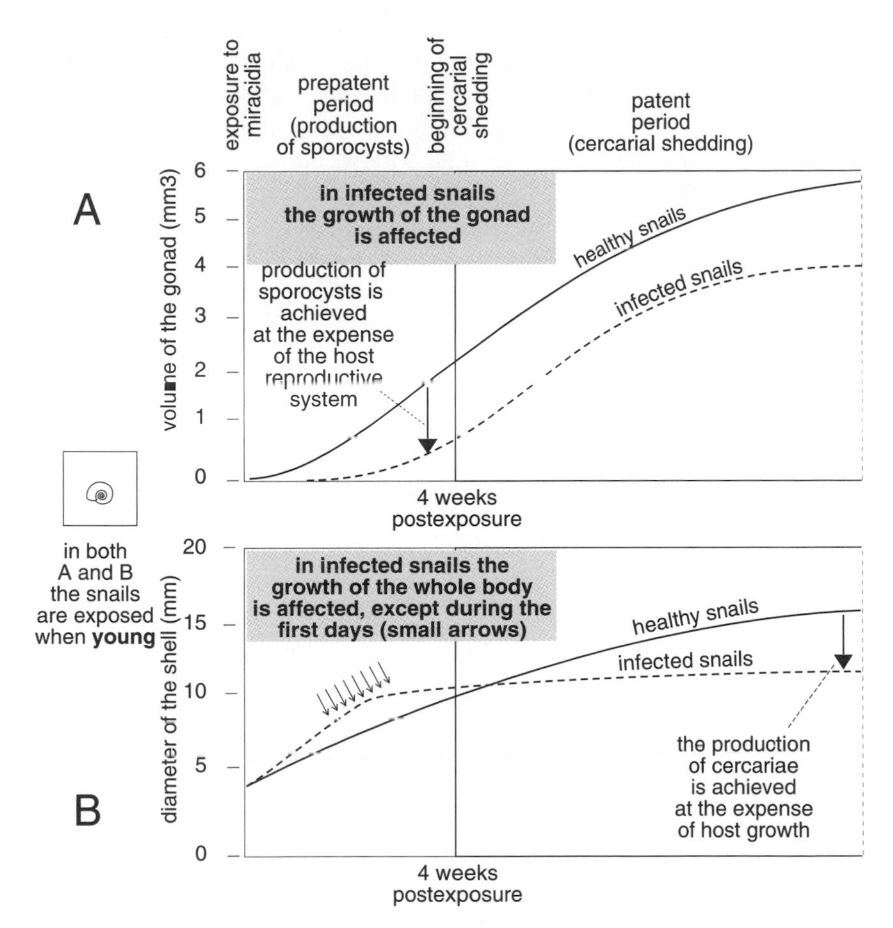

Figure 14.2. The two compartments from which *Schistosoma mansoni* takes most of its energy while developing in juvenile mollusks (modified after Gérard and Théron 1997).

snail's physiology. For instance, at the very beginning of the prepatent period growth is stimulated (favorable to the parasite, since habitat dimensions increase), and fecundity is also stimulated (which may be a kind of compensation to the host for its later castration). Along these lines note that infected snails prefer warmer habitats (Lefcort and Bayne 1991). This effect has been attributed to the increased production of cytokines to fight infection in the snail (ibid.), and in the field this preference would tend to place snails shedding cercariae closer to the shore, increasing the probability of infecting the mammalian host (Vincent A. Connors, personal communication).

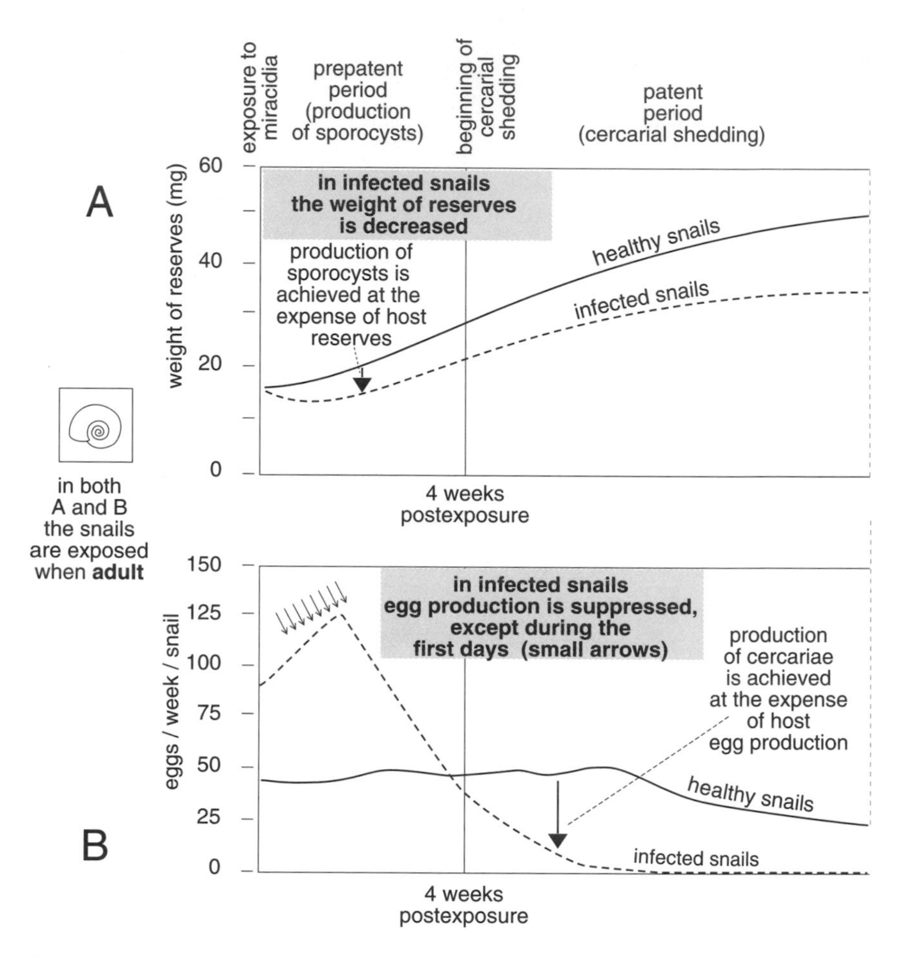

Figure 14.3. The two compartments from which *Schistosoma mansoni* takes most of its energy while developing in adult mollusks (modified after Gérard and Théron 1997).

The stimulation of fecundity following schistosome infection was first discovered by Minchella and LoVerde (1981) and has been the object of some debate since then. This stimulation may be construed as advantageous to either the parasite (in which case it would be selected in the schistosome genome) or the host (in which case it would be selected in the mollusk genome). It would be advantageous for the parasite to allow the compatible host to transmit its genes; otherwise castration would only decrease the number of receptive hosts. In this case the initial stimulation of fecundity may also be interpreted as having a compensatory effect for the host, letting it transmit a maximum number of its genes before be-

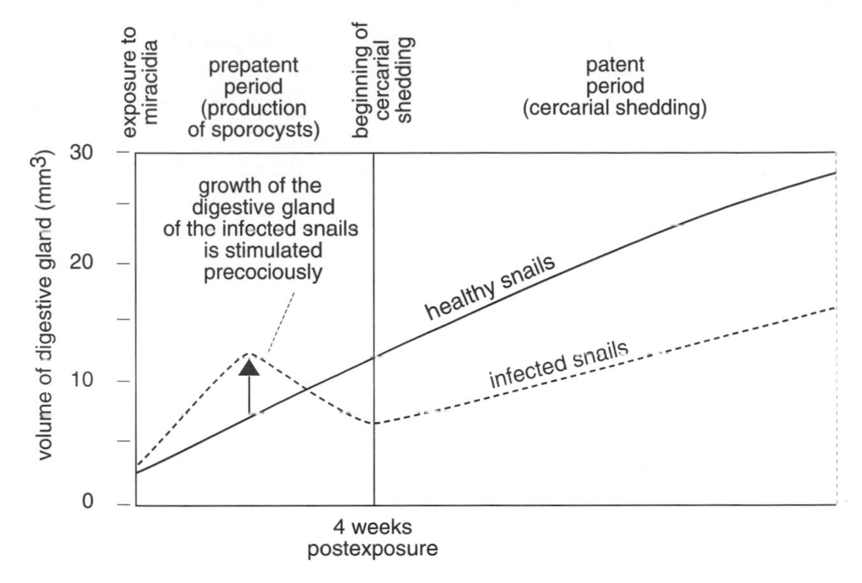

Figure 14.4. Early stimulation of the digestive gland in *Biomphalaria glabrata* infected by *Schistosoma mansoni* as juveniles.

coming genetically dead as the result of infection. It is interesting that fecundity is stimulated in mollusks exposed to miracidia even when the parasites have not developed.

Another alteration of the snail's physiology is the stimulation of digestive gland development (fig. 14.4). If this is a true manipulation by the parasite, it may favor the installation of sporocysts in this organ (about the fourth week after infection). Such an early growth in digestive gland volume might be "preparation" for the parasite and may increase reserves for later use (Théron, Gérard, and Moné 1992a).

When we summarize manipulation of the mollusk by the schistosome (which likely will be found with variation in other trematodes), it appears that immature and adult mollusks are environments with different natures for the parasite and that they require two distinctly different reallocations of resources to ensure cercarial production (Théron and Gérard 1994). It is the energy normally allocated to growth and development that is commandeered in immature mollusks and the energy normally allocated to reproduction that is rerouted in adults.

Overall, exploitation of space or energy by the schistosome is eased because the parasite takes control of the mollusk's physiology. Assorted mechanisms have been proposed to allow this control, and Joosse and Van Eik (1986), Schallig et al. (1991), and Hordijk et al. (1991) have obtained interesting results from studies on a closely related parasite-host system similar to that of *S.mansoni/B. glabrata: Trichobilharzia ocellata* in the snail *Lymnea stagnalis*. In the latter case it has been hypothesized that a not yet identified signal from parasites in the cercarial stage would be recognized by endocrine cells (dorsal bodies) in the mollusk's central nervous system. In response, these cells would then secrete a substance referred to as schistosomine. This molecule, or perhaps another one, would then inhibit the gonad or accessory sexual organs (albumin gland) through an antagonistic effect on gonadotropic hormones. This manipulation, however, has not yet been confirmed.

A large portion of the ensemble of mechanisms of the extended phenotype necessary for the parasite to control the allocation of resources and to limit its invasion (probably by feedback mechanisms) in the molluskan host are still unknown, and it seems that each parasite may have selected its own mode of controlling its host's physiology. For example, the trematode *Prosorhynchus squamatus* does not act on the sexual hormones of its host (the bivalve mollusk *Mytilus edulis*) to reroute host metabolites to its own benefit but instead releases a factor that stimulates (apparently directly) the mobilization of glycogen reserves and inhibits (either directly or indirectly) the division of host's sex cells (Coustau, Renaud, Delay, et al. 1991). In some instances the growth of the molluskan intermediate hosts is not slowed but stimulated. Such an effect leads to gigantism, whose adaptive value is either accepted (Baudoin 1975; Minchella 1985) or argued against (Sousa 1983; Ballabeni 1995).

Nematodes also know how to interfere with their hosts' metabolism. For instance, in scolytid coleopterans numerous proteins synthesized by adipose tissue are stimulated as soon as the parasite *Contortylenchus diplogaster* establishes itself (Lieutier 1984).

In both mollusk/trematode and insect/nematode systems the hosts and parasites belong to relatively distantly related taxonomic groups. When parasitism occurs between members of the same group, however, the molecular interactions become somewhat easier. This is true, for example, of those insects (parasitoids) that parasitize the larvae of other insects (see review by Beckage 1991). Harwood et al. (1998), writing about the wasp *Cotesia congregata,* a parasitoid of the sphingid *Manduca sexta,* note: "Larvae of *M. sexta* parasitized by *C. congregata* exhibit a pleiotropic physiological reprogramming that supports successful development of the parasitoid's progeny." Here the hormones of the parasitoid and its hosts are, if not the same, at least very similar. In the case of the *M. sexta/C. congregata* system, juvenile hormone (JH) titer often rises in parasitized hosts so that the host larval stage is prolonged to the advantage of the parasite. The procedures parasitoids use to manipulate host

JH levels are variable and include the direct secretion of JH by parasitoid larvae, the secretion of molecules stimulating the synthesis of host JH, and the inhibition of molecules inhibiting JH-esterase (an enzyme that inhibits JH). The insect parasite, reminding one of trematodes, is also able to inhibit the early development of its host's gonads. Finally, Geneviève Prévost and Michel Bouletreau (personal communication) have shown that larvae of *Drosophila melanogaster* parasitized by the wasp *Leptopilina boulardi* (which will kill its host in the end as all parasitoids do) are more competitive than nonparasitized larvae when conditions (overpopulation) become difficult, a fact that can be explained only by the parasitoid's affecting its host's physiology.

Lernaeodiscus and the Crab

The rhizocephalid crustacean *Lernaeodiscus porcellanae* is a parasite of the crab *Petrolisthes cabrilloi* in the North Sea. The life cycle involves a swimming larval stage that enters the crab at the level of the gills, migrates to the abdomen, then develops into an amorphous blob toward the outside that is, in fact, the female parasite. At this point a free-swimming male larva throws itself onto the blob, injects its testis, and dies. The parasite at this time is functionally hermaphroditic as the result of a testis graft and can produce several thousand fertile eggs.

Next this parasite's manipulation employs a truly interesting strategy. Female crabs in general tend to take great care of their own eggs, which form an attached mass under the abdomen. During the entire period of egg development the crab uses its abdominal appendages to protect and aerate the eggs. Normal male crabs, of course, do not have eggs under their abdomens and do not show this behavior. If the parasite (a barnacle) ends up under the abdomen of a female crab (thus occupying the regular site of its own brood), the manipulation is easy: there is only the job of sterilizing the host, preventing it from producing its own eggs, and "ordering" it to ventilate the parasite's brood as if it were its own. The female crab will continue this cooperative behavior up through the time when the eggs of *L. porcellanae* hatch, at which time it will climb on a rock and "wave" its abdomen to create a current that favors the dispersion of the young parasites. However, if the host crab is a male this elaborate manipulation is not possible, because the crab does not possess the appropriate ventilating machinery under the abdomen and also does not know how to take care of a brood. Such details do not stop the parasites, though, and when they parasitize a male crab they simply transform it into a female. The parasite has selected all the genes necessary for this manipulation.

This information on the biology of *L. porcellanae* was borrowed from the work of Ritchie and Hoeg (1981), and the same type of relationship exists in comparable parasite-host systems such as *Sacculina,* another rhyzocephalid parasite of crabs that behaves exactly like *Lernaeodiscus* in its host crab.

Besides the "classical" exploitation of metabolic resources, rhyzocephalid parasites of crustaceans also preempt care normally given to the host's own young. This care belongs to the organism's extended phenotype, and as noted at the very beginning of this book, any such extended phenotype may be exploited by parasitism.

In the interactions cited above the procedures of manipulation are ingenious, and the parasite is able to take over without direct modification of the host genome. However, it is just such a direct modification that allows a bacterium to obtain energy at the expense of some of its plant hosts.

Agrobacterium and Plants

Agrobacterium tumefasciens causes tumors called crown galls in various plants. If some cells (without bacteria) are recovered from gall tissues, several surprising things can happen: these cells continue to multiply in culture even in the absence of the normally necessary plant hormones, and when grafted onto a healthy plant they induce tumors. The explanation lies in the transfer of a Ti plasmid (Ti for "tumor inducing") from the bacterial genome to the plant cell's genome. It is not the entire bacterial genome that is transferred, however, but only the plasmid, which is a smaller piece of DNA about 150,000 to 250,000 base pairs long. The plasmid is inserted into the DNA of the host cell like a dancer into a chorus line. There are several varieties of Ti plasmids, and all have genes coding for specific proteins. Thus, infected cells express an "abnormal" phenotype and the bacteria feed on the proteins synthesized from material the plasmid makes available in the cell at the expense of the host. Although the bacteria are parasitic in the metabolic sense, however, they do not use the host cell as a habitat—it is only the plasmid that is inserted into the genome of the plant cell.

Manipulation up to Where?

If a parasite's genome can take control of an organism as in the systems described above, could there be other hidden and unsuspected manipulations? When we see a trematode's genome extend its phenotype into the nervous system, endocrine system, or various organs such as the digestive glands or accessory parts of the genital system, and when one sees trematodes manipulate their molluskan hosts' growth, fecundity, and behavior, one may wonder whether this situation is exceptional or whether it merely illustrates a more general phenomenon that simply is not well documented.

In other words, one must admit that, based on the data currently available, we do not know at what point a host expresses, physiologically and ethologically, its parasite's phenotype as much as, or more than, its own. It is very possible that parasitoses, including those in humans, may have consequences beyond the classical pathogenic effects. That is, there is nothing to indicate that being a vertebrate, even a human, protects one from subtle but deep manipulations that influence both behavior and health more generally.

Exploitation at the Level of Superorganisms

Colonies of social insects (bees, ants, and termites) are considered superorganisms. Although the genetic similarity of the individuals forming an ants' nest is not as great as that of the cells forming a classic metazoan, everything else indicates that the colony should be considered a single entity. Since organisms parasitize other organisms, one can then wonder if some superorganisms parasitize other such superorganisms. Indeed, this is what occurs in "social" parasitism, which has been known and studied for some time. Since the relationship between the two superorganisms lasts for a prolonged period, it is also a durable interaction.

In ants social parasitism is frequent and occurs in several ways.

The first possibility: only the queen is parasitic, and the workers she produces work within the parasitized nest. In this case the parasitic queen does everything to become accepted by the host workers and either becomes the favored queen or kills the host queen.

The second possibility: only the queen is parasitic because workers do not exist in her species. In this case the queen parasite eliminates the host queen and is indefinitely served by the host's workers.

The third possibility: both the queen and her workers are parasitic, because the workers of the parasite species do not work or do very little work. In this case the parasitic colony forces the collaboration of host workers captured as cocoons during "raids."

The mechanisms of parasitic exploitation between superorganisms like ant colonies are very similar to those of parasitism in general, and after obtaining acceptance thanks to a variety of subterfuges involving mimicry, parasitic ants then manipulate their victims' behavior through biochemical processes. Jaisson (1993) considers that colony function in social insects is based on absolute trust (each individual works for the well-being of others without worrying about obtaining something "personally") and notes that social parasitism consists of diverting this trust. Exploited ants serve their exploiters exactly as a mollusk serves its trematode—just as the mollusk host's activity leads to the fabrication of trematode cercariae, the activity of the ant host leads to the production of parasitic ant larvae.

Parasitism of this sort may occur between social insects belonging to distinctly different taxonomic groups. For example, the ant *Hyponeura eduardi* nests in the peripheral galleries of termites of the genus *Reticulitermes* in the south of France. Here the ants are protected by cuticular hydrocarbons that mimic those of the termite. As a result, every day the ants devour some of the hosts, which allow themselves to be approached without ever mounting a defense (Lemaire et al. 1986). This is not a predator-prey system as might at first be thought, because parasitism between the two superorganisms constitutes a durable interaction. Moreover, the virulence of the ant colony is limited so that it does not destroy the termite colony, just as a parasite does not kill the host it needs for continued survival.

Colonies of social insects do not attract only other social insects; they can be parasitized by nonsocial insects or other arthropods that also use biochemical ruses to make themselves tolerated or served. The number of such interactions may be far greater than once imagined.

15 Parasites and Host Populations

Parasitic species are capable of regulating the growth of host populations, even in the complete absence of other influences such as predation or intraspecific competition.
May and Anderson 1978

Parasites should be more speciose (having higher gamma-diversity) than free-living organisms.
Toft 1991b

Effect on Demography

The Addition/Compensation Debate

The innate capacity of a population to increase or decrease in size is characterized in mathematical terms by its net reproductive rate (R_0) a measure of the mean number of descendants per individual in the population. If R_0 equals one the population remains stable, since each individual is, on average, replaced by only one other individual able to reproduce. Similarly, if R_0 is greater than one the population increases, and if it is less than one it decreases. However, there may be a considerable difference between the number of births and the number of individuals surviving to maturity.

As Darwin, and before him Malthus, noted, all living organisms have an innate capacity to increase their number from generation to generation. This capacity is, however, confronted with the dual realities that the available environment (the earth) has discrete and limited dimensions and that all other species have a comparable innate capacity to increase their own populations. As a result, all populations are constrained or somehow regulated in their growth.

Numerous processes such as intra- and interspecific competition for space and resources, predation, and parasitism have the potential to play a role in regulating populations. The question of interest for parasitologists, and more recently for ecologists in general, is, What portion of regulation is due to parasitism? If the role of wolves in regulating ungulate population is not questioned, what role do liver flukes, intestinal nematodes, larval cestodes, nasal bots, and other parasites play in regulating the same ungulate population?

Parasites are in competition with their hosts for resources, so host fecundity and survival are affected even if, in certain cases, there is no visible pathology and even though their effects may at times be blended into the total of all factors affecting fecundity and survival.

For some parasitologists the distinction between microparasites (viruses, bacteria, protists) and macroparasites (helminths, arthropods, etc.) involves a difference in pathogenicity and therefore a difference in their potential to regulate host populations. Microparasites would cause serious damage because they multiply in their host after infection, and macroparasites would cause less because, except in rare cases, they do not multiply after infection. The accepted idea is that in parasitoses "of accumulation" (i.e., caused by macroparasites), infrapopulation recruitment does not usually reach sufficient levels for a pathogenic effect to be detected. In reality though, numerous parasitic metazoans exert heavy enough pressure on their hosts to play as much of a role in population regulation as do microparasites.

When studying a parasite's effect on a host population one must first know if the mortality induced by the parasite (when in fact it does occur), is additive or compensatory. This distinction must be made clear in order to determine if the parasite can regulate the host population.

As noted by Holmes (1982), some epizootics or epidemics are triggered when the host population reaches very high densities, at which point there is a sudden decrease in the population owing to the pathogenic agent. This decrease in the host population would not have occurred in the absence of the parasite and thus is additive—that is, it adds to other causes of mortality in the population.

For many parasitoses, however, such additive mortality is more difficult to confirm. For instance, one can imagine a parasitized and weakened individual that dies under unfavorable conditions or is captured by a predator. If this death takes the place of another death that would otherwise have occurred, parasitism does not add to host population mortality but instead only orients mortality toward certain individuals. This mortality is compensatory.

It is important to recognize the distinction between additive and compensatory mortality because they have different consequences on the host population. If there is additive mortality the host population is regulated at a lower level in the presence of the parasite than when parasites are absent. On the contrary, if there is compensatory mortality, the presence of the parasite does not modify or change the host population size, and the mortality induced by parasitism is not synonymous with regulation.

Experimental studies bring valuable information to bear on the problem. For example, the study of Keymer (1980) on the cestode *Hymenolepis diminuta* showed that this parasite has the potential to regulate natural populations of its intermediate host, the coleopteran *Tribolium confusum.*

Similarly, studies under "seminatural" conditions by Gregory (1991b) on the rodent *Apodemus sylvaticus,* and its nematode parasite, *Heligmosomoides polygyrus* showed that survival of parasitized individuals is lower than that of healthy individuals. Even the best laid out experimental studies, however, cannot replace observations in the field.

In nature the best argument for additivity is temporal correlations between parasite abundance and decline in host population sizes. However, such observations over a long period are few, and after reviewing the studies available at the time Holmes (1982) noted: "In none of them, . . . has disease-induced mortality been shown to be additive."

The difficulty of distinguishing between additive and compensatory mortality was aptly shown by Holmes (1982) using an imaginative scenario involving four players: wolves, mountain lions, bighorn sheep, and protostrongylid nematodes.

Bighorn sheep *(Ovis canadensis)* are parasitized by two species of lungworms (nematodes), both of which destroy a large amount of pulmonary tissue and considerably weaken their hosts. The wolf is one of the sheep's predators and hunts them by chasing, while the mountain lion is another sheep predator but hunts them by stalking.

The scenario of mortality induced by protostrongyles could be as follows: Wolves mainly capture Rocky Mountain sheep that are heavily parasitized because their ability to flee is diminished compared with that of healthy individuals. That is, wolves devour sheep that would have died anyway from parasitism. Mountain lions stalk and capture sheep whether parasitized or not, since the capacity to flee is of little use in escaping stalking cougars. Mountain lions devour whatever individuals they can, and parasites kill some others.

Thus, according to Holmes, the mortality induced by parasitism in this scenario is compensatory where there are wolves and additive where there are mountain lions.

Host Population Regulation in the Field

Arctic Eiders

Based on his own work as well as that of other Russian authors, Galaktionov (1996) presents striking population data on eider ducks *(Somateria mollissima)* from the Arctic regions. In the Arctic, eider chicks are often the victim of mortality so heavy that in some years it can reach 90% (as was the notable case at the end of the 1970s). Examining dead chicks has shown a positive correlation between mortality and parasite load, which was mostly the result of infection by small helminths from three genera: *Paramonostomum, Microphallus* (both trematodes), and *Microsomacanthus* (cestode).

The intensities of these worms may reach extremely high levels, and in the intestine of a single chick up to 50,000 *Paramonostomum,* 60,000 *Microphallus,* and 55,000 *Microsomacanthus* may be found (the absolute record is held by *Microphallus,* with 639,540 found in a single bird). How can such "explosive" parasite loads be possible? To answer this question it is sufficient to examine where the infective stages are found and what the chicks are fed. For *Paramonostomum,* metacercariae are stuck to the shell of very small mollusks of the genus *Hydrobia,* with one single *Hydrobia* capable of carrying as many as 260 metacercariae. For *Microphallus,* metacercariae are encysted in mollusks of the genus *Littorina,* which may carry up to 7,600 of them. For *Microsomacanthus,* cysticercoids are found in amphipods of the genus *Gammarus,* which may harbor up to 250 specimens.

Knowing that *Hydrobia, Gammarus,* and *Littorina* form a large portion of the food eaten by eider chicks, one easily understands that infection may reach record numbers in just a few days' time. If in some years transmission success is a little better than usual, the consequences may be immediate and catastrophic for the young eiders. Galaktionov (1996) writes: "If the host-parasite balance is disturbed, for example by an increase in final host numbers leading to an extensive increase in infection of intermediate hosts, this can lead to a drastic decrease in final host numbers in subsequent years." Thus, there is clearly additive mortality owing to parasitism in this case.

Scottish Red Grouse

Two teams have carried out detailed studies to explain the historically well documented fluctuations in populations of the Scottish red grouse, *Lagopus lagopus scoticus.* Hudson's team has focused mostly on bringing forward evidence of the effect of the nematode *Trichostrongylus tenuis* on the host population, whereas Moss's team has studied an ensemble of grouse population parameters without much consideration of parasitism.

Lagopus lagopus scoticus is a bird much appreciated by hunters—so much so that it is an important element of the local economy in the northern part of the British Isles. As early as the eighteenth century it was noticed that numerous birds were often found dead or dying, and at the end of the nineteenth century the first possible link between disease and the occurrence of a roundworm in the digestive tract of the birds was suggested. In the twentieth century, beginning in the 1930s, certain populations of grouse were shown to have cyclical fluctuations in their numbers. These cycles often consisted of several years of growth followed by a rapid fall and varied in length depending on the region (Moss, Watson, and Parr 1996): four to five years in northern England, six to eight years in certain regions of Scotland (Kerloch), and ten to eleven years in others (Rickarton).

Trichostrongylus tenuis adults locate in the cecum of the red grouse, which is their only important host in Scotland even though they are known to parasitize various other birds in both Europe and America. Adult females lay eggs that are evacuated through the intestine with the feces, after which the L1 larvae quickly hatch from the eggs and molt rapidly to the L2 and then L3 stages. L1 and L2 larvae feed in the feces, while the L3 larvae leave and migrate to the base of heather, then climb to the extremities of the buds. Saunders, Tompkins, and Hudson (1999) have shown experimentally that the phase involving invasion of the plants leads to heavy losses in the parasite population and that larvae do not colonize all plants equally: "One mechanism by which *T. tenuis* appears to compensate for this low migration success, in attempting to maximize their transmission to grouse, is the increased efficiency of larval migration up heather plants compared with other plant species." There they wait, several months if necessary, for grouse to eat the tips, which are the main food for these birds. Once in the digestive tract, the larvae reach the cecum and mature (fig. 15.1). As in many nematodes, a temporary stop in larval development (diapause) may occur, although under the right temperature conditions the time between egg expulsion and the appearance of the infective L3 larvae lasts no more than a week.

Grouse have two ceca, each up to seventy centimeters long and each capable of harboring 10,000 to 15,000 worms (sometimes up to 30,000). The pathogenic effects are lesions and hemorrhages of the walls of the digestive ceca as well as a reduction in enzymatic activity in the absorptive brush border. The result is weight loss, reduction in musculature, loss of appetite, and a decrease in social status for the most heavily infected birds (Delahay, Speakman, and Moss 1995).

Hudson and his team looked to determine the impact of infection on the birds (see in particular Hudson and Dobson 1991; Hudson, Dobson, and Newborn 1992; Hudson, Newborn, and Dobson 1992; Dobson and Hudson 1992) by comparing parasite loads in grouse killed by hunters,

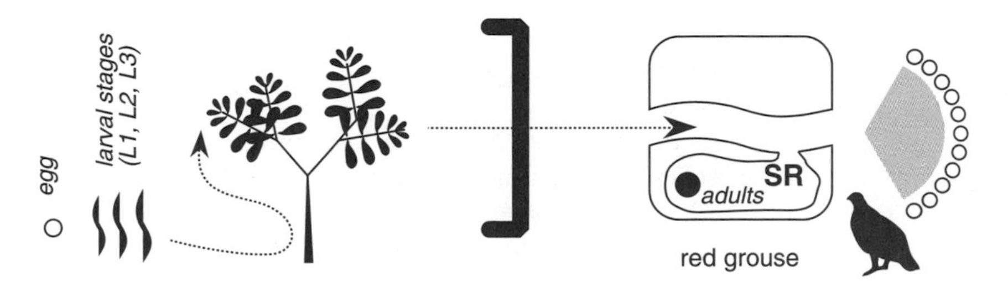

Figure 15.1. Life cycle of *Trichostrongylus tenuis*. SR = sexual reproduction.

those killed by predators (foxes and ravens), and those found dead in the field (in the last two cases determining the number of nematodes was possible only if the digestive ceca were found intact). The result of this comparison showed that, overall, heavy parasite loads occurred in animals found dead and in those captured by predators, whereas those killed by hunters were less parasitized (fig. 15.2). Thus, statistically speaking, ani-

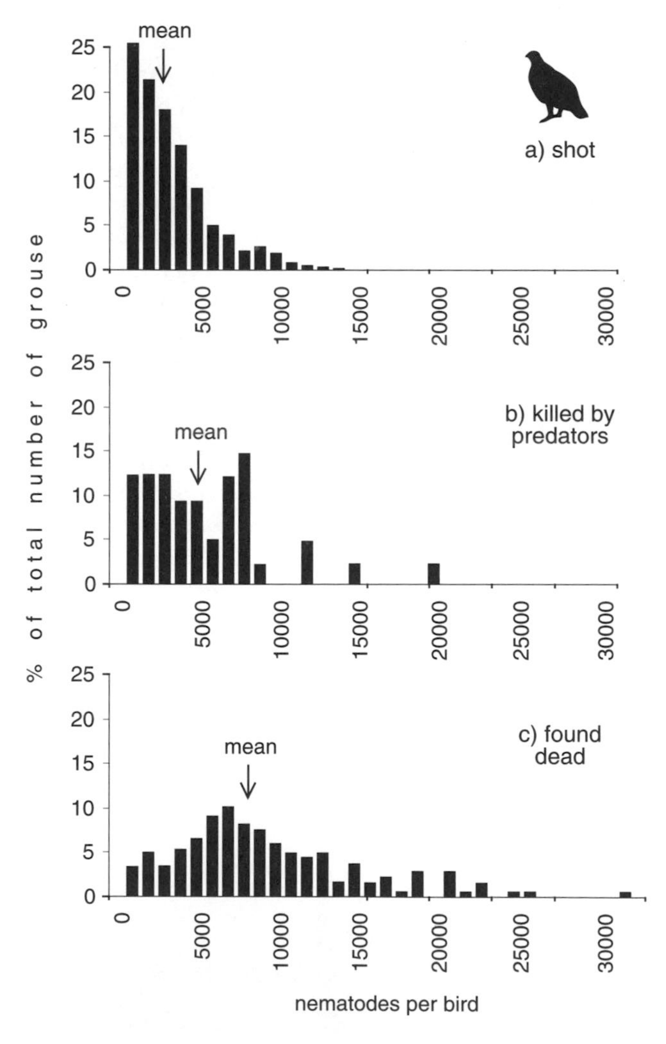

Figure 15.2. Parasitism of the Scottish red grouse by *Trichostrongylus tenuis*. Comparisons of parasitic loads of birds (a) killed by hunters, (b) captured by predators, and (c) found dead from natural causes (modified from Hudson and Dobson 1991).

mals that were the most parasitized were most exposed to predation or most likely to die a natural death.

In addition to an increased risk to predation and death, parasitosis also affected fecundity (the number of eggs and the percentage hatching). This effect was shown in a study where samples of birds captured in a defined area were divided into two subgroups, one treated with an anthelmintic that eliminated at least some of the parasites and the other not. The fecundity of untreated females was compared with that of treated individuals, which served as controls. This part of the study showed that heavy parasite loads were directly correlated with a significant reduction in females' reproductive success.

Without a doubt the most important observation on this system has been that over a period of about ten years the fluctuation of host populations largely was negatively correlated with parasite abundance. Despite the difficulty of recovering this type of data with exactitude, the graph obtained (fig. 15.3) clearly shows that the years when parasites are very abundant coincided with reductions in the grouse population. It is this result that best supports the existence of additive mortality in the system: those grouse whose death is induced by the parasites are added to those that die for other reasons. Last, Hudson, Dobson, and Newborn (1998) have been able to drastically reduce the extent of cyclical population crashes by lightening parasite burdens thanks to treatment with an oral

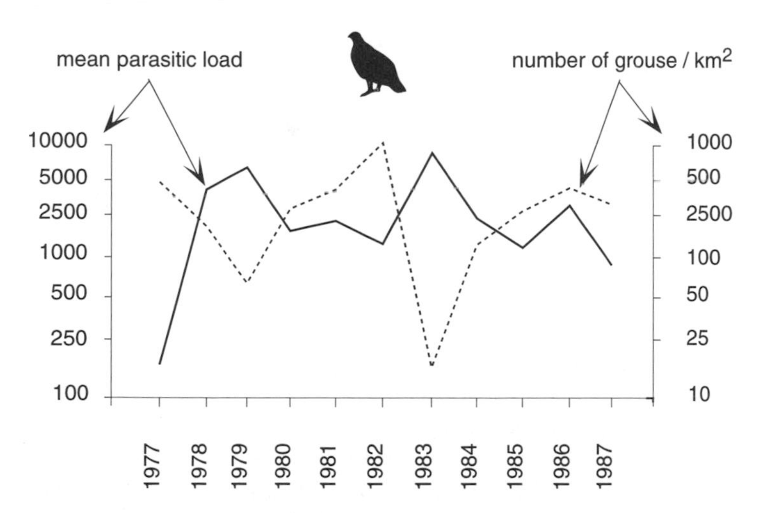

Figure 15.3. Parasitism of the Scottish red grouse by *Trichostrongylus tenuis*. Relation between parasitism and host population over ten years (modified after Hudson and Dobson 1991).

anthelmintic, and Tompkins and Begon (1999), who reviewed the current state of knowledge of the impact of parasites on the fecundity or survival of wild animals, considered this result to be the best proof extant that parasites can regulate wildlife populations.

Moss's team (Moss et al. 1993; Moss, Watson, and Parr 1996) also studied the Scottish red grouse. Based on parasite egg counts over several years, these workers refute the idea that infection by *T. tenuis* explains the observed population fluctuations of the grouse because they did not always find a correlation between parasite load and bird density. These authors considered summer rainfall the main determinant of the parasite load and believed "the pathological impact of worms on grouse might bear little relation to burdens of relatively benign, long-established worms."

It is particularly interesting to note through this controversy how difficult it is to separate the impact of parasitism in natural populations from other factors associated with mortality.[1] Although researchers have a tendency to focus on only one "couple" within a particular system, that is, on one parasite species and one host species, in reality parasites are never by themselves. In the case of the grouse there is also a virus transmitted by ticks that affects both the grouse and sheep. Hudson and his collaborators have shown that although grouse fecundity is strongly diminished because of the nematode, the impact on survival is mostly due to the virus.

Soay Sheep from Saint Kilda

As illustrated by the plagues of Europe (e.g., 1348, 1493, 1518, 1665), which killed up to half the people of some towns or villages, epidemic situations are the best proof that parasites induce additive mortality that may in some cases be particularly destructive.

The occurrence of an arms race places many parasite-host systems in a fragile equilibrium that is easily disturbed. That is, not much is needed—for example, a subtle modification in the environment—to upset the balance. We should not forget that the opening of the encounter and compatibility filters shows a large phenotypic plasticity and that anything that increases the probability of encounter or, even better, weakens a host's immune system can tilt the balance in favor of the parasite.

Soay sheep live on Saint Kilda (Hirta in Gaelic), a small island in the Outer Hebrides, about one hundred kilometers west of the northern tip of Scotland, that is routinely beaten by Atlantic storms. These sheep are supposedly the closest descendants of wild European sheep *(Ovis aries),*

---

1. Thirgood et al. (2000) emphasize the importance of raptor predation in limiting grouse populations.

even though they were transported there by humans about the beginning of our era and then left on their own for some two thousand years. The sheep harbor a community of twelve nematode species and one species of cestode, all living in the digestive tract except for one lung nematode. About four-fifths of the parasites are species of *Ostertagia*, among which *O. circumcincta* dominates (85%). The sheep's population is known to fluctuate greatly over time, with drastic falls occurring every three or four years. One of these drops, which led to the death of two-thirds of the sheep in a three-month period at the beginning of 1989, was closely monitored in a herd of up to 1,600 head.

Field data complemented by experiments involving treatment and infection (Gulland and Fox 1992; Gulland 1992) indicated that the fundamental cause for the observed drops in host number was malnutrition linked to increases in the number of sheep. However, the results also showed that when malnutrition occurred, the animals' natural defenses were weakened, leading to an increase in parasitism that in turn caused additive mortality.

Hares from North America

Field experiments by Murray, Cary, and Keith (1997) and Ives and Murray (1997) demonstrated at what point parasites increase the risk of predation even if the pathogenic effect is not clearly visible. The work was carried out on a hare population *(Lepus americanus)* harboring the nematode *Obeliscoides cuniculi* (up to one hundred per hare). In hares, 95% of natural mortality is due to predation. In this case the researchers tracked, using transmitters, a sample of 612 hares—of which half had been treated with ivermectin (a potent anthelmintic) and the other half with a placebo. The experiment showed that the risk of being caught by a predator over a period of 180 days was 2.4 times larger for untreated animals (those harboring parasites) than for animals that were treated (ivermectin greatly decreased their parasite loads).

Other observations on the same model showed that parasitism has a destabilizing effect on the predator-prey system and that it may induce population cycles involving periods of both low and high density (Ives and Murray 1997).

Regulation and Favorization

We know that processes of favorization very often involve manipulation of upstream hosts, that this manipulation increases the probability of these hosts' being eaten by downstream hosts, and that such a manipulation occurs only when the parasite circulates in a predator-prey system.

Do such manipulations affect the regulation of manipulated host populations?

Thomas's team (Thomas, Renaud, Derothe, et al. 1995; Thomas, Renaud, Rousset, et al. 1995; and Thomas, Renaud, and Cézilly 1996) made such a demonstration while studying parasitism in gammarids in the lagoons of Camargue in the south of France. In these studies the life cycles analyzed were those of microphallid trematodes whose metacercariae encyst in the gammarids and whose definitive hosts are birds. In each case the worms have a life cycle of the same type as that of *Microphallus papillorobustus* (see fig. 7.15).

Three systems were analyzed: metacercariae of *Microphallus papillorobustus* in *Gammarus insensibilis;* metacercariae of *Microphallus papillorobustus* in *Gammarus aequicauda;* and metacercariae of *Microphallus hoffmanni* in *Gammarus aequicauda.*

Gammarid populations were divided into eight age-classes based on body size and sex, and the abundance of metacercariae in each age-class was determined.

The results, summarized in figure 15.4, show that in the *M. papillorobustus/ G. insensibilis* system the oldest age-classes are less parasitized than are other age-classes, a strong indication that infected individuals in populations are selectively depleted by predators because of their manipulated behavior (see chapter 7). However, when the same parasite is in *G. aequicauda* infection regularly increases with age, and the same is true in the *M. hoffmanni/ G. aequicauda* system, suggesting that in both cases predation is independent of parasitism. Significantly, in neither of the two latter systems is there behavioral manipulation as with *M. papillorobustus* in *G. insensibilis,* leading Thomas to conclude that infection intervenes in the regulation of the gammarid population only in the *M. papillorobustus/ G. insensibilis* system, where manipulation occurs.

Some Subtle or Unexpected Effects of Parasites

In some cases the "parasitic disease" is difficult to see and its effects are hard to measure even though the parasite still influences the host—sometimes in subtle yet surprising ways. The regular and apparently innocuous presence of numerous such parasites may even have more important consequences than more readily observed and spectacular epidemics (see Gregory and Keymer 1989). For example:

> – Malaria caused by *Plasmodium mexicanum* in the western fence lizard *Sceloporus occidentalis* in California seems not to be damaging at first sight. In reality,

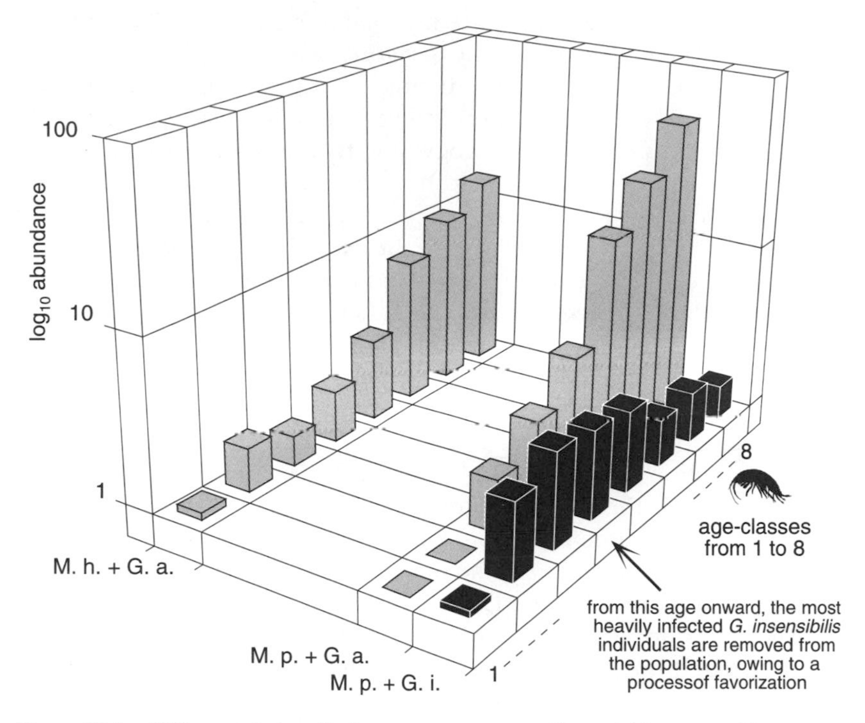

Figure 15.4. Differential mortality between two gammarid species because of the trematode *Microphallus papillorobustus* (after data of Thomas, Renaud, Derothe, et al. 1995). M.h = *Microphallus hoffmanni;* M.p = *Microphallus papillorobustus;* G.a. = *Gammarus aequicauda;* G.i. = *Gammarus insensibilis.*

however, it prevents fat storage and reduces female fecundity by half (Schall 1983).

– The surface-dwelling aquatic insect *Hydrometra myrae* from Florida (which no one suspected was vulnerable to disease) lives half as long when hosting some parasitic mites (Lanciani 1975).

– The clutch size and reproductive period in *Daphnia obtusa* from Texas drop as soon as these freshwater planktonic crustaceans harbor more than two trematode metacercariae (Schwartz and Cameron 1993).

Parasites may even influence their hosts' fitness by directly interfering with their sexual behavior, either by decreasing their capacity to reproduce or reducing their ability to choose the best mate.

The former case is illustrated in the surprising results of Morales et al.

(1996), which show that male mice infected by cysticerci of *Taenia crassiceps* (ten cysticerci injected into the peritoneal cavity) stop ejaculating five weeks after infection, stop introducing their penes into their female mates seven weeks after infection, and cease mounting at thirteen weeks postinfection. In short, there is progressive feminization of the male (owing to a strong increase in female hormones and a drop in male hormones) without any other pathological symptoms. Such modified males are "genetically dead"—a nice case of the extended phenotype.

The latter case (direct interference in the host's sexual behavior) is illustrated by the fish *Culaea inconstans* and two of its parasites, the trematode *Bunodera inconstans* and the acanthocephalan *Neoechinorhynchus rutili*. Females of *C. inconstans* that harbor these parasites are not as good at competing for mates and are less capable of exploring a wide range of territorial males than are healthy females (McLennan and Shires 1995).

The induction of increased mortality may have unexpected aspects where the cost to the host population has nothing to do with "classical" pathology. An example is salmon, which are born in fresh water (often very far upstream), grow up in the sea, and then return to the rivers they were born in to reproduce. The young, called smolts, migrate to the sea when they are one to two years old. In British Columbia (western Canada), smolt populations live in large lakes and thus must find their way out to the rivers that allow them access to the sea. To find their way the smolts use both visual and magnetic signals (Brannon et al. 1981; Quinn and Brannon 1982). These smolts are parasitized by an intestinal cestode of the genus *Eubothrium* found in their pyloric ceca, by larval cestodes of the genera *Diphyllobothrium* and *Proteocephalus* that localize in their muscles, and by a nematode of the genus *Philonema* that localizes in the swim bladder. An original experiment by Garnick and Margolis (1990) analyzed the movement of smolt in artificial open-air circular tanks with eight regularly spaced exits around the periphery. The salmon that were captured in tank exits after twenty-four hours were then examined for parasites, In each case parasitized fish, particularly those infected with *Eubothrium,* showed a greatly reduced variation in their distribution in the various tank exits compared with healthy fish. That is, the presence of parasites increased the concentration of salmon in certain directions, which may correspond to modification of their ability to orient, decreasing their survival.

Such diffuse pathology may perhaps be linked to the enigma of cetacean beaching. Cetaceans (whales, dolphins, etc.) travel long distances in the open ocean and sometimes beach themselves en masse as if they had lost the capacity to orient toward the open ocean. The reason for such beachings is not yet understood with certainty, but some authors have correlated them with the presence of trematodes and other helminths in various regions of the brain (see, for example, Ridgway and Dailey 1972; Dailey and Walker 1978; Morimitsu et al. 1986). For instance, trematodes of the genus *Nasitrema,* which live in the cerebral sinuses, are notably associated with more frequent brain lesions in beached animals than in those captured in the open sea.

Ecologists most often suggest that in predator-prey systems it is not so much that predators regulate the prey (i.e., the availability of resources) as that the prey regulates the predators. Transposing this point of view to parasite-host systems, the hosts would regulate parasites and not the opposite. This is certainly not wrong, though few nowadays doubt that parasites may also regulate their host populations.

The Use of Models

Crofton's Model

In parasitology mathematical modeling has its most useful application in analyzing the relationships between parasite and host populations. Such models are as much analytical tools as are microscopes and electrophoresis equipment, and their equations or ensembles of equations help to describe observed phenomena and to explain or eventually predict the influence of different variables on specific outcomes. The role of models in parasitology is to suggest and test hypotheses, identify those factors that most influence life cycles, and understand particular situations or occurrences (see May 1991; Poulin 1998a). Although it is true that some models produce answers we already know, others clearly contribute new information and knowledge. As Thom (1979) notes, the more surprising the result or prediction a model provides, the more useful it is: "The fecund characteristics of a model mainly appear in the ratio of *a posteriori* justification to *a priori* justification. It is the ratio of what can be gained to what is bet."

Like any other tool, however, modeling has its limits. These limits come from the necessary simplification of variables and the fact that such simplification is often overdone. In terms of demography, for example, the genetic diversity of hosts and parasites is rarely taken into account.[2] Further, it is difficult to assess the quality of the variables (birthrate, mortality rate, etc.) introduced or to validate the model with observations that

2. There is no model that by nature does not simplify. This is not necessarily a weakness, however, since a model must simplify to justify its name. For instance, a geographical map is nothing other than a model of the surface of the earth. Any map that did not simplify would show all the topographic details—it would be impossible to use and therefore of no value.

are sufficiently informative. Last, in nature hosts harbor not one but multiple species of parasites whose pathogenic effects add up, often creating conditions very different from those described in simple models.

Even though models have such limits, they are nevertheless valuable in providing insight into host-parasite relationships, precisely because of their imposed simplicity. For instance, Crofton (1971) proposed the first mathematical model describing a correlation between parasite and host populations, and this work has deeply influenced the development of ideas in parasite ecology ever since. Crofton's model, with its few and simple variables, serves as an ideal starting place for understanding the interest parasitologists have had in modeling population dynamics—variations in parasite and host populations over time.

Crofton built his model by taking into account five variables: the reproductive rate of the host; the reproductive rate of the parasite; the probability of individual transmission; the degree of parasite aggregation; and the lethal threshold of parasitism. He linked these variables in a relatively simple equation, making it possible to give them different values and see how the host and parasite populations would simultaneously develop.

Crofton's model was the first to mathematically formulate the relationship between host and parasite populations, and its only weakness was its oversimplifying nature (May 1977).

Anderson and May's Models

Anderson and May (1978) and May and Anderson (1978) reexplored Crofton's mathematical approach to the relationship between host and parasite populations. These authors first built a basic model describing a simple parasite-host association in which the parasite influenced its hosts but did not multiply on site, then they constructed a more complicated model involving a parasite that did. Their basic formulation improved on Crofton's model by incorporating several new elements.

1. Variation in the time and number of hosts and parasites in populations, calculated as a "rate" (each rate given per individual and per time unit).

2. The change in the host population (H) in a given time period, dH/dt, a measure that was itself dependent on the birthrate of the host, the "natural" mortality rate of the host, and the parasite-induced mortality rate of hosts.

3. The change in the parasite population (P) during the same time period, dP/dt, which depended on the rate of appearance of infective stages, the proportion of infective stages that infect a host with success, the natural mortality rate of the parasite in the host, the mortality rate of

the parasite owing to the natural mortality of the host, and the mortality rate of the parasite owing to parasite-induced host mortality.[3]

By a series of successive corrections to the basic model, notably eliminating certain initial simplifying assumptions, Anderson and May (1978) developed a more advanced model and showed that some processes have a stabilizing effect on parasite-host systems and others have a destabilizing effect.

Factors that tend to stabilize the system are the aggregated distribution of parasites in the host population (i.e., certain hosts disproportionately concentrate part of the parasite population); the existence of a nonlinear relationship between parasite load and host mortality (i.e., mortality increases faster than parasite load); and the density-dependent regulation of parasites in infrapopulations (i.e., constraints appear when density increases).

Factors that tend to destabilize the association are the reproduction of the parasite on site (i.e., the infrapopulation increases in the absence of recruitment from the outside); a parasite-mediated decrease in the fecundity of the host (i.e., there is the induction of decreased fecundity in addition to an induction of mortality); and waiting stages in the parasite life cycle (i.e., transmission is not instantaneous as was previously supposed in the basic model).

Overall, then, the stability of a parasite-host system under natural conditions depends on the influence of each of the stabilizing and destabilizing factors.

Thus, by altering the values of the variables, the models of Anderson and May allow one to more realistically simulate changes in the relationship between parasite and host populations, with one outcome of this more realistic approach being the unequivocal demonstration that parasites can regulate host populations. These models thus allow one to formulate propositions concerning the control of human parasitoses (see Anderson and May 1985).

In general, the regulation of a host population by a parasite requires only that one of these conditions be satisfied: the distribution of parasite individuals must be aggregated (there is host mortality induced by the heaviest parasite loads), or the pathogenic effect must increase more rapidly than the parasite load (even in the absence of an aggregated distribution the heaviest parasite loads induce mortality).

An important distinction in building these mathematical models is that in the case of microparasites hosts may be classified simply as positive

3. The values of dH/dt and dP/dt may be calculated using two differential equations based on the variables listed above.

or negative relative to infection (i.e., they are either infected or not), whereas in the case of macroparasites this distinction makes little sense, since it is the parasite load that determines pathogenicity.

The Graphic Representation

Various graphical representations of the models above are possible and serve as the inspiration for figures 15.5 and 15.6, which illustrate the demographics of transmission of microparasites and macroparasites, respectively. Both figures adopt these conventions:

1. The numbers of individuals (parasites and hosts) at different stages are shown as a series of "boxes," some of which (those of the hosts and the free-living stages of the parasites) occur in the outside environment and others of which (those of the parasitic stages) are nested within boxes that already represent the number of hosts.
2. Each box may increase or decrease in size according to birth, mortality, immigration, and emigration rates, which are depicted by arrows on the diagrams: the open arrows are the "flows" between boxes, while the solid arrows represent mortality.
3. Rates (birth, mortality, etc.) are indicated with letters.

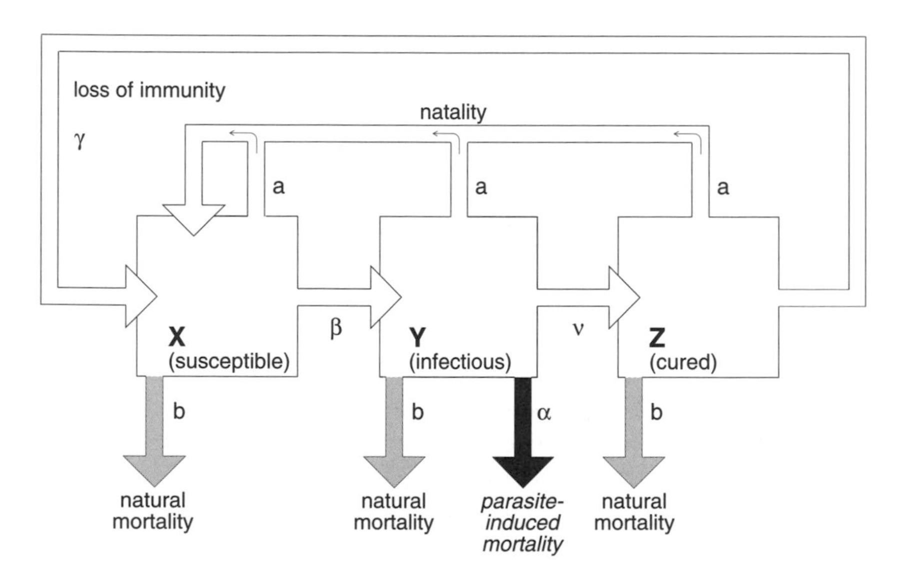

Figure 15.5. Mathematical model of "microparasite" transmission. Lowercase Latin and Greek letters represent the fluxes (see text). This type of model shows that the level of appearance of new cases is related both to the number of susceptible or not yet infected individuals and to the number of infected individuals, which are the source for the infective stages of the parasites (modified after Anderson and May 1979).

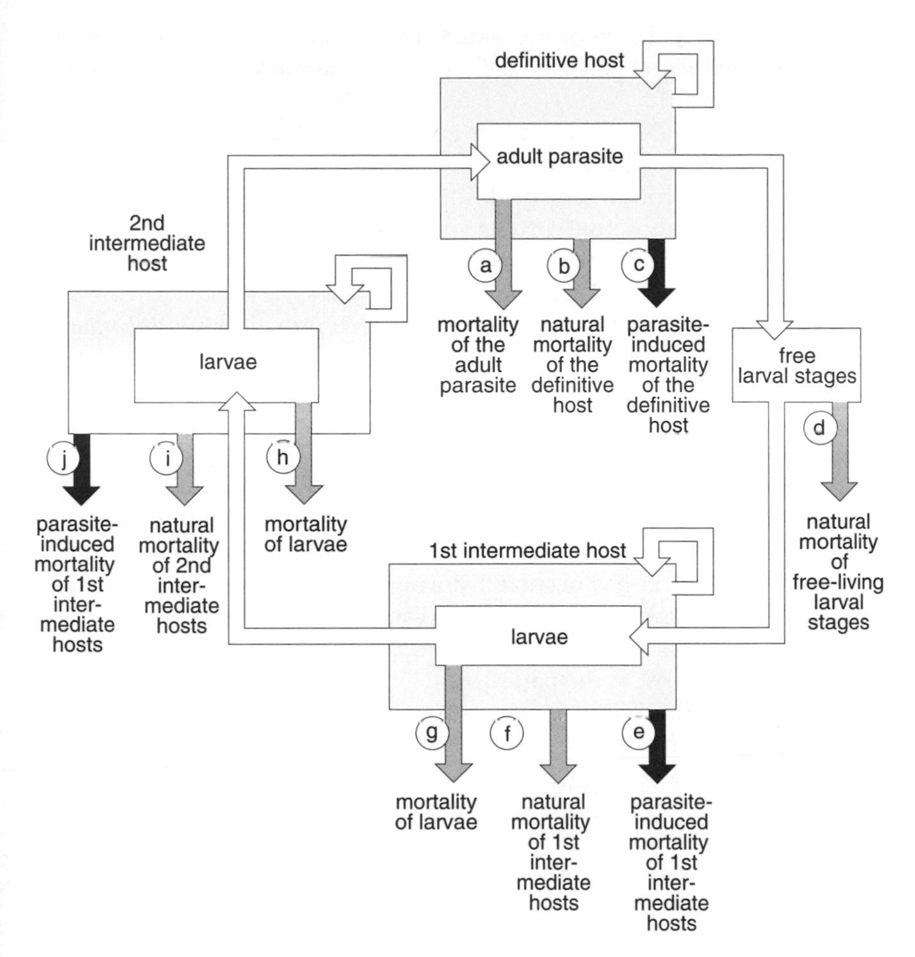

Figure 15.6. Mathematical model of "macroparasite" transmission. Letters in circles represent the fluxes (see text).

Comparing the two figures emphasizes the differences between microparasites (fig. 15.5) and macroparasites (fig. 15.6), taking into account the presence or absence of microparasites on the one hand and parasite loads in the case of macroparasites on the other.

Effects on Sex Ratio

We have already encountered the numerous processes whereby parasites manipulate their hosts' behavior. In the case of bacteria of the genus *Wolbachia,* discussed here, the advantage the pathogenic agent finds in the manipulation is a change not in behavior but rather in the structure of the population of one of its hosts, the pill bug *Armadillidium vulgare* (a terrestrial crustacean).

Wolbachia belongs to a particular group of bacteria (the Proteobacteria) closely related to *Rickettsia* and are parasites (often qualified as "mutualists") of numerous insects and isopods. They are remarkably ancient organisms, with ribosomal DNA sequencing indicating that *Wolbachia* probably separated from their closest bacterial relatives about 700 million years ago (Rousset et al. 1992).

Rigaud and Juchault (1995) have shown that infection by *Wolbachia* is possible if the blood of two injured individuals comes in contact (injury by predators or during molting occurs in up to 10% of the population). Transfer may happen between isopods of the same species or of different but closely related species. On the other hand, *Wolbachia* in insects may be transferred between unrelated hosts, such as between mosquitoes of the genus *Aedes* and fruit flies (Braig et al. 1994), and the phylogenetic comparison of *Wolbachia* and hosts confirms that the horizontal transfer of these bacteria has occurred during evolution (Schilthuizen and Stouthamer 1997). As demonstrated by Vavre et al. (1999) parasitoid-host associations are particularly favorable to the horizontal transfer of *Wolbachia* (from the host to the parasitoid).

Species of *Armadillidium* are terrestrial isopods, and one aspect of their sexuality is particularly unusual. Normally the sexes are genetically determined, with males being ZZ, meaning they have identical sex chromosomes, and females being WZ, meaning they have different sex chromosomes. The W chromosome is responsible for feminization and has a gene that represses the expression of a masculinizing gene on the Z chromosome. Normally the same number of males and females are born, and up to this point there is nothing extraordinary about isopods, since sex determination of this kind is known for numerous organisms.

What is less common (although pill bugs are by no means the only ones to exhibit this trait) is that there is a third type of individual that is genetically ZZ and thus should be male. However, these individuals are functionally females and are referred to as "genetically feminized males."

The question is, What is the feminizing agent? The response is *Wolbachia* (Martin, Juchault, and Legrand 1973), which possesses a sexual factor of extrachromosomal origin (hereafter designated as F). In infected pill bugs the bacteria exist in all tissues but are particularly concentrated in the ovaries.

In summary, there are three types of individuals in populations of *Armadillidium:* WZ individuals that are genetically and functionally female; ZZ individuals that are genetically and functionally male; and ZZ + F individuals, which are genetically male and functionally female.

Armadillidium populations that have been invaded by the bacteria thus have a sex ratio biased in favor of females (Rigaud, Mocquard, and Juchault 1992) because *fake* females (infected feminized males) are added to the *real* females. These feminized males play their role very well, since they lay eggs as well as any real female, and in most populations where *Wolbachia* has spread true females (WZ) have in fact disappeared and all individuals are genetically male but functionally female. This elimination of WZ females occurs because they produce descendants whose sex ratio is balanced (one female:one male), whereas ZZ + F individuals produce *almost* twice as many females. We shall shortly see the significance of "almost" and also see that both "real" and "fake" females are born infected by the bacteria.

If we were to stop at this point in the scenario we would note that we are in the presence of a bacterium whose effect is to alter the sex ratio in favor of females by feminizing a portion of the males. But two questions remain: How does it do it? And why does it do it?

How? Simply, the bacteria possess a gene that inhibits the expression of those male genes that control the differentiation of the androgen gland, thus allowing a genetically male gonad to differentiate into an ovary—another example of the extended phenotype. Further, researchers have observed that in certain strains of *Armadillidium* males are feminized even in the absence of the bacteria, most likely because of the transfer of a small DNA fragment (called *f*) from the bacterium into their genome (Juchault and Mocquard 1993).

Why? The bacteria are transmitted vertically via the cytoplasm. Because the bacteria infect oocytes, such transcytoplasmic transmission means that *Armadillidium* embryos are born harboring the bacteria. Thus the more infected females there are in a pill bug population, the more bacteria are transmitted. That is, there is a distinct advantage for the bacteria in unbalancing the sex ratio in favor of females.

However, it comes naturally to mind that if all the males in a population were feminized, then the eggs of the females (both real and fake) would not be fertilized and that ultimately the population would disappear. This does not happen because a mechanism has been selected in *Armadillidium* to counter feminization—resistance genes that in this precise case have evolved to resist each of the two feminizing factors in the bacteria: against *Wolbachia* itself, a complex of genes *(R)* controls transmission of the bacteria to the isopod's oocytes. Thus a few embryos are born without being infected; and against the DNA fragment inserted into the genome a gene *(M)* exerts a masculinizing effect. This gene may enter into conflict with *Wolbachia,* and in this case it causes the production of intersexed individuals that are not completely feminized males. Thus a male that carries the feminizing DNA *(f)* is not always converted to a female if it has the *M* gene. It seems that *M* inhibits the expression of *f*, which then allows expression of the genes that code for the differentiation of the androgynous gland.

The combined effects of these two sex-altering systems in *Armadillidium* thus ensures that enough males are born to guarantee the continuation of the population (Rigaud and Juchault 1992). Significantly, there appears to be no other explanation for the distortion of the sex ratio in *Armadillidium* populations other than manipulation by *Wolbachia*.

Are *Wolbachia* parasites or mutualists? There is nothing to indicate that *Wolbachia* has a favorable effect on the host, either at the individual or the population level. Thus the qualification of *Wolbachia* as either a "symbiont" in the European sense or a "mutualist" in the American sense seems not to be justified. On the other hand, nothing indicates that *Wolbachia* are unfavorable to their pill bug hosts, either. In *Drosophila*, for instance, *Wolbachia* induces cytoplasmic incompatibility,[4] although no unfavorable effect of *Wolbachia* on individual carriers has been observed (Poinsot and Merçot 1997). These authors note, however, that "[*Wolbachia*] do not induce any detectable deleterious effect on host fitness, [but] . . . it is always possible that a deleterious effect would only be apparent under conditions of stress."

Significantly, the meddling of pathogenic agents in the sex lives of their hosts is not reserved for bacteria, and certain protists and viruses exert the same sort of influence (see Ginsburger-Vogel, Carre-Lecuyer, and Fried-Montaufier 1980; Juchault et al. 1991).

Effects on Social Life

Life in a group has several acknowledged advantages for free-living animals, including decreased risk of predation and increased hunting efficiency. It also has inconveniences, such as increased competition and cheating. Relative to parasitism, two hypotheses are possible concerning hosts living in a group: living in groups either favors parasites or protects individuals from parasitism. Côté and Poulin (1995) performed a metanalysis based on a series of investigations published by a variety of authors.[5] The predictions were that parasites transmitted by contact or having no mobile

4. When an infected male is crossed with an uninfected female, there is high mortality in the embryos.

5. To know if there is a correlation between two variables (for example, parasite prevalence and host group dimension), we may look at investigations done by various authors on different parasite-host systems. If a hypothesis is formulated (for instance, that parasitism increases as the host group becomes larger), there are two ways to test it. The first way is to compare the number of studies that support this hypothesis with the number of studies that

stages should be favored by having hosts living in groups and that mobile parasites, whose behavior may be compared to that of predators, should be disfavored in hosts that live in groups. Although the distinction between only two categories of parasites (mobile and nonmobile) lacks nuance and though transmission via an intermediate host was not examined, the results obtained were consistent with the predictions. The authors concluded that "[parasites] have the potential to act as a force that can select for either larger or smaller group sizes, or even for solitary behavior" (Côté and Poulin 1995).

Hypothesis 1: Living in a group increases parasitism

Sociality is synonymous with either the permanent or the periodic grouping of individuals. "Social animals have a greater chance of acquiring and accumulating contact-transmitted ectoparasites, because of the greater proximity and number of physical contacts among group members than among solitary individuals" (Poulin 1991; see also Poulin 1994a). Freeland (1976) had already shown that transmission of ectoparasites was greater in primates living in groups than in those that were solitary, and in humans parasitism is even sensitive to purely occasional increases in density. For instance, scabies epidemics among humans correspond to increases in all types of close contact: wars, hippie gatherings, rock concerts, and so on (Fain 1992).

The cost of sociality has been particularly well studied in birds, where it has been shown that social behaviors have negative consequences, at least some of them attributable to parasitism. There is an increased transmission of holoxenous parasites and even some heteroxenous ones, such as blood protists transmitted by ectoparasites (Brown and Brown 1986; Poiani 1992). Intraspecific parasitism of several types occurs: laying eggs in a neighbor's nest, theft, and rape (Emlen and Wrege 1986). Interspecific parasitism takes place: birds of a different species find refuge in colonies, and their presence, even if nonaggressive, decreases the reproductive success of the "host" species (Groom 1992).

That holoxenous parasites are more sensitive to the grouping of host individuals than are heteroxenous parasites was confirmed by an investigation carried out in Florida by Moore, Simberloff, and Freehling (1988),

---

reject it; depending on the outcome, the hypothesis will be validated or not. The second way is to carry out a metanalysis, which calls for a statistical process (Hunter and Schmidt 1990) that takes into account the size of the samples analyzed by the various authors and the strength of the correlation. Metanalysis allows one to test the coherence between various studies and to evaluate the strength of the correlation in cases conforming to the starting hypothesis. One of the difficulties with this sort of analysis, however, is that usually only net correlations are published (so that absence of correlation is underrepresented).

comparing the intestinal helminths of young bobwhite quail, *Colinus virginianus,* in different-sized groups. Overall, all helminths had a tendency to be more abundant in young bobwhites belonging to larger groups, while the nematode *Trichostrongylus tenuis,* the sole helminth in the investigation with a direct life cycle (with no intermediate or paratenic hosts), was the most sensitive to an increase in the size of the group.

The difference between parasites that are highly mobile and those that are only slightly mobile has been noted in several investigations. For example, in an analysis using data on forty-five species of California passerines covering more than 43,000 individuals, Poulin (1991) showed that group living in birds increases parasitism with feather mites (slightly mobile) but does not affect parasitism with hippoboscid flies (very mobile).

A positive correlation between the density of cleaner fish (various species of *Labroides*) and the gregariousness of fish "clients" (the species they clean) has been demonstrated by Arnal, Morand, and Kulbicki (1999) in the Pacific. This finding can be taken as an indirect proof that gregariousness increases ectoparasitc loads, because cleaner fish principally ingest ectoparasites they remove from the skin and gills of their "clients."

An occasional form of density increase occurs on farms, which are characterized by the continuous close contact of individuals. This closeness is enough to induce an inflation in transmission and can make a parasite that is only slightly pathogenic under natural conditions the cause of very high mortality. According to Thoney and Hargis (1991), "Monogeneans, usually well accommodated to their fish hosts in nature and causing few easily detectable effects, frequently cause severe epizootics in cultured or aquarium-held populations." A classic example is the monogenean *Diplectanum aequans,* a gill parasite of the sea bass *Dicentrarchus labrax,* which is not dangerous to its host under natural conditions but can induce heavy mortality in tank-reared fish on fish farms around the Mediterranean (Silan and Maillard 1986). While sea bass captured in the sea harbor at the most a few dozen individuals, farm-raised animals can harbor from 10,000 to 15,000. Significantly, malnourished fish may die when they reach a threshold of only about a hundred parasites. A comparable example is *Amyloodinium ocellatum,* a holoxenous protist of slight specificity that lives on the gills of numerous tropical fish that may induce heavy mortality in marine aquariums after a population explosion.

The species barrier itself may be broken when animals are maintained under artificial conditions, such as occurs in aquaculture. For instance, the isopod *Nerocila orbignyi* and the copepod *Caligus pageti* are both ectoparasites of Mediterranean mullet and are only rarely reported on the sea bass *Dicentrarchus labrax.* However, in sea bass raised in cages both arthropods can induce heavy mortality. The same effect occurs for the monogenean *Polylabris tubicirrus,* which is normally a parasite of several *Diplodus* but proliferates on farm-raised gilthead seabream, *Sparus aurata.* Although this loss of specificity may be the consequence of a modification of host immune defenses, it seems more likely that it results from density (for other examples of parasite proliferation in aquaculture, see Euzet and Raibaut 1985).

Hypothesis 2: Living in a group provides protection from parasitism

When we say that sociality makes parasite transmission easier by bringing together prospective hosts, we suppose we are dealing with transmission within the group. But if we consider parasite "attacks" to come from outside the group (such as in predator-prey systems), then the group may have a protective effect. Poulin and Fitzgerald (1989) studied the probability of young sticklebacks *(Gasterosteus)* being parasitized by the infective stages of the crustacean *Argulus canadensis,* depending on whether they swim by themselves or travel in schools of from five to twenty individuals. This study showed that the number of attacks per minute by *A. canadensis* did not differ between isolated and schooling fish, although the larger the group the smaller the number of attacks per minute per fish. They interpreted this in the following way: when a stickleback joins a group, it becomes one target among many others, some more vulnerable to attack by these parasites. The experiment also showed that the fish have an increased tendency to form schools when *A. canadensis* is present, which confirms the value of grouping. Significantly, the outcome of these experiments, in which intragroup transmission is not of concern, do not contradict the frequently observed positive correlations between host density and parasitism.

O'Donnell (1997) defends the original point of view, that in some cases the deleterious effects of parasitism may favor social behavior. According to this author, the reduction in fecundity or even castration that may result from parasitism may favor the expression of altruistic behaviors within populations. For instance, in a parasitized female insect the decision to reproduce may be replaced by the decision to help other reproducers, particularly if they are related. Such a process may have played a role in the passage from "presociality" to sociality, with the conditions that parasite transmission was not favored by the grouping of individuals and that parasite prevalence was low enough.

16 Parasites and Host Ecosystems

A non-specific parasite to which partial immunity has been acquired is a powerful competitive weapon.
Haldane 1949

It is a truth almost universally acknowledged . . . that antagonistic interactions make the world go round. Competition, predation, and increasingly parasitism are all recognized as crucial determinants of the abundance and distribution of individual species and as important factors shaping the structure of ecological communities.
Douglas 1999

Although parasitologists have been interested in the ecology of parasitism for a long time (Baer 1952), it is only recently—as evinced by the many previous books written by mainstream ecologists where the word "parasite" does not appear—that general ecologists and evolutionary biologists alike have begun to be interested in this area. As Dawkins (1990) noted: "Eavesdrop morning coffee at any major centre of evolutionary theory today, and you will find 'parasite' to be one of the commonest words in the language."

In the ecosystems parasites belong to, they influence processes as varied as competition, energy flow, some mechanisms of speciation, nonhost species fitness, and global stability.

Parasitism and Interhost Competition

The Differential Susceptibility of Sympatric Host Species

"One particular interaction is the association between two species which do not compete for resources but share a common natural enemy. . . . It is parasitism that may exert the strongest indirect effects" (Bonsall and Hassell 1998). As early as 1983, Freeland considered that parasite species are distributed among potential host species according to four parameters—phyletic origin, size, morphology, and dietary behavior. In short, parasites cannot usually infect, at the same stage of development, phyletically distant host species, hosts of very different sizes and morphologies, or hosts of very different diets. Freeland (1983) also noted that these four parameters are "parasite barriers" that prevent or limit the transfer of parasites from one species to another and that coexistence between free-living species depends on the efficacy of these barriers (the more efficient the barriers, the more possible coexistence is, and vice versa). Knowing that in general a parasite species has different frequencies and intensities in the different species of its host spectrum, Freeland deduced that differences in susceptibility to parasites can determine what host species can coexist.

Within an ecosystem it has often been demonstrated that the presence of a predator could either ease the coexistence of two or more prey species (as long as the predator preferentially attacked the most abundant prey species) or, in contrast, that it could destroy an already established coexistence (see Schoener 1982). As for predators, parasites are also capable of modifying competition between free-living species, with the difference being that this effect, here called parasitic arbitrage, is probably grossly underestimated in nature.

If two or more sympatric species of free-living animals that compete in the ecosystem "share" the same parasite species, and if one of them is more sensitive to it than the other, this parasite can then modify the combative relationships between them. Although the parasite is disadvantageous (in absolute value) for both species, it is relatively advantageous for the one it damages least.

Knowing that in any competition there is often a "stronger" and a "weaker" combatant, there are two possible ways the parasite can intervene (Combes 1995; Yan 1996): either the parasite runs with the stronger to ensure victory (it further weakens the weaker), in which case it can be responsible for the disappearance of the weaker competitor, or the parasite flies to the rescue of the weaker, allowing it to compete better and survive (contrary to the previous case, parasitism can in this way contribute to the maintenance of diversity, since it weakens the stronger and allows the "weaker" to exist).

The first and probably most renowned demonstration of parasitic arbitrage was by Park (1948), who showed that the flour beetle *Tribolium castaneum* generally dominates the related species *T. confusum,* but that the outcome of the competition is reversed if a parasitic protist, *Adelina tribolii,* is present. Since then other examples have shown that parasites may modify or even reverse the outcome of interspecific competition and thus influence the structure of free-living animal communities (see Price et al. 1986; Holt and Lawton 1994; Hudson and Greenman 1998).

Drosophila melanogaster and *D. simulans* are two species of fruit flies that are related but reproductively well separated. *Leptopilina boulardi* is a small parasitoid wasp (cypinid) that lays its eggs in the larvae of fruit flies, which may be easily studied in caged populations that allow one to follow the

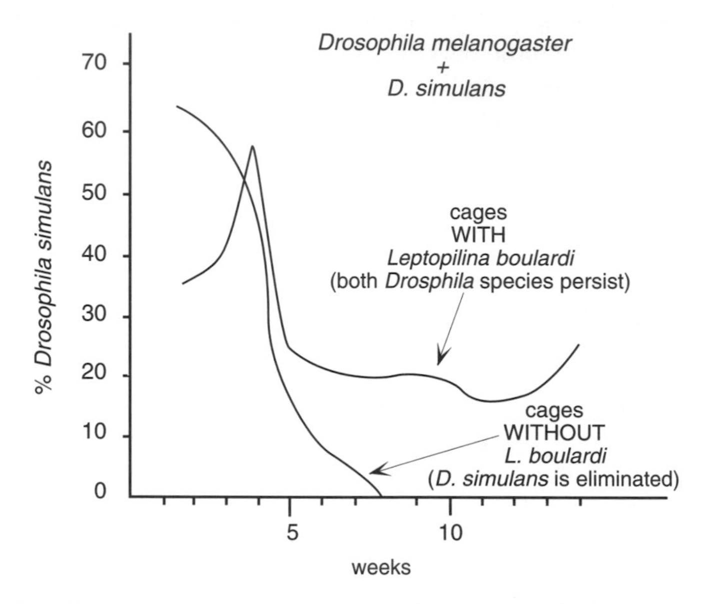

Figure 16.1. Frequency variation between two species of fruit flies in the presence or absence of a parasitoid (after Boulétreau, Fouillet, and Allemand 1991).

population dynamics of one or several species under variable conditions over numerous generations in a short time. In particular, it is easy to introduce parasitoid insects into cages and thus simulate the pressure they exert on natural populations. Under these conditions Boulétreau, Fouillet, and Allemand (1991) showed (fig. 16.1) that when *D. melanogaster* and *D. simulans* are associated in about equal proportions the former eliminates the latter within eight weeks and that when *D. melanogaster* and *D. simulans* are associated in about equal proportions but with *L. boulardi* added, a balance between the two fruit fly populations is quickly established and the two species coexist. Further, when the temperature is dropped to 22°C in the presence of *L. boulardi* there is even elimination of *D. melanogaster* by *D. simulans*—just the opposite of what happens in uninfected cages.

The parasitoid *L. boulardi* thus modifies the relative competitiveness of both species of *Drosophila* in a way that allows *D. simulans* to "resist" *D. melanogaster.* The explanation for this reversal of fortune is in a slight preference by the parasitoid for *D. melanogaster,* which decreases the reproductive success of this species, since the parasitoid larvae systematically kill their hosts.

In nature (for example, in Tunisia), both species of these fruit flies and the parasitoid wasp coexist. In oases in this area it has been observed that the more abundant the parasite, the more *D. melanogaster* decreases in proportion to *D. simulans*—very good support for the idea that in nature, as in caged populations, the parasitoid aids the survival of *D. simulans* (Boulétreau, Fouillet, and Allemand 1991).

Similarly, parasitic arbitrage has been reported in ants in Texas. In this case the ant *Pheidole dentata* is typically dominant over *Solenopsis texana*. This dominance can be observed in the field by offering bait to the ants: *P. dentata* prevents *S. texana* from taking the bait. However, during certain times of the year a dipteran parasitoid (a phorid of the genus *Apocephalus*) is present and reverses the dominance in favor of *S. texana*. The explanation is that the parasitoid, which oviposits on its victims, has a strict specificity for *P. dentata* and attacks only the soldiers of this species. However, these soldiers are precisely those responsible for the dominance of *P. dentata*, because they attack in large numbers when *S. texana* shows up at the bait. Since soldiers hide (under leaves) for about an hour during each attack of the parasitoid, *S. texana* can reach the bait more easily (Feener 1981).

Both these examples illustrate how parasitic arbitrage favors the dominated and thus reestablishes an equilibrium between two species where one would normally tend to eliminate the other. The parasite thus favors one gene pool (that of the less competitive population) while decreasing the success of the other. Obviously such interventions play a role in maintaining species diversity in ecosystems.

The choice of parasitoids to illustrate parasitic arbitrage should not lead one to believe that other types of parasitism do not also intervene in host competitions, and other examples of such parasites are easily cited. For example, certain paramecia *(Paramecia aurelia)* harbor bacteria *(Caedobacter taeniaspiralis)* that produce "toxins" the paramecia are insensitive to. However, other strains of *P. aurelia* that do not have the bacteria are sensitive to these toxins. If conjugation (contact between cytoplasms) occurs between the two types of paramecia, the sensitive partner is then killed by the infected partner.

We saw in the previous chapter that two amphipod species living in the same lagoons along the Mediterranean are affected differently by metacercariae of the trematode *Microphallus papillorobustus*. In *Gammarus insensibilis* part of the metacercariae locate in the brain (see chapter 7), and the total number of metacercariae per host reaches a plateau. Both these facts suggest that "heavily infested individuals are severely removed from the population" (Thomas, Renaud, Rousset, et al. 1995). In *Gammarus aequicauda* there is no migration of metacercariae to the brain, and new cercariae are regularly acquired throughout the life of the gammarid without its survival being directly affected. This differential mortality most likely intervenes in the competition between the two gammarids.

The weight of the young and the clutch size of the great tit, *Parus major,* are negatively correlated with the density of blue tits, *Parus caeruleus,* so the blue tit and the great tit probably are in competition for food. However, the blue tit appears to be the main source of the flea *Ceratophyllus gallinae,* which attacks both species. Thus the decrease of reproductive success of the great tits possibly is related to an increase in the density of fleas, which would act as allies of the blue tit in the competition (Richner, Oppliger, and Christe 1993).

Mathematical modeling also confirms, at least in theory, that competition between species may be highly modified as soon as the species "share" certain parasites (Holt and Pickering 1985; Yan 1996; Hudson and Greenman 1998).

The Differential Susceptibility of Allopatric Host Species

Differential susceptibility to a parasite may be evident in hosts living in different areas. In this case it plays a role only if one of the host species tries to invade the other's territory or is introduced, either purposely or accidentally, by humans. Parasitic arbitrage may then either prevent or promote such invasions.

Freeland (1983) affirms that a species is capable of invading an ecosystem because its susceptibility to parasites may be different from that of the resident species. However, if the immigrant species is susceptible to the same parasites as the resident species, the invader may not be able to establish itself. This is explained by the immigrant species' lack of experience with the parasite; it cannot protect itself because it has not selected the necessary behaviors or immune defenses to limit parasite loads. Further, selecting resistance genes is difficult both because of the small size of the immigrant population, which limits genetic diversity (the founder effect)., and because of the regular arrival of new immigrants, which dilutes the gene pool of the first immigrants and further delays selection for resistance.

In general, the invasion of a new free-living species into an ecosystem will be successful only if one of these two conditions is fulfilled: the immigrant species is insensitive to the parasites of resident species, or the immigrant species carries a parasite that some of the resident species are sensitive to.

According to Freeland, species coexistence should owe more to the presence or absence of shared parasites than to more classical factors such as resource availability and interspecific competition. Price, Westoby, and Rice (1988) furthered these ideas by predicting what would make a winner or a loser during such invasions. According to these authors, winners must be organisms that have developed an efficient immunity and thus would be species with broad geographic ranges (since their parasite richness is larger) and species that are physically small (since their time between generations is shorter).

In either case the collision between two hosts can have different outcomes depending on whether it is the susceptible or the resistant population that moves. The movement of a population that is resistant (but that carries the parasite) into an encounter with a sensitive population of the same species cannot be better illustrated than by the movement of humans. The conquerors of Latin America owed their victory over Native Americans as much to the smallpox, chicken pox, and measles that they carried as to their weapons, horses, and courage. It is estimated than more than 50 million people died as the result of this holocaust alone (Black 1992), and the history of the colonization of the Americas strongly indicates that contact between "disease-experienced" Europeans (because of their travel throughout Eurasia) and "disease-inexperienced" Amerindians (because of their isolation) was catastrophic for the latter.

Along these lines humans have also provoked serious perturbations in ecosystems by introducing into new environments host species that carry parasites unfamiliar to the local inhabitants. Examples of these perturbations are numerous in fish, which often have parasites that are only slightly pathogenic in their area of origin but create serious economic problems when introduced outside their normal range. For instance, European eels are affected by two monogeneans, *Pseudodactylogyrus anguillae* and *P. bini,* that were introduced about 1977 from Asia (Buchmann 1993). Both parasites are known to cause locally high mortality in Europe.

Similarly, the introduction into Norway, starting in 1954, of Baltic salmon carrying the fin-dwelling monogenean *Gyrodactylus salaris* had catastrophic consequences for salmon farming in this area (Malmberg 1993), and the introduction into the Aral Sea of the sturgeon *Acipenser stellatus,* a carrier of the gill monogenean *Nitzchia sturionis* from the Caspian Sea, induced the passage of this parasite onto the local sturgeon, *A. nudiventris.* This passage dramatically reduced the population of *A. nudiventris* for more than twenty years (Bauer and Hoffman 1976).

The introduction into western European rivers of the fish *Lucioperca lucioperca* also introduced its parasite, the trematode *Bucephalus polymorphus,* into the same waters. Cercariae of this parasite are produced in the freshwater mussel *Dreissena polymorpha* and encyst on the fins and opercula of numerous species of small fish that are then eaten by *L. lucioperca. Bucephalus polymorphus* thus induced a heavy mortality in the small fish in these areas (see Combes and Le Brun 1990).

On the other hand, when a population sensitive to a particular parasite penetrates the territory of a population carrying but resistant to a shared parasite, the usual result is that the new population cannot invade. This may happen for populations of either the same or different species.

Once again human populations provide a very good example. On an interspecific scale, the lateral transfer of parasitic diseases to humans from other animals may be seen as parasitic arbitrage in favor of the animal hosts into whose territory humans have penetrated. Undoubtedly many humans have died after moving into new areas as the result of this arbitrage. At an intraspecific scale it is likely that sub-Saharan Africa was protected from European invasion for a number of years by malaria, to which Europeans were extremely sensitive until the advent of quinine treatment at the beginning of the nineteenth century.[1]

In short, then, the intrusion of a sensitive population into a population of resistant hosts may be catastrophic for the immigrants. According to Golvan and Rioux (1961, 1963), certain plague epidemics in Iranian Kurdistan were due to the transfer of the causative agent of the plague, the bacillus *Yersinia pestis,* from a resistant rodent to a sensitive one and then from this sensitive rodent to humans. Two species of gerbils (small rodents) coexist in this region, although each occupies a distinct biotope (fig. 16.2). *Meriones persicus* lives in rocky zones away from crop cultures, is resistant to the bacillus, and constitutes the "reservoir" of the disease. *Meriones vinogradovi* lives in cereal fields and is sensitive to the bacillus. When the population of *M. vinogradovi* becomes very abundant it has a tendency to "overflow" from its regular ecological niche, and a few individuals may become contaminated because of an exchange of fleas with *M. persicus.* At this point a brutal epizootic occurs in *M. vinogradovi,* leading to an almost total destruction of the population. The disease spreads like a forest fire (although at different speeds depending on population densities), and only the few individuals that escape death remain behind to rebuild the *M. vinogradovi* population, which takes a few years. Golvan and Rioux determined that the "plague epizootic ineluctably concluded the period of explosive population growth of *M. vinogradovi* in the Iranian Kurdistan wheat plains." When the epizootic occurred, the possibility for human contamination increased because of imprudent handling of dead gerbils.

Similarly, the plague bacillus allows the resistant rodent *Peromyscus*

1. We see here that the invasions of South America and Africa by Europeans had opposite effects. South America was very vulnerable because of native sensitivity to imported pathogens, whereas Africa was protected by at least one pathogenic agent to which the Europeans were very sensitive.

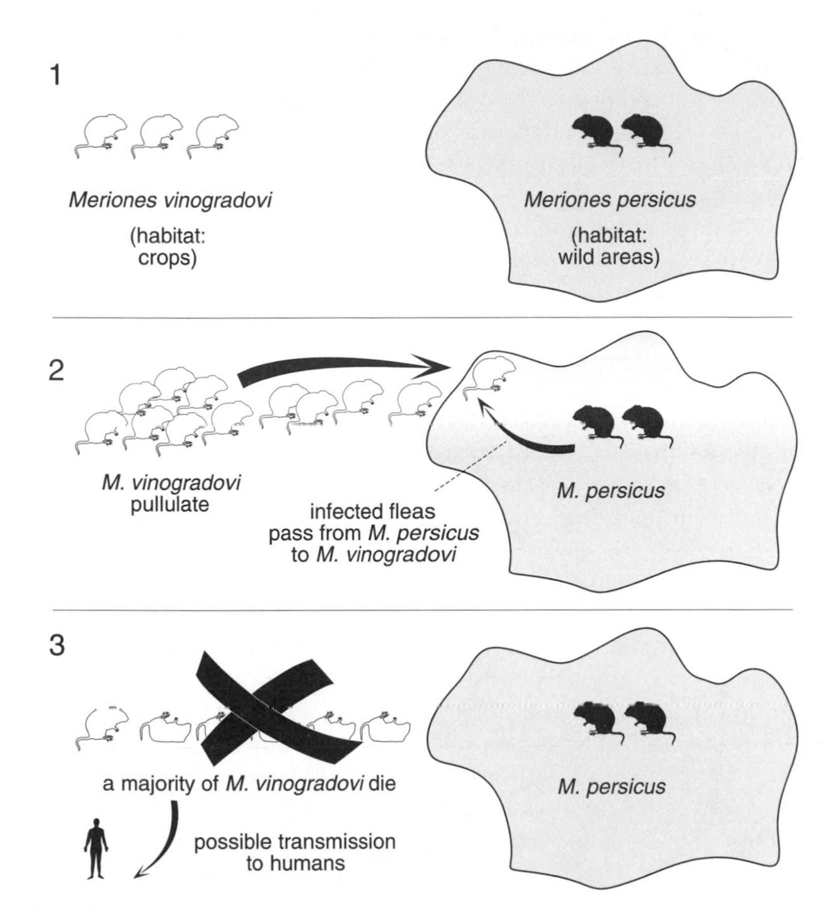

Figure 16.2. Gerbils and plague in Iran. Note the possibility of transfer to humans from crop-dwelling *Meriones vinogradovi* (after Golvan and Rioux 1963).

maniculatus to "defend itself" against invasion by the sensitive rodent *Neotoma cinerea,* which once in a while is subjected to a lethal epidemic when in contact with the former in California (Nelson and Smith 1980).

Examples of the failed movement of sensitive populations are abundant in the history of species introduced into new environments. For instance, *Odocoileus virginianus* (the white-tailed deer) is an American cervid that carries a nematode, *Parelaphostrongylus tenuis,* to which it is "well adapted," so disease is almost never seen. Although whitetails may occur with other ungulates on a regional scale, at the local level everything indicates that their presence excludes other ungulates. For example, the whitetail and the moose *(Alces alces)* have distinct niches characterized by the depth of snow during the winter, with the deer living in regions of low altitude where snow is not too abundant. However, if during the summer moose penetrate into the deer's zone, they can become infected with *P. tenuis* larvae, which induce serious neurological problems that quickly kill them. Numerous attempts have been made to introduce other cervids (the caribou, *Ranger caribou,* the wapiti, *Cervus canadensis,* etc.) into regions occupied by whitetails, and all have failed because the introduced species are sensitive to *P. tenuis* (Anderson 1972). Notably, more recent research has demonstrated that even though some moose may persist in zones of low whitetail density (fewer than four per square kilometer), and even if there are causes of mortality other than the nematode, the density of moose in these areas is inversely correlated with the abundance of *P. tenuis* eggs in the whitetails' feces (Whitlaw and Lankester 1994a, 1994b).

In general, the distribution of animals on the planet is certainly not unrelated to parasitism, as is clear from the observation of species recently introduced into new areas by humans. For instance, the zebu's presence in the Sahel zone of Africa reflects both its resistance to ticks, which are abundant in the Sahel, and its sensitivity to trypanosomes (abundant south of the Sahel). Besides a few particular cases, such as *ndama* bulls, few domestic ungulates resist African trypanosomes. Significantly, horses are absolutely excluded from these regions (Philippe Dorchies, personal communication). Of course in certain cases, after a "difficult" period, invaders or introduced hosts may select resistant individuals.

The examples above support the idea that parasites and disease mediate who wins and who loses, even in human populations. Thus parasites have both stabilizing and destabilizing effects on ecosystem diversity (see Combes 1996). Which of the parasites—the "defensive" ones (those allied to a species on site) or their "offensive" counterparts (those allied to the invaders)—most affect the landscape of evolution? Nobody knows the answer, although it may be essential to understanding the evolutionary history of the biosphere (fig. 16.3). One can only venture the hypothesis

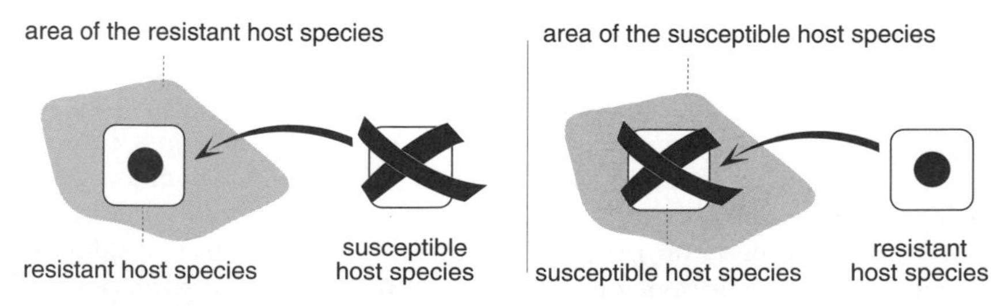

Figure 16.3. Parasites' defensive (left) and offensive (right) arms.

that the defensive side would probably dominate, simply because the parasites are by definition available on site whereas they are often lost during the emigration of hosts, so that the parasites would have more difficulty being exported to confront potentially new host populations.

Up to a certain point it is clear that parasites may mediate either war or peace in the biosphere and that the collision of hosts may then induce host extinctions. For instance, a *Plasmodium* of birds introduced into the Hawaiian Islands ravages the local avifauna even today (see Van Ripper et al. 1986). Such extinctions caused by invading species' virulent parasites have most likely occurred frequently throughout the history of the earth, for instance, when continental bridges such as the one linking North and South America formed. This land bridge is relatively recent: after being separated for about 50 million years, both continents were united only about 3 million years ago. At this time massive exchanges in fauna occurred (until that time there had been only limited exchanges). In this case the consequences were much more dramatic for South America than for North America. During the long separation before linkage, South America had developed its own notable fauna of marsupials, of which 50% were annihilated following the development of a permanent land bridge through Panama. This phenomenon is generally explained by the stronger competitiveness of placental mammals from the north (Bakker 1983), but one wonders whether these massive extinctions may also be due, at least in part, to parasitism (see Barbehenn 1969).

One of the conclusions drawn from such host collisions is that parasites that are often forgotten or ignored when "nothing happens" suddenly become very visible when "things change." In an ecosystem in equilibrium, parasite-host relationships often appear to be in peaceful coexistence, but as soon as the system is disturbed and moves out of equilibrium, conflicts appear. Does this mean we should encourage the conventional wisdom according to which an "ancient" parasite is benign but a newly acquired one is virulent (see chapter 11)? Certainly not, but host collisions clearly allow us to state that more recent associations may be characterized by stronger virulence. This subject will be discussed again when we look at the spread of humans across the planet over only a few hundred thousand years.

Parasitism and Food Webs

Globally, parasite-host systems involve a much smaller transfer of energy than occurs in predator-prey systems. It is not, however, by this direct energy transfer that parasites influence ecosystems. Instead, it is in their influence on predator-prey systems that parasites play their greatest role. Although the influence of predator-prey systems on parasite-host systems has been understood for some time,[2] the influence of parasite-host systems on predator-prey systems has been recognized only of late.[3]

The Fox and the Mite

Predator-prey systems exist in complex and fragile equilibriums. For instance, one may expect that when a prey species becomes rare its predators will exploit another species and that when a predator becomes rare the population of its usual prey will grow explosively. The laws that govern these equilibriums are not all known, and the hypotheses proposed to explain them are often contradictory. This uncertainty may arise in part because it is difficult and dangerous to attempt some "experiments" on natural populations, so prolonged observations are rare.

Given these restrictions, it is interesting to use occasional epizootics induced by parasites as "natural experiments." An epizootic of sufficient magnitude and type in Scandinavian foxes allowed just such a study to be done (Lindstrôm et al. 1994). The outbreak of disease in this case was caused by the sarcoptic mite *Sarcoptes scabiei,* an acarid that burrows in galleries beneath the skin of mammals and induces hair loss and itching. *Sarcoptes scabiei* appeared in foxes in Sweden in 1975 and spread across the country within a few years. Studies carried out in the region of Grimso (60° north latitude) showed that the disease appeared in that area in 1982, reaching maximum intensity between 1983 and 1990. During this period the population of foxes was heavily depressed, which was directly correlated with large increases in almost all their natural prey items, such as deer, hares, and grouse (a total of six prey species). Further, the usually observed cyclical fluctuations in each fox-prey system were also altered, and only voles *(Clethrionomys glareolus)* did not respond with a population increase. Thus the study showed both that foxes exerted the main pressure in structuring the prey community and that a single species of para-

2. As soon as the first heteroxenous life cycles were discovered about the beginning of the twentieth century.

3. Free-living stages of parasites sometimes enter food webs as prey. For instance, certain larval stages, such as trematode cercariae, contribute to the formation of plankton communities and are the prey for planktonophagous organisms. That is, what would otherwise have become benthic organic matter (mollusks) becomes planktonic organic matter (cercariae). Nothing quantitative is known about what this shift in energy in the food web represents: the subject is ignored by ecologists. However, it has been shown that some predators do not neglect this manna. For instance, the oligochaetes *Chaestogaster limnaei* are more numerous on the shells of infected mollusks that emit thousands or cercariae daily than on the shells of healthy mollusks (Fernandez, Goater, and Esch 1991).

site may have multiple consequences in the overall function and energy flow in an ecosystem.

Numerous parasite life cycles have selected genes that influence their intermediate hosts' behavior so they can be captured by definitive hosts. Changing a host's behavior causes certain individuals and even species to be preferentially eaten. A parasite's "choice" of an intermediate host may, because of such favorization, influence the predator's choice of prey. Whatever the choice in question, such an occurrence decreases the energy the predator must expend in capturing its prey. Not only do parasites thus orient the flow of energy, they encourage this flow in an ecosystem. Thus, parasites play a role in directing and easing the flow of energy down specific tracks in food webs, and in this sense they are similar to enzymes in metabolic pathways: they direct energy flows down certain pathways by increasing the rate of predation while regenerating themselves to do it again.

This being said, one may have the feeling that favorization is peculiar to just a few parasites and that it would be incorrect to attribute a more general role to it. This may not be wrong, but parasites may still play much the same role outside favorization processes, and it is interesting to note that the list of parasites that exhibit favorization grows virtually every time it is looked for in parasite-host systems.

The Cat, the Mouse, and the Parasite

In Tom and Jerry cartoons, Jerry (the mouse) is startled each time Tom (the cat) begins his well-prepared aggression. However, Jerry always makes the proper decision, which ultimately makes Tom's plan fail.

Among factors that allow mice to make the right decision in the face of the cat in nature are the processes of neuromodulation (see Thompson and Kavaliers 1994). Stimuli associated with the sight of a predator activate analgesic mechanisms (resistance to pain), which then prevent panic reactions and provide a greater chance for making the appropriate decision. This stress analgesia is the result of the release of opioid peptides at the brain synapses (see Lewis, Cannon, and Libeskind 1980). "These stress- or environmentally-induced analgesic responses are considered fundamental components of an animal's defensive repertoire, facilitating the coordinated expression of other adaptive, defensive, and behavioural responses" (Thompson and Kavaliers 1994).

Kavaliers and Colwell (1993a, 1993b, 1994) studied the influence of mouse parasitism on stress analgesia in the presence of a cat. The parasite was the coccidian *Eimeria vermiformis,* which develops in the mouse intestine. The parasites multiply in the duodenum and mice pass new oocysts (infective stages of coccidia) after eight days, with maximum production about ten days after infection, followed by a decline until the twentieth day, when oocyst emission spontaneously stops. Mice were contaminated with known doses of oocysts, and the quality of stress analgesia was studied by putting their paws on a hot metal plate (50°C) and measuring the time it took for a paw to be removed. Such a measure provides an index of environmentally induced neuromodulatory responses: the longer the latency period, the more efficient the analgesic (resistance to pain) effect and the longer the mouse has to make the right choice.

Two groups of mice in cages, one parasitized for six days and the other not, were set in the presence of either a cat ("an experienced, predatory cat") or an inoffensive guinea pig for thirty seconds. Immediately before and after each experiment, latency time was measured (the experiment also included a group of mice stressed by a factor other than the cat's presence, but this will not be considered here).

There is not much need to comment on figure 16.4, which is very explicit: the latency time for uninfected mice remained the same in the presence of the guinea pig but significantly increased in the presence of the cat; and the latency time for parasitized mice showed no difference when mice were faced with the guinea pig and with the cat.

Thus, under the influence of parasitism infected mice lost one of their main adaptive response mechanisms—a response that might otherwise allow them to make the right decision when under stress and avoid useless fleeing behavior that would, overall, make an infected mouse more likely to be noticed and captured. Significantly, mice infected with *E. vermiformis* also spend more time within range of the cat's smell than do healthy individuals, indicating that infection may also reduce the state of anxiety and increase the risk of finding oneself face to face with the cat. The biochemical mechanisms of this effect will likely be found at the level of the receptors for the neurotransmitter gamma-aminobutyric acid (GABA) in the nervous system, because GABA antagonists induce, in part, a similar reduction in anxiety (Kavaliers and Colwell 1995).

This work demonstrates that even if there is no favorization (*Eimeria*'s life cycle is holoxenous) and even if there is no visible pathogenic effect, parasites can "involuntarily," invisibly, and silently work to the advantage of the predator.

The obvious conclusion is that Jerry is healthy; otherwise Tom would have eaten him long ago.

The Stickleback and the Cichlid

Reported here are a series of experimental studies carried out to analyze the compromises in risk taking made by parasitized animals compared

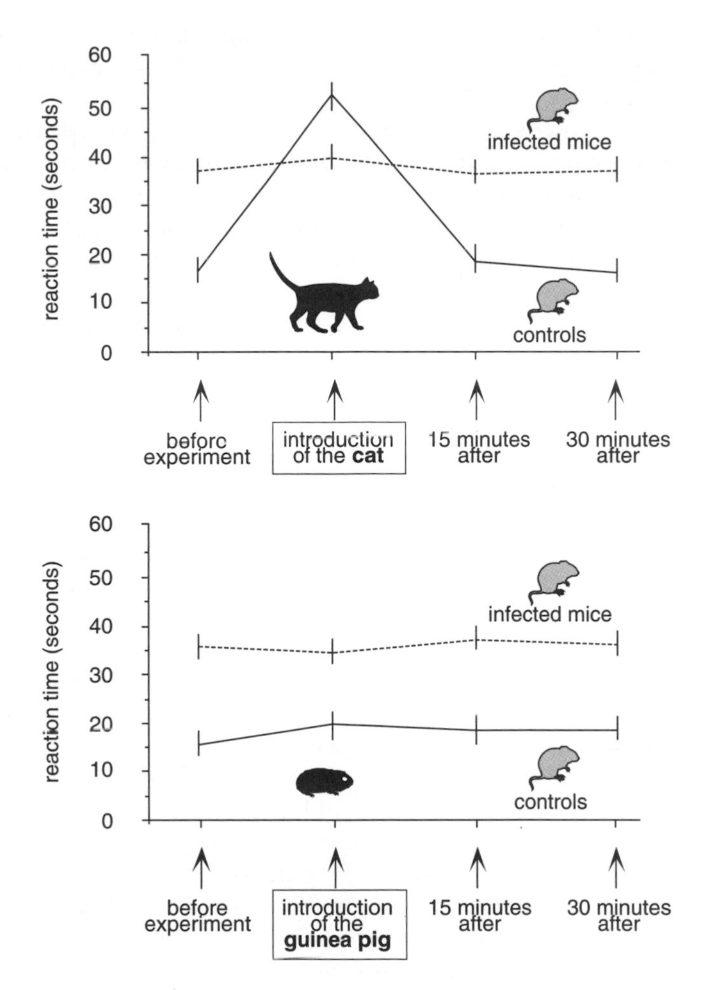

Figure 16.4. Time-lag response of nonparasitized, stressed, or parasitized mice in the presence of either a cat or a guinea pig. When seeing a guinea pig (bottom), healthy, stressed, or parasitized mice show no analgesic reactions. When seeing a cat (top), parasitized mice do not react normally (modified from Kavaliers and Colwell 1994).

with healthy ones (see Milinski 1990; Lima and Dill 1990). Wild animals live perilous lives, and their survival depends on the decisions they make about what risks they take to fulfill their needs, most notably those associated with acquiring food. Just as a pickpocket will evaluate and balance the risks of being caught by the police against the benefits of stealing a wallet, an animal that searches for food "evaluates" and weighs the expected benefits and risks of being captured by a predator. Any decision taken is the result of a compromise, and it is here that the parasite intervenes: if its presence modifies the data input into the decision, the outcome will likely be different.

The situation was studied using the following design. An aquarium was divided into two parts separated by a glass wall. Sticklebacks *(Gasterosteus aculeatus)* were placed in one of the compartments, and a predatory fish (a cichlid) could be introduced into the second. Some *Tubifex* worms were offered to the sticklebacks as food at variable distances from the predator's compartment. The behavior of some sticklebacks harboring plerocercoids of a bird cestode, *Schistocephalus solidus,* was then compared with the behavior of healthy sticklebacks in capturing *Tubifex* in the presence and absence of the predator in the accompanying compartment.

In the absence of predators behind the glass wall, nonparasitized sticklebacks randomly captured *Tubifex* anywhere in the aquarium. When the predator was present, however, significantly more *Tubifex* were captured farther from the predator than close to it. Significantly, parasitized sticklebacks showed no difference in their behavior whether the predator was present or not: captures of *Tubifex* were just as numerous close to or away from the predator.

Since it is unlikely that the sticklebacks understood the presence of a "protective" glass window in one case and not the other, this change in behavior can be attributed to the parasite's modifying the compromise made between hunger and fear. In the healthy stickleback, hunger was not strong enough to justify the risk of approaching the predator. But in the parasitized stickleback the plerocercoid hijacks metabolites to the point where it weighs as much as the fish itself, so the fish is perpetually starved and its hunger dominates its fear. That is, the parasitized stickleback "accepts" the higher risk of being eaten. In this case the word "starved" is not an exaggeration, since healthy sticklebacks that do not eat even for four days do not take the same risks as do parasitized individuals.

There is not much doubt that this change in behavior favors the transmission of the parasite, since it increases the probability that the stickleback will be eaten by the bird that is the definitive host. However, the parasite facilitates the capture of the parasitized stickleback by *all* its predators, host or otherwise.

In general, any metabolic modification leads to multiple consequences, notably the need for more food, so parasitized animals not only are weakened but must spend more time searching for food. Under these conditions infection with any parasite would make parasitized animals easier to catch even if there was no link with transmission. Given the development of a "search image" (predators become used to searching for the prey items that are most abundant or easiest to capture), any parasite that would promote the capture of one specific prey relative to another may reorient a predator's prey preference and thus modify the working of the food web: "It may be the case that in the absence of the parasites, the predator cannot survive on the prey . . . the parasites are an obligate mutualist of the predator . . . survival of the predator population is in large part dependent on the presence of the parasites in the prey" (Freedman 1990). In other words, the cost imposed on the predator by the parasite (owing to its pathogenic effect) is less than the benefit it provides—increased energy consumption because of easier predation. The same idea was developed by Lafferty and Morris (1996) in the conclusion of their work on favorization in the life cycle of the trematode *Euhaplorchis californiensis.* In this work the authors showed that fish harboring metacercariae are much easier to catch than are healthier ones. "If the cost of parasitism is less than the energy gained from capturing more fish, parasites might benefit birds by acting as a delivery service that enables birds to eat fish that are otherwise difficult to capture."

From Parasitism to the Food Web, or Vice Versa?

The structure and function of food webs is one of the most difficult fields of research in ecology, and Marcogliese and Cone (1998b) attracted attention to how studying parasitism can increase our knowledge in the field: "Parasites are a useful means of tracking pathways within food webs. . . . Including parasites not only increases the amount of information on web structure and trophic pathways, but also allows the reassessment of fundamental dynamic web properties." These authors emphasize that heteroxenous parasites that circulate between prey and predator may give more reliable information than the simple examination of stomach contents because cycles reveal permanent characteristics of diets and not just the food the animal ate a few moments earlier. These authors cite as an example the work of Kennedy et al. (1992), who showed from analysis of their parasite communities that eels are plankton feeders. These workers also indicated that studying parasite communities clearly shows that omnivory in food webs (feeding at more than one trophic level) may be more frequent than usually thought.

Marcogliese and Cone are absolutely right in the content of their ideas. In reality, however, one must also take into account that knowledge about parasite life cycles is probably as difficult to acquire as is direct knowledge about food webs. Moreover, parasites may provide only a partial image of food webs because of parasite specificity: if a prey harboring an infective stage is eaten by several predators, it will show the trophic link only between that prey and its predator host—other nonhost predators in which the infective stages will be digested will be ignored. Finally, we may reason in parallel with Marcogliese and Cone: parasitologists are interested in predator-prey relationships because these relationships help them understand parasite life cycles. The logical conclusion is that parasitologists and ecologists can only gain by working together in this area.

Parasites and Nonhost Organisms

Jones, Lawton, and Shachak (1994, 1997) called organisms that modify the availability of other organisms' spatial or energy resources "ecosystem engineers." All living beings are probably ecosystem engineers to one degree or another, but some are more so than others. Where is the parasites' place in this engineering firm?

Thomas, Renaud, de Meeus, et al. (1998) and Poulin (1999a) have carried out the only work that details how the presence of a parasite affects a community of free-living nonhosts. The consequences are related to the manipulation of a host by the parasite and thus prolong the favorization effects above the level of the parasite-host system itself.

The cockle *Austrovenus stutchburyi* is a bivalve that inhabits the sheltered shores off New Zealand. In this area this mollusk constitutes, in many cases, the only available hard surface where an intertidal community of benthic invertebrates (limpets, anemones, bryozoans, barnacles, etc.) can be established. The cockles' behavior is altered (see chapter 7) by metacercariae of the trematode *Curtuteria australis*. In this case local prevalence often reaches 100%, with the number of metacercariae per cockle sometimes exceeding one thousand. Parasitized cockles cannot bury themselves, and they lie exposed at the surface of the mud. Thomas, Renaud, de Meeus, et al. (1998) wrote: "It is not unrealistic to consider that manipulated hosts can be equivalent to new organisms in the ecosystem, involved in new direct and/or indirect interactions with other species." From a study carried out on almost 1,500 harvested cockles from three locations, the authors demonstrate that the limpet *Notoacmea helmsi* and the anemone *Anthopleura aureoradiata* colonize infected and healthy individuals differently. Limpets were thus significantly more frequent, and anemones less frequent, on manipulated (parasitized) cockles than on nonmanipulated ones (fig. 16.5). Further, some experiments showed that limpets preferred manipulated cockles even when anemones were absent, whereas anemones were more susceptible to desiccation at low tide than limpets. The authors concluded that the trematode was an ecosystem engineer, since it modifies the availability and quality of habitats for nonhost species: limpets benefit from parasites, whereas anemones experience less favorable conditions as the result of infection.

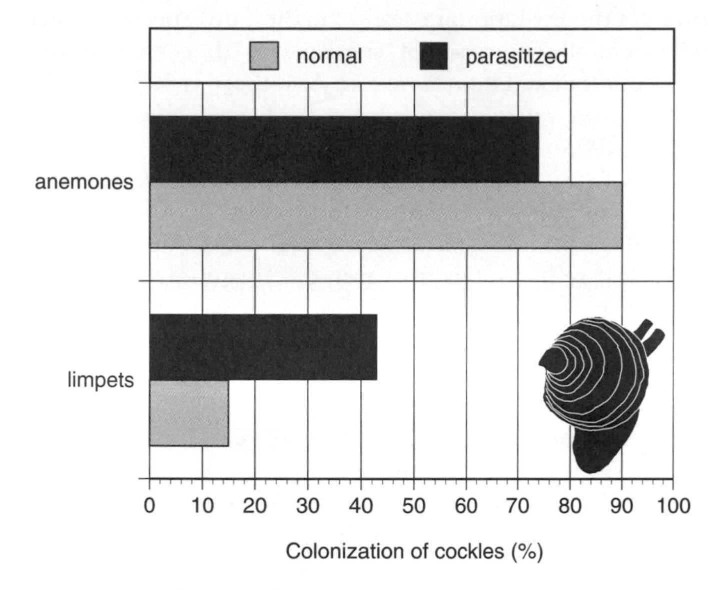

Figure 16.5. Differential colonization by limpets and anemones on parasitized and non-parasitized cockles in New Zealand (adapted from Thomas et al. 1998).

These same workers (Thomas et al. 1999) also noted another possible example of engineering by parasites: when decapod crustaceans are parasitized by other crustaceans (rhizocephalids) such as *Lernaeodiscus* and *Sacculina,* they often fail to molt. The carapace then becomes a more permanent substrate for colonization by various invertebrates in parasitized crustaceans than in those having a normal molting cycle.

Research of this type, at the junction between parasitology and ecology, holds much promise in deciphering the nature and structure of food webs and thus of communities and ecosystems in general.

Parasitism and Speciation

If reproductive isolation did not occur, there would be only one living species on the planet earth. That is, it is only because reproductive barriers exist that all genomes do not mix and that there are humans, dogs, rabbits, mice, and parasites.

Can parasitism play a role in host speciation and thus in the species composition of ecosystems?

Today, European mice are distributed throughout the world. In fact the mouse species *Mus musculus* originated in northern India and has spread out from this focus only within the past ten thousand years—a very short time on the evolutionary scale. Further, this species is peculiar because it is essentially composed of "semispecies" that can hybridize among themselves but that still do not mix: in Asia there is *Mus castaneus* and in Europe both *Mus musculus musculus* and *Mus musculus domesticus* (Bonhomme et al. 1983).

The zone of contact between *M. m. musculus* and *M. m. domesticus* follows a curve extending from Denmark up through Bulgaria, with *M. m. musculus* in the east and *M. m. domesticus* in the west (fig. 16.6). Where the two subspecies are in contact, research on the genotypes has found that there is a very narrow zone of hybridization between the two (always less than forty kilometers). In the hybridization zone individuals come from successive interbreedings between the two parental groups, so individuals in the hybrid zone are a mosaic of parental genes. The word "hybrid" in this sense is thus not taken in the classic sense of individuals' being the progeny of a first-generation cross.

Research on the parasitism of mice has been carried out both within the hybrid zone itself and along each of its sides (first in Germany and then in Denmark).

The study in Germany (Sage et al. 1986) showed that mice captured in the hybrid zone were clearly more parasitized by nematodes and cestodes than were mice on either side. Nothing, however, indicated that the environment in the hybrid zone was particularly favorable to the transmission of helminths, and the authors formulated the hypothesis that a genetic peculiarity could be responsible for the heavy parasite loads they observed.

Studies carried out in Denmark (Moulia, Aussel, et al. 1991; Moulia, Le Brun, et al. 1993) using isozyme analysis, infection rates, and breeding allowed progress toward understanding the mechanisms involved.

In the isozyme studies a total of 120 mice were analyzed individually for their genetic structure (ten diagnostic allozyme loci for each of the two subspecies) and for their parasite loads in terms of intestinal nematodes with holoxenous life cycles *(Aspiculuris tetraptera* and *Syphacia obvelata).*

The genetic analysis allowed the classification of mice into three groups, each corresponding to the location where they were captured: M = those mice carrying between 0 and 20% *domesticus* alleles; H = those carrying between 20% and 80% *domesticus* alleles; and D = those carrying between 80% and 100% *domesticus* alleles.

The abundances of nematodes (average number per mouse) were as follows: mice from M locations (fifty-one individuals), 58.1; mice from H locations (thirty-five individuals): 135.3; and mice from D locations (thirty-four individuals): 55.1.

These results confirm that the hybrid zone is a "privileged" place for

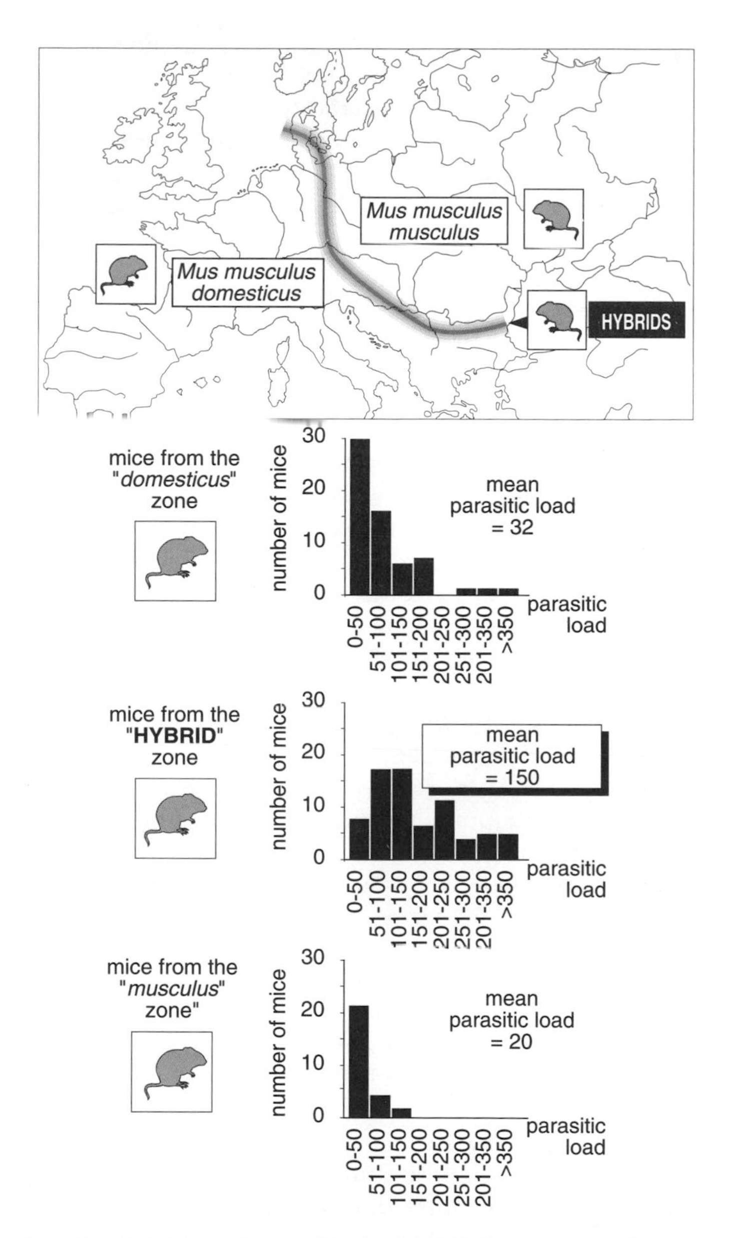

Figure 16.6. Top: hybrid zone between the "territories" of *Mus musculus domesticus* and *M. m. musculus*. Bottom: experimental results showing that the mice from the hybrid zone (center) are more sensitive to parasitism than mice of the pure type *domesticus* or *musculus* (top and bottom, respectively) (modified from Moulia et al. 1991).

parasitism and that there is a "concentration" of the nematodes in this area.

In infection studies purebred mice (both *M. m. musculus* and *M. m. domesticus*) and descendants of mice captured in the hybrid zone in Denmark were then experimentally infected with identical doses of one of the nematodes *(A. tetraptera)* whose life cycle was easy to maintain in the laboratory. These experiments showed that although the mice were all maintained under identical conditions (which eliminated environmental influences), parasitism was significantly higher in the hybrids—proof that susceptibility in the hybrids has a genetic and not an ecological basis in the system.

The explanation for the high susceptibility of the hybrids is that the genomes of the two subspecies are sufficiently different for the recombinants not to be able to correctly express natural resistance to parasitism.

Last, first-generation (F_1) hybrid mice obtained in the laboratory from purebred *M. m. musculus* and *M. m. domesticus* were, in fact, found to be more resistant than their parents. In the F_1's the parental genomes are both present (and one may think that their effects are summed), but the further recombination that would be responsible for the loss of resistance has not yet occurred between parental genomes (Moulia, Le Brun, et al. 1995).

If, as one might suspect based on all available evidence, heavy parasite loads (several hundred nematodes) more often accumulate in mice from the hybrid zone than in *M. m. musculus* and *M. m. domesticus,* and if this accumulation affects the reproductive success of recombinant individuals, it means that parasitism does not favor hybrids and thus that parasites restrict the genetic mixing of the two subspecies in nature. That is, the parasites are very likely a strong factor contributing to the maintenance of the two subspecies. This research indicates how parasitism may play a role both in the genetic structuring of a population and in the process of isolation. It also demonstrates how a hybrid zone would function as a "genetic sink" into which the carriers of recombinant genomes would disappear (Moulia, Le Brun, and Renaud 1996).

Other examples of hybrid susceptibility are reviewed by Moulia (1999). For instance, hybrids between mallards *(Anas platyrhynchos)* and black ducks *(Anas rubripes)* are significantly more frequently infected by protists of the genus *Sarcocystis* than are their parents, and these diseased offspring are more easily caught by predators. Since carnivores are the definitive hosts of *Sarcocystis,* the higher susceptibility of the hybrids might at the same time be both a barrier between duck species and a favorization process for the parasite.

It is interesting that hybrid susceptibility to a particular species of parasite does not mean susceptibility to other species of parasites. Derothe et al. (1999) report that the hybrid mice mentioned above (hybrids between *Mus musculus musculus* and *Mus musculus domesticus*), which are highly susceptible to nematodes, present the same level of susceptibility to *Trypanosoma musculi* as do their parents.

Comparable processes have been described for the insect parasites of plants, within which hybrid zones appear to play a particularly important role. For example, in northern Utah hybrids of the Fremont cottonwood, *Populus fremontii,* and the narrowleaf cottonwood, *P. angustifolia,* are extremely susceptible to the aphid *Pemphigus betae.* This aphid produces galls to such an extent that parasite densities are more than one hundred times higher in hybrid zones than in surrounding areas. This hybrid zone represents only about 3% of the potential tree hosts but harbors at least 85% of the total aphid population. This discrepancy is so great that in some years "pure" trees may be without galls while only hybrids shelter the aphids. Similarly, Floate, Kearsley, and Whitham (1993) showed that hybrids between *Populus fremontii* and *P. angustifolia* are more susceptible to attacks by the beetle *Chrysomela confluens* than are the parent species. However, the results obtained in *Populus* populations were questioned by Paige and Capman (1993), who think that more detailed genetic studies are necessary "to discern the true relationship between hybridization and pest resistance." One should note here that results like these are at the origin of the idea that it may be possible to fight certain crop pathogens by introducing some particularly susceptible plants into cultured fields or near them, so the pathogens would adapt to the nearby crop only imperfectly or not at all (Whitham 1989).

Hybrid zones are not exceptional occurrences in nature, and of course their maintenance is not necessarily always related to parasitism. Numerous hybrid zones exist in birds, and in Europe one of the best-known hybrid zones separates *Corvus corone* (the carrion crow) and *C. cornix* (the hooded crow). This zone in large part follows the Alpine arch, and it is most curious that in Denmark it goes through the same hybrid zone as that of the mice previously described.

Parasitism and Stability

Most ecosystems show astonishing stability over time, and at least in appearance, the composition and function of a forest, savanna, or shore appears not to change over years, centuries, or even millennia (at least in the absence of climatic or cataclysmic events). Stability from this perspective is only a question of scale, and at the global level apparent stability may cover or even hide numerous local perturbations that reflect the underlying dynamic processes resulting in stability at the ecosystem level. That is, ecological stability should not be taken in the sense of indefinite imperturbability. It is clear that throughout the history of the biosphere marked destabilization events have occurred (collisions with comets or asteroids, the fusion of continental plates, climatic chaos, and the emergence of the desertifying species *Homo sapiens*), and it is only at a more modest level that ecosystem stability should be analyzed.

The explanation for ecosystem stability has principally been searched for in complexity-stability relationships, which have been the object of numerous debates. The question can be divided into several parts (see May 1972, 1973; Pimm 1984; McNaughton 1988): Is the stability of the ecosystem real or apparent? If it is real, what are the mechanisms that generate it? and Among the generating mechanisms, what is the role of complexity? Are stable systems the simplest or the most complex? To these the parasitologist might add: In the complexity-stability relationship, what is the role of parasites?

Density-dependent and frequency-dependent processes are considered to be the most important factors determining stability because they have a tendency to correct lags ("perturbations") relative to the current situation. Parasites play a role in stability because they participate in these very same processes, even though this role has been widely ignored by classical ecologists.

Freeland and Boulton (1990) were among the pioneer authors to consider the possible role of parasites in the stability of ecosystems. Their idea was that a disagreement often existed between model predictions (for instance, "the tropical forest has numerous destabilizing processes") and ecological reality ("the tropical forest seems stable") because parasite-host relationships and their regulatory effects have not been taken into account. These authors note that of the over one hundred food webs described by various authors, "none includes a single parasitic species: this has occurred in spite of all organisms being subject to some form of parasitism, and in spite of there being more species of parasite than any other trophic class."

The connection is in the relation between the "realized" and "possible" links that unify the species of an ecosystem (see Frontier 1977; Margalef and Gutierez 1983). A link may result from consumption, predation, parasitism, and so forth. This connection plays an important role in the ecosystems' stability: in general we consider that there is an "optimal" connection that ensures maximum stability. Freeland and Boulton (1990) show that as soon as parasites are taken into consideration in food webs, these connections are deeply modified. How the connection is modified depends on parasite specificity: it is decreased by specific parasites and greatly increased by generalist parasites—especially those having heteroxenous life cycles.

Perhaps one day ecologists will have to consider parasite protection in conservation biology as not necessarily either a paradox or a provocation (Combes 1995). Poulin (1999a) concluded a review on the role of parasites as ecosystem engineers like this: "The challenge for parasitologists is now to show that parasites play important, if often subtle, roles in maintaining biodiversity, and that their study (and conservation!) is essential."

Someone once noted with insight that when we move a bottle of wine on the table, the trajectory of the companion to Sirius is altered because the gravitational field is universal.[4] Similarly, we may write that each time a parasite affects the fecundity or survival of its host it influences the entire ecosystem and perhaps even the biosphere. For example, the fluctuations in red grouse populations discussed earlier cannot fail to have consequences on the heather lands of Scotland and hence on the entire ecosystem (variations in grouse populations influence the populations of their competitors and predators, which in turn influence the populations of decomposers, which then influence the populations of plant producers, etc.). The more the target species of the parasite itself plays a key role in the ecosystem, the more important the consequences to the entire ecosystem. Further widening Dawkins's original idea, the impact of parasites on ecosystems represents the broadest extension of the phenotype, passing the limits of both the host cell and the host organism to reach what can appropriately be called the host ecosystem.

In short, then, the role of parasites in ecosystems should not be conceived of as only the fuzzy sum of multiple elementary effects exerted by each species. As each of the examples above has shown us, parasitism has a precise and deep echo at each of the functioning levels of the biosphere.

---

4. Sirius is the bright star in the constellation Canis Major, eight light-years distant from us. The companion to Sirius is a tiny white dwarf, and its luminosity is less than one four hundredth that of the sun. It was discovered in January 1862 by Alvan G. Clark at the Dearborn Observatory of Northwestern University in Illinois.

17 Parasites and Environment

As the proportion of relatives declines with increasing distance from the heartland and the stranger inhabits an increasingly strange land, so does the importance of the phylogenetic element of the community decline whilst that of the ecological element correspondingly increases.
Kennedy and Bush 1994

The presence and abundance of parasites in an individual host or that of an individual host population is influenced by many factors, some biotic (for instance, the composition of communities of free-living organisms) and others abiotic (for instance, temperature or pollutants). That is, it is not just the influence of the parasite on its ecosystem host that is important in determining a parasite's number but also the influence of the ecosystem on the parasite.

Biotic Factors

Gulls from the Black Sea

The characteristics of infracommunities come from those of xenocommunities, which themselves come in turn from communities.[1] This hierarchical relationship can best be illustrated with an imaginary example (see fig. 17.1). For instance, in any given locality a community of marine birds can harbor a parasite community (top of figure); each bird species in the marine bird community (for example the gull, *Larus argentatus*) has its own xenocommunity of parasites derived from some of the members of the larger parasite community (center on figure); and each individual gull harbors its own infracommunity of parasites, itself an incomplete sampling of the larger xenocommunity (see the bottom of the figure).

As indicated in the figure, not all parasite species forming the xenocommunity of a specific host population have the same "value," and each

1. A parasitologist who performs an investigation does not decide in advance to work either on infracommunities, xenocommunities, or communities. The data recovered in each case are essentially the same, and it is only the later treatment (analysis) and interpretation that are different.

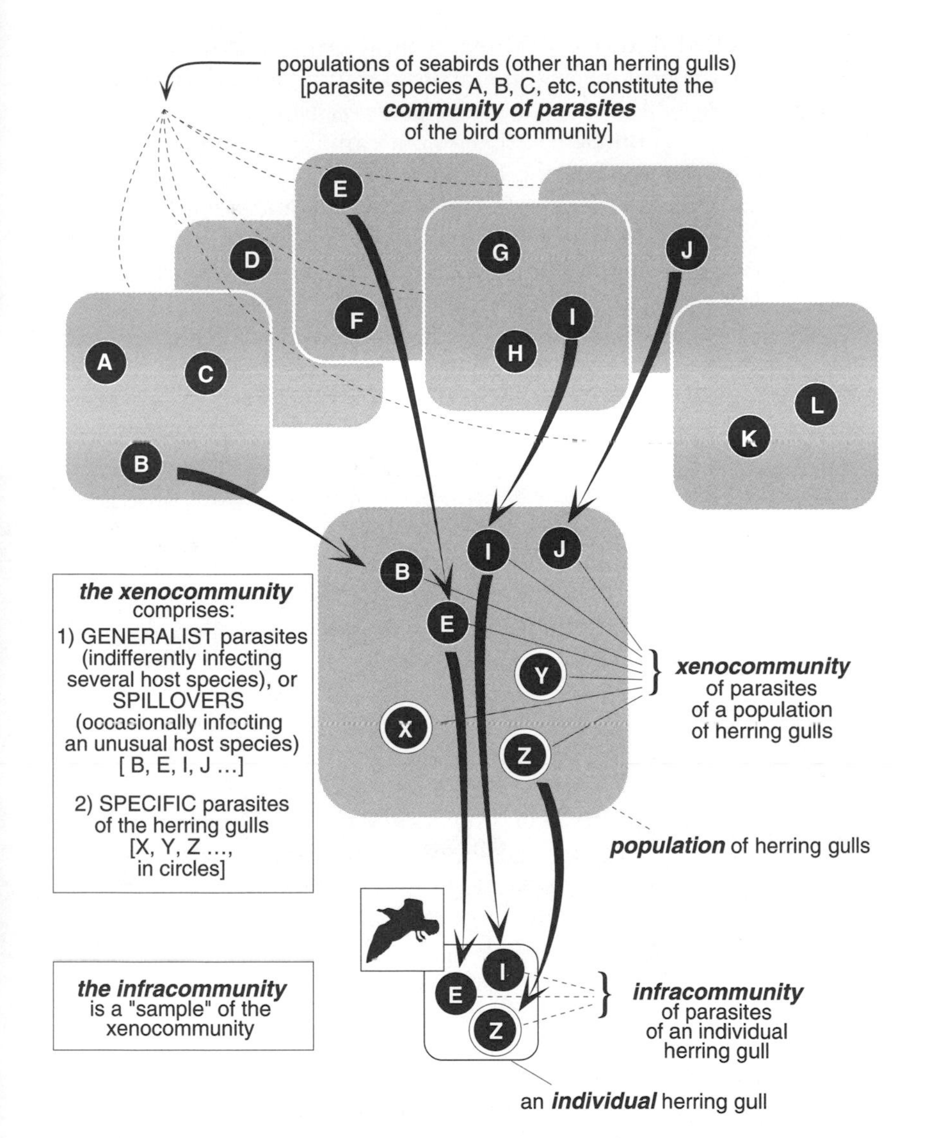

Figure 17.1. Diagram demonstrating how the infracommunity of an individual host (the gull is taken as an example; see the bottom of the drawing) is a sampling of the xenocommunity, which is itself a sampling of the community.

may be classified into one of three general groups: *specialists* that are found only in the host species under consideration; *generalists* that live indifferently in several host species; and *spillovers,* which mainly complete their life cycle in another host species but can also be found in the observed host species and develop normally there.

If, for example, two species of waterfowl (scaup and grebes) share the same environment, their parasites may be distributed according to the following diagram (fig. 17.2), inspired by the body of work of John Holmes.

In short, the xenocommunity makeup is strongly dependent on the

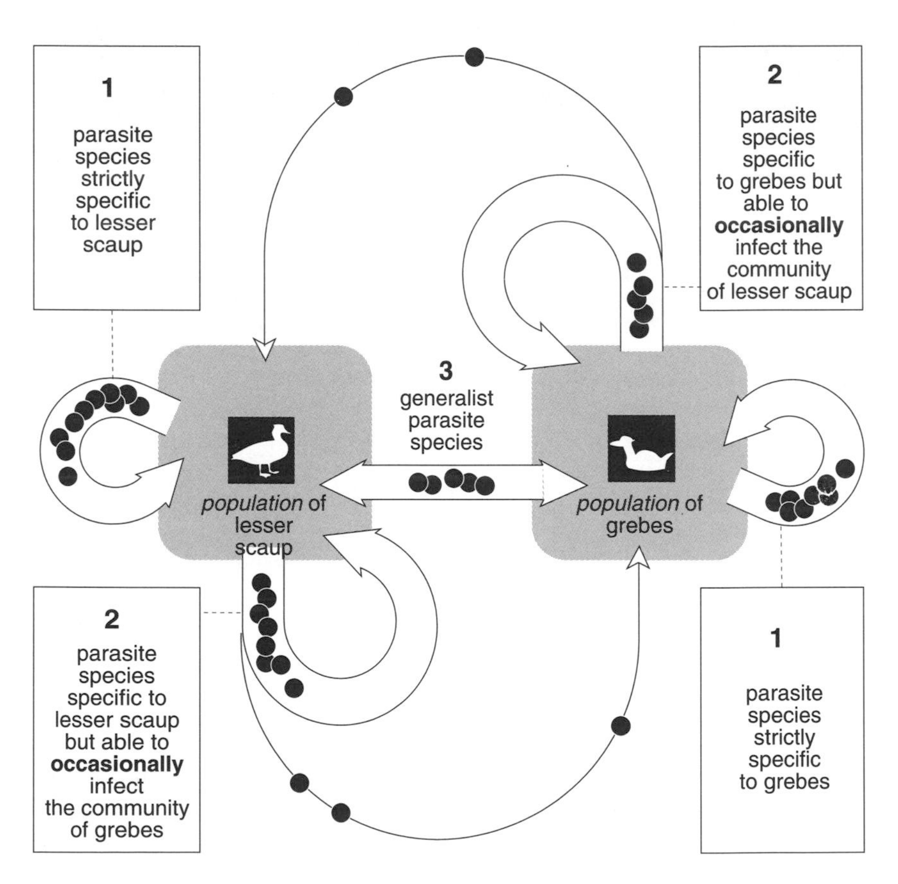

Figure 17.2. Diagram showing that the spectrum of parasites of a host species involves three components: (1) specific parasites found only in the host species being considered; (2) spillover parasites normally harbored by other host species; and (3) generalist parasites indifferently parasitizing several host species. Black circles representing species are only imaginary numbers that do not reflect reality.

presence of those other host species in the environment that potentially make available to any one host both generalist and spillover parasites. Still other causes of change and variation among and within xenocommunities include the presence of intermediate hosts that allow the completion of heteroxenous life cycles, the occurrence of physical and chemical environmental factors affecting transmission (pH, expansion of the aquatic milieu, etc.), and the presence of other organisms preventing or at least limiting the completion of the life cycle. It is at this local scale that what Kennedy (1990) called "parasitic ambiance" is created.

An example of the biotic relationships between the various community levels is supplied by Aneta Kostadinova, who studied the communities of intestinal trematodes of piscivorous birds in the Black Sea (gulls and grebes) and compared her results with those from other investigations done in Europe (particularly Russia and other countries in eastern Europe).[2]

Whereas the earlier studies of larid parasitism (notably the previously cited 1989 investigation by Kennedy and Bakke) may have left one thinking that this group of marine birds is characterized by systematically poor and noninteractive parasite communities, through a detailed analysis of gulls, in particular *Larus argentatus,* Kostadinova shows that this affirmation is inexact at the local level. For example, in populations sampled on the coasts of the Black Sea and in the Ukraine, numerous signs of interactions appear: the existence of core species, a high number of trematode species per host individual, and a large number of infrapopulations. Based on several environment-parasite correlations, Kostadinova thinks that community richness and intra- and interspecific competition (in a word, interactivity) are part of the stability and homogeneity of the environment and that, on the contrary, poor and isolationist communities would characterize perturbed or heterogeneous environments.

The occurrence of parasitic ambiance in parasite communities in Ukrainian grebes is even clearer. Unlike Canadian grebes, whose communities are dominated by cestodes, communities of Ukrainian grebes are largely dominated by trematodes, which constitute up to 98% of the parasite individuals recorded in the grebe *Podiceps cristatus.*

The search for correlations between the environment frequented by grebes and their parasite communities shows that the two species that mainly feed in fresh water *(P. cristatus* and *P. ruficollis)* have communities dominated by specialist trematodes, whereas the two species that mainly frequent the seashore *(P. nigricollis* and *P. grisigena)* have communities

---

2. I thank A. Kostadinova for communicating her results, for the most part not yet published.

that are both richer in species and dominated by generalist trematodes borrowed from a "pool" of parasites circulating in coastal or lagoon ecosystems. One of these generalists is *Cryptocotyle concavum,* a parasite of numerous marine birds that "floods" the ecosystem with infective stages, invading hosts (e.g., grebes) in which they are usually not encountered.

Thus, we see generalist species taking over from specialist species in communities as soon as their infective stages circulate abundantly in the ecosystem.

In the cited case the parasite community structure of a specific grebe species is in the end determined by the species' behavior, in particular its capacity to live alongside other bird species *(P. nigricollis* and *P. grisigena,* but not *P. cristatus* or *P. ruficollis)* and to mix easily with other marine bird colonies, as well as by the composition both of the community of other marine birds in the ecosystem and of their own parasite communities.

The Voles of Poland

The work of Kisielewska (1970a, 1970b, 1970c, 1970d, 1970e) on the helminths of the small rodent *Clethrionomys glareolus* reveals an astonishing variation among the xenocommunities of the same host species, showing that it is not easy to predict xenocommunities either in space or in time. Nevertheless, it is possible to define typical locations that are characterized by relatively homogeneous prevalences and intensities, with some locations being exceptional because of environmental peculiarities; and it is also possible to define typical years, even though certain years are highly aberrant because of climatic irregularities.

Figure 17.3 shows a series of locations in Poland characterized both qualitatively and quantitatively by very different parasite spectra for *C. glareolus.*[3]

A Stranger in a Strange Land

In an article subtitled "A Stranger in a Strange Land," Kennedy and Bush (1994) raised a difficult question: What part of a parasite community encountered in a host is due to its specific location, what part to the identity of the host (its phylogeny), and what part to its environment? To answer this question these authors analyzed the parasite communities of the salmonid fish *Oncorhynchus mykiss* "at different hierarchical taxonomic and spatial scales." The "heartland" of this fish is the north Pacific coast

3. These parasites' spectra, when analyzed in greater detail, may reveal certain factors in the environment notably due to human disturbance.

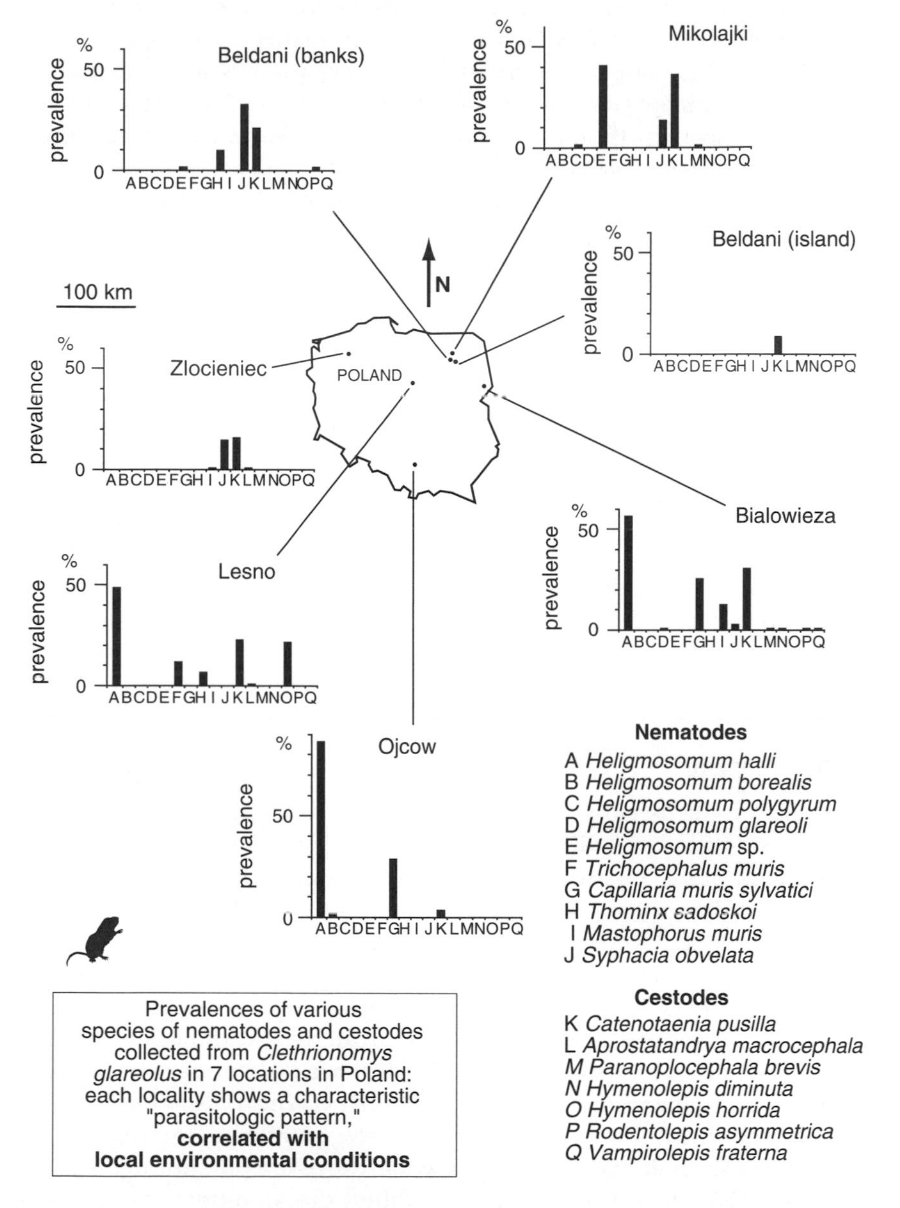

Figure 17.3. Variation of parasitism in the rodent *Clethrionomys glareolus* according to location in Poland, showing the sensitivity of helminths to environmental factors (after Kisielewska 1970c).

of North America and East Asia, from which it was introduced by man throughout most of the rest of the world. The analysis shows:

1. In its heartland *O. mykiss* parasite communities are dominated (about 55%) by species that are specialists of salmonids, some specific and others shared with other salmonids in the same area. It is this group of helminths that must be considered the "phylogenetic element" that has coevolved with its hosts. The remaining 45% of helminths present were generalists borrowed from other nonsalmonid fish.

2. In regions where *O. mykiss* was introduced and salmonids were already present (for example, in eastern North America and northern Europe), helminth communities of *O. mykiss* are dominated by generalist species borrowed from nonsalmonid fish. They still, however, include a notable proportion of salmonid specialist species borrowed from local salmonids.

3. In regions where *O. mykiss* was introduced but where no salmonid fish were originally present, the helminth communities are formed only from generalist species borrowed from local fish.

The result is that the predictability of *O. mykiss* helminth communities differs depending on the location analyzed: it is highly predictable in the heartland, but it becomes more and more dependent on the local environment as one gets farther and farther away. That is, local variation becomes "ecological noise" that obscures the fundamental pattern of the heartland community if one looks at the system only as a whole.

Other Strangers . . .

The study of Bush and Holmes (1986) on the helminths of lesser scaup showed subtle but real differences in community structure relative to environment. In this case differences were dependent on the lakes investigated, with xenocommunities having similar richness in the thirteen lakes investigated. Nevertheless, two variations could be attributed to environmental factors related to diet and the presence of other hosts.

Lesser scaup are large consumers of amphipods, so it is not surprising that they acquire many of their parasites from these crustaceans. In fact two of the commonly ingested amphipods share most of the responsibility for bringing parasites to the birds: *Hyalella azteca* and *Gammarus lacustris*. Bush and Holmes call the "*H. azteca* suite" and the "*G. lacustris* suite" those sets of helminths whose intermediate hosts are one or the other of these two amphipod species. They also show that in different lakes, the relative availability of the two amphipods influences the composition of the xenocommunity.

In addition, part of the parasites encountered in the scaup are spillovers from other birds. That is, their regular hosts are not lesser scaup but they can occasionally be found in these birds. Thus the presence of other waterfowl is another factor affecting the composition of xenocommunities.

Along these lines an investigation by Edwards and Bush (1989) on the intestinal helminths of avocets *(Recurvirostra americana)* in North America showed that those parasite species that were encountered close to temporary ponds were specialists that had little interaction with each other. On the other hand, when close to lakes, avocets were colonized by species that were normally specialists of lesser scaup; they all were added to the community of avocet specialists and then interacted with them, saturating empty niches.

Similarly, the common eider *(Somateria mollissima)* lives mainly in Arctic regions, but isolated colonies are also found farther south. Kornyushin et al. (1995) analyzed parasitism in one of these more southern colonies on an island in the Black Sea and found eleven helminth species common to both this population and those found in Arctic regions, while seventeen species were acquired as spillovers from local birds.

Holmes's research (1990a) on helminth infection in the marine fish *Sebastes nebulosus* showed that these rather territorial fish live on reefs more or less isolated from one another and that they also harbor somewhat different parasites in each community. When several locations are compared—that is, several reefs that differ from each other by environmental factors (notably exposure to currents)—several differences are observed between xenocommunities. These differences are not very predictable and obviously depend on the permanent or transitory presence of other fish that could lend generalist parasites to *Sebastes.*

As noted earlier, Kennedy's work on eels also revealed marked differences in xenocommunities depending on location. In this case xenocommunities had richness varying from zero to nine helminth species depending on location, but most often they were rather poor, usually harboring only one to three species. Further, xenocommunities were also dominated by one or another species (but only one!), either a specialist or a generalist. Most frequently the dominant parasites were found to be acanthocephalans, such as *Acanthocephalus lucii* and *A. clavula* (both generalist species that parasitize numerous other fish as well). Curiously, specialists (eel-specific parasites) such as the cestode *Bothriocephalus claviceps* and the nematode *Paraquimperia tenerrima* were dominant only in the absence of acanthocephalans. The qualitative and quantitative composition of these communities thus reflects the richness of the generalist parasites available in the ecosystem. That is, parasite richness depends on the richness of other fish: where generalists are present (acanthocephalans), they are the dominant species. Where they are absent (and only where they are absent), eel specialists are dominant.

Captivity deeply modifies parasitism, and this is particularly true for all animals found in zoos. For instance, one need only cite a few sentences from Müller-Graf (1995) in his conclusion to a study of 112 lions in Tanzania: "The parasite community in wild lions in this study differed from that of zoo lions. *Spirometra* was the most common parasite in wild lions but has not been reported in zoo lions. *Toxascaris* was not found in wild lions but appears frequently in zoo lions. . . . Taeniids were very common in the wild lions, but few records from zoo lions exists." Müller-Graf adds: "Differences in parasites between wild lions and lions in zoos may be explained by the differences in the meat lions are fed and their close proximity to a wide variety of other host species."

That local communities with only specialist parasites may in other places become richer in generalist species leads to an increased interspecific competition in these other communities. We may then wonder if, at least in some cases, specialist communities are those where competition has previously occurred, whereas communities rich in generalists are those where competition is "current."

That their composition may change seasonally is also a strong indication that environment plays a role in structuring xenocommunities. For example, Bush (1990) showed that in semipalmated plovers *(Catoptrophorus semipalmatus)* those captured on their summer breeding grounds (a freshwater site) have parasite communities dominated by "large" cestodes and are poor in trematodes, whereas those captured on their wintering grounds (saltwater sites) have communities dominated by trematodes and are poor in "large" cestodes.

Since "small" cestodes and nematodes showed little variation between the two sites, Bush deduced that environment plays a fundamental role in the formation of parasite communities: "I believe that environments with their varied habitats, determine the 'supply' of helminths that are available for colonization." Of course behavior (the encounter filter) and immunity (the compatibility filter) both then sort the parasites available in these different environments.

Based on the geographic variability of xenocommunities one may conclude the following:

Interactive communities occur frequently but seem to characterize certain host groups, including most birds or certain unusual species such as the fish *Sebastes nebulosus.*

The existence of empty niches in a given organ does not mean interspecific competition does not occur in a defined region of that same organ. For instance, the posterior portion of the intestine theoretically offers sites, but this availability often coincides with heavy competition for microhabitats in the much more favorable anterior intestine (such as in the case of grebes).

Xenocommunities "feed" infracommunities in many different ways depending on the hosts—if in some cases (lesser scaup) the infracommunity correctly reflects the xenocommunity, in others (eels) the infracommunity is much poorer than the xenocommunity.

Specific parasites that have coevolved with the host usually constitute the core species, but not always.

Communities of related hosts play an important role when they contain generalist parasites, which may then dominate infracommunities (*S. nebulosus,* British eels, and Black Sea grebes) and become core species instead of the specialist species.

Communities that are invaded by generalists may be the place where current competition occurs more frequently than in specialist communities, where only the ghosts of competition past may be observed.

Islands in the Gulf of Lion

Mas-Coma and Esteban (1988) carried out detailed studies of parasitism on Mediterranean Islands. Here we may try to apply the theory of island biogeography to parasites (talking not about the hosts considered as islands, but about the real islands and their parasite populations). In this case we may wonder if an island's parasite richness is derived from particular rules relating to the "sampling" of the nearby continent, which may create communities totally different from those of the mainland. The sampling of parasite species by an island, as for free-living species, must logically be influenced by its dimensions, its distance from the mainland, and its environment in the broadest possible sense.

Research on variation in helminth parasitism in small mammals (mostly *Rattus rattus, Mus musculus domesticus, Eliomys quercinus,* and *Apodemus sylvaticus*) as a function of continental island size was carried out in the Balearic Islands, Medas, the Hyères Islands, Corsica, Cervicales, and Lavezzi (Mas-Coma et al. 1988). For the same host species, parasitism is different between the islands and the neighboring continental populations as well as between islands that differ in size. Differences were found to occur in the number and identity of the parasite species observed as well as in their prevalence, intensity, and even specificity. The consequence was that the pressures parasites exert on host populations on islands change relative to those they exert on the mainland. This may be of the greatest importance in understanding island colonization, either because the processes of regulation of host species are altered or because competition between different host species is modified.

If there is an increase in richness in terms of parasite species with the size of the island, as may be expected, this global tendency may still hide other peculiarities according to group:

Cestodes and nematodes appear on the smallest islands, and the smaller the island, the higher the prevalence. Thus it is on the most minuscule islets that parasites are suspected to exert the strongest pressures on their hosts.

Trematodes and acanthocephalans, on the contrary, appear and more or less take over the preceding parasites as island surface area increases. The nature and most likely also the intensity of parasite pressures thus vary with the size of the island.

The type of life cycle plays an important role, the best "colonizers" being not only helminths with direct life cycles but, curiously, also heteroxenous helminths whose intermediate host is an arthropod. The worst colonizers are heteroxenous helminths with three hosts that require aquatic environments for their transmission.

Thus, when we analyze the representation of a specific parasite species relative to the size of the islands there is an increase or decrease in importance depending on the case. For example, prevalences and intensities of the cestode *Hymenolepis diminuta* and the nematode *Mastophorus muris* for both their definitive host *(Rattus rattus)* and their arthropod intermediate hosts are as high as the surface area of the island is small.

Knowing that populations of insular animals are characterized by behavioral shifts (higher densities, smaller territories, diminished aggressivity, increases in interaction between individuals of the same species, etc.), it would be particularly interesting to someday analyze relationships between these changes in behavior and modifications in parasitism on the islands. Sadly, the role of parasitism has not yet been taken into account in the theory of island biogeography.

Shrinking Ecosystems

When an ecosystem shrinks, for whatever reason, it has a tendency to become more colonized by neighbors (Holmes 1996). This is true, for example, of a forest fragment in the middle of a prairie zone. Humans are creating more and more such shrinking ecosystems, and Holmes shows that two processes occur along with this shrinking that can then lead to the modification of its fauna.

One of these processes is an increase in local parasitism following the increase in host density. Transmission of some parasite species is then made easier, and the resulting imbalance may threaten free-living species. That is, there is a kind of "colonization from the inside."

The other process involves the more and more frequent invasion by parasites belonging to neighboring ecosystems, which is made easier because exchanges are more frequent. The free-living species of the shrinking ecosystem may then become hosts of spillover parasites, itself a new cause of imbalance. This is "colonization from the outside," and Holmes cites the case of the parasitic cowbird *Molothrus ater,* which has a tendency to leave its eggs in the nests of birds in shrinking ecosystems where defense mechanisms are not yet selected for, making the hosts even more susceptible to nest parasitism.

Any parasite species requires an ensemble of environmental conditions (in the broadest sense of the term) in order to develop and thus constitutes a marker (or "tag") for those particular conditions. For example, Speare (1995) has shown that in the Whitsunday Islands close to the Australian coast adult fish *(Istiophorus platypterus)* come from two populations of immature fish, one derived from the north and the other from the south. Some individuals captured in the Whitsundays had parasites "from the north" and others "from the south." Parasites that allow discrimination between the two originating populations are two trypanorhynchid cestodes *(Callitetrarhynchus gracilis* and *Otobothrium dipsacum),* a copepod *(Penella instructa),* and a sanguinicolid trematode *(Cardiocola grandis).*

Natural Abiotic Factors

In the investigations reported above, the specific abiotic factors in the environment that in the end explain the variation in parasite communities have rarely been identified, illustrating just how difficult it is for parasite ecology to go from a descriptive to a causative approach. The parasite-environment relationship thus offers real challenges for study, and the solution to the problem of environmental influences on parasite community structure remains far away. Three studies illustrate the current state of affairs in this realm: one on the coastal mollusks of northern Europe, one on the eels of Nova Scotia, and one on the lizards of Bulgaria.

Galaktionov (1996) performed an impressive investigation on infection in marine mollusks of the genus *Littorina.* This study examined 35,000 individuals, one by one, over several hundred kilometers of coast between Novaya Zemlya in the east and the extreme north of Scandinavia to the west. Environmental conditions over the area of study are described as "extreme" and the more easterly, the more extreme. Galaktionov concludes that in this type of environment those trematodes with two host life cycles (life cycles in which cercariae remain in the mollusk) are favored over those with three hosts (those in which cercariae disperse into the environment. This author has two convincing arguments to support this proposition: trematodes in the genus *Imasthla,* which have cercarial dispersion, are abundant in the west, disappear in the east where the conditions are the most extreme, and are still present in intermediate regions in sites sheltered from the waves; and trematodes in the genus *Microphallus,* which do not have cercarial dispersion, are present throughout the entire zone studied. Overall, the author thinks the disappearance of numerous marine bird trematode species as one goes from temperate Europe toward Arctic regions occurs because these species have free-living stages (miracidia and cercariae) that are inept at encountering hosts in the extreme conditions of the Far North.

Marcogliese and Cone (1996) analyzed parasitism in 1,041 eels *(Anguilla rostrata)* in twenty-eight locations in Nova Scotia and noted that component community diversity (measured by parasite richness or by an index such as Shannon-Weaver) was significantly larger in waters that were slightly acidic (pH > 5.4). For example, the monogenean *Pseudodactylogyrus anguillae* was found only from sites with a pH over 5.4, and trematodes were totally absent in waters with a pH of less than 4.7. The authors suggest that the swimming oncomiracidia of *P. anguillae* could be directly sensitive to acidity, while the disappearance of digeneans in acidic waters could be the indirect consequence of the disappearance of the molluskan intermediate hosts.

Biserkov and Kostadinova (1998) have analyzed the parasites of the green lizard *Lacerta viridis* from four Bulgarian sites. Although the helminth communities of this host proved to be poor, the study led to a surprisingly original conclusion (fig. 17.4). Two of the studied habitats (Bacheco and Golak) were considered to be typical for the green lizard in that they were partially forested ecosystems at an average altitude of between 700 and 800 meters. Two other habitats (Petrich and Burgass) were considered "marginal" and consisted of meadows at low altitude (100 to 150 meters). In typical habitats the dominant parasite is a specialist nematode, *Spauligodon extenuatus,* while in marginal habitats the dominant parasite is a generalist nematode, *Oswaldocruzia filiformis.* Although the link between the dominance of either a specialist or a generalist species and the quality of the environment has yet to be confirmed and the factors responsible need to be identified, the results present an interesting lead in the study of communities.

Abiotic Factors Linked to Pollution

Humans currently impose brutal and often deep changes in all environments, and disturbance ecology studies the way communities of living organisms respond to these perturbations. However, few studies in disturbance ecology have paid attention to parasites. When a perturbation is imposed on an ecosystem, some parasite-host associations may react as single entities (host and parasite are affected or favored as "a whole"),

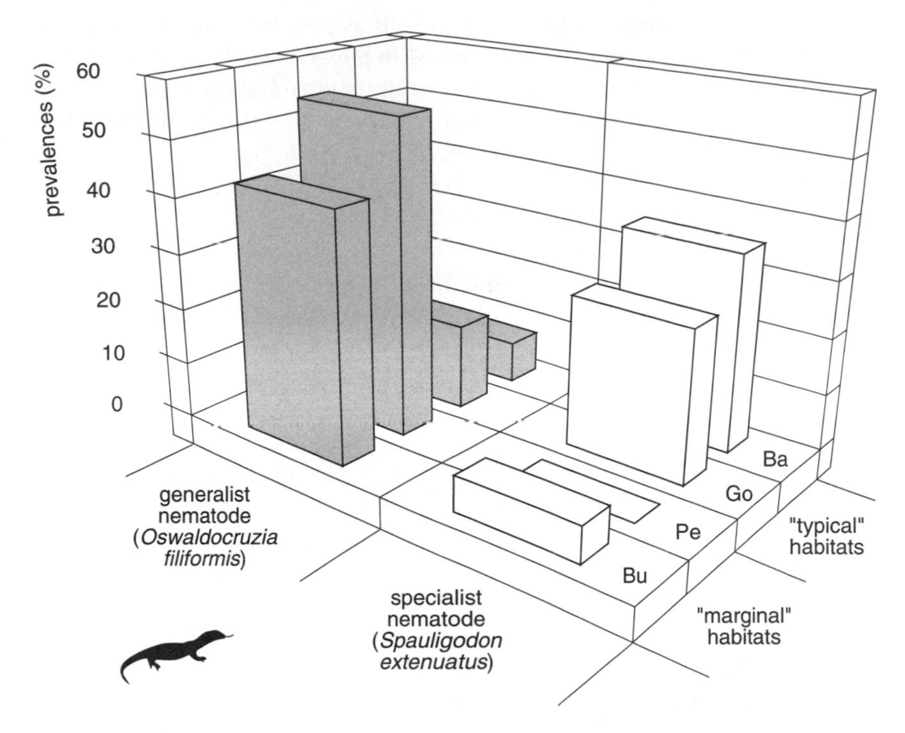

Figure 17.4. Relation between the type of environment and parasite species dominance in green lizards in Bulgaria. Ba, Bu, Go, and Pe refer to locations studied (after Biserkov and Kostadinova 1998).

whereas others become unbalanced (the perturbation either prevents or favors parasite transmission or survival). Undoubtedly some changes linked to pollution are probably due to decreases in the efficacy of host immune systems in polluted environments.

The review of MacKenzie et al. (1995) clearly demonstrates the contradictory influences that perturbation via pollution may have on fish parasitism.

Exposure of fish to oil pollution may *reduce* the abundance of intestinal parasites (for instance, trematodes of the plaice *Pleuronectes platessa* after the *Amoco Cadiz* oil spill) but may *increase* infection with gill parasites (for instance, trichodinid gill ciliates of the cod *Gadus morhua* after the *Exxon Valdez* oil spill and gill monogeneans of the whiting *Merlangus merlangus* in the polluted areas of the North Sea).

Exposure to heavy metals may *reduce* the abundance of certain parasites (for instance, the cestode *Bothriocephalus acheilognathi* in *Gambusia af-*

finis exposed to selenium in North Carolina) and may *increase* infection by others (for instance, the protozoan skin parasite *Ichthyophthirius multifiliis* in carp *(Cyprinus carpio)* experimentally exposed to cadmium. It must be noted that certain parasites accumulate heavy metals to a much higher degree than the surrounding tissues of their hosts (up to one hundred times more!) (Sures, Jürges, and Taraschewski 1998; Sures et al. 1999; Sures and Siddall 1999; Sures, Siddall, and Taraschewski 1999; Zimmermann, Sures, and Taraschewski 1999).

Exposure to thermal pollution caused by hot water ejected from thermal or nuclear plants can reduce parasite richness (for instance, the monogeneans, trematodes, and cestodes of bream *Abramis brama* in Poland) or increase the abundance of a particular parasite. The invasion of the marine fish *Gadus morhua* by a freshwater parasite on Swedish coasts is an exceptional example of the latter case (Thulin 1984). Here examining 614 *G. morhua* close to the nuclear power plant at Oskarshamn showed that 90% of them harbored metacercariae of *Diplostomum,* which are freshwater parasites, in the lens of the eye (see the life cycle of *Diplostomum* in chapter 7). Some of the cods' eyes harbored up to two hundred metacercariae of this trematode, and the lenses showed distinct signs of pathogenicity (progressive opaqueness). Although the infection was not directly linked to the increased temperature of the water by the plant, this troubling observation nevertheless must be added to the file of the relationship between parasite communities and environmental change.

Despite these contradictory findings, Marcogliese and Cone (1998b) consider that "parasite communities may be good indicators of pollution and other environmental stresses that can affect food web structure." Pollution may favor one parasite species while disfavoring an apparently related species, which then may modify the outcome of any competition between the parasites. As such, in Italy Galli et al. (1998) showed that the acanthocephalan *Pomphorhynchus laevis* parasitizes eels only in unpolluted or slightly polluted locations, whereas another acanthocephalan, *Acanthocephalus anguillae,* is more abundant where pollution (nitrates, nitrites, phosphorus, and bacteria) is greater.

The relation of disturbance ecology and parasitism clearly represents an area where much more study is needed.

18 Parasites and Sex

How could animals choose resistant mates? The methods used should have much in common with those of a physician checking eligibility for life insurance.
Hamilton and Zuk 1982

The peacock's tail and the stag's antlers are not mere disabilities; rather, they are handicaps in this very special sense: they allow an individual animal to demonstrate its quality. . . . The cost—the handicap that the signaler takes on—guaranties that the signal is reliable.
Zahavi and Zahavi 1997

The Origin of Sex

What was the first sexed organism on our planet? Nobody knows. Why did this organism appear? Nobody knows.

Even though the answers to these questions may forever remain unknowable, various hypotheses have nevertheless been proposed to explain the origin of sex. Many posit cannibalism between two related cells followed by separation when living conditions became particularly good. During a period of promiscuity between the two genomes, genetic exchanges may have occurred that could be considered the first genetic recombinations.

One interesting hypothesis along these lines postulates that fertilization itself (penetration of the female gamete by the male gamete) is nothing other than parasitism, while yet another postulates that sexuality may be the indirect consequence of the existence of parasitic DNA.

The former hypothesis does not require much of an explanation once one realizes that nothing more resembles the penetration of an oocyte by a spermatozoid than an intracellular parasite penetrating its host cell (e.g., *Plasmodium* entering a hemocyte or *Theileria* entering a lymphocyte).

Gouyon (1994), presents a striking synopsis of this hypothesis:

> Sex is first the encounter of the other, then the production of other "others." . . . When male and female functions are separated into different individuals, then by necessity the offspring are produced by a male and a female. In effect, the different models that allow one to explain the differentiation between male and female give the males a parasitic origin and a strategy that could have consisted of producing numerous small gametes instead of playing the game of giving all the necessary resources to the descendants. The others, the large gametes, overwhelmed by this flow of small mobile gametes, would, out of necessity and not "knowing" how to

> manage by themselves, have ended up waiting for the other, the parasite. As such, the differentiation of sexes, on which so many living organisms have been built, may have been created.

The latter hypothesis—that calling for mobile DNA elements—is also very appealing, however.

Hickey and Rose (1988) and then Hickey (1993) provide the following reasoning: The most important consequence of sexuality is that during meiosis, paternal and maternal genomes exchange genes or fractions of genes thanks to genetic recombination. In each generation, recombination makes new arrangements of genes appear and hence strongly contributes to an organism's variability, which then forms the basis for the evolution of organismic diversity. Hickey and Rose, however, write, "Even though recombination rates may be currently very important, we cannot believe that the early phases evolved in a goal-oriented fashion determined by the adaptive advantages of the end result." Given this, Hickey and Rose noted that mobile DNA elements are numerous in the "higher" sexual eukaryotes. Such elements go from one genome to another during the fusion of the nuclei that characterizes fertilization. As such, the formation of a zygote allows parasitic DNA to garner or colonize a new set of chromosomes (fig. 18.1). For instance, experiments show that the 2m plasmid of the yeast *Saccharomyces cerevisiae* does not spread itself in a yeast population that reproduces asexually but does so very quickly when there is sex.

The logical result is to propose that sexuality is the product of genes that would first belong to parasitic DNA. In prokaryotes, parasitic DNA may have easily passed from genome to genome through conjugation. But in eukaryotes, where the genome is contained within the nucleus, such a passage would not have been possible, and parasitic DNA may have contributed to the selection of genes to induce cell fusion. Since repeated cycles of cell fusion would ultimately lead to cell gigantism, meiosis would then have been selected for to compensate for fertilization.

In conclusion, Hickey (1993) proposes that molecular sequences of an exogenous origin should not be called either parasitic or selfish DNA but rather should be considered "genomic weeds, characterized by a certain replicative exuberance that permits them to affect their host genomes in a variety of novel ways."

Maintenance of Sex

Sex—What For?

Sexuality is so familiar to us that sometimes we forget to wonder about its importance to living organisms. There are, for instance, metazoans that do

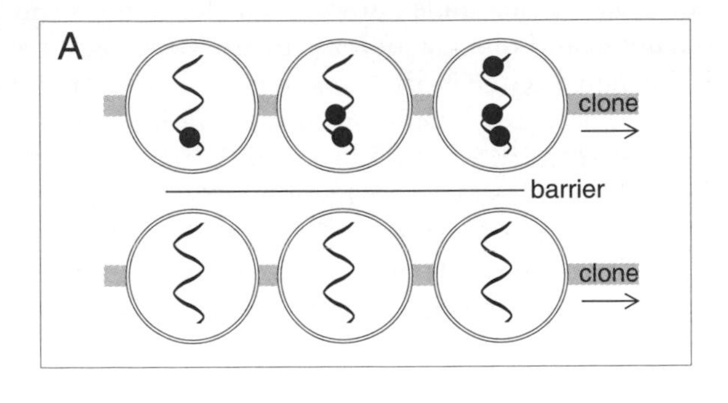

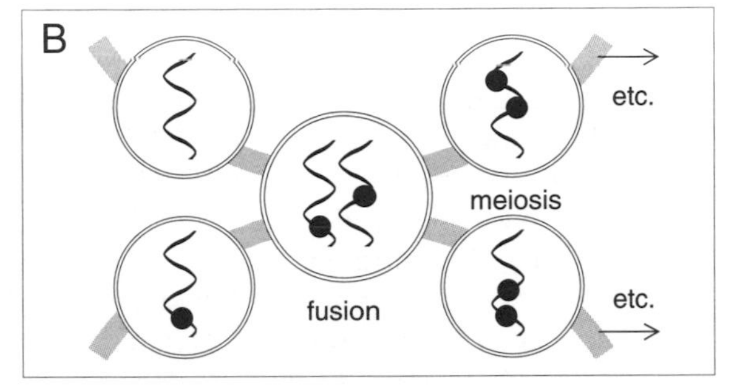

Figure 18.1. Hickey's hypothesis. (A) Clonal reproduction of host cells: the mobile element cannot go from one clone to another. (B) Sexual reproduction of host cells: the mobile element can invade the entire population.

not reproduce sexually, such as the bdelloids, a group of rotifers that live in humid mosses. Bdelloids have uncommon properties, such as entering diapause if the environment becomes dry and returning to activity when conditions improve. These rotifers can also resist temperatures as high as the boiling point (100°C) or as low as near absolute zero (–273°C) and do not seem to be any less competitive than other species of rotifers living in identical environments that have both males and females—all indications are that they are an ancient group that abandoned sexuality a long time ago (maybe as long as 40 to 80 million years ago; see Ridley 1994). Maynard Smith (1986) qualified bdelloids as an evolutionary scandal, and only a very few other metazoans are known to not reproduce sexually.

What, then, are the advantages of sexuality? The surprise is that it is difficult to find any. On the other hand, it has the major disadvantage that half the population is allocated "for nothing," since only the females will actually produce offspring. This is the twofold disadvantage of sex, and it is only when males invest energy in reproduction, for instance, by protecting or feeding the young, that this disadvantage is somewhat reduced.

It is easy to show that an allele allowing females to reproduce without the help of males would quickly increase its frequency and fix itself in a population. The coexistence in the same population of "asexuality" and "sexuality" genes necessarily results in the selection of the former if all other characters are adaptively neutral and if males contribute only sperm to reproduction (see Maynard Smith 1989a and Hamilton, Axelrod, and Tanese 1990 for discussions of the general problem of the possible advantages of sexuality). That is, all other things being equal, "asexual" females have a fitness twice that of sexual females. Ayala (1998) summarizes the difficulty in these terms: "An organism that reproduces asexually passes on all its genes to each one of its individual progeny, whereas one that reproduces sexually passes on only half to each."

In short, theorists usually search for the adaptive value of sex in its amplification of genetic variation. During meiosis, chromosomes of paternal and maternal origin exchange genes or gene fragments because of crossing over, which does not in itself change the richness of alleles in the population but does create new combinations of genes. This event, coupled with other aspects of recombination, such as the alignment of tetrads yielding new combinations of paternal and maternal chromosomes, makes all the descendants of one couple different from each other: "Once allelic variation has been introduced by mutation or immigration, the total array of genotypes may be considerably increased by recombination" (Lewontin 1974).

This being established, what would be the advantage of having genetic variability increased by recombination? Various responses have been proposed. One such hypothesis postulates that if the environment is homogeneous, genetic diversity would reduce competition between descendants when the young disperse (Hamilton and May 1977). An alternative hypothesis states that if the environment is heterogeneous, being genetically diverse would allow the offspring to occupy more niches (Bell 1982). Still another hypothesis postulates that, thanks to recombination, sexual species evolve faster than species that reproduce exclusively by asexual means. According to this hypothesis, sexed organisms would be at a disadvantage at the scale of the individual, whereas at the species level those populations that are sexed would be dominant and eliminate the asexual ones.[1] We can also suppose that by increasing variability sexuality would increase a species' chances to survive sudden or otherwise unexpected ecological changes.

---

1. Species selection is not like group selection. In group selection, the interest of the group opposes the individual interest, as in the parasitized caterpillars that sacrifice themselves for their kin (see chapter 8). In species selection, the interest of the group does not oppose the interest of the individual.

Still other hypotheses evoke the role of sexuality in the repair of DNA and the elimination of unfavorable mutations. However, the most fascinating hypothesis on the origin of sex involves parasitism and the logical extension of the Red Queen hypothesis discussed earlier (Van Valen 1973). In a parasite-host relationship that lasts for a number of generations, each partner is confronted nonstop with the other's new adaptations. In the absence of host genetic variability, parasites would easily select the adaptations necessary to successfully infect them. In contrast, in the absence of parasite genetic diversity hosts would ultimately select adaptations allowing them to eliminate the parasites. This is an application of the precept (Fisher 1930) that if unexpected changes occur between two generations it is an advantage for one's genetic variability to be large so that at least part of your offspring can survive. As Ladle (1992) noted, to become or to remain common is fatal negligence that allows the adversary to adapt. One might add that being genetically unvarying when one has a determined parasite for an adversary is like taking the same route home every day when threatened by terrorists.

The question is not, then, whether one should select a momentarily ideal genetic combination that might bring down one's adversary but whether one should generate continued genetic variation in an effort to stay at least one step ahead of one's foe. The interest of recombination, and thus of sexuality, is that it generates in hosts new genotypes that can oppose even unsuspected weapons selected by the adversary. Producing such diversity via mutation and then recombination thus opposes the homogenizing influence of natural selection.[2] Clearly, variation still exists without sexuality (the proof is that asexual species show genetic diversity), but it derives only from mutation; all offspring derived from the same reproductive individual are, at least in principle, identical.

What is important to long-term survival, then, is resistance against homogenizing selection even if this seems to be a paradox. Too narrow an adaptation renders a population fragile, as much in evolution as on the job market. As P. H. Gouyon notes (personal communication) Noah's ark, which took on board only one male and one female of each species, would not have ensured the survival of species because of insufficient genetic diversity.

The advantage given by the constantly renewed variability sex produces is explained in genetic terms by frequency-dependent selection. In talking about natural selection we often have the impression that in a given environment a particular allele imparts a fixed selective value to the

2. Note that recombination is of interest for parasite virulence and host resistance only if virulence and resistance involve several genes.

individual carrying it. If this were so, selection between several alleles of different value would always end up retaining only the allele giving the best selective advantage. When this effectively happens we talk about directional selection.

Often, however, the value of an allele depends not on a fixed selective value but rather on how frequently it occurs in the population. The classic example for free-living organisms is a mutant allele that determines a particular coloration in a population of prey animals. The predator develops a search image, which means that predation is triggered by a certain image corresponding to the prey most often encountered. The mutant with an unusual color is protected because it does not correspond to that image, and the frequency of the mutant allele then increases in the prey population. However, as mutants become more and more abundant the predators become more and more used to encountering them and thus modify their search image: the mutant is no longer at an advantage. That is, the new allele was advantageous not because of its intrinsic value but simply because it was rare—as soon as it is no longer rare it stops being advantageous.

The process is in all ways comparable to parasite-host systems: rare alleles are at an advantage. A rare allele may code, for example, for a new surface molecule that the host immune system does not recognize or for an immune mechanism the parasite has not yet confronted. Frequency-dependent selection is a factor in maintaining genetic variability for the parasite as well as for the host, since the alleles concerned in both cases have high adaptive value when they are rare and low adaptive value when they are frequent. In other words, *monotony cannot be selected.* The mode of immune evasion of trypanosomes is an extreme example of frequency-dependent selection occurring on a time scale of just a few days: each time some antibodies are directed against an abundant variable surface glycoprotein (VSG) (see chapter 10), the few trypanosomes out of several billion that express a different and unrecognized VSG freely multiply to form a new parasitic wave.

The sum of the discussion above allows one to understand why the pressure of parasitism is, on its own, often considered capable of explaining the maintenance of sexuality (Hamilton 1980).

That is, we may expect for parasites as well as for hosts that all methods capable of enriching genetic diversity may be conserved if not accentuated. Parasites and hosts are both confronted with the exigency of requiring strong interindividual variation to ensure long-term survival. Hamilton, Axelrod, and Tanese (1990) have shown that it is possible to build population genetic models that explain the advantage of sexuality by coevolution with parasites: "The essence of sex in our theory is that it stores genes that are currently bad but have promise for reuse."

Of course, maintaining variability as a way of fighting parasites is not reserved solely for animals. Plants also know how to generate variability, notably in their chemical defenses against the organisms that attack them—either parasites sensu stricto or phytophagous insects. Linhart (1991) gives particular importance to the "advantage of the rare" in the genetic variability of forest groupings where herbivores are the principal selective pressure. Reciprocally, parasites cannot then attack an ensemble of a plant species population with a single "ideal" gene combination and are constrained, as are animal parasites, to generate some variability of their own.

The Sex of the Hosts

Demonstrating that the Red Queen hypothesis explains the maintenance of host sexuality is not easy because there are few models that offer the choice between sexual reproduction and asexual reproduction or allow one to study populations subjected to pressures other than parasitism.

If the hypothesis that parasitism maintains (or contributes to the maintenance of) sexuality is correct, the following predictions may be advanced (fig. 18.2).

1. If *in a given host-species* both sexual populations and parthenogenic *populations* coexist, then the former should be the most heavily infected, since the disadvantage of sexuality is compensated for by the production of genetically diverse offspring capable of fighting the parasites. Where parasitism is not intense, the disadvantage of sexuality takes its toll and parthenogenesis may be selected.

2. If *within the same population* of a species both sexual and parthenogenic *individuals* coexist, then sexual individuals must be the least parasitized, since they supposedly benefit the most from the variability generated by sexuality. Asexual individuals become "trivial" and must have allowed the easy adaptation of the parasites.

The results of rare field investigations are available, with one investigation supporting prediction 1 and two supporting prediction 2.

Lively (1987) has studied populations of the gastropod mollusk *Potamopyrgus antipodarum* in fresh water in New Zealand. Populations of this mollusk offer the peculiarity of having "normal" males and females as well as parthenogenic females in variable proportions according to location (fig. 18.2, top). On the other hand, parasite pressure, represented by various species of trematode larvae (notably *Microphallus*), also vary according to location. According to prediction 1, we may expect a certain coincidence between high parasite prevalence and a higher proportion of

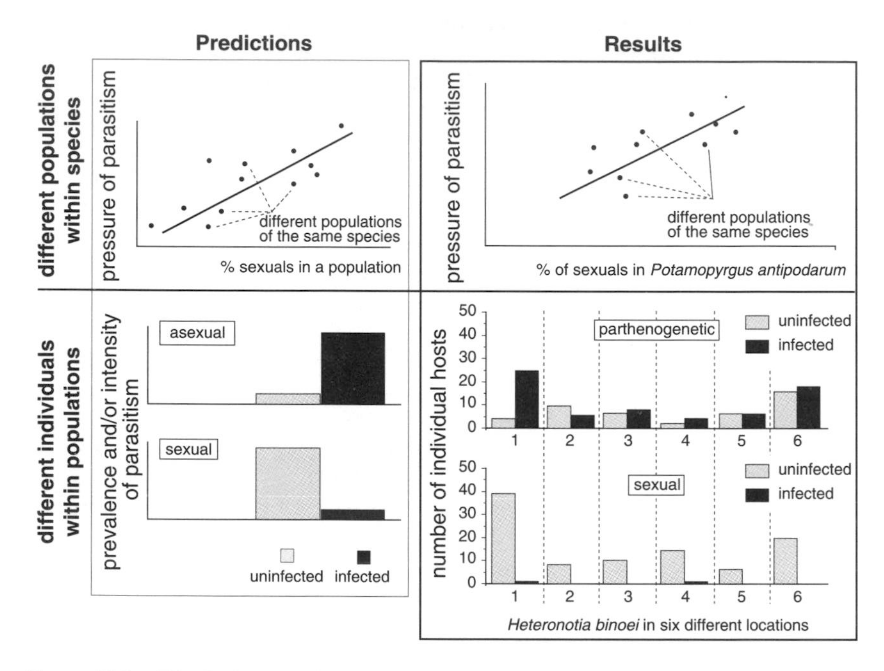

Figure 18.2. The Red Queen hypothesis applied to parasite-host systems. Left: predictions. Top: when parasitic pressure increases, the number of sexual individuals must increase in the population. Bottom: Within the same population, sexual individuals must be more resistant to parasitism; Right: results. Top: parasitism in populations of the gastropod *Potamopyrgus antipodarum* (modified from Lively 1987). Bottom: difference in load of ectoparasitic acarids between sexual and parthenogenetic individuals in a population of the Australian gecko *Heteronotia binoei* (modified from Moritz et al. 1991).

normal males (since the presence of males can only increase genetic diversity, itself supposedly necessary to resist the parasites).

It is, indeed, exactly in populations where parasitism is the most frequent that males of *P. antipodarum* are the most abundant and parthenogenesis is rarest. Lively later enforced these findings (Lively 1992) by showing that although there are differences in environments between more or less "parasitized" locations, these differences were not responsible for the observed variations in the mode of reproduction. In particular, the hypothesis of "the assurance of reproduction," according to which parthenogenesis increases when population density decreases (there is then a purely numerical incentive for the isolated females to reproduce on their own), can be set aside, since it is not consistent with the field data.

The second investigation was carried out in Mexico on freshwater fish and their trematodes (Lively, Craddock, and Vrijenhoek 1990). *Poeciliop-*

sis monacha is found in ponds as both sexual fish and parthenogenic clones. The investigation has shown that metacercariae of the trematode genus *Uvulifer* (whose definitive hosts are birds) accumulate with age more in asexual fish than in sexual ones. These results support prediction 2. The investigation includes a remarkable exception: a pond where sexual fish were highly inbred (with a genetic structure close to that of a clone because of the founder effect) but was also very susceptible. However, Weeks (1996) was not able to confirm these results.

The third investigation consisted of comparing sexual and parthenogenic populations of the gecko *Heteronotia binoei* in Australia for their hematophagous acarids (Moritz et al. 1991). The results are the most surprising (and convincing) of all, since the number of mites is 0.59 on average per sexual individual and 21.6 per asexual individual. Prevalences, which were calculated separately in sexual and parthenogenic individuals, are also significantly different (fig. 18.2, bottom). The most plausible interpretation is summarized by Barbault (1994): "The parthenogenic geckos are the most vulnerable to infection by mites because of their clonal mode of reproduction. The result is an accumulation of identical genotypes in the population, genotypes to which parasites are adapted." This is a second validation of prediction 2 as well as a nice illustration of the reflection of Ladle (1992) cited above.

All three investigations give weight to the Red Queen hypothesis as applied to the maintenance of sexuality in hosts because of parasites; the small number of investigations is the only obstacle to a definitive conclusion.

An appealing variation on the "need for genetic diversity" in host populations concerns social insects. Geneticists easily demonstrate that a gene that determines strict altruistic behavior cannot be selected in natural populations because it inadvertently increases the success of "nonaltruist genes." If nonsexual individuals, such as the workers or soldiers, work for sexual individuals such as a queen bee, it is because they are narrowly related to the sexual individual, and they in fact work for the success of their own genes by supporting these other individuals. However, there are species where individuals of the colony are not as closely related to each other as they would be if they came from a single couple, since the queen mates with several males and not one (polyandry) or because there are several queens (polygyny). Laboratory experiments show that infection by protists (trypanosomes and microsporidians) increases more rapidly in colonies of *Bombus* where individuals are closely related (Shykoff and Schmid-Hempel 1991), from which derives the suggestion of Schmid-Hempel (1994) that polyandry and polygyny may have been selected because they introduce a certain dose of genetic diversity into the colonies—the result being greater resistance to parasites. A comparative analysis (Schmid-Hempel and Crozier 1999) suggests that the level of relatedness within ant colonies, a quantity affected by both polyandry and polygyny, is also significantly correlated with parasite load (the parasites taken into account in the analysis ranged from viruses to parasitoids but excluded mites).

Even when hosts reproduce sexually, they are still at a disadvantage when faced with parasites. A first disadvantage is that while a certain parasite species may specialize on only one or several host species, the host has to confront multiple parasite species with very different genomes (viruses, bacteria, protists, and metazoans). A second disadvantage is that the time between generations is longer for the hosts: "Assuming 30 minutes per bacterial generation, there are 1,226,400 bacterial generations in a human lifetime of 70 years. In the 7 million years since we shared a common ancestor with chimpanzees there have been just over 200,000 'human' generations of 30 years each" (Ridley 1994). Thus hosts have, a priori, less opportunity to generate variability and to select the adaptations necessary for their defense. Could the fact that their genomes are richer in coding sequences compensate for this? Perhaps this is also the root of selection for a mechanism for somatic DNA reshuffling in vertebrates, since this process allows them to produce some tens of millions of different antibodies with great savings in the actual number of genes involved.[3]

The Sex of the Parasites

Curiously, the Red Queen hypothesis applied to the maintenance of parasite sexuality under the influence of the pressure of host resistance has only rarely been studied. However, Harvey and Keymer (1987) wrote: "If sex is maintained among most eucaryotes as a parasite-defence mechanism, as is often suggested, then perhaps it is not surprising that parasites have joined the race by evolving sexuality to keep one step ahead of the defence mechanisms of their host[s]."

If acquired immunity is at least in part genotype specific (if it is more effective against the genotype that elicited it than against other genotypes of the same parasite), one prediction would be that when selective pressures exerted by the hosts' immune system increase, sexual reproduction in the parasite should be favored because it increases the chances that novel genotypes will appear in the offspring and thus allow them to evade

3. This process of somatic DNA rearrangement is an internal genetic recombination in which the individual produces a much greater diversity of defense molecules than is produced during recombination in meiosis alone.

the immune response directed against parental genotypes. The prediction can be tested in parasites that can reproduce either parthenogenetically or sexually.

To my knowledge, only one experimental study has attempted to verify that the investment in sexual reproduction is greater when parasites invade hosts with strong immune responses than when they invade hosts without protective responses. This study was carried out by Gemmil, Viney, and Read (1997, 2000) on *Strongyloides ratti.* This nematode parasitizes the small intestine of rats, which are infected by the nematode's skin-penetrating larvae. Rats harbor only parthenogenetic females, which produce eggs mitotically. The larvae hatch in the external environment and develop either as free-living sexuals (heterogonic development) or as infective L3 larvae that can penetrate rats (homogonic development). Free-living males and females mate, and the females then lay eggs that produce larvae that develop into infective L3s. When sexuality is involved, meiosis and recombination occur. When sexuality is not involved the life cycle is equivalent to clonal multiplication, and the proportion of larvae that develop into sexuals or directly into infective L3s is under both environmental and genetic control.

Gemmill, Viney, and Read (1997, 2000) manipulated the immune system of rats *(Rattus norvegicus)* in various ways (using hypothymic mutants, corticosteroids, irradiation, and previous exposure to the parasite)[4] and demonstrated that larvae from normal rats or rats that have acquired immune protection were more likely to develop into sexuals than larvae from naive or immunosuppressed rats. In other words, when hosts develop immunity against parthenogenetic females, the heterogonic cycle is favored (fig. 18.3). The authors were very cautious in interpreting their results and examined various other hypotheses that also might account for them. However, they concluded that this immune-determined sexuality is consistent with the idea that sexual reproduction by parasites is adaptive and consistent with the Red Queen sex model. In fact, the parasites whose free-living ancestors were sexual forms do not seem to have ever lost their ability for sexual reproduction, even though they have lost many other functions.

Sometimes the objection is raised that numerous metazoan parasites (most notably almost all trematodes, cestodes, and monogeneans) are hermaphroditic because self-fertilization would ultimately reduce genetic variation. However, simultaneous hermaphroditism is not the same as loss of sexuality: the hermaphroditic individual has distinct male and

4. These various techniques were used to minimize the chance that results were due to factors other than immune alteration.

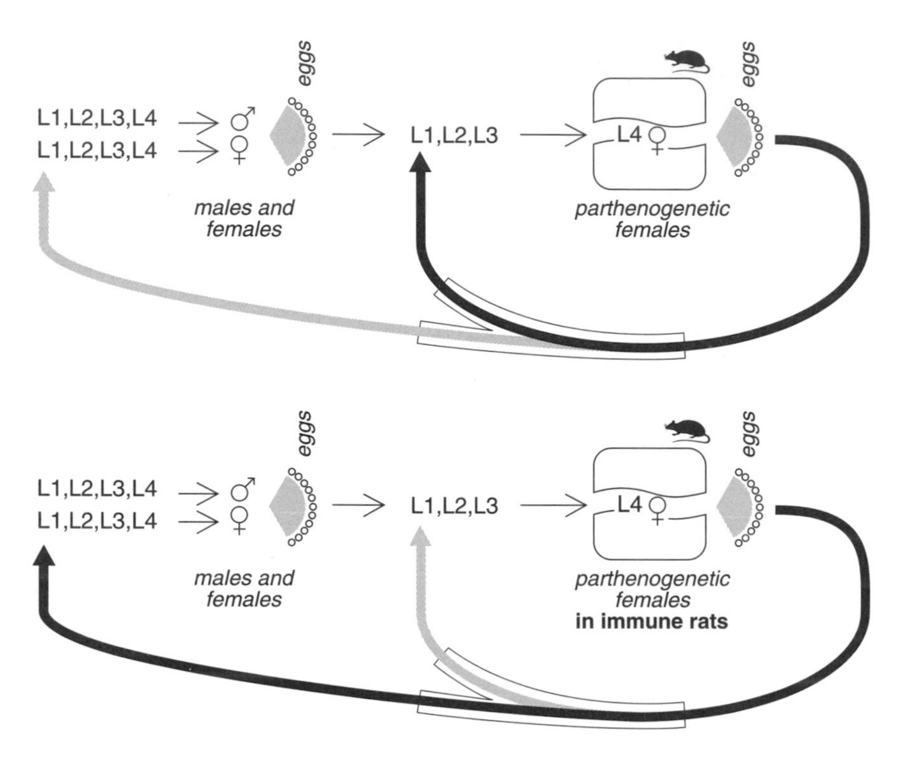

Figure 18.3. Influence of host immunity on parthenogenetic or sexual reproduction of the nematode *Strongyloides ratti* (inspired by Gemmil, Viney, and Read 1997). Dark arrows indicate preferred modes of reproduction.

female genitalia, the gametes are produced via meiosis, and the eggs are the result of fertilization. Simultaneous hermaphroditism is an adaptation, sometimes said to ensure reproduction "in a pinch," that reduces "the risk of rarity" since it allows reproduction via self-fertilization even when an individual is isolated. Simultaneous hermaphroditism outside the world of parasitism characterizes numerous pioneer species.

The only inconvenience of hermaphroditism is that if self-fertilization ends up predominating, the number of homozygous loci in the population increases and recessive defects eventually appear.[5] It may be because

5. Note that while hermaphroditism reduces diversity within populations, it also has a tendency to increase diversity between populations of the same species (Jarne and Charlesworth 1993), which is not automatically a disadvantage if selection occurs at scales other than that of the individual.

of this inconvenience that certain parasites belonging to groups where hermaphroditism predominates have taken the risk of "reselecting" separate sexes (gonochorism). This is the case for schistosomes, the only trematodes living in the blood of homeothermic vertebrates, which may have regained gonochorism in order to have a form of sexuality that guarantees maximum variability and thus an extra chance to oppose the highly sophisticated immune systems of their host (Combes 1991c). Nevertheless, gonochorism still remains costly in terms of the number of unmated, and thus nonreproductive, individuals in the system (Morand et al. 1993).

Few studies have looked into self-fertilization and cross-fertilization in parasites experimentally. Trouvé et al. (1996) infected mice with metacercariae of three strains of the hermaphroditic trematode *Echinostoma caproni*, each identifiable by allozyme markers. Results indicate that when only one parasite is present it uses self-fertilization to ensure reproduction. When there are two parasites (each belonging to a given strain), each both self-fertilizes and cross-fertilizes. When three parasites are present (each belonging to a given strain), each one mates with each possible partner, acting from time to time either as a male or as a female.

To say that sexuality is necessary in order to give parasites a high degree of variability makes them akin to pioneer species; it is therefore instructive to note the circumstances in which free-living organisms use asexual reproduction versus sexual reproduction (see Williams 1975; Ridley 1994).

For example, coral reproduces asexually by budding as long as it needs to conquer the nearby environment. On the other hand, the multitude of larvae that corals let loose in marine currents to be carried away toward the conquest of unknown environments come from sexual reproduction. Numerous plants do the same: they produce shoots or rhizomes (asexual reproduction) on site and use seeds (sexual reproduction) to conquer new areas.

When the conquest of nearby environments is not worthwhile, such as when trees are very close to one another in a forest or mussels are packed on a rock, producing individuals identical to the parents and capable of populating that environment has no particular benefit for the species, and only sexual reproduction exists—it yields variable offspring capable both of dispersing and of occupying at least some new habitats.

Parasite hosts, because of their own variability, are environments even more diverse than the nearshore areas where corals disperse or the soils where seeds land. We can therefore easily understand the need for genetic variability in parasites.

Clones versus Sex

If it is true that parasites need to maintain a high degree of genetic variability to fight against their hosts' immune systems, then it is also true that this variation may be obtained by means other than classic sexuality, such as high mutation rates.

Viruses can't be beaten at this game, and the resulting modification in genetic information, particularly because of nucleotide substitutions, provides an exceptional degree of genetic variability in numerous types of viruses. It is thought that the mutation rate is 1/10,000 nucleotides/replication for HIV1 and that patients may pass from a seropositive state to immunodeficiency when they reach a threshold of "internal viral diversity" where the immune system can no longer respond efficiently (Nowak et al. 1991). The role of mutator alleles in bacteria has also been mentioned previously (see chapter 11).

High rates of mutations are costly, since certain vital mechanisms may not function as well, although Ewald (1994b) thinks these costs are compensated for by the enormous advantage of escaping the immune system. If, as is likely, there is heritable variation in the mutation rates of pathogens, then nothing prevents these pathogens from selecting different rates of mutations depending on the aggressiveness of the immune system they confront. Significantly, those enzymes involved in DNA replication or in regulating this replication are probably the targets of selection.

Thus sexual reproduction is not as essential for the parasite as it is for the host. This observation explains why there is a continuum in the parasite world between the exclusive use of asexual reproduction (which yields the advantage of rapid multiplication) and the exclusive use of sexual reproduction (which ensures constantly renewable variability). Between these two extremes, numerous parasite species can be found that use both.[6]

Asexual reproduction has been known for some time to occur in parasitic protists as well as in numerous other organisms. In all cases asexual reproductive periods, whether short or long, produce clones. The question is then, Which form wins, sexuality or cloning? Determining whether cloning is dominant is important, since therapeutic and vaccination strategies differ depending on whether populations regularly exchange genes.

---

6. Some free-living organisms (planktonic crustaceans, aphids, most plants) show alternating phases of sexual and asexual reproduction, which also allows them to accumulate the advantages of both.

The occurrence or predominance of sexuality in organisms may be recognized because it leads to genetic structures that can be interpreted by the Hardy-Weinberg principle, which allows one to accurately predict allelic frequencies when there is sexual reproduction and panmixia.

The predominance of asexual reproduction in a population, on the other hand, may be recognized by the existence of persistent linkage disequilibrium or fixed heterozygotes.[7]

The predominance of cloning has been demonstrated in several human parasites. In *Trypanosoma brucei* laboratory experiments show that sex may occur during transfer to the tsetse fly (see Schweizer et al. 1994; Gibson and Stevens 1999), although allozyme and DNA analyses show a strong linkage disequilibrium for certain genes, which supports the hypothesis of the dominance of cloning in natural populations. The difficulty, then, is to evaluate the respective importance of sexuality and cloning in *T. brucei*. There is evidence that the clonal expansion of particular genotypes may induce linkage disequilibrium and that recombination exists outside these periods of imbalance. As Mathieu-Daudé et al. (1995) note: "There is no doubt that an explosion in representation of some particular genotypes during epidemic outbreaks produces linkage disequilibrium." In other words, there would be cloning in the short term and sexuality over the long haul. Further, one cannot exclude the possibility that the occurrence of cloning and sexuality may vary with the ecological conditions associated with transmission. According to Hide et al. (1994), clonal explosion would be characteristic of sleeping sickness in humans, whereas genetic exchanges may be more frequent in cattle—the difference being linked to the more recent appearance of this parasitosis in humans. Cloning is also dominant in *T. cruzi* (the causative agent of Chagas' disease) and most likely is the exclusive mode of reproduction in *Entamoeba coli* (amebiasis). In *Candida albicans* (candidiasis), a fungus often involved as an opportunistic parasite in AIDS, Pujol et al. (1993) showed that thirteen enzyme loci are linked (do not recombine) and that it is unlikely this linkage is due to their proximity on the same chromosome.

In *Plasmodium* the alternation between a clonal asexual phase in humans and a sexual phase in mosquitoes is obligatory. However, based on various field observations (linkage disequilibrium and fixed heterozygotes) Tibayrenc, Kjellberg, and Ayala (1990) and Tibayrenc and Ayala

7. There is imbalance in linkage if the ensembles of alleles (or haplotypes) inherited from each parent have frequencies that disagree with the hypothesis of random recombination of alleles. Fixed heterozygotes have a frequency of 100%, also disagreeing with the occurrence of recombination.

(1991) have questioned the importance of sex in the life cycle, opposing the ideas to the contrary defended by Walliker (1991) and Dye (1991). After several years of debate, it now seems certain that it is possible to conclusively demonstrate the occurrence of fertilization and genetic recombination by experimentally infecting mosquitoes with two genetically different clones of mosquitoes; that notable allelic variation occurs and heterozygotes are found in natural populations; that in most cases patients are infected by several different clones; and that the number of different clones in a local population of *Plasmodium* is very high. Most natural populations of *Plasmodium* would then be close to panmixia, if we admit that perfect panmixia almost never exists.

In what cases would the clonal phase be of particular importance? Paul and Day (1998) write: "Clearly, because of the structuring of the parasite population into discrete hosts, the parasite population as a whole does not interbreed randomly."[8]

Each time an infected mosquito bites a human, it establishes a separate infrapopulation of *Plasmodium,* often injecting only a small number of sporozoites (Beier 1993) that may be produced by only one ookinete. If the rate of transmission is low (few infective bites), the genetic diversity of the clones that develop in a patient is also low if not nil, particularly if some clones are eliminated (the proportions of different clones in the same infection are probably always uneven). However, the cloned blood forms produce both male and female gametes. Under these conditions the founder effect following the mosquito bite takes the shape of a total bottleneck, and even though there are billions of gametocytes in the blood, they all have the same genome. Walliker (1991) stated: "It is certainly possible that a single clone could give rise to an outbreak of malaria in a region where the disease occurs infrequently. Mosquito transmission of such a clone would not generate recombinant forms, since the parasite is haploid, and so the same genotype would be found in all infected people . . . if the zygote is homozygous at all loci, as would occur following mating between identical gametes, then recombination events at meiosis would not normally be expected to have any detectable genetic consequences" (fig. 18.4).

Thus it is only in human populations where the level of transmission is very low that the clonal phase may be of particular importance (Babiker and Walliker 1997; Paul and Day 1998): "The extent to which *P. falciparum*

---

8. In reality this remark applies to all parasites. It is only within a given infrapopulation that panmixia might eventually occur, and it never happens in parasite populations (the liver flukes of cow X do not exchange genes, in a given generation, with the liver flukes of cow Y).

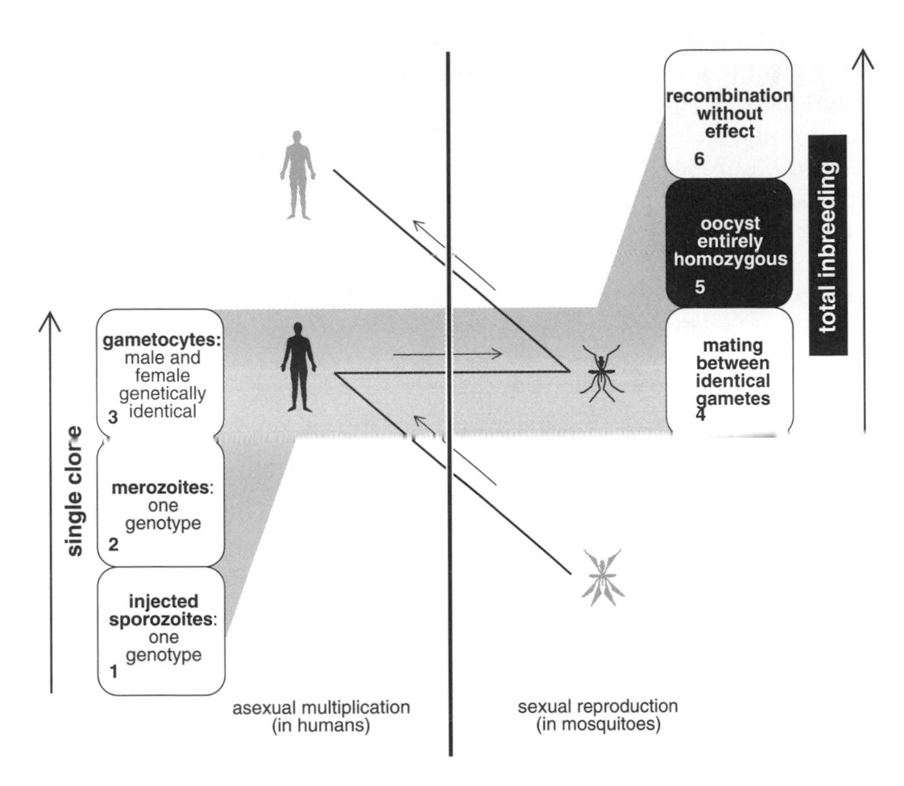

Figure 18.4. If there is a single clone in a patient, fertilization followed by meiosis only reconstitutes the clone from which gametes are produced.

has a clonal mode of reproduction will depend upon the number of different parasite genotypes within a human host" (Paul and Day 1998). In fact, if biological vocabulary is correctly used there is clonal reproduction nowhere in *Plasmodium* (except of course in the human phase of the life cycle); there is only total inbreeding with genetic consequences that mimic clonal reproduction.

Overall, the large majority of field investigations support the "normal" sexuality of *Plasmodium*. For example, Babiker et al. (1994, 1995) have shown that a strong polymorphism occurs in patients from a Tanzanian village where 10% to 20% of the mosquitoes are infected and where inhabitants are bitten by infected mosquitoes more that three hundred times a year. A study of the genetic diversity of the oocytes shows that "an extensive degree of crossing between clones occurs" (Babiker et al. 1995), that "crossing between clones was taking place frequently, following uptakes of mixtures of gametocytes by the mosquitoes" (Babiker et al. 1994), that "random mating events probably occurred within mosquito bloodmeals between gametes belonging to different parasite clones" (Babiker et al. 1994) and that "the findings illustrate emphatically the inappropriateness of the idea that *P. falciparum* in such communities consists of a collection of distinct strains"(Babiker et al. 1995). The authors observed that numerous clones existed within the village inhabitants as a whole but that each patient was infected by only a few. Significantly, only if each patient were infected with all the population clones in the village would the Hardy-Weinberg principle apply exactly. Thus there is an almost necessary deficit in heterozygotes because in general mosquitoes take gametocytes from only one patient during a blood meal. As a result there is "nonrandom mating," caused by the subdivision of the population into infrapopulations. It has also been suggested (Blaineau, Bastien, and Pagès 1992; Bastien, Blaineau, and Pagès 1992) that the characterization of protists into zymodemes—that is, profiling based on allozymes—therefore cannot reveal polymorphism in these organisms.

The controversy on the role of sexuality in parasitic protists shows how poor our knowledge of the genetics of natural populations of parasites is. Even in *Plasmodium,* for example, we do not know (see Read and Day 1992) the rate of inbreeding in mosquitoes, the degree of isolation of *Plasmodium* populations, or if the genes of *Plasmodium* circulate with mosquitoes or with humans.

What we need to recognize, without taking anything away from sexuality, is that clonal multiplication is not an obstacle to variation in these systems. When I talked about antigenic variation (chapter 10), I mentioned that even in clones the antigens exposed by *Plasmodium* on the hemocyte membrane vary rapidly even in the absence of immune system pressures (Roberts et al. 1992). Thus it is possible that after a short time different genotypes may be present in the patients' blood (even if at the origin of the infection the mosquito injected only one sporozoite). Howard (1992) writes: "The combined effects of antigenic diversity in the population and antigenic variation in clonal progeny make it likely that there will always be some parasites able to circumvent pre-existing immunity in the host."

In the case of helminths, we also have very limited knowledge of the role that clonal phases may play in maintaining genetic diversity in populations.

Trematodes are a fascinating group to consider, since at each generation only one miracidium, after developing in the molluskan intermediate host, can produce thousands of cercariae—all copies of the same genome. From the point of view of population biology, if this phase contributes to the success of transmission, there is then the problem of its probable negative impact on variability.[9]

We may certainly consider the idea that the dispersion of cercariae would mix clones and thus restore variability via recruitment in the next host, since numerous cercariae are planktonic in both freshwater and marine systems (see Combes et al. 1994). If we acknowledge that this occurs, then it should also be no surprise that dispersal mechanisms between the mollusk and vertebrate sometimes do not operate. For instance, the cercariae of some species become encysted on site in the mollusk and do not disperse, so that the downstream host that eats the mollusk (often a bird) recruits a stock of parasites that are genetically identical. Further, the cercariae of some other species leave the mollusk but remain united in large numbers (several hundred) by their tails, so that the downstream host (a fish) is attracted by the swirling sphere. As in the previous example, the fish recruits a stock of genetically identical parasites (Beuret and Pearson 1994). Such situations are likely to produce a high degree of inbreeding in the parasite and generate only weak variability. It is possible that the pathogenic effect of the few parasite species that use this "nondispersion" is small and that the defenses of the hosts are themselves weak so that there is no need for a high degree of variation in the parasite. This would then demonstrate a contrario the correlation between high genetic variation and the need to escape host defense mechanisms. In chapter 1 I mentioned that in some trematodes the miracidia may be produced asexually by sporocysts in snails (sexuality disappears totally from the life cycle). Tom Cribb (personal communication) thinks this may be a response to ephemeral definitive host populations. However, he has also suggested that it might be because the snail hosts of these particular trematodes reproduce asexually that the asexual reproduction of miracidia also evolved. If this hypothesis ends up being correct, it constitutes an excellent indirect argument in favor of the Red Queen hypothesis.

Remember that trematodes are not the only helminths that use asexual reproduction. For example, some cestodes, such as *Echinococcus granulosus* and *E. multilocularis,* have larval forms (hydatic cysts and alveolar cysts, respectively) that bud off thousands of genetically identical scolices.

Understanding the nature of genetic variation in parasites is an essen-

---

9. The diversity of cercariae emitted by a single mollusk is influenced by the number of miracidia that infect it. In the *Schistosoma mansoni/Biomphalaria glabrata* association Sire et al. (1999), using molecular markers, reported that 88% of snails harbor a single parasite genotype, whereas Minchella, Sollenberger, and Pereira de Sousa (1995) showed that 57% of the snails harbored multiple infections (up to nine parasite genotypes per snail). Such different situations, the reasons for them yet unexplained, might have very different consequences on the genetic heterogeneity of adult worm infrapopulations.

tial component in the research and control of human parasites. Cox (1991) notes that "this variation is extremely important in both drug therapy and in susceptibility to immune attack." It is the capacity to renew variation that explains the rapid selection and evolution in parasites of strains resistant to the drugs commonly used in therapy. The classical example is that of Nivaquine (a form of chloroquine), a drug the protist *Plasmodium falciparum* became resistant to in numerous parts of the world.[10] The genetic diversity of parasites is one of the most difficult constraints to take into account in making efficient vaccines. Speaking about the antigenic polymorphism of the circumsporozoite molecule (a membrane protein on the sporozoites of *Plasmodium falciparum*), Mazier (1993) writes: "If lymphocytes of immunized patients . . . do not recognize natural variants, it seems difficult, even maybe impossible, to produce a synthetic vaccine that contains enough epitopes to induce protection in endemic zones."

Sexual Selection

The Quest for "Good Genes"

Selective pressures impart different reproductive success to individual carriers of "good genes" (giving them a higher adaptive value) and "bad genes" (giving them a reduced adaptive value). However, note that it is the global adaptive value of the individual that counts and not that of genes taken one by one. *Sexual selection* is one of the selective pressures responsible for the differential success of the transmission of genes by genetically different individuals. In gonochoric species (= dioecious = heterosexual), sexual selection involves the choice of males by females or, more rarely, the choice of females by males (in what follows, only the choice of males by females will be considered).

Sexual selection is a form of true selection only if it results in the retention of genes that give progeny the best possible fitness. For the genes of a given female to be transmitted, her descendants must have the best possible reproductive success. However, the genes of her descendants will include only half of her genes, with the other half coming from her mate. From this the logical conclusion is that to ensure a good future for her own genes, the female serves her own best interest by choosing a mate with the best possible genes (and vice versa if the male does the choosing).

---

10. But strains sensitive to chloroquine have not disappeared. Koella (1994) explains this by a process of frequency-dependent selection: host resistance may be different for different genotypes of the parasite. That is, host resistance develops against the most common genotypes. For instance, when chloroquine resistance started to appear, the most common strains were those sensitive to it, and today the most common strains are those that are resistant. It is against the latter that host resistance develops, and thus strains that are sensitive to chloroquine may end up being favored.

What is a good gene for the potential host of a parasite? Good genes are those that tend to close the encounter filter or the compatibility filter (or both).[11] Thus, in order for a female host to transmit her genes it is to her greatest advantage to select a father that also brings "good genes" capable of avoiding infectious and parasitic diseases. This can be done, however, only if "good genes" can be detected in the candidate *before mating occurs.*

The Pretty and the Useful

"The tails of the peacocks, the song repertoires of nightingales, and the size dimorphism of elephant seals call for an explanation," writes Maynard Smith (1991). It is astonishing that some birds (but not only birds) express a high degree of sexual dimorphism in both their morphology (ornamentation and colors) and their courtship behavior (strutting or parading), whereas other species have males that are morphologically identical to females and exhibit little obvious differential behavior.[12] Two classical examples of the latter are magpies and blue jays: Who knows how to distinguish males from females in these species?

The showy colors and parades of some birds have been known for a long time and intrigued Darwin himself. For him, ornamentation evolved because there was a reproductive reward for brightness in males, since females preferred the shiniest candidates in mating.

For a male, brightness (the "showiness" of Hamilton and Zuk 1982) has one advantage and several disadvantages. The advantage is the increased probability of being chosen as a mate.

The disadvantages are an increased risk of predation (being shiny or having cumbersome feathers increases vulnerability to predators); an energy cost ("building" ornamentation and strutting expend energy that must be diverted from other organs or functions); and immune deficiency (the androgens involved in developing secondary sexual characters in males have a negative effect on the immune system).

Significantly, it is these disadvantages that make brightness an "honest"

---

11. For the parasites, good genes are those that open the encounter filter or the compatibility filter (or both). Nobody knows if sexual selection occurs in parasites with separate sexes, which does not mean it does not. Numerous parasites detect mates through pheromones, and if these pheromones inform about characteristics linked to encounter or compatibility, that would be sufficient to permit sexual selection.

12. The classic examples are the birds of paradise of New Guinea, Australia, and Indonesia. Although all the males of most of the forty-two species known are characterized by bizarre and colorful feathers, about ten species have no sexual dimorphism. Dimorphic species are polygamous, with fathers that take no care of the young, whereas species without dimorphism are monogamous, with fathers that participate in the raising of the young.

signal in the sense of Zahavi (1975, 1977).[13] Honest signals must be costly in order to avoid cheating (see Folstad and Karter 1992).

For a female, there are also a few disadvantages (such as an increased risk in predation during mating with a bright male). Is there an advantage? As far as we know the notion of beauty is purely human, and it is not very serious to say that female birds mate with certain males only because they are "handsomer" than others.

Reflection on this paradox developed in the 1980s and started with the idea that in a given population males that have the best overall genetic qualities—for example, resistance to diseases—develop their colors and ornamentation better than males that are more sensitive to diseases. Thus, by choosing males based on "ornamentation," females would be making the "right genetic choice" (Thornill 1980). Freeland (1976) clearly enunciated the hypothesis that the dominance of certain male mammals reflects their capacity to resist parasites. The progeny of females fertilized by these dominant males would then have the genes necessary for resistance. Thus was born the idea that some traits related to sexuality in a species may be linked to the fight against parasites.

In 1982 Hamilton and Zuk showed that if the hypothesis that females use secondary sexual characteristics to evaluate males' capacity to resist parasites was correct, then a correlation should also exist between the brightness of the males and parasite pressure in the species: logically, only when parasitism is particularly important should brightness in males be selected. To support this hypothesis Hamilton and Zuk assembled the results of various investigations involving the group of protozoan parasites known as hemosporidians. These parasites of blood cells and of some tissues are sporozoans, and three genera are common in birds: *Plasmodium, Leucocytozoon,* and *Haemoproteus.* All are transmitted by biting dipterans (mosquitoes, phlebotomines, etc.), and their pathogenic effect, at least for some species, is serious. Hamilton and Zuk showed that there was, effectively, a correlation between the showiness of the male and the importance of parasitism by hemosporidians in a total of 109 species of passerines in North America.

The Basis for Hamilton and Zuk's Hypothesis

Let's first rephrase the idea and the hypothesis in clearer terms:

If a male contributes to reproduction only with its sperm (if it provides no protection for the female and no care of young), its "quality" is then due only to its sperm (to its genes).

Because individuals are unequal in the face of selection, certain males

---

13. Note that an ornament is not necessarily a signal. It is a signal only if it conveys information from a sender to a recipient (see Zuk 1991).

have genes that impart high quality ("good genes") while other males have genes that impart low quality ("bad genes"). Of course, intermediates of all types exist.

The question then is, Do females choose males randomly, or can they distinguish the carriers of "good genes" from the carriers of "bad genes"?

Riley (1993) writes: "A male's genetic quality does not come written on him." But the hypothesis of Hamilton and Zuk suggests that in certain cases the males' genetic quality is in fact written on them in living color and in a language females clearly know how to interpret.

Trials to assess the validity of the Hamilton and Zuk hypothesis have focused on the selection of males by females.

For this hypothesis to be accepted, four conditions must be fulfilled (in what follows the terms "brightness" and "showiness" are used to designate either color or parade quality): the characteristics of both resistance and showiness are heritable, and there is variation in these traits; within a species individuals with the heaviest parasite load are less bright or showy than are healthy individuals (fig. 18.5, top); females must choose the most showy males; and in an ensemble of host species those that harbor the most parasites must be the brightest, since the selection of showiness must have occurred with the most vigor where parasitism is the most intense (fig. 18.5, bottom).

Resistance and Showiness Are Heritable

If we set apart simple Mendelian traits (the yellow and green colors of Mendel's peas, for instance), we find that the traits of most organisms are both polygenic (they depend on several loci) and multifactorial (they depend on both genetic and environmental factors). The resemblance between parents and offspring, which is studied through quantitative genetics, is thus not always easily predictable. When we study the resemblance between parents and offspring for a quantitative trait, the first question is, "Is there a correlation between parent and offspring for this trait?" If the answer is yes, the second question is then, "Does this correlation have a genetic or environmental cause?" Nonenvironmental covariance is called "inheritance" *(h2)* (fig. 18.6), and the possibility of a response to natural selection is determined by the magnitude of this inheritance.[14]

In practice the inheritance of a quantitative trait is never equal to one because the quality of the phenotype almost always owes something to the

14. Population genetic models demonstrate that fitness traits, like secondary sexual traits, have zero inheritance in a population at equilibrium. This difficulty does not exist, however, if there is continual renewal of genetic variability, as occurs in arms races. As long as the arms race lasts, sexual selection of "good genes" can be sustained indefinitely.

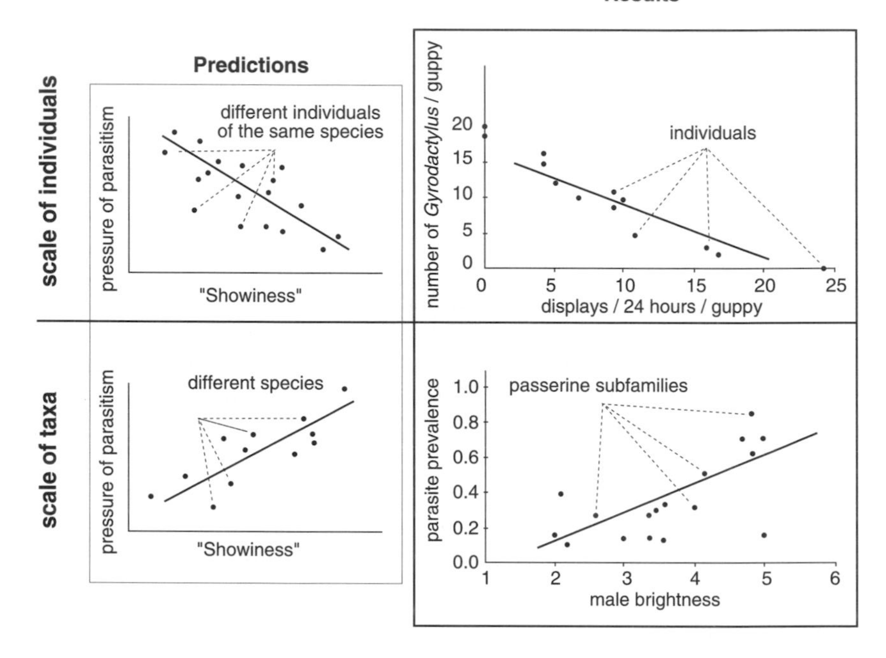

Figure 18.5. The Hamilton and Zuk hypothesis. Left: predictions. Top: within a population there must be a negative correlation between how "showy" individuals are and their parasitic load (the most parasitized individuals must be less showy). Bottom: if several taxa of the same host group are compared, there must be a positive correlation between parasitic pressure and how showy the feathers are (the most parasitized species must be the most showy). Right: results. Top: within a group of guppies the most parasitized individuals are the least capable of parading for the females (after the data of Kennedy et al. 1987). Bottom: within a group of passerine subfamilies in North America, the taxa richest in parasites are the most showy (modified from Read 1988).

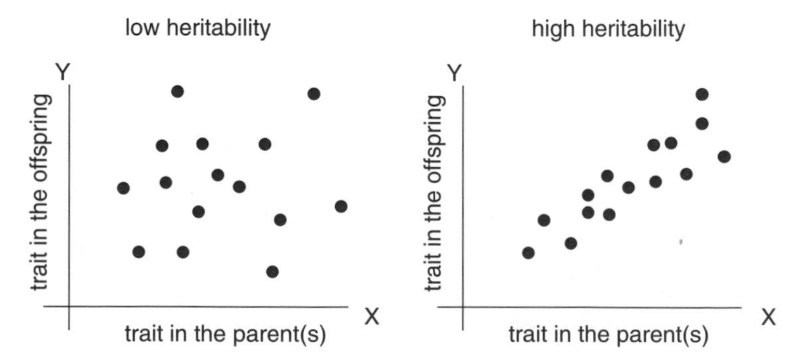

Figure 18.6. No correlation (left) and positive correlation (right) between quantitative traits of parents and offspring.

environment and because combinations of genes producing phenotypes of high quality are disrupted by gene segregation and recombination (see, for example, Read 1988d; Siva-Jothy and Karstein 1998).

Moreover, in the case of complete overdominance (if the best adaptive value is that of heterozygotes) there is no correlation between parents and offspring in the trait considered.

To validate Hamilton and Zuk's proposition, resistance also must be inherited.[15]

The inheritance of resistance has been the object of debate for some time, primarily because this character is strongly influenced by the environment, particularly nutritional condition.[16] A demonstration of the inheritance of resistance in relation to the brightness characteristics of males was done by Møller (1990a), who related the length of the tail feathers of male barn swallows, *Hirundo rustica,* to parasitism by the acarid *Ornithonyssus bursa,* which live on the birds' head. Males with long feathers have significantly fewer parasites, and the descendants of males with long feathers resist experimental infections better than descendants of males with short feathers, even when they are transported to nests other than those of their parents. The "search" for resistance genes by females (for their offspring) is possible only if a diversity of resistance is maintained in the population, and this diversity finds its origins in the mechanisms of coevolution.

Parasitism Makes Affected Males Dull

It has been demonstrated several times that parasitism detracts from the expression of characters as energetically costly as that of the growth of extra-long feathers or makes the performance of complicated parading less perfect. For example, Houde and Torio (1992) analyzed how the monogenean *Gyrodactylus turnbulli* affects the expression of carotenoid pigments (orange spots) on male guppies—ovoviviparous fish characterized by strong sexual dimorphism. *G. turnbulli* is an ectoparasite that mainly attaches to the fins but also attaches to the fish's gonopodium, the organ that allows internal fertilization. These workers compared the color of healthy and parasitized males under the following protocol: Certain males were first set in contact with other fish already infected until

---

15. For a more complete review of the notion of inheritance see, for example, Hartl and Clark (1989).

16. The term "resistance" is here taken in the broadest sense—larger than the one I adopted when describing the functioning of the compatibility filter. In effect, I admitted that an organism may be unsuitable for a parasite (the environment offered does not fit it) or resistant to the parasite (it actively creates a hostile environment for the parasite). Here it is not important for the genes to be unsuitable or truly resistant, and often authors talk about "resistance" in this more general sense.

they carried two to four parasites, after which they were maintained in isolation from one another for ten days. Then fish were cleaned of their parasites using an anthelmintic, with controls receiving the same treatment. The orange spots of previously parasitized and healthy fish were then compared using colored disks (Munsell disks), and indexes of clarity and saturation were determined. The results were eloquent: previously parasitized fish had spots significantly less colored and not as saturated as the spots of healthy fish.

Similarly, Kennedy et al. (1987) showed that parasitism by *Gyrodactylus* also affects the number of sexual displays the male guppy can make in twenty-four hours. In healthy fish this number is about twenty-five, but it drops to as low as zero as fish are parasitized with an increasing number of gyrodactyls (fig. 18.5, top right).

Significantly, such effects are not restricted to fish. For instance, Mulvey and Aho (1993) carried out a large investigation on almost 1,300 male white-tailed deer *(Odocoileus virginianus)* in South Carolina. These authors found a significant negative correlation between the size of the body and the development of the antlers, on the one hand, and parasitism with the liver fluke *Fascioloides magna* on the other.

Wedekind (1992), who studied the tubercles that ornament the head and body of male fish in the roach *Rutilus rutilus,* went further than the previous authors in his conclusions. His reasoning was as follows: the development of different ornaments is related to hormones or mixtures of hormones; different hormones weaken some immune functions in specific ways; parasites of different species are attacked differently by the immune system; hormone mix may reveal the actual use of an animal's immune system, which depends on the presence and burden of, or even susceptibility to, different parasites; and the different species of parasites may be "read" in the ornaments of the host. Wedekind showed that the number of tubercles of *R. rutilus* allows one to distinguish which males are parasitized by gill monogeneans *(Diplozoon)* and which are parasitized by nematodes in the body cavity. The results led him to conclude that "a female roach could potentially decode a male's ornamentation to gather a sort of clinical picture of him and use this information in her choice of mate."

Some studies show positive correlations between parasite load and "sexual display" in surprising ways; that is, they report the opposite of what is predicted by the Hamilton and Zuk hypothesis. For example, Müller and Ward (1995) observed that trematode load in the European minnow *Phoxinus phoxinus* is negatively correlated with the red coloration of the abdomen and fins, which supports the hypothesis, but that it is also positively correlated with the size of the cephalic tubercles, weapons used in combat between males. The authors suggest that males that cannot develop the red coloration (caused by carotenoid pigments) compensate for this handicap by developing tubercles that are useful during intermale competition. This explanation is not impossible, but it seems surprising that a change in resource allocation may occur precisely under the conditions where resources are supposedly already adversely affected by defending against parasites.

Females Choose Bright Males

Numerous studies have demonstrated that females do choose the brightest males. For instance, using the *Ascaridia*/red jungle fowl *(Gallus gallus)* system, Zuk, Johnson, et al. (1990) and Zuk, Thornhill, et al. (1990) showed that infected roosters had duller combs and shorter tail feathers than did controls and that the females preferred unparasitized mates.

Milinski and Bakker (1990) carried out an experiment that allowed them to confirm that in the three-spined stickleback *(Gasterosteus aculeatus)* there is a positive correlation between the color of the males and mate choice by the females. Females make the "good" choice in normal light but choose males randomly if the aquarium is lit with green light.

Houde and Torio (1992), whose work on the orange spots of guppies was described above, also tested females' interest in males that became duller because of parasites. These authors used an aquarium split into three compartments: a central one for females and two side ones—one for healthy males and the other for unparasitized males that had been previously parasitized. The central compartment was itself divided into three virtual zones so that researchers could then note if females were near one or the other of the male compartments. Females showed a statistically significant interest in males with intact colored spots, which then led the authors to conclude that colored spots, not displaying, stimulated the differences observed in female behavior, since the number of displays was very similar in both male samples.

Boyce (1990) studied sexual selection in natural populations of the sage grouse, *Centrocercus urophasianus,* in North America. Sexual selection is particularly well characterized in this bird, and males do their courtship by assembling in leks where they parade in groups in front of the females, which then have the opportunity for precise comparisons before making their choice. In this way females impose a severe selective pressure on male candidates, and the variation in reproductive success is tremendous: one male was observed to mate with twenty-four females in ninety minutes. *C. urophasianus* harbors numerous parasites (twenty-seven species have been recorded), among which is the protist *Plasmodium pediocetii. Plasmodium* causes malaria, and parasitized hemocytes burst in the morning, the time when the birds are at the lek. Males infected with *Plasmodium* frequent leks less regularly and thus are among the less good "reproducers," since assiduous attendance at leks is one of the factors motivating female choice. Because we know there are susceptibility genes to malaria, it is very likely that males that have resistance genes in their genomes transmit all their genes more successfully than those that do not.

Male choice by females may be represented by several models. Zuk, Johnson, et al. (1990) speak of "relative preference" when the female selects from among the males presenting to her the one whose color (or any other trait) is the best developed. These authors also discuss a "threshold preference" model where the female accepts any male whose trait (color or otherwise) passes a certain threshold level and conclude that threshold preference should be favored by selection.

In ungulates the female's role is more limited, because antlers serve first in preliminary fights between males. Thus in these organisms parasitism is involved in sexual selection at the level of both male competition and female choice. Nevertheless, the results of the transmission of "good genes" are exactly the same as in the previous cases.

One may wonder why, as a general rule, males are the target of sexual selection and not females. The explanation is that a brighter livery in females most likely pinpoints them for predators as they care for their young, leading to predation on them as well as the young.

The Hamilton and Zuk hypothesis does not exclude the possibility that the choice of "good genes" for the progeny may be based on signals other than bright colors. Female fruit flies, like birds, can recognize parasitized males and avoid mating with them (Jaenike 1988). In this instance the female's choice appears to have little effect on the selection of particular ornaments, since male fruit flies are (at least to the human eye) as dull as females.

Host Taxa That Are Subjected to Strong Parasite Pressure Develop Ornaments or Displays

The basic work of Hamilton and Zuk mentions, as we have seen, a positive correlation between parasite pressure and the occurrence of bright feathers in male birds. However, it is precisely within this domain that confirmation of the hypothesis remains most uncertain (see Møller 1990b; Read 1990; Zuk 1992). As Møller, Dufva, and Erritzoe (1998) noted, "comparative studies of the relationship between the evolution of bright sexual coloration and parasite loads have provided equivocal results that are consistent with either the hypothesis of parasite-mediated sexual selection or the lack of a general role for parasites in sexual selection." The main difficulty is that the richness in parasite species (which may be benign or virulent, abundant or rare) gives only an imperfect approximation of the real pressure of parasites on a particular host species.

As mentioned, some analyses clearly support the hypothesis of Hamilton and Zuk. Read (1988), for example, tried to neutralize taxonomic, ecological, and ethological variables that might bias conclusions by using mathematical methods. In this way he showed (fig. 18.5, bottom right) that the correlation between color and parasites exists relative to the subfamilies of passerines, thus confirming the ideas of Hamilton and Zuk. Similarly, Zuk (1991) found the same correlation in birds from tropical America, basing herself on previous investigations of the hemosporidians of 35,000 birds belonging to 526 species. Using freshwater fish in Great Britain, Ward (1988) also showed that there was a correlation between the intensity of sexual dichromatism and the number of parasite genera described.

Møller, Dufva, and Erritzoe (1998), instead of seeking a direct correlation between coloration and parasitism, substituted the variable "immune response" for the variable "parasite load." Their notable study, which took into account the size of both the spleen and the bursa of Fabricius, provides evidence that sexual selection is related to immunodefense across a wide range of bird species, thus also supporting the Hamilton and Zuk hypothesis: "If male coloration reflects the size of the spleen or bursa, then choosy females will on average raise offspring with better immune defences and hence obtain an indirect benefit from their mate choice."

On the contrary, however, Chandler and Cabana (1991) do not find any evidence of this type of correlation in North American freshwater fish. Injecting caution into the debate, these authors note that the precision of parasite inventories may influence the conclusions drawn from these types of investigations.

Problems with Hamilton and Zuk's Hypothesis

Hamilton and Zuk's hypothesis initiated numerous debates, posturing, and the taking of positions varying from Read (1987) to John (1997), and it is clear that there are several problems with the proposal.

First Problem: How Does One Distinguish Transmission Avoidance from the Quest for Good Genes?

I noted in chapter 8 that in some cases females avoid choosing parasitized males to prevent their own contamination and not necessarily to provide resistance genes to their progeny. This is a simple antiencounter mechanism—the "transmission-avoidance model" of Clayton (1990, 1991b) and the "contagion indicator" of Able (1996). This makes sense, of course, only for parasites with a direct life cycle, and it is sometimes difficult to choose between the two hypotheses in explaining outcomes.[17] For example, in the previously described work of Houde and Torio (1992) on pigmentation in guppies it is not certain that female behavior was in fact the quest for resistance genes. In the earlier described (chapter 8) work of Clayton (1990) on bird "lice" it was clear that the simple avoidance of transmission was the cause, and it could be the same for the guppies—the gyrodactyls are transmitted by contact, and the parasites may even be found on the copulatory organs. Thus the guppy carriers of gyrodactyls (wormy fish) would simply be the equivalent of Clayton's lousy pigeons. This question is in fact discussed both by Borgia and Collis (1990) and by Howard and Minchella (1990). Able (1996) thinks that if parasites with a direct life cycle are pathogenic, then the hypothesis that ornaments serve as contagion indicators would have as much value as the Hamilton and Zuk hypothesis in explaining sexual selection without the two being mutually exclusive.

In other cases mating with a bright male, and thus one that is probably in good health, may have another direct advantage: if the male participates in raising the young, it is advantageous for the female to have a partner able to carry out this difficult task. Møller (1990a), for example, has shown that males of *Hirundo rustica* infected by *Ornithonyssus bursa* bring less food to their progeny than do healthy males.

Just as it is difficult to separate the Hamilton and Zuk hypothesis from the simple acquisition of direct advantages, it is also difficult to separate the selection of males by females from simple competition between males. Howard and Minchella (1990) think that "unparasitized males fare better than parasitized males in various aspects of mate competition" and suggest that the "Hamilton and Zuk hypothesis has limited applicability, and that the major influence of parasites on host sexual selection is due to their effect on mate competition which may in turn influence mate choice."

Second Problem: Can Males Reconcile Brightness and Immunity?

Selection leads to trade-offs in the allocation of resources in organisms. For example, the energy and nutrients allocated to immune defense cannot be allocated to fecundity.

---

17. However, according to the correlated infection model, "females are selected to avoid males with ectoparasitic infection because the presence of ectoparasites serves as an indicator of low overall male resistance to disease" (Borgia and Collis 1990). This supposes that there is then a positive correlation between the levels of endo- and ectoparasitic infection.

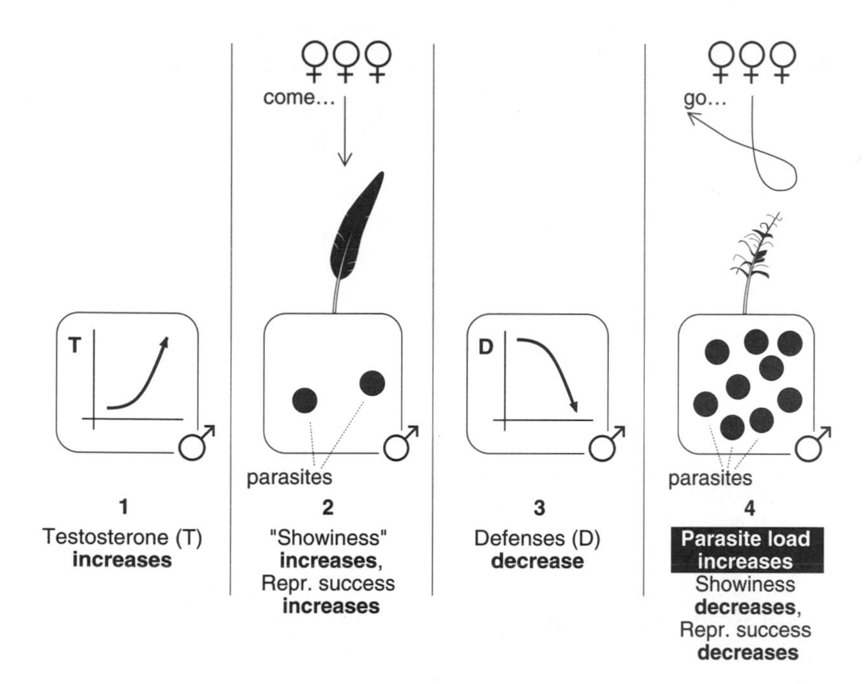

Figure 18.7. The immunocompetence handicap hypothesis (inspired by Folstad and Karter 1992).

Folstad and Karter (1992) focused on the difficult choice organisms face owing to natural selection: the bright ornaments of males develop under the influence of various androgens (testosterone), yet these androgens are known for their immunosuppressive effects, producing an immunocompetence handicap.

Thus, if a given male is particularly bright it may become less resistant to parasites (fig. 18.7). Folstad and Karter (1992) note: "Although a high testosterone titer confers the benefits of exaggerated secondary sexual development and potentially increased mating success, such a condition simultaneously impairs the functioning of the immune system. This double-edged sword creates a real and potential dilemma for males." Similarly, Zuk and McKean (1996) developed the following idea: "Sexual selection has put males in a cruel bind: make yourself vulnerable to disease by developing and maintaining the ornament, or risk lowered mating success by reducing it." They add, "Presumably only males with immune systems robust enough to withstand the compromise associated with high levels of testosterone can maintain elaborate secondary sex characters."

It is here that an explanation to the dilemma probably lies. If a male in-

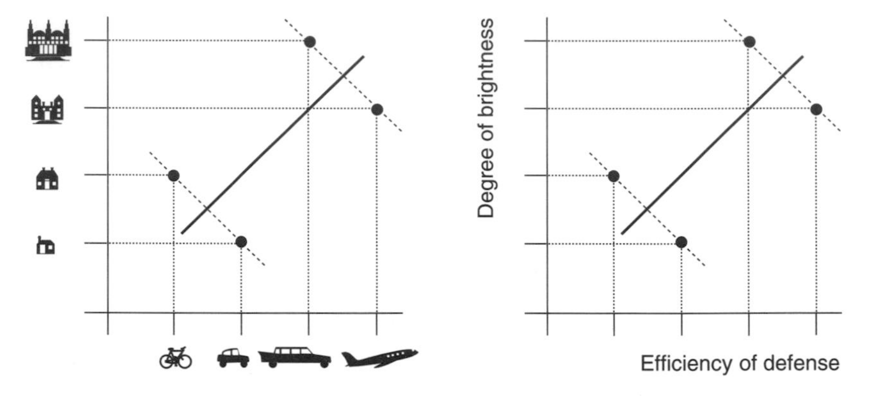

Figure 18.8. The car-house paradox: where negative correlations could be expected (dotted lines), there could in reality be global positive correlations (full line) because the amount of expendable income is highly variable (left), just as is the amount of energy available in hosts (right) (inspired by Zuk et al. 1996).

vests in exaggerated brightness it may increase its immediate reproductive success but compromise its future success because of the immunosuppressive consequences of androgens. Only if it is a carrier of "*very* good genes," then, may it reconcile brightness and immunity.

Here is an image that illustrates this proposition: In humans the person who invests in a nice house might be able to afford only a small car, and vice versa. This is the principle of resource allocation. However, exactly the opposite often occurs (see van Noordwijk and de Jong 1986). This is the "car-house" paradox (fig. 18.8). In fact, this paradox is easily explained if we consider that the basic resources of individuals are very different and that positive correlations are often found between life-history traits (here house and car) whereas theory would predict otherwise.

In parasitology, brightly colored feathers and an efficient immunodefense capability are both costly in terms of resource allocation. But if individuals are truly "superior" (the equivalent of "very rich"), investing in colored ornaments does not prevent their also having a very good defense against parasites. As Zuk noted, "The role of trade-offs in parasitology is an unexplored and potentially exciting area of research" (Zuk et al. 1996). Perhaps an illustration of the car-house paradox can be found in the study of Bronseth and Folstad (1997), who found a positive correlation between the size of the pectoral fins (a dimorphic sexual characteristic) on three-spined sticklebacks, *Gasterosteus aculeatus,* and the intensity of parasitism with various helminths in males. These authors do not believe this finding contradicts the Hamilton and Zuk hypothesis if one admits that "fin size can be indicative of an individual's ability to tolerate the costs of increased parasite exposure and susceptibility."

Third Problem: Are All Possible Variables Taken into Account in Hamilton and Zuk's Hypothesis?

Cox (1989) did not hesitate to qualify the hypothesis as "weak in basic parasitology," noting the difficulty in correctly interpreting investigations about bird blood parasites, which have been the basis of several studies supporting the Hamilton and Zuk hypothesis; the insufficient knowledge of the pathogenic effect of the parasites involved; and how ignorant we most often are about less visible parasites such as bacteria and viruses.

Yezerinac and Weatherhead (1995) evoked the possibility that the parasite "richness" of a particular host species may in fact depend on how attractive it is to arthropod vectors: "Variation among avian species in the prevalence of haematozoa may result from some plumage colours attracting more parasite vectors." Perhaps more serious, however, is the objection of Poulin and Vickery (1993). Based on mathematical models, they think that some of the conditions that allow validation of the Hamilton and Zuk hypothesis occur together only rarely in parasite-host systems. Notably, parasites must be at the same time both common (otherwise females would confuse susceptible but uninfected males with resistant males) and distributed in a slightly aggregated way (otherwise males that are heavily parasitized would be excluded from reproduction even before females could make a choice).

What's the Conclusion?

Despite the problems just noted, the Hamilton and Zuk hypothesis clearly has more supporters than detractors.

The critical synthesis of John (1997) refutes most of the arguments opposing the hypothesis, such as those of Cox (1989). John considers that sexual selection was an important source of innovation throughout evolution and that it is at the crossroads of two coevolutions, that between the parasite and the host and that between the signal emitter (the male) and the signal receiver(s) (the female or predator).[18]

Further, the Hamilton and Zuk hypothesis conforms to one of the two models proposed by Fisher and by Zahavi to explain sexual selection. In

---

18. We are reminded that "parasite-mediated" sexual selection has implications for the evolution of parasite virulence: if parasites affect their hosts' behavior or appearance to the point that the hosts are rejected, then there must be selection for reduced virulence in parasites that are directly contagious, or their transmission will be negatively affected (Møller and Saino 1994).

Fisher's model (1930) females choose one character that was initially advantageous on its own (for example, the length of the feathers), and this choice continues in a kind of runaway selection even though the character in question no longer has a selective advantage.

In Zahavi's model, females choose a character that is disadvantageous as soon as the process starts. However, this disadvantage is compensated for because the character in question indicates the presence of good genes at other loci in the genotype. Clearly the Hamilton and Zuk hypothesis fits the Zahavi model (see Zahavi and Zahavi 1997).

On the question of exactly what "good genes" the females provide to their progeny by choosing the "right fathers," it is clear that not much is known. Von Schantz et al. (1996) have shown that in male pheasants the major histocompatibility complex is associated with variation in both spur length and viability. These data support the "good genes" hypothesis of Hamilton and Zuk, since they show that the same cluster of genes plays a role in both secondary sexual characteristics and resistance. However, according to Siva-Jothy and Skarstein (1998), "In general, the immune system and the relevant good genes are treated as a black box." Clearly much more work is needed on this fascinating subject.

19 Mutualism

It is clear that, by bringing together genetic material from distantly related ancestors, symbiosis provides a source of evolutionary novelty that is additional to mutation and homologous genetic recombination.
Maynard Smith 1989a

From the algae that help power reef-building corals, to the diverse array of pollinators that mediate sexual reproduction in many plant species, to the myriad nutritional symbionts that fix nitrogen and aid digestion, and even down to the mitochondria found in nearly all eucaryotes, mutualisms are ubiquitous, often ecologically dominant, and profoundly influential at all levels of biological organization.
Herre et al. 1999

As opposed to parasitic systems, in a mutualistic system each partner takes advantage of the association between the two protagonists. However, to understand the nature and importance of this type of durable interaction one must first distinguish between internal or intracellular mutualisms, also termed endosymbiosis, and external mutualisms where the partners remain spatially distinct. In parallel with endoparasitism and ectoparasitism, one can speak of ectomutualism and endomutualism in describing the latter pairings. Significantly, internal and external mutualists play similar roles in the biosphere, but endomutualists have also played a major and important role in its evolution.

Numerous books have treated the subject of mutualism (see Cheng 1970; Smith and Douglas 1987; Douglas 1994), and this chapter asks three somewhat different questions: What is the relation between parasitism and mutualism? How do mutualistic associations function in regard to selection? What role has mutualism played in evolution?

What Is the Relation between Parasitism and Mutualism?

The Example of the Phoretic Mites

I noted previously that innovative genes (genes brought to the association by the parasite and whose expression has a positive effect on host fitness) were like "flowers offered to the hosts," and phoretic mites exemplify just how some related species can be either parasitic or mutualistic depending on whether they bring innovative genes to the association.

Most mites are free-living animals that are usually found in detritus or organic waste. Some of them, however, also colonize more discontinuous and temporary habitats such as decaying wood, cow pies, and small pools of water. These acarids, like all mites, are small, are not too mobile, and have difficulty dispersing to colonize new locations that may be separated by distances almost impossible for them to traverse. Alongside these organisms, however, and exploiting similar habitats, are other organisms that can fly or walk and thus can traverse such long distances easily. Obviously, if the acarids could attach to these more motile compatriots and be transported, their difficulty in dispersal would be overcome. The association between mites and their more mobile companions, termed "phoresy," was apparently selected to solve just such a problem (see Athias-Binche 1990, 1991, 1993, 1994). Here are two examples.

Allodinychus flagelliger is a mite that lives in decaying wood where various xylophagous coleopteran larvae also live. During the spring adult beetles emerge out of the wood, mate, and look for new dead wood to lay eggs in. When one looks closely at the legs of these coleopterans, one can see very tiny arthropods—larval mites. These larvae attach to the insects as they leave the wood, remain perfectly still during flight, then drop off as the beetles arrive at their destination.

Naiadacarus arboricola live in small pools of water that accumulate in holes in old trees. Members of the genus *Eristales,* which are aquatic flies, lay their eggs in just such sites. As female flies come to oviposit in the water, the acarids attach to their legs and travel as clandestine stowaways, to drop undetected into newly discovered tree holes filled with water.

In these two examples the association is advantageous for the transported mites (in most cases it is a must for survival) and slightly disadvantageous for the transporter. In effect, the host has no advantage in transporting the acarids; it may at the very least be slightly handicapped by its passengers' weight (as small as they are), and its aerodynamics may be affected. If one defines parasitism (as I do) by the cost induced, then this is a form of parasitism, and the acarid is an uninvited hitchhiker.

Some species of acarids, however, have become more clearly parasitic than those just described. For instance, *Kennethiella trisetosa* is phoretic on the wasp *Ancistrocerus antilope,* itself a parasitoid that deposits its eggs beside paralyzed caterpillars that then serve as a food source. As the wasp oviposits the mites drop off, and before laying their own eggs they feed not only on part of the caterpillars (which is stealing) but also on the hemolymph of the wasp's larvae. Here phoresy leads to true parasitism during the sedentary phase of the mite's life cycle: it is a hitchhiker that not only lets itself be transported but also steals the contents of the glove compartment along the way.

In none of the previous examples is a benefit provided to the host, and only the mites gain some advantage; yet other mites clearly do bring an advantage to their hosts during their association.

Parapygmephorus costaricanus is phoretic on the bee *Agapostemon nasatus*. The female mite detaches from its host as it builds its cells and oviposits beside the bee larva. Later the female mite's daughters attach to the bee as it takes off. These mites are particular because they are present in the cell, which they clean up by eating the bee larva's feces, which in turn prevents fungal or bacterial contamination. In effect they "pay for" the transport by cleaning the transporter's house, and this is referred to as a cleaning symbiosis. In this case the hitchhiker pays for the gas used by washing the car.

Dinogamasus braunsi in Africa is phoretic on the yellow-banded carpenter bee, *Mesotricha caffra*. In autumn and winter the female mites pile up into an acarinarium on the bee's underside. When the host reproduces, the phoretic females leave and also oviposit. The result is a trophic relationship, since the mite family consumes the exudate (substances secreted by the host larvae tegument). However, and as in the previous case, the acarids also clean house by actively eating various waste products that accumulate in the bee's cell.

Dendroleaps neodisetus goes a step further: this species is phoretic on the scolytid coleopteran *Dendroctonus frontalis*, where it devours the free-living stages of a nematode parasite of the beetle and thus reduces its level of parasitism.

In these three associations, Athias-Binche (1990, 1991, 1993, 1994) concludes, the outcome of the association is fair and what the mite takes has "a negligible parasitic effect relative to the positive aspect of the association."

An even more intriguing example of a parasitic mite-host relationship is provided by the mite *Unionicola ypsiliphora*, which parasitizes freshwater mussels. The mite's immature forms leave the bivalves in search of the aquatic pupae of chironomids (for instance, *Paratrichlodius rufiventris*). On encountering a pupa the mite "settles in" and transfers to the adult insect when it first emerges, then engorges itself with its host's hemolymph. Mites thus obtain both a meal and a means of transport to a new aquatic habitat. Interestingly, mites are bright red, and both sexes of the chironomid are infected. However, after mating male chironomids return to their swarms and only the females fly off in search of an aquatic site to oviposit. This means, then, that only those mites attached to ovipositing females will survive, so mites transfer from males to females during copulation. In this case intuition suggests that these ectoparasites, which are relatively large, should impair the aerial performance of parasitized males. Mating success is determined by competition: a male captures a mate only after an aerial duel with other males. However, McLachlan (1999) has demonstrated that exactly the opposite occurs and that infection in fact improves the males' mating success. Although the proportion of infected males in swarms is only about 4%, in mated pairs it increases to about 15%. This counterintuitive result is not totally understood, and McLachlan suggests that either females prefer males exhibiting the red color of the parasites or that the mites manipulate the males' behavior by some still unknown mechanism. Whatever the real explanation, it is clear that from the parasite's perspective fitness is gained when their male hosts mate, since only mites transferred to females survive. From the male host's perspective, carrying parasites increases the probability of transmitting its genes. Thus, although *U. ypsiliphora* is clearly a parasite of females, it is also a very helpful mutualist of males.

The experiments of Jeon (1989) demonstrate that the passage from parasitism to mutualism may result in rapid selective processes if innovative genes are available in the parasite genome. In these studies amoebas of the species *Amoeba proteus* became parasitized by an unknown "bacteria x." After about two hundred generations in culture (a few years), these bacteria became mutualistic and ended up constituting up to 10% of the volume of their host amoebas, which then could not survive without them. In this case it is likely that the cascade of selective processes was as follows: at the beginning of the infection some amoebas carrying resistance genes survived and took over the population; bacteria that carried one or several innovative genes that coded for one or more products (most likely proteins) became useful to the amoebas; and the amoebas lost one or several expensive double-use genes, thus becoming dependent on the bacteria.

Although the latter example provides excellent support for the endosymbiotic origin of eukaryotic organelles, note that although both examples—acarids and amoebas—may indicate otherwise, parasitism will not necessarily always evolve toward mutualism (see chapter 11).

The Example of the Insect Pollinators of Fig Trees

Seven Hundred Fig Trees, Seven Hundred Agaonids

There are about seven hundred species of fig trees *(Ficus)*, for the most part distributed in tropical areas. Significantly, these figs and their associated insect assemblages offer all possible examples of parasitic and mutualistic associations.

The fruits of the fig trees are in fact receptacles (syconia) that were initially flat but over the course of evolution closed themselves upward leaving only a small apical hole, called the ostiole, at the tip. The fig resembles a small cognac glass whose opening would be too narrow to sip through—almost virtual. The inside of the glass is lined by flowers,[1] however. The evolution from a flat receptacle to a fig may have occurred about 100 million years ago during the Cretaceous period.

Although fig trees have developed chemical defenses (notably a form of latex) against bothersome insects, the figs themselves are a resource exploited by numerous organisms ranging from insects that lay their eggs in the flowers to birds and mammals that eat the ripe fruit. The fruit from a single species of fig tree may feed small hymenopterans (chalcidians), coleopterans (weevils), hemipterans (stinkbugs), lepidopterans (butterflies), and dipterans (fruit flies). Such a fig-based community (said to be "sycophilic") commonly contains dozens of insect species (Yves Rasplus, personal communication).

What do these insects do in the figs? Many of these interlopers are parasites that spend their entire larval life in a single fig. For example, the chalcidian hymenopterans (about eight thousand species are estimated to exist) oviposit in the female flowers either by entering the fig through the ostiole or by penetrating the fig wall from the outside. In many cases the chalcidian's larva induces the formation of a gall, and the larva feeds on the resulting proliferation of plant tissues. This proliferation is the insect's extended phenotype into the plant host's phenotype, since it results from substances originating with the parasite's genome. Still other species do not themselves induce galls but instead oviposit in galls already formed by other insects.

Some of the chalcidian hymenopterans, the agaonids, "offer flowers" to their hosts by aiding them in pollination. These small wasps oviposit in the fig trees' ovules but are at the same time indispensable auxiliaries to their reproduction.[2]

These wasp/fig tree systems are remarkably specific, and except for a very few there is only one wasp species per fig tree species (some seven hundred or so species of wasps for some seven hundred or so fig tree species forming some seven hundred or so noninterchangeable mutu-

---

1. Some may say that a good cognac lines the glass with flowers.

2. These "seed-for-seed mutualisms" are known in three associations: the fig/fig wasps, the yucca/yucca moths, and the *Trollius/Chiastocheta* system (see Pellmyr 1989). In each case, the cost for the plant consists of lost seeds (eaten by the insects), and the benefit is expressed in the number of seeds fertilized (because of pollination by these same insects).

alisms). In each case the association is indispensable to both partners: each species of blastophagous wasp can oviposit only in "its" fig tree, and each fig tree is fertilized only in the presence of "its" agaonid.[3]

Female Wasps and the Male Function

Fig trees are either monoecious or dioecious. Monoecious fig tree species have figs in which male and female flowers are found side by side. In dioecious species half the trees are bisexual, with figs that have both female and male flowers, and half are females whose figs have only female flowers.

Parasitism by the wasps modifies the function of some of the female flowers such that these flowers effectively only aid in the dissemination of pollen. These parasitized female flowers fail to produce seed but instead produce wasps that then can carry pollen. As such, part of the female flowers of monoecious species and all the female flowers of the bisexual trees of the dioecious species are diverted from their initial function when parasitized.

Thus there are two distinct situations, which can be summarized as follows (fig. 19.1): In monoecious fig trees one finds within a single fig male flowers that produce pollen, some female flowers that produce wasps that carry pollen, thus making them secondarily part of the "male apparatus" of the plant, and other female flowers that produce seeds. That is, the same tree produces both the pollen and the wasps that carry the pollen to its own ovules. In dioecious fig trees one finds in a fig from a bisexual tree male flowers that produce pollen and female flowers that, without exception, produce wasps that carry pollen and then aid in the "male function."

At the same time, however, the all-female trees from the dioecious species have figs with female flowers that produce seeds. In short, in the dioecious fig species half the trees produce pollen and the wasp carriers of that pollen, and the other half produce seeds.

Let's detail the pollination of *Ficus microcarpa,* a monoecious species that reproduces naturally throughout Asia and that was introduced and is pollinated in North Africa, America, Tahiti, Hawaii, and Bermuda. In this case male and female wasps of the species *Eupristina verticillata* are born in the flowers that line the inside of the fig (fig. 19.2).

The male wasps in this system have two important functions: the first one, as expected, is to fertilize the females, and usually males appear slightly before females so they can do this before the females leave the parasitized ovules. The second function of the males (or at least of some of them) is to make a hole in the fig wall so that the fertilized females can exit and fly to another fig (after which the males in the fig die).

---

3. There are, as noted, a few exceptions to this otherwise ideal scheme.

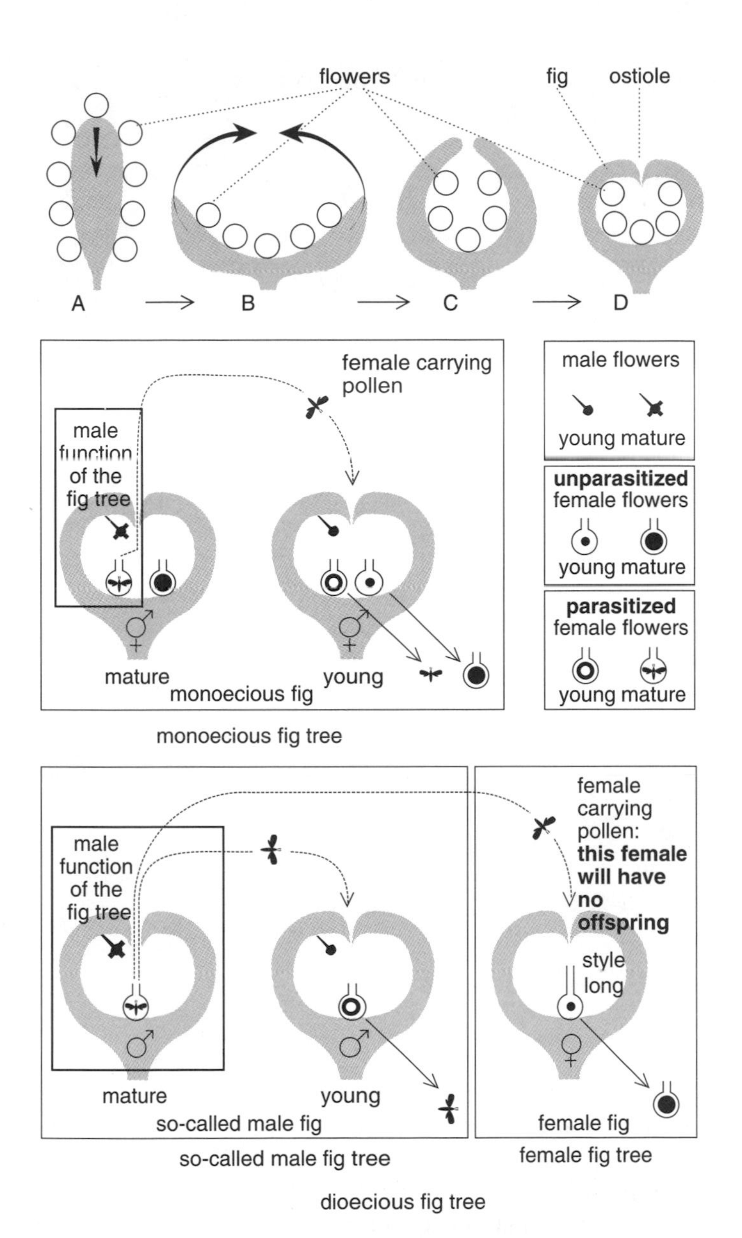

Figure 19.1. Mutualism between fig trees and their pollinators. In a monoecious fig tree (middle), the fig produces both seeds and pollinator insects, which are part of the male function of the tree. In a dioecious fig tree (bottom) half the trees are females that are pollinated—in these female trees the insects are not permitted to lay eggs. Since each insect can visit only one fig, those who enter a female fig on these trees die on site without transmitting their genes.

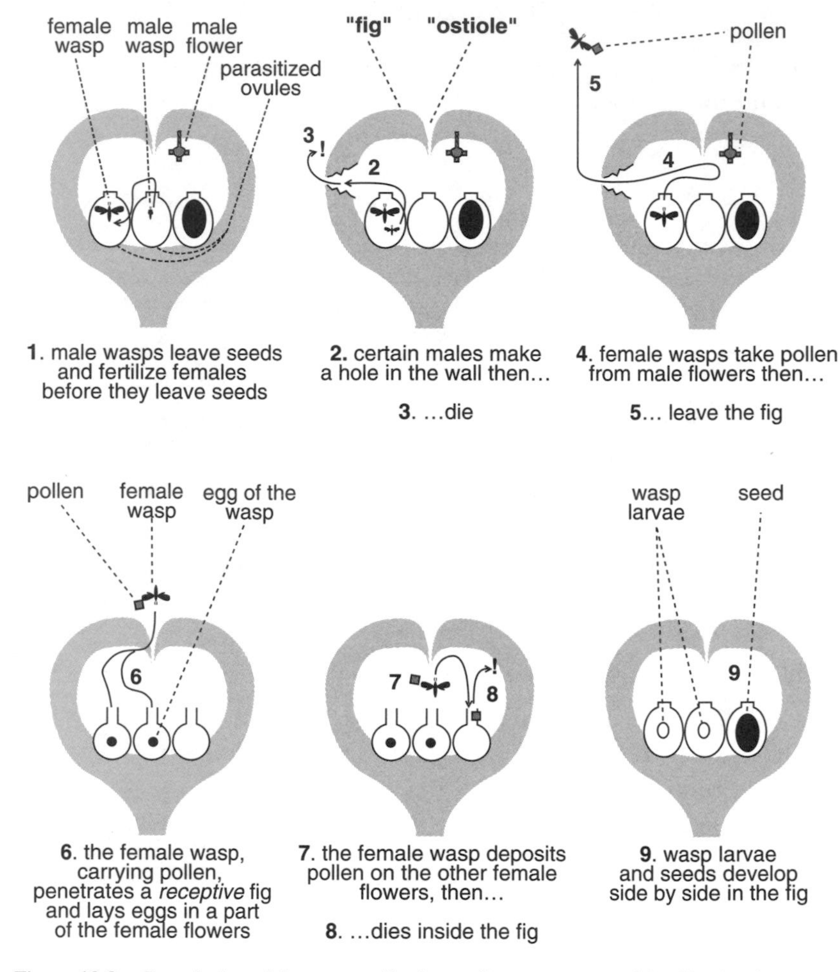

Figure 19.2. Description of the events allowing pollen transport and fertilization in a monoecious fig tree.

The females do not, however, rush to the opening made by the males. Instead, they first visit male flowers in the same fig, where they fill small pockets on the thorax with pollen. Only after this do the females fly out the holes made by the males (not out the ostiole, since it is closed in ripe figs). During their short life females seek out receptive young figs, enter through the ostiole, oviposit in some of the female flowers, and in doing so deposit pollen on the stigmata of others (after which they die). Each flower that receives pollen but no wasp egg then produces a *Ficus* seed, and each flower that receives an egg produces a wasp, either male or female.

In this system there cannot be synchronous development between the plant and the insect—if there were, emerging female wasps would find no flowers to fertilize several weeks after the initial oviposition. The system persists because of the "nonsynchrony" of fig tree blossoming, and it cannot last in areas where the climate precludes this spreading out of the bloom. It also cannot exist below a certain fig tree density, since the wasps, whose life span is only a few days, would have too few chances to find a tree with figs in the right state of bloom (with female flowers receptive to both pollen and insect eggs; Kjellberg et al. 1987; Kjellberg and Maurice 1989).

Fragile Equilibriums

Agaonids are thus both fig seed parasites and indispensable auxiliaries in the production of fig seeds. Such associations are an excellent model to study the conflicts that exist in mutualistic systems, notably to understand why, despite these conflicts, the systems persist (Anstett, Hossaert-MacKey, and Kjellberg 1997; Patel et al. 1995). These associations raise several questions.

Why don't wasps oviposit in all the female flowers in a monoecious fig tree?

Any flower that produces a wasp does not produce a seed, and any flower that produces a seed does not produce a wasp. Thus there is potential competition for the flowers.

It was thought for a long time that there were two types of flowers, one with a short style allowing the wasps to oviposit and others with a long style preventing them from doing so. In fact, the explanation is simpler: female flowers are stacked up on the fig wall in several layers, and acccss to the deeper layers is more difficult for the insect. Moreover, other nonmutualistic wasp species also oviposit in the same flowers through the fig wall and thus have easier access to the deeper layers of flowers. By ovipositing only in the superficial layers, the mutualistic wasps thus avoid interspecific competition, emerge more easily in the fig cavity, and are themselves more easily fertilized.

Why don't wasps oviposit in the female flowers of female trees in dioecious fig trees?

Here the explanation is easier: there has been selection in the female flowers of the female trees for long styles that in fact do prevent the ovipositors of the female wasps from reaching the ovary, thus effectively stopping the wasps from ovipositing there. That is, even if the female wasps that have penetrated the female flowers of the female trees try to oviposit, the structure of the flower prevents them from doing so.

Why do wasps visit the female figs whose style length prevents oviposition in dioecious fig trees?

This question is truly an evolutionary enigma. Let's consider the different situations that the wasps are confronted with depending on whether the tree is monoecious or dioecious.

In an association with a monoecious fig tree, the female wasps visit the figs and are selected for this behavior because here they find female flowers in which they can oviposit. In an association with a dioecious tree, part of the female wasps visit the figs of "male" trees. As I noted previously, this is logical, since they then seek out and find female flowers to oviposit in (let's not forget, however, that these trees are males because the female flowers they produce are diverted from the female function by wasp larvae). Also, another part of the wasps visit the figs of the female trees, which then ensures the pollination of the female flowers and thus the reproduction of the fig tree. This seems illogical, however, since the length of the styles does not allow them to oviposit in these flowers, and wasps that act this way die without leaving any descendants. Logically, female wasps in this case should be selected to visit the flowers of the male trees while avoiding the flowers of the female trees. The first of these two traits was selected. The second one was not. What are the possible explanations?

In this case everything happens as if the fig trees determine which of two strategies to prevent the wasps from avoiding female figs is selected.

The first strategy consists of desynchronizing the receptivity of male and female flowers so that when the wasps emerge only female figs are available and the wasps die without leaving offspring whether or not they visit female figs. In this way entering a female fig cannot be counter-selected, because this behavior is neutral in regard to selection.

The second strategy consists of giving exactly the same characteristics (odor and color) to both male and female figs so that no selection allowing wasps to discriminate between them may be possible.

Why, in some associations, does pollination imply an active behavior for the wasps?

In insect-plant mutualisms pollination is often "a by-product of insect visit for a reward" (Anstett, Hossaert-MacKey, and Kjellberg 1997). Insects become "involuntarily" covered by pollen by going through or past a flower's stamen and then ensure pollination by contacting the pistil of another flower. Such a procedure is termed passive pollination. In some associations, however (for instance, in monoecious or dioecious fig trees), active pollination is observed. In this case insects actively gather pollen and put it in specialized pockets. Finn Kjellberg thinks that such a characteristic and the associated behavior could have been established and maintained only because they were giving an immediate advantage to the insect (personal communication). He suggests that fertilization of the female flowers by the pollen allowed better development of the wasp's larvae, so that the insects would then have an "interest" in transporting pollen, which then explains the selection of the necessary structures and behaviors.

Who parasitizes whom?

"Mutualism can also be viewed as reciprocal slavery; over their prolonged common evolutionary history, each species has been selected to exploit its obligate partner without being able to avoid being exploited" (Anstett, Hossaert-MacKey, and Kjellberg 1997). With this in mind, fig tree mutualism may leave us perplexed about reciprocal benefits; that is, Who parasitizes whom? Of course blastophagous insects are fundamentally parasites because they inhabit and feed on the figs during their larval life. They are secondarily mutualists because they bring genes for mobility to the association, and it is the fig tree that is phoretic on the insect with its pollen grains. As transporters, insects have the tremendous advantage that the "next stop" is strictly written into the biology of their specialism. Thus, whereas wind would normally disperse pollen every which way, phoresy on a *specialist* insect guarantees that pollen will arrive on the female flower of a fig tree of the same species. If the insects were to visit just any species of fig, part of the pollen would be lost. It is therefore possible that the only agaonid/fig tree systems that have subsisted are those on which the insect was narrowly specialized. That is, everything happens as if the tree is willingly "paying" for the transport of its pollen by losing some of its flowers and as if the "male function" of the tree is to produce not only pollen but also the vehicles for the pollen—the insect pollinators themselves.[4]

Such blastophagous pollinators illustrate the passage from a situation in which the parasite does not bring anything positive (and thus is really nothing but a parasite) to a situation where it makes itself useful and ultimately even indispensable (a mutualist). This is, then, one of the mechanisms that

4. In order for their genes to survive, the plant's male gametes (pollen) must encounter female gametes as imperatively as parasites must encounter their hosts. Thus they need "encounter genes." In the agaonid/fig tree association these "encounter genes" for the male gametes are, in essence, the insect's genome.

has allowed parasites to make themselves accepted—by bringing a selective advantage to host individuals that do not defend themselves.[5]

As in the case of mites and amoebas, it is necessary to add that not all insect parasites of fig trees evolve toward mutualism. For this evolution to occur, the parasites must undoubtedly have some preadaptations, that is, some innovative genes capable of ensuring that a plant would function better with them than without them (with only its own genes).

How Do Mutualistic Associations Function in Regard to Selection? Nongradual Gradualism

During evolution new genes are usually acquired by "internal growth" (here I use a vocabulary that is more familiar to economists than to biologists), that is, by mutations, duplications, and recombinations, all of which produce innovations in place.

Mutualism, on the other hand, represents an acquisition of genes by "external growth," which has the characteristic of being nongradual. Whereas acquiring one or several genes by internal growth usually is a slow evolutionary process, any external growth is a more sudden evolutionary process similar to a step along the staircase of evolution.

As soon as an association between two genomes becomes indispensable, the two partners function as a sole entity and are confronted by selection as a whole even if the two remain physically distinct. When the association of the two genomes brings innovative genes, these genes are then the equivalent of a major mutation not acquired by the usual small steps of selection. Thus acquiring these genes greatly accelerates the move toward complexity. That is, mutualism entails a positive added phenotypic value (APA) that improves the host's fitness (fig. 19.3).

Significantly, it is probably true in numerous such instances that genes can and do pass from one genome to the other in these associations. When any type of parasite (protist, helminth, etc.) comes to inhabit a host, it brings its own genetic information. The difference (the only difference!) with parasitic DNA is that the DNA of the parasitic protist or the parasitic helminth does not insert itself directly into the host's DNA. When the association is of an obligatory mutualistic type, the resemblance to parasitic DNA is even more striking: the superorganism formed by the host and its symbiont is now coded for by a supergenome. As such, how important is it for the genes of this supergenome to be grouped into two ensembles that are only topologically distinct (fig. 19.4)?

5. The situation in which a consumer brings an advantage to the organism it eats exists not only in parasite-host systems but in predator-prey systems as well: some predators, by consuming nectar or pollen, participate in plant reproduction and dissemination.

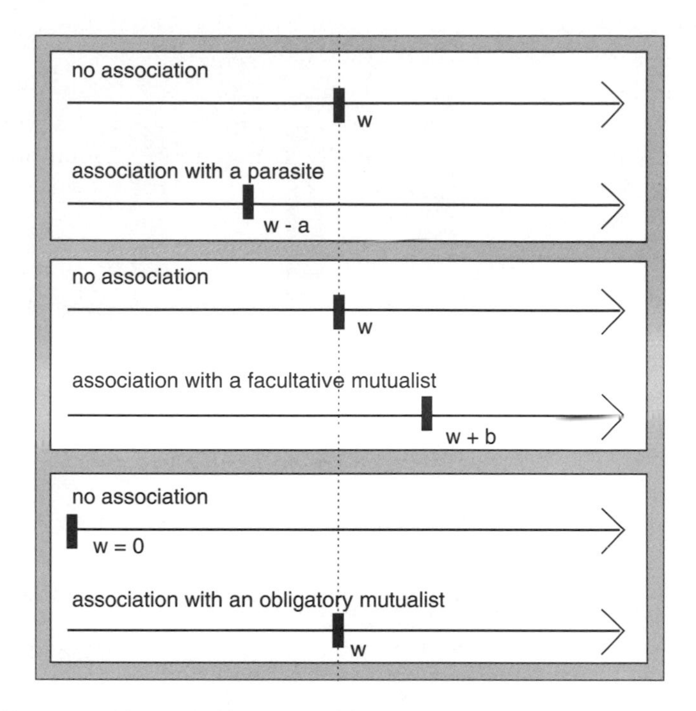

Figure 19.3. Parasitism and obligatory and facultative mutualism: w = host fitness; a, b = changes in fitness.

Let's take an example. In order to produce light, an organism is supposed to have five genes: two for luciferin and three for an aldehyde-reductase (luciferase) that will react with luciferase to induce the emission of photons. Bioluminescence of this type is common in deep-sea animals, since sunlight is sparse at one thousand feet and absent farther down. Some of these deep-dwelling animals, particularly fish, can produce light themselves because they have the genes coding for the substrates and enzymes of the luciferin/luciferase complex. Others, however, emit light only because of a mutualistic association with bacteria of the genus *Photobacterium* or *Vibrio.*[6]

Maynard Smith (1989b) emphasizes that it is highly improbable, if not

6. Several hypotheses may be proposed to explain why light is sometimes produced by the organisms themselves and sometimes as the result of durable interactions. One may think either that fish with bioluminescent bacteria have lost their own genes for bioluminescence after acquiring bacteria that had genes for a similar or better function (the elimination of double-use genes) or that acquiring bioluminescent bacteria allowed the fish to save on the selection of genes for bioluminescence (the acquisition of innovative genes).

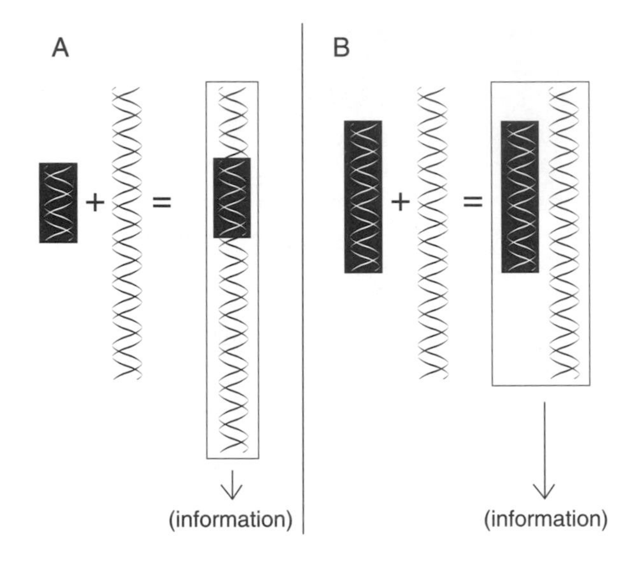

Figure 19.4. Whether a DNA molecule gets inserted into another DNA molecule (A) or whether two distinct DNA molecules are associated without insertion (B), these molecules carry information that constitutes one entity in regard to selection. In the case of B the association often ends up with sequence exchanges occurring later on.

impossible, for a fish to acquire the five genes necessary to produce light all "at once" through a single event occurring in its genome. However, if the fish becomes associated with a bacterium that already has the five genes, then they can be acquired in a single event. This does not, however, alter the gradual acquisition of the five genes in the bacteria, in which multiple stages of mutation and selection must have occurred previously.

The acquisition of adaptations via mutualism thus overlays a saltatory process on a gradualist process—mutualism is a lateral transfer of previously acquired advantages, and once the mutualistic association becomes indispensable the system functions as "a whole." In essence, the host of a mutualist does not make a distinction between the genes of the mutualist and its own. In the case of the fish and the bioluminescent bacteria, the fish genome evolves to make the best of the genes it has acquired by such external growth. For instance, the leiognathids (ponyfish) that live in the Pacific Ocean, Indian Ocean, and Mediterranean Sea all have bioluminescent bacteria in their esophagus and build a true light fixture around this light source: the muscles lining the esophagus have become transparent, the swim bladder has been transformed into a reflector, and chromatophores (pigmented cells that can spread out or retract themselves) are used as interrupters (MacFall-Ngai 1989).

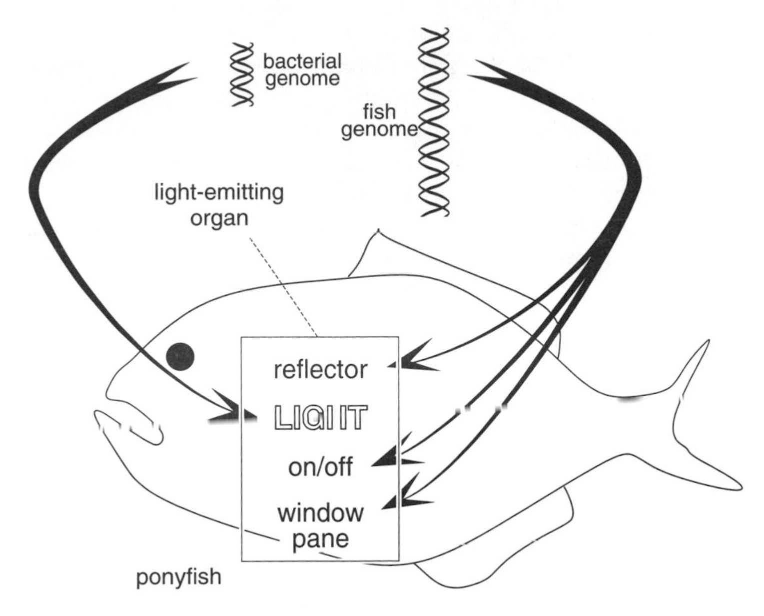

Figure 19.5. Ponyfish have a light producing system with light production being the result of the expression of mutualistic bacterial genes and an optical system derived from the expression of the fish's genes.

This example demonstrates the functional incorporation of the mutualist's genome into the host's genome: although the bacterial DNA is not physically incorporated into the fish genome, the fish acquire the bacteria by ingestion each generation (19.5).

When a host benefits from innovative genes brought by parasites, it saves on what is ensured by its mutualist. For example, many benthic marine organisms are associated with autotrophic microorganisms that obtain their own energy from the oxidation of sulfur. This the case for the bivalve mollusks *Calyptogena* and *Bathymodiolus,* as well as for the vestimentiforan *Riftia pachyptila,* which lives in hydrothermal vent ecosystems (Fiala-Medioni and Métivier 1985; Fiala-Medioni 1988; Fiala-Medioni and Le Pennec 1989). In these animals the organs normally associated with food capture and digestion have regressed, and they have saved on the expression of genes that would normally be used to build the digestive tract and ensure food capture. The resources saved in this way are redistributed to other functions.[7]

7. The host bivalves have lost their digestive tracts just as the tapeworms of vertebrates have and thus behave as parasites of their bacteria!

Sometimes very little is needed for such associations to fall on either the parasitic or the mutualistic side of the continuum.

A unique example of such an ambiguous situation is provided by the *Drosophila* C virus (DCV), a "pathogen" of the fruit fly *Drosophila melanogaster.* This virus belongs to the same group as poliovirus and causes some mortality in the fly, which becomes contaminated while ingesting virion-containing food (Thomas-Orillard 1984). However, contaminated adult female fruit flies grow faster, have larger ovaries, and reproduce more than do "healthy" fruit flies. That is, their reproductive success is increased, which is a characteristic of a mutualistic association. The explanation for the advantage provided by the viruses likely resides in the production of proteins that may be profitably used by the fruit flies. Thomas-Orillard states the following hypothesis: "The host would consider viral proteins as its own proteins and might use them for reproduction; this would limit DCV multiplication and consequently the pathological effect of the parasite." That is, if larval mortality wins, then the cost for the host is higher than the benefit. But if the increase in reproductive success in adults does better than compensate for this mortality, the benefit wins. By itself this example shows very well all the subtleties of the relationship that characterizes a durable interaction.

What Role Has Mutualism Played in Evolution?

The Eukaryotic Cell

Mutualism was undoubtedly associated with a marked and important event in evolution: the "assembling" or "building" of the eukaryotic cell.

It is important to remember that having a nucleus is not the only characteristic that differentiates the eukaryotic cell from the prokaryotic cell. Two other differences are particularly important. One is size (a eukaryotic cell is on average a thousand times bigger than a prokaryotic cell), and the other is that there are mitochondria in the cytoplasm of a typical animal cell and both mitochondria and chloroplasts in a typical plant cell. Both of these—mitochondria and chloroplasts—have their own DNA, and the hypothesis that these two organelles were originally distinct living organisms has not stopped gaining ground in the latter portion of the twentieth century.

In the 1970s and 1980s Margulis (see Margulis 1993) championed the exogenous origin of these organelles and patiently accumulated more and more information in its support. The idea has clearly been bolstered by molecular biology, which shows that most of the characteristics of these organelles' DNA make them closer to bacterial than to nuclear DNA. Price (1989) has concluded: "The evolution of eucaryotes, the acquisition of organelles, and the evolution of biotic complexity have depended on the initial invasion of parasitic organisms, an invasion which became mutualistic secondarily."

Both mitochondria and chloroplasts play a fundamental role in eukaryotic cellular metabolism.[8] From an evolutionary point of view, there are three fundamentally different scenarios outlining what might have occurred had mutualism not been attained with the bacterial precursors of what are today mitochondria and chloroplasts.

Hypothesis 1: The evolution of complex organisms would have been the same, because the genes whose function is today ensured by bacteria existed in the eukaryotic cell and were lost only because they had a double use after association with the bacteria.

Hypothesis 2: The evolution of complex organisms would have been delayed because the genes whose function is today ensured by the bacteria would have ended up appearing in the eukaryotic cell's own genome over a longer period.

Hypothesis 3: Evolution would not have produced the more complex life forms to which we belong.

Most biologists favor hypothesis 3, and in any case the history of life would have been deeply altered.

Clearly, the vast majority of data indicates that the cell from which the eukaryotic branch of the evolutionary tree arose came from the encounter and marriage of several living organisms that were originally independent of one another. This parasitic origin of the eukaryotes is summarized in figure 19.6, which was inspired by Katz (1998).

It is thought today that the flagella of eukaryotic cells also have a mutualistic origin (among bacteria from the spirochete group), as do both the centrioles and microtubules (see Maynard Smith 1989b; Katz 1998). However, an explanation for the origin of the nucleus remains difficult, and its evolution has yet to be satisfactorily explained.

Theoretically, the plant cell has an advantage over the animal cell in that it is the only one of the two to possess mutualistic photosynthetic

8. The mutualistic association that gave rise to the eukaryotic cell was certainly not the only one that played a major role in evolution. Mobile elements themselves must have been important, not only by increasing genetic diversity by the mutations they induced but also by effecting the selection of repressive mechanisms. Some "silencing mechanisms" repressing the expression of certain regions of the genome may have been necessary at some key moments of evolution, and it has been suggested that these mechanisms "arose as adaptive responses to the selfish drive of transposable elements to expand in number within host genomes" (McDonald 1998).

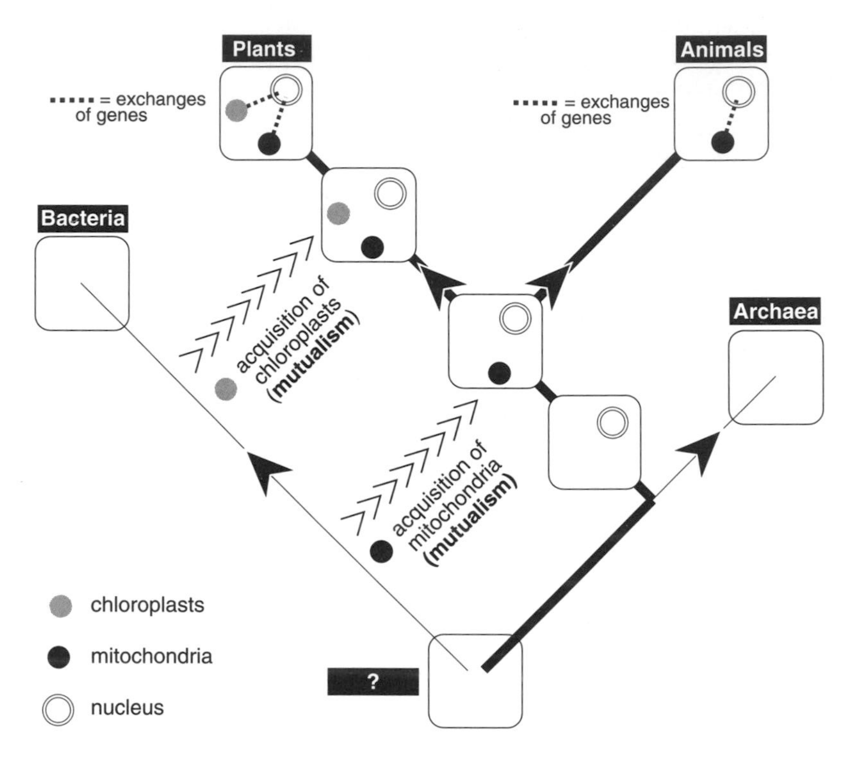

Figure 19.6. Model for the origin of eukaryotes showing how important endomutualists are.

bacteria, the chloroplasts. This advantage derives from the fact that chloroplasts make the cell autotrophic, that is, capable of using sunlight to power organic synthesis.[9]

Figure 19.6 diagrams the existence of gene transfers between the associated organisms. The genetic code being, except for a few details, universal, a DNA fragment that "arrives" within another even taxonomically distant genome can be read by the enzymes whose function is to replicate their own DNA before each cell division. Thus nothing prevents genes from circulating between one genome and another if the interaction

9. The existence of chloroplasts is not just a characteristic of "green" plants in the traditional sense. Protozoans from the group Apicomplexa, such as *Plasmodium, Toxoplasma,* and *Eimeria,* also contain DNA of the chloroplast type that codes for ribosomal RNA (Köhler et al. 1997). During each division this DNA divides to be shared among the daughter cells, supporting the idea that it must have a function.

is sufficiently durable, and today some genes present in the nuclear genome of eukaryotic cells clearly result from transfers from mitochondrial or chloroplast genomes.[10]

Sometimes, genes that belong to a host and others that belong to a mutualist collaborate in the making of the same product. This is the case of a molecule present in the *Rhizobium*-containing nodules of legumes—leghemoglobin, an oxygen transporter that plays a key role in nitrogen fixation: part of the molecule is coded for by the plant and part by *Rhizobium.* Similarly, the mitochondria of today's organisms are composed of functional gene products from both the host's nuclear DNA and the DNA of the mitochondrion itself.

In order to obtain energy resources, heterotrophic organisms that are not associated with chloroplasts must consume organic matter made by others. They do this by becoming either predators, parasites, or decomposers. It is therefore intriguing that heterotrophs have sometimes acquired partial or even total autotrophy through the secondary acquisition of autotrophic mutualists. For instance, some animals, principally protozoans, cnidarians, free-living platyhelminths, and mollusks have secondarily associated themselves with photosynthetic algae and in this way benefit from the bacterial chloroplast within the algal cell. Interestingly, in some green unicellular algae it appears that what was once thought to be a chloroplast is in reality a eukaryote made of a cell associated with a chloroplast (see McFadden and Gilson 1995). Still other heterotrophs consume unicellular algae and digest their essential components (cytoplasm, nucleus, etc.) while conserving the intact and functional chloroplasts for their own use. This is the case of the ciliated oligotrich protozoan *Laboea strobila,* which lives in the surface waters of the Atlantic Ocean and has discriminating digestive vacuoles that allow it to adopt chloroplasts instead of digesting them (Stoecher, Michaels, and Davis 1987). This is also the case of the marine mollusk *Elysia viridis,* as well as other related species that sequester recovered chloroplasts for months and then expose themselves to light to activate photosynthesis.

Figure 19.7 depicts the various ways one can become autotrophic, either by a "primary" association with the chloroplast-bacterium, by a "secondary" association with an already photosynthetic organism, or by the recouping of prey chloroplasts.

---

10. Contrary to mitochondria and chloroplasts, the flagella, whose origin is sought for in a mutualistic relationship with spirochetes, do not have their own DNA. Margulis (1993) supposes that the entire "useful" genome of the spirochete was transferred into the nuclear genome during evolution and that the rest was lost.

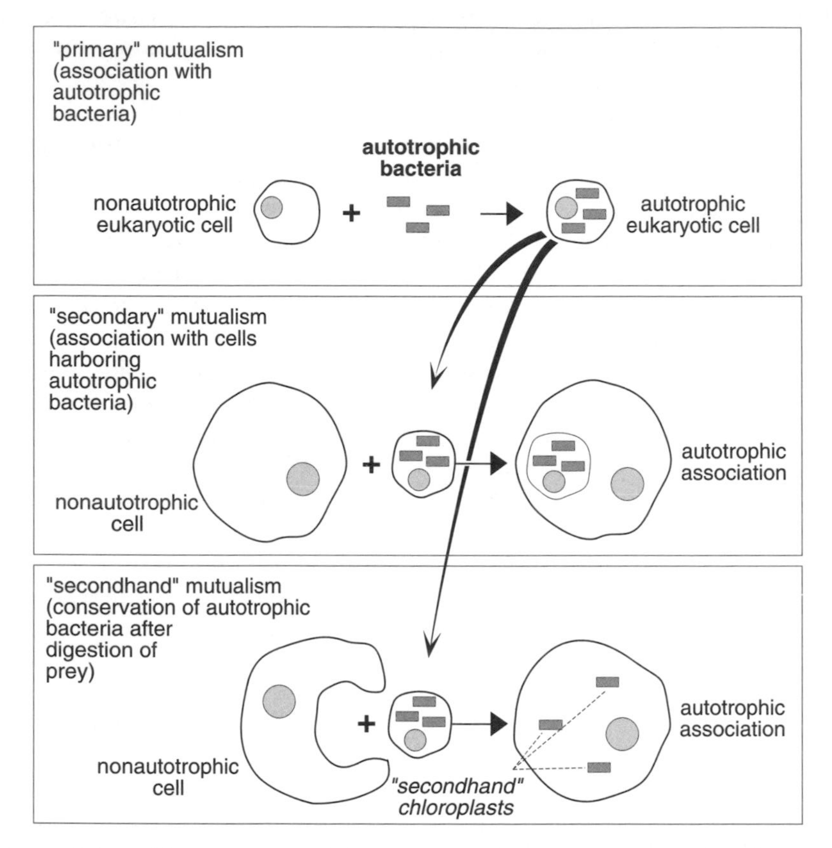

Figure 19.7. Different processes of the acquisition of autotrophy via mutualism. Top: mutualism between the primitive eukaryotic cell and an autotrophic prokaryotic cell. Center: mutualism between a heterotrophic cell and an "already" autotrophic cell. Bottom: acquisition of choloroplasts after consumption of an autotrophic cell.

Prokaryote-Eukaryote Associations

Associations between prokaryotic microorganisms and eukaryote macroorganisms hold an essential place in the exchange of energy in the living world. In such associations the host of the prokaryote is no longer just a cell but is in fact a complex organism.

For example, larvae of the weevil *Sitophilus oryzae* house mutualistic bacteria in specialized cells called bacteriocytes that together form a bacteriome. The bacteria in these cells are transmitted vertically by the female gametes, and at the time of host metamorphosis the bacteriocytes disintegrate, leaving the bacteria free to migrate to cells close to the insect's digestive tract. These bacteria receive nutrients from the host and in turn participate in several important processes: they provide vitamins (panthotenic acid, biotin, and riboflavin) and degrade excess methionine brought in with ingested wheat. Weevils experimentally cleansed of their bacteria (aposymbiotic weevils) have slower growth and decreased fertility than controls, and it is likely that in nature weevils would not be able to survive if they were to lose their bacteria. In this case each genome penetrates the other's functions: insect genes limit bacterial multiplication (Nardon and Grenier 1988), and compared with related free-living bacteria of the genus *Erwinia, Sitophilus*-associated bacteria have lost about one-third of their DNA—a loss most likely corresponding to the disappearance of genes that became useless after the association with the insect. "The symbiotic bacteria of *S. oryzae* seem to have lost a number of genes which prevents their survival outside the bacteriome. The symbiotic bacteria were completely domesticated by the weevil and behave as permanent organelles" (Charles et al. 1997). Genetic analysis of the mutualistic bacteria of *Sitophilus* suggests that this association may be very ancient—about 50 to 100 million years old (Heddi et al. 1998). Such bacteria/insect associations are numerous, and many are even more ancient than the *Sitophilus*/bacteria association discussed here (from 100 to 200 million years in certain cases).[11]

Other associations that are metabolic in character can also easily be cited. These include, for instance, the numerous flowering plants (probably 70% to 80% of the known species) that are associated with the mycelium of soil-dwelling fungi (mycorrhizae). In these cases, fungal mycelia wind tightly around the lateral roots and root hairs of the host plant and insinuate themselves between the cells, sometimes even branching into the cells themselves. The fungus receives carbohydrates from the plant host, and in exchange the mycorrhizal network penetrates the soil better and farther than the plant roots can, bringing a greater supply of minerals back to the plant. It is possible to experimentally grow plants in the absence of their mycorrhizal associates (a facultative associate), but the plants produced are small. The advantage provided by mycorrhizae may thus be critical when soils are poor, and in some instances a particular environment may be conquered only by species associated with the right fungus. This is the case, for instance, for heathers (Ericascidae), which are able to grow in acidic soils as a result of their association.

Curiously, the nature of a mutualism-parasite relationship may change depending on the partner—here exemplified by the association between

---

11. These associations may also involve insect parasites: for example, phthiropterans that feed on bird feathers digest this less than appetizing material thanks to symbiotic bacteria.

orchids and their mycorrhizal fungi. Orchid seeds are able to germinate and grow only if they are infected with a fungus that penetrates the cells. In this type of mycorrhizae it is not the plant that feeds the fungus but the opposite. This can occur to such an extent that some orchids never develop photosynthetic abilities and are then totally dependent on the fungus, even as adults. In this case it does not look as if the orchid gives anything indispensable to the fungus, and it is the orchid that has all the characteristics of being a parasite of the fungus. However, when the fungus *Armillairia mellea* (normally a destroyer of herbaceous plants or bushes) becomes associated with the developing young of certain orchids, it feeds the plant and does not damage it at all. Thus in the "armillairid/nonorchid plant" association the fungus exploits the plant genome, whereas in an "armillairid/orchid" association the orchid exploits the fungus. That is, *Armillairia* is sometimes a parasite, sometimes a host.

Nitrogen is one of the four essential components of living organisms, and no eukaryote has the nitrogenase genes that would allow it to fix nitrogen (i.e., to transform gaseous nitrogen, or N_2, into ammonia, or NH_3, which can then be incorporated into proteins). This lack explains the occurrence of plant associations with either bacteria *(Rhizobium),* actinomycetes *(Frankia),* or blue-green algae (the exceptional case of the aquatic fern *Azolla* associated with the alga *Anaboena*), all of which are capable of fixing gaseous nitrogen into ammonia.

The association of *Rhizobium* with various legumes of economic interest (beans, peas, soybeans, peanuts, etc.) occurs in root nodules whose formation is under the control of bacterial genes (the expression of these genes leads to the synthesis of lipo-oligosaccharides called "nod factors"). Associations of this type illustrate well the phenomenon of crossed phenotypes, since there are exchanges not only of products but also of activation signals (for example, the genes for *Rhizobium* that induce the formation of nodules are themselves activated by substances issued from the host plant).

According to Price (1989), the digestion of cellulose by bacteria, protozoans, and fungi involves some two thousand species of termites, ten thousand coleopterans, and two hundred species of mammalian herbivores (elk, antelopes, camels, etc.). As such, the trophic relationships between the world of the herbivore and that of the plant are essentially due to durable interactions of the mutualistic type. The posterior region of the digestive tract of termites harbors, depending on the case, either bacteria or various species of flagellated protists that digest the cellulose in ingested wood, and the termites infect each other by exchanging their intestinal contents. Similarly, marine boring clams make wrecked wooden boats disappear in a few dozen years—also by digesting wood because of bacteria. Ruminants, which are heavy consumers of cellulose, harbor both ciliated protists and bacteria in their rumens. In fact, butter is made from particular fatty acids that are produced by the bacteria and then passed into protozoans as they eat the bacteria in the cow's stomach. These fatty acids become part of the milk, from which the cream is skimmed and churned into butter.

Bacteria do more than digest intestinal contents; they also provide vitamins that their eukaryote host cannot synthesize. All vertebrates have an intestinal flora, and not having to synthesize vitamins of the B group saves the expense of producing the hundred or so enzymes involved in fabricating these compounds.[12]

Parasites themselves can also use mutualistic bacteria, such as for metabolizing their host's tissues. This is the case for *Steinernema*, nematodes that parasitize insects (see chapter 10).

In fact, numerous parasites may use mutualistic bacteria to improve their fitness and thus to invade and exploit their hosts.

A good example of this is provided by the intracellular bacteria found in filarial nematodes, including the human parasites *Wuchereria bancrofti, Brugia malayi,* and *Onchocerca volvulus* (see Taylor and Hoerauf 1999). These bacteria, which are genetically related to the *Wolbachia* found in insects and isopods, are present in the reproductive tissues of the female worms and are transmitted vertically through the cytoplasm of the eggs. Further, they appear to have no pathological effects on their host nematode's organs. Hoerauf et al. (1999) have shown that eliminating them by antibiotic treatment (tetracycline) inhibited development and caused infertility in the rodent filarial worm *Litomosoides sigmodontis* (treatment with antibiotics not affecting the intracellular bacteria had no effect on the nematodes, and filarial species that lacked mutualistic bacteria were insensitive to the treatment). Although it cannot be totally excluded that the bacteria killed by tetracycline might have had a pathogenic effect, the most likely explanation is that the bacteria provide the worms with innovative genes coding for some useful (and possibly necessary) molecules. A phylogenetic analysis suggests that, although related to the *Wolbachia* of arthropods, the bacteria of the filarial nematodes diverged from them at least 50 million years ago. Thus this particular mutualistic association is also very ancient.

12. Numerous mutualistic relationships are most likely ignored, and numerous organisms considered as strict parasites may also bring useful molecules to their hosts (see Lincicome 1971). To illustrate these lesser known situations, we may cite the example of trypanosomes, which are traditionally classified among the pure parasites but might also provide their mammalian hosts with some vitamin B-6 (Munger and Holmes 1988).

Conflicts between Mutualists

Viewed from afar, a mutualistic association presents two characteristics of the perfect couple: faithfulness and getting along well (especially if it is an obligatory match). Is this really what happens?

First let's test *faithfulness*. A good way to test the faithfulness of mutualistic associations is to compare phylogenies. In chapter 5 we saw that by comparing the phylogeny of a parasite with the phylogeny of a defined group of hosts we can test the congruence of the phyletic trees. If the congruence is excellent, it means that parasites and hosts have evolved and cospeciated together. If the phylogenies do not look alike, it means that lateral transfers have occurred.

Logically and if, as we expect, mutualistic systems are more "peaceful" than parasitic systems, then the congruence of the trees must be better.

The study of Hinkle et al. (1994) supports this hypothesis. This work was carried out on ants of the group *Atta* and the fungi they culture for their subsistence. This is a durable interaction at the scale of the superindividual formed within the ants' nest, which live constantly in a very tight relationship with the fungus. In fact there is even vertical transmission of the mutualist in the system—queens take with them a little fraction of the fungus's mycelium when they leave to form new colonies. This is an obligatory type of mutualism, and the ants are never encountered without the fungus and vice versa. The ants "feed" the fungus with freshly cut leaves and then feed on it. Hinkle et al. (1994) studied the comparative phylogeny of five species of "fungalist" ants and that of the five species of basidiomycete fungi associated with them using cladistic analysis of both morphological and sequence data derived from the gene coding for the small ribosomal subunit. The number five may seem small, but the result was astonishing. It seems that, for a period estimated to be between 25 and 40 million years, the evolutionary relationship between the ants and their fungi is *entirely congruent*. There was no unfaithfulness during all that time, and the authors discuss a "mutualism of considerable antiquity." In this example the mutualistic system functions as a true evolutionary unit; if diversification occurs, it does so in both partners at once. However, molecular data suggest that not all such ant/fungus associations are so ancient (for a brief synthesis, see Herre et al. 1999). This is not surprising, since the fungus-gardening ants constitute over two hundred species, all dependent on cultivating fungi.

Now let's test them for *getting along*. Even within the same genome, genes are in competition. Such intragenomic conflicts (see Hurst, Hurst, and Johnstone 1992) are, like intergenomic competitions, arbitrated by selection simply because a gene that multiplies exaggeratedly within a genome without adaptive advantage for the phenotype quickly compromises the individual harboring that genome. In other words, the advantage acquired at the level of the DNA molecule by an intensive and selfish multiplication of one gene may be counterselected by the disadvantage resulting at the level of the phenotype.

Similarly, conflicts exist between the two genomes in a mutualist association even when it is obligatory. This happens for at least two reasons: the first is that the host genome must make sure the mutualist genome is not too zealous in its multiplication. With endomutualists, for example, host genes regulate the mutualist population, as in the coleopteran *Sitophilus*. Similarly with ectomutualism, plants must select mechanisms that moderate pollinator insects with too invasive an attitude. "Is it always better to attract more mutualists? Apparently not" (Bronstein 1994a).

The second reason conflicts exist in mutualistic associations is that the "reward" each partner gives the other for the "services" rendered has a cost and that service should never be overpaid: "Mutualisms are clearly not altruistic" (Bronstein 1994b). It is certain also that the equilibrium between the service offered and the price paid for it may fluctuate in space and time depending on environmental conditions.

Such conflicts between mutualists and hosts perpetuate themselves even in the most ancient associations, as shown by the interference of mitochondrial mutants with the sexual reproduction of certain plants. In quite a number of flowering plants there is a curious phenomenon—the coexistence in natural populations of hermaphroditic individuals and female individuals (or more exactly, those whose male function has been abrogated). Male sterility in at least some of these cases is interpreted as resulting from cytoplasmic genes, that is, from the influence of certain mitochondria that enhance their own transmission (Couvet et al. 1990). This conflict between mitochondrial and nuclear DNA is easy to understand: since mitochondrial DNA is transmitted solely by ovules (female gametes), mitochondria are better served by favoring their own production to the detriment of host pollen grains (male gametes). In this case nuclear DNA then must select genes that restore the male function, and in that way it opposes the process engaged in by the mitochondria. Thus, despite the great age of the system, there are conflicts of interest and confrontation between both the plant and the mitochondrial genomes.[13]

That such conflicts between mutualists do in fact exist simply implies

---

13. As emphasized by Atlan and Gouyon (1994), the often inappropriately labeled "harmony" observed in nature is not due to the harmony between species, between individuals of the same species, between nuclear and mitochondrial DNA of the same cell, or even between sequences of the same genome.

that the classic interpretation of mutualism may be inexact. Although very little is known about the transfer of metabolites between two associated organisms (even in models as well studied as lichens), we may wonder with Smith (1992) if in numerous cases the host might exploit the mutualist much more intensively than suspected, just as I proposed in the introduction to this book.

Last, I should add that parasitism and mutualism may simply be durable interactions where the advantages of "habitat" and "energy" are reversed. In true parasitism the intruder benefits from the host habitat and exploits the energy of the host source, whereas in mutualism the intruder benefits from the host habitat but is in turn exploited and becomes a source of benefit or energy for the host.

In light of the importance of understanding the mechanisms that govern durable interactions of this type, it is reassuring that in the past decade general ecologists have demonstrated increasing interest in mutualisms (see Bronstein 1994a).

20 Parasitism and Humanity

A mobile band of primates would be exposed to numerous new infections.
Cockburn 1971

There is a great danger that in the course of thousands and hundreds of thousands of years, occasionally there is a particular genetic or ecological constellation which permits the transfer to a new host.
Mayr 1957

A Different Kind of Host

Humans are a "different kind of host." By this I mean that hominids constitute the group of free-living organisms who, above all others, have most influenced the evolutionary history of parasitism by offering exceptional opportunities for the transfer and spread of parasitic diseases around the globe. The reasons for this difference are simple: the ecological and ethological diversification of our ancestors over time has opened their encounter filter to a variety of parasites that have never before had the ability to make contact with hominids.

When the infective stages of any nonhuman parasite encounter the human body they may either be destroyed by the immune system or simply be unable to exploit the offered environment. Alternatively, some may survive long enough to induce a pathogenic effect without ever completing their life cycle, while yet others may be successful in their attempts at infection.

The latter possibility obviously broadens a parasite's host spectrum and may even be followed eventually by speciation. Significantly, the equivalent of these human-related conquests had never occurred to such an extent in the history of biological evolution simply because they took place over the very short period during which mankind spread around the world.

There were several causes for the opening of the hominids' encounter filter to parasites. Humans abandoned the arboreal mode of life of their ancestors, and in doing so they invaded all or almost all of the available terrestrial environments on the planet. Second, in concert with the acquisition of a terrestrial mode of life they diversified their behavior in terms of hygiene, food acquisition, habitation, and social structure. Last, they established large sedentary populations (villages and cities), and they domesticated animals that harbored their own suites of parasites. In doing so, hominids doubtless greatly modified both their own behavior and the ecosystems they occupied over the short period of their evolution, thus greatly increasing their exposure to parasites they had never before encountered.

The passage from an arboreal existence to life on the ground was obviously a decisive factor in the acquisition of new parasitic diseases. Living at ground level increased contact with ground-dwelling hematophagous arthropods, brought hominids into contact with the contaminated feces of both arboreal and nonarboreal vertebrates, and led them to visit water more often. Clearly, humans owe the privilege of having intercepted parasites from all of the major food webs on the planet to these factors as well as to their having become exceptionally eclectic omnivores after their descent from the trees—the most spectacular example of this today being anisakiasis. Nematodes of the genus *Anisakis* and their relatives (*Heterakis,* etc.) are mainly whale and seal parasites whose infective larvae live in marine fish; when these infected fish are eaten raw by an omnivorous human instead of a marine carnivore, the larvae quickly become highly pathogenic in the intestine (fig. 20.1).

It is relatively straightforward to see how becoming sedentary, the emergence of agriculture, animal domestication, and the creation of cities all constituted new phases in the relationship between humans and parasitism. As a general rule, large numbers of individuals living close together almost always favors the transmission of pathogenic agents, and some agricultural practices, such as using feces for fertilizer, even make the transmission of some newly acquired parasites easier. Significantly, and perhaps most important, the new mode of life presented by the combination of these factors created new niches for a variety of organisms that

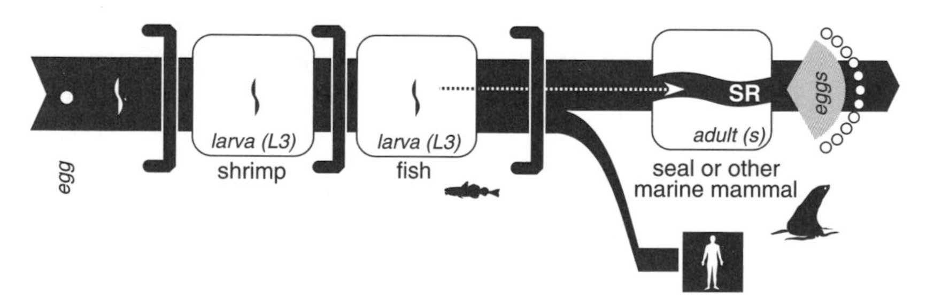

Figure 20.1. Life cycle of *Anisakis* and related genera (nematodes): the “normal” (regular?) definitive hosts are marine mammals; humans become infected by eating raw fish. A = adult; SR = sexual reproduction.

carried parasites. For instance, under such conditions it becomes easier for vectors such as mosquitoes to select for anthropophilic strains preferentially adapted to humans, and other vectors find ideal living conditions in the habitations used by humans, such as the thatched houses favored by the hemipteran vectors of Chagas' disease in South America.

Domestication has had the effect of constraining the animal carriers of some parasites to live close to humans where they can more easily spread their infective stages. Still other animals closely associated with domestication, such as rodents, found the more closely packed human habitations perfectly suited for their proliferation. Southwood (1987) made an interesting comparison between the presumed long history of animal domestication and the number of transfers of parasites between humans and these domesticates. The curve obtained shows a positive correlation between the two variables (fig. 20.2). The longer ago domestication oc-

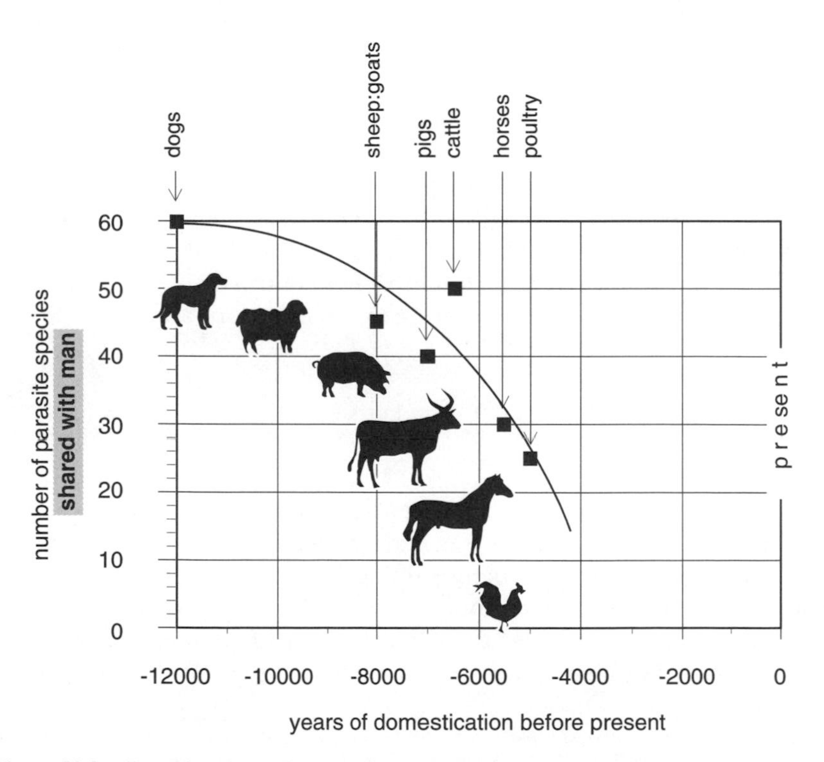

Figure 20.2. Possible relation between how ancient domestication is and the number of lateral transfers to humans. Note that the relation does not change if only mammals are taken into consideration (modified after Southwood 1987).

curred, the higher the number of parasites shared with humans. Although this analysis does not take into account other important variables, such as the taxonomic proximity of the hosts, Southwood's work nevertheless shows that transfers do not occur immediately upon domestication but only after some time spent in close association.

Without doubt the changes humans have wrought to the parasites' natural environment often favor their transmission. Dams and irrigation networks offer ideal conditions for the explosive growth of mollusks (grassy shores), mosquito larvae (puddles of water), and sometimes even blackfly larvae (the production of stream sections with strong currents at the base of impoundments). In short, these and other transformations, such as the fragmentation of habitats or deforestation, may greatly favor the transfer of some parasites by concentrating vectors, life-cycle stages, or reservoir hosts. However, note also that in some circumstances the evolution of human behavior and activities may close the encounter filter: preventing contamination, draining wetlands, cooking food, and migrating from tropical environments to temperate climates all make parasite transmission less easy in some cases. Further, humans are also the first species with specialist subpopulations, such as medical doctors and parasitologists, whose job it is to fight parasites. Such undertakings, however, exert huge selective pressures that themselves can lead to the emergence of new genetic structures and pathologies in the offending parasites. In this regard it is truly worrisome that at the start of this twenty-first century we humans, who have been able to eliminate or control all of our predators, still pay a heavy and often devastating toll to infections and parasitic diseases.

Where Do Human Parasites Come From?

The Parasites of Adam and Eve

The authors of the Renaissance had an apparently unsolvable problem (see de Witt 1994): if paradise was really paradise, Adam and Eve should not have had any parasites before biting the apple; and if animals had been created before humans, then human parasites did not have hosts before Adam and Eve. To solve the problem it is sufficient to look into lateral transfers (and eventually see in these transfers a punishment inflicted on the first couple).

Numerous human parasitoses were acquired during the evolution of our ancestors following the changes in ecology and ethology mentioned above—that is, as the result of transfers (Cockburn 1971). In the face of any such disease, the question in determining the source of infection is simple (McKeown 1988; Combes 1991e): Is this disease coming from our primate ancestors, or was it was derived by lateral transfer from a nonprimate?

Some human parasites almost certainly were inherited from our primate ancestors: the classic examples include *Plasmodium* (except *P. falciparum*), pinworms *(Enterobius)*, filarial worms of the genus *Loa*, body lice *(Phthirus)*, head lice *(Pediculus)*, various intestinal protozoans, and numerous viruses. In fact oxiurids (pinworms) and primates provide one of the best examples of parallel radiation between parasites and hosts, with an overwhelming congruence of their phylogenies and only a few minor discordances (Hugot 1999).

Among the human parasites almost certainly resulting from lateral transfers we may cite *Plasmodium falciparum*, the schistosomes, *Onchocerca volvulus*, *Leishmania*, the trypanosomes, and fleas.

One indication that human parasites are only in part inherited from our direct ancestors is given by the construction of a primate phylogeny based on the list of all known parasites in orangutans, gorillas, chimpanzees, gibbons, and Old World monkeys. The tree obtained is not congruent with that of the primates. Humans are lumped in the vicinity of the Old World monkeys, while relationships among the other groups remain logical (Glen and Brooks 1986)—not an acceptable alternative based on all other evolutionary data.

The Hominization of Parasitoses

The passage of a disease to humans may lead to two distinct situations: either "sharing" the parasite with the host of origin or speciation. Here I apply to human populations what I said about parasite transfers in chapter 5.

Physicians qualify as "zoonotic" a disease that continues to affect the animal host of origin after it has been acquired by humans, in which case the animal hosts are then called reservoirs.[1] An example of such a zoonosis is sleeping sickness in eastern Africa, which is caused by *Trypanosoma brucei rhodesiense:* the disease affects numerous wild ungulates, and humans are infected through the bite of an infected tsetse fly when they enter the environment where the parasite's life cycle is maintained (for instance, when they go hunting).

When combating diseases in which reservoir hosts are involved, it has been very tempting to simply get rid of the reservoirs, such as the ungulates that might harbor *T. b. rhodesiense*. For instance, with the objective of

1. The World Health Organization defined zoonoses in 1959 as "those diseases that are naturally transmitted between vertebrate animals and man."

eradicating sleeping sickness in Rhodesia, 41,000 wild animals were killed in 1955 alone (Garnham 1958). Of course, such a strategy ignores the impact of removal on potentially delicate ecosystems and in fact should have put the dog first for extirpation, since it is a "reservoir of innumerable zoonoses" (Garnham 1958).

If isolation occurs (see chapter 5) the ancestral parasite may divide into two species, one of which continues to infect the animal while the other becomes specific to humans. This often happens with lateral transfers, and such a mode of speciation requires, besides a long enough time, a mechanism of isolation (gene flow must be greatly reduced between the population infecting the animal and the population infecting humans). Additionally, this process also requires that human groups have become large enough to maintain the survival of a parasite species they can call their own.

Since only a limited time has passed since the appearance of the first hominids (australopithicines lived a little more than 3 million years ago), not many successful speciations have occurred as the result of transfer over the course of hominid history. As such, it is expected that the recently formed species found in humans may not be perfectly separated from their sister species in animal reservoirs and that in some cases speciation may still "be occurring." For example, the filarial nematodes *Loa* and *Onchocerca* appear to illustrate two stages in such an ongoing process.

Loa loa has a life cycle comparable to that of *Wuchereria bancrofti* (chapter 1). The insect vectors in this case, however, are horseflies in the genus *Chrysops,* and there is good epidemiological evidence for two life cycles that are in large part distinct (Duke 1955; Noireau and Gouteux 1989). One cycle involves Old World monkeys *(Papio, Cercocebus, Cercopithecus)* and "zoophilic" *Chrysops (C. langi and C. centurionis),* and the other involves humans and anthropophilic *Chrysops (C. silicea* and *C dimidiata).* As evidenced by a much higher microfilaremia in these hosts, the parasite seems to like it better in the monkeys than in humans. Remarkably, microfilaremia peaks are nocturnal in monkeys and diurnal in humans, and the existence of different vectors along with the chronobiology of microfilaremia (periodicity) leads to the conclusion that isolation is ongoing but not necessarily total between the *Loa* of monkeys and that of humans. Thus human loaiasis is originally a disease of primates whose causative agent is still in the process of speciation.

Onchocerciasis is another disease caused by a filarial nematode *(Onchocerca volvulus).* In this case the microfilariae may induce lesions in the cornea, the choroid, the iris, and the retina of humans (river blindness).

The blackfly *Simulium damnosum* is the main vector in Africa,[2] whereas in America *O. volvulus* is adapted to local blackflies, the main ones being *S. guianense, S. metallicum,* and *S. ochraceum.*

Onchocerca volvulus is very closely related to several species in the same genus, such as *O. lienalis, O. gibsoni,* and *O. ochengi,* which are all parasites of African ungulates. A large body of morphoanatomical data indicates that the human *Onchocerca* is most likely the result of a lateral transfer followed by speciation. That there are apparently no other *Onchocerca* specific for primates, other than humans, supports this hypothesis. Further, ribosomal DNA analyses shows that the human parasite *O. volvulus* and the ungulate parasite *O. ochengi,* have exactly the same sequence of four hundred base pairs in the 5S spacer region, normally a highly variable noncoding region usually used to separate closely related species. Xie, Bain, and Williams (1994) conclude: "It is possible that *O. volvulus* evolved from the captured *O. ochengi* of the herbivore reservoir too recently for *O. volvulus* to show genetic distinctions from *O. ochengi.*" Despite this very large congruence with the animal parasite species, all observations so far indicate that *O. volvulus* is specific to humans even though both the gorilla and chimpanzee have been found to be locally infected (but only rarely). Such specificity has, in fact, been a large obstacle to research because no viable animal models are available as surrogates for study in the laboratory. The posttransfer evolution of *Onchocerca* has thus gone further than that of *Loa.*

Note that the lateral transfer of infectious or parasitic diseases to humans is not restricted to prehistoric times, and one of the hypotheses to explain the occurrence of the virus causing AIDS is the very recent lateral transfer of a related primate retrovirus. The SIVs (simian immunodeficiency viruses) of primates do not usually cause serious problems in these hosts, and evidence suggests that two distinct transfers to humans may have occurred, one giving rise to HIV1 (the more pathogenic) and the other to HIV2. The primate viruses that are closest to human HIV1 are

---

2. In the savanna ocular lesions are both serious and frequent, while in the forest lesions leading to blindness are rare even though parasite load seems to be comparable. However, other factors might be responsible for these differences. One possible explanation resides in a "polymorphism of virulence" of *Onchocerca volvulus,* if not in the existence of distinct species, since some DNA sequences seem specific for the "forest forms" (Erttmann et al. 1987). A huge control program (OCP = Onchocerciasis Control Program) has been led in West Africa by the World Health Organization, mainly based on the destruction of the aquatic larvae of the blackflies. The exceptionally positive results make this eradication program one of the most successful in the fight against human endemic diseases (see Hougard et al. 1997).

specific for the chimpanzee subspecies *Pan troglodytes troglodytes* (Gao et al. 1999), and it is this host that is most likely at the origin of the AIDS virus in humans.

On Death Row

Evolutionarily, the recent acquisition of an infectious or parasitic disease may be positively correlated with how serious it is (see the example of rabbit myxomatosis in chapter 11), and numerous parasitoses are more serious in humans than the same or related diseases in animals. The explanation for this seriousness is known: if both the encounter and compatibility filters are open wide when the transfer occurs, the initial virulence might be greater than the optimal virulence. If conditions are then such that selection may occur, the parasite becomes less pathogenic (because of the selection of less pathogenic individuals) and the parasite load comes under control (because of the selection of more resistant hosts). As Crompton (1991) notes, "Conflict is a debatable and less secure concept, no doubt influenced by the daily, wide scale occurrence of parasitic diseases in humans and their domestic animals."

While one can argue that a decrease in virulence over time has not always followed a transfer, it is no less true that there is a troubling coincidence between the seriousness of human parasitoses and their relatively recent acquisition.

There is one situation, however, in which a parasitic disease not only does not seem to evolve toward a "minimal pathology" but *cannot* evolve in a such way. We know, for instance, that parasites do not bring gifts to their hosts when a high degree of pathology favors their transmission. That is, when transmission requires that one organism be eaten by another, selection favors a pathogenic effect in the "eaten" that makes it easier to catch.

Because humans are omnivore/carnivores at the top of the food web, we have a tendency to take the place of the "eater" when we substitute ourselves into other parasite-host systems. Except for rare exceptions, those parasitoses where infection occurs through consumption and where humans become definitive hosts are not extremely grave. This is the case, for example, with the cestode *Hymenolepis diminuta,* whose normal definitive host is a rodent and that infects humans when they accidentally eat the mealworm (coleopteran) intermediate host (fig. 20.3A).

There are, however, instances where humans take the place of the "eaten," such as with several cestodes other than *H. diminuta* (fig. 20.3B). Here the gravity of the situation is quite different. For example, *Echinococcus multilocularis* is a cestode with the following natural life cycle: adults

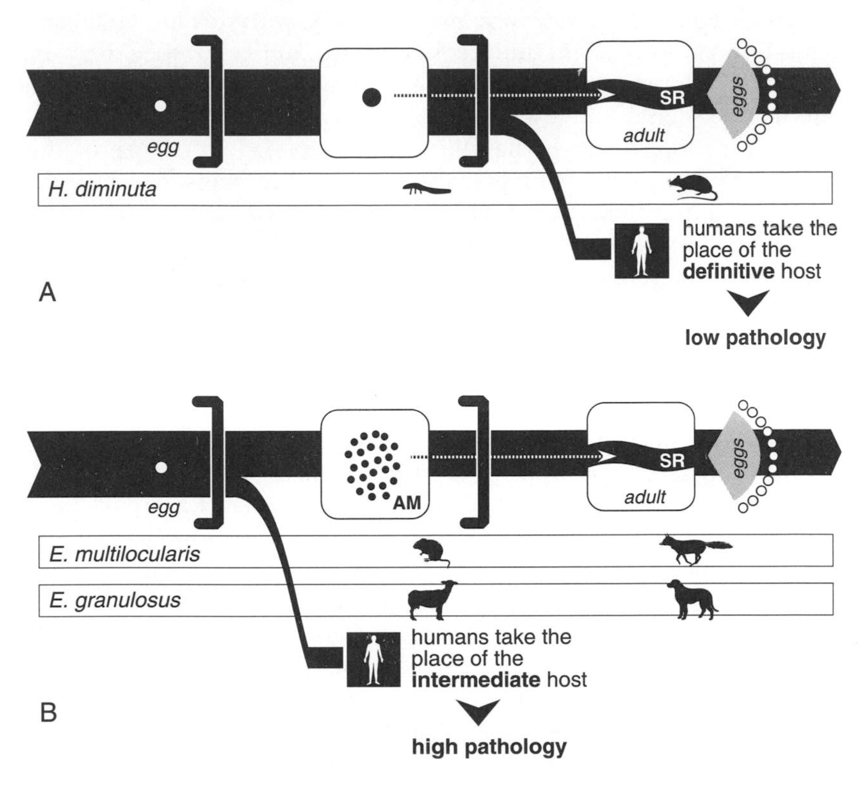

Figure 20.3. Diagram showing the very different place humans occupy in the life cycle of the cestodes *Echinococcus multilocularis* and *E. granulosus* compared with their place in the life cycle of a cestode such as *Hymenolepis diminuta:* the diseases caused by *Echinococcus multilocularis* and *E. granulosus* are extremely serious because humans take the place of the intermediate host (the herbivore or the rodent) and not that of the definitive host (the carnivore). AM = asexual multiplication; SR = sexual reproduction.

live in the fox intestine, where they can reach impressive densities of several thousand individuals owing to their very small size (only two to three proglotids). The gravid proglotids, packed with eggs, are released outside with the feces of the fox, and if a vole *(Microtus)* or a related rodent swallows even one egg, a larva (the cysticercus) can develop in the liver. In numerous cestodes this is not a major problem because the cysticercus is not a very active stage, essentially only prefiguring the adult with its sole scolex. Here, however, the cysticercus buds and grows, producing numerous scolices, and this enlargement results in the invasion of other organs in addition to the liver. This invasion has an obvious pathogenic effect, and it is highly likely that it makes infected voles easier for the fox to capture. Humans usually have nothing to do with this life cycle, and even if humans were to eat undercooked voles (unlikely), specificity opposes the development of the adult in the human intestine. However, humans may accidentally ingest eggs of *E. multilocularis* (for example, opening the encounter filter by handling dead foxes). In the case of this parasite, the compatibility filter is also open, and humans then take the place of the intermediate host. This occurs, of course, with no future for the parasite, but it causes a very serious pathology in the human because of cyst growth and development. Unfortunately for those infected with this parasite, there is little hope for cure aside from surgical intervention.

Echinococcus granulosus, whose definitive hosts are dogs, jackals, or other carnivores, has a similar fate in humans, who in this case take the place of the normal sheep intermediate host when they eat released eggs. The budding cysticercus develops mostly in the liver in the form of an "hydatid cyst," and as in the previous example, it can form secondary cysts, causing serious pathology. This disease is frequent in animal-raising regions where dogs are fed sheep viscera, which then ensures continued transmission of the parasite. The life cycles of *Multiceps multiceps* and *Taenia coenurus* are also similar to that of *E. granulosus,* with the localization of the budding cysticerci (in the brain) the only difference.

We thus see that when a new host is involved in a life cycle it is better for this host to take the place of the "eater" than that of the "eaten," since the latter is like being placed on death row without much chance of a pardon.

Two Important Parasitic Endemics

Human infectious and parasitic diseases are extremely numerous. Among the "important endemics" are parasitoses caused by a wide range of protozoans and helminths that, when taken together, constitute the major health problems facing the world today.

Protozoans induce numerous diseases in humans, in a variety of forms. The most important of these afflictions (according to the number of infected individuals) are malaria, trypanosomiasis (sleeping sickness in Africa and Chagas' disease in South America), amebiasis, and leishmaniasis. Helminths in turn are responsible for schistosomiasis (also called bilharziasis), hepatic trematodiasis (fascioliasis, clonorchiasis), pulmonary trematodiasis (paragonomiasis), filariasis (onchocerciasis, wuchereriasis, loaiasis, dracunculosis), hydatidosis, and intestinal nematodiasis (anguillulosis, ancylostomiasis, ascariasis, etc.). The description of these parasitoses can be found in numerous books (Olsen 1974; Cheng 1986; Melhorn 1988; Roberts and Janovy 2000), and the total number of humans afflicted is staggering to the point of disbelief (over 5 billion).

To help readers appreciate the magnitude of the problem presented by human parasites, here I summarize some of the important characteristics of two human endemics, focusing on their evolutionary origins—one caused by microparasites and the other by macroparasites.

Microparasitosis: The Malarias

The human malarias (with probably 800 million people infected today) are caused by protozoans in the genus *Plasmodium.* This genus belongs to the Apicomplexa, which are all parasites of vertebrates and do not have free-living forms. The genus includes species parasitic of amphibians, reptiles, birds, and mammals (rodents and primates) and four species that parasitize humans. Two of the human forms are common *(Plasmodium falciparum* and *P. vivax),* two are rarer *(P. malariae* and *P. ovale),* and the seriousness of the diseases each causes is very different. The malady caused by *P. falciparum* is often lethal, whereas those caused by the three other species are almost always benign but not without danger. *Plasmodium falciparum* is found spread throughout the tropical zones; *P. vivax* is common only in North Africa, Southeast Asia, Madagascar, and certain regions of South America. *Plasmodium malariae* is present in several countries of tropical Africa, South America, and India, whereas *P. ovale* occurs only in the western half of Africa and in the Middle East.

The historical impact of this disease has been immense, and its effects have been far-reaching throughout both historical and recent times. Malaria was previously present in Italy, France, central Europe, and even England, where its transmission was seasonal up to the nineteenth century. This malaria was almost certainly caused by *P. vivax,* which is known to have existed as far north as Moscow, most likely as the result to its greater cold tolerance (development in the mosquito is possible down to 16°C).

The life cycle of *Plasmodium* is heteroxenous and involves an obligatory passage through both a vertebrate (the intermediate host) and a mosquito (the definitive host). The possible but rare existence of transplacental infections in humans (contamination of the fetus by the mother during pregnancy) or by transfusion (contamination of the transfused by the donor's blood) notwithstanding, there is an absolute necessity for this alternation between the mosquito and vertebrate in order to maintain the disease at an endemic level. Even though the cycle is complex, it can be easily divided into four parts: three in the human and one in the mosquito (fig. 20.4).

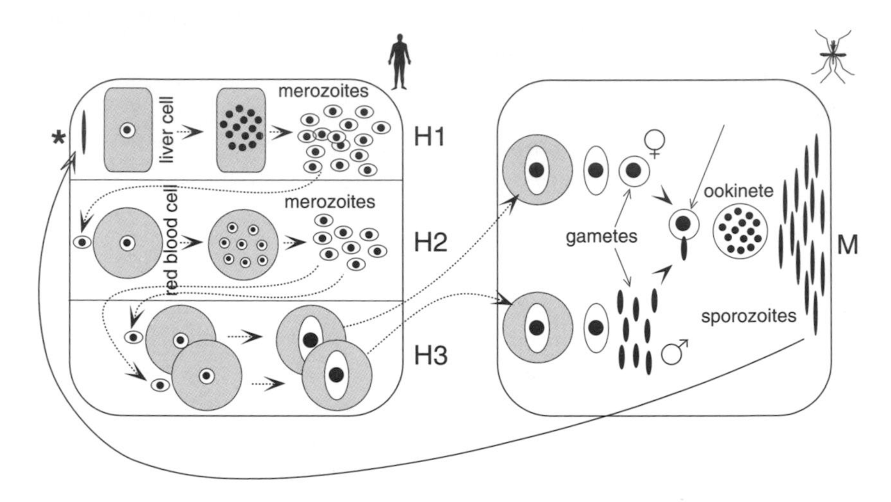

Figure 20.4. Life cycle of human *Plasmodium.* H1, H2, H3: the three phases of the life cycle that occur in humans. M: the phase that occurs in the mosquito. See text for details. Note that H2 is repeated (not shown).

When an infected *Anopheles* mosquito bites a human, it injects with its saliva parasites in the form of sporozoites, which then enter the vertebrate's circulatory system (see the asterisk on the figure). As sporozoites pass through the liver, they quickly penetrate hepatic cells (hepatocytes), and within just a few minutes after the bite they disappear from the circulatory system. It is estimated that the salivary glands of an infected mosquito can contain up to six thousand sporozoites at any one time and that the number of sporozoites inoculated during a blood meal usually ranges between twenty and thirty (although sometimes up to a thousand may make the passage; see Rosenberg et al. 1990). Once in the hepatocytes, the parasite divides many times (preerythrocytic schizogony) to give about twenty thousand merozoites (H1 on the figure). After bursting from hepatocytes, the merozoites then reach the blood and spread throughout the circulatory system.

Once in the bloodstream, the released merozoites penetrate red blood cells (most often one per cell) and divide (a new round of schizogony), each producing a few dozen other merozoites that in turn are able to penetrate other red blood cells after bursting free. The result of this repeated penetration, division, and release is a continuous cycling in the bloodstream (H2 on the figure) that proceeds unabated on a defined schedule every few days. The number of erythrocytes that can be parasitized as the result of this cyclical multiplication process is impressive: up to 20% of the number of red blood cells (equivalent to many trillions) in a single human.

Significantly, not all merozoites are created equal, and a few that have penetrated into erythrocytes do not divide but instead transform into gametocytes, some male and others female (H3 on the figure). When a mosquito takes a blood meal on an infected human it ingests red blood cells, some of which carry *Plasmodium* in the form of these gametocytes. Once in the digestive tract of the mosquito, the blood cells are digested and the parasites released. Female gametocytes then transform into female gametes and male gametocytes divide to give several male gametes in a process called exflagellation. Fertilization occurs in the gut to form zygotes, called ookinetes because they are mobile, and each ookinete then passes through the stomach wall of the mosquito (penetrating between cells) and installs itself on the gut's outer edge. The ookinete then divides and gives rise to several thousand sporozoites that in turn migrate to the salivary glands of the mosquito (M on figure), ready for another bite (for a detailed outline of the development of the malaria parasite in the mosquito, see Ghosh et al. 2000).

Of the four phases (H1, H2, H3, and M), only H2 (infection of the erythrocytes and the formation of new merozoites) is repeated cyclically; thus the succession of events in the cycle is H1, ($n \times$ H2), H3, and M.

A complication in the life cycle may occur with the formation of a particular form of cell called a hypnozoite, which is a sporozoite injected with the mosquito's saliva that becomes "dormant" in a hepatocyte and does not undergo the usual immediate rounds of intrahepatic schizogony. Months or even years later hypnozoites may "wake up," divide to give merozoites that then infect erythrocytes, and continue the life cycle as described above. In this case people then become sick again, sometimes even after years without recurrence (relapse). Hypnozoites are known to occur in *P. vivax* and *P. ovale* but not in *P. falciparum* or *P. malariae* (for which there is no relapse). However, in *P. malariae* erythrocytic stages can survive for years, resulting in sporadic recurrence even years later (recrudescence).

The specificity of *Plasmodium* for its sites within hosts is narrow, since at each stage the parasite penetrates only well-defined cells. It is also narrow for the invertebrate host, and the parasites use only mosquitoes in the genus *Anopheles* as vectors (about thirty out of four hundred known mosquito species in this genus). By far the most important members of the genus for transmission are those from the *Anopheles gambiae* complex, and there is good reason to believe that modifications of the environment induced by humans and the development of dense human populations linked to the development of agriculture have been to the good fortune of both *A. gambiae* and *P. falciparum.*

Contrary to that of the invertebrate host, specificity for the vertebrate host is variable. It is narrow in the case of both *P. falciparum* and *P. vivax*, which seem to be specific to humans, while *P. malariae* and *P. ovale* also parasitize the chimpanzee in Africa and several other monkeys in South America.[3] The compatibility of *Plasmodium* within specific populations and individuals varies strongly within the human species and is a function of both assorted genetic characteristics and acquired immunity (see chapter 4). Within endemic regions human immune defenses produce a kind of equilibrium between people and parasite. At birth, newborns have antibodies transmitted by the mother that protect them during the first few months of life. After about eighteen months, babies begin to acquire their own defenses in response to sporozoites injected when infected mosquitoes bite them. There is, however, a critical period (between about one and two years old) during which the mother's antibodies decrease and children do not yet have much protection of their own. Significantly, protection is never long lasting, and it decreases if the body stops being stimulated by constant reinfection, such as when one travels outside an endemic area and then returns.

The origin of the four human malarias is probably not the same. *Plasmodium malariae, P. vivax,* and *P. ovale* induce diseases that are usually not too serious, and they either also infect monkeys *(P. ovale* and *P. malariae)* or are closely related to those *Plasmodium* found in monkeys (e.g., *P. vivax* is very closely related to *P. fragile*).

Plasmodium falciparum, on the other hand, causes serious disease in humans, does not infect monkeys, and is not closely related to those found infecting monkeys.

Based on these peculiarities, it was suspected for a long time that *P. falciparum* in humans may have resulted from a recent evolutionary transfer (from which comes its pronounced pathology) whereas the other three would have been transmitted to us some time ago through our primate ancestors.

Molecular techniques have since confirmed this conjecture, and studies have also revealed the surprising host from which *P. falciparum* was borrowed. The sequence of the circumsporozoite protein, which completely surrounds the invasive sporozoite when it first enters the vertebrate (and thus is of great interest in the making of a vaccine), first showed that *P. falciparum* was clearly separate from the other *Plasmodium* found in mam-

3. That humans are the only vertebrates parasitized by *P. falciparum* has had two major consequences. One is positive: there is no wild reservoir. The other is negative, since laboratory studies were difficult up to the time in vitro cultures were made possible in the late 1970s.

mals (Di Giovanni, Cochrane, and Enea 1990). Later, analysis of the DNA coding for the small ribosomal subunit brought a surprising result: an astonishingly close relation between *P. falciparum* and the bird parasites *P. gallinaceum* and *P. lophorae*.[4] Further, the work also revealed that there is no close relation to either the *Plasmodium* of monkeys *(P. fragile)* or the parasites of other mammals, such as those found in rodents *(P. berghei, P. vinckei,* and *P. yoleii)*. In short, the lateral transfer from a bird is the likely origin of the current human *P. falciparum* (Waters, Higgins, and MacCutchan 1991), which appears to be very recent and linked to the development of agriculture and the consequent formation of large groups of sedentary humans. That is, it is not the behavior of individuals but rather a shift away from a more solitary mode of existence toward group living that would ultimately have made possible the transfer and maintenance of this particular disease in humans.

In general, the epidemiological characteristics of malaria depend greatly on how humans relate to the ecology of the vector, itself tightly linked to water because mosquito larvae are aquatic. Thus malaria is typically a water-dependent parasitosis, and it is therefore not surprising to find more than one malaria epidemiology, each corresponding to a different set of environmental parameters (climate, the nature of the aquatic environment, habitat management, habitat density, the quality of life, etc.). That is, transmission may look very different a few kilometers down the road (Robert et al. 1985), and it is stable wherever there is a good vector—one with both an open encounter filter (long mosquito life span, strong attraction to humans, presence for a prolonged period during the year) and an open compatibility filter (high receptivity to *Plasmodium*).

On the other hand, malaria is unstable wherever the vector does not present these qualities, and paradoxically, the more unstable malaria is, the more serious it becomes as people lose their acquired immunity. Infections thus may be more severe in regions where malaria occurs episodically than in regions where people's immune systems are regularly stimulated, and it is in just such unstable zones that murderous epidemics occur.

To make readers realize what an endemic parasitic disease can be, here I present a short summary of some of the transmission data for *P. falciparum* in the region of Brazzaville, Congo. In this region the environment is that of a degraded forest with high humidity and a variety of water impoundments, ensuring continuous year-round transmission (Carnevale et al. 1985). The researchers obtained data by capturing mosquitoes

4. *Plasmodium falciparum* has elongated gametocytes, which had already suggested that the species was better adapted to the oval form of the bird's red blood cell than to the disk-like mammalian red blood cell.

(Anopheles gambiae) on their bare legs several nights each month throughout the year. *Anopheles* were found to be essentially nocturnal, with the maximum number of bites occurring around 1:00 A.M. (the bite rate was about the same inside and outside houses). Captured *Anopheles* were dissected and examined under the microscope, and the distinction between healthy and parasitized mosquitoes was made either by identifying ookinetes around the stomach or sporozoites in salivary glands or by using parasite-specific DNA probes.

The results illustrated in the figure 20.5 do not require much comment: depending on the month, the number of bites per person per night varies on average from one to one hundred. Based on the percentage of infected mosquitoes, it is then easy to calculate that the average number of infective bites per person per night varies from zero to almost four. As hard as it is to believe for those of us who live in industrialized countries, the inhabitants of a village in Congo, and thus those of thousands of similar tropical African villages, therefore receive close to a thousand infective mosquito bites each year.

In such circumstances the classical remark that during each generation parasites are continually confronted with host defenses but not all hosts are confronted with the parasite is clearly not valid. All host individuals get the disease each generation, and there is no chance for people living in such endemic areas to escape.

Clearly, human activity continues to play a major and key role in the transmission of malaria in endemic regions, particularly through the need for water. Hydrologic intervention in order to irrigate, produce energy, or attain any other economic objective often increases transmission of malaria in endemic regions simply because it produces more mosquito breeding habitat. More rarely, human-designed impoundments or alterations reduce transmission. For example, the urbanization of Africa disrupts the conditions necessary for malaria transmission. In the town of Bobo Dioulasso in Burkina Faso, for example, transmission is estimated to be one hundred times lower than in neighboring villages (0.14 bites/person/year versus 50 to 130; Robert et al. 1986). However, transmission remains high in some parts of Brazzaville that have recently been urbanized but still have small open-water streams, though in other parts of the town, where aquatic environments have become rare, the transmission rate has dropped to almost zero (Trape 1987).

Macroparasitosis: The Schistosomes

Human schistosomiasis is caused by a relatively small group of trematodes in the genus *Schistosoma,* which itself belongs to a "small" family, the Schis-

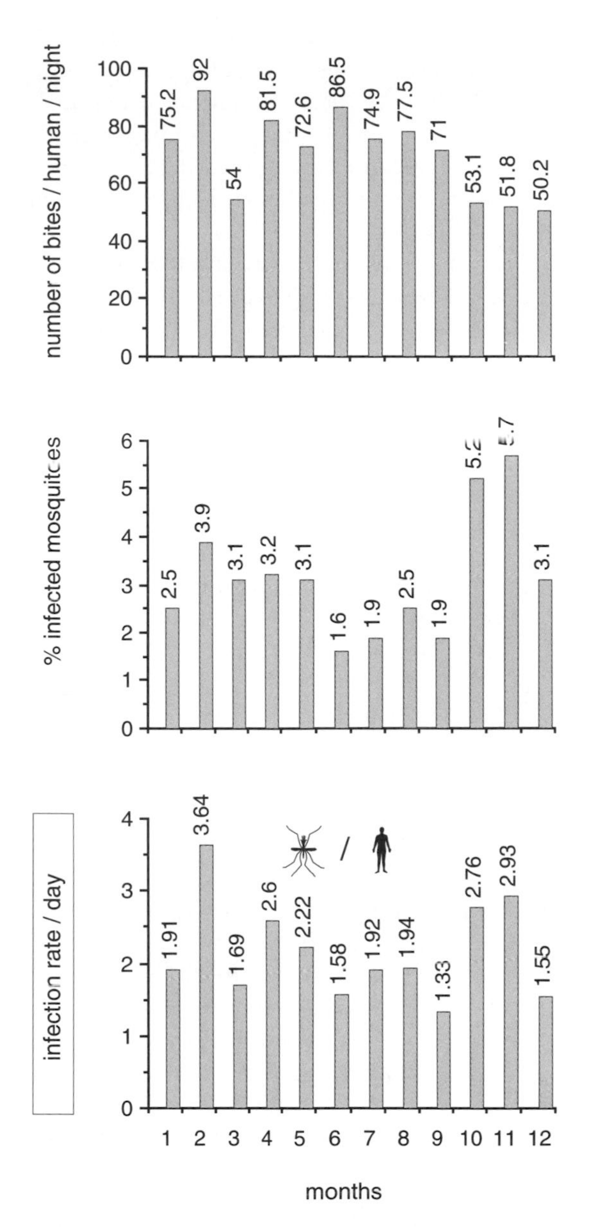

Figure 20.5. Malaria in equatorial Africa: in some regions each person receives several infective bites every day throughout the year (after the data of Carnevale et al. 1985).

tosomatidae. Included in the family are only one hundred or so species, virtually all of them parasites of warm-blooded vertebrates (birds or mammals) with sufficiently aquatic behavior to encounter the waterborne infective cercariae (though one species has been described in an Australian crocodile; see Platt et al. 1991). Although the group may be small, its impact is anything but: in the world today probably close to 300 million people are infected.

The heteroxenous life cycle of *Schistosoma* is familiar from the discussion in chapter 2. It contains two successive obligatory hosts, a freshwater mollusk within which larval stages develop and a vertebrate within which sexual reproduction occurs. Seven species of schistosomes, three major and four minor, are known to parasitize humans; all occur in tropical or subtropical climates.

The three "major" species are *Schistosoma mansoni* (Africa, Madagascar, the Middle East, and South America), *S. haematobium* (Africa, Madagascar, the Middle East, and one controversial focus in India), and *S. japonicum* (Japan, China, the Philippines, and Sulawesi).

The four "minor" species known to infect humans are *S. mattheei* (austral Africa), *S. intercalatum* (Zaire, Cameroon, and dispersed foci in central Africa and Mali), *S. mekongi* (Laos and Cambodia), and *S. malayensis* (Malaysia).

The adults of all of these species show a strong sexual dimorphism and are found in their vertebrate hosts' blood vessels, where they live as permanent couples: the relatively stout male measures about one centimeter and carries the slightly longer and thinner female in a ventral-facing arched groove (the gynecophoral canal) in a state of perpetual copulation. Schistosomes feed on blood, and one couple can ingest about 350,000 erythrocytes per hour (Lawrence 1973). The couples are not randomly located in the body, since they need to be close to an "exit" so their eggs can be released (see chapter 2). As a result, most species live near the intestine, where the eggs can be evacuated with the feces. However, *S. haematobium,* a parasite of humans in Africa, lives in the blood vessels surrounding the bladder, and its eggs are passed in the urine.

Regardless of the species, once the eggs reach water they release miracidia—ciliated swimming larvae about 0.4 millimeter long. Once these miracidia recognize and enter a compatible molluskan host, they transform into primary sporocysts; these in turn produce secondary sporocysts that then produce numerous cercariae. These cercariae are small swimming larvae with a long forked tail that escape through the soft tissues of the mollusk about four weeks after initial miracidial penetration. Since 1980, however, we have known that the basic textbook scheme of miracidia–primary sporocyst–secondary sporocyst–cercaria is incomplete (Jourdane, Théron, and Combes 1980; Jourdane and Théron 1987). Surprisingly, secondary sporocysts stop producing cercariae after a time and instead give rise to a new generation of tertiary sporocysts. This tertiary production of sporocysts is not a marginal process; on the contrary, it is essential to increasing and maintaining the production of cercariae in the mollusk. In fact, after only a few weeks of cercarial production most of the active sporocysts in the mollusk are tertiary, and experiments have shown that potentially they could give rise to new generations of sporocysts indefinitely as long as there is virgin "terrain" to occupy. This situation has allowed the cloning of schistosomes in the snail host by microsurgical transplantation (Jourdane and Théron 1980). As the result of this process sporocyst "replication" can be obtained indefinitely through transplantation, and continuous sporocyst production can be maintained through asexual reproduction after grafting sporocysts from infected mollusks into healthy ones every four to five weeks.

Although sporocysts can live at least as long as their molluskan host, the cercariae have only a few hours to encounter their vertebrate definitive host once they leave the snail. When this encounter occurs the cercariae adhere to the skin, and in a few minutes they penetrate the tegument, leaving their forked tails behind. The part that has penetrated, now called a schistosomule, then reaches a lymphatic vessel or vein and starts its long migration toward its final destination.

It was believed for a long time that the schistosomule migrated toward its final location through the tissues. Today, however, we know from the studies of Wilson (1987) on *Schistosoma mansoni* that it migrates through the circulatory system, as outlined below (fig. 20.6).

After penetrating a vessel, the schistosomules (about 0.3 millimeter long) are passively carried by the circulation to the right side of the heart, where they are then propelled through the pulmonary arteries to the lungs. There the schistosomules are stopped because of their size, and over the course of three to four days they metamorphose into slimmer organisms capable of passing through the capillaries. Interestingly, this metamorphosis occurs without cell division and without the schistosomule's feeding. Once the schistosomules pass the obstacle of the lung they are led through the pulmonary veins first to the left side of the heart and then through the aorta to the systemic circulation. As long as a schistosomule is not led to the arteries reaching the intestine (mesenteric arteries), it continues to circulate in the bloodstream and may even "turn" through the body dozens or hundreds of times. When schistosomules are by chance propelled into the mesenteric arteries, however, they migrate to the capillaries that surround the intestine, cross them, arrive in the portal vein, and then reach the capillaries of the liver. These capillaries

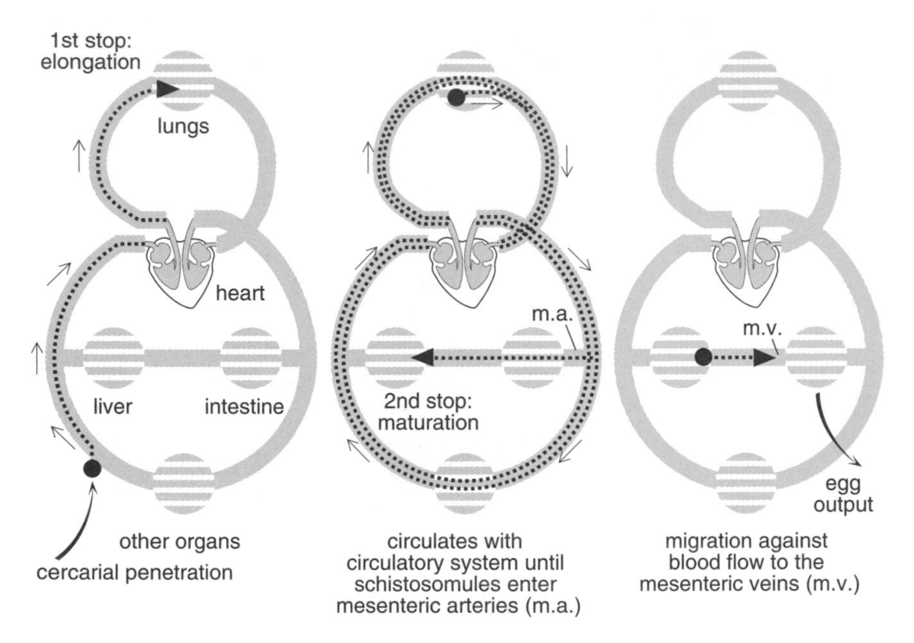

Figure 20.6. Migration of *Schistosoma mansoni* in the human.

are special for the schistosomule, and here they stop for the second time, begin ingesting blood, and grow. Here also they reach sexual maturity, and within about three weeks they form permanent couples. At this point the paired worms start their last migration: going against the current, they migrate up the portal vein, enter the intestinal veinules, and start laying their eggs as close as possible to the intestinal lumen.[5]

Given this treacherous and chancy route, it is a wonder that any schistosomes at all make it to adulthood, and in fact more than 50% cannot find the proper final site, for reasons that may or may not be linked to immunity (the "leaky liver" hypothesis of Wilson 1990 and of Establet and Combes 1992; see chapter 12, note 4).

Even though the information on migration in the vertebrate body outlined above comes from experiments carried out in mice, numerous clues indicate that the migration in humans is very similar for each of the other species of schistosomes (though the migration data for *S. haematobium* are incomplete). Overall, then, the migration pathways taken by the schisto-

5. Male and female schistosomes are generally assumed to form "faithful" monogamous pairs throughout life (years or tens of years). Experiments with two strains differing by genetic markers have shown that this is not always true (Pica-Mattoccia et al 2000).

somes as a group appear to be alike, although the duration of the migration varies with the species. For instance, experiments in mice have shown that schistosomules of *S. japonicum* spend less time in the lungs than do those of *S. mansoni,* allowing them to better escape the host's potent defenses at this site (Gui et al. 1995).

Although much is now known about the migration and physiology of schistosomes, determining the other parameters needed for a full understanding of the disease remains a problem. For instance, estimating the worm burden in infected hosts has been the object of numerous discussions; it is obviously not easy either to count adults directly or to determine their number by fecal egg counts. Relatively recent work in this area includes that of Gryseels and de Vlas (1996), who estimated that a female *S. mansoni* may lay as many as three hundred eggs a day, that 50% of these are passed with the feces, and that one egg per gram of stool corresponds to about one pair of schistosomes in the host.

Based on these numbers, and taking into account the aggregated distribution of adult worms in human populations, Gryseels and de Vlas then built a model that was tested and validated in several endemic zones. Figure 20.7 shows the distribution suggested by the model, which is based on counting eggs obtained from infected individuals in a village in eastern Zaire. The authors estimate that in this endemic zone 8% of adult humans harbor over 10,000 schistosomes, or some 5,000 couples each.

Although as many as 30,000 to 40,000 eggs per gram of stool have sometimes been reported, it is not likely that such cases correspond to loads of 60,000 to 80,000 schistosomes (remember that these worms are each about one centimeter long). Instead, these heavy loads are most probably due to a high degree of variability in the worms' ability to successfully pass eggs to the outside as well to the efficacy of the individual host's immune system. Interestingly, the passage of eggs to the outside across the intestine is decreased in immunodeficient patients: Karanga et al. (1997) have shown in a population of Kenyans heavily infected by *S. mansoni* that HIV seropositive individuals excreted significantly fewer eggs (650 ± 600 eggs per gram of feces) than did seronegative ones (1,900 ± 1,800) even though the levels of circulating antigens were similar. This observation alone confirms just how difficult it is to deduce parasite load based on fecal egg concentrations alone.[6]

As noted in earlier chapters, the presence of density-dependent competition in schistosomes allows one to better understand some aspects of the pathology and epidemiology associated with this parasite. However,

---

6. This also indicates that granulomas, that is, the immune response, promotes the passage of schistosome eggs to the outside.

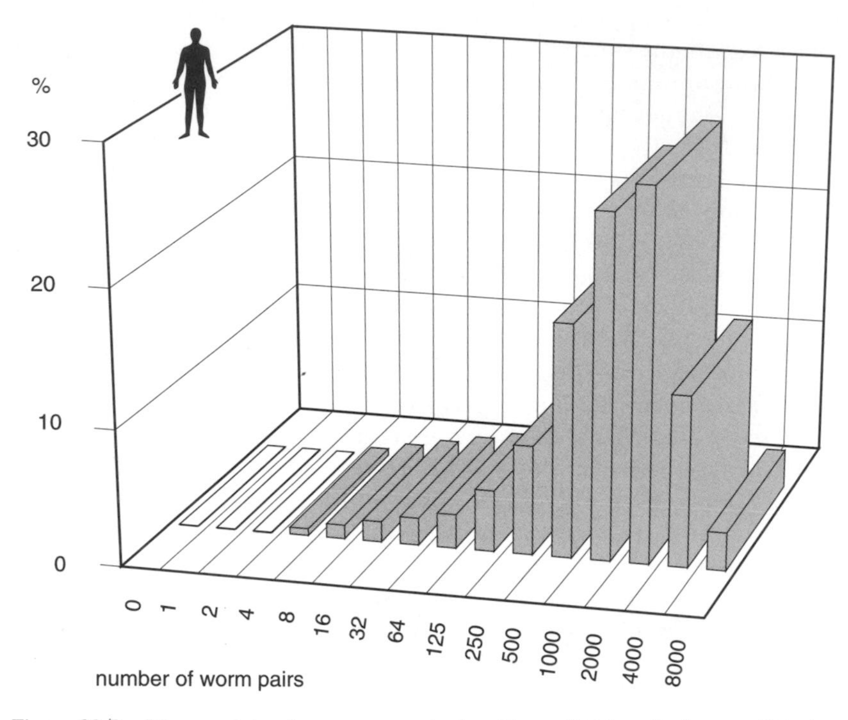

Figure 20.7. Worm pair burden per person in the village of Makundju (eastern Zaire). The numbers of pairs are derived from egg counts: the distribution is supposed to be aggregated with the parameter $k = 1.2$.

note that just because established schistosomes can "protect" themselves against overpopulation (see chapter 12), this does not mean serious cases of schistosomiasis do not exist (theoretically, a certain threshold number of adult worms should never be surpassed). In fact, in areas where cercarial recruitment is intense, schistosomiasis does become a serious problem, which can be explained only if new parasites penetrate and survive in already infected people despite an existing infrapopulation. The explanation for this paradox resides in the variable nature and efficacy of host immunity—some host individuals simply allow the establishment of heavy parasite loads whereas others do not. That is, the threshold where competition-dependent regulation occurs is determined not solely by parasite infrapopulation density, but also by the type of host and the variable nature and effectiveness of immunity in individuals of the same host species.

Several other characteristics are also important to consider as we attempt to understand the effect of schistosomiasis on human biology.

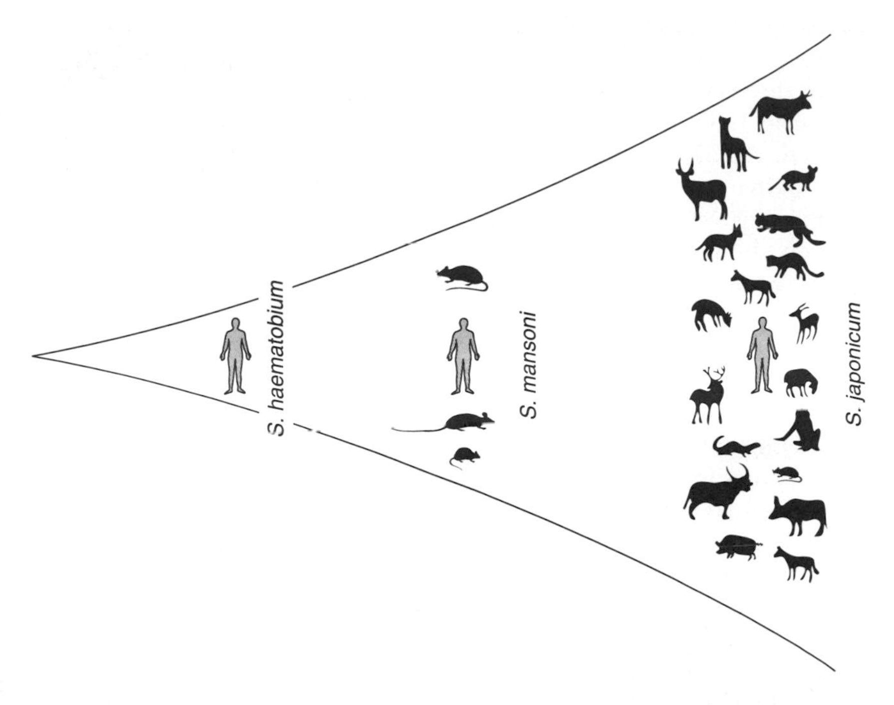

Figure 20.8. Occurrence in the genus *Schistosoma* of species with very different host spectra.

Among these are how long the adult worms can live and reproduce and the specificity of the worms for both their invertebrate and vertebrate hosts. The latter can be evaluated each time eggs containing live miracidia are passed by a patient whose date of infection (last stay in an endemic area) is known with certainty. The "record" is held by a patient who was infected for over thirty-seven years (Chabasse et al. 1985; see also Wilkins 1987).

Relative to the former, it is clear that the specificity of schistosomes for vertebrates varies considerably according to species (fig. 20.8). For instance, *Schistosoma haematobium* develops only in humans (and experimentally in a few other primates), whereas *S. mansoni* develops in humans and, in some regions (mostly in the New World), also in rodents. This is the case in Guadeloupe for the black rat *Rattus rattus* (see chapter 7) as well as in Brazil, where two rodents living in wetlands and marshes *(Holochilus* and *Nectomys)* may even be the main hosts. *Schistosoma japonicum,* on the other hand, develops in any mammal that visits aquatic habitats often enough to come in contact with its cercariae, and in mainland

China alone it has been reported from more than thirty definitive hosts besides humans. These include three species of insectivores, fifteen rodents, three artiodactyles, nine carnivores, and two other primates (Mao and Shao 1982; Xia 1990). Further, specificity varies depending on the endemic region: the main host of *S. japonicum* in the Philippines is the rat, whereas in most of the known foci in mainland China rats are not infected but humans are. Moreover, in Taiwan only a "zoophilic strain" exists.

Even though variation in vertebrate hosts exists for the different species of human schistosomes, these same worms are nevertheless specialists for their intermediate hosts. In fact, each schistosome uses only a small group of gastropods within a specific genus as its invertebrate hosts: *Bulinus* for *S. haematobium; Biomphalaria* for *S. mansoni;* and *Oncomelania* for *S. japonicum.*

Based on all available evidence, it is clear that the origin of the human schistosomes is among those related species that parasitize other mammals and not birds or reptiles. This is supported by the fact that in humans the bird schistosomes cause only a temporary although highly annoying dermatitis, or "swimmer's itch." Members of the Schistosomatidae are known from rodents, ungulates, carnivores, hippopotamuses, and elephants, and no other schistosome specific to primates besides those of humans has ever been described. In fact, all the species that parasitize humans belong to the same genus—*Schistosoma*—and though there is some mention in the literature of members of the genus *Schistosoma* parasitizing African primates (chimpanzees and baboons), the reports invariably end up involving human parasites likely to have been acquired at foci where humans are the usual hosts. For example, Müller-Graf and colleagues (Müller-Graf, Collins, and Woolhouse 1996; Müller-Graf et al. 1997) analyzed *S. mansoni* infections in a natural population of olive baboons in Tanzania and noted that the most heavily infected troops were those that lived close to humans: "The range of the troop with the highest infection rate includes the park village. As a consequence, baboons in this troop have the highest contact rate with water also used by humans; this may indicate that the initial source of parasite originates from the people."

It is this ensemble of data that supports the hypothesis of a lateral transfer to our nearest relatives rather than of a lineage of schistosomes that evolved with the primates.

Clearly, two factors, one ethological (the encounter filter) and the other immunological (the compatibility filter), must have played a role in the successful transfer of schistosomes from nonprimates to humans. In short, a vertebrate can be the potential host of a schistosome only if it has regular contact with infected water; as I noted earlier, hominids have increased their visits to such habitats during their recent evolution. As such, humans and their immediate ancestors multiplied their chances of contracting the swimming cercariae of the schistosomes from other mammals.[7] At the same time the pressure exerted by the human immunodefense system on the newly contacted schistosomes must have been intense, resulting in the selection of parasite genotypes capable of avoiding these defenses.[8] That is, other characters, such as the cercarial emission profile previously discussed in the context of the parasite-host encounter, must then have been selected for, and this is how the lateral transfers of schistosomes into hominids from nonprimates must have occurred.

Thus both logic and evidence support the hypothesis of a lateral transfer of schistosomes from nonprimates to humans. However, note that the human schistosomes of Africa, Asia, and America do not all have the same animal hosts of origin (Combes 1991e). In fact, two lineages of schistosomes have been recognized for some time in Africa—each characterized by the shape of the egg and the nature of the molluskan intermediate hosts.

The first lineage contains those schistosomes whose eggs show a lateral spine and whose intermediate host is a pulmonate snail in the genus *Biomphalaria.* This lineage includes a parasite of rodents, *S. rodhaini,* two lesser-known parasites of the hippopotamus, and the causative agent of human intestinal schistosomiasis, *S. mansoni.*

The second lineage is that of the schistosomes whose eggs show a terminal spine and whose intermediate host is also a pulmonate snail, this time in the genus *Bulinus.* This lineage includes several ungulate parasites *(S. bovis, S. matthei, S. curassoni, S. leiperi,* and *S. margrebowiei)* and two human parasites, *S. haematobium* and *S. intercalatum*—the culprits in human urinary and rectal schistosomiasis, respectively.

Within these two lineages the close relation between species is easily demonstrated by the fact that fertile hybrids can be obtained in the laboratory as well as by the occurrence of natural hybrids identified from sev-

---

7. This process continues today, and human behavior in regard to water remains the main risk of contamination by schistosomes. If water resources (dams, irrigation networks, etc.) are not managed correctly, the probability of the parasite's succeeding increases greatly. On the other hand, transmission may be greatly reduced through education, well-conceived management plans, or making clean water available.

8. The boundary between compatible and noncompatible hosts for schistosomes is very small: Malek and Armstrong (1967) infected monkeys and a human volunteer with the cercaria of the raccoon, *Heterobilharzia americana,* and showed that schistosomules not only could cross the skin but also could survive for the relatively long time it took them to reach the lungs and the liver. In these experiments the only stage that was missing was the final stage of sexual maturation.

eral locations in the wild (Pitchford 1961; Wright et al. 1974; Rollinson et al. 1990). Thus the sum total of the data strongly suggests that humans have borrowed their intestinal schistosome, *S. mansoni,* from a lineage that evolved in rodents, whereas *S. haematobium* and *S. intercalatum* have been borrowed from a lineage that evolved in ungulates.

Clearly, the transfer of schistosomes into humans in Africa has been followed by speciation, since the "human" species of today are obviously distinct from those found in animals.[9] This hypothesis has been confirmed using DNA analysis (Després et al. 1992; Barral et al. 1993). The sequencing of two regions, one from the nuclear genome and the other from the mitochondrial genome, both showed concordance in the placement of the divergence between the human's schistosomes and their related species at about 2 million years ago, a time that coincides with the emergence of hominids (fig. 20.9).[10] Bowles, Blair, and McManus (1995a) later confirmed the existence of the two African clades, the "*haematobium*" group and "*mansoni*" groups, using similar techniques.

In Asia, people are infected by the other major human schistosome, *S. japonicum,* which also cycles in numerous domestic animals (notably water buffalo) as well as in wild animals. In this case the broadening of the host spectrum has not been followed by schistosome speciation, either because humans populated Asia long after Africa or because living in close proximity with domesticated water buffalo has prevented the isolation necessary for speciation to occur. There may, however, be a very different reason. Chilton et al. (1999) believe, based on allozyme analysis, that there is in fact a complex of species confounded under the name *S. japonicum.* Not only would the strains from Taiwan, Japan, the Philippines, and Sulawesi be different species, but there would also be two cryptic species present in mainland China. DNA analysis has shown that *S. japonicum* has been separated from the African schistosomes for a long time (tens of millions of years), as has been suspected from its use of a prosobranch rather than a pulmonate intermediate host and from various morphoanatomical characteristics. *Schistosoma mekongi* in Laos and *S. malayensis* in Malaysia, whose intermediate hosts are also prosobranchs, are also closely related to *S. japonicum,* and both appear to be the result of more recent speciations. Bowles, Blair, and McManus (1995a) confirmed the monophyly of *S. japonicum* and *S. mekongi* by sequencing two portions of the nuclear and mitochondrial genomes as discussed above.

---

9. When there is speciation, species parasitic of current animals cannot be considered the ancestors of species parasitic of humans; instead, they all have a common ancestor.

10. Snyder and Loker (2000) consider that the genus *Schistosoma* might have an Asian origin, based on molecular studies (sequence of the 28S ribosomal subunit).

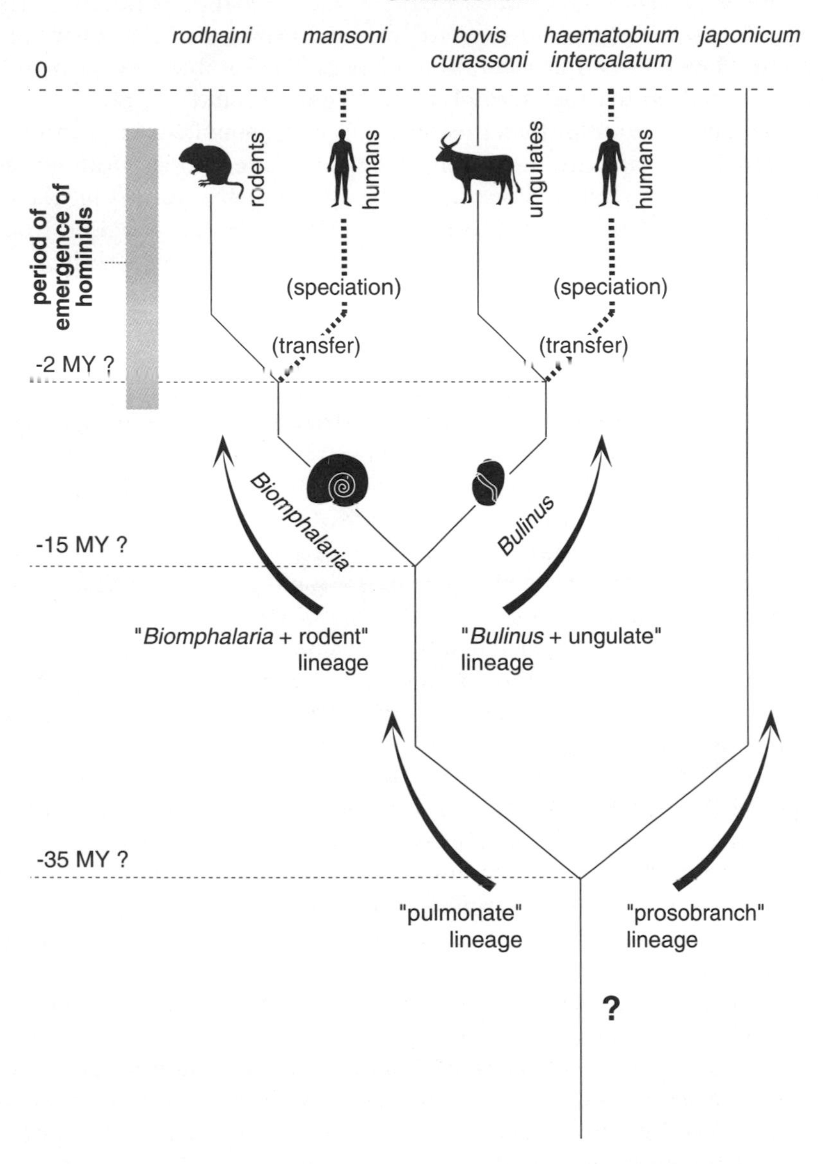

Figure 20.9. Hypothesis of the evolution of schistosomes based on analyses of DNA sequences in relation to hominid emergence (after the data of Després et al. 1992; Barral et al. 1993; Barral and Combes 1995).

Several schistosomes from India and Southeast Asia, either parasites of ungulates (*S. spindale, S. indicum,* and *S. incognitum*) or parasites of rodents (*S. sinensium*), do not seem to parasitize humans and thus have apparently not followed an evolutionary path comparable to that of the species cited above. Interestingly, a phylogenetic study by Snyder and Loker (2000) suggests that the family Schistosomatidae comprises a "mammalian" clade and a "nonmammalian" clade and that the genus *Schistosoma* has an Asian (not African!) origin. Further, Johnston, Kane, and Rollinson (1993) have shown that *S. spindale* (a strain coming from Sri Lanka) is related to African schistosomes but not to *S. japonicum,* based on 18S ribosomal DNA analysis.

When we look at the distribution maps of African schistosomes, it is easy to see that *S. mansoni* is the only one of the group also present on the other side of the Atlantic (in Brazil, Venezuela, and the Antilles). In fact, *S. mansoni* from America and those from Africa are morphologically very similar, and molecular biology has shown (Després et al. 1992) that *S. mansoni* is a very recent addition to the Americas, brought from Africa with its human hosts during the slave trade of the sixteenth, seventeenth, and eighteenth centuries (fig. 20.10).

Schistosoma mansoni was able to establish its life cycle in the New World because it encountered local species of *Biomphalaria* capable of acting as intermediate hosts, in particular the excellent vector *B. glabrata.* Although it is possible that this species had been present in the Americas for a very long time, in the end it is genetically more closely related to modern African species than to its American congeners (Woodruff and Mulvey 1997). Thus it is highly likely that *B. glabrata* may also have been imported to the New World quite recently. As for *S. mansoni,* both *S. haematobium* and *S. intercalatum* were in all likelihood introduced to the New World under the same circumstances as *S. mansoni.* However, these worms could not establish themselves because of the absence of *Bulinus* in the Americas—a nice illustration of how important specificity is as an obstacle to parasitic invasion.

Even though the species of schistosomes are both morphologically and biochemically distinct, they still nevertheless appear to be very closely related and thus "not well separated" from each other. From this arises, as I noted above, the possibility of hybridization between the various species (Jourdane and Southgate 1992). That such genetic exchanges may occur between human parasites and closely related animal parasites is truly worrisome. Such exchanges thus have the potential to locally modify important epidemiological and immunological characteristics associated with the disease. If, for example, the HIV-mediated immunosuppression at present sweeping through parts of the world were to break the immuno-

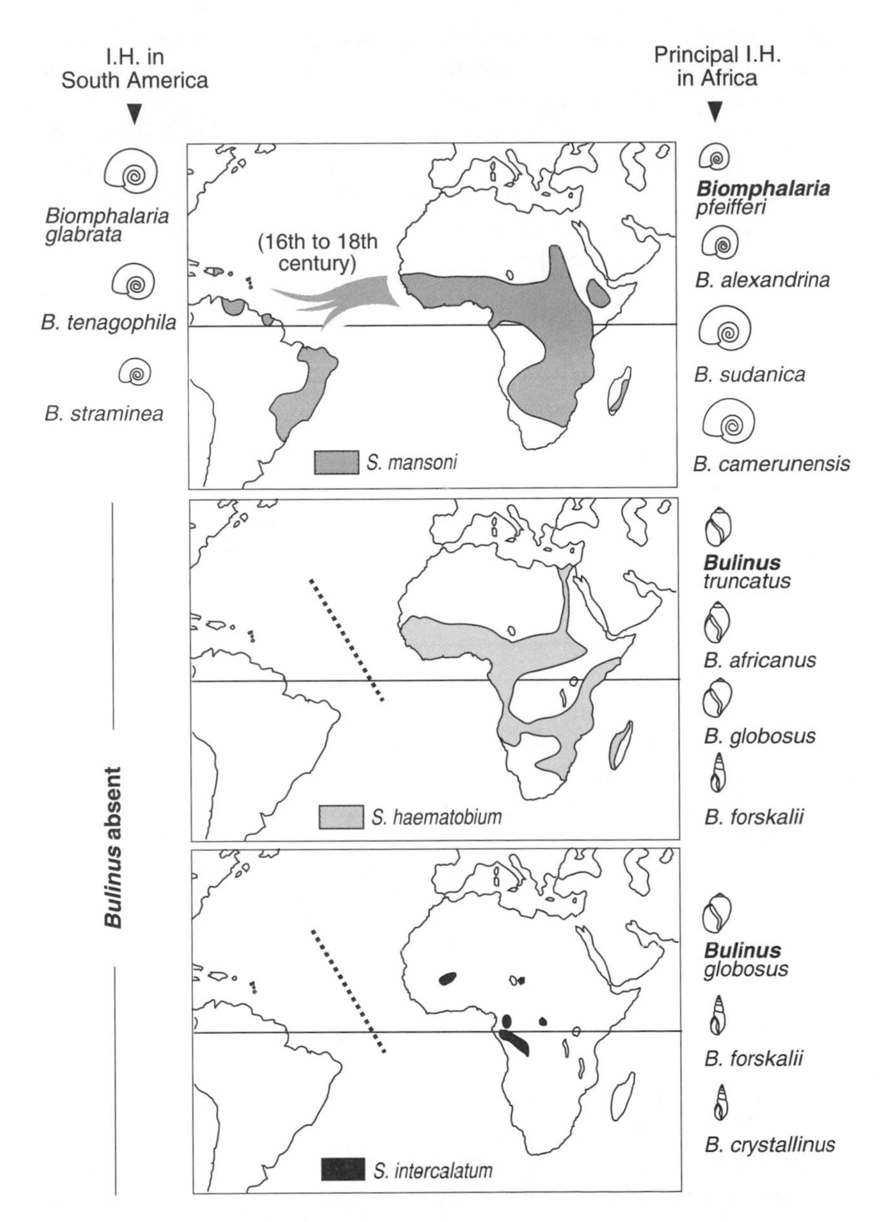

Figure 20.10. Approximate expansion of the human schistosomiasis of African origin relative to worm transport in historical times and the geographical distribution of appropriate vectors. Only *Schistosoma mansoni* has maintained itself in South America because of the presence of mollusks of the genus *Biomphalaria*. I.H. = intermediate host.

logical barrier of specificity, the introgression of foreign genes from animal schistosomes into human schistosomes could, as feared by Combes and Jourdane (1991), become amplified with potentially devastating consequences to humans who are not immunocompromised. Significantly, evidence exists that this introgression is already occurring. For example, *Schistosoma haematobium,* a human parasite, and *S. bovis,* an ungulate parasite, are found at the same transmission sites and readily produce fertile hybrids under experimental conditions (i.e., when the encounter filter is circumvented). In nature, however, this hybridization is unlikely because, as with the cercariae of bird schistosomes, *S. bovis* cercariae penetrating humans are normally destroyed by the immune system (the compatibility filter). However, in immunocompromised people some *S. bovis* may reach sexual maturity and form couples with *S. haematobium.* Brémond et al. (1993) have shown, by allozyme analysis, that such hybridizations, though rare, already occur in nature: some schistosomes recovered from naturally infected ungulates do in fact show alleles usually present only in *S. haematobium.* It is safe to say, then, that such gene flow has the ability to greatly modify, at the same time, several major disease-related parameters, including transmission (the use of new molluskan vectors), pathology, and drug resistance.

Since schistosomes do not multiply in their vertebrate hosts, (unlike *Plasmodium*), the parasite load in these hosts depends on both the number of cercariae that penetrate and the number of those penetrating that survive. Thus the worm burden of a particular vertebrate host depends on both the degree of exposure at the transmission site and the efficacy of a host's immunodefense system. Field investigations indicate that parasitism usually peaks at adolescence. This age-dependent prevalence is shown in an investigation done in Africa in the south of Benin (fig. 20.11), which clearly shows that prevalence is comparable, within a population, for *S. mansoni* and *S. haematobium.* For both species infection "peaks" at about age eighteen, and a rapid decrease is observed thereafter. Even though the time a person spends in contact with water may vary with age, it is certain that the human immunodefense system becomes more efficient at about this age (eighteen to twenty; see Butterworth 1990, 1994; Hagan et al. 1991), and several studies on different schistosome species suggest that host IgEs are one of the key players in the development of protective immunity.[11]

The various schistosomiases are diseases for which there is remarkably

11. Rats and mice have been used as models for the study of the immunoregulation of schistosomiasis. Interestingly, the rat model appears to be relevant for human immunity, whereas the mouse model is inappropriate (see Capron, Dombrowicz, and Capron 1999).

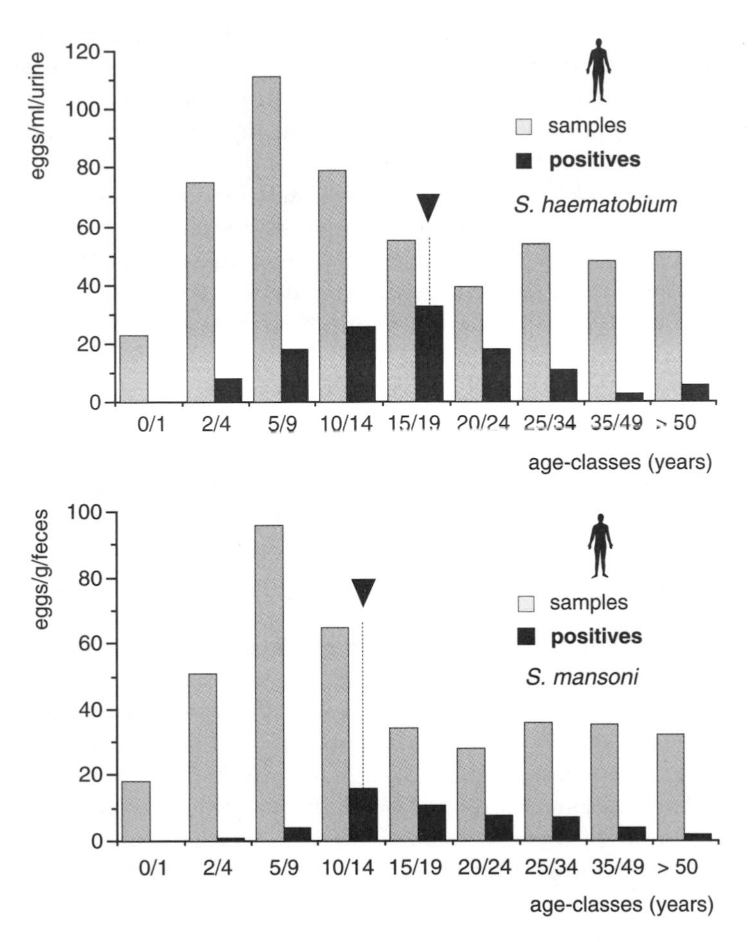

Figure 20.11. Age dependence of infection with schistosomes in an African locality. Note the same increase in the prevalence until the teenage years for the two types of schistosomiasis in the same location, followed by a decrease. The bottom graphs are the same data expressed in percentages (modified after Chippaux et al. 1990).

ancient historical information, and it is now normal to remind every parasitology class that on several occasions eggs of *S. haematobium* have been found in Egyptian mummies. Significantly, more recent studies have shown that a particular protein produced by schistosomes, the circulating anodic schistosomal antigen, has been identified in a mummy dated at 5200 B.P. It has even been possible to determine, by searching for this antigen in a series of twenty-three mummies from the fifth century B.C., that at least sixteen of them were infected, a proportion that corresponds to a prevalence on the same order of magnitude as commonly reported today in endemic regions. Along these lines it is also very interesting that this antigen has not been observed in South American mummies interred before the conquest of the Americas by Europeans (see Capron 1993; Deelder et al. 1990; Miller et al. 1992).

In general, schistosomiasis, like malaria, is one of the major human parasitoses whose transmission is also closely linked to human practices. Significantly, this was already true in ancient times when people first started to irrigate crops and thus to expand the breeding habitats of the snail intermediate hosts. More recently, the construction of large dams (the Akosombo on the Volta in Ghana, the Aswan on the Nile in Egypt, and the Richard-Toll on the Senegal) have created huge human health problems (owing to increased snail habitat) that are extremely difficult to control. Similarly, the creation of irrigation canals and the continued development and farming of rice continue to favor the appearance of new transmission sites and disease foci for schistosomiasis as well.

Parasitic Endemics and Hominid Evolution

According to most anthropologists, hominids originated in Africa beginning with the appearance of *Australopithecus afarensis* (about 3.5 million years ago) followed by *Homo habilis* (about 2 million years ago) and then *Homo erectus* (about 1.2 million years ago). From the estimations of Cann, Stoneking, and Wilson (1987) based on the study of mitochondrial DNA,[12] modern humans *(Homo sapiens sapiens)* come from a small group that lived in Africa no more than 200,000 years ago. That there is more genetic diversity between human populations south of the Sahara than between all other populations in the world attests to the rapid and recent expansion of modern humans "out of Africa."

The wealth of data available on hominid evolution and sociality allows us to speculate on the role parasites might have played in the evolution and expansion of humans (Combes 1991b). The cost of sociality is easily demonstrated in both birds and mammals and leaves little doubt that the evolution of society in those primates that gave rise to the hominid lineage came at a cost paid to those parasites capable of profiting from this social behavior. Obviously, human sociality must have provided a benefit larger than this cost, or it would not have been selected and maintained over the course of hominid evolution.

In this regard I must emphasize that the region of the planet where hominids emerged was not the most healthful one. Notwithstanding that there are serious parasitoses and diseases in the more temperate parts of

12. For a critique of this study, see Excoffier and Roessli (1990).

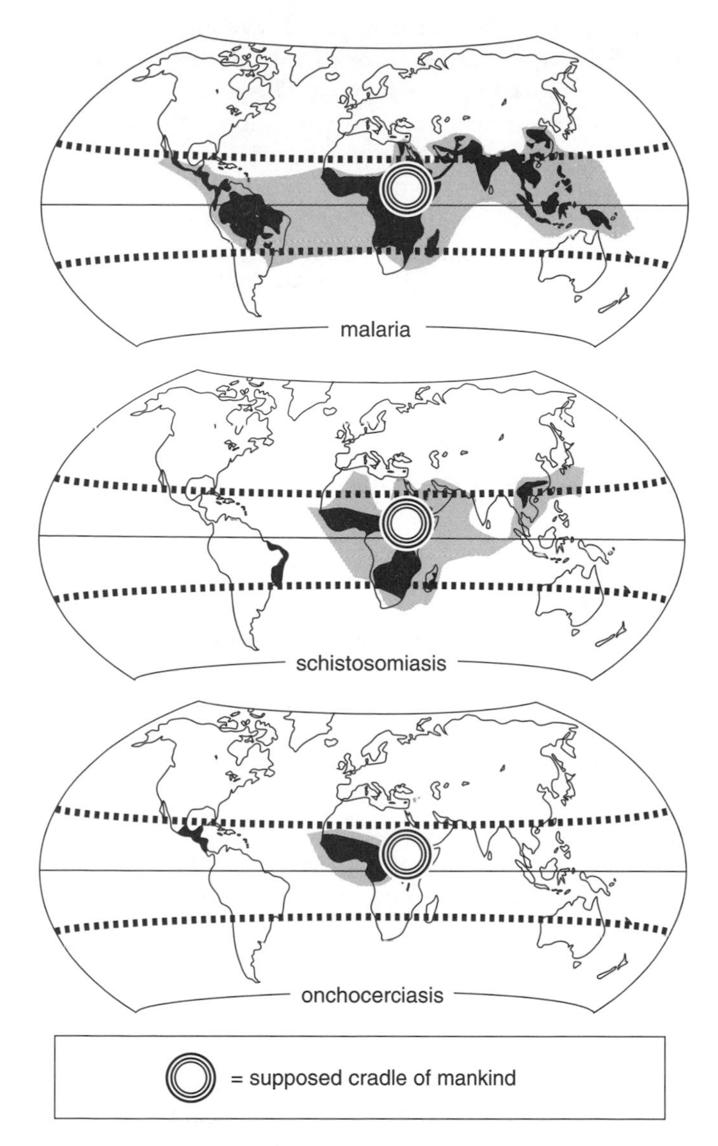

Figure 20.12. Maps of the approximate distribution of a few parasitic endemics showing that they rarely pass out of the intertropical zone (after various sources).

the globe, the intertropical areas are today, and probably also were in the past, the areas most heavily affected by parasites and other disease-causing organisms (fig. 20.12). This pattern follows the general north-south gradient of richness in free-living species on the planet, with the warmer latitudes having the greatest diversity compared with the more temperate climates. There have always been many more free-living species close to the equator than away from it and there is no reason there would not also be more parasitic species. This is true today, and it is unlikely the situation was any different a few hundred thousand years ago.

It is in precisely these warm regions, where all evidence indicates there were at least as many parasites then as there are today, that hominids emerged. It is then, in a sense, only singularly bad luck that exposed the start of humanity to these parasitic dangers, of which malaria and schistosomiasis are just two examples. As early as 1949 Haldane postulated that parasites had historically exerted a considerable selective pressure on humans, and the extreme diversity of our immunodefense-related genes bears witness to an evolutionary past where confrontations with diverse pathogenic agents were both frequent and common.

Add to this diverse array of infective worms and germs the inequality of humans in the face of parasites, and it is quite likely that the first transfers of animal parasites to our ancestors ravaged human populations until the genes for resistance appeared and were selected for in large enough numbers. May and Anderson (1978) write: "Today, for all but a few primitive tribes, it is unlikely that parasites play a major role in determining the growth of human populations." Although this may have been true in 1978, following early chemical-based victories against parasites and their vectors, it is clearly no longer true today, nor was it true in the distant past. That is, it is not unreasonable to think that the additive effects of various parasitic diseases, one after the other over the millennia, may have slowed down and influenced the development and expansion of our African ancestors. It is also not unreasonable to conclude, on the contrary, that human expansion and development outside the tropical areas benefited when humans arrived in areas where the toll paid to parasitism, though certainly present, was not nearly as heavy. Extending the thought further, it is even likely that glaciations may have made some parasitoses less of a burden to human development simply because cooler climates would have clearly disrupted the development of the external life-cycle stages of holoxenous parasites.

We know that in historical times the roles parasites have played differed greatly depending on the part of the global stage occupied. Current human populations have been decimated when confronted with diseases "new" to them, either from the mixing of populations or because of increased contact with animal reservoirs. One need only think of the conquest of the Americas, already invoked several times in this book, as an illustration of such parasitic arbitrage. Still other examples include the decimation of French troops, with their five thousand dead in Madagascar at the end of the nineteenth century (five died in combat, the rest were felled by malaria; Aubry 1979); the cost Allied troops paid to malaria in the South Pacific during World War II; the limitation of Arabic expansion into Africa because of the extreme sensitivity of their horses to sleeping sickness; and last, the death of Alexander the Great, whose global conquest was stopped when he fell mortally ill in Babylon at age thirty-three with what was in all likelihood *falciparum* malaria.

There is no doubt that malaria prevented, for many years, the conquest of much of sub-Saharan Africa by Caucasians, while the Africans themselves resisted the pathogens brought by Europeans because of the gift of their large and innate genetic diversity. In contrast, endemic North and South American pathogenic agents did not protect the Amerindians, who were rendered vulnerable by their own lack of genetic diversity (a founder effect as the result of limited migration across the Bering Strait). The pathogens brought by Europeans caused a massive and profound decimation that ultimately allowed the conquering of the New World, whereas the conquest of Africa was made possible only after the discovery of quinine (a product that came from the Americas).

Curiously, a similar contrast is also found with cattle: all attempts at importing European cattle into Africa have failed because of sleeping sickness (African trypanosomiasis), whereas European cows were imported into South America without much difficulty because of a lack of any serious locally endemic pathogens. However, the ease of introduction in the latter case has had catastrophic ecological consequences, as in some elevated parts of Peru where overgrazing has led to almost desertlike conditions.

Thus, although predators have the most obvious role in the ongoing drama of life, the more cryptic and behind-the-scenes work of infectious and parasitic diseases has clearly played the more important part in the evolution of humans, their resources, and life in general on the planet. Parasites wrote the script and direct the play; predators are simply the most visible actors.

Reflections

The Evolution of Ideas about Communities

It is only since the mid-1970s—really the early 1980s—that parasitic communities have been the object of serious investigation and conceptual reflection by ecologically minded parasitologists.

Krystina Kisielewska (1970a–e, 1971, 1983; see also Kisielewska et al. 1973) was the first to consider the need to take a community approach to parasite ecology. Most of the ideas that would later be developed by other workers are already found in her pioneering works (the idea of parasite communities being either predictable or not, the notion of dominant species, the value of parasites as environmental indicators, etc.). In fact Kisielewska went so far as to propose a series of terms for the analysis of communities that were more coherent than those later used and whose only defect was their not being widely distributed. Additionally, her ideas may have came too early for Western researchers.

Peter W. Price (1980) considers that, with some exceptions, parasite populations are not in balance and that parasite communities are not interactive. The basis for his idea of imbalance is the difficulty parasites may have, particularly if they have complex life cycles, in finding the conditions necessary to maintain their existence in a durable manner. Thus populations would only exceptionally attain levels above the threshold where regulation would be triggered. The idea of non-interactivity is essentially based on the observation that potential habitats are not saturated (there are available niches) and that the realized niche in the presence of a potentially competitive species is not different from the fundamental niche. Price does not, however, deal with overlapping distributions (such as in the digestive tract), and he invokes other forms of niche separation to explain lack of overlap, most notably the idea of specialization for different resources under which niche separation would owe nothing to competition. Price considers that parasites are very specialized by necessity (because of their small size and limited mobility) and believes this specialization implies niche reduction, including its spatial dimension, even in the absence of any competition. Thus competition would play only a very minor role in the adjustment of parasite communities, and then only in particular situations. Price considers that the organization of a parasite community depends, before anything else, on its basic resource (its host) and essentially denies any role for interparasite coevolution as long as the basic resource is not limited. The opinion that parasite population sizes rarely reach levels where competitive pressures can be exerted is shared by various ecologists outside the field of parasitology, such as J. A. Wiens, D. R. Strong, and D. S. Simberloff.

Using as a model monogeneans on fish gills, Klaus Rohde (1979, 1992, 1993, 1994) has shown that numerous niches are vacant and that each species of monogenean is specialized for a particular portion of the gill. In other words, the absence of competition does not mean species extend their niches to all the space that is apparently available. That is, it doesn't seem that there is a change in the restriction of the niche relative to species number, which may also be explained by the occurrence of past competition with species that have since disappeared ("the ghost of competitions past"). Rohde has proposed an ingenious explanation for the restricted niches he observed based on the idea that specialization—and thus concentration in a restricted space—has an adaptive value for genetic exchange: if the populations are not saturated, the adaptation common to an environment that is narrowly defined strongly increases the probability of encounter.

John C. Holmes, opposed to such isolationist theses, championed interactivity in parasite communities and published numerous studies showing that certain communities are in fact interactive (see for instance Holmes 1971, 1987; Bush and Holmes 1986). He considers that, as long as the cooccurrences are frequent enough, two or several species of parasites may exert reciprocal selective pressures on each other and thus that they will coevolve. In fact the disagreement between the two schools does not hinge on knowing if certain communities are interactive or not, since Price admits that competition may affect "a minority of species" and Holmes recognizes that interaction is not the general rule. Holmes instead focuses on the interpretation of cases where niches are separated without demonstrable current competition, such as in the Aporocotylidae of fish and *Calliobothrium* in sharks. Overall, Price considers parasite communities to be systematically young and pioneers, whereas Holmes considers them the fruit of a long evolution. Holmes and Price (1986) have agreed that, depending on the community, parasite niches may originate by interspecific competition, in the concentration of individuals for mating, or in specialization for resources. Further, both authors also agree that the objective of research into parasite community structure should be to discover where, when, and why these processes or others play a dominant role.[1]

Parasite Sex Ratios

The sex ratio is about fifty-fifty in most species and thus is approximately balanced (there are about as many females as males).

In bisexual species males are in part useless mouths to feed, because a single male is usually able to fertilize numerous females. A good way for species to reduce the "twofold disadvantage of sex" would thus be to produce fewer males and more females. In fact, Fisher (1930) has shown that the value of the fifty-fifty split between the sexes is the occurrence of an equilibrium whereby male and female genes have about the same success. That is, it is an example of frequency-

1. Simberloff (1990), who is a specialist in community ecology in general and who has also shown an interest in parasitism, is quite critical of the work of parasitologists who, in his opinion, do not do enough experiments and analyze their data with insufficiently rigorous statistical tests.

dependent selection: when males are in the minority the male genes are favored and vice versa, so that the sex ratio is always brought back to an equilibrium value.

Some parasites, however, have peculiar sex ratios. Hamilton (1967) has shown that if competition between males occurs between brothers that fertilize their sisters (local mate competition), then the sex ratio can become unbalanced in favor of females. This occurs mostly in hymenopteran parasitoids that lay several eggs in the same host individual. Among these eggs, some will give rise to males and others to females. However, female hymenopterans can choose whether their eggs are fertilized or not and can give birth to either females (fertilized eggs) or males (haploid, unfertilized eggs) depending on whether the seminal vesicle is opened. Two situations may then result: if the host individual is not already parasitized, it is to the advantage of the female wasp to lay many "female" eggs and fewer "male" eggs, since a small number of males will be sufficient to fertilize their sisters. However, if a female of the same species has already parasitized the host individual, the second female has an advantage in producing more males, which will then fertilize the females from the first clutch. These predictions have been verified in nature: in the blastophagous hymenopterans of figs, where the number of females visiting the same fig may vary from one to five, Herre (1985) has shown that the later a female arrives the more male eggs she oviposits. Similarly, in the acarid *Acarophenax tribolii,* considered by Gould (1980) to be among the "enigmas of evolution," the young are parasites in the body of their own mother, and in each clutch there is only one male that fertilizes its sisters (about fifteen) before they are born, after which it dies.

The case of schistosomes is more nuanced. Although the sex ratio is fifty-fifty at the miracidial stage (at the beginning of the life cycle), there are significant differences throughout the rest of the life cycle. For instance, female cercarial production in mollusks is higher than male cercarial production, and female cercariae live longer than males. On the contrary, however, a male-biased adult sex ratio is always detected in vertebrate hosts. Boissier, Morand, and Moné (1999) think that female and male strategies are different: females invest in quantity (greater multiplication in the intermediate host), whereas males invest in quality (greater infectivity and survival in the definitive host). The final balance is in favor of males and causes the observed adult male-biased sex ratio. The possible adaptive consequences of this approach are a higher probability that each female will find a male and the sexual selection of males by females, as Read and Nee (1990) have suggested.

Family Planning among Parasites

I have already mentioned that when an infectious agent encounters a host individual it does not "know," in general, whether this host is already occupied by conspecific organisms. Thus, if regulation becomes necessary it must be *postinvasive.* I emphasized, however, that things are different in the parasitoid world, since in these organisms adult females oviposit in hosts and can thus distribute eggs according to the best strategy possible (the one that gives their progeny the best chance of survival and that minimizes fratricidal competition).

Wajnberg (1994) (from whom I borrowed the title and content of this reflection) explains that each parasitoid female with a stock of eggs to oviposit during its life also has the ability to make very different decisions during ovipositing. These include laying all its eggs in the first host encountered or, at the opposite end of the spectrum, laying only one egg in each host individual (with all choices in between being possible). Specialists call this range of options selection for optimal clutch size. The optimal clutch size depends on factors such as the total quantity of eggs available to the female and the probability of her survival between the discovery of two successive hosts. This discovery depends in turn on other factors, including host density in the environment. Research has shown that females' decisions were genetically determined and that there are important variations from one individual to another. Thus natural selection appears to operate on the range of behaviors available in these systems.

Such preinvasive strategies do not exclude the possibility of *postinvasive* regulation (see chapter 12). As Wajnberg notes, the descendants may "judge" inappropriate the mother's family planning decision and then modify it in the form of an implacable competition between brothers and sisters.

Do Parents Pay the Costs of Parasitism in Their Offspring?

Bloodsucking ectoparasites are extremely frequent in birds' nests. Some of these parasites attack the nestlings but not the parents, and it has been experimentally demonstrated that they can reduce the offspring's body mass and even cause death (Richner, Oppliger, and Christe 1993). As such they may clearly lower the value of a brood, stimulating behavioral responses to these parasites. Christe, Richner, and Oppliger (1996) made two predictions about parental behavior relative to nestling infection. Either parasitized nestlings would beg less because they are weakened and so parents might reduce the rate of food provisioning, or parasitized nestlings would beg more in order to obtain more food and parents should increase the rate of food provisioning.

Tripet and Richner (1997) investigated the hypothesis of parental food compensation in a blue tit *(Parus coerulus)* population in Switzerland. By comparing nests infected with the bird flea *Ceratophyllus gallinae* and those uninfected by the flea, they found that there was no apparent cost of parasitism for nestlings: the fleas had no significant effect on body mass, and no mortality was attributable to the fleas. These workers also found that blue tit pairs from infested nests provisioned their young at a higher rate than pairs of parasite-free nests (fig. R3.1). Nestlings were thus able to cope with parasites when especially well fed, whereas they would suffer growth reduction and possibly death if otherwise "normally" nourished. In short, the parents compensated for the effects of parasites by providing more food.

But what effect did increased feeding activity have on the parents? Tripet and Richner found that the body condition of both males and females was not different between the parasite-free and the infected group, although it is not known whether they fed themselves more while provisioning the nest.

However, life history theory predicts a trade-off between current and future re-

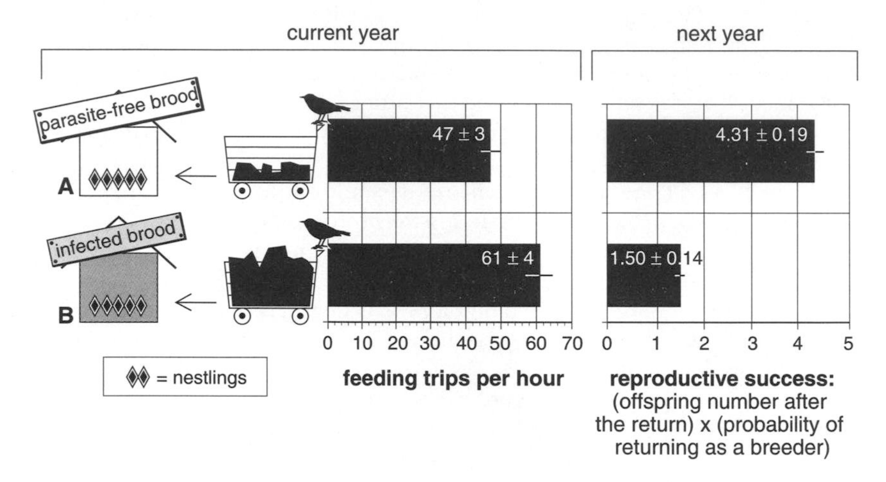

Figure R3.1. Parents pay the cost of parasitism: when the brood is parasitized by fleas, parents increase food provisioning but suffer lower future reproductive success (inspired by the work of Tripet and Richner 1997; and Richner and Tripet 1999).

production, and we can expect that future reproductive success will decrease with current reproductive effort. Since fleas were responsible for a greater parental investment in feeding, the prediction was that a decrease would occur in one or both of the components of the parents' future reproductive success—survival and residual reproductive value. Experimental manipulations of parasitism (Tripet and Richner 1997; Richner and Tripet 1999) demonstrated that the prediction was true: parasitism had a significant negative effect on the likelihood that tits would return as breeders as well as on the mean number of young raised the following year. That is, the parents rather than their offspring paid the cost of parasitism.

The extent to which chick parasitism can be compensated for by more food depends on numerous factors, and studies on different parasite-host associations have given results comparable to those of Tripet and Richner. For instance, Hurtez-Boussès et al. (1998) showed that blue tits whose nests were parasitized by blowflies *(Protocalliphora)* also increased their feeding rate. But in this case the parents did not fully compensate for the effects of the parasites, and the chicks from infested nests had lower body masses than uninfected broods. In a study of great tits *(Parus major)* parasitized by fleas *(Ceratophyllus gallinae)*, Christe, Richner, and Oppliger (1996) demonstrated that it was principally the males that increased their food provisioning.[2]

2. Why do males respond more than females? Christe, Richner, and Oppliger (1996) suppose that this is because in great tits and blue tits, females divorce males if breeding success is low. As a result, males reduce the probability of divorce by making a heavier investment in the current brood.

Although some interesting work has been done in this area, two questions nevertheless remain.

1. How is the current increase in parental effort translated into future reproductive success? Richner and Tripet (1999) discuss various hypotheses along this line; one is that the mediation could operate by reducing immunodefense ability. Various studies have shown that there is often a positive correlation between the parents' effort and the prevalence of endoparasites. For instance, Richner, Christe, and Oppliger (1995) demonstrated that if *Parus major* broods were experimentally enlarged, males increased their feeding rates by 50% while the prevalence of bird malaria in these males increased from 35% to 76%. There was also a positive correlation between natural clutch size and the prevalence of malaria in females (Oppliger, Christe, and Richner 1997).

2. Since food processing also carries a cost for future reproduction, why don't parents choose to reduce their investment in the current brood in order to increase their potential to invest in future broods? Little work has been done on this, but Perrin, Christe, and Richner (1996) have constructed a mathematical model indicating that parents lose less in terms of lifetime reproductive success if they increase their food provisioning when their current brood is infected than if they reduce their effort.

AGV and APV

Parasites introduce into their host populations an extra element of genetic heterogeneity that is due to their own genes. Briefly, parasite genomes are equivalent to massive DNA insertions (although not necessarily directly into the host's genome). All these insertions have comparable effects, thanks to the durable interaction, which are then visible via the expression of the parasites' genes in the hosts' genotypes.

In a host population the following coexist: individuals that possess only the host's genome and thus the "standard" phenotype; and individuals that possess the host's genome plus the parasite's genome and thus have a modified phenotype.

Although these two ensembles of host individuals belong to the same species, they often have distinct phenotypes within an ecosystem: they may occupy different strata in the environment, they may have different behaviors, they may be the prey of different predators, they may have different chances of survival, and so on. These phenotypes (parasitized and nonparasitized) in the same species may be more distinct than are two closely related species.

This added genetic value (AGV) imported to the host's genome thus results in an added phenotypic value (APV) in a parasitized individual.

For instance, in a population of *Gammarus* from a lagoon in Camargue (France), the heterogeneity due to *Microphallus papillorobustus* separates the individuals with a "*Gammarus*" genome and "*Gammarus*" behavior from individuals with a "*Gammarus* plus *Microphallus*" genome and a *Microphallus*-modified "*Gammarus*" behavior.

The consequences of the APV are numerous for *Gammarus:* there is preferential predation by birds (Helluy 1984; see chapter 7), and an assortative pairing by parasitic prevalence occurs (Thomas, Renaud, Derothe, et al. 1995; Thomas, Renaud, and Cézilly 1996). This is probably due to the vertical segregation between healthy gammarids (which mostly stay deep) and the infected gammarids (which mostly swim just beneath the surface). There is also a modification of competition between the two *Gammarus* species (*G. aequicaudata* and *G. insensibilis*), to the benefit of *G. aequicaudata*, as the result of behavioral changes in one species but not the other (Thomas, Renaud, Rousset, et al. 1995). Further, cercariae of *Maritrema subdolum* mainly encyst in the "crazy" gammarids simply because these cercariae swim at the surface, thus opening the encounter filter only to the "crazy" gammarids.

Because of the large number of parasites and the prevalences reached in natural populations of free-living animals, the influence of the AGV thus exerts itself on fundamental processes such as predation, competition, and adaptation.[3]

The Renewal of Diversity

If we go back to the origin of life, it is likely that molecules similar to current RNAs, but made of only a few nucleotides and capable of self-replication, must have been the first to form. Either these molecules remain to be discovered or they do not exist any more for several possible reasons. Perhaps the environmental conditions did not allow them to survive; or they were eliminated by more evolved entities; or they found refuge in other living beings and thus became parasites themselves.

Associations between primitive genomes may very well have occurred very early in the evolution of life, "as soon as the first self-replicating systems emerged from the primeval soup" (Bremerman 1983), if for no other reason than as the result of collisions. If, after such an "encounter," one of the two genomes seized the chance to exploit the other genome's replication machinery (which conferred an advantage on the former genome since it could save on something), then the first nucleic acid parasitic of another nucleic acid was born. Nothing would have kept this relationship from being mutualistic if the newcomer harbored an adaptive "plus." If we allow ourselves to push this hypothesis further, we may ask whether the primitive genomes mainly "fed on" exogenous newcomers before gene duplication took over the job of replication. Our genome would then be a "metagenome" in the same sense as an organism is a metazoan. This stage in life, however, is no longer accessible to us.

However, DNA and RNA insertions go on playing a role in nature in the form of mobile elements. Their insertion into coding sequences and their capacity to transport genes from one genome to another add up to mutations that increase the diversity and complexity of the genomes.

One may suppose that mobile elements that use a reverse transcriptase, that is, that synthesize an "intermediate RNA," are more interesting in terms of evolution

3. The term "value" in AGV has a very general sense and is not synonymous with benefit for the host individual.

than those using transposases. This is because, unlike DNA, RNA can leave the nucleus of cells,[4] although we should not forget that DNA transfers may be frequent even between living organisms from phyletically distant groups (see Heinemann 1991).

If parasitic DNA were only borrowing material to build itself and exploit enzymes that allowed replication in the host's genome, then the pathogenic effect would limit itself to a very small hijacking of energy. Since genomic parasites are usually very well "contained," their importance could be easily limited—recall that their abusive multiplication is limited either by frequency-dependent mechanisms (multiplication decreases when the number of copies increases) or by counterselection (cells or individuals that harbor too many genomic parasites disappear).

The importance of parasitic DNAs is of a very different order. It is not the energy they hijack that counts but rather the perturbations they induce in the hosts' genomes. Indeed, insertions of mobile elements may have played a prime role in evolution, since such insertions induce mutations or chromosomal modifications. Many mutations are unfavorable, but others are favorable and thus positively selected. That is, there may have been "explosions of transposition" (Biémont, Arnault, and Heizmann 1990), which then would result in "the possibility for a population to be able to respond to new selective pressures, and thus to be able to adapt to new environmental conditions." It is not impossible that a very long time ago mobile elements may have been at the origin of introns that today split up the eukaryotes' genes.

Thus DNAs that are parasitic or are thought of as such most likely have numerous positive consequences. While this may not be true at the individual scale (in humans, mobile elements are at the basis of the activation of certain oncogenes and thus of cancers), it is so at the scale of biological evolution, and the main consequence of these positive aspects is the increase of genetic diversity.

Who Is a Parasite?

Mutualism, in a sense, may be considered upside-down parasitism where it is the host that exploits the parasite. Extending this concept to its fullest means it is the host that is parasitic (see Smith 1992: "The contrast between symbiosis and parasitism is a contrast between hosts which exploit their associates, and hosts which are being exploited by their associates"). This apparent paradox arises because in these associations the "habitat" functions and "energy source" functions are separate and not summed—which needs some explanation.

In a classic parasite-host system an organism (the parasite) accrues the three advantages of room, board, and transportation by associating itself with another organism (the host). In a mutualist parasite-host system the room and transportation continue to be provided by the one that generally is still called the host.

4. The ability of mobile elements to insert themselves into genomes is widely used in making transgenic organisms.

In this case, however, the table is provisioned by the one we often qualify as a parasite. If we go even further in the analysis, we often see that "the landlord goes shopping while the tenant cooks."

Let's take, for example, the durable interaction between molluskan marine invertebrates from the deep ocean floors and the chemosynthetic bacteria they harbor.

Who are the landlords (and eventually the transportation in the case of a mobile mollusk)? The mollusks. Who are the tenants? The bacteria. Who provide the raw nutrients (sulfur)? The mollusks. Who transforms these nutrients into usable organic matter? The bacteria.

In terms of cost and benefits the system functions much like a rabbit farm: the farmer provides a home for the rabbits, the rabbits eat the feed the farmer provides, and the farmer eats the rabbits. The comparison is even better when you know that chemosynthetic bacteria are also digested by the mollusk's cells, which, like the farmer, always leave enough behind not to deplete the source so that a new crop can be raised.

These remarks show that the question of who is the parasite is interesting only anecdotally. It is the tenant if the criterion used is the use of habitat and the landlord if it is the use of energy.

Who Is "Inferior"?

It is striking that, between them all, the "inferior" organisms (mostly prokaryotes, but also protists and fungi) know how to make many more things than do the "superior" organisms: from photosynthesis to chemosynthesis to the digestion of cellulose or the making of vitamins and antibiotics, these organisms literally do it all.

It is, then, somewhat of a paradox that the genome of the "superior" organisms, because they do not have the right genes, cannot ensure those functions that are otherwise perfectly ensured by bacterial genomes, such as the production and storage of energy as the result of either photosynthesis or chemosynthesis; the total oxidation of sugars and fatty acids to CO_2 and H_2O (the genome of the eukaryotes practice only anaerobic glycolysis, which is only one-twentieth as efficient as the energy transformations that occur in the mitochondria); and the transformation of nitrogen gas (N_2) into ammonia NH_3 (the form of nitrogen needed for incorporation into proteins).

Then again, one can always believe that "superior" organisms have lost their own genes for these vital processes because they had a double use in the association, though it is more likely that they never had them in the first place.

If the hypothesis is correct, any "superiority" exhibited by the "more advanced" organisms in fact comes from their association with their "inferiors."[5]

---

5. A. Lambert (personal communication) has pointed out that prokaryotes are characterized by a wide variety of metabolic pathways while eukaryotes are characterized by a wide diversity of structures, and that the marriage of the two is a logical outcome of this complementarity.

Preadaptation

Wondering about the origin of a mutualistic or parasitic association, we can ask, What traits are necessary in a free-living organism to make the passage to a durable interaction? Such traits, selected for another function but necessary in order for the association to occur, are generally termed preadaptations. Mutualism in particular allows one to reflect on this question.

Davidson and McKey (1993) insist on the importance of preadaptations and cite as an example those associations between plants and ants in which the benefit for the ant is to be protected by the special structures of the plant (domatia), while the benefit for the plant is to be defended by the ants against phytophagous insects. According to Davidson and McKey, these associations would have occurred only if, before the association, the plants had some hollows between their knots where the ants could settle. These hollows would be an example of a preadaptation.

Death as an Advantage

Certain authors (Kroemer 1997; Rudel and Meyer 1999) consider not only that the acquisition of the aerobic mitochondrion allowed the ancestral anaerobic cell to live in an environment rich in oxygen, but also that the ancestral mitochondrion could kill the host cell if it was not providing the mitochondrion with the right conditions. These authors believe it was only later that the host cell would use and control the killing mechanisms of the mitochondria and thus generate the capacity for apoptosis or programmed cell death—which these authors consider to perhaps be "a major prerequisite for the development of multicellular organisms" (Rudel and Meyer 1999) and thus a critical event in biological evolution.

The Cheaters

A cheater is an individual that benefits without paying the price. To maintain themselves, cheaters must be less abundant than the species they take from and must not exercise too strong a selective pressure on their victims. If these conditions did not occur, there would be selection for countermeasures among the victims.

Significantly, there are both interspecific cheaters (in a community, certain species take advantage of processes selected by other species) and intraspecific cheaters (in a population, certain individuals do not pay the price normally paid for a service).

An example of interspecific cheaters may be found in some ants that live in association with certain plants. In general the association is in the form of a classical mutualism in which the plants provide shelter to the ants (the domatia mentioned earlier) in exchange for protection against phytophagous insects. The ant workers patrol the young leaves and chase away potentially harmful insects. For example, this is the case in Africa for the ant *Petalomyrmax phylax* and its host tree *Leonardoxa africana*. However, when we examine a population of *L. africana*, about a quarter of the trees are colonized by another species of ant, *Cataulacus mckeyi,* which behaves like a parasite of the *P. phylax/L. africana* mutualistic association. *Cataulacus mckeyi* expels *P. phylax,* takes its place, and benefits from the plant's domatia but does not protect the plant at all against phytophagous insects (Gaume and McKey 1998, 1999). As a result, the trees that harbor the cheater ants are probably at a distinct disadvantage compared with those that harbor the mutualist ants.

An example of an intraspecific cheater exists in the yuccas and their pollinators. There originally occurred in North and Central America about forty species of yuccas that are today used as decorative plants that gardeners have dispersed in parks and gardens throughout the world. The inflorescences of these plants are parasitized by caterpillars (this is where the durable interaction occurs), while the adult moths in exchange allow cross-fertilization. The reproduction of yuccas in their area of origin occurs in the following way (see Powell 1992; Addicott, Bronstein, and Kjellberg 1993).

At the end of each spring, moths of the genera *Tegeticula* and *Parategeticula* fly toward the yuccas and mate on the flowers. Right after mating, the females gather pollen using special tentacles on their maxillaries that allow them to form and hold a ball of pollen. The female then flies to another flower, puts the pollen ball on its pistil (allowing cross-fertilization), and at the same time oviposits. When the young caterpillars hatch, they feed on the forming seeds for a time (they eat about 15% of the seeds), drop to the ground, enter diapause, and become adults the next spring.

Exactly like the blastophage/fig tree mutualism, the yucca/moth mutualism is necessary for both partners, and the exchange is food for mobility.

However, there are cheaters on both the yucca side and the moth side. In the moths certain females do not harvest pollen, even if there is a lot of it, while still others do harvest the pollen and form the ball but do not put it on the pistil of the next flower visited. In both cases the females nevertheless oviposit in the ovary of the flower and thus behave as parasites. Tyre and Addicott (1993), who report these facts, believe such cheaters may exist in more or less constant proportion in this and other pollinator species.

Significantly, in the yuccas *(Yucca baccata)* the apex of the ovary contains flowers with few viable ovules, and moth larvae that fail to encounter viable ovules perish (Bao and Addicott 1998). As a result, females who have oviposited in these flowers have transported the pollen but will have no descendants. Thus, although the association is of a mutualist type, there is also defensive cheating on the part of the plant, which likely limits the insect population.

Why Are Humans Naked?

Attempts have been made to explain the link between nudity and parasitism by considering the following points.

On the one hand, making a nude primate in tropical regions appears to be a real provocation, since the myriad biting arthropods that serve as vectors for numerous parasites couldn't ask for a better opportunity to get their blood meal.

Even though the fur of an animal is not very efficient protection against such bites, bare skin would be bound to encourage biting, and those mutations that allowed nudity probably arose well before the invention of clothing.

On the other hand, one may propose the contrary idea that the fur of wild animals promotes the settlement of true ectoparasites—those that establish a durable relationship with their host, such as lice, ticks, and fleas. Supporting this hypothesis, hair, beards, and pubic hair remain the favorite domain of ectoparasites.

The confrontation of these two contradictory ideas shows just how difficult it is to understand the adaptive value of some characters.

The Enigmas of Parasitism

The study of parasitism provides numerous enigmas whose evolutionary explanations challenge the logical reasoning of scientists. Here are two examples.

1. Sexual grafting consists of an intimate association (and thus is itself a durable interaction) between a male and a female that are initially distinct organisms. In this case it is always the male that grafts onto/into the female, and the problem is to decide, as usual, if this type of association is clearly parasitism or constitutes mutualism.

Sexual grafting of the "first degree" is encountered in several fish families from deep ocean waters, such as the Ceratiidae, Neoceratiidae, Linophrynidae, and Caulophrynidae (Pietsch 1976). In about thirty-five species the males, after a free-living phase, attach to females by biting them, after which their tissues fuse and the blood vessels of both partners come to communicate directly. At this stage the males receive their nutrients only from the female's blood, and both male and female develop their gonads only after this association has occurred.

A second degree of perfectionism in grafting is illustrated by what has been referred to as "cryptogonochorism" (see Raibaut and Trilles 1993). Diverse mollusks and crustaceans have long been considered hermaphroditic because one ovary and several testes have been found in the same individual. In fact, thorough studies on the reproductive biology of these species have shown that the testes in question are the remains of males that injected themselves into the females' bodies.

Cryptogonochorism is known to occur in the following:

Rhizocephalid crustaceans such *Sacculina, Peltogaster,* and *Lernaeodiscus,* all parasites of other crustaceans (crabs). These species have free-swimming larvae (cyprids), some male and others female. Females inject themselves into their hosts and there become a "visceral mass" into which the males later inject themselves as the mass grows to the outside of the host crab on the ventral side (see chapter 14);

Copepod crustaceans, such as *Gonophysema* (parasites of ascidians), *Xenocoeloma* (parasites of polychaetes), and *Aphanodomus* (parasites of bivalves). In these organisms the organs that were initially considered to be testes are in fact dwarf males grafted onto the females' bodies;

Last, we also know of cryptogonochorism in parasitic mollusks such as *Entoconcha* and *Enteroxenos,* both of which are parasites of holothuroidians.

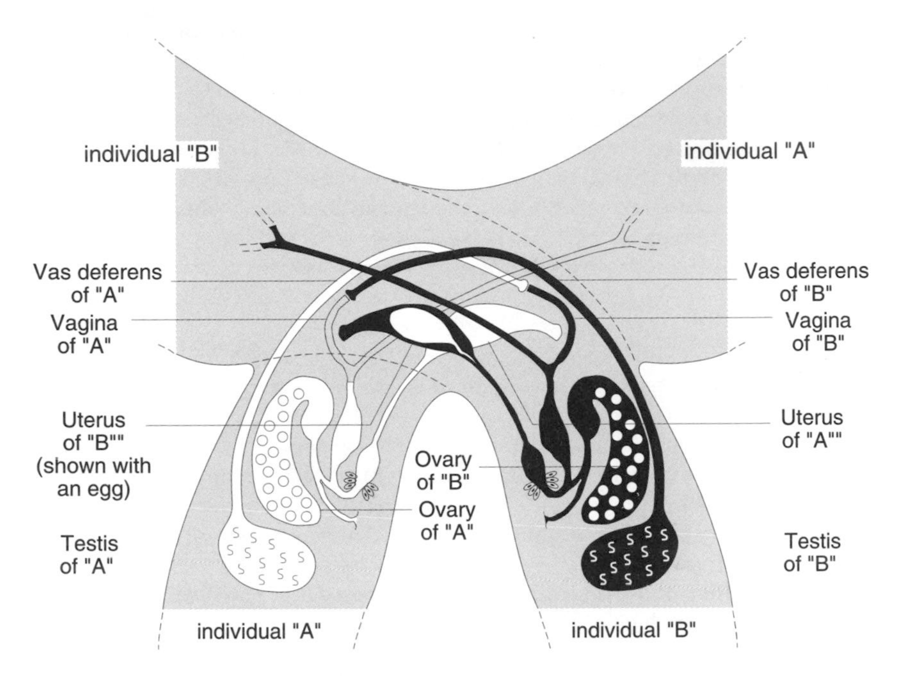

Figure R3.2. This figure represents the area of fusion of double monogeneans *(Diplozoon)* where the genitalia of the two hermaphroditic individuals are in permanent communication. The main regions of the genitalia are labeled for one individual (individual A). An egg is shown in the uterus of individual B. Members of the genus *Diplozoon* are candidates for the title "most enigmatic animal on the planet" (modified from Euzet and Combes 1980).

Relative to such "sexual parasitism," the question Why be parasitic? cannot be answered using any of the usual responses (habitat, mobility, energy) discussed at the beginning of this book. Instead, in these cases the advantage provided by the parasitic mode of life is most certainly that it permits continued encounter between the sexes. In abyssal waters, for example, the scarcity of energy allows predatory fish to exist only as populations with very low densities. As a result, individuals are far from each other and encounters are not frequent. The permanent attachment or fixation of the male(s) onto the female guarantees that when the time comes for reproduction the problem of encountering a mate is already taken care of. When females lay their eggs, the males simply expel their sperm (whatever their position on the female), and fertilization is thus ensured. Some authors believe that males, which are small, are also protected from predators by being attached to the much larger females.

In invertebrates the adaptive value of sexual grafting is most likely the same as in vertebrates. That parasitic invertebrates among all other invertebrates selected this mode of reproduction indirectly indicates that organisms that are simultaneously hermaprodítíc (for which the encounter of two individuals is not necessary for reproduction) are the best candidates for becoming parasitic. In effect, cryptogonochorism is a reconstitution of hermaphroditism (functionally but *not genetically*) from an initial state of gonochorism (separate sexes).

This type of association must be considered mutualism, since the reproductive success of the host female depends on the presence of the parasitic male. The cost the male induces by being fed is then largely, if not entirely, compensated for by his precious "luggage": his gametes.

Perhaps the most extraordinary animals in the world are those monogeneans belonging to the genus *Diplozoon*. In these organisms two hermaphroditic individuals fuse their tissues and form a lifelong connection between their genitalia that results in an inseparable double individual. Figure R3.2 shows that canals originating in each individual end up in the vagina of the other so that cross-fertilization is permanent. Compared with other Platyhelminthes, the members of the genus *Diplozoon* have thus conserved two advantages of hermaphroditism (e.g., sexuality is conserved and all individuals reproduce themselves) but have lost the third—self-fertilization is not possible, and thus there is no limitation to the cost of being rare. One probable but not clearly explained consequence of this mode of life is that the spermatozoa of *Diplozoon* are unique among all monogeneans (Justine, Le Brun, and Mattei 1985).

The way the two organisms fuse is as unusual as the fusion itself: larvae have on their backs the "male" part of what is essentially a snap fastener, with the female part on the ventral side. When the snap closes, the fusion between two larvae starts.

Too Much of a Good Thing

The eukaryote plus mitochondrion plus chloroplast association has produced the plant world; the eukaryote plus mitochondrion association has produced the animal world. In other words, the autotrophy that allows organisms to make a living from the available light, water, and minerals in their environment appears to be antagonistic to the development of intelligence. It is only in those organisms where autotrophy was not acquired that nervous systems developed (probably because of the nonstop quest for food), which then led to the evolution of the brain and ultimately to the level of consciousness observed in humans today.

Conclusion

> A walk through the woods or along the creekside reveals the living world as we commonly think about it—the flash of a colorful bird, all the shades of photosynthesis, the near-frantic activity of insects as the day warms up. . . . Dwelling inside (and on) most of these organisms are other populations, other communities.
> *Moore 1993b*

It is only in the later years of the twentieth century that we have come to realize that parasites and mutualists intervene in both the evolution and the functioning of the living world. It is also only recently that we have come to recognize that a living being is not solely the expression of an autonomic and isolated genome. The phenotype as we observe it is the result of the expression of genetic information that has been modified by the expression of multiple parasite and mutualist genes from both within and outside itself. That is, each host is a singularity wherein different genomes cross pathways, either in conflict with each other or in cooperation. When we see a pigeon fly overhead we rarely realize that it may carry more parasitic and mutualistic genetic information than information of its own.

Because consciousness appeared on earth at the individual level (neither the genome, the species, or an ecosystem thinks—only human beings do), for a long time attention was focused on the individual cost of pathogens, and it has been only in the past few years that parasites and mutualists have began to be included in debates about genetics, sexuality, speciation, adaptation, ecosystems, and conservation.

It is also only recently that the concept of durably interacting organisms has emerged and that these organisms have begun to be recognized as major contributors to biological processes and not just isolated and oddball effectors of their individual host's health and well-being.

Even when a balance had been recognized between free-living and pathogenic organisms, it was understood that such an equilibrium was extremely fragile and that factors such as malnutrition, climatic events, or environmental perturbation could increase the success of pathogen transmission and jeopardize host populations.

Although humans have known for some time how to use their native intelligence to counter their more visible predators, it is only very recently that this ability has been put to use against parasites that could not be seen but that in the end cause the greater mortality. The result has been that within a few decades the advantage in the arms race, which until then had been with the pathogens because of their innately greater genetic variability, may have shifted.

In humans the Lamarckian transmission of knowledge has come to the rescue of Darwinian mechanisms, and the host genome is no longer alone in the battle. In effect, the resulting cultural heritage involves both behavioral modifications (closing the encounter filter by providing clean water, cooking food, limiting contagion, disinfecting, avoiding vectors) and the ability to fight directly against those parasites that manage to contact and settle on or in their hosts (closing the compatibility filter by vaccination, chemotherapy, or gene therapy).

Culture has made the human being a bioengineer who can now count more on intelligence than on genes to fight against parasites.

References

Abel, L., Demenais, F., Baule, M. S., et al. 1989. Genetic susceptibility to leprosy on a Caribbean island: Linkage analysis with five markers. *Int. J. Lepr. Other Mycobact. Dis.* 57:465–471.

Abel, L., Demanais, F., Prata, A., et al. 1991. Evidence for the segregation of a major gene in human susceptibility/resistance to infection by *Schistosoma mansoni. Am. J. Hum. Genet.* 48:959–970.

Able, D. J. 1996. The contagion indicator hypothesis for parasite-mediated sexual selection. *Proc. Natl. Acad. Sci. U.S.A.* 93:2229–2233.

Adamson, M. L., and Caira, J. N. 1994. Evolutionary factors influencing the nature of parasite specificity. *Parasitology* 109:S85–S95.

Adamson, M. L., and Noble, S. 1992. Structure of the pinworm (Oxyurida: Nematoda) guild in the hindgut of the American cockroach, *Periplaneta americana. Parasitology* 104:497–507.

Addicott, J. F., Bronstein, J., and Kjellberg, F. 1993. Evolution of mutualistic life-cycles: Yucca moths and fig-wasps. In *Insect life-cycles: Genetics, evolution and coordination,* ed. F. Gilbert, 143–161. London: Springer-Verlag.

Adema, C. M., and Loker, E. S. 1997. Specificity and immunobiology of larval digenean-snail associations. In *Advances in trematode biology,* ed. B. Fried and T. K. Graczyk, 229–263. Boca Raton, Fla.: CRC Press.

Adlard, R. D., and Lester, R. J. G. 1994. Dynamics of the interaction between the parasitic isopod, *Anilocra pomacentri,* and the coral reef fish, *Chromis nitida. Parasitology* 109:311–324.

Aeby, G. S. 1991. Behavioral and ecological relationships of a parasite and its hosts within a coral reef system. *Pac. Sci.* 45:263–269.

Aeschlimann, A. 1991. Ticks and disease: Susceptible hosts, reservoir hosts, and vectors. In *Parasite-host associations: Coexistence or conflict?* ed. C. A. Toft, A. Aeschlimann, and L. Bolis, 148–156. New York: Oxford University Press.

Agnew, P., and Koella, J. C. 1997. Virulence, parasite mode of transmission, and host fluctuating asymmetry. *Proc. Roy. Soc. Lond.,* ser. B, *Biol. Sci.* 264:9–15.

Aguilar, J. d', Dommanget, J. L., and Préchac, R. 1985. *Guide des libellules d'Europe et d'Afrique du Nord.* Paris: Delachaux, and Niestlé.

Aho, J. M., Bush, A. O., and Wolfe, R. W. 1991. Helminth parasites of bowfin *(Amia calva)* from South Carolina. *J. Helminthol. Soc. Wash.* 58:171–175.

Akurst, R. J. 1983. *Neoplectana* species: Specificity of association with bacteria of the genus *Xenorhabdus. Exp. Parasitol.* 55:258–262.

Alexander, J., and Stimson, W. H. 1988. Sex hormones and the course of parasitic infection. *Parasitol. Today* 4:189–193.

Allison, A. C. 1954. Protection afforded by sickle-cell trait against subtertian malarial infection. *Br. Med. J.* 1:290–294.

Alvarez, F. 1993. Proximity of trees facilitates parasitism by cuckoos *Cuculus canorus* on rufous warblers *Cercotrichas galactotes*. *Ibis* 135:331.

Ameisen, J. C., Idziorek, T., Billaut-Mulot, O., et al. 1995. Apoptosis in a unicellular eukaryote *(Trypanosoma cruzi):* Implications for the evolutionary origin and role of programmed cell death in the control of cell proliferation, differentiation and survival. *Cell Death Differ.* 2:285–300.

Amiri, P., Locksley, R. M., Parslow, T. G., et al. 1992. Tumour necrosis factor α restores granulomas and induces parasite egg-laying in schistosome-infected SCID mice. *Nature* 356:604–607.

Anderson, R. C. 1972. The ecological relationships of meningeal worm and native cervids in North America. *J. Wildl. Dis.* 8:304–310.

Anderson, R. M. 1979. The influence of parasitic infection on the dynamics of host population growth. In *Population dynamics,* ed. R. M. Anderson, B. D. Turner, and L. R. Taylor, 245–281. Oxford: Blackwell Scientific Publications.

Anderson, R. M., and Gordon, D. M. 1982. Processes influencing the distribution of parasite numbers within host populations with special emphasis on parasite-induced host mortalities. *Parasitology* 85:373–398.

Anderson, R. M., and May, R. M. 1978. Regulation and stability of host-parasite population interactions. I. Regulatory processes. *J. Anim. Ecol.* 47:219–247.

———. 1979. Population biology of infectious diseases. Part 1. *Nature* 280:361–367.

———. 1982. Coevolution of hosts and parasites. *Parasitology* 85:411–426.

———. 1985. Helminth infections of humans: Mathematical models, population dynamics, and control. *Adv. Parasitol.* 24:1–101.

———. 1991. *Infectious diseases of humans: Dynamics and control.* New York: Oxford University Press.

Anderson, T. J. C., Blouin, M. S., and Beech, R. N. 1998. Population biology of parasitic nematodes: Applications of genetic markers. *Adv. Parasitol.* 41:219–283.

Anderson, T. J. C., and Jaenike, J. 1997. Host specificity, evolutionary relationships and macrogeographic differentiation among *Ascaris* populations from humans and pigs. *Parasitology* 115:325–342.

Anstett, M. C., Hossaert-MacKey, M., and Kjellberg, F. 1997. Figs and fig pollinators: Evolutionary conflicts in a coevolved mutualism. *Trends Ecol. Evol.* 12:94–99.

Arnal, C., Morand, S., and Kulbicki, M. 1999. Patterns of cleaner wrasse density among three regions of the Pacific. *Mar. Ecol. Prog. Ser.* 177:213–220.

Arnold, E. N. 1986. Mite pockets of lizards, a possible means of reducing damage by ectoparasites. *Biol. J. Linn. Soc.* 29:1–21.

Artois, M., Aubert, M., and Stahl, P. 1990. Organisation spatiale du renard roux (*Vulpes vulpes* L., 1758) en zone d'endémie de rage en Lorraine. *Rev. Ecol. Terre Vie* 45:113–134.

Artois, M., Aubert, M., Blancou, J., et al. 1991. Écologie des comportements de transmission de la rage. *Ann. Rech. Vet.* 22:163–172.

Ataev, G. L., and Dobrovolskij, A. A. 1992. Development of microhemipopulation of *Philophthalmus rhionica* rediae in molluscs naturally infected with other species of trematodes. *Parazitologiya* 26:227–233.

Athias-Binche, F. 1990. Sur le concept de symbiose: L'exemple de la phorésie chez les acariens et son évolution vers le parasitisme ou le mutualisme. *Bull. Soc. Zool. Fr.* 115:77–98.

———. 1991. Ecology and evolution of phoresy in mites. In *Modern acarology*, ed. F. Dusbabek and V. Bukva, 27–41. Prague: Academia.

———. 1993. Dispersal in varying environments: The case of uropodid mites. *Can. J. Zool.* 71:1793–1798.

———. 1994. *La phorésie chez les acariens: Aspects adaptatifs et évolutifs.* Perpignan, France: Editions du Castillet.

Athias-Binche, F., and Morand, S. 1993. From phoresy to parasitism: The example of mites and nematodes. *Res. Rev. Parasitol.* 53:73–79.

Athias-Binche, F., Schwarz, H. H., and Meierhofer, I. 1992. Phoretic association of *Neoseius novius* (Ouds, 1902) (Acari: Uropodina) with *Necrophorus* spp. (Coleoptera: Silphidae): A case of sympatric speciation? *Int. J. Acarol.* 19:75–86.

Atlan, A., and Gouyon, P.-H. 1994. *Les conflits intra-génomiques.* Paris: Société Française de Génétique.

Aubry, P. 1979. L'expédition française de Madagascar de 1895: Un désatre sanitaire. Pourquoi? *Med. Armées* 7:745–752.

Ayala, F. J. 1998. Is sex better? Parasites say "no." *Proc. Natl. Acad. Sci. U.S.A.* 95:3346–3348.

Aznar, F. J., Balbuena, J. A., and Raga, J. A. 1994. Helminth communities of *Pontoporia blainvillei* (Cetacea: Pontoporiidae) in Argentinian waters. *Can. J. Zool.* 72:702–706.

Ba, C. T., Wang, X. Q., Renaud, F., et al. 1993. Diversity and specificity in cestodes of the genus *Moniezia*: Genetic evidence. *Int. J. Parasitol.* 23:853–857.

Babiker, H. A., Charlwood, J. D., Smith, T., et al. 1995. Gene flow and cross-mating in *Plasmodium falciparum* in households in a Tanzanian village. *Parasitology* 111:433–442.

Babiker, H. A., Ranford-Cartwright, L. C., Currie, D., et al. 1994. Random mating in a natural population of the malaria parasite *Plasmodium falciparum. Parasitology* 109:413–421.

Babiker, H. A., and Walliker, D. 1997. Current views on the population structure of *Plasmodium falciparum:* Implications for control. *Parasitol. Today* 13:262–267.

Baer, J. G. 1952. *Ecology of animal parasites.* Urbana: University of Illinois Press.

———. 1957. Répartition et endémicité des cestodes chez les reptiles, oiseaux et mammifères. In *Premier symposium sur la spécificité parasitaire des parasites de vertébrés,* 270–292. Neuchâtel: Attinger.

———. 1972. Transmission d'helminthes larvaires par le lait. *Parassitologia* (Rome) 14:11–14.

Baer, J. G., and Euzet, L. 1961. Les monogènes. In *Traité de zoologie,* ed. P. P. Grassé, 4:561–577. Paris: Masson.

Bakker, R. T. 1983. The deer flees, the wolf pursues: Incongruencies in predator-prey coevolution. In *Coevolution,* ed. D. J. Futuyma and M. Slatkin, 350–382. Sunderland, Mass.: Sinauer.

Balbuena, J. A., and Raga, J. A. 1993. Intestinal helminth communities of the long-finned pilot whale *(Globicephala melas)* off the Faroe Islands. *Parasitology* 106:327–333.

Ball, G. H. 1941. Parasitism and evolution. *Am. Nat.* 77:345–364.

Ballabeni, P. 1995. Parasite-induced gigantism in a snail: A host adaptation? *Funct. Ecol.* 9:887–893.

Bansemir, A. D., and Sukhdeo, M. V. K. 1994. The food resource of adult *Heligmosomoides polygyrus* in the small intestine. *J. Parasitol.* 80:24–28.

Bao, T., and Addicott, J. F. 1998. Cheating in mutualism: Defection of *Yucca baccata* against its yucca moths. *Ecol. Lett.* 1:155–159.

Barbault, R. 1994. *Des baleines, des bactéries et des hommes.* Paris: Odile Jacob.

Barbehenn, K. R. 1969. Host-parasite relationships and species diversity in mammals: An hypothesis. *Biotropica* 1:29–35.

Barker, D. E., Marcogliese, D. J., and Cone, D. K. 1996. On the distribution and abundance of eel parasites in Nova Scotia: Local *versus* regional patterns. *J. Parasitol.* 82:697–701.

Barker, S. C., and Cribb, T. H. 1993. Sporocysts of *Mesostephanus haliasturis* (Digenea) produce miracidia. *Int. J. Parasitol.* 23:137–139.

Barker, S. C., Cribb, T. H., Bray, R. A., et al. 1994. Host-parasite associations on a coral reef: Pomacentrid fishes and digenean trematodes. *Int. J. Parasitol.* 24:643–647.

Barral, V., and Combes, C. 1995. Coévolution et transferts chez les Schistosomes, d'après plusieurs marqueurs moléculaires. Colloque INRA 72, Techniques et utilisation des marqueurs moléculaires, 177–185.

Barral, V., This, P., Imbert-Establet, D., et al. 1993. Genetic variability and evolution of the *Schistosoma* genome analysed by using random amplified polymorphic DNA markers. *Mol. Biochem. Parasitol.* 59:211–222.

Bartoli, P. 1973a. La pénétration et l'installation des cercaires de *Gymnophallus fossarum* P. Bartoli, 1965 (Digenea, Gymnophallidae) chez *Cardium glaucum* Bruguière. *Bull. Mus. Natl. Hist. Nat.,* Sect. A, *Zool. Biol. Ecol. Anim.* 117:319–334.

———. 1973b. Les microbiotopes occupés par les métacercaires de *Gymnophallus fossarum* P. Bartoli, 1965 (Trematoda, Gymnophallidae) chez *Tapes decussatus* L. *Bull. Mus. Natl. Hist. Nat.,* sect. A, *Zool. Biol. Ecol. Anim.* 117:335–349.

Bartoli, P., and Combes, C. 1986. Stratégies de dissémination des cercaires de trématodes dans un écosystème marin littoral. *Acta Oecol. Oecol. Gen.* 7:101–114.

Bartoli, P., Morand, S., Riutort, J.-J., et al. 2000. Acquisition of parasites correlated with social status and behavioural changes in a fish species. *J. Helminthol.* 74:289–293.

Bartoli, P., and Prévot, G. 1978. Recherches écologiques sur les cycles évolutifs de trématodes dans une lagune de Provence (France). II. Le cycle de *Maritrema misenensis* (A. Palombi, 1940). *Ann. Parasitol. Hum. Comp.* 53:181–193.

Basch, P. F., Lie, K. J., and Heynemann, D. 1969. Antagonistic interaction between strigeid and schistosome sporocysts within a snail host. *J. Parasitol.* 55:753–758.

———. 1970. Experimental double and triple infections of snails with larval trematodes. *Southeast Asian J. Trop. Med. Public Health* 1:129–137.

Bastien, P., Blaineau, C., and Pagès, M. 1992. *Leishmania*: Sex, lies and karyotype. *Parasitol. Today* 8:174–177.

Bates, R. M., and Kennedy, C. R. 1990. Interactions between the acanthocephalans *Pomphorhynchus laevis* and *Acanthocephalus anguillae* in rainbow trout: Testing an exclusion hypothesis. *Parasitology* 100:435–444.

———. 1991. Potential interactions between *Pomphorhynchus laevis* and *Acanthocephalus anguillae* in their natural hosts chub, *Leuciscus cephalus,* and the European eel, *Anguilla anguilla. Parasitology* 102:289–297.

Batra, L. R., and Batra, S. W. T. 1985. Floral mimicry induced by mummy-berry fungus exploits host's pollinators as vectors. *Science* 228:1011–1013.

Baudoin, M. 1975. Host castration as a parasitic strategy. *Evolution* 29:335–352.

Bauer, A. M., Russel, A. P., and Dollakon, N. R. 1990. Skin folds in the gekkonid genus *Rhacodactylus:* A natural test of the damage limitation hypothesis of mite-pocket function. *Can. J. Zool.* 68:1196–1201.

Bauer, O. N., and Hoffman, G. L. 1976. Helminth range extension by translocation of fish. In *Wildlife diseases,* ed. L. A. Page, 163–172. New York: Plenum.

Bayne, C. J. 1990. Phagocytosis and non-self recognition in invertebrates. *BioScience* 40:723–731.

———. 1991. Invertebrate host immune mechanisms and parasite escapes. In *Parasite-host associations: Coexistence or conflict?* ed. C. A. Toft, A. Aeschlimann, and L. Bolis, 299–315. New York: Oxford University Press.

Bayne, C. J., and Loker, E. S. 1987. Survival within the snail host. In *The biology of schistosomes: From genes to latrines,* ed. D. Rollinson and A. J. G. Simpson, 321–346. New York: Academic Press.

Bayne, C. J., and Yoshino, T. 1989. Determinants of compatibility in mollusc-trematode parasitism. *Am. Zool.* 29:399–407.

Bayssade-Dufour, C. 1980. L'appareil sensoriel des cercaires et la systématique des trématodes digénétiques. *Mem. Mus. Natl. Hist. Nat.,(r)MDRV* ser. A, *Zool.* 113:1–81.

Beckage, N. E. 1991. Host-parasite hormonal relationships: A common theme? *Exp. Parasitol.* 72:332–338.

Behnke, J. M., Barnard, C. J., and Wakelin, D. 1992. Understanding chronic nematode infections: Evolutionary considerations, current hypotheses and the way forward. *Int. J. Parasitol.* 22:861–907.

Beier, J. C. 1993. Malaria sporozoites: Survival, transmission and disease control. *Parasitol. Today* 9:210–215.

Bell, G. 1982. *The masterpiece of nature.* Berkeley: University of California Press.

Bell, G., and Burt, A. 1991. The comparative biology of parasite species diversity: Internal helminths of freshwater fish. *J. Anim. Ecol.* 60:1047–1063.

Benassi, V., Frey, F., and Carton, Y. 1998. A new specific gene for wasp cellular immune resistance in *Drosophila. Heredity* 80:347–352.

Berrada Rkhami, O., and Gabrion, C. 1986. Synchronisation par la lumière de l'éclosion des larves de deux espèces de bothriocéphales. *Ann. Parasitol. Hum. Comp.* 61:255–260.

Besansky, N. J., and Collins, F. H. 1992. The mosquito genome: Organization, evolution and manipulation. *Parasitol. Today* 8:186–192.

Bethel, W. M., and Holmes, J. C. 1973. Altered evasive behaviour and responses to light in amphipods harboring acanthocephalan cystacanths. *J. Parasitol.* 59:945–956.

Betschart, B., Marti, S., and Glaser, M. 1990. Antibodies against the cuticlin of *Ascaris suum* cross-react with epicuticular structures of filarial parasites. *Acta Trop.* 47:331–338.

Betterton, C. 1979. The intestinal helminths of small mammals in the Malaysian tropical rain forest: Patterns of parasitism with respect to host ecology. *Int. J. Parasitol.* 9:313–320.

Beuret, J., and Pearson, J. C. 1994. Description of a new zygocercous cercaria (Opisthorchiodea: Heterophyidae) from prosobranch gastropods collected at Heron island (Great Barrier Reef, Australia) and a review of Zygocercariae. *Syst. Parasitol.* 27:105–125.

Beveridge, I., and Spratt, D. M. 1996. The helminth fauna of Australian marsupials: Origins and evolutionary biology. *Adv. Parasitology* 37:135–254.

Biémont, C., Arnault, C., and Heizmann, A. 1990. Massive changes in genomic locations of P elements in an inbred line of *Drosophila melanogaster. Naturwissenschaften* 77:485–488.

Binns, E. S. 1982. Phoresy as migration: Some functional aspects of phoresy in mites. *Biol. Rev.* 57:571–620.

Biserkov, V., and Kostadinova, A. 1998. Intestinal helminth communities in the green lizard, *Lacerta viridis,* from Bulgaria. *J. Helminthol.* 72:267–271.

Black, F. L. 1992. Why did they die? *Science* 258:1739–1738.

Blaineau, C., Bastien, P., and Pagès, M. 1992. Multiple forms of chromosome I, II and V in a restricted population of *Leishmania infantum* contrasting with monomorphism in individual strains suggest haploidy or automixy. *Mol. Biochem. Parasitol.* 50:197–204.

Blair, D., Agatsuma, T., and Watanobe, T. 1997. Molecular evidence for the synonymy of three species of *Paragonimus, P. ohirai* Miyazaki, 1939, *P. iloktsuenensis* Chen, 1940 and *P. sadoensis* Miyazaki *et al.,* 1968. J. Helminthol. 71:305–310.

Blair, D., Agatsuma, T., Watanobe, T., et al. 1997. Geographical genetic structure within the human lung fluke, *Paragonimus westermanni,* detected from DNA sequences. *Parasitology* 115:411–417.

Blankespoor, C. L., Pappas, P. W., and Eisner, T. 1997. Impairment of the chemical defence of the beetle, *Tenebrio molitor,* by metacestodes (cysticercoids) of the tapeworm, *Hymenolepis diminuta. Parasitology* 115:105–110.

Blaxter, M. L., De Ley, P., Garey, J. R., et al. 1998. A molecular evolutionary framework for the phylum Nematoda. *Nature* 392:71–75.

Blondel, J. 1986. *Biogéographie évolutive.* Paris: Masson.

Blondel, J., Perret, P., and Maistre, M. 1990. On the genetical basis of the laying date in an island population of blue tit. *J. Evol. Biol.* 3:469–475.

Blondel, J., Perret, P., Maistre, M., et al. 1992. Do harlequin Mediterranean environments function as source sink for blue tits (*Parus coeruleus* L.)? *Landscape Ecol.* 6:213–219.

Blouin, M. S., Dame, J. B., Tarrant, C. A., et al. 1992. Unusual population genetics of a parasitic nematode: mtDNA variation within and among populations. *Evolution* 46:470–476.

Boemare, N., Givaudan, A., Brehélin, M., and Laumond, C. 1997. Symbiosis and pathogenicity of nematode-bacterium complexes. *Symbiosis* 22:21–45.

Boissier, J., Morand, S., and Moné, H. 1999. A review of performance and pathogeneicity of male and female *Schistosoma mansoni* during the life-cycle. *Parasitology* 119:447–454.

Bonas, U., and Van den Ackerveken, G. 1997. Recognition of bacterial avirulence proteins occurs inside the plant cell: A general phenomenon in resistance to bacterial diseases? *Plant J.* 12:1–7.

Bonhomme, F., Catalan, J., Gerasimov, S., et al. 1983. Le complexe d'espèces du genre *Mus* en Europe Centrale et Orientale. *Z. Säugetierkd.* 48:78–85.

Bonsall, M. B., and Hassell, M. P. 1998. Population dynamics of apparent competition in a host-parasitoid assemblage. *J. Anim. Ecol.* 67:918–929.

Booth, D. T., Clayton, D. H., and Block, B. A. 1993. Experimental demonstration of the energetic cost of parasitism in free-ranging hosts. *Proc. Roy. Soc. Lond.*, ser. B, *Biol. Sci.* 253:125–129.

Borgia, G., and Collis, K. 1990. Parasites and bright male plumage in the satin bowerbird *(Ptilonorhynchus violaceus). Am. Zool.* 30:279–285.

Boucot, A. J. 1990. Host-parasite and host-parasitoid relations. In *Evolutionary paleobiology of behaviour and coevolution,* ed. A. J. Boucot, 59–125. New York: Elsevier.

Bouix-Busson, D., Rondelaud, D., and Combes, C. 1985. L'infestation de *Lymnaea glabra* Müller par *Fasciola hepatica* L. Les caractéristiques des émissions cercariennes. *Ann. Parasitol. Hum. Comp.* 60:11–21.

Boulard, C. 1988. Hypodermose bovine. *Point Vet.* (Maisons-Alfort) 20:17–30.

———. 1992. Effect of hypodermin A, an enzyme secreted by *Hypoderma lineatum* (Insect Oestridae) on the bovine immune system. *Vet. Immunol. Immunopathol.* 31:167–177.

Boulétreau, M., Fouillet, P., and Allemand, R. 1991. Parasitoids affect competitive interactions between the sibling species, *Drosophila melanogaster* and *D. simulans. Redia* 84:171–177.

Boulinier, T., Ives, A. R., and Danchin, E. 1996. Measuring aggregation of parasites at different host population levels. *Parasitology* 112:581–587.

Boulinier, T., and Lemel, J.-Y. 1997. Spatial and temporal variations of factors affecting breeding habitat quality in colonial birds: Some consequences for dispersal and habitat selection. *Acta Oecol.* 17:531–552.

Boulinier, T., McCoy, K. D., and Sorci, G. 2001. Dispersal and parasitism. In *Dispersal,* ed. J. Clobert, J. D. Nichols, E. Danchin, and A. Dhondt. Oxford: Oxford University Press. In press.

Bowles, J., Blair, D., and McManus, D. P. 1995a. A molecular phylogeny of the human schistosomes. *Mol. Phylogenet. Evol.* 4:103–109.

———. 1995b. A molecular phylogeny of the genus *Echinococcus. Parasitology* 110:317–328.

Bowman, A. S., Dillwith, J. W., and Sauer, J. R. 1996. Tick salivary prostaglandins: Presence, origin and significance. *Parasitol. Today* 12:388–395.

Boyce, M. S. 1990. The Red Queen visits sage grouse leks. *Am. Zool.* 30:263–270.

Bradley, D. J. 1980. Host genes as determinants of the natural history of parasitic infection. *Colloq. INSERM* 80:525–533.

Braig, H. R., Guzman, H., Tesch, R. B., et al. 1994. Replacement of the natural *Wolbachia* symbiont of *Drosophila simulans* with a mosquito counterpart. *Nature* 367:453–455.

Branch, G. M. 1984. Competition between marine organisms: Ecological and evolutionary implications. *Oceanogr. Mar. Biol. Ann. Rev.* 22:429–593.

Brannon, E. L., Quinn, T. P., Lucetti, G. L., et al. 1981. Compass orientation of sockeye salmon *(Onchorhynchus nerka)* smolts from Great Central Lake, British Columbia. *Can. J. Zool.* 59:1548–1553.

Brehélin, M., Drif, L., Baud, L., et al. 1989. Insect hemolymph: Cooperation between humoral and cellular factors in *Locusta migratoria. Insect Biochem.* 19:301–307.

Bremerman, H. J. 1983. Parasites at the origin of life. *J. Math. Biol.* 16:165–180.

Brémond, P., Sellin, B., Sellin, E., et al. 1993. Arguments en faveur d'une modification du génome (introgression) du parasite humain *Schistosoma haematobium* par des gènes de *S. bovis,* au Niger. *C. R. Acad. Sci.,* ser. 3, *Sci. Vie* 316:667–670.

Breyev, K. A. 1968. On the distribution of cattle grubs (Diptera, Hypodermatidae) in the herds of cattle. I. Negative binomial distribution as a model of the distribution of the cattle grubs. *Parasitology* 2:322–333.

Brodeur, J., and McNeil, J. 1989. Seasonal microhabitat selection by an endoparasitoid through adaptive modification of host behavior. *Science* 244:226–228.

———. 1992. Host behaviour modification by the endoparasitoid *Aphidius nigripes*: A strategy to reduce hyperparasitism. *Ecol. Entomol.* 17:97–104.

Bronseth, T., and Folstad, I. 1997. The effect of parasites on courtship dance in threespine sticklebacks: More than meets the eye. *Can. J. Zool.* 75:589–594.

Bronstein, J. L. 1994a. Our current understanding of mutualism. *Q. Rev. Biol.* 69:31–51.

———. 1994b. Conditional outcomes in mutualistic interactions. *Trends Ecol. Evol.* 9:214–216.

Brooke, M. de L., and Davies, N. B. 1988. Egg mimicry by cuckoos *Cuculus canorus* in relation to discrimination by hosts. *Nature* 335:630–632.

Brooker, L. C., and Brooker, M. G. 1990. Why are cuckoos host specific? *Oikos* 57:301–309.

Brooks, D. R. 1979. Testing the context and extent of host-parasite coevolution. *Syst. Zool.* 28:299–307.

———. 1980. Allopatric speciation and non-interactive parasite community structure. *Syst. Zool.* 29:192–203.

———. 1985. Historical ecology: A new approach to studying the evolution of ecological associations. *Ann. Mo. Bot. Gard.* 72:660–668.

Brooks, D. R., and McLennan, D. A. 1993. *Parascript: Parasites and the language of evolution.* Washington, D.C.: Smithsonian Institution Press.

Brown, C. R., and Brown, M. B. 1986. Ectoparasitism as a cost of coloniality in cliff swallows *(Hirundo pyrrhonota)*. *Ecology* 67:1206–1218.

———. 1992. Ectoparasitism as a cost of natal dispersal in cliff swallows. *Ecology* 73:1718–1723.

Brown, J. H. 1995. *Macroecology.* Chicago: University of Chicago Press.

Brown, W. L., and Wilson, E. O. 1956. Character displacement. *Syst. Zool.* 5:49–64.

Buchmann, K. 1993. Epidemiology and control of *Pseudodactylogyrus* infections in intensive eel culture systems: Recent trends. *Bull. Fr. Pêche Piscic.* 328:66–73.

Bucknell, D., Hoste, H., Gasser, R. B., et al. 1996. The structure of the community of strongyloid nematodes of domestic equids. *J. Helminthol.* 70:185–192.

Bukva, V. 1985. *Demodex flagellurus* sp. n. (Acarii: Demodicinae) from the preputial and clitoral glands of the house mouse, *Mus musculus* L. *Folia Parasitol.* 32:73–81.

Bullini, L., Biocca, E., Nascetti, G., et al. 1978. Ricerche cariologiche ed elettroforetiche su *Parascaris univalens* e *Parascaris equorum. Rend.—Ist. Lomb., Accad. Sci. Lett., A Sci. Mat., Fis., Chim. Geol.* 65:151–156.

Buron, I. de, and Beckage, N. E. 1992. Characterization of a Polydnavirus (PDV) and virus-like filamentous particles (VLFP) in the braconid wasp *Cotesia congregata* (Hymenoptera: Braconidae). *J. Invertebr. Pathol.* 59:315–327.

Buron, I. de, and Maillard, C. 1987. Transfert expérimental d'helminthes adultes chez les poissons par ichtyophagie et cannibalisme. *Ann. Parasitol. Hum. Comp.* 62:188–191.

Buron, I. de, Renaud, F., and Euzet, L. 1986. Speciation and specificity of acanthocephalans: Genetics and morphological studies of *Acanthocephaloides geneticus* sp. nov., parasitizing *Arnoglossus laterna* (Bothidae) from the Mediterranean littoral. *Parasitology* 92:165–171.

Bursten, S. N., Kimsey, R. B., and Owings, D. H. 1997. Ranging of male *Oropsylla montana* fleas via male California ground squirrel *(Spermophilus beecheyi)* juveniles. *J. Parasitol.* 83:804–809.

Bush, A. O. 1990. Helminth communities in avian hosts: Determinants of pattern. In *Parasite communities: Patterns and processes,* ed. G. W. Esch, A. O. Bush, and J. M. Aho, 197–232. London: Chapman, and Hall.

Bush, A. O., Aho, J. M., and Kennedy, C. R. 1990. Ecological versus phylogenetic determinants of helminth parasite community richness. *Evol. Ecol.* 4:1–20.

Bush, A. O., Heard, R. W., and Overstreet, R. M. 1993. Intermediate hosts as source communities. *Can. J. Zool.* 71:1358–1363.

Bush, A. O., and Holmes, J. C. 1986. Intestinal helminths of lesser scaup ducks: An interactive community. *Can. J. Zool.* 64:142–152.

Bush, A. O., Lafferty, K. D., Lotz, J. M., et al. 1997. *Parasitology* meets ecology on its own terms: Margolis *et al.* revisited. *J. Parasitol.* 83:575–583.

Bush, G. L. 1975. Sympatric speciation in phytophagous parasitic insects. In *Evo-*

lutionary strategies of parasitic insects and mites, ed. P. W. Price, 187–206. New York: Plenum.

Butterworth, A. E. 1990. Studies on human schistosomiasis: Chemotherapy, immunity and morbidity. *Ann. Parasitol. Hum. Comp.* 65 (suppl.), 1:53–57.

———. 1994. Human immunity to schistosomes: Some questions. *Parasitol. Today* 10:378–380.

Butterworth, A. E., Bensted-Smith, R., Capron, A., et al. 1987. Immunity in human *schistosomiasis mansoni*: Prevention by blocking antibodies of the expression of immunity in young children. *Parasitology* 94:281–300.

Butterworth, E. W., and Holmes, J. C. 1984. Character divergence in two species of trematodes (Pharyngostomoides: Strigeoidea). *J. Parasitol.* 70:315–316.

Byrne, P. J. 1992. On the biology of *Rhabdochona rotundicaudatum* and *R. cascadilla* (Nematoda: Thalazioidea) in stream fishes from southern Ontario, Canada. *Can. J. Zool.* 70:485–494.

Cabaret, J. 1982. Polymorphisme de *Euparypha pisana* et réceptivité à l'infestation par les Protostrongylides. *Malacologia* 22:49–50.

———. 1990. Sheep and goats: Epidemiology of protostrongylid lungworm infections. *Int. Goat Sheep Res.* 2:142–152.

Cabaret, J., and Vendrous, P. 1986. The response of four terrestrial molluscs to the presence of herbivore feces: Its influence on infection by protostrongylids. *Can. J. Zool.* 64:850–854.

Caira, J. N. 1992. Verification of multiple species of *Pedibothrium* in the Atlantic nurse shark with comments on the Australasian members of the genus. *J. Parasitol.* 78:289–308.

Caira, J. N., Benz, G. W., Borucinska, J., et al. 1997. Pugnose eels, *Simenchelys parasiticus* (Synaphobranchidae) from the heart of a shortfin mako, *Isurus oxyrhynchus* (Lamnidae). *Environ. Biol. Fishes* 49:139–144.

Cann, R. L., Stoneking, M., and Wilson, A. C. 1987. Mitochondrial DNA and human evolution. *Nature* 325:31–36.

Canning, E. U., and Hollister, W. S. 1987. Microsporidia of mammals: Widespread pathogens or opportunistic curiosities. *Parasitol. Today* 3:267–273.

Capron, A., Capron, M., Grangette, C., et al. 1989. IgE and inflammatory cells. *CIBA Found. Symp.* 147:153–160.

Capron, A., Dombrowicz, D., and Capron, M. 1999. Regulation of the immune response in experimental and human schistosomiasis: The limits of an attractive paradigm. *Microbes Infect.* 1:485–490.

Capron, L. 1993. 5000 ans de bilharziose en Egypte. *Rev. Prat.* 43:448.

Carmichael, L. M., Moore, J., and Bjostad, L. B. 1993. Parasitism and decreased response to sex pheromones in male *Periplaneta americana* (Dictyoptera: Blattidae). *J. Insect Behav.* 6:25–31.

Carnevale, P., Bosseno, M.-F., Zoulani, A., et al. 1985. La dynamique de la transmission du paludisme humain en zone de savane herbeuse et de forêt dégradée des environs nord et sud de Brazzaville, R. P. du Congo. *Cah. O. R.S. T.O.M, Ser. Entomol. Med. Parasitol.* 23:(r)MDBO95–115.

Carney, J. P., and Brooks, D. R. 1991. Phylogenetic analysis of *Alloglossidium* Simer,

1929 (Digenea: Plagiorchiiformes: Macroderoididae) with discussion of the origin of truncated life-cycle patterns in this genus. *J. Parasitol.* 77:890–900.

Carney, W. P. 1969. Behavioral and morphological changes in carpenter ants harboring *Dicrocoeliid metacercaria. Am. Midl. Nat.* 82:605–611.

Caro, A. 1998. Effets de facteurs stressants sur la cultivabilité, la viabilité cellulaire et le pouvoir pathogène de *Salmonella typhimurium* dans l'environnement aquatique. Thesis, Université Montpellier II, France.

Caro, A., Combes, C., and Euzet, L. 1997. What makes a fish a suitable host for Monogenea in the Mediterranean? *J. Helminthol.* 71:203–210.

Carter, R., Schofield, L., and Mendis, K. 1992. HLA effects in malaria: Increased parasite-killing immunity or reduced immunopathology. *Parasitol. Today* 8:41–42.

Carton, Y., and David, J. R. 1985. Relation between variability of digging behavior of *Drosophila* larvae and their susceptibility to a parasitic wasp. *Behav. Genet.* 15:143–154.

Carton, Y., Frey, F., and Nappi, A. 1992. Genetic determinism of the cellular immune reaction in *Drosophila melanogaster. Heredity* 69:393–399.

Carton, Y., Haouas, S., Marrakchi, M., et al. 1991. Two competitive parasitoid species coexist in sympatry. *Oikos* 60:222–230.

Carton, Y., and Nappi, A. 1991. The *Drosophila* immune reaction and the parasitoid capacity to evade it: Genetic and coevolutionary aspects. *Acta Oecol.* 12:89–104.

———. 1997. *Drosophila* cellular immunity against parasitoids. *Parasitol. Today* 13:218–226.

Carton, Y., and Sokolowski, M. 1992. Interactions between searching strategies of *Drosophila* parasitoids and the polymorphic behavior of their hosts. *J. Insect Behav.* 5:161–175.

———. 1994. Parasitization of embedded and nonembedded *Drosophila melanogaster* (Diptera: Drosophilidae) pupae by the hymenopteran parasitoid *Pachycrepoideus vindemniae* (Hymenoptera: Pteromalidae). *J. Insect Behav.* 7:129–131.

Caswell, H. 1978. Predator-mediated coexistence: A nonequilibrium model. *Am. Nat.* 112:127–154.

Cavalier-Smith, T. 1985. *The evolution of genome size.* New York: Wiley.

Chabasse, D., Bertrand, G., Leroux, J. P., et al. 1985. Bilharziose à *Schistosoma mansoni* évolutive découverte 37 ans après l'infestation. *Bull. Soc. Pathol. Exot.* 78:643–647.

Chabaud, A. G. 1965. Particularités physiologiques des nématodes: Spécificité parasitaire. In *Traité de zoologie,* vol. 4, part 2, ed. P.-P. Grassé, 548–557. Paris: Masson.

Chabaud, A. G., and Durette-Desset, M.-C. 1978. Parasitisme par plusieurs espèces congénériques. *Bull. Soc. Zool. Fr.* 103:459–464.

Challier, A., and Laveissière, C. 1973. Un nouveau piège pour la capture des glossines (Glossina: Diptera, Muscidae): Description et essais sur le terrain. *Cah. O.R.S.T.O.M., Ser. Entomol. Med. Parasitol.* 11:251–262.

Champagne, D. E. 1994. The role of salivary vasodilatators in bloodfeeding and parasite transmission. *Parasitol. Today* 10:430–433.

Chandler, M., and Cabana, G. 1991. Sexual dichromatism in North American freshwater fish: Do parasites play a role? *Oikos* 60:322–328.

Channon, J. Y., and Kasper, L. H. 1995. Parasite subversion of the host cell endocytic network. *Parasitol. Today* 11:47–48.

Chappell, L. H. 1969. Competitive exclusion between two intestinal parasites of the three-spined sticleback, *Gasterosteus aculeatus* L. *J. Parasitol.* 55:775–778.

———. 1995. The biology of diplostomatid eyeflukes of fishes. *J. Helminthol.* 69:97–101.

Charles, H., Condemine, G., Nardon, C., et al. 1997. Genome size characterization of the principal endocellular symbiotic bacteria of the weevil *Sitophilus oryzae,* using pulse field gel electrophoresis. *Insect Biochem. Mol. Biol.* 27:345–350.

Charlesworth, B., Jarne, P., and Assimiacopoulos, S. 1994. The distribution of transposable elements within and between chromosomes in a population of *Drosophila melanogaster.* III. Element abundances in heterochromatin. *Genet. Res.* 64:183–197.

Charlesworth, B., Sniegowski, P., and Stephan, W. 1994. The evolutionary dynamics of repetitive DNA in eucaryotes. *Nature* 371:215–220.

Chassé, J. L., and Théron, A. 1988. An example of circular statistics in chronobiological studies: Analysis of polymorphism in the emergence rhythms of *Schistosoma mansoni* cercariae. *Chronobiol. Int.* 5:433–439.

Cheng, T. C. 1970. *Symbiosis, organisms living together.* New York: Pegasus.

———. 1986. *General parasitology.* Orlando, Fla.: Academic Press.

———. 1991. Is parasitism symbiosis? A definition of terms and the evolution of concepts. In *Parasite-host associations: Coexistence or conflict?* ed. C. A. Toft, A. Aeschlimann, and L. Bolis, 15–36. New York: Oxford University Press.

Chernin, E. 1970. Behavioral responses of miracidia of *Schistosoma mansoni* and other trematodes to substances emitted by snails. *J. Parasitol.* 56:287–296.

Chilton, N. B., Bao-Zhen, Q., Bogh, H. O., et al. 1999. An electrophoretic comparison of *Schistosoma japonicum* (Trematoda) from different provinces in the People's Republic of China suggests the existence of cryptic species. *Parasitology* 119:375–383.

Chippaux, J.-P. 1993. Dracunculose: La fin d'un fléau. *Cah. Santé* (Montrouge) 3:77–86.

Chippaux, J.-P., Massougbodji, A., Zomadi, A., et al. 1990. Étude épidémiologique des schistosomes dans un complexe lacustre cotier de formation récente. *Bull. Soc. Pathol. Exot.* 83:498–509.

Choe, J. C., and Kim, K. C. 1987. Microhabitat preference and coexistence of ectoparasitic arthropods on Alaskan seabirds. *Can. J. Zool.* 66:987–997.

Christe, P., Møller, A. P., and de Lope, F. 1998. Immunocompetence and nestling survival in the house martin: The tasty chick hypothesis. *Oikos* 83:175–179.

Christe, P., Oppliger, A., and Richner, H. 1994. Ectoparasite affects choice and use of roost sites in the great tit, *Parus major. Anim. Behav.* 47:895–898.

Christe, P., Richner, H., and Oppliger, A. 1996. Begging, food provisioning, and nestling competition in great tit broods infested with ectoparasites. *Behav. Ecol.* 7:127–131.

Christensen, B. M., and Severson, D. W. 1993. Biochemical and molecular basis of mosquito susceptibility to *Plasmodium* and filarioid nematodes. In *Parasites and pathogens of insects*, ed. N. E. Beckage, S. N. Thompson, and B. A. Federici, 245–266. New York: Academic Press.

Cioli, D., and Dennert, G. 1976. The course of *Schistosoma mansoni* infection in thymectomized rats. *J. Immunol.* 117:59–65.

Cioli, D., Knopf, P. M., and Senft, A. W. 1977. A study of *Schistosoma mansoni* transferred into permissive and nonpermissive hosts. *Int. J. Parasitol.* 7:293–297.

Cislo, P. R., and Caira, J. N. 1993. The parasite assemblage in the spiral intestine of the shark *Mustelus canis*. *J. Parasitol.* 79:886–899.

Clarebout, G., Gamain, B., Slomianny, C., et al. 1996. The course of *Plasmodium berghei, P. chabaudi* and *P. yoelii* infections in β-thalassaemic mice. *Parasitology* 112:269–276.

Clark, C. G., and Diamond, L. S. 1994. Pathogenicity, virulence and *Entamoeba histolytica*. *Parasitol. Today* 10:46–47.

Clark, L. 1991. The nest protection hypothesis: The adaptive use of plant secondary compounds by European starlings. In *Bird-parasite interactions: Ecology, evolution, and behaviour*, ed. J. E. Loye and M. Zuk, 205–221. Oxford Ornithology Series 2. Oxford: Oxford University Press.

Clatworthy, A. L. 1998. Neural-immune interactions—an evolutionary perspective. *Neuroimmunomodulation* 5 (3–4): 136–142.

Clayton, D. H. 1990. Mate choice in experimentally parasitized rock doves: Lousy males lose. *Am. Zool.* 30:251–262.

———. 1991a. Coevolution of avian grooming and ectoparasite avoidance. In Bird-parasite Interactions: Ecology, evolution, and behaviour, ed. J. E. Loye and M. Zuk, 258–289. Oxford Ornithology Series 2. Oxford: Oxford University Press.

———. 1991b. The influence of parasites on host sexual selection. *Parasitol. Today* 7:329–334.

Clayton, D. H., Gregory, R. D., and Price, R. D. 1992. Comparative ecology of Neotropical bird lice (Insecta: Phthiroptera). *J. Anim. Ecol.* 61:781–795.

Clayton, D. H., and Tompkins, D. M. 1994. Ectoparasite virulence is linked to mode of transmission. *Proc. Roy. Soc. Lond.*, ser. B, *Biol. Sci.* 256:211–217.

———. 1995. Comparative effects of mites and lice on the reproductive success of rock doves *(Columba livia)*. *Parasitology* 110:195–206.

Cockburn, T. A. 1971. Infectious diseases in ancient populations. *Curr. Anthropol.* 12:45–62.

Coleman, F. C. 1993. Morphological and physiological consequences of parasites encysted in the bulbus arteriosus of an estuarine fish, the sheepshead minnow, *Cyprinodon variegatus*. *J. Parasitol.* 79:247–254.

Colette, J., Sellin, B., Garrigue, G., et al. 1982. Étude épidémiologique de la substitution de *Schistosoma haematobium* par *Schistosoma mansoni* dans une zone d'endémie bilharzienne d'Afrique de l'Ouest (Haute-Volta). *Med. Trop.* 42:289–296.

Combes, C. 1968. Biologie, écologie des cycles et biogéographie de digènes et monogènes d'amphibiens dans l'est des Pyrénées. *Mem. Mus. Natl. Hist. Nat.* 51:1–195.

———. 1971a. Influence of the behaviour of amphibians in helminth life-cycles. In *Behavioural aspects of parasite transmission,* ed. E. U. Canning and C. A. Wright, 151–170. London: Academic Press.

———. 1971b. Über die Rolle des Verhaltens der zweiten zwischen Wirtes im biologischen Zyklus der Gorgoderiden. *Parasitol. Schriftenr.* 21:183–185.

———. 1972. Écologie des Polystomatidae (Monogenea): Facteurs influençant le volume et le rythme de la ponte. *Int. J. Parasitol.* 2:233–238.

———. 1980. Les mécanismes de recrutement chez les métazoaires parasites et leur interprétation en termes de stratégies démographiques. *Vie Milieu* 30:55–63.

———. 1982. Trematodes: Antagonism between species and sterilizing effects on snails in biological control. *Parasitology* 84:151–175.

———. 1983a. Les parasites et leurs cibles vivantes. *Pour Sci.* 70:64–72.

———. 1983b. Application à l'écologie parasitaire des indices d'association fondés sur le caractère présence-absence. *Vie Milieu* 33:203–212.

———. 1987. Préface. In *L'adaptation,* 4–8. Paris: Bibliothèque pour la Science.

———. 1991a. Evolution of parasite life-cycles. In *Parasite-host associations: Coexistence or conflict?* ed. C. A. Toft, A. Aeschlimann, and L. Bolis, 62–82. New York: Oxford University Press.

———. 1991b. Did parasites interfere with the evolution of hominids? Proceedings of the eighth International Congress of Human Genetics, Washington, D.C. *Am. J. Hum. Genet.* 49 (suppl.): 459.

———. 1991c. The schistosome scandal. *Acta Oecol.* 12:165–173.

———. 1991d. Ethological aspects of parasite transmission. *Am. Nat.* 138:866–880.

———. 1991e. Where do human schistosomes come from? *Trends Ecol. Evol.* 5:334–337.

———. 1992. Etre un parasite et transmettre ses gènes. *Pour Sci.* 174:71.

———. 1995. *Interactions durables: Écologie et évolution du parasitisme.* Paris: Masson.

———. 1996. Parasites, biodiversity and ecosystem stability. *Biodivers. Conserv.* 5:953–962.

———. 1997. Fitness of parasites: Pathology and selection. *Int. J. Parasitol.* 27:1–10.

———. 2000. Natural selection and virulence/resistance processes. *Nova Acta Leopoldina.* In press.

Combes, C., Fournier, A., Moné, H., et al. 1994. Behaviours in trematode cercariae that enhance parasite transmission: Patterns and processes. *Parasitology* 109:S57–S67.

Combes, C., and Imbert-Establet, D. 1980. Compared infectivity in rodents of *Schistosoma mansoni* cercariae of human origin and *Schistosoma mansoni* cercariae of murine origin. *J. Helminthol.* 54:167–171.

Combes, C., and Jourdane, J. 1991. Immunodeficiencies in humans: A possible cause of hybridizations and gene introgressions in bisexual parasites. *Acta Oecol.* 12:829–883.

Combes, C., and Knoepffler, L.-P. 1977. Parasitisme d'une population de *Pelobates cultripes* (Cuvier, 1829) à la sortie de l'eau par les postlarves de *Polystoma pelobatis* Euzet and Combes, 1965. *Vie Milieu* 28:215–219.

Combes, C., and Le Brun, N. 1990. Invasions by parasites in continental Europe. In *Biological invasions in Europe and the Mediterranean basin,* ed. F. Di Castri, A. J. Hansen, and M. Debussche, 285–296. Dordrecht: Kluwer Academic Publishers.

Combes, C., and Moné, H. 1987. Possible mechanisms of the decoy effect in *Schistosoma mansoni* transmission. *Int. J. Parasitol.* 17:971–975.

Combes, C., and Nassi, H. 1977. Metacercarial dispersion and intracellular parasitism in a strigeid trematode. *Int. J. Parasitol.* 7:501–503.

Combes, C., and Théron, A. 2000. Metazoan parasites and resource heterogeneity: Constraints and benefits. *Int. J. Parasitol.* 30:299–304.

Connell, J. H. 1980. Diversity and the coevolution of competitors, or the ghost of competition past. *Oikos* 35:131–138.

Connors, V. A., Buron, I. de, and Granath, W. O. 1995. *Schistosoma mansoni:* Interleukin-1 increases phagocytosis and superoxide production by hemocytes and decreases output of cercariae in schistosome-susceptible *Biomphalaria glabrata. Exp. Parasitol.* 80:139–148.

Connors, V. A., Buron, I. de, Jourdane, J., Théron, A., Agner, A., and Granath, W. O. 1998. Recombinant Interleukin-1 mediated killing of *Schistosoma mansoni* primary sporocysts in *Biomphalaria glabrata. J. Parasitol.* 84:920–926.

Connors, V. A., M. J. Lodes, and T. P. Yoshino. 1991. Isolation of a *Schistosoma mansoni* larval excretory-secretory product and its effect on superoxide production by *Biomphalaria glabrata* hemocytes. *J. Invertebr. Pathol.* 58:387–395.

Connors, V. A., and Nickol, B. B. 1991. Effects of *Plagiorhynchus cylindraceus* (Acanthocephala) on the energy metabolism of adult starlings, *Sturnus vulgaris. Parasitology* 103:395–402.

Connors, V. A., and Yoshino, T. P. 1990. *In vitro* effect of larval *Schistosoma mansoni* excretory-secretory products on phagocytosis-stimulated superoxyde production in hemocytes from *Biomphalaria glabrata. J. Parasitol.* 76:895–902.

Conway Morris, S. 1981. Parasites and the fossil record. *Parasitology* 82:489–509.

Cooper, E. L., Zhang, Z., Raftos, D. A., Habicht, G. S., Beck, G., Connors, V., Cossarizza, A., Franceschi, C., Ottaviani, E., and Scapigliati, G. 1995. When did communication in the immune system begin? *Int. J. Immunopathol. Pharmacol.* 7:203–217.

Cortesero, A. M., and Monge, J. P. 1994. Influence of pre-emergence experience on response to host and host plant odours in the larval parasitoid *Eupelmus vuilleti. Entomol. Exp. Appl.* 72:281–288.

Cortesero, A. M., Monge, J. P., and Huignard, J. 1993. Response of the parasitoid *Eupelmus vuilleti* to the odours of the phytophagous host and its host plant in an olfactometer. *Entomol. Exp. Appl.* 69:109–116.

———. 1995. Influence of two successive learning processes on the response of

Eupelmus vuilleti Crw (Hymenoptera: Eupelmidae) to volatile stimuli from hosts and hosts plants. *J. Insect Behav.* 8:751–762.

Côté, I. M., and Poulin, R. 1995. Parasitism and group size in social animals: A meta-analysis. *Behav. Ecol.* 6:159–165.

Cotgreave, P., and Clayton, D. H. 1992. Comparative analysis of time spent grooming by birds in relation to parasite load. *Behaviour* 120:1–12.

Coulson, P. S., and Wilson, R. A. 1989. Portal shunting and resistance to *Schistosoma mansoni* in 129 strain mice. *Parasitology* 99:383–389.

Coustau, C., Carton, Y., Nappi, A. J., et al. 1996. Differential induction of antibacterial transcripts in *Drosophila* susceptible and refractory to parasitism by *Leptopilina boulardi. Insect Mol. Biol.* 3:167–172.

Coustau, C., Renaud, F., Delay, B., et al. 1991. Mechanisms involved in parasitic castration: *In vitro* effects of the trematode *Prosorhynchus squamatus* on the gametogenesis and the nutrient storage metabolim of the marine bivalve *Mytilus edulis. Exp. Parasitol.* 73:36–43.

Coustau, C., Renaud, F., Maillard, C., et al. 1991. Differential susceptibility to a trematode parasite among genotypes of the *Mytilus edulis/galloprovincialis* complex. *Genet. Res.* 57:207–212.

Couvet, D., Atlan, A., Belhassen, E., et al. 1990. Co-evolution between two symbionts: The case of cytoplasmic male-sterility in higher plants. *Oxford Surv. Evol. Biol.* 7:225–249.

Cox, F. E. G. 1989. Parasites and sexual selection. *Nature* 341:289.

———. 1991. Variation and vaccination. *Nature* 349:193.

Coyne, J. A., and Barton, N. H. 1988. What do we know about speciation? *Nature* 331:485–486.

Crampton, J., Morris, A., Lycett, G., et al. 1990. Transgenic mosquitoes: A future vector control strategy? *Parasitol. Today* 6:31–36.

Crampton, J. M. 1994. Molecular studies of insect vectors of malaria. *Adv. Parasitol.* 34:1–31.

Crofton, H. D. 1971. A model of host-parasite relationships. *Parasitology* 63: 343–364.

Croll, N. A., Anderson, R. M., Gyorkos, T. W., et al. 1982. The population biology and control of *Ascaris lumbricoides* in a rural community in Iran. *Trans. Roy. Soc. Trop. Med. Hyg.* 76:187–1970.

Crompton, D. W. T. 1973. The sites occupied by some parasitic helminths in the alimentary tract of vertebrates. *Biol. Rev.* 48:27–83.

———. 1991. Nutritional interactions between hosts and parasites. In *Parasite-host associations: Coexistence or conflict?* ed. C. A. Toft, A. Aeschlimann, and L. Bolis, 228–257. New York: Oxford University Press.

Crowden, A. E., and Broom, D. M. 1980. Effects of the eye-fluke, *Diplostomum spathaceum,* on the behaviour of dace *(Leuciscus leuciscus). Anim. Behav.* 28:287–294.

Cruz, A., and Wiley, J. W. 1989. The decline of an adaptation in the absence of a presumed selection pressure. *Evolution* 43:55–62.

Curtis, L. A. 1987. Vertical distribution of an estuarine snail altered by a parasite. *Science* 235:1509–1511.

———. 1990. Parasitism and the movements of intertidal gastropod individuals. *Biol. Bull.* 179:105–112.

———. 1993. Parasite transmission in the intertidal zone: Vertical migrations, infective stages, and snail trails. *J. Exp. Mar. Biol. Ecol.* 173:197–209.

———. 1997. *Ilyanassa obsoleta* (Gastropoda) as a host for trematodes in Delaware estuaries. *J. Parasitol.* 83:793–803.

Czaplinski, B. 1975. Hymenolepididae parasitizing wild mute swans *Cygnus olor* (Gm.) of different age, in Poland. *Acta Parasitol. Pol.* 23:305–327.

Dailey, M. D., and Walker, W. A. 1978. Parasitism is a factor (?) in single strandings of Southern California cetaceans. *J. Parasitol.* 64:593–597.

Dame, J. B., Blouin, M. S., and Courtney, C. H. 1993. Genetic structure of populations of *Ostertagia ostertagi. Vet. Parasitol.* 46:55–62.

Damian, R. T. 1964. Molecular mimicry: Antigen sharing by parasite and host and its consequences. *Am. Nat.* 98:129–149.

Danchin, E. 1992. The incidence of the tick parasite *Ixodes uriae* in kittiwake *Rissa tridactyla* colonies in relation to the age of the colony, and a mechanism of infecting new colonies. *Ibis* 134:134–141.

Danchin, E. T., Boulinier, T., and Massot, M. 1998. Conspecific reproductive success and breeding habitat selection: Implications for the study of coloniality. *Ecology* 79:2415–2428.

David, P. H., Hudson, D. E., Hadley, T. J., et al. 1985. Immunization of monkeys with a 140 kilodalton merozoite surface protein of *Plasmodium knowlesi* malaria: Appearance of alternate forms of this protein. *J. Immunol.* 134:4146–4152.

Davidson, D. W., and McKey, D. 1993. The evolutionary ecology of symbiotic antplant relationships. *J. Hym. Res.*3 2:13–83.

Davies, N. B., Brooke, M., and Kacelnik, A. 1996. Recognition errors and probability of parasitism determine whether reed warblers should accept or reject mimetic cuckoo eggs. *Proc. Roy. Soc. Lond.*, ser. B, *Biol. Sci.* 263:925–931.

Davies, N. B., Kilner, R. M., and Noble, D. G. 1998. Nestling cuckoos, *Cuculus canorus,* exploit hosts with begging calls that mimic a brood. *Proc. Roy. Soc. Lond.*, ser. B, *Biol. Sci.* 265:673–678.

Dawes, B., and Hughes, D. L. 1970. Fascioliasis: The invasive stages in mammals. *Adv. Parasitol.* 8:259–274.

Dawkins, R. 1982. *The extended phenotype: The gene as the unit of selection.* Oxford: Oxford University Press.

———. 1990. Parasites, desiderata lists and the paradox of the organism. *Parasitology* 100:S63–S73.

Dawkins, R., and Krebs, J. R. 1979. Arms races between and within species. *Proc. Roy. Soc. Lond.*, ser. B, *Biol. Sci.* 205:489–511.

Day, J. F., and Edman, J. D. 1983. Malaria renders mice susceptible to mosquito feeding when gamatocytes are most active. *J. Parasitol.* 69:163–170.

Dean, D. A., Minard, P., Murrel, K. D., et al. 1978. Resistance of mice to secondary infection with *Schistosoma mansoni.* II. Evidence for a correlation between egg deposition and worm elimination. *Am. J. Trop. Med. Hyg.* 27:957–965.

Deelder, A. M., Miller, R. L., de Jonge, N., et al. 1990. Detection of schistosome antigen in mummies. *Lancet* 335:724–725.

Delahay, R. J., Speakman, J. R., and Moss, R. 1995. The energetic consequences of parasitism: Effects of a developing infection of *Trichostrongylus tenuis* (Nematoda) on red grouse *(Lagopus lagopus scoticus)* energy balance, body weight and condition. *Parasitology* 110:473–482.

de Lope, F., Gonzalez, G., Pérez, J. J., et al. 1993. Increased detrimental effects of ectoparasites on their bird hosts during adverse environmental conditions. *Oecologia* 95:234–240.

de Lope, F., and Møller, A. P. 1993. Effects of ectoparasites on reproduction of their swallow hosts: A cost of being multi-brooded. *Oikos* 67:557–562.

de Meeus, T., Michalakis, Y., Renaud, F., et al. 1993. Polymorphism in heterogeneous environments, evolution of habitat selection and sympatric speciation: Soft and hard selection models. *Evol. Ecol.* 7:175–198.

de Meeus, T., Renaud, F., and Gabrion, C. 1990. A model for studying isolation mechanisms in parasite populations: The genus *Lepeophtheirus* (Copepoda, Caligidae). *J. Exp. Zool.* 254:207–214.

Derothe, J.-M., Loubes, C., Perriat-Sanguinet, M., et al. 1999. Experimental trypanosomiasis of natural hybrids between house mouse subspecies. *Int. J. Parasitol.* 29:1011–1016.

Des Clers, S. A., and Wootten, R. 1990. Modelling the population dynamics of the sealworm *Pseudoterranova decipiens. Neth. J. Sea Res.* 25:291–299.

Despommier, D. D. 1993. *Trichinella spiralis* and the concept of niche. *J. Parasitol.* 79:472–482.

Després, L., Imbert-Establet, D., Combes, C., et al. 1992. Molecular evidence linking hominid evolution to recent radiation of schistosomes (Platyhelminthes: Trematoda). *Mol. Phylogenet. Evol.* 1:295–304.

Després, L., and Maurice, S. 1995. The evolution of dimorphism and separate sexes in schistosomes. *Proc. Roy. Soc. Lond.*, ser. B, *Biol. Sci.* 262:175–180.

Dessein, A. 1988. Human resistance to *Schistosoma mansoni* is associated with IgG reactivity to a 37-kDa larval surface antigen. *J. Immunol.* 140:2727–2736.

Dessein, A. J., Hillaire, D., El Wali, N. E., et al. 1999. Severe hepatic fibrosis in *Schistosoma mansoni* infection is controlled by a major locus that is closely linked to the interferon-gamma receptor gene. *Am. J. Hum. Genet.* 65:709–721.

Dessein, A. J., Marquet, S., Henri, S., et al. 1999. Infection and disease in human *Schistosomiasis mansoni* are under distinct major gene control. *Microbes Infect.* 1:561–567.

de Vries, R. R. 1991. Genetic control of immunopathology induced by *Mycobacterium leprae. Am. J. Trop. Med. Hyg.* 44:12–16.

de Witt, H. C. D. 1994. *Histoire du développement de la biologie.* Lausanne: Presses Polytechniques et Universitaires Romandes.

Dias, P. C., and Blondel, J. 1996. Local specialization and maladaptation in the Mediterranean blue tit *(Parus caeruleus). Oecologia* 107:79–86.

Dickman, C. R. 1992. Commensal and mutualistic interactions among terrestrial vertebrates. *Trends Ecol. Evol.* 7:194–197.

Di Giovanni, L., Cochrane, A. H., and Enea, V. 1990. On the evolutionary history of the circumsporozoite protein in Plasmodia. *Exp. Parasitol.* 70:373–381.

Dissous, C., and Capron, A. 1995. Convergent evolution of tropomyosin epitopes. *Parasitol. Today* 11:45–46.

Dissous, C., Grzych, J. M., and Capron, A. 1986. *Schistosoma mansoni* shares a protective oligosaccharide epitope with freshwater and marine snails. *Nature* 323:443–445.

Dissous, C., Torpier, G., Duvaux-Miret, O., et al. 1990. Structural homology of tropomyosins from the human trematode *Schistosoma mansoni* and its intermediate host *Biomphalaria glabrata. Mol. Biochem. Parasitol.* 43:245–256.

Dobbelacre, D. A. E., Coquerelle, T. M., Roditi, I. J., et al. 1988. *Theileria parva* infection induces autocrine growth of bovine lymphocytes. *Proc. Natl. Acad. Sci. U.S.A.* 85:4730–4734.

Dobson, A. P. 1985. The population dynamics of competition between parasites. *Parasitology* 91:317–347.

———. 1988. The population biology of host-induced changes in host behaviour. *Q. Rev. Biol.* 63:139–165.

Dobson, A. P., and Hudson, P. J. 1992. Regulation and stability of a free-living host-parasite system, *Trichostrongylus tenuis* in red grouse. II. Population models. *J. Anim. Ecol.* 61:487–500.

Dobson, A. P., and Merenlender, A. 1991. Coevolution of macroparasites and their hosts. In *Parasite-host associations: Coexistence or Conflict?* ed. C. A. Toft, A. Aeschlimann, and L. Bolis, 83–101. New York: Oxford University Press.

Dobzhansky, T. 1973. Nothing in biology makes sense except in the light of evolution. *Amer. Biol. Teacher* 35:125–129.

Doenhoff, M. J. 1997. A role for granulomatous inflammation in the transmission of infectious disease: Schistosomiasis and tuberculosis. *Parasitology* 115:S113–S125.

———. 1998. Granulomatous inflammation and the transmission of infection: Schistosomiasis and TB too. *Immunol. Today* 19:462–467.

Dogiel, V. A. 1964. *General parasitology.* Trans. Z. Kabata. Edinburgh: Oliver, and Boyd.

Donelson, J. E., and Fulton, A. B. 1989. The pushy ways of a parasite. *Nature* 342:615–616.

Donges, J. 1972. Double infection experiments with echinostomatids (Trematoda) in *Lymnaea stagnalis* by implantation of rediae and exposure to miracidia. *Int. J. Parasitol.* 2:409–423.

Donnelly, R. E., and Reynolds, J. D. 1994. Occurrence and distribution of the parasitic copepod *Leposphilus labrei* on corkwing wrasse *(Crenilabrus melops)* from Mulroy Bay, Ireland. *J. Parasitol.* 80:331–332.

Dorchies, P., and Guitton, C. 1993. Les ascaridioses des carnivores domestiques. *Recl. Med. Vet.* 169:333–343.

Douglas, A. E. 1994. *Symbiotic interactions.* Oxford: Oxford University Press.

———. 1999. Mutualisms, ecology and ecologists. *Bull. Br. Ecol. Soc.* 30:14–15.

Dubois, G. 1929. Les cercaires de la région de Neuchâtel. *Bull. Soc. Neuchatel. Sci. Nat.* 53:1–177.

Dufour, V., and Galzin, R. 1993. Colonization patterns of reef fish larvae to the lagoon at Moorea Island, French Polynesia. *Mar. Ecol. Prog. Ser.* 102:143–152.

Dugatkin, L. A., FitzGerald, G. J., and Lavoie, J. 1994. Juvenile three-spined sticklebacks avoid parasitised conspecifics. *Environ. Biol. Fishes* 39:215–218.

Duke, B. O. L. 1955. The development of *Loa* in flies of the genus *Chrysops* and the probable significance of the different species in the transmission of loiasis. *Trans. Roy. Soc. Trop. Med. Hyg.* 49:115–121.

Dumont, H. J., and Hinnekint, B. O. N. 1973. Mass migration in dragonflies, especially in *Libellula quadrimaculata* L.: A review, a new ecological approach and hypothesis. *Odonatologica* 2:1–2.

Dupas, S., and Boscaro, M. 1999. Geographic variation and evolution of immunosuppressive genes in a *Drosophila* parasitoid. *Ecography* 22:284–291.

Dupas, S., Brehélin, M., Frey, F., et al. 1996. Immune suppressive virus-like particles in a *Drosophila* parasitoid: Significance of their intraspecific morphological variations. *Parasitology* 113:207–212.

Dupas, S., Frey, F., and Carton, Y. 1998. A single parasitoid segregating factor controls immune suppression in *Drosophila. J. Hered.* 89:306–311.

Durette-Desset, M.-C. 1985. Trichostrongyloid nematodes and their vertebrate hosts: Reconstruction of the phylogeny of a parasitic group. *Adv. Parasitol.* 24:239–306.

Duvaux-Miret, O., Stefano, G. B., Smith, E. M., et al. 1992. Immunosuppression in the definitive and intermediate hosts of the human parasite *Schistosoma mansoni* by release of immunoactive neuropeptides. *Proc. Natl. Acad. Sci. U.S.A.* 89:778–781.

Dye, C. 1991. Population genetics of nonclonal, nonrandomly mating malaria parasites. *Parasitol. Today* 7:236–240.

Dye, C., and Davies, C. R. 1990. Glasnot and the great gerbil: Virulence polymorphisms in the epidemiology of leishmaniasis. *Trends Ecol. Evol.* 5:237–238.

Dzika, E. 1999. Microhabitats of *Pseudodactylogyrus anguillae* and *P. bini* (Monogenea: Dactylogyridae) on the gills of large-size European eel *Anguilla anguilla* from Lake Gaj, Poland. *Folia Parasitol.* 46:33–36.

Ebert, D. 1994. Virulence and local adaptation of a horizontally transmitted parasite. *Science* 265:1084–1086.

———. 1995. The ecological interactions between a microsporidian parasite and its host *Daphnia magna. J. Anim. Ecol.* 64:361–369.

———. 1998. Infectivity, multiple infections, and the genetic correlation betxeen within-host growth and parasite virulence: A reply to Hochberg. *Evolution* 52:1869–1871.

Ebert, D., and Hamilton, W. D. 1996. Sex against virulence: The coevolution of parasistic diseases. *Trends Ecol. Evol.* 11:79–82.

Ebert, D., and Herre, E. A. 1996. The evolution of parasitic diseases. *Parasitol. Today* 12:96–101.

Ebert, D., and Mangin, K. L. 1997. The influence of host demography on the evolution of virulence of a microsporidian gut parasite. *Evolution* 51:1828–1837.

Eckwalanga, M., Marussig, M., Tavares, M. D., et al. 1994. Murine AIDS protects mice against experimental cerebral malaria: Down-regulation by interleukin 10 of a T-helper type 1 CD4+ cell-mediated pathology. *Proc. Natl. Acad. Sci. U.S.A.* 91:8097–8101.

Edwards, D. D., and Bush, A. O. 1989. Helminth communities in avocets: importance of the compound community. *J. Parasitol.* 75:225–238.

Elmes, G. W., Thomas, J. A., and Wardlaw, J. C. 1991. Larvae of *Maculinea rebeli,* a large-blue butterfly, and their *Myrmica* host ants: wild adoption and behaviour in ant-nests. *J. Zool.* (London) 223:447–460.

Emlen, S. T., and Wrege, P. H. 1986. Forced copulations and intra-specific parasitism: Two costs of social living in the white-fronted bee-eater. *Ethology* 71:2–29.

Emson, R. H., and Mladenov, P. V. 1987. Brittlestar host specificity and apparent host discrimination by the parasitic copepod *Ophyopsyllus reductus. Parasitology* 94:7–15.

Erttmann, K. D., Unnasch, T. R., Greene, B. M., et al. 1987. A DNA sequence specific for forest form *Onchocerca volvulus. Nature* 327:415–417.

Esch, G. W., and Fernandez, J. C. 1993. *A functional biology of parasitism.* London: Chapman and Hall.

Esch, G., Gibbons, J. W., and Bourque, J. E. 1975. An analysis of the relationship between stress and parasitism. *Amer. Midl. Nat.* 93:339–353.

Esch, G. W., Hazen, T. C., and Aho, J. M. 1977. Parasitism and r- and K-selection. In *Regulation of parasite populations,* ed. G. W. Esch, 9–62. New York: Academic Press.

Eslin, P., Giordanengo, P., Fourdrain, Y., et al. 1996. Avoidance of encapsulation in the absence of VLP by a braconid parasitoid of *Drosophila* larvae: An ultrastructural study. *Can. J. Zool.* 74:2193–2198.

Establet, D., and Combes, C. 1992. Relocation of *Schistosoma mansoni* in the lungs and resistance to reinfection in *Rattus rattus. Parasitology* 104:51–57.

Euzet, L. 1956. Recherches sur les cestodes tetraphyllides des sélaciens des côtes de France. *Nat. Monspel., Ser. Zool.* 3:1–265.

———. 1989. Ecologie et parasitologie. *Bull. Ecol.* 20:277–280.

Euzet, L., and Combes, C. 1980. Les problèmes de l'espèce chez les animaux parasites. *Bull. Soc. Zool. Fr.* 40:239–285.

———. 1998. The selection of habitats among the Monogenea. *Int. J. Parasitol.* 28:1645–1652.

Euzet, L., and Raibaut, A. 1985. Les maladies parasitaires en pisciculture marine. *Symbiosis* 17:51–68.

Euzet, L., Renaud, F., and Gabrion, C. 1984. Le complexe *Bothriocephalus scorpii* (Mueller, 1776): Différenciation à l'aide des méthodes biochimiques de deux espèces parasites du turbot *(Psetta maxima)* et de la barbue *(Scophthalmus rhombus). Bull. Soc. Zool. Fr.* 109:84–88.

Ewald, P. W. 1994a. *The evolution of infectious disease.* Oxford: Oxford University Press.

———. 1994b. Evolution of mutation rate and virulence among human retroviruses. *Philos. Trans. Roy. Soc. Trop. Med. Hyg.* 346:333–343.

——— 1995. The evolution of virulence: A unifying link between parasitology and ecology. *J. Parasitol.* 81:659–669.

———. 1996. Guarding against the most dangerous emerging pathogens: Insights from evolutionary biology. *Emerg. Infect. Dis.* 2:245–257.

Excoffier, L., and Roessli, D. 1990. Origine et évolution de l'ADN mitochondrial humain: Le paradigme perdu. *Bull. Mem. Soc. Anthropol. Paris* 2:25–42.

Fain, A. 1992. Progrès et lacunes dans nos connaissances de la gale sarcoptique chez l'homme. *Bull. Soc. Fr. Parasitol.* 10:295–310.

Faliex, E., and Morand, S. 1994. Population dynamics of the metacercarial stage of the bucephalid trematode *Labratrema minimus* (Stossich, 1887) from Salses-Leucate lagoon (France) during the cercarial shedding period. *J. Helminthol.* 68:35–40.

Feener, D. H., Jr. 1981. Competition between ant species: Outcome controlled by parasitic flies. *Science* 214:815–816.

Feliu, C., Renaud, F., Catzeflis, F., et al. 1997. A comparative analysis of parasite species richness of iberian rodents. *Parasitology* 115:453–466.

Felsenstein, J. 1985. Phylogenies and the comparative method. *Am. Nat.* 125:1–15.

Ferdig, M. T., Beerntsen, B. T., Spray, F. J., et al. 1993. Reproductive costs associated with resistance in a mosquito-filarial worm system. *Am. J. Trop Med. Hyg.* 49:756–762.

Fernandez, J., and Esch, G. W. 1991a. Guild structure of larval trematodes in the snail *Helisoma anceps:* Patterns and processes at the individual host level. *J. Parasitol.* 77:528–539.

———. 1991b. The component community structure of larval trematodes in the pulmonate snail *Helisoma anceps. J. Parasitol.* 77:540–550.

Fernandez, J., Goater, T. M., and Esch, G. W. 1991. Population dynamics of *Chaetogaster limnaei limnaei* (Oligocheta) as affected by a trematode parasite in *Helisoma anceps* (Gastropoda). *Am. Midl. Nat.* 125:195–205.

Fiala-Medioni, A. 1988. Synthèse sur les adaptations structurales liées à la nutrition des mollusques bivalves des sources hydrothermales profondes. *Oceanologia Acta* 8:173–179.

Fiala-Medioni, A., and Le Pennec, M. 1989. Adaptive features of the bivalve molluscs associated with fluid venting in the subduction zones of Japan. *Palaeogeogr. Palaeoclimatol. Palaeoecol.* 71:161–167.

Fiala-Medioni, A., and Métivier, C. 1985. Ultrastructure of the gill of the hydrothermal bivalve *Calyptogena magnifica* with a discussion on its nutrition. *Mar. Biol.* 90:215–222.

Fiorillo, R. A., and Font, W. F. 1996. Helminth community structure of four species of *Lepomis* (Osteichthyes: Centrarchidae) from an oligohaline estuary in southeastern Louisiana. *J. Helminthol. Soc. Wash.* 63:24–30.

Fisher, R. A. 1930. *The genetical theory of evolution.* Oxford: Oxford University Press.

Flegr, J., and Havlicek, J. 1999. Changes in personnality profile of young women with latent toxoplasmosis. *Folia Parasitol.* 46:22–28.

Flegr, J., and Hrdy, I. 1994. Influence of chronic toxoplasmosis on some human personality factors. *Folia Parasitol.* 41:122–126.

Flegr, J., Zitkova, S., Kodym, P., et al. 1996. Induction of changes in human behaviour by the parasitic protozoan *Toxoplasma gondii. Parasitology* 113:49–54.

Fleming, J. A. 1992. PolyDNAviruses: Mutualists and pathogens. *Ann. Rev. Entomol.* 37:401–425.

Fleury, F. 1993. Les rythmes circadiens d'activité chez les hyménoptères parasitoïdes de drosophiles: Variabilité, déterminisme génétique, signification écologique. Thesis, Université Claude Bernard Lyon I, France.

Fleury, F., Pompanon, F., Mimouni, F., et al. 1991. Daily rhythmicity of locomotor activity in adult hymenopteran parasitoids. *Redia* 74:287–293.

Floate, K. D., Kearsley, M. J. C., and Whitham, T. G. 1993. Elevated herbivory in plant hybrid zones: *Chrysomela confluens populus* and phenomenological sinks. *Ecology* 74:2056–2065.

Foley, M., and Tilley, L. 1995. What makes a malaria host? *Parasitol. Today* 11:111–112.

Folstad, I., and Karter, A. J. 1992. Parasites, bright males, and the immunocompetence handicap. *Am. Nat.* 139:603–622.

Folstad, I., Nilssen, A. C., Halvorsen, O., et al. 1991. Parasite-avoidance: The cause of post-calving migrations in *Rangifer*? *Can. J. Zool.* 69:2423–2429.

Fons, R., and Mas-Coma, S. 1990. Système hôte/parasite et "syndrome insulaire": Les micromammifères et leurs helminthes parasites dans les îles méditerranéennes. I Congreso Internacional de las Asociaciones Sudoccidental-Europeas de Parasitologia, Valencia.

Forbes, M. R. L. 1993. Parasitism and host reproductive effort. *Oikos* 67:444–450.

Fournier, A., Clément, P., Mimouni, P., et al. 1993. Swimming patterns of schistosome miracidia analysed by computer aided trajectometry. *Ethol. Ecol. Evol.* 4:477–487.

Fraile, L., Escouffier, Y., and Raibaut, A. 1993. Analyse des correspondances de données planifiées: Étude de la chémotaxie de la larve infestante d'un parasite. *Biometrics* 49:1142–1153.

Frank, S. A. 1992. Models of plant-pathogen coevolution. *Trends Genet.* 8:213–218.

———. 1993. Coevolutionary genetics of plants and pathogens. *Evol. Ecol.* 7:45–75.

———. 2000. Polymorphism of attack and defense. *Trends Ecol. Evol.* 15:167–171.

Freedman, H. I. 1990. A model of predator-prey dynamics as modified by the action of a parasite. *Math. Biosci.* 99:143–155.

Freeland, W. J. 1983. Parasites and the coexistence of animal species. *Am. Nat.* 121:223–236.

Freeland, W. J., and Boulton, W. J. 1990. Coevolution of food webs: Parasites, predators and plant secondary compounds. *Biotropica* 24:309–317.

Freeland, W. T. 1976. Pathogens and the evolution of primate sociality. *Biotropica* 8:12–24.

Frevert, U., Sinnis, P., Cerami, C., et al. 1993. Malaria circumsporozoite protein binds to heparan sulfate proteoglycans associated with the surface membrane of hepatocytes. *J. Exp. Med.* 177:1287–1298.

Friedman, M. J. 1978. Erythrocytic mechanisms of sickle-cell resistance to malaria. *Proc. Natl. Acad. Sci. U.S.A.* 80:5421–5424.

Friedman, M. J., Roth, E. F., Nagel, R. L., et al. 1979. *Plasmodium falciparum*: Physiological interactions with the human sickle cell. *Exp. Parasitol.* 47:73–80.

Fritz, R. S. 1982. Selection for host modification by insect parasitoids. *Evolution* 36:283–288.

Frontier, S. 1977. Réflexions pour une théorie des écosystèmes. *Bull. Ecol.* 8:445–464.

Funch, P., and Kristensen, R. M. 1995. Cycliophora is a new phylum with affinities to Entoprocta and Ectoprocta. *Nature* 378:711–714.

Futuyma, D. J. 1986. *Evolutionary biology.* Sunderland, Mass.: Sinauer.
———. 1998. *Evolutionary biology.* 3d ed. Sunderland, Mass.: Sinauer.
Galaktionov, K. V. 1993. Trematode life cycles as components of ecosystems. *Proc. Russ. Acad. Sci. Ser. Biol.* 2:1–19.
———. 1996. Life cycles and distribution of seabird helminths in Arctic and subarctic regions. *Bull. Scand. Soc. Parasitol.* 6:31–49.
Galat-Luong, A., and Galat, G. In press. Chimpanzees and baboons drink filtered water. *Folia Primatol.*
Galli, P., Mariniello, L., Crosa, G., et al. 1998. Populations of *Acanthocephalus anguillae* and *Pomphorhynchus laevis* in rivers with different pollution levels. *J. Helminthol.* 72:331–335.
Galtier, N., and Gouy, M. 1995. Inferring phylogenies from DNA sequences of unequal base compositions. *Proc. Natl. Acad. Sci. U.S.A.* 92:11317–11321.
Gandon, S., Michalakis, Y., and Ebert, D. 1996. Temporal variability and local adaptation. *Trends Ecol. Evol.* 11:431.
Gao, F., Bailes, E., Robertson, D. L., et al. 1999. Origin of HIV1 in *Pan troglodytes troglodytes. Nature* 397:436–441.
Garnham, P. C. C. 1958. Zoonoses or infections common to man and animals. *J. Trop. Med. Hyg.* 61:92–94.
Garnick, E., and Margolis, L. 1990. Influence of four species of helminth parasites on orientation of seaward migrating sockeye salmon *(Oncorhynchus nerka)* smolts. *Can. J. Fish. Aquat. Sci.* 47:2380–2389.
Gaume, L., and McKey, D. 1998. Protection against herbivores of the myrmecophyte *Leonardoxa africana* (Baill.) Aubrev-T3 by its principal ant inhabitant *Aphomomyrmex afer* Emery. *C. R. Acad. Sci.,* ser. 3, *Sci. Vie* 321:593–601.
———. 1999. An ant-plant mutualism and its host-specific parasite: Activity rhythms, young leaf patrolling, and effects on herbivores of two specialist plant ants inhabiting the same myrmecophyte. *Oikos* 84:130–144.
Gelnar, M. 1987. Experimental verification of the effect of physical condition of *Gobio gobio* (L.) on the growth rate of micropopulations of *Gyrodactylus gobiensis* Gläser, 1974 (Monogenea). *Folia Parasitol.* 34:211–217.
Gemmill, A. W., and Read, A. F. 1998. Counting the cost of disease resistance. *Trends Ecol. Evol.* 13:8–9.
Gemmill, A. W., Viney, M. E., and Read, A. F. 1997. Host immune status determines sexuality in a parasitic nematode. *Evolution* 51:393–401.
———. 2000. The evolutionary ecology of host-specificity: Experimental studies with *Strongyloides ratti. Parasitology* 120:429–437.
Gérard, C. 1993. Écologie d'une interaction durable: *Schistosoma mansoni/Biomphalaria glabrata.* Dynamique de l'infrapopulation parasite dans l'écosystème mollusque, interactions spatiales et énergétiques. Thesis, Université de Paris VI.
Gérard, C., Moné, H., and Théron, A. 1993. *Schistosoma mansoni–Biomphalaria glabrata*: Dynamics of the sporocyst population in relation to the miracidial dose and the host size. *Can. J. Zool.* 71:1880–1885.
Gérard, C., and Théron, A. 1997. Age/size and time-specific effects of *Schistosoma*

mansoni on energy allocation patterns of its snail host *Biomphalaria glabrata. Oecologia* 112:447–452.
Ghosh, A., Edwards, M. J., and Jacobs-Lorena, M. 2000. The journey of the malaria parasite in the mosquito: Hopes for the new century. *Parasitol. Today* 16:196–201.
Gibson, W., and Stevens, J. 1999. Genetic exchange in the Trypanosomatidae. *Adv. Parasitol.* 43:1–46.
Giles, N. 1987. Predation risk and reduced foraging activity in fish: Experiments with parasitized and non-parasitized three-spined sticklebacks, *Gasterosteus aculeatus* L. *J. Fish Biol.* 31:37–44.
Ginsburger-Vogel, T., Carre-Lecuyer, M., and Fried-Montaufier, M. C. 1980. Transmission expérimentale de la thélygénie liée à l'intersexualité chez *Orchestia gammarellus* (Pallas): Analyse des phénotypes sexuels dans les descendances de femelles normales transformées en femelles thélygènes. *Arch. Zool. Exp. Gen.* 122:261–270.
Giraudoux, P. 1991. Utilisation de l'espace par les hôtes du taenia multiloculaire (*E. multilocularis*): Conséquences épidémiologiques. Thesis, Université de Dijon, France.
Glen, D. R., and Brooks, D. R. 1986. Parasitological evidence pertaining to the phylogeny of the hominoid primates. *Biol. J. Linn. Soc.* 29:331–354.
Goater, C. P., and Bush, A. O. 1988. Intestinal helminth communities in long-billed curlews: The importance of congeneric host-specialists. *Holarct. Ecol.* 11:140–145.
Goater, T. M., Esch, G. W., and Bush, A. O. 1987. Helminth parasites of sympatric salamanders: Ecological concepts at infracommunity, component and compound community levels. *Am. Midl. Nat.* 118:289–300.
Godfrey, D. G., Baker, R. D., Rickman, L. R., et al. 1990. The distribution, relationships and identification of enzymic variants within the subgenus *Trypanozoon. Adv. Parasitol.* 29:1–74.
Goff, L. J. 1991. Symbiosis, interspecific gene transfer, and the evolution of new species: A case study in the parasitic red algae. In *Symbiosis as a source of evolutionary innovation,* ed. L. Margulis and R. Fester, 341–363. Cambridge: MIT Press.
Golvan, Y. J., and Rioux, J. A. 1961. Écologie des mérions du Kurdistan Iranien: Relation avec l'épidémiologie de la peste rurale. *Ann. Parasitol. Hum. Comp.* 36:449–558.
———. 1963. La peste, facteur de régulation des populations de mérions au Kurdistan iranien. *Rev. Ecol. Terre Vie* 1:3–34.
Gordon, D. M., and Rau, M. E. 1982. Possible evidence for mortality induced by the parasite *Apatemon gracilis* in a population of brook sticklebacks *(Culaea inconstans). Parasitology* 84:41–47.
Gould, S. J. 1980. *The panda's thumb: More reflections in natural history.* New York: Norton.
Gouyon, P.-H. 1994. Le sexe, ce choix de l'évolution biologique. *Turbulence* 1:8–15.

Grabda-Kazubska, B. 1976. Abbreviation of life-cycles in plagiorchid trematodes: General remarks. *Acta Parasitol. Pol.* 24:125–141.

Granath, W. O., Connors, V. A., and R. L. Tarleton. 1994. Interleukin-1 activity from strains of the snail *Biomphalaria glabrata* varying in susceptibility to the human blood fluke, *Schistosoma mansoni:* Presence, differential expression, and biological function. *Cytokine* 6:21–27.

Granath, W. O., and Esch, G. W. 1983. Temperature and other factors that regulate the composition and infrapopulation densities of *Bothriocephalus acheilognathi* (Cestoda) in *Gambusia affinis* (Pisces). *J. Parasitol.* 69:1116–1124.

Granovitch, A. I. 1999. Parasitic systems and the structure of parasite population. *Helgol. Meeresunters.* 53:9–18.

Granovitch, A. I., and Sergievskii, S. O. 1989. The use of acorn barnacle colonies by molluscs *Littorina saxatilis* (Gastropoda, Prosobranchia) depending on their infections with parthenitae of trematodes. *Zool. Zh.* 68:39–47 (in Russian).

Grant, P. 1975. The classical case of character displacement. *Evol. Biol.* 8:237–337.

Gray, C. A., Gray, P. N., and Pence, D. B. 1989. Influence of social status on the helminth community of late-winter mallards. *Can. J. Zool.* 67:1937–1944.

Gregory, R. D. 1991a. Parasites and host-geographic range as illustrated by waterfowl. *Funct. Ecol.* 4:645–654.

———. 1991b. Parasite epidemiology and host population growth: *Heligmosomoides polygyrus* (Nematoda) in enclosed wood mouse populations. *J. Anim. Ecol.* 60:805–821.

Gregory, R. D., and Keymer, A. E. 1989. The ecology of host-parasite interactions. *Sci. Prog.* 73:67–80.

Gregory, R. D., Keymer, A. E., and Harvey, P. H. 1991. Life history, ecology and parasite community structure in Soviet birds. *Biol. J. Linn. Soc.* 43:249–262.

———. 1996. Helminth parasite richness among vertebrates. *Biodivers. Conserv.* 5:985–997.

Groom, M. J. 1992. Sand-colored nighthawks parasitize the antipredator behavior of three nesting bird species. *Ecology* 73:785–793.

Gryseels, B., and de Vlas, S. J. 1996. Worm burdens in schistosome infections. *Parasitol. Today* 12:115–119.

Guégan, J.-F., and Agnèse, J.-F. 1990. Parasite evolutionary events inferred from host phylogeny: The case of *Labeo* species (Teleostei, Cyprinidae) and their dactylogyrid parasites (Monogenea, Dactylogyridae). *Can. J. Zool.* 69:595–603.

Guégan, J.-F., and Hugueny, B. 1994. A nested parasite species subset pattern in tropical fish: Host as major determinant of parasite infracommunity structure. *Oecologia* 100:184–189.

Guégan, J.-F., and Kennedy, C. R. 1993. Maximum local helminth parasite community richness in British freshwater fish: A test of the colonization time hypothesis. *Parasitology* 106:91–100.

———. 1996. Parasite richness/sampling effort/host range: The fancy three-piece jigsaw puzzle. *Parasitol. Today* 12:367–369.

Guégan, J.-F., Lambert, A., Lévêque, C., et al. 1992. Can host body size explain the parasite species richness in tropical freshwater fishes? *Oecologia* 90:197–204.

Guégan, J.-F., and Morand, S. 1996. Polyploid hosts: Strange attractors for parasites. *Oikos* 77:366–370.

Gui, M., Kusel, J. R., Shi, Y. E., et al. 1995. *Schistosoma japonicum* and *S. mansoni*: Comparison of larval migration patterns in mice. *J. Helminthol.* 69:19–25.

Gulland, F. M. D. 1992. The role of nematode parasites in Soay sheep (*Ovis aries* L.) mortality during a population crash. *Parasitology* 105:493–503.

Gulland, F. M. D., Albon, S. D., Pemberton, J. M., et al. 1993. Parasite-associated polymorphism in a cyclic ungulate population. *Proc. Roy. Soc. Lond.*, ser. B, *Biol. Sci.* 254:7–13.

Gulland, F. M. D., and Fox, M. 1992. Epidemiology of nematode infections of Soay sheep (*Ovis aries* L.) on St Kilda. *Parasitology* 105:481–492.

Guyatt, H. L., and Bundy, D. A. P. 1990. Are wormy people parasite prone or just unlucky? *Parasitol. Today* 6:282–283.

Haas, W., and Haberl, B. 1997. Host recognition by trematode miracidia and cercariae. In *Advances in trematode biology,* ed. B. Fried and T. K. Graczyl, 197–227. Boca Raton, Fla.: CRC Press.

Haas, W., Haberl, B., Kalbe, M., et al. 1995. Snail host-finding by miracidia and cercariae: Chemical host cues. *Parasitol. Today* 12:468–472.

Hacker, J., Blum-Oehler, G., Mühldorfer, I., et al. 1997. Pathogenicity islands of virulent bacteria: Structure, function and impact on microbial evolution. *Mol. Microbiol.* 23:1089–1097.

Haenen, O. L. M., Grisez, L., de Charleroi, D., et al. 1989. Experimentally induced infections of European eel *Anguilla anguilla* with *Anguillicola crassus* (Nematoda, Dracunculoidea) and subsequent migration of larvae. *DAO (Dis. Aquat. Org.)* 7:97–101.

Hafner, M. S., and Nadler, S. A. 1988. Phylogenetic trees support the coevolution of parasites and their hosts. *Nature* 332:258–259.

Hagan, P., Blumenthal, U. J., Dunn, D., et al. 1991. Human IgE, IgC4 and resistance to reinfection with *Schistosoma haematobium. Nature* 349:243–245.

Hair, J. D., and Holmes, J. C. 1970. Helminths of Bonaparte's gulls, *Larus philadelphia,* from Cooking Lake, Alberta. *Can. J. Zool.* 48:1129–1131.

Haldane, J. B. S. 1949. Disease and evolution. *Ric. Sci.* 19 (suppl.): 68–76.

Halvorsen, O. 1985. On the relationship between social status of host and risk of parasitic infection. *Oikos* 47:71–74.

Hamilton, W. D. 1967. Extraordinary sex-ratios. *Science* 156:477–488.

———. 1976. The genetical theory of social behaviour. *J. Theor. Biol.* 7:1–52.

———. 1980. Sex *versus* non-sex versus parasite. *Oikos* 35:282–290.

Hamilton, W. D., Axelrod, R., and Tanese, R. 1990. Sexual reproduction as an adaptation to resist parasites (a review). *Proc. Natl. Acad. Sci. U.S.A.* 87:3566–3573.

Hamilton, W. D., and May, R. M. 1977. Dispersal in stable habitats. *Nature* 269:578–581.

Hamilton, W. D., and Zuk, M. 1982. Heritable true fitness and bright birds: A role for parasites? *Science* 218:384–386.

Hammond-Kosack, K. E., and Jones, J. D. G. 1996. Resistance gene-dependent plant defense responses. *Plant Cell* 8:1773–1791.

Handunnetti, S. M., Mendis, K. N., and David, P. H. 1987. Antigenic variation of cloned *Plasmodium fragile* in its natural host *Macaca sinica*. *J. Exp. Med.* 165:1269–1283.

Hanley, K. A., Biardi, J. E., Greene, C. M., et al. 1996. The behavioral ecology of host-parasite interactions: An interdisciplinary challenge. *Parasitol. Today* 12:371–373.

Hanski, I. 1982. Dynamics of regional distribution: The core and satellite species hypothesis. *Oikos* 38:210–221.

———. 1998. Metapopulation dynamics. *Nature* 396:41–49.

Haraguchi, Y., and Sasaki, A. 1996. Host-parasite arms races in mutation modifications: Indefinite escalation despite a heavy load. *J. Theor. Biol.* 183:121–137.

Harris, P. D. 1988. Changes in the site specificity of *Gyrodactylus turnbulli* Harris, 1896 (Monogenea) during infections of individual guppies (*Poecilia reticulata* Peters, 1859). *Can. J. Zool.* 66:2854–2857.

Hart, B. J., Hart, L. A., Mooring, M. S., et al. 1992. Biological basis of grooming behaviour in antelope: The body-size, vigilance and habitat principles. *Anim. Behav.* 44:615–631.

Hart, B. L. 1994. Behavioural defense against parasites: Interaction with parasite invasiveness. *Parasitology* 109:S139–S151.

Hartl, D. L., and Clark, A. G. 1989. *Principles of population genetics*. Sunderland, Mass.: Sinauer.

Harvey, P., and Pagel, M. 1991. *The comparative method in evolutionary biology*. Oxford: Oxford University Press.

Harvey, P. H. 1989. Parasitological teeth for evolutionary problems. *Nature* 342:230.

Harvey, P. H., and Keymer, A. E. 1987. Sex among the parasites. *Nature* 330:317–318.

Harvey, P. H., Read, A. W. F., and Nee, S. 1995. Why ecologists need to be phylogenetically challenged. *J. Ecol.* 83:535–536.

Harwood, S. H., McElfresh, J. C., Nguyen, A., et al. 1998. Production of early expressed parasitism-specific proteins in alternate sphingid hosts of the braconid wasp *Cotesia congregata*. *J. Invertebr. Pathol.* 71:271–279.

Hawdon, J. M., and Schad, G. A. 1991. Developmental adaptations in nematodes. In *Parasite-host associations: Coexistence or conflict?* ed. C. A. Toft, A. Aeschlimann, and L. Bolis, 274–298. New York: Oxford University Press.

Hawking, F. 1962. Microfilaria infestation as an instance of periodic phenomena seen in host-parasite relationships. *Ann. N.Y. Acad. of Sci.* 98:940–953.

———. 1967. The 24 hour periodicity of microfilariae: Biological mechanisms responsible for its production and control. *Proc. Roy. Soc. Lond.*, ser. B, *Biol. Sci.* 169:59–76.

———. 1975. Circadian and other rhythms of parasites. *Adv. Parasitol.* 13:123–182.

Hawkins, B. A., and Lawton, J. H. 1987. Species richness patterns: Why do some insects have more parasitoids than others? *Colloq. INRA* 48:131–136.

Heddi, A., Charles, H., Khatchadourian, C., et al. 1998. Molecular characterization of the principal symbiotic bacteria of the weevil *Sitophilus orizae:* A peculiar G+C content of an endocytobiotic DNA. *J. Mol. Evol.* 47:52–61.

Heeb, P., Werner, I., Mateman, A. C., et al. 1999. Ectoparasite infestation and sex-biaised local recruitment of hosts. *Nature* 400:63–65.

Heesemann, J., Schubert, S., and Rakin, A. 1999. Ecological aspects of evolutionary development of bacterial pathogens. *Nova Acta Leopold.* 307:23–38.

Heinemann, J. A. 1991. Genetics of gene transfer between species. *Trends Genet.* 7:181–185.

Helluy, S. 1981. Relations hôte parasite du trématode *Microphallus papillorobustus* (Rankin, 1940). I. Pénétration des cercaires et rapports des métacercaires avec le tissu nerveux des *Gammarus* hôtes intermédiaires. *Ann. Parasitol. Hum. Comp.* 57:263–270.

———. 1982. Relations hôte parasite du trématode *Microphallus papillorobustus* (Rankin, 1940). II. Modifications du comportement des *Gammarus* hôtes intermédiaires et localisation des métacercaires. *Ann. Parasitol. Hum. Comp.* 58:1–17.

———. 1983. Un mode de favorisation de la transmission parasitaire: La manipulation du comportement de l'hôte intermédiaire. *Rev. Ecol. Terre Vie* 38:211–223.

———. 1984. Relations hôtes-parasites du trématode *Microphallus papillorobustus* (Rankin, 1940). III. Facteurs impliqués dans les modifications du comportement des Gammarus hôtes intermédiaires et tests de prédation. *Ann. Parasitol. Hum. Comp.* 59:41–56.

Helluy, S. 1988. On the mechanisms by which a larval parasite (*Polymorphus paradoxus,* Acanthocephala) alters the behaviour of its intermediate host (*Gammarus lacustris,* Crustacea). Thesis, University of Alberta.

Helluy, S., and Holmes, J. C. 1990. Serotonin, octopamine, and the clinging behavior induced by the parasite *Polymorphus paradoxus* (Acanthocephala) in *Gammarus lacustris* (Crustacea). *Can. J. Zool.* 68:1214–1222.

Hernandez, A. D., and Sukhdeo, V. K. 1995. Host grooming and the transmission strategy of *Heligmosoides polygyrus. J. Parasitol.* 81:865–869.

Herre, E. A. 1985. Sex-ratio adjustment in fig wasps. *Science* 228:896–898.

———. 1993. Population structure and the evolution of virulence in nematode parasites of fig wasps. *Science* 259:1442–1445.

———. 1995. Factors affecting the evolution of virulence: Nematode parasites of fig wasps as a case study. *Parasitology* 111:S179–S191.

Herre, E. A., Knowlton, N., Mueller, U. G., et al. 1999. The evolution of mutualisms: Exploring the paths between conflict and cooperation. *Trends Ecol. Evol.* 14:49–53.

Hesselberg, C. A., and Andreassen, J. 1975. Some influences of population density on *Hymenolepis diminuta* in rats. *Parasitology* 71:517–523.

Heussler, V. T., Eichhorn, M., Reeves, R., et al. 1992. Constitutive IL-2 mRNA expression in lymphocytes infected with the intracellular parasite *Theileria parva. J. Immunol.* 149:562–567.

Hickey, D. A. 1993. Molecular symbionts and the evolution of sex. *J. Hered.* 84:410–414.

Hickey, D. A., and Rose, M. R. 1988. The role of gene transfer in the evolution of eukaryotic sex. In *The evolution of sex,* ed. R. E. Michod and B. R. Levin, 161–175. Sunderland, Mass.: Sinauer.

Hide, G. 1994. The exclusive trypanosome. *Parasitol. Today* 10:85–86.

Hide, G., Welburn, S. C., Tait, A., et al. 1994. Epidemiological relationships of *Trypanosoma brucei* stocks from south east Uganda: Evidence for different population structures in human infective and non-human infective isolates. *Parasitology* 109:95–112.

Hill, A. V. S. 1992. Malaria resistance genes: A natural selection. *Trans. Roy. Soc. Trop. Med. Hyg.* 86:225–226.

Hill, A. V. S., Allsopp, C. E., Kwiatkowski, D., et al. 1991. Common West African HLA antigens are associated with protection from severe malaria. *Nature* 352:595–600.

Hill, A. V. S., Yates, S. N., Allsopp, C. E., et al. 1994. Human leukocyte antigens and natural selection by malaria. *Philos. Trans. Roy. Soc. Trop. Med. Hyg.* 346:379–385.

Hinkle, G., Wetterer, J. K., Schultz, T. R., et al. 1994. Phylogeny of the attine fungi based on analysis of small subunit ribosomal RNA gene sequences. *Science* 266:1695–1697.

Hita, M., Poirié, M., Leblanc, N., et al. 1999. Genetic localization of a *Drosophila melanogaster* resistance gene to a parasitoid wasp and physical mapping of the region. *Genome Res.* 9:471–481.

Hochberg, M. E. 1998. Establishing genetic correlations involving parasite virulence. *Evolution* 52:1865–1868.

Hochberg, M. E., Clarke, R. T., Elmes, G. W., et al. 1994. Population dynamic consequences of direct and indirect interactions involving a large blue butterfly and its plant and red ant hosts. *J. Anim. Ecol.* 63:375–391.

Hochberg, M. E., and Hawkins, B. A. 1992. Refuges as a predictor of parasitoid density. *Science* 255:973–976.

Hochberg, M. E., Michalakis, Y., and de Meeus, T. 1992. Parasitism as a constraint on the rate of life-history evolution. *J. Evol. Biol.* 5:491–504.

Hoerauf, A., Nissen-Pähle, K., Schmetz. C., et al. 1999. Tetrachycline therapy targets intracellular bacteria in the filarial nematode *Litomosoides sigmodontis* and results in filarial infertility. *J. Clin. Invest.* 103:11–17.

Hohorst, W., and Lämmler, G. 1962. Experimentelle dicrocoeliose-studien. *Z. Tropenmed. Parasitol.* 13:377–397.

Holland, C. V., Crompton, D. W. T., Asaolu, S. O., et al. 1992. A possible genetic factor influencing protection from infection with *Ascaris lumbricoides* in Nigerian children. *J. Parasitol.* 78:915–916.

Holmes, J. C. 1971. Habitat segregation in Sanguinicolid blood flukes (Digenea) of scorpaenid rockfishes (Perciformes) on the Pacific Coast of North America. *J. Fish. Res. Board Can.* 28:903–909.

———. 1976. Host selection and its consequences. In *Ecological aspects of parasitology,* ed. C. R. Kennedy, 21–39. Amsterdam: North-Holland.

———. 1982. Impact of infectious disease agents on the population growth and geographical distribution of animals. In *Population biology of infectious diseases,* ed. R. M. Anderson and R. M. May, 37–51. New York: Springer-Verlag.

———. 1987. The structure of helminth communities. *Int. J. Parasitol.* 17:203–208.

———. 1990a. Helminth communities in marine fishes. In *Parasite communities: Patterns and processes,* ed. G. Esch, A. O. Bush, and J. Aho, 101–113. New York: Chapman and Hall.

———. 1990b. Competition, contacts, and other factors restricting niches of parasitic helminths. *Ann. Parasitol. Hum. Comp.* 65 (suppl.) 1:69–72.

———. 1996. Parasites as threats to biodiversity in shrinking ecosystems. *Biodivers. Conserv.* 5:975–983.

Holmes, J., and Bartoli, P. 1993. Spatio-temporal structure of the communities of helminths in the digestive tract of *Sciaena umbra* L. 1758 (Teleostei). *Parasitology* 106:519–525.

Holmes, J. C., and Bethel, W. M. 1972. Modification of intermediate host behavior by parasites. *Zool. J. Linn. Soc.* 51 (suppl.) 1:123–149.

Holmes, J. C., and Price, P. W. 1986. Communities of parasites. In *Community ecology: Pattern and process,* ed. J. Kittawa and D. J. Anderson, 187–213. New York: Blackwell Scientific Publications.

Holmes, J. C., and Zohar, S. 1990. Pathology and host behaviour. In *Parasitism and host behaviour,* ed. C. J. Barnard and J. M. Benke, 34–63. London: Taylor and Francis.

Holt, R. D., and Lawton, J. H. 1994. The ecological consequences of shared natural enemies. *Ann. Rev. Ecol. Syst.* 25:495–520.

Holt, R. D., and Pickering. J. 1985. Infectious diseases and species coexistence: A model of Lotka-Volterra form. *Am. Nat.* 126:196–211.

Hoogland, C., and Biémont, C. 1996. Chromosomal distribution of transposable elements in *Drosophila melanogaster:* Test of the ectopic recombination model for maintenance of isertion site number. *Genetics* 144:197–204.

Horak, P., Dvorak, J., Kolarova, L., et al. 1999. *Trichobilharzia regenti,* a pathogen of the avian and mammalian central nervous systems. *Parasitology* 119:S77–S81.

Hordijk, P. L., Van Loenhout, H., Ebberink, R. H. M., et al. 1991. Neuropeptide schistosomin hormonally-induced ovulation in the freshwater snail *Lymnaea stagnalis. J. Exp. Zool.* 259:268–271.

Houde, A. E., and Torio, A. J. 1992. Effect of parasitic infection on male color pattern and female choice in guppies. *Behav. Ecol.* 3:346–351.

Hougard, J.-M., Yaméogo, L., Sékétéli, A., et al. 1997. Twenty-two years of blackfly control in the Onchocerciasis Control Programme in West Africa. *Parasitol. Today* 13:425–431.

Howard, R. D., and Minchella, D. J. 1990. Parasitism and mate competition. *Oikos* 58:120–122.

Howard. R. J. 1992. Asexual deviants take over. *Nature* 357:647–648.

Hudson, D. E., Wellems, T. E., and Miller, L. H. 1988. Molecular basis for mutation in a surface protein expressed by malaria parasites. *J. Mol. Biol.* 203:707–714.

Hudson, P. J., and Dobson, A. P. 1991. The direct and indirect effects of the caecal nematode *Trichostrongylus tenuis* on red grouse. In *Bird parasite interactions,* ed. J. E. Loye and M. Zuk, 49–68. Oxford: Oxford University Press.

Hudson, P. J., Dobson, A. P., and Newborn, D. 1998. Prevention of population cycles by parasite removal. *Science* 282:2256–2258.

———. 1992. Do parasites make prey vulnerable to predation? Red grouse and parasites. *J. Anim. Ecol.* 61:681–692.

Hudson, P. J., and Greenman, J. 1998. Competition mediated by parasites: Biological and theoretical progress. *Trends Ecol. Evol.* 13:387–390.

Hudson, P. J., Newborn, D., and Dobson, P. 1992. Regulation and stability of a free-living host-parasite system, *Trichostrongylus tenuis* in red grouse. I. Monitoring and parasite reduction experiments. *J. Anim. Ecol.* 61:477–486.

Hugot, J.-P. 1999. Primates and their pinworm parasites: The Cameron hypothesis revisited. *Syst. Biol.* 48:523–546.

Hugueny, B., and Guégan, J.-F. 1997. Community nestedness and the proper way to assess statistical significance by Monte-Carlo tests: Some comments on Worthen and Rohde's (1996) paper. *Oikos* 80:572–574.

Hunter, J. E., and Schmidt, F. L. 1990. *Methods of meta-analysis: Correcting error and bias in research findings.* Newbury Park, Calif.: Sage.

Hurd, H. 1990. Physiological and behavioural interactions between parasites and invertebrates hosts. *Adv. Parasitol.* 29:271–318.

Hurst, G. D. D., Hurst, L. D., and Johnstone, R. A. 1992. Intranuclear conflict and its role in evolution. *Trends Ecol. Evol.* 7:373–378.

Hurtez-Boussès, S., Blondel, J., Perret, P., et al. 1998. Chick parasitism by blowflies affects feeding rates in a Mediterranean population of blue tits. *Ecol. Lett.* 1:17–20.

Ives, A. R., and Murray, D. L. 1997. Can sublethal parasitism destabilize predator-prey population dynamics? A model of snowshoe hares, predators and parasites. *J. Anim. Ecol.* 66:265–278.

Jackson, J. A., and Tinsley, R. C. 1994. Infrapopulation dynamics of *Gyrdicotylus gallieni* (Monogenea: Gyrodactylidae). *Parasitology* 108:447–452.

———. 1998. Reproductive interference in concurrent infections of two *Protopolystoma* species (Monogenea: Polystomatidae) *Int. J. Parasitol.* 28:1201–1204.

Jaenike, J. 1988. Parasitism and male mating success in *Drosophila testacea. Am. Nat.* 131:774–780.

———. 1996. Population-level consequences of parasite aggregation. *Oikos* 76:155–160.

Jaisson, P. 1993. *La fourmi et le sociobiologiste.* Paris: Odile Jacob.

Janovy, J., Jr., Clopton, R. E., and Percival, T. J. 1992. The roles of ecological and evolutionary influences in providing structure to parasite species assemblages. *J. Parasitol.* 78:630–640.

Jarne, P. 1993. Resistance genes at the population level. *Parasitol. Today* 9:216–217.

Jarne, P., and Charlesworth, D. 1993. The evolution of the selfing rate in functionally hermaphrodite plants and animals. *Ann. Rev. Ecol. Syst.* 24:441–466.

Jennings, J. B., and Calow, P. 1975. The relationship between high fecundity and the evolution of entoparasitism. *Oecologia* 21:109–115.

Jeon, K. W. 1989. Amoeba and x-bacteria: Symbiont acquisition and possible species change. In *Symbiosis as a source of evolutionary innovation,* ed. L. Margulis and R. Fester, 118–131. Cambridge: MIT Press.

John, J. L. 1997. The Hamilton-Zuk theory and initial test: An examination of some parasitological critiscisms. *Int. J. Parasitol.* 27:1269–1288.

Johnston, D. A., Kane, R. A., and Rollinson, D. 1993. Small subunit (18S) ribosomal RNA gene divergence in the genus *Schistosoma. Parasitology* 107:147–156.

Jones, C. G., Lawton, J. H., and Shachak, M. 1994. Organisms as ecosystem engineers. *Oikos* 69:373–386.

———. 1997. Positive and negative effects of organisms as physical ecosystem engineers. *Ecology* 78:1946–1957.

Joosse, J., and Van Elk, R. 1986. *Trichobilharzia ocellata*: Physiological characterization of giant growth, glycogen depletion and absence of reproductive activity in the intermediate snail host, *Lymnaea stagnalis. Exp. Parasitol.* 62:1–13.

Jourdane, J. 1974. Découverte de l'hôte vecteur de *Nephrotrema truncatum* (Leuckart, 1842) (Trematoda) et mise en évidence d'une phase hépatique au cours de la migration du parasite chez l'hôte définitif. *C. R. Acad. Sci.*, ser. 3, *Sci. Vie* 278:1533–1536.

Jourdane, J., and Mounkassa, J. B. 1986. Topographic shifting of primary sporocysts of *Schistosoma mansoni* in *Biomphalaria pfeifferi* as a result of coinfection with *Echinostoma caproni. J. Invertebr. Pathol.* 48:269–274.

Jourdane, J., and Southgate, V. 1992. Genetic exchanges and sexual interactions between species of the genus *Schistosoma. Res. Rev. Parasitol.* 52:21–26.

Jourdane, J., and Théron, A. 1980. *Schistosoma mansoni*: Cloning by microsurgical transplantation of sporocysts. *Exp. Parasitol.* 50:349–357.

———. 1987. Larval development: Eggs to cercariae. In *The biology of schistosomes*, ed. D. Rollinson and A. J. G. Simpson, 83–113. London: Academic Press.

Jourdane, J., Théron, A., and Combes, C. 1980. Demonstration of several sporocysts generation as a normal pattern of reproduction of *Schistosoma mansoni. Acta Trop.* 37:177–182.

Jousson, O., Bartoli, P., Zaninetti, L., et al. 1998. Use of ITS rDNA for elucidation of some life-cycles of Mesometridae (Trematoda, Digenea). *Int. J. Parasitol.* 28:1403–1411.

Jouy-Avantin, F., Combes, C., de Lumley, H., et al. 1999. Helminth eggs in animal coprolites from a middle Pleistocene site in Europe. *J. Parasitol.* 85:376–379.

Juchault, P., Louis, C., Martin, G., et al. 1991. Masculinization of female isopods (Crustacea) correlated with non-mendelian inheritance of cytoplasmic viruses. *Proc. Natl. Acad. Sci. U.S.A.* 88:10460–10464.

Juchault, P., and Mocquard, J. P. 1993. Transfer of a parasitic sex factor to the nuclear genome of the host: A hypothesis on the evolution of sex-determining mechanisms in the terrestrial isopod *Armadillidium vulgare* Latr. *J. Evol. Biol.* 6:511–528.

Justine, J. L., Le Brun, N., and Mattei, X. 1985. The aflagellate spermatozoon of *Diplozoon* (Platyhelminthe: Monogenea: Polyopisthocotylea): A demonstrative case of relationship between sperm ultrastructure and biology of reproduction. *J. Ultrastruct. Res.* 92:47–54.

Kalbe, M., Haberl, B., and Haas, W. 1997. Miracidial host-finding in *Fasciola hepatica* and *Trichobilharzia ocellata* is stimulated by species-specific glycoconjugates released from the host snails. *Parasitol. Res.* 83:806–812.

Kaltz, O., and Shykoff, J. A. 1998. Local adaptation in host-parasite systems. *Heredity* 81:361–370.

Karanga, D. M., Colley, D. G., Nahlen, B. L., et al. 1997. Studies on schistosomiasis in western Kenya: I. Evidence for immune-facilitated excretion of schistosome eggs from patients with *Schistosoma mansoni* and human immunodeficiency virus coinfections. *Am. J. Trop Med. Hyg.* 56:515–521.

Kasper, L. H., and Mineo, J. R. 1994. Attachment and invasion of host-cells by *Toxoplasma gondii. Parasitol. Today* 10:184–187.

Katz, L. A. 1998. Changing perspectives on the origin of eukaryotes. *Trends Ecol. Evol.* 13:493–497.

Kavaliers, M., and Colwell, D. 1993a. Multiple opioid system involvment in the mediation of parasitic-infection induced analgesia. *Brain Res.* 623:316–320.

———. 1993b. Aversive responses of female mice to the odors of parasitized males: Neuromodulatory mechanisms and implications for mate choice. *Ethology* 95:202–212.

———. 1994. Parasite infection attenuates nonopioid mediated predator-induced analgesia in mice. *Physiol. Behav.* 55:505–510.

———. 1995. Decreased predator avoidance in parasitized mice: Neuromodulatory correlates. *Parasitology* 111:257–263.

Kearn, G. C. 1973. An endogenous circadian rhythm in the monogenean skin parasite *Entobdella soleae* and its relationship to the activity rhythm of the host *(Solea solea)*. *Parasitology* 66:101–122.

Kechemir, N. 1978. Démonstration expérimentale d'un cycle à quatre hôtes obligatoires chez les trématodes hémiurides. *Ann. Parasitol. Hum. Comp.* 53:75–92.

Kechemir, N., and Théron, A. 1997. Intraspecific variability of *Schistosoma haematobium* in Algeria. *J. Helminthol.* 71:29–33.

Kennedy, C. E. J., Endler, J. A., Poynton, S. L., et al. 1987. Parasite load predicts male choice in guppies. *Behav. Ecol. Sociobiol.* 21:291–295.

Kennedy, C. R. 1990. Helminth communities in freshwater fish: Structured communities or stochastic assemblages? In *Parasite communities: Patterns and processes,* ed. G. W. Esch, A. O. Bush, and J. M. Aho, 130–156. London: Chapman, and Hall.

———. 1994. The distribution and abundance of the nematode *Anguillicola australiensis* in eels *Anguilla reinhardtii* in Queensland, Australia. *Folia Parasitol.* 41:279–285.

———. 1995. Richness and diversity of macroparasite communities in tropical eels *Anguilla reinhardtii* in Queensland, Australia. *Parasitology* 111:233–245.

Kennedy, C. R., and Bakke, T. A. 1989. Diversity patterns in helminth communities in common gulls, *Larus canus. Parasitology* 98:439–445.

Kennedy, C. R., Bates, R. M., and Brown, A. F. 1989. Discontinuous distributions of the fish acanthocephalans *Pomphorhynchus laevis* and *Acanthocephalus anguillae* in Britain and Ireland: An hypothesis. *J. Fish Biol.* 34:607–619.

Kennedy, C. R., Berrilli, F., Di Cave, D., et al. 1998. Composition and diversity of helminth communities in eels *Anguilla anguilla* in the River Tiber: Long-term changes and comparison with insular Europe. *J. Helminthol.* 72:301–306.

Kennedy, C. R., and Bush, A. O. 1992. Species richness in helminth communities: The importance of multiple congeners. *Parasitology* 104:189–197.

———. 1994. The relationship between pattern and scale in parasite communities: A stranger in a strange land. *Parasitology* 109:187–196.

Kennedy, C. R., Bush, A. O., and Aho, J. M. 1986. Patterns in helminth communities: Why are birds and fish different? *Parasitology* 93:205–215.

Kennedy, C. R., and Guégan, J.-F. 1994. Regional versus local helminth parasite richness in British freshwater fish: Saturated or unsaturated parasite communities. *Parasitology* 109:175–185.

———. 1996. The number of niches in intestinal helminth communities of *Anguilla anguilla:* Are there enough spaces for parasites? *Parasitology* 113:293–302.

Kennedy, C. R., Nie, P., Kaspers, J., et al. 1992. Are eels (*Anguilla anguilla* L.) planktonic feeders? Evidence from parasite communities. *J. Fish Biol.* 41:567–580.

Kennedy, M. J., and MacKinnon, J. D. 1994. Site segregation of *Thelazia skrjabini* and *Thelazia gulosa* (Nematoda: Thelazioidea) in the eyes of cattle. *J. Parasitol.* 80:501–504.

Kerr, P. J., and Best, S. M. 1998. Myxoma virus in rabbits. *Rev. Sci. Tech.—Off. Int. Epizoot.* 17:256–268.

Keymer, A. E. 1980. The influence of *Hymenolepis diminuta* on the survival and fecundity of the intermediate host, *Tribolium confusum. Parasitology* 81:405–421.

———. 1981. Population dynamics of *Hymenolepis diminuta* in the intermediate host. *J. Anim. Ecol.* 50:941–95.

———. 1982. Density-dependent mechanisms in the regulation of intestinal helminth populations. *Parasitology* 84:573–587.

Keymer, A. E., and Read, A. F. 1991. Behavioural ecology: The impact of parasitism. In *Parasite-host associations: Coexistence or conflict?* ed. C. A. Toft, A. Aeschlimann, and L. Bolis, 37–61. New York: Oxford University Press.

Kidwell, M. G., and Ribeiro, J. M. C. 1992. Can transposable elements be used to drive disease refractoriness genes into vector populations? *Parasitol. Today* 8:325–329.

Kim, A. I., Terzian, C., Santamaria, P., et al. 1994. Retroviruses in invertebrates: The gypsy retrotransposon is apparently an infectious retrovirus of *Drosophila melanogaster. Proc. Natl. Acad. Sci. U.S.A.* 91:1285–1289.

King, K. A., Keith, J. O., Mitchell, C. A., et al. 1977. Ticks as a factor of nest desertion of California brown pelicans. *Condor* 79:507–509.

Kingsolver, J. G. 1987. Mosquito host choice and the epidemiology of malaria. *Am. Nat.* 130:811–827.

Kisielewska, K. 1970a. Ecological organization of intestinal helminth groupings in *Clethrionomys glareolus* (Schreb.) (Rodentia). I. Structure and seasonal dynamics of helminth groupings in a host population in the Bialowieza Park. *Acta Parasitol. Pol.* 18:121–147.

———. 1970b. Ecological organization of intestinal helminth groupings in *Clethrionomys glareolus* (Schreb.) (Rodentia). II. An attempt at an introduction of helminths of *C. glaerolus* from the Bialowieza National Park into an island of the Beldany Lake (Mazurian Lakeland). *Acta Parasitol. Pol.* 18:149–162.

———. 1970c. Ecological organization of intestinal helminth groupings in *Clethrionomys glareolus* (Schreb.) (Rodentia). III. Structure of helminth groupings in *C. glaerolus* populations of various forest biocoenoses in Poland. *Acta Parasitol. Pol.* 18:163–176.

———. 1970d. Ecological organization of intestinal helminth groupings in *Clethrionomys glareolus* (Schreb.) (Rodentia). IV. Spatial structure of a helminth grouping within the host population. *Acta Parasitol. Pol.* 18:177–196.

———. 1970e. Ecological organization of intestinal helminth groupings in *Clethrionomys glareolus* (Schreb.) (Rodentia). V. Some questions concerning helminth groupings in the host individuals. *Acta Parasitol. Pol.* 18:197–208.

———. 1971. Intestinal helminths as indicators of the age structure of *Microtus arvalis* Pallas 1778 population. *Bull. Acad. Pol. Sci. Ser. Sci. Biol.* 19:275–282.

———. 1983. Ecological characteristics of parasitic worm (helminth) communities. *Acta Theriol.* 28 (suppl.) 1:73–88.

Kisielewska, K., Fraczak, K., Krasowska, I., and Zubczeweska, Z. 1973. Structure of the intestinal helminthocoenosis in the population of *Microtus arvalis* Pallas, 1778, and the mechanism of its variability. *Acta Parasitol. Polonica* 21:71–83.

Kjellberg, F., Gouyon, P. H., Ibrahim, M., et al. 1987. The stability of the symbiosis between dioecious figs and their pollinators: A study of *Ficus carica* L and *Blastophaga psenes* L. *Evolution* 41:693–704.

Kjellberg, F., and Maurice, S. 1989. Seasonality in the reproductive phenology of *Ficus*: Its evolution and consequences. *Experientia* 45:653–660.

Koella, J. C. 1994. Linking evolutionary ecology with epidemiology. *Rev. Suisse Zool.* 101:865–874.

Koella, J. C., and Packer, M. J. 1996. Malaria parasites enhance blood-feeding of their naturally infected vector *Anopheles punctulatus*. *Parasitology* 113:105–109.

Kôhler, S., Delwiche, C. F., Denny, P. W., et al. 1997. A plastid of probable green algal origin in apicomplexan parasites. *Science* 275:1488–1491.

Kontrimavicius, V. L. 1987. Basic aspects of population biology of helminths. In *Parasite-host-environment,* 141–156. Varna, Bulgaria: First International Autumn School.

Körner, M., and Haas, W. 1998. Chemo-orientation of echinostome cercariae towards their snail-hosts: Amino acids signal a low host-specificity. *Int. J. Parasitol.* 28:511–516.

Kornyushin, V. V., Iskova, N. L., Smogorzhevskaya, L. A., et al. 1995. The helminth fauna of an isolated Black Sea population of the common eider *Somateria mollissima*. *Proc. Jubilee Conf. Ukr. Soc. Parasitol., Kiev* (16–17 May 1995): 39–49 (in Russian).

Koslowski, J. 1992. Optimal allocation of resources to growth and reproduction: Implications for age and size at maturity. *Trends Ecol. Evol.* 7:15–19.

Kraaijeveld, A. R., and Godfray, H. C. J. 1997. Trade-off between parasitoid resistance and larval competitive ability in *Drosophila melanogaster*. *Nature* 389:278–280.

Krause, J., and Godin, J.-G. 1996. Influence of parasitism on shoal choice in the banded killifish (*Fundulus diaphanus,* Teleostei, Cyprinodontidae). *Ethology* 102:40–49.

Krause, J., Ruxton, G. D., and Godin, J.-G. 1999. Distribution of *Crassiphiala bulboglossa,* a parasitic worm, in shoaling fish. *J. Anim. Ecol.* 68:27–33.

Kroemer, G. 1997. Mitochondrial implication in apoptosis towards an endosymbiont hypothesis of apoptosis evolution. *Cell Death Differ.* 4:443–456.

Kuris, A. M. 1974. Trophic interactions: Similarity of parasitic castrations to parasitoids. *Q. Rev. Biol.* 49:129–148.

———. 1990. Guild structure of larval trematodes in molluscan hosts: Prevalence, dominance and significance of competition. In *Parasite communities: Patterns and processes,* ed. G. Esch, A. Bush, and J. Aho, 69–100. London: Chapman and Hall.

Kuris, A. M., and Lafferty, K. D. 1994. Community structure: Larval trematodes in snail hosts. *Ann. Rev. Ecol. Syst.* 25:189–217.

Kwo, E. H., Lie, K. J., and Owyang, C. K. 1970. Predation of sporocysts of *Fasciola gigantica* by rediae of *Echinostoma audyi. Southeast Asian J. Trop. Med. Public Health* 1:429.

Lachaise, D. 1994. Fig tree-fig wasp assemblage history: the cospeciation versus empty barrel hypotheses. *Proc. Thirteenth Plenary Meet. AETFAT Cong, Malawi* 1:461–466.

Ladeveze, V., Galindo, I., Chaminade. N., et al. 1998. Transmission pattern of *hobo* transposable element in transgenic lines of *Drosophila melanogaster. Genet. Res.* 71:97–107.

Ladle, R. J. 1992. Parasites and sex: Catching the Red Queen. *Trends Ecol. Evol.* 7:405–408.

Lafferty, K. D. 1992. Foraging on prey that are modified by parasites. *Am. Nat.* 140:854–867.

———. 1993. The marine snail, *Cerithidea californica,* matures at smaller sizes where parasitism is high. *Oikos* 68:3–11.

Lafferty, K. D., and Morris, A. K. 1996. Altered behavior of parasitized killifish greatly increases susceptibility to predation by bird final hosts. *Ecology* 77:1390–1397.

Lafferty, K. D., Sammond, D. T., and Kuris, A. M. 1994. Analysis of larval trematode communities. *Ecology* 75:2275–2285.

Lambert, M. 1976. Cycle biologique de *Parasymphylodora markewitschi* (Kulakovskaya, 1947) (Trematoda Digenea, Monorchiidae). *Bull. Mus. Natl. Hist. Nat.,* sect. A, *Zool. Biol. Ecol. Anim.* 407:1107–1114.

Lanciani, C. A. 1975. Parasite-induced alterations in host reproduction and survival. *Ecology* 56:689–695.

Landau, I., Chabaud, A., Cambie, G., et al. 1991. Chronotherapy of malaria: An approach to malaria chemotherapy. *Parasitol. Today* 7:350–352.

Lawrence, J. D. 1973. The ingestion of red blood cells by *Schistosoma mansoni. J. Parasitol.* 59:60–63.

Le Brun, N., Renaud, F., and Lambert, A. 1988. The genus *Diplozoon* in southern France: Speciation and specificity. *Int. J. Parasitol.* 18:395–400.

Lebedev, B. I. 1978. Some aspects of monogenean existence. *Folia Parasitol.* 25:131–136.

Lecomte, C., and Thibout, E. 1986. Analyse dans deux types d'olfactomètres du comportement de quête des femelles de *Diadromus pulchellus* en présence d'odeurs du phytophage hôte et du végétal attaqué ou non. *Entomophaga* 31:69–78.

Lefcort, H., and Bayne, C. J. 1991. Thermal preferences of resistant and susceptible strains of *Biomphalaria glabrata* (Gastropoda) exposed to *Schistosoma mansoni* (Trematoda). *Parasitology*. 103 3:357–362.

Lemaire, M., Lange, C., Lefebvre, J., et al. 1986. Stratégie de camouflage du prédateur *Hyponeura eduardi* dans les sociétés de *Reticulitermes* européens. *Actes Colloq. Insectes Sociaux, Vaison la Romaine* (September 1985, France), 3:97–101.

Lenski, R. E., and May, R. M. 1994. The evolution of virulence in parasites and pathogens: Reconciliation between two competing hypotheses. *J. Theor. Biol.* 169:253–266.

Levin, B. R. 1996. The evolution and maintenance of virulence in microparasites. *Emerg. Infect. Dis.* 2:93–102.

Levin, B. R., and Lenski, R. E. 1985. Bacteria and phage: A model system for the study of the ecology and co-evolution of hosts and parasites. In *Ecology and genetics of host-parasite interactions,* ed. D. Rollinson and R. M. Anderson, 227–242. London: Academic Press.

Lewin, B. 2000. *Genes VII.* New York: Oxford University Press.

Lewis, J. W., Cannon, J. T., and Libeskind, J. C. 1980. Opioid and nonopioid mechanisms of stress analgesia. *Science* 208:623–625.

Lewontin, R. C. 1974. *The genetic basis of evolutionary change.* New York: Columbia University Press.

Li, W.-H., and Graur, D. 1991. *Fundamentals of molecular evolution.* Sunderland, Mass.: Sinauer.

Lie, K. J., Basch, P. F., and Heyneman, D. 1968. Direct and indirect antagonism between *Paryphostomum segregatum* and *Echinostoma paraensei* in the snail *Biomphalaria glabrata. Z. Parasitenkd.* 31:101–107.

Lie, K. J., and Richards, C. S. 1977. Studies on resistance in snails: Interference by nonirradiated echinostome larvae with natural resistance to *Schistosoma mansoni* in *Biomphalaria glabrata. J. Invertebr. Pathol.* 29:118–125.

Lieutier, F. 1984. Ovarian and fat body protein concentrations in *Ips sexdentatus* (Coleoptera ; Scolytidae) parasitized by nematodes. *J. Invertebr. Pathol.* 43:21–31.

Lim, H. K., and Heyneman, D. 1972. Intramolluscan inter-trematode antagonism: A review of factors influencing the host-parasite system and its possible role in biological control. *Adv. Parasitol.* 10:191–268.

Lima, S. L., and Dill, L. M. 1990. Behavioral decisions made under the risk of predation: A review and prospective. *Can. J. Zool.* 68:619–640.

Lincicome, D. R. 1971. The goodness of parasitism: A new hypothesis. In *The biology of symbiosis,* ed. T. C. Cheng, 139–227. Baltimore: University Park Press.

Lindholm, A. K. 1999. Brood parasitism by the cuckoo on patchy reed warbler populations in Britain. *J. Anim. Ecol.* 68:293–309.

Lindstrôm, E. R., Andrén, H., Angelstram, P., et al. 1994. Disease reveals the predator: Sarcoptic mange, red fox predation, and prey populations. *Ecology* 75:1042–1049.

Linhart, Y. B. 1991. Disease, parasitism and herbivory: Multidimensional challenges in plant evolution. *Trends Ecol. Evol.* 6:392–396.

Little, T. J., and Ebert, D. 1999. Associations between parasitism and host genotype in natural populations of *Daphnia* (Crustacea: Cladocera). *J. Anim. Ecol.* 68:134–149.

Littlewood, D. T. J., Rohde, K., and Clough, K. A. 1997. Parasite speciation within or between host species? Phylogenetic evidence from site-specific polystome monogeneans. *Int. J. Parasitol.* 27:1289–1297.

Lively, C. M. 1987. Evidence from a New Zealand snail for the maintenance of sex by parasitism. *Nature* 328:519–521.

———. 1992. Parthenogenesis in a freshwater snail: Reproductive assurance *versus* parasitic release. *Evolution* 46:907–913.

Lively, C. M., Craddock, C., and Vrijenhoek, R. J. 1990. Red Queen hypothesis supported by parasitism in sexual and clonal fish. *Nature* 344:864–866.

Llewellyn, J. 1984. The biology of *Isancistrum subulatae* n. sp., a monogenean parasitic on the squid, *Alloteuthis subulata,* at Plymouth. *J. Mar. Biol. Assoc. U.K.* 64:285–302.

Lo, C. M., Morand, S., and Galzin, R. 1998. Parasite diversity/host age and size relationship in three coral-reef fishes from French Polynesia. *Int. J. Parasitol.* 28:1695–1708.

———. 1999. Le parasitisme des poissons coralliens: Reflet de l'habitat? *C. R. Acad. Sci.,* ser. 3, *Sci. Vie* 322:281–287.

Loker, E. S. 1994. On being a parasite in an invertebrate host: A short survival course. *J. Parasitol.* 80:728–747.

Loker, E. S., and Adema, C. M. 1995. Schistosomes, echinostomes and snails: Comparative immunobiology. *Parasitol. Today* 11:120–124.

Loker, E. S., Cimino, D. F., and Hertel, L. A. 1992. Excretory-secretory products of *Echinostoma paraensei* sporocysts mediate interference with *Biomphalaria glabrata* hemocyte functions. *J. Parasitol.* 78:104–115.

Loos-Frank, B., and Zimmermann, G. 1976. Über eine dem *Dicrocoelium*-Befall analoge Verhaltensänderung bei Ameisen der Gattung *Formica* durch einen Pilz der Gattung *Entomophtora. Z. Parasitenkd.* 49:281–289.

Lotem, A., and Rothstein, S. I. 1995. Cuckoo-host coevolution: From snapshots of an arms race to the documentation of microevolution. *Trends Ecol. Evol.* 11:436–437.

Lotz, J. M., Bush, A. O., and Font, W. F. 1995. Recruitment-driven, spatially discontinuous communities: A null model for transferred patterns in target communities of intestinal helminths. *J. Parasitol.* 81:12–24.

Lotz, J. M., and Font, W. F. 1985. Structure of enteric helminth communities in two populations of *Eptesicus fuscus* (Chiroptera). *Can. J. Zool.* 63:2969–2978.

LoVerde, P. T. 1979. Defense reactions in snails against the intramolluscan stages of schistosomes. *Zentralbl. Bakteriol., Parasitenkd., Infektionskr. Hyg. 1 Abt. Ref., Med. Mikrobiol., Parasitol., Hyg., Präv. Med.* 263:207–209.

Lowenberger, C. A., and Rau, M. E. 1994. *Plagiorchis elegans*: Emergence, longevity and infectivity of cercariae, and host behavioural modifications during cercarial emergence. *Parasitology* 109:65–72.

Loye, J. E., and Carroll, S. P. 1991. Nest ectoparasite abundance and cliff swallow colony site selection, nestling development, and departure time. In *Bird-*

parasite interactions. ecology, evolution, and behaviour, ed. J. E. Loye and M. Zuk, 222–241. Oxford: Oxford University Press.

Lozano, G. A. 1991. Optimal foraging theory: A possible role for parasites. *Oikos* 60:391–395.

Lysne, D. A., Hemmingsen, W., and Skorping, A. 1997. Regulation of infrapopulations of *Cryptocotyle lingua* on cod. *Parasitology* 114:145–150.

McAllister, M. K., and Roitberg, B. D. 1987. Adaptive suicidal behaviour in pea aphids. *Nature* 328:797–799.

MacArthur, R. H. 1958. Population ecology of some warblers of northeastern coniferous forests. *Ecology* 39:599–619.

MacArthur, R. H., and Wilson, E. O. 1967. *The theory of island biogeography.* Princeton: Princeton University Press.

McCarthy, A. M. 1990. Speciation of echinostomes: Evidence for the existence of two sympatric sibling species in the complex *Echinoparyphium recurvatum* (von Linstow 1873) (Digenea: Echinostomatidae). *Parasitology* 101:35–42.

McCoy, K. D., Boulinier, T., Chardine, J. W., et al. 1999. Dispersal and distribution of the tick *Ixodes uriae* within and among seabird host populations: The need for a population genetic approach. *J. Parasitol.*85:196–202.

McDonald, J. F. 1998. Transposable elements, gene silencing and macroevolution. *Trends Ecol. Evol.* 13:94–95.

MacDonald, S., and Combes, C. 1978. The hatching rhythm of *Polystoma integerrimum,* a monogenean from the frog *Rana temporaria. Chronobiologia* 5:277–285.

MacDonald, S., and Jones, A. 1978. Egg laying and hatching rhythms in the monogenean *Diplozoon homoion gracile* from the southern barbel *(Barbus meridionalis). J. Helminthol.* 52:23–28.

McFadden, G., and Gilson, P. 1995. Something borrowed, something green: Lateral transfer of chloroplasts by secondary endosymbiosis. *Trends Ecol. Evol.* 10:12–17.

MacFall-Ngai, M. J. 1989. Luminous bacterial symbiosis in fish evolution: Adaptive radiation among the leiognathid fishes. In *Symbiosis as a source of evolutionary innovation,* ed. L. Margulis and R. Fester, 381–409. Cambridge: MIT Press.

McGrew, W. C., Tutin, C. E. G., Collins, D. A., et al. 1989. Intestinal parasites of sympatric *Pan troglodytes* and *Papio* spp at two sites: Gombe (Tanzania) and Mt. Assirik (Senegal). *Am. J. Primatol.* 17:147–155.

McGrew, W. C., Tutin, C. E. G., and File, S. K. 1989. Intestinal parasites of two species of free-living monkeys in far western Africa, *Cercopithecus (Aethiops) sabaeus* and *Erythrocebus patas patas. Afr. J. Ecol.* 27:261–262.

McInnes, F. J., Crompton, D. W. T., and Ewald, J. A. 1994. The distribution of *Centrorhynchus aluconis* (Acanthocephala) and *Porrocaecum spirale* (Nematoda) in tawny owls (*Strix aluco*) from Great Britain. *J. Raptor Res.* 28:34–38.

McKenzie, A. A. 1990. The ruminant dental grooming apparatus. *Zool. J. Linn. Soc.* 99:117–128.

McKenzie, A. A., and Weber, A. 1993. Loose front teeth: Radiological and histocorrelation with grooming function in the impala *Aepyceros melampus. J. Zool. (Lond.)* 231:167–174.

MacKenzie, K., Williams, H. H., Williams, B., et al. 1995. Parasites as indicators of water quality and the potential use of helminth transmission in marine pollution studies. *Adv. Parasitol.* 35:85–114.

McKeown, T. 1988. *The origins of human diseases.* New York: Blackwell.

McLachlan, A. 1999. Parasites promote mating success: The case of a midge and a mite. *Anim. Behav.* 57:1199–1205.

McLachlan, A., Ladle, R., and Bleay, C. 1999. Is infestation the result of adaptive choice behaviour by the parasite? A study of mites and midges. *Anim. Behav.* 58:615–620.

McLennan, D. A., and Shires, V. L. 1995. Correlation between the level of infection with *Bunodera inconstans* and *Neoechinorhynchus rutili* and behavioral activity in female brook sticklebacks. *J. Parasitol.* 81:675–682.

McManus, D. P. 1985. Enzyme analysis of natural populations of *Schistocephalus solidus* and *Ligula intestinalis. J. Helminthol.* 59:323–332.

McNaughton, S. J. 1988. Diversity and stability. *Nature* 333:204–205.

Maillard, C. 1970. Distomatoses de poissons en milieu lagunaire. Thesis, Université de Montpellier II, France.

Maillard, C., and Aussel, J.-P. 1988. Host specificity of fish trematodes investigated by experimental ichthyophagy. *Int. J. Parasitol.* 18:493–498.

Malek, E. A., and Amstrong, J. C. 1967. Infection with *Heterobilharzia americana* in primates. *Am. J. Trop Med. Hyg.* 16:708–714.

Malmberg, G. 1993. Gyrodactylidae and gyrodactylosis of Salmonidae. *Bull. Fr. Pêche Piscic.* 328:5–46.

Manning, S. D., Woolhouse, M. E. J., and Ndamba, J. 1995. Geographic compatibility of the freshwater snail *Bulinus globosus* and schistosomes from the Zimbabwe highveld. *Int. J. Parasitol.* 25:37–42.

Mao, S. P., and Shao, B. R. 1982. Schistosomiasis control in the People's Republic of China. *Am. J. Trop Med. Hyg.* 31:92–99.

Marchetti, K., Nakamura, H., and Lisle Gibbs, H. 1998. Host-race formation in the common cuckoo. *Science* 282:471–472.

Marcogliese, D. J., and Cone, D. K. 1996. On the distribution and abundance of eel parasites in Nova Scotia: Influence of pH. *J. Parasitol.* 82:389–399.

——— 1998a. Comparison of richness and diversity of macroparasite communities among eels from Nova Scotia, the United Kingdom and Australia. *Parasitology* 116:73–83.

———. 1998b. Food webs: A plea for parasites. *Trends Ecol. Evol.* 12:320–325.

Margalef, R., and Gutierez, E. 1983. How to introduce connectance in the frame of an expression for diversity? *Am. Nat.* 121:601–607.

Margolis, L., Esch, G. W., Holmes, J. C., et al. 1982. The use of ecological terms in *Parasitology* (report of an ad hoc committee of the American Society of Parasitologists). *J. Parasitol.* 68:131–133.

Margulis, L. 1993. *Symbiosis in cell evolution.* New York: Freeman.

Marquet, S., Abel, L., Hillaire, D., et al. 1996. Genetic localization of a locus controlling the intensity of infection by *Schistosoma mansoni* on chromosome 5q31–q33. *Nat. Genet.* 14:181–184.

Martin, G., Juchault, P., and Legrand, J. J. 1973. Mise en évidence d'un microorganisme intracytoplasmique symbiote de l'oniscoïde *Armadillidium vulgare* dont la présence accompagne l'intersexualité ou la féminisation totale des mâles génétiques de la lignée thélygène. *C. R. Acad. Sci.*, ser. 3, *Sci. Vie* 276:2313–2316.

Martin, P. M. V., and Combes, C. 1997. Emerging infectious diseases and the depopulation of French Polynesia in the 19th century. *Emerg. Infect. Dis.* 2:359–360.

Mas-Coma, S., and Esteban, J. G. 1988. La evolucion de una fauna parasitaria en islas "continentales": El caso de los helmintos de micromamiferos en las Baleares y Pitiusas. *Bull. Ecol.* 19:211–218.

Mas-Coma, S., Galan-Puchades, M. T., Fuentes, M. V., et al. 1988. Sobre la composicion cuantitativa de las parasitofaunas insulares: Posible efecto regulador de las especies parasitas sobre las poblaciones de sus hospedadores. In *Mamiferos y helmintos,* ed. V. Sans-Coma, S. Mas-Coma, and J. Gosalbez, 217–251. Barcelona: Ketres.

Mas-Coma, S., and Montoliu, I. 1978. Sobre la biolog°a de los trem todos del liron careto, *Eliomys quercinus ophiusae* Thomas, 1925 (Rodentia: Gliridae), en Formentera (Islas Pitiusas). Res. Rev. Parasitol. 38:95–109.

———. 1987. The life-cycle of *Dollfusinus frontalis,* a brachylaimid trematode of small mammals (Insectivora and Rodentia). *Int. J. Parasitol.* 17:1063–1079.

Mathieuhid-Daudé, F., Stevens, J., Welsch, J., et al. 1995. Genetic diversity and population structure of *Trypanosoma brucei*: Clonality versus sexuality. *Mol. Biochem. Parasitol.* 72:89–101.

Mauël, J. 1996. Intracellular survival of protozoan parasites with special reference to *Leishmania* spp., *Toxoplasma gondii* and *Trypanosoma cruzi. Adv. Parasitol.* 38: 1–51.

May, R. M. 1972. Will a large complex system be stable? *Nature* 238:413–414.

———. 1973. *Stability and complexity in model ecosystems.* Monographs in population biology 6. Princeton: Princeton University Press.

——— 1977. Dynamical aspects of host-parasite associations: Crofton's model revisited. *Parasitology* 75:259–276.

———. 1986. How many species are there? *Nature* 324:514–515.

———. 1991. The dynamics and genetics of host-parasite associations. In *Parasite-host associations: Coexistence or conflict?* ed. C. A. Toft, A. Aeschlimann, and L. Bolis, 102–128. New York: Oxford University Press.

May, R. M., and Anderson, R. M. 1978. Regulation and stability of host-parasite population interactions. II. Destabilizing processes. *J. Anim. Ecol.* 47:249–267.

May, R. M., and Nowak, M. A. 1995. Coinfection and the evolution of parasite virulence. *Proc. Roy. Soc. Lond.*, ser. B, *Biol. Sci.* 261:209–215.

Maynard Smith, J. 1986. Contemplating life without sex. *Nature* 324:300–301.

———. 1989a. *Evolutionary genetics.* Oxford: Oxford University Press.

———. 1989b. A Darwinian view of symbiosis. In *Symbiosis as a source of evolutionary innovation,* ed. L. Margulis and R. Fester, 26–39. Cambridge: MIT Press.

———. 1989c. Generating novelty by symbiosis. *Nature* 341:284–285.

———. 1991. Theories of sexual selection. *Trends Ecol. Evol.* 6:146–156.

Mayr, E. 1957. Concluding remarks. In *First symposium on host specificity among parasites of vertebrates,* 7–14. Neuchâtel: Attinger.

Mazars, E., and Dei-Cas, E. 1998. Epidemiological and taxonomic impact of *Pneumocystis* biodiversity. *FEMS Immunol. Med. Microbiol.* 22:5–13.

Mazier, D. 1993. Épitopes plasmodiaux exprimés en surface des hépatocytes infectés: Leur rôle dans la protection. *Cah. Santé* (Montrouge) 3:267–273.

Meiners, T., and Hilker, M. 1997. Host location in *Oomyzus gallerucae* (Hymenoptera: Eulophidae), an egg parasitoid of the elm leaf beetle *Xanthogaleruca luteola* (Coleoptera luteola) (Coleoptera: Chrysomelidar). *Oecologia* 112:87–93.

Melhorn, H. 1988. *Parasitology in focus: Facts and trends.* Berlin: Springer-Verlag.

Michel, M. 1955. Parasitological significance of bovine grazing behaviour. *Nature* 175:1088–1089.

Mikhailova, N. A., Granovitch, A. I., and Sergievskii, S. O. 1988. Influence of trematodes on the microbiotopical distribution of the molluscs *Littorina obtusata* and *L. saxatilis. Parazitologiya* 22:398–407 (in Russian).

Milinski, M. 1990. Parasites and host decision making. In *Parasitism and host behaviour,* ed. C. J. Barnard and J. M. Behnke, 95–116. London: Taylor and Francis.

Milinski, M., and Bakker, T. C. M. 1990. Female sticklebacks use male coloration in mate choice and hence avoid parasitized males. *Nature* 344:330–333.

Miller, G. C. 1981. Helminths and the transmammary route of infection. *Parasitology* 82:335–342.

Miller, L. H. 1994. Impact of malaria on genetic polymorphism and genetic diseases in Africans and African Americans. *Proc. Natl. Acad. Sci. U.S.A.* 91:2415–2419.

Miller, R. L., Armelagos, G. J., Ikram, S., et al. 1992. Palaeoepidemiology of *Schistosoma* infection in mummies. *Br. Med. J.* 304:555–556.

Miller, R. S. 1967. Pattern and process in competition. *Adv. Ecol. Res.* 4:1–74.

Minchella, D. J. 1985. Host life-history variation in response to parasitism. *Parasitology* 90:205–216.

Minchella, D. J., and LoVerde, P. T. 1981. A cost of increased early reproduction effort in the snail *Biomphalaria glabrata. Am. Nat.* 118:876–881.

Minchella, D. J., Sollenberger, K. M., and Pereira de Sousa, C. 1995. Distribution and schistosome genetic diversity within molluscan intermediate hosts. *Parasitology* 11:217–220.

Møller, A. P. 1990a. Effects of a haematophagous mite on the barn swallow (*Hirundo rustica*): A test of the Hamilton and Zuk hypothesis. *Evolution* 44:771–784.

———. 1990b. Parasites and sexual selection: Current status of the Hamilton and Zuk hypothesis. *J. Evol. Biol.* 3:319–328.

———. 1995. Bumblebee preference for symmetrical flowers. *Proc. Natl. Acad. Sci. U.S.A.* 92:2288–2292.

———. 1996. Parasitism and developmental instability of hosts: A review. *Oikos* 77:189–196.

———. 1997. Developmental stability and fitness: A review. *Am. Nat.* 149:916–932.

Møller, A. P., Dufva, R., and Erritzoe, J. 1998. Host immune function and sexual selection in birds. *J. Evol. Biol.* 11:703–719.

Møller, A. P., and Erritzoe, J. 1996. Parasite virulence and host immune defense: Host immune response is related to nest reuse in birds. *Evolution* 50:2066–2072.

———. 1998. Host immune defence and migration in birds. *Evol. Ecol.* 12:945–953.

Møller, A. P., and Saino, N. 1994. Parasites, immunology of hosts, and sexual selection. *J. Parasitol.* 80:850–858.

Molloy, S., Holland, C., and Poole, R. 1995. Metazoan parasite community structure in brown trout from two lakes in western Ireland. *J. Helminthol.* 69:237–242.

Molnar, K., Szekely, C. S., and Baska, F. 1991. Mass mortality of eel in Lake Balaton due to *Anguillicola* infection. *Bull. Eur. Assoc. Fish Pathol.* 11:211–212.

Molyneux, D. H., and Jenni, L. 1981. Mechanoreceptors, feeding behaviour and trypanosome transmission in *Glossina. Trans. Roy. Soc. Trop. Med. Hyg.* 75:160–161.

Monconduit, H., and Prévost, G. 1994. Avoidance of encapsulation by *Asobara tabida,* a larval parasitoid of *Drosophila* species. *Norw. J. Agric. Sci.* 16 (suppl.): 301–309.

Moné, H., and Combes, C. 1986. Analyse expérimentale de l'effet "decoy" (ou effet leurre) exercé par les mollusques non cibles sur le système hôte-parasite *Biomphalaria glabrata* (Say, 1818)—*Schistosoma mansoni* Sambon, 1907. *Acta Oecol. Oecol. Appl.* 7:281–286.

Monge, J. P., and Cortesero, A. M. 1996. Tritrophic interactions among larval parasitoids, bruchids and leguminosae seeds: Influence of pre- and post-mergence learning on parasitoids' response to host and host-plant cues. *Entomol. Exp. Appl.* 80:293–296.

Monod, R. 1977. Étude de l'helminthofaune des couleuvres du genre *Natrix* dans le sud de la France: Biologie et écologie de la progénèse chez les digènes du genre *Paralepoderma.* Thesis, Université de Montpellier II, France.

Moore, J. 1983a. Responses of an avian predator and its isopod prey to an acanthocephalan parasite. *Ecology* 64:1000–1015.

———. 1983b. Altered behavior in cockroaches *(Periplaneta americana)* infected with an archiacanthocephalan, *Moniliformis moniliformis. J. Parasitol.* 69:1174–1176.

———. 1993a. Parasites and the behavior of biting flies. *J. Parasitol.* 79:1–16.

———. 1993b. Worthy animals. *Science* 262: 124.

Moore, J., and Gotelli, N. J. 1990. A phylogenetic perspective on the evolution of altered host behaviours: A critical look at the manipulation hypothesis. In *Parasitism and host behaviour,* ed. C. J. Barnard and J. M. Behnke, 193–233. London: Taylor and Francis.

———. 1992. *Moniliformis moniliformis* increases cryptic behaviors in the cockroach *Supella longipalpa. J. Parasitol.* 78:49–53.

Moore, J., and Lasswell, J. 1986. Altered behavior in isopods *(Armadillidium vulgare)* infected with the nematode *Dispharynx nasuta. J. Parasitol.* 72:186–189.

Moore, J., and Simberloff, D. 1990. Gastrointestinal helminth communities of bobwhite quail. *Ecology* 71:344–359.

Moore, J., Simberloff, D., and Freehling, M. 1988. Relationships between bobwhite quail social-group size and intestinal helminth parasitism. *Am. Nat.* 131:22–32.

Morales, J., Larrade, C., Arteaga, M., et al. 1996. Inhibition of sexual behavior in male mice infected with *Taenia crassiceps* cysticerci. *J. Parasitol.* 82:689–693.

Morand, S. 1988. Cycle évolutif de *Nemhelix bakeri* Morand et Petter (Nematoda, Cosmocercidae) parasite de l'appareil génital de *Helix aspersa* (Gastropoda, Helicidae). *Can. J. Zool.* 66:1796–1802.

———. 1996a. Biodiversity of parasites in relation to their life-cycle. In *The genesis and maintenance of biological diversity,* ed. M. Hochberg, J. Clobert, and R. Barbault, 243–260. Oxford: Oxford University Press.

———. 1996b. Life-history traits in parasitic nematodes: A comparative approach for the search of invariants. *Funct. Ecol.* 10:210–218.

———. 1997. Comparative analyses of continuous data: The need to be phylogenetically correct. *Mem. Mus. Natl. Hist. Nat.* 173:73–90.

Morand, S., Manning, S. D., and Woolhouse, M. F. J. 1996. Parasite-host coevolution and geographic patterns of parasite infectivity and host susceptibility. *Proc. Roy. Soc. Lond.,* ser. B, *Biol. Sci.* 263:119–128.

Morand, S., and Müller-Graf, C. D. M. 2000. Muscles or testes? Comparative evidence for sexual competition among dioecious blood parasites (Schistosomatidae) of vertebrates. *Parasitology* 120:45–56.

Morand, S., and Petter, A. 1986. *Nemhelix bakeri* n. gen., n. sp. (Nematoda: Cosmocercidae) parasite de l'appareil génital de *Helix aspersa* (Gastropoda: Helicidae) en France. *Can. J. Zool.* 64:2008–2011.

Morand, S., Pointier, J.-P., Borel, G., et al. 1993. Pairing probability of schistosomes related to their distribution among the host population. *Ecology* 74:2444–2449.

Morand, S., and Poulin, R. 1998. Density, body mass and parasite species richness of terrestrial mammals. *Evol. Ecol.* 12:717–727.

Morand, S., and Rivault, C. 1992. Infestation dynamics of *Blatticola blattae* Graeffe (Nematoda: Thelastomatidae), a parasite of *Blattella germanica* L. (Dictyoptera: Blattellidae). *Int. J. Parasitol.* 22:983–989.

Morand, S., Robert, F., and Connors, V. 1995. Complexity in parasite life-cycles: Population biology of cestodes in fish. *J. Anim. Ecol.* 64:256–264.

Moravec, F. 1992. Spreading of the nematode *Anguillicola crassus* (Dracunculoidea) among eel populations in Europe. *Folia Parasitol.* 39:247–248.

Morimitsu, T. T., Nagai, T., Ide, M., et al. 1986. Parasitic octavus neuropathy as a cause of mass stranding in Odontoceti. *J. Parasitol.* 72:469–472.

Moritz, C., MacCallum, H., Donnellan, S., et al. 1991. Parasite loads in pathenogenetic and sexual lizards *(Heteronotia binoei):* Support for the Red Queen hypothesis. *Proc. Roy. Soc. Lond.,* ser. B, *Biol. Sci.* 244:145–149.

Moritz, R. F. A. 1994. Selection for varroatosis resistance in honeybees. *Parasitol. Today* 10:236–238.

Moss, R., Watson, A., and Parr, R. 1996. Experimental prevention of a population cycle in red grouse. *Ecology* 77:1512–1530.

Moss, R., Watson, A., Trenholm, I. B., et al. 1993. Caecal threadworm *Trichostrongylus tenuis* in red grouse *Lagopus lagopus scoticus*: Effects of weather and host density upon estimated worm burdens. *Parasitology* 107:199–209.

Mosser, D. M., and Brittingham, A. 1997. Leishmania, macrophages and complement: A tale of subversion and exploitation. *Parasitology* 115:S9–S23.

Mott, K. E., Desjeux, P., Moncayo, A., et al. 1991. Parasitoses et urbanisation. *Bull. Org. Mond. Santé* 69:9–16.

Mouahid, A., Moné, H., Chaib, A., et al. 1991. Cercarial shedding patterns of *Schistosoma bovis* and *Schistosoma haematobium* from single and mixed infections. *J. Helminthol.* 65:8–14.

Mouchet, F., Théron, A., Brémond, P., et al. 1992. Pattern of cercarial emergence of *Schistosoma curassoni* from Niger and comparison with three sympatric species of schistosomes. *J. Parasitol.* 78:61–63.

Mouchet, J., Faye, O., Julvez, J., et al. 1996. Drought and malaria retreat in the Sahel, West Africa. *Lancet* 348:1735–1736.

Moulder, J. W. 1979. The cell as an extreme environment. *Proc. Roy. Soc. Lond.*, ser. B, *Biol. Sci.* 204:199–210.

Moulia, C. 1999. Parasitism of plant and animal hybrids: Are facts and fates the same? *Ecology* 80:392–406.

Moulia, C., Aussel, J. P., Bonhomme, F., et al. 1991. Wormy mice in a hybrid zone: A genetic control of susceptibility to parasite infection. *J. Evol. Biol.* 4:679–687.

Moulia, C., Le Brun, N., Loubès, C., et al. 1995. Hybrid vigour against parasites in interspecific crosses between two mice species. *Heredity* 74:48–52.

Moulia, C., Le Brun, N., Orth, A., et al. 1993. Experimental evidence of genetic determinism in high susceptibility to pinworms infections in mice: A hybrid zone model. *Parasitology* 106:387–393.

Moulia, C., Le Brun, N., and Renaud, F. 1996. Mouse-parasite interactions: From gene to population. *Adv. Parasitol.* 38:119–165.

Mueller, J. F. 1974. The biology of *Spirometra*. *J. Parasitol.* 60:3–14.

Müller, C. B., and Schmid-Hempel, R. 1992. To die for host or parasite? *Anim. Behav.* 44:177–179.

Müller, G., and Ward, P. I. 1995. Parasitism and hetrozygosity influence the secondary sexual characters of the European minnow, *Phoxinus phoxinus* (L.) (Cyprinidae). *Ethology* 100:309–319.

Müller-Graf, C. D. M. 1995. A coprological survey of intestinal parasites of wild lions *(Panthera leo)* in the Serengeti and the Ngorongoro Crater, Tanzania, East Africa. *J. Parasitol.* 81:812–814.

Müller-Graf, C. D. M., Collins, D. A., Packer, C., et al. 1997. *Schistosoma mansoni* infection in a natural population of olive baboons *(Papio cynocephalus anubis)* in Gombe Stream National Park, Tanzania. *Parasitology* 115:621–627.

Müller-Graf, C. D. M., Collins, D. A., and Woolhouse, M. E. J. 1996. Intestinal parasite burden in five troops of olive baboons *(Papio cynocephalus anubis)* in Gombe Stream National Park, Tanzania. *Parasitology* 112:489–497.

Mulvey, M., and Aho, J. M. 1993. Parasitism and mate competition: Liver flukes in white-tailed deer. *Oikos* 66:187–192.

Munger, J. C., and Holmes, J. C. 1988. Benefits of parasitic infection: A test using a ground squirrel-trypanosome system. *Can. J. Zool.* 66:222–227.

Murdoch, W. W., Luck, R. F., Walde, S. J., et al. 1989. A refuge for red scale under control by *Aphytis*: Structural aspects. *Ecology* 70:1707- 1714.

Murray, D. L., Cary, J. R., and Keith, L. B. 1997. Interactive effects of sublethal nematodes and nutritional status on snowshoe hare vulnerability to predation. *J. Anim. Ecol.* 66:250–264.

Nadler, S. A. 1995. Microevolution and the genetic structure of parasite populations. *J. Parasitol.* 81:395–403.

Nadler, S. A., Lindquist, R. L., and Near, T. J. 1995. Genetic structure of midwestern *Ascaris suum* populations: A comparison of isozyme and RAPD markers. *J. Parasitol.* 81:385–394.

Nardon, P., and Grenier, A. M. 1988. Genetical and biochemical interactions between the host and its endocytobiotes in the weevils *Sitophilus* (Coleoptera, Curculionidae) and other related species. In *Cell to cell signals in plant, animal and microbial symbiosis,* ed. S. Scannerini et al., 255–270. NATO ASI Series 17. Berlin: Springer-Verlag.

Nascetti, G., Cianchi, R., Mattiucci, S., et al. 1993. Three sibling species within *Contracaecum osculatum* (Nematoda, Ascaridida, Ascaridoidea) from the Atlantic Arctic-Boreal region: Reproductive isolation and host preferences. *Int. J. Parasitol.* 23:105–120.

Nasincova, V., and Scholz, T. 1994. The life-cycle of *Asymphylodora tincae* (Modeer, 1790) (Trematoda: Monorchiidae): A unique development in monorchiid trematodes. *Parasitol. Res.* 80:192–197.

Nelson, B. C., and Smith, C. R. 1980. Ecology of sylvatic plague in lava caves at Lava Beds National Monument, California. In *Fleas: Proceedings of the International Conference on Fleas,* ed. R. Traub and H. Starcke, 273–275. Rotterdam: Balkema.

Nelson, G. S. 1990. Human behaviour and the epidemiology of helminth infections: Cultural practices and microepidemiology. In *Parasitism and host behaviour,* ed. C. J. Barnard and J. M. Behnke, 234–263. London: Taylor and Francis.

Newbold, C. I., and Marsh, K. 1990. Antigens on the *Plasmodium falciparum* infected erythrocyte surface are parasite-derived. *Parasitol. Today* 6:320–322.

Newhouse, J. R. 1990. Chesnut blight. *Sci. Am.* 263:106–111.

N'Goran, E., Brémond, P., Sellin, E., et al. 1997. Intraspecific diversity of *Schistosoma haematobium* in West Africa: Chronobiology of cercarial emergence. *Acta Trop.* 66:35–44.

Nilsson, L. A. 1988. The evolution of flowers with deep corolla tubes. *Nature* 334:147–149.

———. 1998. Deep flowers for long tongues. *Trends Ecol. Evol.* 13:259–260.

Noble, E. R., Noble, G. A., Schad, G. A., et al. 1989. *Parasitology: The biology of animal parasites.* Philadelphia: Lea and Febiger.

Noble, S. J. 1991. Factors influencing the pinworm community (Oxyurida; Nematoda) parasitic in the hindgut of the American cockroach *Periplaneta americana.* M.Sc. thesis, University of British Columbia.

Noireau, F., and Gouteux, J. P. 1989. Current considerations on a *Loa loa* simian reservoir in the Congo. *Acta Trop.* 46:69–70.

Nokes, C., and Bundy, D. A. P. 1994. Does helminth infection affect mental processing and educational achievement? *Parasitol. Today* 10:14–18.

Noldus, L. 1989. Chemical espionage by parasitic wasps. Thesis, Université de Wageningen, Hollande.

Nollen, P. M. 1983. Patterns of sexual reproduction among parasitic platyhelminths. *Parasitology* 86:99–120.

———. 1993. *Echinostoma trivolvis:* Mating behavior of adults raised in hamsters. *Parasitol. Res.* 79:130–132.

Norris, K., Anwar, M., and Read, A. F. 1995. Reproductive effort influences the prevalence of haematozoan parasites in great tits. *J. Anim. Ecol.* 63:601–610.

Nowak, M. A., Anderson, R. M., MacLean, A. R., et al. 1991. Antigenic diversity thresholds and the development of AIDS. *Science* 254:963–969.

Obrebski, S. 1975. Parasite reproductive strategy and evolution of castration of hosts by parasites. *Science* 188:1314–1316.

O'Donnell, S. 1997. How parasites can promote the expression of social behaviour in their hosts. *Proc. Roy. Soc. Lond.,* ser. B, *Biol. Sci.* 264:689–694.

Oien, I. J., Honza, M., Moksnes, A., et al. 1996. The risk of parasitism in relation to the distance from reed warbler nests to cuckoo perches. *J. Anim. Ecol.* 65:147–153.

Ojcius, D. M., and Dautry-Varsat, A. 1995. Les mille et une ruses des microbes intracellulaires. *Recherche* 273:142–148.

Olsen, O. W. 1974. *Animal parasites: Their life-cycles and ecology.* Baltimore: University Park Press.

Oppliger, A., Christe, P., and Richner, H. 1996. Clutch size and malaria resistance. *Nature* 381:565.

———. 1997. Clutch size and malarial parasites in female great tits. *Behav. Ecol.* 8:148–152.

Orgel, L. E., and Crick, F. H. C. 1980. Selfish DNA: The ultimate parasite. *Nature* 284:604–607.

Overath, P., Chaudhri, M., Steverding, D., et al. 1994. Invariant surface proteins in bloodstream forms of *Trypanosoma brucei. Parasitol. Today* 10:53–58.

Pagès, J. R., and Théron, A. 1990a. Analysis and comparison of cercarial emergence rhythms of *Schistosoma haematobium, S. intercalatum* and *S. bovis* and their hybrid progeny. *Int. J. Parasitol.* 20:193–197.

———. 1990b. *Schistosoma intercalatum* from Cameroon and Zaire: Chronobiological differenciation of cercarial emergence. *J. Parasitol.* 76:743–745.

Paige, K. N., and Capman, W. C. 1993. The effects of host-plant genotype, hybridization, and environment on gall-aphid attack and survival in cottonwood: The importance of genetic studies and the utility of RFLPS. *Evolution* 47:36–45.

Pariselle, A. 1996. Diversité, spéciation et évolution des monogènes branchiaux de cichlidae en Afrique de l'Ouest. Thesis, Université de Perpignan, France.

Park, T. 1948. Experimental studies of interspecific competition. I. Competition between populations of flour beetles, *Tribolium confusum* Duval and *Tribolium castaneum* Herbst. *Ecol. Monogr.* 18:265–308.

Pasteur, G. 1981. A classificatory review of mimicry systems. *Ann. Rev. Ecol. Syst.* 13:169–199.

Patel, A., Anstett, M. C., Hossaert-MacKey, M., et al. 1995. Pollinators entering female dioecious figs: Why commit suicide? *J. Evol. Biol.* 8:301–313.

Patrick, M. J. 1991. Distribution of enteric helminths in *Glaucomys volans* L. (Sciuridae): A test for competition. *Ecology* 72:755–758.

Patterson, B. D., and Atmar, W. 1986. Nested subsets and the structure of insular mammalian faunas and archipelagos. *Biol. J. Linn. Soc.* 28:65–82.

Paul, R. E. L., and Day, K. P. 1998. Mating patterns of *Plasmodium falciparum. Parasitol. Today* 14:197–202.
Pays, E., and Steinert, M. 1988. Control of antigen gene expression in African trypanosomes. *Ann. Rev. Genet.* 22:107–126.
Peeters, M., Liégeois, F., Torimiro, N., et al. 1999. Characterization of a highly replicative intergroup M/O human immunodeficiency virus type 1 recombinant isolated from a Cameroonian patient. *J. Virol.* 73:7368–7375.
Pellmyr, O. 1989. The cost of mutualism: Interactions between *Trollius europaeus* and its pollinating parasites. *Oecologia* 78:53–59.
Pellmyr, O., Leebens-Mack, J., and Huth, C. J. 1996. Non-mutualistic yucca moths and their evolutionary consequences. *Nature* 380:155–156.
Perez-Maluf, R., Kaiser, L., Wajnberg, E., et al. 1998. Genetic variability of conditioned probing responses to a fruit odor in *Leptopilina boulardi* (Hymenoptera: Eucolidae) a *Drosophila* parasitoid. *Behav. Genet.* 28:67–73.
Perrin, N., Christe, P., and Richner, H. 1996. On host life-history response to parasitism. *Oikos* 75:317–320.
Petter, A. 1966. Équilibre des espèces dans les populations de nématodes parasites du colon des tortues terrestres. *Mem. Mus. Natl. Hist. Nat.,* ser. A. *Zool.* 39:1–252.
Phares, C. K. 1987. Plerocercoid growth factor: A homologue of human growth hormone. *Parasitol. Today* 3:346–349.
Phares, K. 1996. An unusual host-parasite relationship: The growth hormone-like factor from plerocercoids of *Spirometra* tapeworms. *Int. J. Parasitol.* 26:575–588.
Philippon, B. 1977. Étude de la transmission d'*Onchocerca volvulus* (Leuckart, 1883) (Nematoda, Onchocercidae) par *Simulium damnosum* Theobald, 1903 (Diptera, Simuliidae) en Afrique Tropicale. *Trav. Doc. ORSTOM* 63:1–308.
Pianka, B. R. 1970. On r- and K- selection. *Am. Nat.* 104:592–597.
Pica-Mattoccia, L., Moroni, R., Tchuem-Tchuenté, L. A. Southgate, V. R., and Cioli, D. 2000. Changes of mate occur in *Schistosoma mansoni. Parasitology* 120:495–500.
Pichelin, S., Whittington, I., and Pearson, J. 1991. *Concinnocotyla* (Monogenea: Polystomatidae), a new genus for the polystome from the Australian lungfish *Neoceratodus forsteri. Syst. Parasitol.* 18:81–93.
Picot, H., and Benoist, J. 1975. Interaction of social and ecological factors in the epidemiology of helminth parasites. In *Biosocial interrelations in population adaptation,* ed. E. Watts, 233–247. The Hague: Mouton.
Pietsch, T. W. 1976. Dimorphism, parasitism and sex: Reproductive strategies among deepsea ceratioid anglerfishes. *Copeia* 1:781–793.
Pimm, S. L. 1984. The complexity and stability of ecosystems. *Nature* 307:321–326.
———. 1991. *The balance of nature.* Chicago: University of Chicago Press.
Pitchford, R. J. 1961. Observations on a possible hybrid between the two schistsosomes *S. haematobium* and *S. matthei. Trans. Roy. Soc. Trop. Med. Hyg.* 55:44–51.
Platt, T. R., Blair, D., Purdie, J., et al. 1991. *Griphobilharzia amoena* n. gen. n. sp. (Digenea: Schistosomatidae), a parasite of the freshwater crocodile *Crocodylus johnstoni* (Reptilia: Crocodylia) from Australia, with erection of a new subfamily, Griphobilharziinae. *J. Parasitol.* 77:65–68.
Platt, T. R., and Brooks, D. R. 1997. Evolution of the schistosomes (Digenea:

Schistosomatidea): The origin of dioecy and colonization of the venous system. *J. Parasitol.* 83:1035–1044.
Poiani, A. 1992. Ectoparasitism as a possible cost of social life: A comparative analysis using Australian passerines. *Oecologia* 92:429–441.
Poinsot, D., and Merçot, H. 1997. *Wolbachia* infection in *Drosophila simulans*: Does the female host bear a physiological cost? *Evolution* 51:180–186.
Poirié, M., Frey, F., Hita, M., Huguet, E., Lemeunier, F., Periquet, G., and Carton, Y. 2000. *Drosophila* resistance genes to parasitoids: Chromosomal location and linkage analysis. *Proc. R. Soc. Lond.*, ser. B, *Biol. Sci.* 267:1417–1421.
Pojmanska, T. 1982. The co-occurrence of three species of *Diorchis* Clerc, 1903 (Cestoda: Hymenolepididae) in the European coot, *Fulica atra* L. *Parasitology* 84:419–429.
Poliansky, Y. I., and Bychowsky, B. E. 1963. Parasite fauna of sea fish. In *Results and perspectives of investigations by Soviet parasitologists on fish parasites in seas of the USSR,* 187–193. Jerusalem: Israel Program of Scientific Translations.
Potts, W. K., and Wakeland, E. K. 1990. Evolution of diversity at the major histocompatibility complex. *Trends Ecol. Evol.* 5:181–187.
Poulin, R. 1991. Group-living and infestation by ectoparasites in passerines. *Condor* 93:418–423.
———. 1992a. Altered behaviour in parasitized bumblebees: Parasite manipulation or adaptative suicide? *Anim. Behav.* 44:174–176.
———. 1992b. Determinants of host-specificity in parasites of freshwater fishes. *Int. J. Parasitol.* 22:753–758.
———. 1994a. Meta-analysis of parasite-induced behavioural changes. *Anim. Behav.* 48:137–146.
———. 1994b. The evolution of parasite manipulation of host behaviour: A theoretical analysis. *Parasitology* 109:S109–S118.
———. 1995a. Phylogeny, ecology, and the richness of parasite communities in vertebrates. *Ecol. Monogr.* 65:283–302.
——— 1995b. Adaptive changes in the behaviour of parasitized animals: A critical review. *Int. J. Parasitol.* 25:1371–1383.
———. 1996. The evolution of life history strategies in parasitic animals. *Adv. Parasitol.* 37:107–134.
———. 1998a. *Evolutionary ecology of parasites.* London: Chapman and Hall.
———. 1998b. Large-scale patterns of host use by parasites of freshwater fishes. *Ecol. Lett.* 1:118–128.
———. 1999a. The functional importance of parasites in animal communities: Many roles at many levels? *Int. J. Parasitol.* 29:903–914.
———. 1999b. The intra-and interspecific relationships between abundance and distribution in helminth parasites of birds. *J. Anim. Ecol.* 68:719–725.
Poulin, R., Brodeur, J., and Moore, J. 1994. Parasite manipulation of host behaviour: Should hosts always lose? *Oikos* 70:479–484.
Poulin, R., and Combes, C. 1999. The concept of virulence: Interpretations and implications. *Parasitol. Today* 15:474–475.
Poulin, R., and Fitzgerald, G. J. 1989. Shoaling as an anti-ectoparasite mechanism in juvenile sticklebacks (*Gasterosteus* spp.). *Behav. Ecol. Sociobiol.* 24:251–255.

Poulin, R., and Rohde, K. 1997. Comparing the richness of metazoan ectoparasite communities of marine fishes: Controlling for host phylogeny. *Oecologia,* 110:278–283.

Poulin, R., and Vickery, W. L. 1993. Parasite distribution and virulence: Implication for parasite-mediated sexual selection. *Behav. Ecol. Sociobiol.* 33:429–436.

Powell, J. A. 1992. Interrelationships of yuccas and yucca moths. *Trends Ecol. Evol.* 7:10–15.

Preston, T. M., and Southgate, V. R. 1994. The species specificity of *Bulinus-Schistosoma* interactions. *Parasitol. Today* 10:69–73.

Prévost, M. C., Lesourd, M., Arpin, M., et al. 1992. Unipolar reorganization of F-actin layer at bacterial division and bundling of actin filaments by plastin correlate with movement of *Shigella flexneri* within HeLa cells. *Infec. Immun.* 60:4088–4099.

Price, E. W. 1989. The web of life: Development over 3.8 billion years of trophic relationships. In *Symbiosis as a source of evolutionary innovation,* ed. L. Margulis and R. Fester, 262–272. Cambridge: MIT Press.

Price, P. W. 1977. General concepts on the evolutionary biology of parasites. *Evolution* 31:405–420.

———. 1980. *Evolutionary biology of parasites.* Princeton: Princeton University Press.

———. 1984. Communities of specialists: Vacant niches in ecological and evolutionary time. In *Ecological communities: Conceptual issues and the evidence,* ed. D. R. Strong, D. Simberloff, L. G. Abele, and A. B. Thistle, 510–523. Princeton: Princeton University Press.

———. 1987. Evolution in parasite communities. *Int. J. Parasitol.* 17:209–214.

Price, P. W., and Clancy, K. M. 1983. Patterns in number of helminth parasite species in freshwater fishes. *J. Parasitol.* 69:449–454.

Price, P. W., Westoby, M., and Rice, B. 1988. Parasite-mediated competition: Some predictions and results. *Am. Nat.* 131:544–555.

Price, P. W., Westoby, M., Rice, B., et al. 1986. Parasite mediation in ecological interactions. *Ann. Rev. Ecol. Syst.* 17:487–505.

Profet, M. 1993. Menstruation as a defense against pathogens transported by sperm. *Q. Rev. Biol.* 68:335–386.

Pujol, C., Reynes, J., Renaud, F., et al. 1993. The yeast *Candida albicans* has a clonal mode of reproduction in a population of infected human immunodeficiency virus–positive patients. *Proc. Natl. Acad. Sci. U.S.A.* 90:9456–9459.

Pulliam, H. R. 1988. Sources, sinks, and population regulation. *American Naturalist* 132:652–661.

Quentin, J.-C. 1989. Prévalence des helminthes de rongeurs myomorphes d'Afrique Centrale: Relations entre l'écologie et l'éthologie de ces rongeurs et leurs peuplements parasitaires. *Bull. Ecol.* 20:311–323.

Quinn, S. C., Brooks, R. J., and Cawthorn, R. J. 1987. Effects of the protozoan parasite *Sarcocystis rauschorum* on open-field behaviour of its intermediate vertebrate host, *Dicrostonyx richardsoni. J. Parasitol.* 73:265–271.

Quinn, T. P., and Brannon, E. L. 1982. The use of celestial and magnetic cues by orienting sockeye salmon smolts. *J. Comp. Physiol.* 147:547–552.

Radomski, A. A., and Pence, D. B. 1995. Persistence of a recurrent group of intestinal helminth species in a coyote population from Southern Texas. *J. Parasitol.* 79:371–378.

Raibaut, A., Combes, C., and Benoit, F. 1998. Analysis of the parasitic copepod species richness among Mediterranean fish. *J. Mar. Syst.* 15:185–206.

Raibaut, A., and Trilles, J.-P. 1993. The sexuality of parasitic Crustaceans. *Adv. Parasitol.* 32:367–455.

Raikova, E. V. 1994. Life cycle, cytology, and morphology of *Polypodium hydriforme,* a coelenterate parasite of the eggs of Acipenseriform fishes. *J. Parasitol.* 80:1–22.

Ralley, W. E., Galloway, T. D., and Crow, G. H. 1993. Individual and group behaviour of pasture cattle in response to attack by biting flies. *Can. J. Zool.* 71:725–734.

Ranta, E. 1992. Gregariousness *versus* solitude: Another look at parasite faunal richness in Canadian freshwater fishes. *Oecologia* 89:150–152.

Ratti, O., Dufva, R., and Alatalo, R. V. 1993. Blood parasites and male fitness in the pied flycatcher. *Oecologia* 96:410–414.

Raymond, K., and Probert, A. J. 1992. The daily cercarial emission rhythm of *Schistosoma margrebowiei* with particular reference to dark period stimuli. *J. Helminthol.* 65:159–168.

Rea, J. G., and Irwin, S. W. B. 1994. The ecology of host-finding behaviour and parasite transmission: Past and future perspectives. *Parasitology* 109:S31–S39.

Read, A. F. 1987. Comparative evidence supports the Hamilton and Zuk hypothesis on parasites and sexual selection. *Nature* 328:68–70.

———. 1988. Sexual selection and the role of parasites. *Trends Ecol. Evol.* 3:97–102.

———. 1990. Parasites and the evolution of host sexual behaviour. In *Parasitism and host behaviour,* ed. C. J. Barnard, and J. M. Behnke, 117–157. London: Taylor and Francis.

Read, A. F., and Day, K. P. 1992. The genetic structure of malaria parasite populations. *Parasitol. Today* 8:239–242.

Read, A. F., and Harvey, P. H. 1993. The evolution of virulence. *Nature* 362:500–501.

Read, A. F., and Kilejian, A. Z. 1969. Circadian migratory behavior of a cestode symbiote in the rat host. *J. Parasitol.* 55:574–578.

Read, A. F., and Nee, S. 1990. Male schistosomes: More than just muscle? *Parasitol. Today* 6:297.

Read, A. F., and Skorping, A. 1995. The evolution of tissue migration by parasitic nematode larvae. *Parasitology* 111:359–371.

Renaud, F., and Gabrion, C. 1988. Speciation in Cestoda: Evidence of two sibling species in the complex *Bothrimonus nylandicus* (Scheider, 1902) (Cestoda, Pseudophyllidea). *Parasitology* 97:1–9.

Rennie, J. 1992. Living together. *Sci. Am.* 266:104–113.

Reversat, J., Renaud, F., and Maillard, C. 1989. Biology of parasite populations: Differential specificity in the genus *Helicometra* Odhner, 1902 (Trematoda,

Opecoelidae) in the Mediterranean Sea demonstrated by enzyme electrophoresis. *Int. J. Parasitol.* 19:885–890.

Reynaud, C. A., Anquez, V., Dahan, A., et al. 1985. A single rearragement event generates most of the chicken immunoglobulin light chain diversity. *Cell* 40:283–291.

Reynaud, C. A., Garcia, C., Hein, W. R., et al. 1995. Hypermutation generating the sheep immunoglobulin repertoire is an antigen-independent process. *Cell* 80:115–125.

Richards, C. S., Knight, M., and Lewis, F. A. 1992. Genetics of *Biomphalaria glabrata* and its effect on the outcome of *Schistosoma mansoni* infection. *Parasitol. Today* 8:171–174.

Richner, H., Christe, P., and Oppliger, A. 1995. Paternal investment affects prevalence of malaria. *Proc. Natl. Acad. Sci. U.S.A.*, 92:1192–1194.

Richner, H., and Heeb, P. 1995. Are clutch and brood size patterns in birds shaped by ectoparasites? *Oikos* 73:435–441.

Richner, H., Oppliger, A., and Christe, P. 1993. Effect of an ectoparasite on the reproduction in great tits. *J. Anim. Ecol.* 62:703–710.

Richner, H., and Tripet, F. 1999. Ectoparasitism and the trade-off between current and future reproduction. *Oikos* 86:535–538.

Ridgway, S. H., and Dailey, M. D. 1972. Cerebral and cerebellar involvement of trematode parasites in dolphins and their possible role in stranding. *J. Wildl. Dis.* 8:33–43.

Ridley, M. 1994. *The Red Queen: Sex and the evolution of human nature.* London: Penguin Books.

Rigaud, T., and Juchault, P. 1992. Genetic control of the vertical transmission of a cytoplasmic sex factor in *Armadillidium vulgare* Latr. (Crustacea, Oniscidea). *Heredity* 68:47–52.

———. 1995. Success and failure of horizontal transfers of feminizing *Wolbachia* endosymbionts in woodlice. *J. Evol. Biol.* 8:249–255.

Rigaud, T., Mocquard, J. P., and Juchault, P. 1992. The spread of parasitic sex factors in populations of *Armadillidium vulgare* Latr. (Crustacea, Oniscidea): Effects on *sex ratio. Genet. Sel. Evol.* (Paris) 24:3–18.

Riley, M. 1993. *Evolution.* New York: Blackwell Scientific Publications.

Rio, D.C. 1991. Regulation of *Drosophila* P element transposition. *Trends Genet.* 7:282–287.

Rioux, J. A., Lanotte, G., Petter, F., et al. 1986. Les leishmanioses cutanées du bassin méditerranéen occidental: De l'identification enzymatique à l'analyse éco-épidémiologique. L'exemple de trois "foyers," tunisien, marocain et français. In *Leishmania: Taxonomie et phylogénèse. Applications éco-épidémiologiques,* 365–395. Montpellier: IMEEE.

Ritchie, L. E., and Hoeg, J. T. 1981. The life-history of *Lernaeodiscus porcellanae* (Cirripedia: Rhizocephala) and co-evolution with its porcellanid host. *J. Crustac. Biol.* 1:334–347.

Ritter, W. 1981. Varroa disease of the honey-bee *Apis mellifera. Bee World* 62:141–153.

Robert, F., Boy, V., and Gabrion, C. 1990. Biology of parasite populations: Population dynamics of bothriocephalids (Cestoda, Pseudophyllidea) in teleostean fish. *J. Fish Biol.* 37:327–342.

Robert, F., Renaud, F., Mathieu, E., et al. 1988. Importance of the paratenic host in the biology of *Bothriocephalus gregarius* (Cestoda, Pseudophyllidea), a parasite of the turbot. *Int. J. Parasitol.* 18:611–621.

Robert, M., and Sorci, G. 1999. Rapid increase of host defence against brood parasites in a recently parasitized area: The case of village weavers in Hispaniola. *Proc. Roy. Soc. Lond.*, ser. B, *Biol. Sci.* 266:1–6.

Robert, V., Gazin, P., Boudin, C., et al. 1985. La transmission du paludisme en zone de savane arborée et en zone rizicole des environs de Bobo Dioulasso (Burkina Faso). *Ann. Soc. Belge Med. Trop.* 65 (suppl.) 2:201–214.

Robert, V., Gazin, P., Ouédraogo, V., et al. 1986. Le paludisme urbain à Bobo-Dioulasso (Burkina Faso). *Cah. ORSTOM,* ser. *Entomol. Med. Parasitol.* 24:121–128.

Roberts, D. J., Craig, A. G., Berendt, A. R., et al. 1992. Rapid switching to multiple antigenic and adhesive phenotypes in malaria. *Nature* 357:689–692.

Roberts, L. S., and Janovy, J. J. 2000. *Gerald D. Schmidt and Larry S. Roberts' Foundations of parasitology.* 6th ed. Boston:McGraw-Hill.

Roberts, T. M., Ward, S., and Chernin, E. 1979. Behavioral responses of *Schistosoma mansoni* miracidia in concentration gradients of snail-conditioned water. *J. Parasitol.* 65:41–49.

Robinson, J., Poynter, D., and Terry, R. J. 1962. The role of the fungus *Pilobolus* in the spread of the infective larvae of *Dictyocaulus viviparus. Parasitology* 52:17–18.

Rodriguez, L., Sokolowski, M. B., and Carton, Y. 1991. Intra- and interspecific variation in pupation behaviours of *Drosophila* from different habitats. *Can. J. Zool.* 69:2616–2619.

Rodriguez-Girones, M. A., and Lotem, A. 1999. How to detect a cuckoo egg: A signal-detection theory model for recognition and learning. *Am. Nat.* 153:633–648.

Roepstorff, A., Eriksen, L., Slotved, H.-C., et al. 1997. Experimental *Ascaris suum* infection in the pig: Worm population kinetics following single inoculations with three doses of infective eggs. *Parasitology* 115:443–452.

Rohde, K. 1979. A critical evaluation of intrinsic and extrinsic factors responsible for niche restriction in parasites. *Am. Nat.* 114:648–671.

———. 1984. Zoogeography of marine parasites. *Helgol. Meersunters.* 37:35–52.

———. 1992. Latitudinal gradients in species diversity: The search for the primary cause. *Oikos* 65:514–527.

———. 1993. *Ecology of marine parasites.* 2d ed. Wallingford, Eng.: CAB International.

———. 1994. Niche restriction in parasites: Proximal and ultimate causes. *Parasitology* 109:S69–S84.

———. 1999. Latitudinal gradients in species diversity and Rapoport's rule revisited: A review of recent work and what can parasites teach us about the causes of gradients? *Ecography* 22:593–613.

Rohde, K., and Heap, M. 1998. Latitudinal differences in species and community richness and in community structure of metazoan endo- and ectoparasites of marine teleost fish. *Int. J. Parasitol.* 28:461–474.

Rollinson, D., Southgate, V. R., Vercruysse, J., et al. 1990. Observations on natural and experimental interactions between *Schistosoma bovis* and *S. curassoni* from West Africa. *Acta Trop.* 47:101–114.

Rosenberg, R., Wirtz, R. A., Schneider, I., et al. 1990. An estimation of the number of malaria sporozoites ejected by a feeding mosquito. *Trans. Roy. Soc. Trop. Med. Hyg.* 84:209–212.

Rossignol, P. A. 1988. Parasite modification of mosquito probing behaviour. *Misc. Publ. Entomol. Soc. Am.* 68:25–28.

Rothstein, S. I., Robinson, S. I., and Robinson, S. K. 1994. Conservation and coevolutionary implications of brood parasitism by cowbirds. *Trends Ecol. Evol.* 9:162–164.

Rotter, J. I., and Diamond, J. M. 1987. What maintains the frequencies of human genetic diseases? *Nature* 329:289–290

Rousset, F., Bouchon, D., Pintureau, B., et al. 1992. *Wolbachia* endosymbionts responsible for various alterations of sexuality in arthropods. *Proc. Roy. Soc. Lond.*, ser. B, *Biol. Sci.* 250:91–98.

Rowland, M., and Boersma, E. 1988. Changes in the spontaneous flight activity of the mosquito *Anopheles stephensi* by parasitization with the rodent malaria *Plasmodium yoelii. Parasitology* 97:221–227.

Roye, O., Delhem, N., Trottein, F., et al. 1998. Dermal endothelial cells and keratinocytes produce IL-7 in vivo after human *Schistosoma mansoni* percutaneous infection. *J. Immunol.* 161:4161–41–68.

Rudel, T., and Meyer, T. F. 1999. Infection of human cells by *Neisseria*—a paradigm of ancestral apoptosis? *Nova Acta Leopold.* 307:71–86.

Rudin, W. 1990. Comparison of the cuticular structure of parasitic nematodes recognized by immunocytochemical and lectin binding studies. *Acta Trop.* 47:255–268.

Ruiz, G. M. 1991. Consequences of parasitism to marine invertebrates: Host evolution? *Am. Zool.* 31:831–839.

Russell, S. W., Baker, N. F., and Raizes, G. S. 1966. Experimental *Obeliscoides cuniculi* infections in rabbits: Comparison with *Trichostrongylus* and *Ostertagia* infections in cattle and sheep. *Exp. Parasitol.* 19:163–173.

Ruttner, F., and Hanel, H. 1992. Active defense against Varroa mites in a carniolan strain of honeybee (*Apis mellifera carnica* Pollman). *Apidologie* 23:173–187.

Saad-Fares, A. 1985. Trématodes de poissons des côtes du Liban: Spécificité, transmission et approche populationnelle. Thesis, Université de Montpellier II, France.

Saad-Fares, A., and Combes, C. 1992. Abundance/host size relationship in a fish trematode community. *J. Helminthol.* 66:187–192.

Sage, R. D., Heyneman, D., Lim, K. C., et al. 1986. Wormy mice in a hybrid zone. *Nature* 324:60–63.

Sakanari, J. A., and Moser, M. 1990. Adaptation of an introduced host to an indigenous parasite. *J. Parasitol.* 76:420–423.

Sanfilippo, D. 1978. Microhabitat des monogènes Dactylogyroidea parasites branchiaux de téléostéens Mugilidae et Sparidae. Thesis, Université de Montpellier II, France.

Sasal, P., Desdevises, Y., and Morand, S. 1998. Host-specialization and species diversity in fish parasites: Phylogenetic conservatism? *Ecography* 21:639–643.

Sasal, P., Niquil, N., and Bartoli, P. 1999. Community structure of digenean parasites of sparid and labrid fishes of the Mediterranean Sea: A new approach. *Parasitology* 119:635–648.

Sasal, P., Trouvé, S., Müller-Graf, C., et al. 1999. Specificity and host predictability: A comparative analysis among monogenean parasites of fish. *J. Anim. Ecol.* 68:437–444.

Sato, T. 1986. A brood parasitic catfish of mouthbrooding cichlid fishes in Lake Tanganyika. *Nature* 323:58–59.

Saunders, M. L., Tompkins, D. M., and Hudson, P. J. 1999. The dynamics of nematode transmission in the red grouse *(Lagopus lagopus scoticus):* Studies on the recovery of *Trichonstrongylus tenuis* larvae from vegetation. *J. Helminthol.* 73:171–175.

Schad, G. A. 1977. The role of arrested development in the regulation of nematode populations. In *Regulation of parasite populations,* ed. G. W. Esch, 111–167. New York: Academic Press.

Schall, J. J. 1983. Lizard malaria: Cost to vertebrate host's reproductive success. *Parasitology* 87:1–6.

Schall, J. J., and Marghoob, A. B. 1995. Prevalence of a malarial parasite over time and space: *Plasmodium mexicanum* in its vertebrate host, the western fence lizard *Sceloporus occidentalis. J. Anim. Ecol.* 64:177–185.

Schallig, H. D., Sassen, J. M., Hordjik, P. L., et al. 1991. *Trichobilharzia ocellata*: Influence of infection on the fecundity of its intermediate host *Lymnaea stagnalis* and cercarial induction of the release of schistosomin, a snail neuropeptide antagonizing female gonadotropic hormones. *Parasitology* 102:85–91.

Schilthuizen, M., and Stouthamer, R. 1997. Horizontal transmission of parthenogenesis-inducing microbes in *Trichogramma* wasps. *Proc. Roy. Soc. Lond.*, ser. B, *Biol. Sci.* 264:361–366.

Schmid-Hempel, P. 1994. Infection and colony variability in social insects. *Philos. Trans. Roy. Soc. Trop. Med. Hyg.* 346:313–321.

Schmid-Hempel, P., and Crozier, R. H. 1999. Polyandry *versus* polygyny *versus* parasites. *Philos. Trans. Roy. Soc. Trop. Med. Hyg.* 354:507–515.

Schmid-Hempel, R., and Müller, C. B. 1991. Do parasitized bumblebees forage for their colony? *Anim. Behav.* 41:910–912.

Schmidt, D. G., and Roberts, L. S. 1977. *Foundations of parasitology.* St. Louis: Mosby.

Schmidt-Rhaesa, A. 1999. Nematomorpha. In *Encyclopedia of reproduction,* 3:333–341. New York: Academic Press.

Schoener, T. W. 1982. The controversy over interspecific competition. *Am. Sci.* 70:586–595.

Schulz-Key, H., and Wenk, P. 1981. The transmission of *Onchocerca tarsicola* (Filarioidea: Onchocercidae) by *Odagmia ornata* and *Prosimulium nigripes* (Diptera: Simuliidae). *J. Helminthol.* 55:161–166.

Schwartz, S. S., and Cameron, G. N. 1993. How do parasites cost their hosts? Preliminary answers from trematodes and *Daphnia obtusa. Limmnol. Oceanogr.* 38:602–612.

Schweizer, J., Pospichal, H., Hide, G., et al. 1994. Analysis of a new genetic cross between two East African *Trypanosoma brucei* clones. *Parasitology* 109:83–93.

Scott, J. S. 1982. Digenean parasite communities in flatfishes of the Scotian Shelf and southern Gulf of St. Lawrence. *Can. J. Zool.* 60:2804–2811.

Scott, M. 1985. Experimental epidemiology of *Gyrodactylus bullatarudis* (Monogenea) on guppies *(Poecilia reticulata):* Short- and long-term studies. In *Ecology and genetics of host-parasite interactions,* ed. D. Rollinson and R. M. Anderson, 21–38. London: Academic Press.

Seureau, C., and Quentin, J.-C. 1986. L'insecte *Locusta migratoria* (Orthoptère, Acrididae): Un hôte intermédiaire expérimental modèle pour l'étude des cycles de nématodes phasmidiens parasites de vertébrés terrestres. *Année Biol.* 25:25–47.

Sevenster, J. G. 1996. Aggregation and coexistence. I. Theory and analysis. *J. Anim. Ecol.* 65:297–307.

Sewell, S. 1922. Cercariae Indicae. *Ind. J. Med. Res.* 10:1–327.

Shapiro, A. 1976. Beau geste! *Am. Nat.* 110:900–902.

Shaw, D. J., and Dobson, A. P. 1995. Patterns of macroparasite abundance and aggregation in wildlife populations: A quantitative review. *Parasitology* 111:S111–S133.

Shaw, D. J., Grenfell, B. T., and Dobson, A. P. 1998. Patterns of macroparasite aggregation in wildlife host populations. *Parasitology* 117:597–610.

Sheldon, B. C., and Verhulst, S. 1996. Ecological immunology: Costly parasite defences and trade-offs in evolutionary ecology. *Trends Ecol. Evol.* 11:317–321.

Sheppard, C. H., and Kazacos, K. R. 1997. Susceptibility of *Peromyscus leucopus* and *Mus musculus* to infection with *Baylisascaris procyonis. J. Parasitol.* 83:1104–1111.

Sher, A. 1992. Parasitizing the cytokine system. *Nature* 356:565–566.

Sherman, I. W., and Winograd, E. 1990. Antigens on the *Plasmodium falciparum* infected erythrocyte surface are not parasite-derived. *Parasitol. Today* 6:317–320.

Shoop, W. L. 1988. Trematode transmission patterns. *J. Parasitol.* 74:46–59.

———. 1991. Vertical transmission of helminths: Hypobiosis and amphiparatenesis. *Parasitol. Today* 7:51–54.

Shostak, A. W., and Esch, G. W. 1990. Photocycle dependent emergence by cercariae of *Halipegus occidualis* from *Helisoma anceps,* with special reference to cercarial emergence patterns as adaptations for transmission. *J. Parasitol.* 76:790–795.

Shykoff, J. A., and Schmid-Hempel, P. 1991. Parasites and the advantage of genetic variability within social insect colonies. *Proc. Roy. Soc. Lond.,* ser. B, *Biol. Sci.* 243:55–58.

Sibley, L. D., and Boothroyd, J. C. 1992. Virulent strains of *Toxoplasma gondii* comprise a single clonal lineage. *Nature* 359:82–84.

Silan, P., Euzet, L., Maillard, C., et al. 1987. Le biotope des ectoparasites branchiaux de poissons: Facteurs de variations dans le modèle bar-monogènes. *Bull. Ecol.* 18:383–391.

Silan, P., and Maillard, C. 1986. Modalités de l'infestation par *Diplectanum aequans,* monogène ectoparasite de *Dicentrarchus labrax* en aquiculture: Éléments d'épidémiologie et de prophylaxie. In *Pathology in marine aquaculture,* ed. C. Vivarès, J.-R. Bonami, and E. Jaspers, 139–152. Special Publication 9. Bredene, Belg.: European Aquaculture Society.

———. 1990. Comparative structures and dynamics of some populations of helminths, parasites of fishes: The sea-bass-*Diplectanum* model. *Acta Oecol.* 11:857–874.

Simberloff, D. 1990. Free-living communities and alimentary tract helminths: Hypotheses and pattern analyses. In *Parasites communities, patterns and processes,* ed. G. Esch, A. Bush, and J. Aho, 288–319. London: Chapman and Hall.

Simberloff, D., and Moore, J. 1997. Community ecology of parasites and free-living animals. In *Host-parasite evolution: General principles and avian models,* ed. D. H. Clayton and J. Moore, 174–197. Oxford: Oxford University Press.

Sire, C., Durand, P., Pointier, J.-P., et al. 1999. Genetic diversity and recruitment pattern of *Schistosoma mansoni* in a *Biomphalaria glabrata* snail population: A field study using random-amplified polymorhic DNA markers. *J. Parasitol.* 85:436–441.

Sire, C., Rognon, A., and Théron, A. 1998. Failure of *Schistosoma mansoni* to reinfect *Biomphalaria glabrata* snails: Acquired humoral resistance or intra-specific larval antagonism? *Parasitology* 117:117–122.

Sirot, E. 1996. The pay-off from superparasitism in the solitary parasitoid *Venturia canescens. Ecol. Entomol.* 21:305–307.

Sirot, E., Ploye, H., and Bernstein, C. 1997. State dependent superparasitism in a solitary parasitoid: Egg load and survival. *Behav. Ecol.* 8:226–232.

Siva-Jothy, M. T., and Skarstein, F. 1998. Towards a functional understanding of "good genes." *Ecol. Lett.* 1:178–185.

Smith, D.C. 1992. The symbiotic condition. *Symbiosis* 14:3–15.

Smith, D.C., and Douglas, A. E. 1987. *The biology of symbiosis.* Baltimore: Edward Arnold.

Smith, J. E., and Dunn, A. M. 1991. Transovarial transmission. *Parasitol. Today* 7:146–148.

Smith, M. L., Bruhn, J. N., and Anderson, J. B. 1992. The fungus *Armillaria bulbosa* is among the largest and oldest living organisms. *Nature* 356:428–431.

Smith, N. G. 1968. The advantage of being parasitized. *Nature* 219:690–694.

Smithers, S. R., and Gammage, K. 1980. Recovery of *Schistosoma mansoni* from the skin, lungs and hepatic portal system of naive mice and mice previously exposed to *S. mansoni:* Evidence for two phases of parasite attrition in immune mice. *Parasitology* 80:289–300.

Smithers, S. R., and Terry, R. J. 1969. The immunology of schistosomiasis. *Adv. Parasitol.* 7:41–93.

Smith Trail, D. 1980. Behavioral interactions between parasites and hosts: Host suicide and the evolution of complex life-cycles. *Am. Nat.* 116:77–91.

Snounou, G., Jarra, W., and Preiser, P. R. 2000. Malaria multigene families: The price of chronicity. *Parasitol. Today* 16:28–30.

Snyder, S. D., and Esch, G. 1993. Trematode community structure in the pulmonate snail *Physa gyrina. J. Parasitol.* 79:205–215.

Snyder, S. D., and Loker, E. S. 2000. Evolutionary relationships among the Schistosomatidae (Platyhelminthes: Digenea) and an Asian origin for *Schistosoma. J. Parasitol.* 86:283–288.

Sokolowski, M. B., and Carton. Y. 1989. Microgeographic variation in a *Drosophila melanogaster* larval behavior. *J. Insect Behav.* 2:829–834.

Soler, J. J., Møller, A. P., and Soler, M. 1998. Mafia behaviour and the evolution of virulence. *Theor. Biol.* 191:267–277.

Soler, M., and Møller, A. P. 1990. Duration of sympatry and coevolution between the great spotted cuckoo and its magpie host. *Nature* 343:748–75.

Soler, M., Soler, J. J., Martinez, J. G., et al. 1995. Magpie host manipulation by the great spotted cuckoo: Evidence for an avian Mafia? *Evolution* 49:770–775.

Solomon, G. B. 1969. Host hormones and parasitic infection. *Int. Rev. Trop. Med.* 3:101–158.

Sorci, G., Massot, M., and Clobert, J. 1994. Maternal parasite load increases sprint speed and philopatry in female offspring of the common lizard. *Am. Nat.* 144:153–164.

Sorci, G., Møller, A. P., and Boulinier, T. 1997. Genetics of host-parasite interactions. *Trends Ecol. Evol.* 12:196–200.

Sorci, G., Morand, S., and Hugot, J.-P. 1997. Host-parasite coevolution: Comparative evidence for covariation of life-history traits in primates and oxyurid parasites. *Proc. Roy. Soc. Lond.,* ser. B, *Biol. Sci.* 264:285–289.

Sorensen, E. 1999. Population dynamics of *Schistosoma japonicum* in pigs. Ph.D. diss., Danish Center for Experimental Parasitology, Copenhagen.

Sousa, W. P. 1983. Host life history and the effect of parasitic castration on growth: A field study of *Cerithidea californica* Haldeman (Gastropoda: Prosobranchia) and its trematode parasites. *J. Exp. Mar. Biol. Ecol.* 73:273–296.

———. 1990. Spatial scale and the processes structuring a guild of larval trematode parasites. In *Parasite communities, patterns and processes,* ed. G. Esch, A. Bush, and J. Aho, 41–67. London: Chapman and Hall.

———. 1994. Patterns and processes in communities of helminth parasites. *Trends Ecol. Evol.* 9:52–57.

Southgate, V. R. 1979. Host-parasite relationship between schistosomes and their intermediate hosts. *Zentralbl. Bakteriol. Parasitenkd. Infektionskr. Hyg. 1 Abt. Ref. Med. Mikrobiol. Parasitol. Hyg. Präv. Med.* 263:195–196.

Southgate, V. R., Brown, D. S., Warlow, A., et al. 1989. The influence of *Calicophoron microbothrium* on the susceptibility of *Bulinus tropicus* to *Schistosoma bovis. Parasitol. Res.* 75:381- 391.

Southgate, V. R., Van Wijk, H. M., and Wright, C. A. 1976. Schistosomiasis at Loum, Cameroun: *Schistosoma haematobium, Schistosoma intercalatum* and their natural hybrid. *Z. Parasitenkd.* 49:145–159.

Southwood, T. R. E. 1987. Species-time relationships in human parasites. *Evol. Ecol.* 1:245–246.

Speare, P. 1995. Parasites as biological tags for sailfish *Istiophorus platypterus* from east coast Australian waters. *Mar. Ecol. Prog. Ser.* 118:43–50.

Spindler, E.-M., Zahler, M., and Loos-Frank, B. 1986. Behavioral aspects of ants as second intermediate hosts of *Dicrocoelium dendriticum*. *Z. Parasitenkd.* 72:689–692.

Spurrier, M. F., Manly, B. J. F., and Boyce, M. S. 1990. Effects of parasites on mate choice by captive sage grouse. In *Bird-parasite interactions*, ed. J. E. Loye, C. V. Riper, and M. Zuk, 389–398. Oxford Ornithology Series. Oxford: Oxford University Press.

Stearns, S. C. 1992. *The evolution of life histories*. Oxford: Oxford University Press.

Steele, J. H. 1985. The zoonoses. *Int. J. Zoonoses* 12:87–97.

Stenseth, N. C., and Maynard Smith, J. 1984. Coevolution in ecosystems: Red Queen evolution or stasis? *Evolution* 38:870–880.

Stock, T. M., and Holmes, J. C. 1987. *Dioecocestus asper* (Cestoda: Dioecocestidae): An interference competitor in an enteric helminth community. *J. Parasitol.* 73:1116–1123.

———. 1988. Functional relationships and microhabitat distributions of enteric helminths of grebes (Podicipedidae): The evidence for interactive communities. *J. Parasitol.* 74:214–227.

Stoecher, D. K., Michaels, A. E., and Davis, L. H. 1987. Large proportion of marine planctonic ciliates found to contain functional chloroplasts. *Nature* 326:790–792.

Storfer, A., and Sih, A. 1998. Gene flow and ineffective antipredator behavior in a stream-breeding salamander. *Evolution* 52:558–565.

Stromberg, P. C., and Crites, J. L. 1973. Specialization, body volume, and geographical distribution of Camallanidae (Nematoda). *Syst. Zool.* 23:189–201.

Stunkard, H. W. 1923. Studies on North American blood flukes. *Bull. Am. Mus. Nat. Hist.* 48:165–221.

Sukhdeo, M. V. K., and Sukhdeo, S. C. 1994. Optimal habitat selection by helminths within the host environment. *Parasitology* 109:S41–S55.

Sures, B., Jürges, G., and Taraschewski, H. 1999. Relative concentrations of heavy metals in the parasites *Ascaris suum* (Nematoda) and *Fasciola hepatica* (Digenea) and their respective porcine and bovine definitive hosts. *Int. J. Parasitol.* 28:1173–1178.

Sures, B., and Siddall, R. 1999. *Pomphorhynchus laevis:* The intestinal acanthocephalan as a lead sink for its fish host, chub *(Leuciscus cephalus)*. *Exp. Parasitol.* 93:66–72.

Sures, B., Siddall, R., and Taraschewski, H. 1999. Parasites as accumulation indicators of heavy metal pollution. *Parasitol. Today* 15:16–21.

Sures, B. Steiner, W., Rydlo, M., and Taraschewski, H. 1999. Concentrations of 17 elements in the zebra mussel *(Dreissena polymorpha)* in different tissues of perch *(Perca fluviatilis)* and in perch intestinal parasites *(Acanthocephalus lucii)* from the subalpine Lake Mondsee, Austria. *Environ. Toxicol. Chem.* 18:2574–2579.

Sutherst, R. W., Floyd, R. B., Bourne, A. S., et al. 1986. Cattle grazing behaviour regulates tick populations. *Experientia* 42:12–15.

Szathmary, E., and Maynard Smith, J. 1995. The major evolutionary transitions. *Nature* 374:227–232.

Szidat, L. 1969. Structure, development and behaviour of new strigeatoid Metacercariae from subtropical fishes of South America. *J. Fish. Res. Board Can.* 26:753–786.

Tabachnick, W. J., and Black, W. C., IV, 1995. Making a case for molecular population genetic studies of arthropod vectors. *Parasitol. Today* 11:27–30.

Taddei, F., Matic, I., Godelle, B., et al. 1997. To be a mutator, or how pathogenic and commensal bacteria can evolve rapidly. *Trends Microbiol.* 5:427–428.

Taddei, F., Radman, M., Maynard Smith, J., et al. 1997. Role of mutator alleles in adaptaive evolution. *Nature* 387:700–702.

Tardieux, I., Webster, P., Ravesloot, J., et al. 1992. Lysosome recruitment and fusion are early events required for trypanosome invasion of mammalian cells. *Cell* 71:1–20.

Taskinen, J., Mäkelä, T., and Valtonen, E. T. 1997. Exploitation of *Anodonta piscinalis* (Bivalvia) by trematodes: Parasite tactics and host longevity. *Ann. Zool. Fenn.* 34:37–46.

Taylor, F. J. R. 1974. Implication of the serial endosymbiosis theory of the origin of eukaryotes. *Taxon* 23:229–258.

Taylor, L. H., Mackinnon, M. J., and Read, A. F. 1998. Virulence of mixed-clone and single-clone infections of the rodent malaria *Plasmodium chabaudi. Evolution* 52:583–591.

Taylor, L. H., and Read, A. F. 1998. Determinants of transmission success of individual clones from mixed-clone infections of the rodent malaria *Plasmodium chabaudi. Int. J. Parasitol.* 28:719–725.

Taylor, L. H., Walliker, D., and Read, A. F. 1997a. Mixed-genotype infections of the rodent malaria *Plasmodium chabaudi* are more infectious to mosquitoes than single-genotype infections. *Parasitology* 115:121–132.

———. 1997b. Mixed-genotype infections of malaria parasites: Within-host dynamics and transmission success of competing clones. *Proc. Roy. Soc. Lond.,* ser. B, *Biol. Sci.* 264:927–935.

Taylor, M. G. 1970. Hybridization experiments on five species of African schistosomes. *J. Helminthol.* 44:253–314.

Taylor, M. J., and Hoerauf, A. 1999. *Wolbachia* bacteria of filarial nematodes. *Parasitol. Today* 15:437–442.

Tchuem-Tchuenté, L. A., Imbert-Establet, D., Delay, B., et al. 1993. Choice of mate, a reproductive isolating mechanism between *Schistosoma intercalatum* and *S. mansoni* in mixed infections. *Int. J. Parasitol.* 23:179–185.

Tchuem-Tchuenté, L. A., Southgate, V. R., Njiokou, F., et al. 1997. The evolution of schistosomiasis at Loum, Cameroon: Replacement of *Schistosoma intercalatum* by *S. haematobium* through introgressive hybridization. *Trans. Roy. Soc. Trop. Med. Hyg.* 91:664–665.

Terry, R. J. 1994. Human immunity to schistosomes: Concomitant immunity? *Parasitol. Today* 10:377–378.

Théron, A. 1975. *Parabascus lepidotus* Looss, 1907 (Trematoda, Lecithodendriidae): Un exemple de parasite transfuge. *Vie Milieu* 25:181–185.

———. 1976. Le cycle biologique de *Plagiorchis neomidis* Brendow, 1970, Digène parasite de *Neomys fodiens* dans les Pyrénées: Chronobiologie de l'émission cercarienne. *Ann. Parasitol. Hum. Comp.* 51:329–34.

———. 1984. Early and late shedding patterns of *Schistosoma mansoni* cercariae: Ecological significance in transmission to human and murine hosts. *J. Parasitol.* 70:652–655.

———. 1985a. Dynamiques de production des cercaires de *Schistosoma mansoni* en relation avec les variations de la dose miracidiale proposée au mollusque vecteur *Biomphalaria glabrata. Ann. Parasitol. Hum. Comp.* 60:665–674.

———. 1985b. Le polymorphisme du rythme d'émission des cercaires de *Schistosoma mansoni* et ses relations avec l'écologie de la transmission du parasite. *Vie Milieu* 35:23–31.

———. 1989. Hybrids between *Schistosoma mansoni* and *S. rodhaini*: Characterization by cercarial emergence rhythms. *Parasitology* 99:225–228.

Théron, A., and Combes, C. 1988. Genetic analysis of cercarial emergence rhythms of *Schistosoma mansoni. Behav. Genet.* 18:201–209.

———. 1995. Asynchrony of infection timing, habitat preference and sympatric speciation of schistosome parasites. *Evolution* 49:372–375.

Théron, A., and Gérard, C. 1994. Development of accessory sexual organs in *Biomphalaria glabrata* (Planorbidae) in relation to timing of infection by *Schistosoma mansoni*: Consequences for energy utilization patterns by the parasite. *J. Molluscan Stud.* 60:25–31.

Théron, A., Gérard, C., and Moné, H. 1992a. Early enhanced growth of the digestive gland of *Biomphalaria glabrata* infected with *Schistosoma mansoni*: Side effect or parasite manipulation? *Parasitol. Res.* 78:445–450.

———. 1992b. Spatial and energy compromise between host and parasite: The *Biomphalaria glabrata–Schistosoma mansoni* system. *Int. J. Parasitol.* 22:91–94.

Théron, A., Pointier, J.-P., Morand, S., et al. 1992. Long-term dynamics of natural populations of *Schistosoma mansoni* among *Rattus rattus* in a patchy environment. *Parasitology* 104:291–298.

Théron, A., Rognon, A., and Pagès, J.-R. 1998. Host choice by larval parasites: A study of *Biomphalaria glabrata* snails and *Schistosoma mansoni* miracidia related to host size. *Parasitol. Res.* 84:727–732.

Théron, A., and Xia, M. Y. 1986. Rythme d'émission des cercaires de *Schistosoma japonicum* de Chine Continentale par des *Oncomelania hupensis. Ann. Parasitol. Hum. Comp.* 61:553–558.

Thibout, E. 1988. La spécificité de *Diadromus pulchellus* (Hyménoptère: Ichneumonidae) vis-à-vis de son hôte *Acrolepiopsis assectella,* la teigne du poireau. *Entomophaga* 33:439–452.

Thibout, E., Guillot, J. F., and Auger, J. 1993. Microorganisms are involved in the production of volatile kairomones affecting the host seeking behaviour of *Diadromus pulchellus,* a parasitoid of *Acrolepiopsis assectella. Physiol. Entomol.* 18:176–182.

Thirgood, S. J., Redpath, S. M., Rothery, P., et al. 2000. Raptor predation and population limitation in red grouse. *J. Anim. Ecol.* 69:504–516.

Thom, R. 1979. Modélisation et scientificité. In *Élaboration et justification des modèles: Applications en biologie,* ed. P. Delattre and M. Thellier, 21–29. Paris: Maloine.

Thomas, F., and Poulin, R. 1998. Manipulation of a mollusc by a trophically transmitted parasite: Convergent evolution or phylogenetic inheritance? *Parasitology* 116:431–436.

Thomas, F., Poulin, R., de Meeus, T., et al. 1999. Parasites and ecosystem engineering: What roles could they play? *Oikos* 84:167–171.

Thomas, F., Renaud, F., and Cézilly, F. 1996. Assortative pairing by parasitic prevalence in *Gammarus insensibilis* (Amphipoda): Patterns and processes. *Anim. Behav.* 52:683–690.

Thomas, F., Renaud, F., de Meeus, T., et al. 1998. Manipulation of host behaviour by parasites: Ecosystem engineering in the intertidal zone? *Proc. Roy. Soc. Lond.*, ser. B, *Biol. Sci.* 265:1091–1096.

Thomas, F., Renaud, F., Derothe, J. M., et al. 1995. Assortative pairing in *Gammarus insensibilis* (Amphipoda) infected by a trematode parasite. *Oecologia* 104:259–264.

Thomas, F., Renaud, F., Rousset, F., et al. 1995. Differential mortality of two closely related host species induced by one parasite. *Proc. Roy. Soc. Lond.*, ser. B, *Biol. Sci.* 260:349–352.

Thomas, F., Ward, D. F., and Poulin, R. 1998. Fluctuating asymmetry in an insect host: A big role for big parasites? *Ecol. Lett.* 1:112–117.

Thomas, J. A., and Elmes, G. W. 1993. Specialized searching and the hostile use of allomones by a parasitoid whose host, the butterfly *Maculinea rebeli,* inhabits ant nests. *Anim. Behav.* 45:593–602.

Thomas, R. J. 1953. On the nematode and trematode parasites of some small mammals from the inner Hebrides. *J. Helminthol.* 28:153–168.

Thomas-Orillard, M. 1984. Modification of mean ovariole number, fresh weight of adult females and developmental time in *Drosophila melanogaster* induced by *Drosophila* C virus. *Genetics* 107:635–644.

Thompson, J. N. 1994. *The coevolutionary process.* Chicago: University of Chicago Press.

———. 1999. The evolution of species interactions. *Science* 284:2116–2118.

Thompson, S. N., and Kavaliers, M. 1994. Physiological basis for parasite-induced alterations of host behaviour. *Parasitology* 109:S119–S138.

Thoney, D. A., and Hargis, W. J., Jr, 1991. Monogenea (Platyhelminthes) as hazards for fish in confinement. *Ann. Rev. Fish Dis.* 1:133–153.

Thornhill, J. A., Jones, J. T., and Kusel, J. R. 1986. Increased oviposition and growth in immature *Biomphalaria glabrata* after exposure to *Schistosoma mansoni. Parasitology* 93:443–450.

Thornill, R. 1980. Competitive, charming males and choosy females: Was Darwin correct? *Fla. Entomol.* 63:5–30.

Thulin, J. 1984. The impact of some environmental changes on the parasite fauna of cod in Swedish waters (abstract). *Abstr. Fourth European Multicolloq. Parasitol.* 239–240.

Tibayrenc, M., and Ayala, F. J. 1988. Isozyme variability in *Trypanosoma cruzi,* the agent of Chagas' disease: Genetical, taxonomical, and epidemiological significance. *Evolution* 42:277–292.

———. 1991. Towards a population genetics of microorganisms: The clonal theory of parasitic protozoa. *Parasitol. Today* 7:228–240.

Tibayrenc, M., Kjellberg, F., and Ayala, F. J. 1990. A clonal theory of parasitic protozoa: The population structures of *Entamoeba, Giardia, Leishmania, Naegleria, Plasmodium, Trichomonas* and *Trypanosoma* and their medical and taxonomical consequences. *Proc. Natl. Acad. Sci. U.S.A.* 87:2414–2418.

Tierney, J. F., and Crompton, D. W. T. 1992. Infectivity of plerocercoids of *Schisto-*

cephalus solidus (Cestoda: Ligulidae) and fecundity of the adults in an experimental definitive host, *Gallus gallus. J. Parasitol.* 78:1049–1054.

Timms, R., and Read, A. F. 1999. What makes a specialist special? *Trends Ecol. Evol.* 14:333–334.

Tinsley, R. C., and Jackson, J. A. 1998a. Speciation of *Protopolystoma* Bychowsky, 1957 (Monogenea: Polystomatidae) in hosts of the genus *Xenopus* (Anura: Pipidae). *Syst. Parasitol.* 40:3–141.

———. 1998b. Correlation of parasite speciation and specificity with host evolutionary relationships. *Int. J. Parasitol.* 28:1573–1582.

Tirard, C., and Raibaut, A. 1989. Quelques aspects de l'écologie de *Lernaeocera lusci* (Bassett-Smith, 1896), copépode parasite de poissons Merluciidae et Gadidae. *Bull. Ecol.* 20:289–294.

Toft, C. A. 1991a. Current theory of host-parasite interactions. In *Bird-parasite interactions: Ecology, evolution and behaviour,* ed. J. E. Loye and M. Zuk, 3–15. Oxford Ornithology Series 2. Oxford: Oxford University Press.

———. 1991b. An ecological perspective: The population and community consequences of parasitism. In *Parasite-host associations: Coexistence or conflict?* ed. C. A. Toft, A. Aeschlimann, and L. Bolis, 148–156. New York: Oxford University Press.

Tompkins, D. M., and Begon, M. 1999. Parasites can regulate wildlife populations. *Parasitol. Today* 15:311–313.

Touassem, R., and Combes, C. 1985. Le choix du mollusque hôte intermédiaire par le miracidium de *Schistosoma bovis. Bull. Soc. Pathol. Exot.* 78:637–642.

Touassem, R., and Théron, A. 1989. *Schistosoma rodhaini*: Dynamics and cercarial production for mono- and pluri-miracidial infections of *Biomphalaria glabrata. J. Helminthol.* 63:79–83.

Trape, J. F. 1987. Malaria and urbanization in Central Africa: The example of Brazzaville. *Trans. Roy. Soc. Trop. Med. Hyg.* 81 Suppl. 2:1–9.

Tregenza, T., and Bridle, J. R. 1997. The diversity of speciation. *Trends Ecol. Evol.* 12:382–383.

Tripet, F., and Richner, H. 1997. Host responses to ectoparasites: Food compensation by parent blue tits. *Oikos* 78:557–561.

Trouvé, S., Renaud, F., Durand, P., et al. 1996. Selfing and outcrossing in a parasitic hermaphrodite helminth (Trematoda, Echinostomatidae). *Heredity* 77: 1–8.

Tyre, A. J., and Addicott, J. F. 1993. Facultative non-mutualistic behaviour by an "obligate" mutualist: "Cheating" by yucca moths. *Oecologia* 94:173–175.

Ulmer, M. J. 1971. Site-finding behavior in helminths in intermediate and definitive hosts. In *Ecology and physiology of parasites,* ed. A. M. Lallis, 123–159. Toronto: University of Toronto Press.

Upeniece, I. 1999. Fossil record of parasitic helminths in fishes. In *Proceedings of the fifth international symposium on fish parasites,* Ceske Budejovice, Czech Republic, 9–13 August.

Urdal, K., Tierney, J. F., and Jakobsen, P. J. 1995. The tapeworm *Schistocephalus solidus* alters the activity and response, but not the predation susceptibility of infected copepods. *J. Parasitol.* 81:330–333.

Vaidova, S. M. 1978. *Helminths of the birds of Azerbaidzhan.* Elm, Baku: Akademy Nauk Azerbaidzhan, SSR (in Russian).

Van Baalen, M., and Sabelis, M. W. 1995. The dynamics of multiple infection and the evolution of virulence. *Am. Nat.* 146:881–910.

Van den Ackerveken, G., and Bonas, U. 1997. Bacterial avirulence proteins as triggers of plant disease resistance. *Trends Microbiol.* 5:394–398.

Van der Knaap, W. P. W., and Loker, E. S. 1990. Immune mechanisms in trematode-snail interactions. *Parasitol. Today* 6:175–182.

Van der Plank, J. E. 1968. *Disease resistance in plants.* New York: Academic Press.

Van Noordwijk, A. J., and de Jong, G. 1986. Acquisition and allocation of resources: Their influence on variation in life history tactics. *Am. Nat.* 128:137–142.

Van Ripper, C., III, Van Riper, S. G., Lee Goff, M., et al. 1986. The epizootiology and ecological significance of malaria in Hawaiian land birds. *Ecol. Monogr.* 56:127–144.

Van Valen, L. 1973. A new evolutionary law. *Evol. Theory* 1:1–30.

Van Vuren, D. 1996. Ectoparasites, fitnes and social behaviour of yellow-bellied marmots. *Ethology* 102:686–694.

Vass, E., Nappi, A. J., and Carton, Y. 1993. Comparative study of immune competence and host susceptibility in *Drosophila melanogaster* parasitized by *Leptopilina boulardi* and *Azobara tabida. J. Parasitol.* 79:106–112.

Vavre, F., Fleury, F., Lepetit, D., et al. 1999. Phylogenetic evidence for horizontal transmission of *Wolbachia* in host-parasitoid associations. *Mol. Biol. Evol.* 16:1711–1723.

Véra, C. 1991. Contribution à l'étude de la variabilité génétique des schistosomes et de leurs hôtes intermédiaires: Polymorphisme de compatibilité entre diverses populations de *Schistosoma haematobium, S. bovis* et *S. curassoni* et les bulins hôtes potentiels en Afrique de l'Ouest. Thesis, Université de Montpellier II, France.

Verneau, O., Catzeflis, F. M., and Renaud, F. 1997. Molecular relationships between closely related species of *Bothriocephalus* (Cestoda: Platyhelminthes). *Mol. Phylogenet. Evol.* 7:201–207.

Vinson, S. B. 1975. Biochemical coevolution between parasitoids and their hosts. In *Evolutionary strategies of parasitic insects and mites,* ed. P. W. Price, 14–48. New York: Plenum.

———. 1976. Host selection by insect parasitoids. *Ann. Rev. Entomol.* 21:109–133.

Von Schantz, T., Wittless, H., Gôransson, G. et al. 1996. MHC genotype and male ornamentation: Genetic evidence for the Hamilton-Zuk model. *Proc. Roy. Soc. Lond.,* ser. B, *Biol. Sci.* 263:265–271.

Waage, J. K. 1981. How the zebra got its stripes: Biting flies as selective agents in the evolution of zebra coloration. *J. Entomol. Soc. South. Afr.* 41:351–358.

Wajnberg, E. 1994. Le planning familial chez les parasites d'insectes. *Pour Sci.* 196:62–68.

Wajnberg, E., Rosi, M. C., and Colazza, S. 1999. Genetic variation in patch time allocation in a parasitic wasp. *J. Anim. Ecol.* 68:121–133.

Wakelin, D. 1978. Genetic control of susceptibility and resistance to parasitic infection. *Adv. Parasitol.* 16:219–308.

———. 1985. Genetic control of immunity to helminth infections. *Parasitol. Today* 1:17–23.

Walker, J. C. 1979. *Austrobilharzia terrigalensis*: A schistosome dominant in interspecific interactions in the molluscan host. *Int. J. Parasitol.* 9:137–140.

Wallet, M., Théron, A., and Lambert, A. 1985. Rythme d'émission des cercaires de *Bucephalus polymorphus* Baer, 1827 (Trematoda, Bucephalidae) en relation avec l'activité de *Dreissena polymorpha* (Lamellibranche, Dreissenidae) premier hôte intermédiaire. *Ann. Parasitol. Hum. Comp.* 60:675–684.

Walliker, D. 1991. Malaria parasites: Randomly interbreeding or "clonal" populations. *Parasitol. Today* 7:232–235.

Walter, G. H. 1983. Divergent male ontogenies in Aphelinidae (Hymenoptera: Chalcidoidea): A simplified classification and a suggested evolutionary sequence. *Biol. J. Linn. Soc.* 19:63–82.

Walther, B. A., Clayton, D. H., Cotgreave, P., et al. 1995. Sampling effort and parasite species richness. *Parasitol. Today* 11:306–310.

Wang, C. C. 1991. Metabolic deficiencies in parasites and their relation to host metabolism. In *Parasite-host associations: Coexistence or conflict?* ed. C. A. Toft, A. Aeschlimann, and L. Bolis, 258–273. New York: Oxford University Press.

Ward, P. I. 1988. Sexual dichromatism and parasitism in British and Irish freshwater fish. *Anim. Behav.* 36:1210–1215.

Wasserthal, L. T. 1997. The pollinators of the Malagasy star orchids *Angraecum sesquipedale, A. sororium* and *A. compactum* and the evolution of extremely long spurs by pollinator shift. *Bot. Acta* 110:343–349.

Wasson, D. L. 1993. Immunoecological succession in host-parasite communities. *J. Parasitol.* 79:483–487.

Waters, A. P., Higgins, D. G., and MacCutchan, T. F. 1991. *Plasmodium falciparum* appears to have arisen as a result of lateral transfer between avian and human hosts. *Proc. Natl. Acad. Sci. U.S.A.* 88:3140–3144.

Webber, R. A., Rau, M. E., and Lewis, D. J. 1987. The effects of *Plagiorchis noblei* (Trematoda: Plagiorchidae) metacercariae on the behavior of *Aedes aegypti* larvae. *Can. J. Zool.* 65:1340–1342.

Webster, J. P., Brunton, C. F. A., and MacDonald, D. W. 1994. Effect of *Toxoplasma gondii* upon neophobic behaviour in wild brown rats, *Rattus norvegicus*. *Parasitology* 109:37–43.

Wedekind, C. 1992. Detailed ornementation about parasites revealed by sexual ornementation. *Proc. Roy. Soc. Lond.*, ser. B, *Biol. Sci.* 247:169–174.

Wedekind, C., and Milinski, M. 1997. Do three-spined sticklebacks avoid consuming copepods, the first intermediate host of *Schistocephalus solidus*? An experimental analysis of behavioural resistance. *Parasitology* 112:371–383.

Weeks, S. C. 1996. A reevaluation of the Red Queen model for the maintenance of sex in a clonal-sexual fish complex (Poeciliidae: Poeciliopsis) *Can. J. Fish. Aquat. Sci.* 53:1157–1164.

Weidner, E. 1985. Early morphogenesis of *Glugea*-induced xenomas in laboratory-reared flounder. *J. Protozool.* 32:269–275.

Westoby, M., Leishman, M. R., and Lord, J. M. 1995. On misinterpreting the "phylogenetic correction." *J. Ecol.* 83:531–534.

Wetzel, E. J., and Esch, G. W. 1996. Influence of odonate intermediate host ecology on the infection dynamics of *Halipegus* spp., *Haematoloechus longiplexus,* and *Haematoloechus complexus* (Trematoda: Digenea). *J. Helminthol. Soc. Wash.* 63: 1–17.

Whitham, T. G. 1989. Plant hybrid zones as sinks for pests. *Science* 244:1490–1493.

Whitlaw, H. A., and Lankester, M. W. 1994a. A retrospective evaluation of the effects of parelaphostrongylosis on moose populations. *Can. J. Zool.* 72:1–7.

———. 1994b. The co-occurrence of moose, white-tailed deer and *Parelaphostrongylus tenuis* in Ontario. *Can. J. Zool.* 72:819–825.

Whittington, I. D. 1996. Benedeniine capsalid monogeneans from Australian fishes: Pathogenic species, site-specificity and camouflage. *J. Helminthol.* 70:177–184.

Whittington, I. D., Cribb, B. W., Hamwood, T. E., et al. 2000. Host-specificity of monogenean (platyhelminth) parasites: Role for anterior adhesive areas? *Int J. Parasitol.* 30:305–320.

Wickler, W. 1976. Evolution-oriented ethology, kin selection, and altruistic parasites. *Z. Tierpsychol.* 42:206–214.

Wilkins, H. A. 1987. The epidemiology of schistosome infections in man. In *The biology of schistosomes,* ed. D. Rollinson and A. J. G. Simpson, 379–397. London: Academic Press.

Williams, G. C. 1975. *Sex and evolution.* Monographs in Population Biology. Princeton: Princeton University Press.

Williams, J. A., and Esch, G. W. 1991. Infra- and component community dynamics in the pulmonate snail *Helisoma anceps,* with special emphasis on the hemiurid trematode *Halipegus occidualis. J. Parasitol.* 77:246–253.

Wilson, A. 1987. Cercariae to liver worms: Development and migration in the mammalian host. In *The biology of schistosomes,* ed. D. Rollinson and A. J. G. Simpson, 115–146. London: Academic Press.

Wilson, D. S. 1982. Genetic polymorphism for carrier preference in a phoretic mite. *Ann. Entomol. Soc. Am.* 75:293–296.

Wilson, E. O. 1992. *The diversity of life.* Cambridge: Harvard University Press.

Wilson, K., and Edwards, J. C. 1986. The effects of parasitic infection on the behavior of an intermediate host, the american cockroach, *Periplaneta americana,* infected with the acanthocephalan *Moniliformis moniliformis. Anim. Behav.* 34:942–944.

Wilson, R. A. 1990. Leaky livers, portal shunting and immunity to schistosomes. *Parasitol. Today* 6:354–358.

Winfree, R. 1999. Cuckoos, cowbirds, and the persistence of brood parasitism. *Trends. Ecol. Evol.* 14:338–343.

Wolowczuk, I., Delacre, M., Roye, O., et al. 1997. Interleukin-7 in the skin of *Schistosoma mansoni*–infected mice is associated with a decrease in interferon-γ production and leads to an aggravation of the disease. *Immunology* 91:35–44.

Wolowczuk, I., Nutten, S., Roye, O., et al. 1999. Infection of mice lacking interleukin-7 (IL-7) reveals an unexpected role for IL-7 in the development of the parasite *Schistosoma mansoni. Infect. Immun.* 67:4183–4190.

Wolowczuk, I., Roye, O., Nutten, S., et al. 1999. Role of interleukin-7 in the relation between *Schistosoma mansoni* and its definitive vertebrate host. *Microbes Infect.* 7:545–551.

Woltz, P., Stockey, R. A., Gondran, M., et al. 1994. Interspecific parasitism in the gymnosperms: Unpublished data on two endemic New Caledonian Podocarpaceae using scanning electron microscopy. *Acta Bot. Gall.* 141:731–746.

Woodruff, D. S., and Mulvey, M. 1997. Neotropical schistosomiasis: African affinities of the host snail *Biomphalaria glabrata* (Gastropoda: Planorbidae). *Biol. J. Linn. Soc.* 60:505–516.

Worthen, W. B., and Rohde, K. 1996. Nested subset analyses of colonization-dominated communities: Metazoan ectoparasites of marine fishes. *Oikos* 75:471–478.

Wright, C. A. 1971. *Flukes and snails.* London: Allen and Unwin.

Wright, C. A., Southgate, V. R., Van Wijk, H. B., et al. 1974. Hybrids between *Schistosoma haematobium* and *S. intercalatum* in Cameroon. *Trans. Roy. Soc. Trop. Med. Hyg.* 68:413–414.

Wright, S. 1931. Evolution in Mendelian populations. *Genetics* 16:97–159.

———. 1932. The roles of mutation, inbreeding, crossbreeding and selection in evolution. *Proc. VI Int. Congr. Genet.* 1:356–366.

———. 1969. *Evolution and the genetics of populations.* Chicago: University of Chicago Press.

Xia, M. 1990. Contribution à l'étude du développement et de la variabilité génétique de *Schistosoma japonicum.* Thesis, Université de Perpignan, France.

Xia, M., Jourdane, J., and Combes, C. 1998. Local adaptation of *Schistosoma japonicum* in its snail host demonstrated by transplantation of sporocysts. *ICOPA* 9:573–576.

Xie, H., Bain, O., and Williams, S. A. 1994. Molecular phylogenetic studies on filarial parasites based on 5S ribosomal spacer sequences. *Parasite* 1:141–151.

Yan, G. 1996. Parasite-mediated competition: A model of directly transmitted macroparasites. *Am. Nat.* 148:1089–1112.

Yasuda, J., Shortridge, K. F., Shimizu, Y., et al. 1991. Molecular evidence for a role of domestic ducks in the introduction of avian influenza viruses to pigs in southern China, where the A/Hong Kong/68 (H3N2) strain emerged. *J. Gen. Virol.* 72:2007–2010.

Yezerinac, S. M., and Weatherhead, P. J. 1995. Plumage coloration, differential attraction of vectors and haematozoa infections in birds. *J. Anim. Ecol.* 64:528–537.

Yoshino, T. P., and Bayne, C. J. 1983. Mimicry of snail host antigens by miracidia and primary sporocysts of *Schistosoma mansoni. Parasite Immunol.* (Oxford) 5:317–328.

Yoshino, T. P., Cheng, T. C., and Renwrantz, L. R. 1977. Lectins and human blood determinants of *Schistosoma mansoni*: Alteration following *in vitro* transformation of miracidium to mother sporocyst. *J. Parasitol.* 63:818–824.

Yoshino, T. P., and Lodes, M. J. 1988. Secretory protein biosynthesis in snail hemocytes: *In vitro* modulation by larval schistosome excretory-secretory products. *J. Parasitol.* 74:538–547.

Zahavi, A. 1975. Mate selection—a selection for a handicap. *J. Theor. Biol.* 53:205–214.

———. 1977. The cost of honesty. *J. Theor. Biol.* 67:603–605.

———. 1979. Parasitism and nest predation in parasitic cuckoos. *Am. Nat.* 113:157–159.

Zahavi, A., and Zahavi, A. 1997. *The handicap principle.* Oxford: Oxford University Press.

Zervos, S. 1988a. Population dynamics of a thelastomatid nematode of cockroaches. *Parasitology* 96:353–368.

———. 1988b. Evidence for population self-regulation, reproductive competition and arrhenotoky in a thelastomatid nematode of cockroaches. *Parasitology* 96:369–379.

Zimmermann, S., Sures, B., and Taraschewski, H. 1999. Experimental studies on lead accumulation in the eel-specific endoparasites *Anguillicola crassus* (Nematoda) and *Paratenuisentis ambiguus* (Acanthocephala) as compared with their host, *Anguilla anguilla*. *Arch. Environ. Contam. Toxicol.* 37:190–195.

Zuk, M. 1991. Parasites and bright birds: New data and a new prediction. In *Bird-parasite interactions: Ecology, evolution, and behaviour,* ed. J. E. Loye and M. Zuk, 317–327. Oxford: Oxford University Press.

———. 1992. The role of parasites in sexual selection: Current evidence and future directions. *Adv. Study Behav.* 21:39–68.

Zuk, M., Bryant, M. J., Kolluru, G. R., et al. 1996. Trade-offs in Parasitology, evolution and behavior. *Parasitol. Today* 12:46–47.

Zuk, M., Johnson, K., Thornhill, R., et al. 1990. Mechanisms of female choice in red jungle fowl. *Evolution* 44:477–485.

Zuk, M., and McKean, K. A. 1996. Sex differences in parasite infections: Patterns and processes. *Int. J. Parasitol.* 26:1009–1024.

Zuk, M., Thornhill, R., and Ligon, J. D. 1990. Parasites and mate choice in red jungle fowl. *Am. Zool.* 30:235–244.

Zuniga, J. M., and Redondo, T. 1992. No evidence for variable duration of sympatry between the great spotted cuckoo and its magpie host. *Nature* 359:410–411.

Index

Italic page numbers refer to illustrations